The Human Side of Managing Technological Innovation

THE HUMAN SIDE OF MANAGING TECHNOLOGICAL INNOVATION

A Collection of Readings

SECOND EDITION

Edited by **RALPH KATZ**

New York   Oxford
OXFORD UNIVERSITY PRESS
2004

Oxford University Press

Oxford   New York
Auckland   Bangkok   Buenos Aires
Cape Town   Chennai   Dar es Salaam   Delhi
Hong Kong   Istanbul   Karachi   Kolkata
Kuala Lumpur   Madrid   Melbourne   Mexico City
Mumbai   Nairobi   São Paulo   Shanghai
Taipei   Tokyo   Toronto

Published by Oxford University Press, Inc.
198 Madison Avenue, New York, New York, 10016
http://www.oup-usa.org

Library of Congress Cataloging-in-Publication Data
The human side of managing technological innovation : a collection
of readings / edited by Ralph Katz.—2nd ed.
     p. cm.
   Includes bibliographical references and index.
   ISBN 0-19-513530-X (hardcover : alk. paper)—
   ISBN 978-0-19-513531-2 (pbk. : alk. paper)
   1. Technological innovations—Management. I. Katz, Ralph.
HD45.H84 2003
658.5′14–dc21   2003007795

Printed in the United States of America
on acid-free paper

This book is dedicated
to my wonderful wife and family
Judy, Dan, Cecille, Gary, and Elana.

CONTENTS

VI.   The Management of Organizational Processes for Innovation

INTRODUCTION

Organizations competing in today's world of rapid technological and market changes are faced with the challenges of "dualism," that is, operating efficiently in the present while innovating effectively for the future. Managers and leaders within these organizations not only have to focus on the market success and profitability of each of their established products and services but must also ensure their capability to introduce into next generation offerings those specific technical advances and product attributes that will sustain and even augment their continued global competitiveness. Technology-based companies, no matter how they are structured or organized, must develop and implement strategies that satisfy *both* of these sets of concerns. The surprisingly fast collapse or near collapse of so many supposedly dominant, world-class, excellent organizations—especially over the last decade—gives ample testimony to the importance of managing this dualism.

Now, it would be somewhat straightforward if everyone working in an organization could agree on the strategies and means by which this dualism could be achieved, but such consensus is rare. Amid the diverse pressures of daily job requirements and rewards, managers representing different business units and departmental areas usually disagree on the relative merits of allocating already scarce resources and talents among the range of activities that might benefit "today's," versus "tomorrow's," products and services. Furthermore, most managerial practices and procedures cannot offer simple, ready-made solutions that are accommodating to what are essentially inherently conflicting demands

and challenges. Classical management theory, for example, deals primarily with the efficient utilization and production of the company's existing assets and product offerings. The principles of high task specialization, unity of command and direction, high division of labor, and equality of responsibility and authority all address the problems of structuring work systems and information flows in clear, repetitive ways. They are geared toward the resolution of familiar problems and the facilitation of productivity and control through formal lines of positional and hierarchical authority or through job standardization. And organizations that are designed and managed for doing the same things well repetitively, as in manufacturing, are not particularly appropriate for doing something well once, as in Research and Development (R&D). Similarly, organizations designed and managed for doing something well once are not going to be appropriate for doing the same things well repetitively. Put simply, operating and innovating organizations represent *opposing* logics.

What is needed, therefore, is some comparable set of concepts and principles that would reveal how to organize research, development, and engineering (RD&E) activities so that the outputs and creative ideas from innovating professionals are utilized effectively even as the organization deals with the pressures of meeting the everyday demands from its intensely competitive marketplace. How does one establish and manage an organizational RD&E setting such that new technical advances and developmental efforts not only take place but ultimately take place in a very timely, effective, and successful manner? If, in fact, "visionary" ideas and "stupid" ideas look the same at the earliest stages of the innovation process, then how does one manage, lead, allocate resources, or make the "right" decisions in the face of this dilemma?

The need to manage technology-based innovation and enhance the commercialization of new technical knowledge is not new. The ability to develop and incorporate research advances and technical know-how into new products and services has long been a cornerstone of economic growth and industrialization. According to recent U.S. National Science Foundation and National Research Council reports, leveraging the effective contributions of technical professionals through cross-functional teams is one of the more important areas in the management of technological innovation. Organizations not only have to improve their abilities to assess, coordinate, and integrate alternative technical developments within their overall business strategies but must also be capable of implementing their plans through the leadership of their technical management and the efforts of their engineering and scientific personnel. Far more often than not, the frustrations identified in the management of innovation are simply not technical in nature, the problems being rooted more in the complex interplay among the many different priorities and motivational efforts of the technical professionals and their cross-functional peers. The management of technological innovation has to take the human side of the equation into consideration!

In most industries today, a variety of economic, technical, and market-driven forces have coalesced to make these people-related problems even more salient. The diversification and decentralization of businesses, the growth of international alliances and competition, the faster rates at which new technologies are being generated worldwide, the increasing demands of more sophisticated customers, the growing use of computerized information and communication networking, and the dramatic reduction of lead times for new product development have all placed greater pressure on the firm to expand both its commitment to and its competence in managing technological innovation and organizational change. In particular, the ways in which a company's professional personnel and resources are allocated, positioned, and generally managed will have a strong bearing on the firm's ability to compete successfully in tomorrow's more global economy. As a result, decision makers in organizations today must have those managerial skills and perspectives necessary to enhance the innovation process and to bring technical advances successfully to the marketplace. They must acquire substantial knowledge and ability not only in managing and directing the technical and marketing developments themselves but also in utilizing and directing the professionals involved in these development efforts. They must be able to communicate, understand, inspire, and guide technical professionals and integrate them with other functional parts of the organization, including finance, marketing, and manufacturing. The need for executive leadership in managing both innovation and people has never been greater.

The main objective of the second edition of this readings book, *The Human Side of Managing Technological Innovation: A Collection of Readings,* is to provide a more recent comprehensive picture of a variety of approaches and perspectives on issues critical to the effective management of technical professionals and cross-functional teams throughout the innovation process. My hope is to bridge the gap between theory and application by selecting those readings that contain concepts and empirical findings that will prove meaningful and useful to both students and practicing managers, executives, and professional staff personnel. The selections are not meant to span the entire gamut of issues involved in the management of technology and innovation. I have, instead, focused on readings that deal with the management and leadership of professional specialists and cross-functional teams, as well as the structuring and coordination of organizational roles during the innovation process. For the most part, the selections discuss how internal work environments can be organized and structured to provide the kinds of opportunities, experiences, and managerial leadership that will enable professional specialists to innovate effectively through their organizational and team-based relationships. The articles included in this second edition represent the current thinking of a large number of researchers and practitioners seeking a richer understanding of the complex interplay between the specialized knowledge and skills of creative professionals and the realistic pressures and constraints required by successful business organizations.

The readings in this book can be used as a text for advanced undergraduate or graduate courses that focus primarily on how organizational managers, individual professionals, project teams, and/or functional groups deal with problems and issues related to the management of technology-based innovation. The collection can also be used as a complementary text for any course that emphasizes product, process, and organizational or technological innovation. It could be used in behavioral courses on issues of leadership, organizational behavior, organizational change, or general management, but more specifically, it could be used for courses broadly related to the management of innovation, courses such as R&D management, new-product development, project management, engineering management, managing professionals, and so forth. Another audience includes all of those managers and staff professionals working on innovative projects or dealing with product development activities, especially when they attend in-house company or university-based workshops and want to have background material on the kinds of problems and concerns included in the various readings. Finally, although many of the selections in this volume focus on technical professionals, the concepts and issues covered apply to a wide range of professionals, including those involved in new products and services, consulting services, organizational or technological change and uncertainty, or any situation involving the development or application of specialized knowledge and abilities.

The second edition of *The Human Side of Managing Technological Innovation* is organized into seven broad parts comprising eighteen sections that include a total of fifty-eight individual selec-

tions. The sections range from issues related to the motivation and leadership of individual technologists to more macro organizational frameworks for innovation and new product development. Part I looks at problems associated with fostering the creative performance and career contributions of individual professionals. In Section 1, we examine the issue of motivation, emphasizing the importance of designing energizing and challenging work environments as well as the strong role that managers can play in influencing these motivational conditions through socialization processes and career development activities. Section 2 contains several readings that discuss the results and implications of important studies dealing with the management and leadership of very creative individual contributors in organizational settings. While Section 2 deals primarily with creativity and managing the contributions of creative individuals, the selections in Section 3 discuss several ideas and frameworks for capturing many of the problems technical professionals face as they transition into management, trying to become effective technical managers and organizational leaders.

Part II covers a number of important issues related to the management and leadership of professionals within groups and cross-functional project teams. Section 4 deals with the formation and management of high-performing technical teams both through case studies of DEC's Alpha design team and Sun Microsystem's development of JAVA and through a selection that summarizes many years of research around exceedingly high-performing groups and project teams. Section 5 then explores some of these cases' issues and problems from a cross-functional team perspective and discusses how to maintain a team's performance over time, especially if the team becomes increasingly stable and complacent. Section 6 rounds out the picture with its discussions of managing R&D groups and project performances within functional, project, and matrix-type organizational structures.

Part III examines the different technical, managerial, leadership, and entrepreneurial roles that professionals can and should play at particular stages of the innovation process. Section 7 investigates the role of the technical manager and describes many of the problem-solving activities he or she needs to orchestrate and manage during the development and design of new products and services. The readings in Section 8 discuss and compare various informal leadership roles that technical professionals and managers have to carry out, especially as they try to start, entrepreneur, and implement critical organizational changes and radically new product development and service initiatives. One of the selections in this section presents a nice roadmap-type model for conceptualizing the innovation process and the critical roles that need to take place to support this process.

The readings in Part IV illustrate just how powerfully organizational structures and technical cultures influence the attitudes and communication behaviors of professionals within the innovation process. Section 9 cautions against generalizing across all types of professionals and R&D activities, pointing out critical differences between science and technology as well as between research and development activities. The section also shows the difficulty of communicating and transferring information effectively across different "technical cultures" but then goes on to reveal how the critical role of "technical gatekeeper" can help overcome this difficulty and how organizations can manage information and technical know-how. Section 10 focuses on an important dilemma within technical settings: How does one measure and assess the performance and productivity of an organization's R&D effort as well as the individual R&D projects within the laboratory? While it may be easy to measure repetitive activities, it is very hard to assess R&D-type activities that are unique and only need to be done once.

Within Part V, Section 11 deals with this same dilemma but does so by discussing the strong effects that cultures have on professional performance, R&D breakthroughs, and organizational innovation; one of the selections illustrates these effects through a detailed case study of the innovation of 3M's Post-it Notes. While Section 11 emphasizes the positive aspects of culture, Section 12 discusses many of the problems of sustaining innovative activities and creative R&D performance within functioning organizations.

Part VI discusses a variety of organizational processes and practices that influence an organization's ability to innovate in a timely and effective manner. Section 13 focuses on the decision-making process, stressing important differences in the management of routine, versus nonroutine, work as well as how managers who make "fast" decisions differ from those who are substantially slower at making key decisions. The last selection in Section 13 illustrates the ever-present tensions between development efforts, testing, quality control, and schedule pressures by retelling the classic tale of the demise of Sweden's warship, the *Vasa,* as it immediately sank upon sailing away from its dock. Section 14 discusses a number of important issues from a human resource management perspective, including issues of diversity, rewards, and dual ladder promotional systems. The issue of reducing product development cycle times as well as the conventional wisdom about being "first to market" are addressed directly in the readings in Section 15 by including the frameworks, experiences, and caveats from several well-known consultants and researchers.

The last part of the book, Part VII, deals with several key dynamics and interrelationships underlying the innovation process. Section 16 concentrates on the critical interface between marketing and R&D by including results from three renowned empirical studies. The section's first reading investigates the characteristics that lead to more effective cross-functional relationships; the second selection in this section summarizes Robert G. Cooper's results surrounding his classic studies of the development activities and product attributes of those products that eventually win in the marketplace; and the third reading explores Clayton Christensen's more recent perspective on disruptive versus sustaining innovations, that is, the "innovator's dilemma." Section 17 contains two selections that capture Eric von Hippel's seminal work on user innovation and how to organize to take advantage of this phenomenon through lead user methodology and through his most recent work on toolkits. The readings portray some of the limitations of trying to come up with new product ideas by identifying key market needs, but then go on to show how organizations can enhance the management of new product developments by actively co-opting ideas and solutions from lead users and customers using toolkits. Section 18 completes the volume by presenting a range of perspectives on the innovation process. Two of the readings in this section present the latest thinking and findings from several very well-known researchers who have been studying the success of using product platform strategies within the innovation process for both products and services. Another selection discusses the innovation dilemma by comparing the patterns of product, versus process, innovation over the course of a product's life cycle and discussing how these patterns affect the dynamics and organizational arrangements between old and new technologies. And finally, the section's last reading presents a broad overview of the strategic management of alliances to augment and support the organization's technological innovation program.

Clearly, no book could hope to cover every topic associated with the human side of managing technological innovation, nor could it possibly include all of the excellent readings related to the many issues and nuances involved in each of the parts and sections. Nevertheless, it is my hope that the fifty-eight selections included in this second edition will be relevant and useful to those readers who are part of an organization's innovation efforts and to those who have to manage the innovation or change process within their organizations. I hope the readings will increase their sensitivity to the kinds of problems and issues they will face and that they will have some new ideas, tools, and insights for handling these problems or—better yet—for preventing them.

R.K.

The Human Side of Managing Technological Innovation

I

THE MOTIVATION AND LEADERSHIP OF TECHNICAL PROFESSIONALS

1

The Motivation of Professionals

Motivating Professionals in Organizations

RALPH KATZ

Although motivation is a critical ingredient for fostering and sustaining high performance and creativity among individual professionals and project teams, it is often one of the most difficult things to understand and manage. While the presence of motivation does not guarantee high performance and success, its absence seems to result in long-term problems. Highly motivated people and teams push themselves to overachieve, stretching their thinking and working arduously to accomplish considerably more than brighter and even more technically competent peers. In fact, as leaders gain experience, they soon realize that to get creative ideas and innovative advances commercialized more quickly and successfully through organizational systems, they are much better off having technical professionals with *A-rated* motivations and *B-rated* capabilities than the other way around. Or, as Thomas Edison once remarked: "I have more respect for the person who gets there than for the brilliant person with a thousand ideas who does nothing."

Reprinted with permission of the author.
Ralph Katz is Professor of Management at Northeastern University's College of Business and Research Associate at MIT's Sloan School of Management.

Although the benefits of motivation are clear, it is still a very perplexing issue within most organizational settings. One might think that with all the cumulative wisdom and knowledge that has been built up through countless past generations of work and organizational experiences that today's executives and managers would be masters of this element of a leader's role and supervisory responsibility. And yet, after more than twenty-five years of teaching, researching, and consulting on issues related to the management of professionals, project teams, and technology-based innovation, I still find individuals at all organizational levels listing the motivation of subordinates, peers, and bosses as one of the most important, albeit frustrating, aspects of their jobs.

What's the secret to getting and keeping people motivated? While there's an abundance of motivational theories and models, it is well acknowledged that a person's true calling comes from the nature of the work he or she is asked to do. Put simply, people in general, and technical professionals in particular, like to do *neat* things—"to boldly go where no man has gone before." Or, as cleverly stated by one pharmaceutical plant geneticist, a revolution was taking shape in human genomics and she just didn't want to miss it! If organizational employees believe their work is challenging, innovative, significant, and exciting, then no demands are so difficult that they eventually can't be met. In his feature story on "techies," for example, Alpert (1992) emphatically points out that in the many dozens of interviews he conducted with engineers and programmers, not a single person mentioned career advancement as a primary goal; instead, they talked about technical challenge, that is, the work itself, as their chief reward.

The degree of motivational potential of any specific job, then, is dramatically influenced by how a person views the job assignment on which he or she is working at a given point in time, including how tasks, information, reward, and decision-making processes are organized, structured, and managed. Even in high-pressure and stressful situations, professionals report that work is truly motivational when they really feel they are having

fun doing it. In some sense, then, highly motivating work assignments should be similar to the kinds of activities individuals choose to do for fun on their own.

To pursue this analogy, let me argue that many professionals enjoy playing golf—some even like to go bowling. Let me further suggest that one of the underlying reasons people like to golf or bowl is that it's a use of their own skills and abilities. They perform the activities by themselves based on their own individual styles and competencies. They can set their own target scores and strive to achieve those scores at their own pace without holding others back or being dependent on others. In researching the career transitions of technical professionals, I have found this job characteristic to be one of the more difficult changes one encounters when one is promoted from functional specialist to manager.

A staff professional is typically working on tasks that have been partitioned so that he or she is relying primarily on his or her own personal skills and capabilities for solving problems and completing work-related requirements. When a professional becomes a manager, however, the person soon discovers that he or she must now count on the motivations and competencies of subordinates and other project-related members for getting the problems solved and for getting the work done. In other words, the professional-turned-manager is now responsible for the outputs of others, and he or she is now dependent on them for how well the overall project is completed and how well success is achieved.

This career transition can be rather problematic, especially for some of the most technically proficient professionals who reluctantly find themselves having to learn to manage and supervise the work of subordinates and/or peers who may not be as technically creative or as capable as themselves. It is one thing to delegate work to someone who is more talented or more motivated; it is often much harder and at times even exasperating to delegate work to someone who is less capable or less motivated. What company education and training programs need to understand is that if the most talented technical contributors within an R&D setting are promoted, at least initially, to first-level supervisory positions, then by definition the ways in which these new leaders learn to work with and manage subordinates who may not be as technically competent as themselves will significantly impact the motivations of their personnel. Of course, such uneasiness can also occur in bowling or golfing if one does not develop the patience and sensitivity for having fun with individuals or teammates who cannot play as well, or who are much less talented, or who play at a much slower or more methodical pace.

Professionals probably also like to bowl or play golf because they have a complete understanding of what the *whole* game is from beginning to end. They see themselves playing an active and genuine role during the game because their contributions are valued and respected. As a result, they feel they are significant and somewhat equal players or colleagues within the overall team effort. Their scores count just as much and are as crucial as the scores of any other team member. In sharp contrast, people would not be excited to play if they were told their scores did not count—or their scores were going to be canceled or shelved—or if everyone else's scores were seen as more pivotal or were weighted more highly.

Even though the goals and objectives of the games are clear, bowlers and golfers are given a lot of autonomy within which to play. They are free to develop their own individual style, movement, and pace within the rules and constraints of the game. They also know exactly how well they are performing while playing. Feedback is quick and unequivocal—and they know how they're doing compared with others in the game. But again, no one would like to bowl or golf if the alleys or fairways were draped so they couldn't see the pins or greens, or if they were told to come back in six months for performance appraisals. Nor would anyone enjoy the game if a boss stood by with constant suggestions such as, "Move to your left, watch the red line, keep your grip steady, your arm straight." And finally, who would want to play if, as soon as one achieved a terrific score, the CEO ended up con-

gratulating the boss rather than the player? It is doubtful that anyone would truly find bowling or golfing *fun* under these kinds of conditions.

MOTIVATIONAL TASK CHARACTERISTICS

If work is to be viewed in the same vein as having fun, then the job's tasks should have some of the same characteristics as golfing and bowling. But what are these task characteristics that create such high levels of intrinsic work motivation? Building on the motivation framework put forth by Hackman and Oldham (1980), Figure 1.1 shows that people are more motivated when they feel their jobs require them to use a wide variety of skills and abilities. All too often, professionals who work in well-defined job or project positions become dissatisfied and even disillusioned when narrowly structured tasks require them to use only a small portion of their overall competencies and educational training. Perhaps even worse, their assigned tasks might involve only existing, older, or more mundane technologies rather than the newer and more exciting ones that are being developed within their disciplines for future product and/or service introductions.

Software designers, for example, are considerably more engaged and energized when task assignments require them to write programs in newer languages such as Perl, Python, XML, Java, or C++ than when they are asked to program in less jazzy languages, such as Cobol or HTML—or when they are simply asked to modify and basically reuse someone else's "old code." The evidence accumulated from many previous research studies seems to indicate that the broader the range of skills and abilities tapped early in a person's career, the more likely it is that the person will remain to become a more effective and successful contributing member of the organization in the future. The key is not how many different projects or tasks individuals are asked to perform, but whether they are able to develop the cumulative knowledge, perspective, and credibility essential for continued success within the organization's particular setting and culture.

Dimensions of Task Characteristics	Definitions
Skill Variety	The degree to which the job requires the use of different skills, abilities, and talents
Task Identity	The degree to which the person feels that he or she is part of the whole job or project activity from beginning to end
Task Significance	The degree to which the job is considered important by and has impact on the lives of others
Autonomy	The degree to which the job provides freedom, independence, and discretion in how the work is carried out
Feedback	The degree to which the person is provided with clear and direct information about the effectiveness of his or her performance

Figure 1.1. A framework for work motivation (adapted from Hackman and Oldham, 1980).

In his seminal study of professionals in technical organizations, for example, Donald Pelz (1988) discovered that scientists and engineers were judged most effective when they devoted their time across the range of activities within the research, development, and technical service continuum rather than concentrating in only one of these domains. Young professionals not only have to solidify their technical reputations by focusing activities and accomplishments in a major project and/or technical area but must also have enough diversity and exposure in their task challenges and networks to come up with the kinds of initiatives and ideas that enable them to become the organization's future high flyers and star performers (Kelley, 1998; McCall, 1998). And of the many job experiences investigated in my own panel studies of technical professionals' early career years, the ones that seem the most predictive for achieving promotions to higher managerial positions within an organization are the opportunities to work at jobs that help familiarize them with the ways in which the organization deals with business and financial information.

In several organizations in which my colleague, Professor Tom Allen, and I were able to conduct follow-up investigations, young technical professionals who had previously indicated that their task assignments required them to work with financial information and budgets were significantly more likely over the next five years to have received higher rates of managerial promotion within their technical settings. One possible explanation is simply that as a result of such fiscally related experiences and activities, these individuals were more likely to have had the chances of developing into and being seen as broader contributors. They were now perceived as more capable of putting together the kinds of *business* plans and strategies that make those managing the organization's existing operations and finances more comfortable—at least in terms of focus, scope, and language—rather than the more one-sided, narrow *technical* arguments that are all too often put forward by those who come to be viewed as individual contributors, specialists, or just one-dimensional thinkers. It is extremely important for technical managers to learn how to frame or "couch" their ideas and efforts into the kinds of business screens and financial terminology that are specifically used by those in power to run and evaluate the organization's day-to-day pressures and alternative interests. Even General Electric's Jack Welch, *Fortune* magazine's most studied and written about CEO in the last twenty-five years, points out in his autobiography that as a Ph.D. chemical engineer, he was sufficiently "green" in the intricacies of finance that he asked his staff to prepare a book translating all of GE's financial jargon into layman's terms so he could be more conversant with the people in the businesses (Welch, 2001).

Two other analogous and important characteristics of work are *task identity* and *task significance*. Work assignments are more fun and motivating when people are given a complete picture of the project and feel as if they are *real* members of the project team. All too often, however, many professionals supporting and contributing to projects complain that they are not really being treated as part of the overall effort because the only time they hear from program managers, project leaders, or core members is when those people need something. Once these individuals have what they want, they are virtually never heard from or seen by the professionals again. In one highly respected electronics company, for example, an expensive professional videotape was made to promote the new product development processes that had been used by the project team to achieve what had become an outstanding technical and market success story. The plan, of course, was to use the tape throughout the company as an exemplary model of how new product development should be organized and conducted in the future. Unfortunately, the videotape only included the hardware design and development project team members who were around at the end of the project's completion. The organization somehow forgot about and failed to include in the video any members of the software design group whose efforts were invaluable in making the product more functionally competitive and user friendly. Needless to say, the software group was a bit upset. And they became even more disgruntled when the organization continued to use the tape as originally filmed rather than redoing it.

Many professionals want jobs that not only let them feel like strong contributors but also let them build credible professional reputations with members of their disciplines. Studies show that kudos received from respected colleagues and peer groups often motivate R&D professionals more than those from management. Much of this peer recognition and reinforcement occurs through the informal contacts and networking opportunities that take place not only during their organizational and project-related activities but also at professional conferences and industry shows. In one of my research sites, a project team of about twenty-five engineers struggled for several years to overcome numerous obstacles to finally accomplish some very noteworthy technical breakthroughs that they had targeted for an important product line. What truly mattered to this research team after all these years of effort was the opportunity to present their "breakthrough" as a team at their profession's annual international conference in order to impress their colleagues and sort of "stick it" to their competitors. In addition, the company was not intending to keep this breakthrough a secret; it was in fact going to promote it aggressively to enhance the product's market attractiveness. Rather than rewarding this team and enhancing its motivational commitment for accomplishing something special, the organization replied to the team's request with its normal policy. Senior management logically explained that budgeting procedures would only allow one or two members of a project to present results at a conference (even though the company would be spending hundreds of millions of dollars to build its future business around this breakthrough). You can, of course, predict how this perfunctory response affected the future enthusiasm and motivation of this team's technical staff to want to continue to work on this product, a team whose members had literally put their lives "on hold" to meet incredibly aggressive deadlines and requirements

Most professionals display motivational problems at work when assigned tasks appear to have little significance or import. People don't want to waste their time and effort. They're motivated most when working on projects they believe will be taken seriously—projects that are considered important and that could significantly affect their organizations, professions, or society. To work on projects that improve the quality of people's lives is the ultimate dream. As Steve Jobs framed it so convincingly to John Sculley when he was enticing him to leave Pepsi and join Apple Corporation as its CEO: "You can either stay and sell colored water for the rest of your life or you can leave to join us at Apple and help change the world." While money and position may be very visible on report cards, they are not the incentives—what really drives most technical innovators is excitement in their work and pride in their accomplishments.

I once worked with a large project team that had been assembled to develop a state-of-the-art fault-tolerant computer (i.e., a computer that supposedly would not *crash* and *go down*). The project manager wanted to share his enthusiasm for this challenging effort with his new team, so he held a kick-off project meeting to explain the demands and requirements of this upcoming opportunity. But he quickly discovered the team didn't share his excitement and wasn't interested in learning about the technical details and specifications. What the team really wanted to hear was whether the company was serious about the new business venture. Was the company really going to try to bring a product to market, or would it eventually give up on the project as it had done on so many other occasions? Only when a very senior-level executive in the organization met with the team and explained that he would seriously sponsor the effort (i.e., essentially putting his own neck on the line) did members of the group switch from being passive spectators to motivated participants.[1] Project teams must be assured that management believes what they're working on is important and supports their work wholeheartedly, infusing them with the kind of can-do spirit that pervades many high-performing teams (Katz, 1994)

A fourth task characteristic that's key for inducing higher levels of motivation is autonomy—the degree of freedom a person has in carrying out work requirements. As autonomy increases, individuals tend to become more reliant on their own efforts, initiatives, and decisions. They begin to feel more personal responsibility and are willing to ac-

cept more personal accountability for the outcomes of their work. There is, however, an important distinction that needs to be made between *strategic autonomy* and *operational autonomy*—between *what* has to be done in terms of goals, expectations, direction, and constraints versus *how* one chooses to accomplish the goals. Management, unfortunately, often confuses these two aspects of autonomy. Rather than clarifying expectations and establishing clear parameters that would encourage people to make decisions and take initiatives within well-defined boundaries, organizations often give too much free reign with aphorisms such as "Take risk" or "Don't be afraid to fail." Then, as they get increasingly nervous, organizations try to control and micromanage the work, imposing all kinds of unanticipated constraints and changes.

People I've met with and taught from the former Bell Labs, for example, recount that Bell Lab's management would often give initial research assignments to young technical professionals with well-intentioned but rather unclear mandates such as "Go be creative" or "Go think great thoughts." Many of these young professionals would then go off on their own trying to be creative. Many would eventually emerge months or even years later with some wonderful new thoughts or ideas only to discover that Bell Lab managers were not really interested in pursuing something in that particular area, something that risky, or something that required that much money or time, and so forth. Managers need to do it the other way around. They need to clarify expectations and conditions as much as possible and then give people the freedom to function within those definitions and constraints.

To sustain motivation, individuals also need to see the results of their work. If professionals cannot determine whether they are performing well or poorly, they have no basis for trying to improve. So another important motivating task characteristic is *feedback*—the degree to which individuals receive clear information and unambiguous evaluations of their performance. Although technologists often feel that as professionals they are excellent judges of their own roles and performances, they still have a need to calibrate their activities and career progress against the expectations of others. They especially want to know how others whom they respect and whose opinions they value judge their abilities, efforts, contributions, and success.

FRAMEWORK FOR DIAGNOSIS

Having developed an initial instrument for measuring the task dimensions within this framework of work motivation, Hackman and Oldham (1980) argue that managers in any organizational environment can evaluate the jobs of their employees to determine whether any of them is relatively weak with respect to each of the five major characteristics, namely, skill variety, task identity, task significance, autonomy, and feedback. Jobs and task assignments could be restructured, redesigned, or managed to enrich them. Fractionalized project tasks, for example, could be combined to form new jobs that would strengthen a worker's interactions and sense of inclusion in the project, and the new jobs might also require using broader skills, abilities, and understanding. Expectations and directions could be clarified so that technical professionals could be given the autonomy they need to make the everyday decisions that would let them complete their work in a more timely and effective manner. The professional's sense of a job's significance could even be increased if he or she had more direct, relevant contact with critical customers rather than feeling isolated from the marketplace in which his or her company competes.

As technical and business managers gain more experience, they eventually learn one of the most important tenets of motivation: It is easier to destroy morale than to create excitement. Only by closely examining how jobs are structured and organized along these five dimensions of task characteristics can a manager hope to foster the kind of setting in which professionals might find themselves having fun as they do their work.

Over the years, the research we've done in professional type settings to quantify and compare these task dimensions seems to show that feedback is the most deficient task characteristic. In our sur-

veys of more than thirty-five hundred technical professionals working in some fifteen different companies, feedback usually had the lowest mean value across the hundreds of project groups studied. It also seemed to be the one around which technologists complained the most. There are probably at least three reasons for this result. First, in most technical or creative-type settings where one is often trying to do something that has never been done before, it is very hard to define or measure exactly what good work is. Trying to explain what constitutes "good code," for example, is very difficult to do ahead of time—only after one sees it can one really say whether or not the code was "neat" or cleverly done. Also, one may not be able to determine for quite some time whether a new idea is truly creative or stupid. Or, to say it another way, the most creative idea and the stupidest idea often look the same at the early stages; one may not know for quite some time whether the idea was brilliant or just plain foolish.

A second reason for feedback being low is simply that most managers and supervisory leaders in technical settings are not particularly comfortable or well trained at giving constructive feedback. And in addition, most technical professionals are not particularly good at receiving feedback, especially feedback that may not be totally positive. It is not uncommon to discover that many of the most creative individuals have never received feedback or been given anything but uniform praise in the past. This can make the transmission of what was intended as "constructive" interaction into a troublesome and unpleasant exchange.

While engineers, scientists, and other professionals may complain the most about the inadequate communication, information, and feedback they get surrounding job performance and career opportunities, it is not the most powerful task dimension for establishing a high level of work motivation. The most critical dimension by far for elevating motivation is task significance. Professionals become more excited and energized when they feel they are working on something important—something that clearly makes a difference within the business unit and is not considered trivial, low priority, or mundane. Interestingly enough, even though task significance seems to be the most influential task characteristic for influencing motivation and performance, the dimension that often surfaces from survey responses is autonomy. Organizations, however, need to be careful. It is easy for technical professionals to say they want a lot of autonomy; it does not necessarily mean they will thrive under autonomy. And autonomy decoupled from a sense of significance can lead to very dissatisfying and unpleasant work experiences.

MULTIDIMENSIONAL TASK DIMENSIONS

As previously described, one can easily examine any job for its motivational potential by trying to determine whether it is relatively strong or weak with respect to each of the five task characteristics. One of the problems of managing and motivating professionals in RD&E (research, development, and engineering) type environments, however, is that there are at least two alternative ways of looking at each of these task characteristics (see Figure 1.2). The previous motivational framework, for example, talked about skill variety. Organizations hire professional employees because they need them—

they want to use their skills, knowledge, and abilities. The individual's prior education, training, and work experiences are all strongly considered in the organization's decision to hire new personnel and in its distribution of critical work and project assignments among its present employees. On the other hand, from the professionals' points of view, they not only want to be *used* but also want to *grow*. They want to learn—to extend their skills, knowledge, and abilities. That's the professional norm! For professionals working in organizations, what makes a job or project especially enticing is the belief that they will learn a heck of a lot as they carry out their task activities and complete their assignments. Put simply, they not only want to utilize what they already know but also want to keep up to date, incorporating relevant leading-edge thinking and advances from their disciplines and areas of expertise into their project's requirements and specifications. In a rapidly changing environment, keeping up with the latest knowledge is essential for sustained success, and some techies are manic about keeping abreast of the latest developments. While executives and professionals would readily agree that *both* skill utilization and extension are crucial elements of any job environment, organizations must realize there are often major differ-

Task Dimensions	Organizational Orientation	Professional Orientation
Skill Variety	To utilize one's skills and abilities	To learn and develop new skills and abilities
Task Identity	To become a contributing member of the organization	To become a contributing member of the profession
Task Significance	To work on projects that are important to the organization	To work on projects that are exciting within the profession
Autonomy	Strategic clarity	Operational autonomy
Feedback	Subjective data and information processes	Objective data and information processes

Figure 1.2. Multidimensional framework for work motivation.

ences in how the two groups rank order the relative priorities of skill utilization and extension. And such priority differences can become de-motivating and counterproductive if disagreements and conflicts materialize and are left unresolved for too long a period.

In a similar vein, organizations and knowledge workers can view the dimensions of task identity and task significance in very different ways. Is the professional employee a member of the organization or a member of the profession? In essence, they have one head but two hats. Are they software or hardware engineers at IBM, for example, or are they IBM employees doing software or hardware engineering? Most managers would contend that professionals who work for them should perceive themselves first as organizational contributors and second as members of their profession. The opposite, however, is often the way in which many professionals prioritize their orientations, scientists and Ph.D.'s in particular (Allen and Katz, 1992).[2] A similar pattern often occurs with respect to task significance. Is it more important to work on projects that are important to the organization or on technical issues that are exciting to the profession? Is it more important to try to incorporate the most sophisticated technological developments and features or to focus on technological advances and features that fit within infrastructure of use, that are manufacturable, reliable, cost effective? And so on. Once again, many professionals might prefer to work on breakthrough solutions for problems defined as important by their fields.[3] Organizations, on the other hand, would prefer that they concentrate on coming up with technical advances that are "good enough"—advances that solve customer problems and can be quickly turned into products, services, or intellectual properties that eventually make money (Steele, 1988).

Many professionals see themselves as having studied and internalized a body of knowledge and a code of conduct that supersedes the companies for which they work. At the same time, however, they want to influence (but not necessarily lead or make) decisions that determine the projects on which they'll work and on how their expertise will be used and applied. These individual contributors often have strong beliefs and personalities and are more motivated when *pulled* rather than *pushed*. They want to know *why* something is asked of them, and they want to mull over ideas, information, assignments, and strategies and then have the opportunities to challenge any with which they're uncomfortable or disagree. These professionals respond best to leaders who have an empathetic understanding of their technical problem-solving worlds and who make their lives easier by (1) respecting their expertise; (2) supporting them in their technical efforts; (3) providing them with the best available tools, equipment, and information; and (4) protecting them from nonproductive hierarchical demands and inflexible bureaucratic constraints.

As previously discussed, there are also two kinds of autonomy: strategic and operational. If organizations want to enrich jobs and motivate employees through empowerment, that is, giving them the freedom to function independently and make decisions based on their own careful professional judgments, then business managers need to establish as much strategic focus and clarity as possible and then let their technical personnel function autonomously within these clearly defined goals and boundary conditions. In developing new products and services, project teams are constantly making critical decisions and tradeoffs. And the clearer the organizational leadership is about its expectations and constraints, its proverbial "lines in the sand," the easier it is to empower the teams and project groups effectively.[4] Managers and leaders have to be careful not to do it the other way around, confusing people with unclear directives and support while simultaneously maintaining rigid control over the means by which they have asked their people to do things differently, that is, to be creative and innovative.

And of course, there are two kinds of feedback: *subjective* and *objective*. While both kinds of information are important and have to be integrated in some way during any organizational decision-making process, when objective data conflict with a manager's subjective gut, that is, his or her more intuitive analysis and understanding, the subjective elements are more likely to dominate. But remember, technical professionals come from perhaps the most objective environments in the world, namely, educational and university type institutions. When students take courses, they are told what to read, when to read, and even how long it should take them to read and study distributed materials. They are told which models and formulas to memorize, and when tested, they are given all information they need to solve carefully crafted technical problems. There is, moreover, a single correct solution to each of the test's problems, and individual answers are separately scored to give each student his or her own objective piece of feedback, called a grade. Finally, if the person takes and passes any prescribed sequence of required courses and electives, he or she will graduate in the normal time frame. All of this programmed structure fits nicely with the educational culture and philosophy propagated within technical professions—cultures in which analyses and decisions that are based on logical thinking, clear discussions, and reliable, unambiguous data are highly valued.

In the real world of work, however, the opposite sets of conditions typically apply. Technical professionals soon discover that problems are not well defined or self-contained but are intertwined with all sorts of organizational politics and behavioral-type issues. Nor are they given all of the formulas, tools, and information needed to come up with appropriate solutions. In fact, they usually have to figure out what it is they need, where to get it, what may or may not be valid information, and what is and is not acceptable. Some problems may not even be solvable. Promotions and rewards, moreover, are not guaranteed. And unlike the university, feedback is no longer based on objective grading from test answers evaluated by professors who have more technical expertise than the students. Feedback is now based on the subjective perceptions of one's supervisor or manager who may actually have less technical knowledge and understanding about the tasks and problems on which the individual has worked. In short, the critical skill for performing effectively in the university is *solving*

problems that have been carefully structured and communicated. The real skill in the world of work, however, is being able to *formulate* (or get *formulated*) comprehensive definitions of what are often incomplete or ill-defined problems, especially in arenas strongly influenced by subjective interpretations, preconceived judgments, and personal emotions and commitments.

COGNITIVE MODELS OF MOTIVATION

Generalized models of the motivation process, as reviewed by Steers and Porter (1995), can be characterized by three basic common denominators. Motivation is primarily concerned with (1) what energizes particular behaviors, (2) what directs or channels these behaviors, and (3) how these behaviors are sustained or altered. The first component concentrates on those needs, drives, or expectations within individuals or work settings that trigger certain behaviors, while the second component emphasizes the goals and visions of the individuals and groups toward which the energized behaviors are directed. The last component of any motivational model has to deal with feedback, focusing on those forces within individuals or their work environments that can either reinforce and intensify the energy and direction of desired behaviors or shift them to another course of action, thereby redirecting their efforts.

Although a wide range of motivational models has been put forth, many of them are psychological in nature, focusing on the willingness of a professional to undertake action in order to satisfy some need. An unsatisfied need creates tension, which stimulates a drive within the individual to identify particular goals that, if attained, will satisfy the need and lead to a reduction of tension. In some sense, motivated employees are in a state of tension, and they undertake activities and behave in ways to relieve this tension. The greater the tension, the greater the drive to bring about and seek relief. Cognitive models of this type require managers to understand the psychological needs of their R&D workforce, for when technical professionals are working hard at some set of activities, they are motivated by a desire to achieve goals they value. Consequently, if organizations want motivated engineers and scientists, they must create the kinds of job assignments, careers, and work-related conditions that allow these professionals to satisfy their individual needs.

Maslow's Hierarchy of Needs

Arguably one of the most venerable models of motivation, Maslow's (1954) hierarchy of needs theory claims that within every individual there exists a hierarchy of five classes of needs:

1. *Physiological:* These involve bodily needs such as hunger, shelter, and sex.
2. *Safety:* These include one's needs for security and protection from physical and and emotional harm.
3. *Social:* These include one's needs for affection and a sense of belonging and acceptance.
4. *Self-esteem:* These include one's internal needs for self-respect, autonomy, and achievement as well as one's external needs for status and recognition.
5. *Self-actualization:* This involves the need for self-fulfillment, to grow and achieve one's full potential.

Based on the hierarchical nature of this model, as each class of needs becomes more satisfied, the next class of needs in the hierarchy becomes more dominant. Maslow separated the five classes of needs into lower (physiological and safety) and higher (social, self-esteem, and self-actualization) orders. He then suggested that lower-order needs are satisfied *externally* through wages, bonuses, job security, and the like, while higher-order needs are satisfied *internally* through the individual's own sense of personal growth and development. Since satisfied needs no longer motivate, individuals tend to move up the hierarchy as their lower-order needs are met. (One can of course move down the hierarchy as lower needs, such as job security, become threatened.) The managerial implications of this motivational model are rather straightforward. RD&E settings should be organized and led to satisfy the higher-order needs of their technical professionals. To be strongly motivated, technologists need to feel that their jobs are both important and meaningful and that their contributions are truly valued by their organizations, their professions, and even by society. Nevertheless, many organizations still have trouble providing job experiences that consistently give their professional employees the opportunities for growth and achievement they desire.

Herzberg's Two-Factor Theory of Motivation

In the belief that an individual's attitude toward work can greatly affect success or failure on the job, Professor Fred Herzberg asked professional employees to write two separate paragraphs: one to describe a situation in which they felt exceptionally satisfied about their jobs and one describing a situation in which they felt especially dissatisfied. After analyzing hundreds of paired comparisons from these critical incident descriptions, Herzberg observed that the kinds of issues professionals mentioned and described when they felt good about their jobs were distinctly different from the replies given when they felt bad. After categorizing these paired comparison responses, Herzberg concluded that certain job characteristics, labeled *motivators,* are more influential with respect to job satisfaction while other characteristics, labeled *hygiene factors,* are more strongly connected with job dissatisfaction. Motivating or intrinsic-type factors included items involving achievement, recognition, responsibility, challenging work, and opportunities for growth and advancement, while hygiene or extrinsic-type factors associated with job dissatisfaction were associated with items such as company policies, administrators, supervisory relationships, working conditions, salary, and peer relationships.

Based on this empirical divergence, Herzberg argued that it may not be very meaningful to measure or think about satisfaction and dissatisfaction as end points along a single dimension or continuum. Instead, one should consider satisfaction and dissatisfaction as two independent factors comprised by very different items. There are intrinsic factors that lead to greater levels of job satisfaction, and consequently more motivated individuals within the workplace. And then there are separate hygiene-type factors that mostly affect one's level of job dissatisfaction. The basic notion underneath this two-factor theory is that managers who focus on improving hygiene-type factors may succeed in reducing employee dissatisfaction, but such efforts will probably not increase job motivation. Most likely, such managers are placating their workers rather than motivating and engaging them. When hygiene factors are adequate, employees may not be dissatisfied but neither will they feel motivated. As in Maslow's theory, if organizations want to enhance the motivations of technical professionals, they have to emphasize the intrinsic aspects of jobs and create the organizational changes and climates that result in positive energies that foster creative and innovative performances rather than instituting policies and changes that lower dissatisfactions but which can also yield complacency in place of excitement.

McClelland's Theory of Needs

Additional psychological models have been developed to look at motivation as a function of the "fit" between the individual and the organizational job setting. McClelland and his colleagues, for example, contend that individuals' needs and motives can be measured along three critical dimensions (McClelland and Boyatzis, 1982). The *need for affiliation* describes an individual's desire for friendly and close interpersonal relationships. Employees with a high need for affiliation want to be well liked and accepted by their colleagues. They prefer job situations that are cooperative rather than competitive, environments in which relationships are built on high levels of mutual trust and understanding. A second dimension depicts an individual's *need for power,* that is, the drive to influence others and have an impact. People with a high need for power strive for control, prefer competitive situations, and seem to enjoy being in charge. The

third dimension captures a person's *need for achievement,* the desire to excel or succeed at some challenging activity or project. Individuals with a high need for achievement seek to overcome obstacles in order to do things better or more efficiently than they were done in the past. They work to accomplish difficult goals that have intermediate levels of risk, and they are willing to accept the personal responsibility for a project's success or failure rather than leaving the outcome to chance or to the actions of others.

Studies in R&D settings have consistently found that technical professionals with a high need for achievement are more motivated and successful in entrepreneurial activities (Roberts, 1991), although a high need to achieve does not necessarily lead to being a good technical manager. In fact, McClelland reported from his most recent research that the best managers have a relatively low need for affiliation but have a relatively high need for power—a need for power, however, that is not unbridled but is carefully kept under restraint by the individual. Since innovation often requires entrepreneurial risk-taking behavior, R&D managers can either select professionals with high achievement needs to lead such efforts or can establish appropriate training programs to stimulate the achievement needs of those technologists undertaking entrepreneurial-type projects and activities.

Schein's Career Anchor Model

As discussed by Schein (1996), a person's career anchor is his or her self-concept, consisting of (1) self-perceived talents and abilities; (2) basic values; and, most importantly, (3) the individual's evolved sense of motives and needs. Career anchors develop as one gains occupational and life experiences; once they are formed, they function as a stabilizing force, making certain job and career changes either acceptable or unacceptable to the individual. The research showed that most people's self-concepts revolve around eight categories reflecting basic values, motives, and needs. And as careers and lives evolve, Schein discovered that most individuals employ one of these eight categories as their decision-making anchor, the thing they will *not* give up, even though most careers permit the fulfilling of several of the needs that underlie the different anchors.

1. *The Managerial Anchor:* The fundamental motivation of individuals in this category was to become competent in the complex set of activities comprising general management. These professionals want to rise in the organization to higher levels of responsibility and authority. Schein argues that there are three component areas of competence that these managers need to develop. First, they must have the analytical competence to identify and solve conceptual problems under conditions of uncertainty and incomplete information. Second, they must have the interpersonal competence to influence and lead people toward the more effective achievement of organizational goals. And third, these professionals in general management positions have to develop emotional competence, that is, the ability to remain strong and make tough decisions in highly pressured and stressful situations without feeling debilitated or paralyzed.[5]

2. *The Technical-Functional Anchor:* Many professionals indicate that their careers are motivated by the technical challenges of the actual work they do. Their anchor lies in using the knowledge and skills of their technical fields or disciplinary areas, not the managerial process itself. If members of this group hold supervisory responsibility, they are usually most comfortable supervising others who are like themselves or who are working in the same disciplinary area. Unlike those in the managerial anchor category, it is the nature of the technical work and not the supervising that excites them. They have a strong preference not to be promoted to job positions that place them outside their technical area, although many admit to having taken positions of greater managerial responsibility in the hope of garnering more influence and rewards.

3. *The Security-Stability Anchor:* This group of professionals seems to have an underlying need for security in that they seek to stabilize their careers by linking them to particular organizations. Professionals with this anchor are more willing than other anchor types to accept the organization's definition of their careers. Regardless of their personal aspirations or areas of technical competences, these individuals come to rely on the organization to recognize their needs and provide them the best possible options and opportunities. Professionals anchored in security/stability often have considerable difficulty in higher levels of managerial responsibility where emotional competence is the prime requisite for effective performance. They will also experience a great deal of stress as their companies eschew policies of no layoffs and guaranteed employment, essentially forcing them to shift their career dependence from the organization to themselves.

4. *The Entrepreneurial Creativity Anchor:* Some professionals have a strong need to create something of their own. This is the fundamental need operating in an entrepreneur. It manifests itself in the individual's desire to commercialize new products or services or to create something with which he or she can be clearly identified. Although these individuals often express a strong desire to be on their own and away from the constraints of established organizations, the decisive factor for leaving their prior organizations was not to achieve autonomy or to make a great deal of money. Instead, technical entrepreneurs typically start their own companies and businesses because they really believe in a given product or service, and the organization in which they had been working would not allow them to move forward with their idea. Schein also reports that entrepreneurially anchored people are often obsessed with the need to create and can easily become bored or restless with the demands and routines of running a business. Other studies (see Roberts, 1991) have drawn a sharp distinction between creative individual contributors who may be good at generating new ideas and those professionals who have the strong desire and capability to grab or exploit good ideas and persevere with them until they have been commercialized in the marketplace.

5. *The Autonomy-Independence Anchor:* Some professionals discover that they strongly prefer not to be bound by the kind of rules, policies, procedures, dress codes, working hours, and other behavioral norms that are present in almost any traditional organization. These individuals are primarily concerned about their own sense of freedom and autonomy. They have an overriding need to do things their own way, at their own pace, and along their own standards. They find organizational life to be restrictive, somewhat irrational, or intrusive into their own private lives. As a result, they prefer to pursue more independent careers, often seeking work as consultants, professors, independent contractors, or researchers in R&D laboratories. Professionals anchored in autonomy are often highly educated and self-reliant, and they will usually decline much better work and job offers if such work significantly impinges on their independence.

6. *The Lifestyle Anchor:* Professionals in this anchor category are focusing on careers that can be integrated with their total lifestyle. They value careers that permit them to stabilize their life patterns by settling into a given region without having to move and relocate. These individuals differ from the security-stability anchor professionals in that they do not define their careers in terms of economic security but rather see their careers as part of a larger life system. Most are in dual career situations, and so they seek to work in organizational settings that will provide them the flexibility and kinds of options they need to integrate two careers and two sets of personal and family concerns into a coherent overall pattern. Such options might include working at home, flexible schedules, part-time work when needed, leaves of absence or sabbaticals, paternity and maternity leaves, day-care coverage, and so on. Although the size of this anchor grouping is probably tied to various social and societal trends, what these professionals want from their employing organizations is an understanding attitude and genuine respect for their lifestyle pressures and needs.

7. & 8. *The Service-Dedication and Pure Challenge Anchors:* A number of individuals reported that they were feeling the need not only to maintain an adequate income but also to do something meaningful in a larger context. As the world's problems became more visible, professionals anchored in service wanted to devote their careers to organizations and activities that allowed them to contribute to these issues in a significant way. And finally, a small group of people defined their careers in terms of overcoming impossible odds, solving the unsolved problem, or winning out over competitors. These professionals seek jobs in which they face tough challenges or difficult kinds of strategic problems, but in contrast to the technical-functional anchored group, they are less concerned about the particular kind of problem or technology that is involved.

Since individual career anchors vary so widely, Schein contends that constructive career management is impossible unless individuals know their own needs and biases. It is therefore critically important that professionals gain more self-insight by analyzing their own career anchors and then managing more proactively their own career courses. Organizations also need to create more flexible career paths and incentive systems in order to meet this wide range of pluralistic needs. If individuals are given a more accurate picture of career patterns and the work that needs to be done, then they will be better able to set a constructive direction for themselves.

Motivation Through the Design of Work

All of these cognitive theories of motivation imply that when professional employees are well matched with their jobs, it is rarely necessary to force or manipulate them into working hard to perform their jobs well. When there is a good *fit* between the individual and the job, the person typically describes a very high level of internal work motivation, feeling good both about themselves and what they are accomplishing. Good performance becomes self-rewarding, which, in turn, serves as an incentive for continuing to do well. In similar fashion, poor performance creates unhappy feelings, which prompts the person to try harder in the future to avoid unpleasant outcomes and regain the intrinsic rewards that good performance can bring. The result is a self-perpetuating cycle of positive work motivation powered through the intrinsically rewarding nature of the work itself. The critical issue is for organizations to structure and design requisite project activities in a way that professionals will also find the work personally rewarding and satisfying.

Several theories have been developed to examine job-related conditions that would affect an employee's level of work motivation. *Equity theory*, for example, asserts that employees' motivations center around *relative* rewards and what they believe is comparatively equitable. Basically, individuals weigh what they put into a job situation, that is, their inputs, against what they get from their jobs, that is, their outcomes, and then compare their input-to-outcome ratios against the input-to-outcome ratios of relative others. If employees perceive their ratios to be equal to that of relevant others with whom they choose to compare themselves, the motivational system is in equilibrium since it is viewed as being fair. However, if the ratios are unequal, then individuals will see themselves as either *underrewarded* or *overrewarded* and will be motivated to reestablish equity either by changing the levels of effort they put into a job, by changing the kinds of outcomes they seek from their jobs, or by altering their comparative others.

Expectancy theory argues that the strength of a tendency to act in a certain way depends on the strength of an expectation that the act will be followed by a given outcome and on the attractiveness of that outcome to the individual. According to this model, the strength of a person's motivation is conditioned on three sets of variables: (1) the *effort-performance linkage,* that is, the probability perceived by the individual that exerting a given amount of effort will lead to high performance, (2) the *performance-reward linkage,* that is, the degree to which the individual believes that this level of performance will result in the attainment of a certain reward or outcome, and (3) the *importance* or *attractiveness* that the individual places on that par-

ticular reward or outcome. This motivational theory requires R&D managers not only to understand what outcomes and rewards professionals value from the workplace, but also to establish the kind of organizational practices and processes that will help professionals build more reassuring reward linkages and expectations.

While equity and expectancy theories emphasize relationships between work motivation and extrinsic rewards, the Hackman and Oldham *job design model* previously discussed is the one most known for showing how *internal* work motivation is linked to the way job tasks are organized and structured. First, an individual must have *knowledge of results,* for, if the employee who does the work never finds out whether it is being performed well or poorly, then he or she has no basis for feeling good about having done well or unhappy about doing poorly. Second, the individual must *experience responsibility* for the results of the work, believing that she or he is personally responsible for the work outcomes. This allows the individual to feel proud if one does well and sad if one doesn't. And third, the individual must *experience the work as meaningful,* something that counts in the person's own system of values. According to Hackman and Oldham, the five task dimensions described in Figure 1.1 underlie these three psychological states, and all three must be represented for strong internal work motivation to develop and persist. High levels in one of the psychological states cannot easily compensate for a deficiency in another. Even individuals with important jobs, for example, can exhibit low levels of internal work motivation if they feel they are being micromanaged or have to follow too many bureaucratic procedures and, consequently, experience little personal responsibility for the outcomes of the work. Individuals display motivational problems when their tasks are designed so that they have little meaning, when they experience little responsibility for work outcomes, or when they are separated from information about how well they are performing. Based on these work design models, motivation in organizational settings has more to do with how project tasks are designed and managed than with the personal dispositions of the professionals who are doing them.

TENETS OF MOTIVATION

After many years of studying behaviors and managerial practices in R&D laboratories, Manners and colleagues (1988) offer a number of interesting tenets of motivation for practitioner managers in technological settings.[6] The authors assert that too many managers confuse the concept of motivation with performance and behavior. Just because a group of professionals is excited about work and is exhibiting lots of activity, it does not necessarily mean that the group is productively or effectively active. Technologists could be highly motivated and end up running amuck perhaps because their motivational spirit is not well focused or well managed.

The first tenet of motivation is simply that generating incremental excitement about work is very difficult. Destroying excitement and morale, on the other hand, is relatively easy. Inexperienced R&D managers need to understand this difficulty so that they don't get discouraged too quickly and give up trying to create a highly motivated work group. Another important observation by the researchers is that "fat, happy rats never run mazes." This does not mean that managers and project leaders should keep their technical professionals deprived. What is implied is that a *positive tension* needs to be present in order to have productive motivation. Rather than focusing only on the technical credentials of professionals, organizations should also consider recruiting and selecting professionals who are capable of generating their own excitement or who have high achievement needs. This is similar to Pelz's (1988) concept of *creative tension*—technical professionals perform best and are most innovative when forces for stability and change are both present at the same time. A third interesting facet of motivation is that emotion has almost no intellectual content. This creates problems for R&D because it is staffed by educated in-

dividuals of high intellect who believe you can intellectualize your way through most motivational problems. People, however, will do things simply because it *feels* good, and so professionals need to have the celebrations and fun times, that is, the emotional opportunities to feel good not only about what they are doing but also with and for whom they are doing it.

Although the concept of *hedonism* is the fundamental principle that people seek pleasure and avoid pain, individuals are very different in what they like and dislike. One of the most critical errors a manager can commit is to make broad generalizations about what motivates all of his or her people. The more managers recognize the salient individual differences among their personnel, the more they can tailor the use of informal rewards, including travel, equipment, assignments, and so on, to motivate their staffs. Managers also need to communicate clearly how they intend to protect their technical professionals if they truly want their people to take more risks, and the authors strongly advise that incremental rewards and status symbols be associated only with eventual success and not the risk-taking act itself. A more controversial tenet is that managers should learn to distinguish between time spent on supervision and time spent on motivation. Although low performers generally require disproportionate amounts of supervisory time, managers should not spend more motivational time on them. Instead they should *invest* this kind of time on the higher performers. The argument is that one moves a group's mean performance to a new plateau by motivating the high performers, not by rescuing the low performers.[7]

Finally, the researchers conclude that effective managers do not rely solely on the organization's formal reward system; instead, they employ a continual stream of informal rewards that they can deliver on a timely basis to generate employee excitement. These managers do not reward incremental performance every time nor do they use or rely on the *same* rewards every time. By keeping the system from getting too stable, they are able to prevent complacency and the expectations that rewards are entitlements. Ultimately, the capacity to motivate is dependent upon the manager's credibility. If the subordinate does not believe in the manager, the manager cannot motivate the subordinate. It is in this ability of the manager to control or get access to rewards and not let them get lost in a bureaucratic system that allows the employees to trust and build continued confidence in their manager. Ultimately, what's critically important is keeping professionals happy and excited about their jobs, projects, contributions, and intellectual "food fights," for when they are having fun with colleagues at work, they tend to churn out all kinds of creative ideas, breakthroughs, and product improvements.

ENDNOTES

1. Studies have shown that one of the most commonly asked questions by teams that have been asked to do something extraordinary is how they can strengthen the nature of their sponsorship (Larson and Lafasto, 1989). And in my experience, unfortunately, teams almost always wait for this sponsorship to walk through the door. It doesn't walk through the door. Teams usually have to build it. A team has to tell its potential sponsor what to do and what kind of sponsorship it needs in order to complete its task demands: What priority conflicts need to be resolved; what authority relationships need to be clarified; what obstacles need to be removed; and what resources need to be freed up or provided.

2. During the technology craze of the 1990s, many technical professionals working in Silicon Valley saw themselves as "technical gypsies." They didn't picture themselves as loyal employees of any specific company; rather, they were Silicon Valley professionals who could easily switch from one company to another without having to relocate and sometimes even without having to change carpools.

3. Seymour Cray, the technical design genius behind almost all Cray Corporation's initial supercomputers, proudly attested that in all of the machines he designed, cost was always a secondary consideration. He would figure out how to build his new platform of supercomputers as fast as possible, completely disregarding the cost of construction. In fact, Seymour's first line of commercialized Cray machines, the Cray-1, was jokingly referred to as the world's most expensive love seat (the computers were ensconced by cushions) while his design of the Cray-2 generation of supercomputers was called the world's most expensive aquarium (these computers were immersed in liquid nitrogen for cooling purposes).

4. Exxon Corporation, for example, use to tell its new business venture managers that they could do anything as long as they did not jeopardize the company's AAA-Bond rating. As a more recent example, Captain Michael Abrashoff describes how he transformed his ship from one of the worst in the U.S. Navy to perhaps the Navy's best by empowering his officers and crew to take actions and make decisions they deemed appropriate as long as they did not endanger the ship, risk people's lives, or demean anyone—his lines in the sand.

5. More recently, researchers such as Goleman, Boyatzis, and McKee (2002) draw a similar conclusion when they claim that it is the emotional intelligences or competences of managers that make them great leaders, not just their IQ's.

6. These tenets have not been derived from research studies but are the cumulative opinions of the authors after carefully observing successful motivational practices in many industrial and government laboratories. The authors have come to view these tenets as fundamental truths of work excitement.

7. Peter Drucker (1999), perhaps the most respected management guru of our day, arrived at a similar conclusion when he argued that the task of an executive is not to change people but to multiply performance capability of the whole by putting to use their strengths and aspirations. According to Drucker: "One should waste as little effort as possible on improving areas of low competence. It takes far more energy and work to improve from incompetence to mediocrity than it takes to improve from first-rate performance to excellence. Rather than concentrating on making incompetent performers into mediocre ones, one should focus more making competent people into star performers." Buckingham and Coffman (2001) make essentially the same point after studying some eighty thousand managers from four hundred companies. The best managers didn't waste their time trying to rewire people; instead, the managers who excelled at getting great performance from their teams spent more of their time drawing out and reveling in the strengths of their teams rather than focusing on their weaknesses.

REFERENCES

Abrashoff, M. (2002). *It's Your Ship: Management Techniques from the Best Damn Ship in the Navy.* New York: Warner Books.

Allen, T., and Katz, R. (1992). Age, Education and the Technical Ladder. *IEEE Transactions on Engineering Management,* 39, 237–245.

Alpert, M. (1992). Engineers. *Fortune,* September 21, 87–95.

Buckingham, M., and Coffman, C. (2001). *First Break*

All the Rules: What the World's Greatest Managers Do Differently. New York: Simon & Schuster.

Drucker, P. (1999). Managing Oneself. *Harvard Business Review*, March–April, 65–74.

Goleman, D., Boyatzis, R. E., and McKee, A. (2002). *Primal Leadership.* Cambridge, MA: Harvard Business School Press.

Hackman, J. R., and Oldham, G. R. (1980). *Work Redesign.* Reading, MA: Addison-Wesley.

Katz, R. (1994). Managing High Performance R&D Teams. *European Management Journal*, 12, 243–252.

Kelley, R. E. (1998). *How to Be a Star Performer.* Random House.

Larson, C., and Lafasto, F. (1989). *Teamwork.* New York: Sage Publications.

Manners, G., Steger, J., and Zimmerer, T. (1988). Motivating Your R&D Staff. In R. Katz (Ed.), *Managing Professionals in Innovative Organizations* (pp. 19–26). New York: Harper Business.

Maslow, A. (1954). *Motivation and Personality.* New York: Harper and Row.

McCall, M. W. (1998). *High Flyers: Developing the Next Generation of Leaders.* Cambridge, MA: Harvard Business School Press.

McClelland, D. C., and Boyatzis, R. E. (1982). Leadership Motive Pattern and Long-Term Success in Management. *Journal of Applied Psychology*, 67, 737–743.

McClelland, D. C., and Boyatzis, R. E. (2002). *Primal Leadership: Realizing the Power of Emotional Intelligence.* Cambridge, MA: Harvard Business School Press.

Pelz, D. (1988). Creative Tensions in the Research and Development Climate. In R. Katz (Ed.), *Managing Professionals in Innovative Organizations* (pp. 37–48). New York: Harper Business.

Roberts, E. (1991). *Entrepreneurs and High Technology: Lessons from MIT and Beyond.* New York: Oxford University Press.

Schein, E. (1996). Career Anchors Revisited: Implications for Career Development in the 21st Century. *Academy of Management Executive*, 10, 80–88.

Steele, L. (1988). Managers' Misconceptions About Technology. In R. Katz (Ed.), *Managing Professionals in Innovative Organizations* (pp. 280–287). New York: Harper Business.

Steers, R. M., and Porter, L.W. (1995). *Motivation and Work Behavior.* New York: McGraw-Hill.

Welch, J. (2001). *Jack: Straight from the Gut.* New York: Warner Books.

2

How Bell Labs Creates Star Performers

ROBERT KELLEY AND JANET CAPLAN

It's a given that today's companies must keep new products and services coming—and respond quickly to continually shifting customer demands. To maintain this competitive pace, managers need to improve the productivity of their knowledge professionals. But while many have expected new technologies like companywide computer networks to boost performance, the real promise lies elsewhere. Changing the ways professionals work, not installing new computers, is the best way to leverage this intellectual capital.

Yet managers have been loath to tackle this kind of productivity improvement. For good reason, companies in high-tech and other creatively driven businesses often avoid direct exhortations to be more productive. Professionals such as engineers, scientists, lawyers, programmers, and journalists already work hard, perhaps 50 to 60 hours a week. To demand more of them is counterproductive because, unlike many manufacturing or service workers, these professionals have options. When pushed, they may withhold their best ideas or simply leave the company.

Although managers know that some professionals excel, few people, including the "stars" themselves, can describe exactly how they do it. But we believe that defining the difference between top performers and average workers is essential for improving professional productivity. For the past seven years, our research has focused on the engineers and computer scientists at AT&T's prestigious Bell Laboratories. This study has led to a successful training program based on the work strategies of star performers. The program has dra-

matically improved productivity, as evaluated by both managers and engineers.

In fields like computer programming, an eight-to-one difference between the productivity of stars and average workers has been reported. As one of the Bell Labs executives observed, "Ten to fifteen percent of our scientists and engineers are stars, while the vast majority are simply good, solid middle performers." When asked why this is so, most managers come up with a variety of plausible explanations. Top performers, they say, have higher IQs. Or they're better problem solvers, or driven by an enormous will to win. In other words, stars are better people in some fundamental sense, while middle performers lack the inborn traits that are necessary for more than solid plodding.

Since such traits are exceedingly difficult to change, a job-training program to become a "better person" sounds hopeless. Our research, however, has revealed a basic flaw in this reasoning. None of the above explanations for the difference between stars and middle performers stands up to empirical testing. Based on a wide range of cognitive and social measures, from standard tests for IQ to personality inventories, there's little meaningful difference in the innate abilities of star performers and average workers.

Rather, the real differences turn up in the strategic ways top performers do their jobs. While it's impossible to get in the door of Bell Labs without technical competence and high-level reasoning abilities, these cognitive skills don't guarantee success. But specific work strategies like taking initiative and networking make for star performance *and* are trainable. When companies promote such

Robert Kelley is an Adjunct Professor of Management at Carnegie Mellon University. **Janet Caplan** is an independent management consultant in Minneapolis, MN.

strategies systematically, individual professionals not only improve but also pass along the benefits to their colleagues and the company's bottom line.

DEMYSTIFYING HIGH PRODUCTIVITY

Let's consider the first major hurdle to a training program: defining productivity for a particular job. Some software companies, for example, use lines of computer code as their productivity measure, based on the assumption that good programmers generate more lines than others. This measure, however, ignores the fact that 4 lines of elegant computer code are better than 100 lines that accomplish the same objective. In addition, few professionals do the exact same job. Two computer-code testers may have the same job title, yet one may test 50 small computer programs in a single day, while the other spends as long as 3 weeks testing 1 large program.

Peter Drucker has discussed the apparently impossible task of understanding the productivity of knowledge professionals. In particular, he has pointed to the difficulties of analyzing the process that produces high-quality results in knowledge work. The best we can do, Drucker says, is ask, "What works?" Implicit in this question is the reality that the work of knowledge professionals happens inside their heads. And managers can't directly observe, let alone accurately evaluate, these mental processes or strategies.

That leaves asking workers to disclose their mental secrets. This is no simple task, however. First of all, many people have a hard time describing what goes on in their minds when they work or even determining whether or not they've been productive (see the box "How Do Professionals Define Their Productivity?"). Second, researchers can fall for nonsensical productivity recipes if their methods aren't sufficiently focused.

In the early 1980s, for example, much ado was made about peak performance. Many researchers interviewed Olympic champions, who dutifully recounted this typical daily regimen: they woke at dawn, stretched out, ate their Wheaties, spent an hour visualizing their success, and practiced their sport for three hours. After enough champions had described the same regimen, a spate of books hit the market on how to become a peak performer in sports, sales, or management.

But what about the Olympic contenders who didn't win? Chances are these athletes also woke at dawn, stretched out, ate their Wheaties, spent an hour visualizing success, and practiced their sport for three hours. In other words, it's not enough to ask the stars what works; researchers must compare the regimens of star performers to those of the also-rans and then target the differences.

In fact, no one has come up with a generally accepted definition of productivity in any knowledge profession, let alone across these professions. In our research at Bell Labs, rather than grappling with a broad definition of productivity, we focused on the practical ways managers can distinguish stars from middle performers. And when it came time to evaluate the training program, we asked managers to tell us what practical changes, such as spotting and fixing problems or pleasing customers, they expected to observe in the engineers whose performances had improved.

When we began our study in 1986, the Bell Labs Switching Systems Business Unit (SSBU) was feeling the pinch of competition from companies like Northern Telecom. Before the breakup of AT&T's Bell system, the Labs felt as much like a university research center as a corporate entity. Top-flight engineers went there for a combination of reasons: the opportunity to work on leading-edge telecommunications projects, the outstanding reputation of Bell Labs as an applied R&D think tank, and the job security that came with working for AT&T.

But Bell Labs executives watched market share drop sharply during the 1980s, and these managers soon realized that recruiting the best and brightest computer engineers and scientists wasn't enough. As it turned out, academic talent was not a good predictor of on-the-job productivity. As in other companies, applied R&D at Bell Labs now means fast, cost-effective product cycles. And job security is tied to value-added contribution, not scholarly performance.

How Do Professionals Define Their Productivity?

While a company's performance-rating system may not identify all high achievers, don't expect individual professionals to have a clearer or more consistent view of their own work. In 1990, for instance, we asked 40 engineers at Bell Labs and 25 engineers at another high-tech telecommunications company to evaluate themselves. Using a short e-mail survey, we queried the engineers every day for two weeks. The four questions in the survey were:

- How productive were you today?
- How did you measure your productivity?
- What caused you to be either productive or unproductive today?
- Did you get feedback about your productivity?

Although you might expect individual workers to rate their own productivity more positvely than their bosses would, the engineers actually turned out to be quite hard on themselves. On average, these self-doubters rated their daily productivity at a rather low 68% given the performance needs of a technical environment like Bell Labs.

One reason for such uncertainty could be unclear performance standards. For instance, the most popular method for measuring personal productivity turned out to be checking off items on a to-do list, which engineers cited 41% of the time. A gut feeling of being productive came in a distant second at 16%, and the actual amount of time spent working trailed at 14%. Only one engineer on one day cited "amount of my work making a direct contribution to the company" as a measure of personal productivity.

Indeed, our work with expert engineers indicates that many of the assumptions about what makes for high productivity in an organization are tacit. More often than not, such measures aren't explicitly or specifically enunciated by top managers. Although the engineers we surveyed preferred concrete accomplishments as personal gauges of productivity, they also complained about having a tough time deciding whether or not the tasks on their to-do lists added any value to the company. And on a day-to-day basis, managers don't seem to help with this problem. Of the 65 engineers, 44% said they received absolutely no productivity feedback from their managers during the period of the survey.

But what interfered with productivity? The engineers cited meetings, meetings, and more meetings. When they weren't in meetings, our survey respondents complained, they were being interrupted. These two factors accounted for 45% of the cited obstacles to productivity on any given day. However, while most professionals, especially top managers, would agree that meetings are the bane of their existence, we suggest there are also other reasons for reduced productivity. Organizations would do well to examine their performance expectations, clearly outlining which activities add value to the company and providing the feedback necessary to keep people on track.

Consider the actual work of an engineer at Bell Labs. The SSBU creates and develops the switches that control telephone systems around the world. These switches entail substantial computer hardware and millions of lines of software code. SSBU engineers spend considerable time simply maintaining the lines of code that run a switch. The jobs of SSBU engineers also call for creativity. For example, engineers write software programs for switching systems in response to customer requests for services like caller ID, which displays an incoming caller's name and phone number on a telephone set before the call is answered.

These engineers usually work in teams because the scale of the work is beyond any one person. It can take anywhere from 5 engineers to 150 to complete a software application, in 6 months or as long as 2 years. According to one experienced engineer, "No one engineer can understand the entire switch or have all the knowledge needed to do his or her job." Individual productivity at the SSBU, then, depends on the ability to channel one's expertise, creativity, and insight into work with other professionals—a formidable job assignment, even for the smartest knowledge worker.

TARGETING THE RIGHT STRATEGIES

To specify how a star engineer does his or her work, we developed an expert model, but we turned the usual approach on its head. Expert models were invented by artificial intelligence researchers in an effort to get computers to mimic the skills of human beings. Researchers have created such models by interviewing expert welders, for example, and asking them to explain in concrete detail how they go about their job. Researchers then used the interview data to construct a computer program that reproduced the experts' skills in the form of a robotic welder. But based on our interviews with the SSBU experts—in this case, star performers in software development—the expert model for engineers was one people could use, not computers.

First we had to identify the experts. Initially, we relied on managers to point out star performers. We looked for those who had received the highest performance ratings and merit awards. We also asked managers, "If you were starting a new company and could hire only ten knowledge professionals from your present staff, whom would you hire?" There was surprising consensus among managers about who these software engineers were.

Yet once we started interviewing the engineers themselves, the picture grew murkier. As we discovered, managers sometimes overlook important components of star performance, like who originates an idea and who helps colleagues the most when it comes to solving critical problems. Being closer to the action, however, knowledge professionals certainly consider these skills when rating their peers.

In addition, the engineers believed that the Bell Labs performance evaluation system was flawed because it turned up too many false negatives, that is, people who were outstanding performers but for reasons of work style or modesty received low ratings from managers. (Later on, we found only a 50% agreement between peer and manager ratings.) The experts selected for our study, therefore, had to be highly valued performers in *both* their managers' and peers' eyes.

What Star Performers Said. We asked each of the expert engineers to define productivity, how they knew when they were productive, and what exactly it was that they did to be productive. For example, one expert told us that networking was crucial to getting his job done. We then asked him how he went about networking with other experts. He explained that networking was a barter system in which an engineer needed to earn his or her own way. From his perspective, that meant first becoming a technical expert in a particularly sought-after area, then letting people know of your expertise, then making yourself available to others. Once an engineer has developed his or her bargaining chips, it's possible to gain access to the rest of this knowledge network. But once in the network, you have to maintain a balance of trade to stay in.

After we met with the experts in groups, they came to a consensus about the two categories—cognitive skills and work strategies—that influence high productivity. Since all Bell Labs engineers score at the top in IQ tests, cognitive abilities neither guarantee success nor differentiate stars from middle performers. However, the Bell engineers identified nine work strategies that do make a difference: taking initiative, networking, self-management, teamwork effectiveness, leadership, followership, perspective, show-and-tell, and organizational savvy (see Figure 2.1, "An Expert Model for Engineers").

Moreover, the engineers ranked the work strategies in order of importance. Taking initiative is the core strategy in this expert model. An engineer must be able to take initiative upon arriving at Bell Labs or develop the ability for doing so soon after. In a competitive technical environment, it's just not possible to survive otherwise. Yet taking initiative is also one of the most elusive strategies and therefore difficult to quantify. As one engineer explained, "I go into my supervisor's office for a performance evaluation, and she tells me that I should take more initiative. I say to myself that I'm already taking initiative, so what exactly is it that she wants me to do?"

Clearly, any training program for improving the productivity of professionals must first target taking initiative. During our discussions with the Bell Labs experts, one proposed creating practical checklists to detail each work strategy. The "Checklist for Taking Initiative" outlines a sample of specific actions and behaviors that define this core strategy.

The second layer of the expert model includes work strategies like networking and self-management. Although the Bell Labs engineers thought these were critical for high productivity, they acknowledged that they could be acquired at a slower pace. The third and final layer contains show-and-tell and organizational savvy, which these star performers considered "icing on the cake." Professionals who develop these work strategies have a leg up for managerial promotions, but giving riveting presentations and playing the correct political games aren't essential to getting the technical job done.

What Middle Performers Said. At the same time that we were defining the expert model with star engineers, we were also interviewing middle performers at the Bell Labs SSBU. When we first compared the interviews, it appeared that stars and

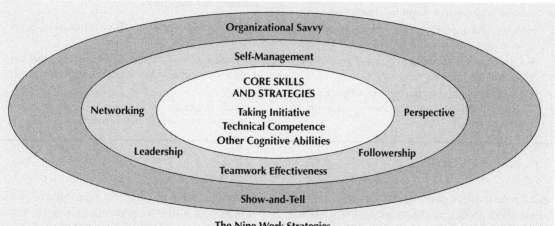

The Nine Work Strategies

Taking initiative: accepting responsibility above and beyond your stated job, volunteering for additional activities, and promoting new ideas.

Networking: getting direct and immediate access to coworkers with technical expertise and sharing your own knowledge with those who need it.

Self-management: regulating your own work commitments, time, performance level, and career growth.

Teamwork effectiveness: assuming joint responsibility for work activities, coordinating efforts, and accomplishing shared goals with coworkers.

Leadership: formulating, stating, and building consensus on common goals and working to accomplish them.

Followership: helping the leader accomplish the organization's goals and thinking for yourself rather than relying solely on managerial direction.

Perspective: seeing your job in its larger context and taking on other viewpoints like those of the customer, manager, and work team.

Show-and-tell: presenting your ideas persuasively in written or oral form.

Organizational savvy: navigating the competing interests in an organization, be they individual or group, to promote cooperation, address conflicts, and get things done.

Figure 2.1. An expert model for engineers.

Checklist for Taking Initiative

Going Beyond the Job

- I make the most of my present assignment.
- I do more than I am asked to do.
- I look for places where I might spot problems and fix them.
- I fix bugs that I notice in programs or at least tell someone about them.
- I look for opportunities to do extra work to help the project move along more quickly.

New Ideas and Follow-Through

- I try to do some original work.
- I look for places where something that's already done might be done better.
- I have ideas about new features and other technical projects that might be developed.
- When I have an idea, I try to make it work and let people know about it.
- I try to document what my idea is and why it's a good idea.
- I think about and try to document how my idea would save the company money or bring in new business.

- I seek advice from people who have been successful in promoting ideas.
- I construct a plan for selling my idea to people in the company.

Dealing Constructively with Criticism

- I tell colleagues about my ideas to get their reactions and criticisms.
- I use their comments and criticisms to make my ideas better.
- I consult the sources of criticisms to help find solutions.
- I continue to revise my ideas to incorporate my colleagues' concerns.

Planning for the Future

- I spend time planning what I'd like to work on next.
- I look for other interesting projects to work on when my present work gets close to the finish line.
- I talk to people to find out what projects are coming up and will need people.

middle performers gave similar answers. For example, both groups identified taking initiative as a useful work strategy. But closer inspection revealed that the answers of stars and average engineers differed in two critical ways: how they ranked the strategies in importance and how they described them.

To begin with, middle performers inverted the expert model's ranking of the nine work strategies. According to these engineers, show-and-tell and organizational savvy were the core strategies and were largely responsible for high performance ratings from managers. It's easy to understand why these nonexpert engineers came to this conclusion. One of the few times senior managers see knowledge professionals in action is when they give pre-

sentations. And in some cases, mediocre professionals with a flair for showmanship are rewarded by top management. But in general, executives use such public presentations to infer the skills and strategies that produce good technical work. Picking up on only the superficial aspect, the Bell Labs middle performers were overly focused on impression management rather than critical strategies like networking.

As for describing the work strategies, the differences between stars and middle performers were even more striking. One middle performer at the SSBU, for instance, told us of gathering and organizing source materials, including documents and software tools, for a project he was beginning with his group. Another described writing a memo to his supervisor about a software bug. Both engineers believed they showed a great deal of initiative in taking it upon themselves to do this work.

Yet when we described these examples to the Bell Labs experts, they were critical. They thought these engineers were barely doing their jobs, let alone taking initiative. For example, one expert explained that by the time a software bug is documented, it is often impossible for the software developers to re-create the problem in order to fix it. For the experts, fixing a bug yourself or preparing for a project is what's expected of you in your job. Real initiative means going above and beyond the call of duty. In addition, such actions must help other people besides yourself and involve taking some risks.

Discussions about networking surfaced equally revealing differences, since both stars and middle performers said networks of knowledgeable people are critical for highly productive technical work. For example, a middle performer at Bell Labs talked about being stumped by a technical problem. He painstakingly called various technical gurus and then waited, wasting valuable time while calls went unreturned and e-mail messages unanswered. Star performers, however, rarely face such situations because they do the work of building reliable networks before they actually need them. When they call someone for advice, stars almost always get a faster answer.

In fact, we found similar differences between stars and middle performers in their definitions of all the work strategies. In particular, some middle-performing engineers clearly lacked perspective. One engineer described the many hours he had spent mastering a software tool for organizing files, which ended up delaying the delivery of a customer's product. From an expert's perspective, of course, the customer comes first. Although star performers agreed that upgrading their knowledge of current software tools is useful, they also emphasized the need to set priorities. And these experts stressed the need to "shift gears" between the narrow focus required for certain tasks and a broad view of how their project may fit into a larger one.

TRAINING KNOWLEDGE PROFESSIONALS

Not surprisingly, knowledge workers don't like off-the-shelf productivity training programs. Our discussions with engineers at Bell Labs and elsewhere show that these people like to make their own choices. Such professionals readily admit that they could do their jobs better, but they're also wary, as at least one Bell Labs participant put it, of "becoming a clone."

Knowledge professionals value the real experts on productivity in their laboratory or law firm, not trainers who breeze in, teach a day-long workshop, and then breeze out. Therefore, once the Bell Labs SSBU training program got underway, respected engineers led the training sessions. In fact, the process of developing the expert model became the foundation for the training program itself. The Bell Labs experts we interviewed reported increases in their own productivity because they had picked up valuable tips from listening to their star colleagues.

The expert model also has a clear advantage over a system like mentoring. While many professionals are experts about their own productivity, no single star performer knows everything. Unlike a mentoring program in which one senior professional advises a junior staff member or a group of new workers, an expert model pools the strategies of many stars. And a training program based on such a model makes those strategies explicit.

Developing the Curriculum. In the spring of 1989, top managers at Bell Labs agreed to a pilot training program for the SSBU. Sixteen engineers chosen by managers participated in two groups that met once a week over the course of ten weeks. These groups included a mix of stars and middle performers but were weighted more heavily with stars, since we wanted them to become trainers later. After the initial pilot sessions, we reversed the ratio of stars to middle performers in the groups.

The training program's primary task was to make the critical work strategies concrete, accessible, and learnable. Each week of the pilot program focused on one of the nine strategies, and the last week was used for a wrap-up. But despite this neat schedule, the first engineers to participate revised the curriculum as they went along, testing ideas out in real time, keeping what worked, and discarding what didn't.

For example, the engineers developed a teamwork exercise based on work-related issues at the SSBU. The group formed a mock task force to focus on a pressing company issue like whether or not the software development process should be standardized. Participants decided to spend part of each remaining session in this mock task force. A few weeks into the exercise, however, one engineer complained that while this was more realistic than most training activities, it still had no real impact on her day-to-day work or that of the company. Within a week, top managers at Bell Labs told the pilot group that they would read and respond to a written report from this no longer "mock" task force. Suddenly, this particular teamwork exercise became more compelling than anything the group had done before.

By the end of the pilot program, the 16 engineers had created a detailed curriculum for each of the 9 work strategies. Each piece of that curriculum included frank discussion, work-related exercises, ratings on the work strategy checklists, and homework that required participants to practice while they learned. As the box "A Day in the Life of Productivity Trainers" indicates, a Bell Labs workshop session involved not only specific case studies and exercises but also active disagreement among all participants.

Eventually, the Bell Labs training program was streamlined to six weeks, with the sessions facilitated by expert engineers who had previously participated. Yet continually reshaping the curriculum in response to critical events on the job is still the current program's most important feature.

For example, during one of the later sessions, top management issued a memo on company quality initiatives. Engineers at the SSBU thought the memo blamed them for poor quality. So participants in the training program decided to respond directly to the memo as part of that session's work.

Up to that point, it was quite unusual for engineers to take such a step because most believed that top managers would not appreciate, let alone respond to, a direct approach. But as it turned out, the engineers got a quick and constructive response. Top managers sent e-mail messages and talked to some of the participants about their concerns.

In addition, if professionals try to analyze their own productivity, they need a clear idea of how others, especially managers, view their performance. Bell Labs trainees received feedback from peers, managers, customers, and fellow participants. They also rated themselves on the work strategy checklists and filled out several other self-evaluations. With such a range of feedback, most participating engineers knew what their strengths were and where they most needed to improve by the end of the program.

Measuring the Bottom Line. Since 1989, more than 600 of the 5,000 engineers at the Bell Labs SSBU have participated in what is now called the Productivity Enhancement Group (PEG). Since these engineers were scattered across many projects and departments, it's difficult to demonstrate the program's effectiveness through measures like fewer person hours spent on a particular project. In their self-evaluations, however, participants reported a 10% increase in productivity immediately after the sessions ended, which grew to 20% after 6 months and 25% after a full year. This steady upward curve is the opposite of what follows most training programs. Typically, effectiveness is greatest on the last day of the program and falls to zero after a year.

But even if PEG participants reported substantial productivity increases, this doesn't prove that the performance of these engineers actually changed. The corporate goal for PEG was not, for example, taking initiative for initiative's sake but adding value to the company. Therefore, we met with managers again, asking them, "What would you look for as indices of increased productivity in a person who worked for you?" Figure 2.2, "What Managers Thought: The Real Test of Productivity" shows that the productivity of PEG participants improved twice as much as nonparticipants over an eight-month pe-

---

A Day in the Life of Productivity Trainers

7:30 A.M. The engineer-facilitators, who train in pairs, meet their partners to get ready for a class on taking initiative. They review the agenda, organize notes, decide who's going to do what, and worry.

8:00 Class begins with a discussion of taking initiative, based on a case study that was part of the homework assignment. The case describes an engineer who volunteered to organize a technical project assigned to her department. Discussion is heated. Did organizing the project go beyond what was expected of this engineer? Did it involve taking any risks? An argument breaks out about whether the definition of initiative should include risk taking. One participant, a recognized technical guru, describes how he recently took initiative by letting his department head know that a technical project was floundering. He ruffled department feathers but succeeded because of how he strategically documented ideas, built a network of allies, and took calculated risks.

9:05 Facilitators sit back and observe the 30-minute mock task force meeting. The task force agrees that today's goal is to concentrate on soliciting input from all members, especially the quiet ones. The task force is working on recommendations for revamping the recognition and reward system at the Labs. There's disagreement over whether or not performance evaluations should be scrapped, and the task force gets bogged down. Members use a round-robin approach to solicit suggestions. Deadlock is broken by addressing the deeper issue of how to define the bottom-line goals of reward systems. Facilitators give feedback about the meeting. Task force members exchange feedback.

10:15 Coffee break. Facilitators huddle briefly to make changes in the next portion of the program based on participants' reactions to the first part.

10:30 The group discusses the checklist for taking initiative, another part of the homework assignment. Almost all the participants rated themselves on the list, but some are clearly overwhelmed: for example, one says, "I never knew I could have been doing these things. What an eye-opener."

11:45 Facilitators summarize the session and go over the new homework assignment. Each participant is expected to use his or her networks to find an answer to the same tough technical question. Next week, participants will compare their networking strategies by discussing how long it took them to get the answer, whether they received the right answer, and how many people they had to contact.

Noon Lunch with ten facilitators from other groups. Some sample comments and questions:

- "What did you do when that initiative discussion bogged down?"
- "We had an engineer who hated the checklists!"
- "I think we had the group from hell. Everybody was quiet except for this one engineer who thought he had the answer to everything. He wouldn't let go of this one idea about initiative, so we took a poll of the other group members. It turned out they disagreed with him, and that diffused it."
- "After this group, I'll never be afraid to run a meeting again."
- "What changes should be made to the program for the groups that will meet this afternoon?"

---

riod. According to the SSBU managers, these engineers improved in seven areas, including spotting and fixing problems, getting work done on time with high quality, pleasing customers, and working well with other departments. And star performers were not alone in benefiting from PEG training. Star and middle performers improved at similar rates.

We also compared our manager surveys with the company's standard performance ratings, which are routinely collected at Bell Labs and are the basis for salary adjustments and promotions. We looked at these ratings before participants began PEG and then eight months after they had finished the program. Interestingly enough, the performance ratings of PEG participants improved at twice the rate of nonparticipants, mirroring the results of our manager surveys.

In addition, PEG had an especially strong impact on women and minority engineers (see the Box "Women and Minorities at Bell Labs"). In traditional organizations, these groups are often excluded from the expert loop. But creating an expert model that demystifies certain productivity secrets, particularly the importance of key work strategies and how to acquire them, makes the loop explicit and accessible to all.

Ultimately, of course, such productivity increases for individual professionals fall to the company's bottom line. If the total compensation package for a knowledge professional is about $62,500 (salary plus fringe benefits), the ROI is $625 each year for every 1% productivity increase. Thus a

10% increase yields $6,250 for each participant, while a 25% productivity increase would pay back $15,625, and so on.

But these ROI numbers don't include the more indirect productivity benefits. PEG participants improved dramatically in the ways they assisted colleagues. These engineers also built stronger ties to customers. While such positive changes are hard to measure, they are essential to a highly productive work team.

MAKING A COMMITMENT TO STAR PERFORMANCE

Some managers still wonder whether high productivity is due only to individual work style and motivation. In many cases, they're searching for a jus-

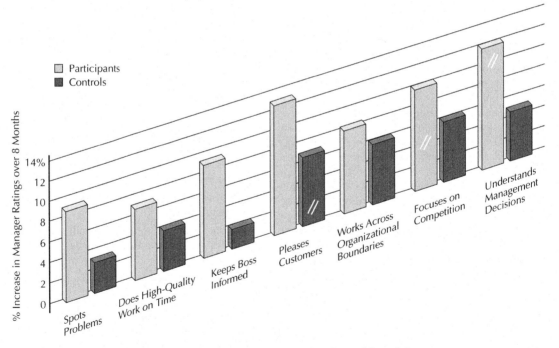

This chart is based on a survey that was used to evaluate the effects of the training program. The study compared 300 participants with 300 nonparticipants (controls). Managers of each group completed the survey before the training sessions began and then again eight months later.

Figure 2.2.   What managers thought, the real test of productivity.

tification for their own style. "Clean desk" people want to believe that being organized leads to higher productivity. "Sloppy desk" managers, however, view their style as evidence of the creativity that translates into high performance.

But we've found no such relationships. Rather, training programs like PEG can help professionals discover the strengths and weaknesses of their individual work styles. Managers gain little by foisting a time-management system, complete with scheduling book and to-do priority tabs, on someone who prefers to keep such information in his or her head. Helping this worker to develop a better strategy for storing information mentally and setting priorities makes more sense.

Motivation, however, is another matter. At Bell Labs, most of the engineers we worked with were eager for productivity tips. They knew their success was tied to high performance, and they saw how easily the work piled up. Since the PEG sessions focused on developing individual work strategies, motivated professionals benefited by improving their own productivity.

But when workers aren't motivated to improve, a program like PEG is of little help. In surveys done outside of Bell Labs, we've found that about one-third of knowledge workers don't feel tied to their company's destiny, nor do they feel that their productivity and good ideas are sufficiently rewarded. For example, teamwork is often touted by corporate headquarters as critical to both individual and company success; however, an employee's ability to work with others often has little to do with annual performance ratings or rewards. Many professionals know this and are right to resent it.

Yet such resentment can lead to serious drops in productivity. A company with unproductive and actively resentful professionals, then, may need to address additional organizational issues, such as revising the reward system or treating its professionals as individuals with individual needs.

Clearly, it's not possible to turn every average worker into a star. Despite PEG or any other training program, there are differences among professionals, much as there are among athletes who follow the same training regimen. It's probably not

Women and Minorities at Bell Labs

At large companies like AT&T, which employ many women and minority professionals, there are often separate support groups for women and each minority. Such groups do foster the sharing of problems and success stories; yet they may also limit the contact of women and minorities with a wider range of company experts.

The effectiveness of an expert-model approach in no way dismisses the fact that women and minorities have traditionally been excluded from productivity secrets in engineering or other high-tech environments. But since the Productivity Enhancement Group (PEG) program focuses on improving individual productivity, it can sidestep some of these organizational barriers, particularly stereotypes about the productivity and work styles of women and minorities. In fact, teaching professionals expert work strategies may be the most pragmatic form of affirmative action.

As the graphs in Figure 2.3 based on the manager surveys show, the productivity of women and minority PEG participants improved at four times the rate of comparable nonparticipants. On the dimensions that most directly affected managers—"does high-quality work on time" and "keeps boss informed"—the productivity of women and minorities who didn't participate in PEG actually *decreased* during the eight-month survey period.

The fact that managers reported that some women engineers lost ground in the crucial area of keeping their bosses informed is particularly telling. Men and women may indeed communicate with their bosses differently. But in PEG, everyone learns the necessity of regular communication with managers.

even desirable or cost-effective to put all professionals through a training program, since not everyone enjoys or is compelled by intense workshop sessions.

However, the PEG participants at Bell Labs

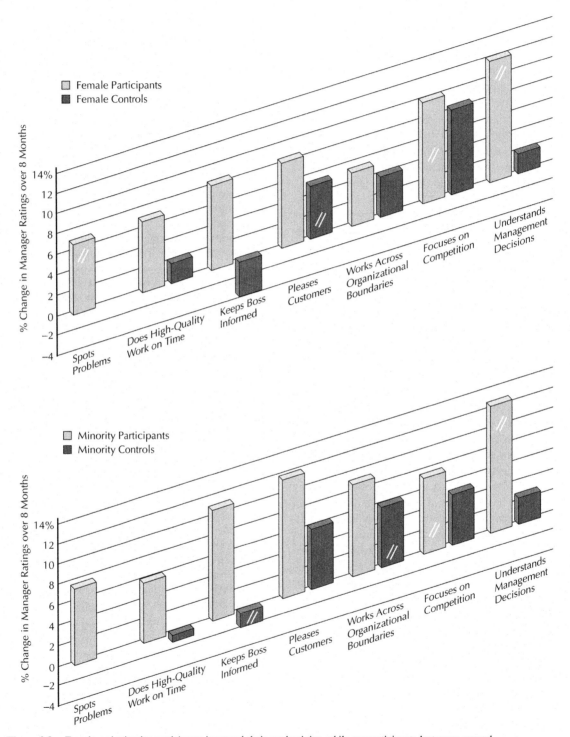

Figure 2.3. Female and minority participants increased their productivity, while nonparticipants lost some ground.

have not only improved their own productivity but also positively affected the productivity of nonparticipating coworkers. The checklists and other materials derived from the expert model have been photocopied, passed around, and incorporated into the everyday functioning of the SSBU. Once such informal dissemination happens on a wide scale, formal training is no longer necessary.

Developing an expert model can provide a powerful platform for leveraging intellectual capital in many professions. The PEG program at the BELL Labs SSBU, however, isn't a blueprint for another company, even a research laboratory with a similar environment. The mix of work strategies may differ from profession to profession; a marketing department, for example, may find that show-and-tell is a core strategy for star performance, along with taking initiative. Yet, regardless of profession, top managers need to focus on people when they address productivity improvement. In the new knowledge economy, it's the performance of knowledge professionals, not just complex technologies, that will make or break a business.

3

Organizational Socialization and the Reduction of Uncertainty

RALPH KATZ

For many professionals, their first year of organizational employment is a very frustrating experience, full of stress, anxiety, and disillusionment. Their struggles to become accepted by others and to function as "true" contributing members within their new work settings are sufficiently dissatisfying that many switch companies within the first couple of years. In fact, it has been estimated that more than 50 percent of college graduates entering industrial corporations leave their initial firms within the first three or four years. Similar high rates of turnover were found in my own longitudinal studies of young technical professionals. Many of them even decided to abandon engineering or science as a career during this period of time. The end result, of course, is a very high rate of organizational turnover among new groups of professional employees—a wasteful outcome, not only economically but also in terms of lost promise and potential, especially if the turnover takes place among the most talented individuals.

For other professionals, however, the first years of work are seen as a marvelously satisfying and challenging experience, full of excitement, achievement, and personal development. Not only are these individuals more likely to remain with their organizations, but it is also likely that they will continue to perform effectively, developing strong commitments to both their project and their organizational settings.

Given this range of difference in the individual experiences of organizational newcomers, what is it that takes place during one's initial work years

that affects the amount of stress one feels and determines one set of outcomes over the other? One explanation lies in the perceptual accuracy with which individuals enter their new professional and organizational work environments. Generally speaking, the more that individuals begin their jobs with unrealistic views and expectations, the more they encounter "reality shock" as they confront the true demands of their everyday task activities. On the other hand, individuals who assume their new organizational positions with a more realistic understanding and perspective feel less surprised and disenchanted, because they possess, at least initially, more compatible relationships and interactions with supervisors, peers, and work-related expectations and pressures.

Based on this argument, if newcomers were given more accurate information about their prospective jobs, they would be able to undertake their new responsibilities with far less discomfort and frustration. They would be, in a sense, better innoculated against the idealistic hopes and expectations that so many young employees form about their upcoming organizational involvements. More "realistic previews" during recruiting, then, can play an important role in preventing disappointments from emerging and disillusioning professional newcomers as they begin to carry out their daily job assignments.

Although much can be done to educate and prepare new hires to meet the demands of their new world of work, one should also realize that the concerns, reactions, and accomplishments of new em-

Reprinted with permission from the author.
Ralph Katz is Professor of Management at Northeastern University's College of Business and Research Associate at MIT's Sloan School of Management.

ployees are eventually shaped by the structure of events and interactions taking place throughout their entire socialization experience within the organization (Katz, 1980; Schein, 1978). A "sink or swim" type of socialization process, for example, evokes considerably more tension and stress than a socialization process that is highly supportive and well structured, even though the actual task demands may be equivalent. If our ultimate objective is to learn how to provide new entrants with a better "joining-up" process—one that is not only less stressful but also more meaningful and personally developmental—then we need to understand more fully how individual needs and concerns should be met throughout this important introductory period of organizational careers.

CONTENT OF SOCIALIZATION

During the socialization phase of a new job, individuals are very uncertain insofar as organizationally relevant attitudes, behaviors, and procedures are concerned. It is in this early job state, therefore, that professionals learn not only the specific technical requirements of their jobs but also the socially acceptable attitudes and behaviors necessary for becoming effective organizational members. If new employees hope to direct and orient their own organizational performance in a meaningful and contributive manner, then they must develop a genuine understanding of the events and activities taking place around them. They must build a "situational perspective" within which ideas and assumptions can be tested, interpreted, and interrelated.

Creating this perceptual outlook is analogous to building a mental or cognitive map of one's organizational surrounding, including its particular cast of characters (Louis, 1980). To come to know a job situation and act within it implies that the professional hire has developed a sufficiently useful scheme for making sense out of the vast array of experiences associated with his or her participation in the new job setting. The newcomer must come to know what others in the organization are about, how they operate, and how he or she should per-

form on the job relative to these others. In time, these perceptions provide the new employee with a meaningful way of classifying events and organizing the many interrelationships that exist within the workplace.

In developing this local organizational perspective, every professional newcomer must accomplish over time at least three important tasks. They must build their own role identities within their new job contexts. They must discover how to deal with peers and certain influential authority figures, especially their boss or bosses, in addition to establishing a more informal network of information and support. And they must decipher the appropriate reward systems and situational norms of acceptable social and task-related behaviors.

1. Establishing One's Organizational Role Identity. As new hires start their organizational careers, they are faced with the problem of developing situational identities that will be viable and suitable both from their own perspectives and from the perspectives of other relevant organizational members. Whether one is aware of it or not, each new professional must find answers to the questions of "Who am I and what are my contributions going to be in this organization?"

This issue exists simply because all of us have a large repertory of possible roles and behavioral styles that can be enacted in any particular situation. The person who is viewed as influential, aggressive, and helpful in one organizational setting, for example, may be seen by others in a different situation as quiet, reserved, and uninvolved. To some extent, therefore, we can be different people in different situations, depending upon the particular sets of perceptions that come to surround and envelop us.

Newcomers are typically hired into their organizations as the result of some valued educational background or some highly specialized training program. But until they are actually working and participating in their specific job contexts, neither they nor their organizations can really be sure how they will fit into their new work environments. The socialization period, therefore, is a time of mutual discovery between the new employee and the employing organization, each learning more and more about the other. With increasing experience and organizational exposure, the new employee gradually acquires enough self-knowledge to develop a clearer image of his or her own strengths and weaknesses, assessing his or her own preferences, values, talents, and abilities. In a similar fashion, other organizational employees also develop their own perceptual views of the individual newcomer. And as these perceptions and expectations become more firmly established, they function to constrain the role behaviors that the individual is allowed to play within the overall work setting. Thus, it is the intersection among many sets of perceptions that eventually defines the specific role or situational identity of each individual employee. During socialization, then, every newcomer is testing his own self-image against the views and reactions of other organizational employees. The greater the fit between these developing perspectives, the less stress experienced by the new individual since the paths through which he is expected to contribute become increasingly well defined and mutually agreeable.

It is very possible, of course, that not everyone in the work environment will have the same reaction to or will develop the same impression of the new employee. Some may come to value and respect his or her particular skills and abilities; others may view such areas of expertise as unnecessary and irrelevant. Coworkers or project colleagues might also develop a picture of the new employee that is vastly different from the one constructed by his or her supervisor, the one formulated by his or her subordinates, or even the one held by those professionals, customers, or clients outside the organization. New professionals should realize that they are building different identities with different parts of the organization, depending on the kinds of interaction and contributions that take place during their socialization period. If there is no time for socialization to take place, then the identity that an individual will have will be based on some preconceived stereotype. Without socialization, all engineers are alike, all marketing people are alike, all Americans or Japan-

ese are alike, and so on. It is the relationships and identities that are built during the socialization process that break down and overcome such indiscriminate stereotypes.

In building these role identities, socialization must take place along both dimensions of interpersonal and task-related activities. The newcomer has a strong need to obtain answers to a number of important underlying interpersonal questions; while only some of these may be conscious, all need to be answered as quickly as possible. Having entered a strange and unfamiliar social arena, newcomers are strongly concerned with inclusion, that is, becoming a necessary and significant part of the overall organization. According to Schein (1978), Graen (1976), and Katz (1978), to become accepted and recognized as an important contributing member within one's work setting is one of the major obstacles with which new employees must struggle and which they must eventually overcome. To what extent, then, will they be considered worthwhile? Will they be liked, supported, and appreciated? Will they be kept informed, included, and be given opportunities to make meaningful contributions? These are some of the key interpersonal issues preoccupying employees as their new situational identities become progressively established.

From the technical or task-performance point of view, the newcomer must also figure out whether or not he or she can do the job effectively. To prove or test themselves on the job is another very important concern of new employees. Having spent most of their lives in an educational environment, which has kept them at arm's length from the "real," industrial world, young employees need to discover just what sort of persons they are and of what they are really capable. They need to see how they function on actual work tasks where the outcomes make a significant difference. For this reason the testing of one's skills and abilities is of critical importance.

An outstanding academic record may be indicative of a professional's willingness to work hard and accept responsibility in a well-structured environment, one in which the examinations, degree requirements, problem information, and schedules are laid out quite clearly. These same skills and abilities, however, may not be sufficient in a less certain and more complex work environment, one in which the professional must learn to work with others in order to define or clarify problems and their corresponding specifications, information needs, and schedule requirements. Much of a technologist's effort is often spent defining the problem in terms of the organization's capabilities, operations, constraints, interests, history of successes and failures, and so forth. This is not to say that educational achievement is not important but rather that there is not a direct mapping between how problem sets are presented and worked on in a classroom setting and how problems are formulated and solved in an organization.

Furthermore, new employees are not only concerned with using their present knowledge, but they also want opportunities that enable them to continue to learn and grow—to extend their talents and areas of expertise. One of the inevitable results of prolonged professional education is the expectation that one should continue to self-develop and be given the opportunities and freedom to do so. In short, what is really important to young employees during socialization is the opportunity to clearly demonstrate their work competence and future promise by being meaningfully utilized in some critical aspect of the organization's activities.

2. Learning to Deal and Network with One's Boss and Other Employees. A first boss plays a disproportionate role in a young person's career. According to the results of many studies, one of the most critical factors influencing the professional and organizational career success of young employees falls within the mentoring domain of one's immediate supervisor. Despite the obvious importance of this supervisor-subordinate relationship, very few newcomers are entirely satisfied with their initial boss. One of the major tasks of socialization, then, is to learn how to relate and get along with this individual. He or she may be too Machiavellian, too unstructured, too busy, too fickle, too competent, or even relatively incompe-

tent in certain technical areas. Nevertheless, young employees must learn to cope with the reality of being dependent upon their particular supervisors.

The newcomer's immediate boss also plays a critical role in sponsoring or linking the new professional to the rest of the organization, in making sure his or her work priorities are consistent with organizational needs, in securing adequate resources, and in providing the additional information and expertise that are necessary to perform well. While some bosses do an excellent job of caring for their new subordinates in these ways, a more reasonable expectation is that only a modest amount of assistance and clarity will be forthcoming. In most instances, therefore, professionals have to assume primary responsibility for their own careers and development, seeking the kind of help, information, and sponsorship they need in order to complete their work activities more effectively. Simply wishing for supervisors to provide the kind of mentorship one would like to have, or simply waiting for stronger sponsorship to "walk through the door," will not work in today's more demanding and busy world.

This is not always an easy undertaking. It is often very stressful simply because of the high "psychological costs" that are involved in seeking help from supervisors who are in evaluative and more powerful positions. As summarized by Lee (1994), much research has shown that individuals tend to use information sources that have the least psychological risk instead of using the most effective sources of information. However, as in any negotiation or conflict situation, it becomes somewhat easier to approach the "other party" as one generates more meaningful information about those individuals. In a similar fashion it should become easier and less stressful for newcomers to deal with their boss as they acquire a more comprehensive picture of that individual. The more new employees gain insight into the characteristics and perspectives of their specific supervisor, the easier it will be for them to interact with him or her. Such insight typically requires a very good understanding of supervisors' goals, expectations, and work-related values; the personal and task-related pressures and demands that confront them; their areas of managerial strength and weakness; their preferences for different work styles, habits, and so on. In general, then, new employees will become less anxious and will have added control and predictability in dealing with their bosses as they generate increasingly more useful information about them.

In any interpersonal situation, individuals are also more likely to get along and work well together when they are more similar to each other and more compatible in their goals, values, and priority systems. Communication and interaction are always facilitated when individuals have common frames of reference and shared experiences. Accordingly, one can speculate that organizational socialization will be smoother for those new employees who have the most in common with their immediate supervisors. Lindholm (1983) for example, clearly demonstrated that the single most important determinant of a supervisor-subordinate mentoring relationship resided in the nature of their interpersonal relationship and attraction and not in the nature of their task association or performance.

It is also likely that the very idea of having a boss is inherently uncomfortable to young professionals who have recently left the autonomy of university student life. New employees, as a result, are likely to experience considerable conflict as they struggle to balance their desire for professional independence with their more immediate sense of dependence upon their new manager(s) for the definition of their work, their information, their resources, their rewards and promotions, and so forth. This conflict may be even more pronounced for the most creative young professionals who often have low tolerances for formal authority, structure, and procedures. Nonetheless, new hires must first clarify and then learn how to relate to the demands and expectations of their supervisors—how to keep them informed and how to seek their support and approval. At the same time, they must also begin to display an ability to function on their own, to take initiative, to define problems accurately by themselves, and to uncover relevant sources of new and useful information. One of the major accom-

plishments of socialization, then, in the ability to cope with the creative tension that stems from being dependent, on one hand, yet demonstrating one's independence on the other.

This trade-off between autonomy and control is often one of the major sources of tension between young professionals and their employing work organizations, according to the personal interviews conducted by Bailyn (1982). Quite often, organizations try to create an atmosphere for their young professionals that is very conducive to creative work, one in which professionals are given as much autonomy as possible in choosing problems. This high level of independence, however, is very frustrating to the relatively new professional. In the rapidly changing world of technology, it is not clear to the new employee just what problems or projects are most relevant to the organization's overall goals and objectives. What the young professional really wants is to be placed in a well-defined project that is central to the organization's mission. Having been given this kind of assignment, he or she then expects to be given the necessary resources, support, and decision-making discretion for carrying out the assignment. Too often, however, the opposite seems to take place. After giving the young professional the autonomous mandate to "be creative," the organization then places a great deal of control on his everyday problem-solving activities. In short, what becomes stressful to young professionals in the early years of work is to be given a very high level of freedom to choose which problems to work on only to be told *how* to work on these problems. In the industrial world of work, young professionals are more likely to welcome the reverse situation; otherwise, they would have remained in university-type settings.

In addition to gaining the acceptance of their boss, newcomers must also learn to deal with other members of the hierarchy and with other peer group members. For those entering with a clear group assignment, the only problem is how to mesh their own needs and abilities with the requirements of the group. For others, however, the problem is to locate the appropriate peer or reference groups with which to align themselves. Much of what goes on in an organization occurs through informal channels and associations that have evolved over time. Thus, one comes to understand and appreciate the political aspects of different reporting relationships and organizational undertakings primarily through the individual and group contacts one has made.

The building of relationships within the organization is also important, simply because they help us form the contacts and connections through which we are able to discover or gather key pieces of key information, make timely and important decisions, and implement project activities and decisions successfully. Very few organizations are ruled through omnipotent hierarchies, and very few function as pure democracies where all members have equal votes and the majority point of view dominates. Most organizations require the skillful building of interpersonal and political relationships, both formal and informal, in order to contribute effectively or to get new things done in new ways.

All too often, it takes much too long to finally convince new technical professionals that the organization is composed of many people with whom they must cultivate a work-related and/or personal relationship. They must build the communication network that not only keeps them informed but also allows them to draw upon the most useful knowledge and information embedded within the organization. Being relatively inexperienced, the new professional needs guidance and learning from more senior colleagues in order to carry out work activities in a more effective and timely manner. It should not be surprising, therefore, that studies of young technologists have consistently found that those professionals with the highest ratings of performance and contribution also have significantly broader information contacts within the organization than their technical counterparts (Allen, 1984). Unfortunately, what is surprising is just how hesitant and difficult it is for many new engineers to seek advice and information from other experienced professionals, especially those who are outside of their immediate work or project grouping. In studying the assimilation of young engineers going through their first six months of socialization,

Lee (1994) discovered that those engineers who reported having social activities with more experienced staff colleagues outside of the normal work context received the highest ratings of performance contribution in their initial appraisals. Such informal, after-work social contacts were even more strongly related to these engineers' performance ratings than their grade-point averages. Lee argues from his research findings that such informal social interaction makes it more comfortable and less threatening for the new technologist to approach and learn from her or his busy professional colleagues. It is also known that "friendlier" colleagues are more likely to provide more extensive background information and a richer context from which the new professional can learn more quickly, not only the organization's technical culture, but also its political one.

3. Deciphering Reward Systems and Situational Norms. As new employees learn to relate to other relevant individuals within the work setting, they must also unravel the customary norms of acceptable social and task related behaviors. If one truly hopes to become a viable, functioning member of the organization and pursue a long-term relationship, then one is required to learn the many attitudes and behaviors that are appropriate and expected within the new job setting. One must discover, for example, when to ask questions, offer new ideas or suggestions, push for change, take a vacation, or ask for a pay raise or promotion. One must learn which work elements or requirements are really critical, which should be given the most attention, and which can be scrimped. Newcomers must also align their own assumptions, values, and behavioral modes of conduct against parallel perspectives that are held by their peer and reference groups. Most likely, as employees are able to adopt the collectively held view of things during their socialization, they will be viewed more positively by the organization and will receive more favorable evaluations.

On the other hand, if the new employee finds it difficult to develop a situational perspective that is consistent with those that already exist within the work environment, a high level of stress is likely to result. This dilemma can come about in at least two ways. First, the new employee may strongly disagree with the collectively held view, operating under a very different set of assumptions, values, or priorities. The new assistant professor, for example, may be very excited about teaching and working with graduate students only to discover that his university colleagues value research output almost exclusively. Or there may be strong disagreement concerning the importance of different areas of research, the value of different research methodologies, or the merits of particular application areas. The specific sources of strain will certainly vary with each particular organizational and occupational setting; nevertheless, the more the individual sees oneself as having a "deviant" perspective within the workplace, the more he or she will experience stress in attempting to gain acceptance and prove oneself as a valuable, contributing member.

A second source of discomfort can occur simply when there is no collectively held view of things to guide the new employee. This situation can come about when there is little consensus within the environment, either because the individuals disagree among themselves, or because there has been insufficient time or insufficient stability to develop a collective viewpoint. In either case new employees who find themselves in these situations will experience considerably more stress because of the vast amount of uncertainty that still exists both within their own roles and within their relevant work environments. Each professional must discover what is *really* expected and what is *really* rewarded. To what extent can they trust the official formal statements of reward practices and policies? To what extent can they rely on information provided by older, more experienced employees, especially if the situation happens to be changing? New employees must determine for themselves how reward systems actually function so they can comfortably decide where to put their efforts and commitments.

Surprisingly enough, in most organizational settings, the criteria surrounding advancement and other kinds of rewards are very ambiguous espe-cially to young employees in the midst of socialization. Moreover, different employees usually see very different things as being important in getting ahead, covering the full spectrum of possibilities from pure ability and performance to pure luck and politics. Part of the reason for all this ambiguity is that organizational careers are themselves highly variable in that one can succeed in many different ways. Nevertheless, as long as newcomers remain uncertain about the relative importance of alternative outcomes, they will experience considerable tension and stress as they execute their daily activities.

Much of the uncertainty that surrounds reward systems in today's organizations can be traced to the local-cosmopolitan distinctions originally discussed many years ago by Gouldner (1957). Having recently been trained in an educational or university-type setting, new professional employees usually enter their organizations with a relatively strong professional orientation. At the same time, however, they must begin to apply this professional knowledge for the good of the organization; that is, they must develop a parallel orientation in which their professional interests and activities are matched against the current and future demands of their functioning organization. This balancing act can lead to a great deal of tension and frustration, particularly if the young employee is forced to allocate his or her time and efforts between two relatively independent sets of interests and rewards. To what extent should they pursue task activities that will be well recognized and rewarded within their profession or within their organization? Can they, in fact, do both relatively easily? All too often, young employees face job situations in which there is too little overlap between the demands, challenges, and rewards of their profession and those of their actual work environments.

By trying to build an accurate picture of the reward system, the young employee is dealing, of course, with only a part of the overall issue. He or she must also begin to question the likelihood of achieving certain results and desired outcomes. An individual's willingness to carry out an action is greatly influenced by whether one feels he or she can perform the action, by one's beliefs concerning the consequences of doing it, and by the attractiveness of the outcomes associated with doing it. To answer these kinds of questions, a reasonable amount of critical performance feedback from supervisors and peers is required. Such feedback provides the newcomer with a clearer sense of how he or she is being viewed and regarded, helping each to find his or her particular "niche" in the overall scheme of things. Unfortunately, in most organizations this kind of useful feedback from supervisors is a rather rare occurrence. In fact, in my own research surveys on engineering professionals, performance feedback was one of the lowest rated behaviors attributed to engineering supervisors in over ten separate RD&E facilities. New employees, therefore, are faced with the problem of obtaining adequate feedback on their own individual performances which can be particularly difficult if one's work is diffused within a larger group or project effort. What becomes most distressing to new employees, then, is that they have entered the organization with an underlying expectation that they would be learning and improving on their first job, yet their supervisors fail to realize that they should act and feel responsible for teaching and helping the employees to accomplish this objective.

Employees need to determine how effectively they are currently performing, how difficult it might be for them to achieve desired outcomes, and how readily they could obtain or develop the various skills and knowledge necessary to meet the demands and expectations of the organization. As long as these critical concerns remain unmet and uncertain to the new professional, the amount of stress he or she experiences will escalate. One of the most important, yet most trying, learning experiences for the newly hired professional undergoing socialization is how to obtain valid feedback in those particular situations in which it does not automatically or effectively take place. And for many young professionals, the ultimate learning experience is to figure out how to become an excellent judge of one's own individual performance. Socialization, as a result, is facilitated to the extent that one's supervisor is able to: (1) make an accurate assessment of the new employee's performance and give useful and valid feedback on this performance; (2) transmit the right kinds of values and norms to the new employee in terms of the long-run contributions that are expected of him or her; and (3) design the right mix of meaningful, challenging tasks that permit the new employee to utilize and extend his or her professional skills and build his or her new situational identity.

PROCESS OF SOCIALIZATION

Underlying the tasks represented by the socialization process is the basic idea that individuals are strongly motivated to organize their work lives in a manner that reduces the amount of uncertainty they must face and that is therefore low in stress (Pfeffer, 1981; Katz, 1982). As argued by Weick (1980), employees seek to "enact" their environments by directing their activities toward the establishment of a workable level of certainty and clarity. As they enter their new job positions, they are primarily concerned with reality construction, building more realistic understandings of their unfamiliar social and task environments and their own situational roles within them. They endeavor, essentially, to structure the world of their experience, trying to unravel and define the many formal and informal rules that steer the workplace toward social order rather than toward social chaos.

One of the most obvious, yet most important and often overlooked aspects of new employee socialization is simply that it must take place. By and large, people will not accept uncertainty. They must succeed over time in formulating situational definitions of their workplace with which they can coexist and function comfortably; otherwise, they will feel terribly strained and will seek to leave the given work organization. Until the new employee has created a situational perspective on himself or herself, constructed guidelines regarding what is expected, and built certain situationally contingent understandings necessary to participate meaningfully within the work setting, the individual cannot act as freely and as fully as he or she would like.

What is very important to recognize here is that stress does not come from the uncertainty itself; it comes from the individual's inability to reduce or lower it. As long as one is making progress in reducing uncertainty, that is, as long as socialization is being facilitated and the individual is making increasing sense out of his or her new work surroundings, stress and anxiety will diminish and satisfaction and motivation will rise. If, however, new employees are somehow prevented from accomplishing any or all of the broad socialization tasks previously discussed, then they are not succeeding in reducing as much of the uncertainty as they need to. This can become highly frustrating and anxiety producing, resulting eventually in higher levels of dissatisfaction and increased levels of organizational turnover. Just as the engineer is highly motivated to reduce technical uncertainty in his or her laboratory activities, the new employee is highly motivated to reduce social and interpersonal uncertainties within his or her new environments. Generally speaking, activity that results in the reduction of uncertainty leads to increasing satisfaction and reduced stress, whereas activity or change that generates uncertainty creates dissatisfaction and higher stress. Thus, it is not change per se that is resisted, it is the increase in uncertainty that usually accompanies it that is so difficult for individuals to accept.

One must also realize that socialization, unlike an orientation program, does not take place over a day or two. It takes a fair amount of time for employees to feel accepted and competent and to accomplish all of the tasks necessary to develop a situational perspective. How long this socialization period lasts is not only influenced by the abilities, needs, and prior experiences of individual workers, but it also differs significantly across occupations. In general, one might posit that the length of one's initial socialization stage varies positively with the level of complexity of one's job and occupational requirements, ranging perhaps from as little as a month or two on very routine, programmed-type jobs to as much as a year or more on very skilled, unprogrammed-type jobs, as in the engineering and scientific professions. It is generally recognized, for example, that a substantial so-cialization phase is usually required before an engineer can fully contribute within the organization, making use of his or her knowledge and technical specialty. Even though one might have received an excellent university education in mechanical engineering principles, one must still figure out how to become an effective mechanical engineer at Westinghouse, Dupont, or General Electric.

Socialization Is a Social Process

Another very important assumption about socialization is that it must take place through interaction—interaction with other key organizational employees and relevant clientele. By and large, new employees can only reduce uncertainty through interpersonal activities and interpersonal feedback processes.

Newcomers' perceptions and responses are not developed in a social vacuum but evolve through successive encounters with their work environments. Their outlooks become formulated as they interact with and act upon different aspects of their job setting. Their development cannot transpire in isolation, for it is the social context that provides the information and cues with which new employees define and interpret their work experiences (Salancik and Pfeffer, 1978).

One of the more important features of socialization is that the information and knowledge previously gathered by employees from their former colleges or other institutional settings are no longer sufficient nor completely appropriate for interpreting and understanding their new organization domains. As a result, they must depend on more-established professionals within their new situations to help them make sense out of the numerous activities taking place around them. The greater their unfamiliarity, the more they must rely on their new situations to provide the kinds of information and interaction by which they can eventually construct their own individual perspectives and situational identities. It is precisely this reliance on situational information and dependence on others for interpretation and meaning that forces new employees to be more vulnerable and more easily influenced during socialization.

Clearly, as employees become increasingly cognizant of their overall job surroundings, they become increasingly capable of relying on their own knowledge and experiences for interpreting organizational events and for executing their daily task activities. They are now freer to operate more self-sufficiently in that they are now better equipped to determine for themselves the importance and meaning of the various events and information flows surrounding them. On the other hand, as long as new employees have to balance their situational perspectives against the views of significant others within the workplace, frequent interaction with those individuals will be required.

New employees absorb the subtleties of local organizational culture and climate and construct their own definitions of organizational reality—and in particular their own role identities—through interactions with other individuals, including peers, supervisors, subordinates, and customers. Verbal and social interaction, in contrast to written documentation, are the predominant means by which new professionals acquire the most pertinent information about their new environments. Since multiple meanings are likely for any particular event, the more individuals with whom the new employee interacts, the more likely she or he is to put together a view that is both comprehensive and realistic. Different individuals will emphasize different aspects of the work setting and will also vary in the way they interpret events. Recent hires, as a result, formulate their concepts and guide their activities around the anticipated reactions and expectations of the many key employees with whom they are connected.

Of all the concepts that each individual newcomer acquires through the plethora of interpersonal contacts that takes place, perhaps the most important is one's self-concept. It has often been argued that one's fellow workers help to define for each newcomer many of the diverse aspects of the new job setting by the way they act and behave toward these aspects, for example, how they deal with absenteeism, budget overruns, schedule slippages, staff reports, or subordinate suggestions. Similarly, fellow employees help newcomers create perspectives on themselves as particular kinds of individuals by the way the fellow employees act and respond toward these organizational newcomers. As a result, a new employee's self-image is largely a social product, significantly affected by the behaviors and attitudes of other employees within his or her organizational neighborhood. In essence the newcomer's situational identity is strongly influenced by the self-concept that is gleaned from the eyes of those significant others whom they come to know and with whom they interact.

If newcomers strive to reduce uncertainty by locating and orienting themselves relative to the views and expectations that emerge from those individuals on whom they are most dependent and with whom they are most interactive, then it should not be surprising that some of the most important and most satisfying experiences for new employees are those which attune them to what is expected of them. There is a strong need for newcomers to identify closely with those colleagues and supervisors who can furnish guidance and reassurance concerning such expectations. If on the other hand, the individual newcomer is precluded from reducing uncertainty and making increasing sense out of his or her organizational surrounding, then he or she will feel stressed and will be unable to act in a completely responsive and undistracted manner.

Many circumstances can arise in any work setting to delay a newcomer's socialization, circumstances that invariably prevent or inhibit the kinds of interpersonal interactions that are essential for uncertainty reduction. Consider, for example, the new employee whose boss is out of town, on vacation, has just quit, or is simply too busy to help with one's integration; or the new employee who is assigned to a job location or given an office far away from the boss or key reference groups. Chances are that the reduction of situational uncertainty under these kinds of conditions will be a much prolonged process, perhaps interfering with the newcomer's potential career success or even his or her willingness to remain in the organization. Research studies have shown, for example, that young engineers who are not well networked internally (so that their stronger communication links are with individuals outside their group or organization) are much less likely to receive a managerial promotion and are much more likely to leave the organization over time (Katz and Tushman, 1983).

Socialization Experiences Are Highly Influential and Long-Lasting

According to the law of primacy expressed by Brown (1963), early socialization experiences are particularly important because they greatly influence how later experiences will be interpreted. The early images and perspectives that are formed in the first year or two of one's organizational career have a strong and lasting influence on one's future task assignments, perceived performances and abilities, and promotional success.

What has become clear from a large number of studies (e.g., McCall and Lombardo, 1989; Bray and Howard, 1988; Lee, 1992) is that the degree to which an employee perceives his or her job as important and challenging by the end of the first year will strongly influence future performance and promotional opportunities. Such studies have shown that young professionals and managers who viewed their job assignments more positively or who were evaluated more positively by their supervisors after only one year of employment were also more likely to have received higher performance ratings and higher rates of promotion some five to ten years later. In their classic study of engineers and scientists, Pelz and Andrews (1976) reported that technical professionals who were able to utilize and demonstrate more diverse skills and abilities in accomplishing their tasks during their early career years were significantly more likely to advance within the organization than professional counterparts who were frequently rotated from project to project. General Electric even discovered from their own career-tracking studies of young professionals that the best predictor of career success at GE was the *number* of different supervisors who had personal knowledge of the task accomplishments of the young professional. Given the general consistency in the pattern of findings from these different studies, it is clear that the newcomer who gets widely known and comes to be seen and

sponsored as a valued high performer gains a considerable long-term advantage over the newcomers not so fortunately viewed—the proverbial self-fulfilling prophecy.

In the process of interpreting their early work experiences, young employees begin to observe their colleagues as well as other members who have been labelled as successful or unsuccessful within the organization. They then begin to assess their own careers relative to these individuals. This process of comparison involves many factors, but temporal comparisons represent some of the most critical. By comparing one's progress against these other individuals, the new employee begins to form an implicit "career benchmark" against which both the individual and the organization can start to determine how well he or she is doing. In their studies of British and American managers respectively, Sofer (1970) and Lawrence (1990) show just how sensitive organizational members can become to their relative career progress. In another example Dalton et al. (1982) strongly argue from their study of R&D professionals that organizations have clearly defined expectations about the behaviors and responsibilities of their more successful engineers at well-defined age-related career stages.

All of these examples emphasize that soon after beginning work, employees gradually become concerned about how their progress fits within some framework of career benchmarks. Where are they—are they on schedule, ahead of schedule, behind schedule? The pressure from these kinds of comparisons can become extremely acute especially as the relative judgments become increasingly salient and competitive and their timing increasingly fixed and inflexible. Such events seem to occur in at least two different ways. First, the comparisons can become highly intense as employees enter an organization as part of a well-defined, well-bounded cohort but are then forced to compete amongst themselves for the best individual evaluations, as in the case of many law firms, public accounting firms, consulting companies, universities, and so on. The directness and clarity of these comparisons make the implicit aspects of the career benchmarking process more explicit and, in general, place the young employee under a great deal of stress as he or she competes for the next level of advancement.

For other young employees the occupation or other organization itself can present a fixed timetable for measuring success and career advancement. The tenure process in universities, standardized professional exams (e.g., registered engineer, CPA), or certain apprenticeship or associate periods are all examples of highly structured, well-defined timetables of career progress. These kinds of explicit benchmarks can also place the young employee under severe stress particularly if the employee loses control over the timing of the process or it becomes more like an "up or out" or a "pass/fail" type of system. While the climate that emerges from these more explicit models of career benchmarking may not be very supportive, they may "energize" a great deal of activity and long hours of work on the part of the new employee, at least during his or her early career years.

Because the employee's immediate supervisor influences and controls so many aspects of the communication, task, and career benchmarking factors during socialization, it becomes clear why so many studies have pinpointed the new employee's first boss as being so critical with respect to his or her successful advancements both organizationally and professionally (Kanter, 1977; Schein, 1978; Henning and Jardim, 1977). While supervisors play a critical role in linking their subordinates to other parts of the organization, they can also assume a broader role within their work groups, becoming actively involved in the training, integration, and socialization of their more recently hired members.

By building close working relationships with young subordinates, supervisors might not only improve their group's performance (Katz and Tushman, 1981), but they might also directly affect the personal growth and development of their young professionals. To the extent that supervisors help their new employees participate and contribute more effectively within their work settings, have clearer working relationships with other key organizational individuals, and communicate more easily with outside customers, clients, or profession-

als, these young professionals will experience less stress and will be less likely to leave the organization. Graen and Ginsburgh (1977) showed, for example, that organizational newcomers who built strong dyadic relationships with their immediate supervisors and who saw a strong relationship between their work and their professional careers were more likely to remain with the organization.

In a longer longitudinal study, Katz and Tushman (1983) found that young engineers who had high levels of interaction with their first and second level supervisors were significantly less likely to leave the organization over the next five years. These supervisors were seen as technically competent and were viewed as valuable sources of new ideas and information. As a result, they became more interactive simply because they were consulted and listened to more frequently on work related matters. At the same time, this high level of interpersonal activity allowed these supervisors to create close working relationships with their younger engineering subordinates, helping them become established and integrated during their early career years. Thus, it may be this high level of interpersonal contact with technically competent supervisors that not only facilitates socialization but also results in more accurate expectations, perceptions, and understandings about one's role in the job and in the larger organization—all of which are important in decreasing turnover and the anxiety levels of new employees.

It has been argued throughout this paper that becoming an integral part of the organization's communication and information processing networks and learning the organization's customs and norms are critically important for reducing stress and fostering more positive attitudes during the early stages of employees' careers. It has also been argued that supervisors play a very direct role in dealing with the initial concerns of young employees, allowing them to reduce uncertainty by helping them understand and interpret the reality of their new settings. In essence, supervisors operate as effective socializing agents and networks builders for their young employees.

In many cases, however, the supervisor is not

the only socializing agent of the new employee. The veteran group as a unit can also affect the attachment of new members to the organization. In line with the findings of Katz (1982) and McCain et al. (1981), for example, the larger the proportion of group members with the same group tenure and shared work history, the more distinct that cluster of individuals might become from other organizational members, in general, and from new entering members, in particular. Young employees, for example, might experience a great deal of stress and frustration in trying to integrate themselves into a well-established, older cohort or vice-versa.

Additional conflicts and power issues are also likely to result when there are larger gaps between cohorts within the overall work group. If the group has been staffed on a regular basis, then the new employee's integration is more likely to proceed in a smooth fashion since socialization can be nurtured through the existence of closer, linking cohorts (e.g., Ouchi and Jaeger, 1978). If there are large gaps between cohorts, however, then it is likely that perceptions and beliefs will differ more, resulting in considerable communication difficulty and impedance. The existence of well-differentiated cohorts, according to McCain et al. (1981), increases the possibility of different intragroup norms and expectations which can result in a group atmosphere that is characterized by severe intragroup conflict—a very stressful experience for new employees.

CONCLUSIONS

Perhaps the most important notion in this paper is that individuals undergoing a transition into a new organization are placed in a high anxiety-producing situation. They are motivated, therefore, to reduce this anxiety by learning the functional and social requirements of their new role as quickly as possible. What must be recognized and understood by organizations is that supervisors and group colleagues of new employees have a very special and important role to fulfill in inducting and socializing the new employee. The careful selection of

these individuals for young professionals should go a long way toward alleviating many of the problems that usually occur during the "joining-up" process. One must recognize that the problems and concerns of young professionals are real and must be dealt with before these young employees can become effective organizational members. Although organizations might want to develop specific training programs to teach managers how to "break in" the young professional more effectively, an alternative strategy would be to make sure that young professionals are integrated and socialized into their job environments only through those particular groups and supervisors who appear especially effective in this function. Rather than allowing all supervisors and groups to recruit and hire new employees as additional staffing needs arise, a more centralized policy of rotating or transferring individuals to some areas in order to hire new employees through other key integrating areas might prove more beneficial to the organization as a whole in the long run. The careful assignment of groups and supervisors to new college recruits, combined with some training for these individuals, should go a long way toward utilizing the great potential in most young professionals, thereby reducing the high levels of frustration, stress, and dissatisfaction that so many of them experience during their initial career years.

REFERENCES

Allen, T. J. *Managing the flow of technology.* Cambridge: MIT Press, 1984.

Bailyn, L. Resolving contradictions in technical careers: or, what if I like being an engineer. *Technology Review,* 1982, November-December, 40–47.

Bray, D. W. and Howard, A. *Managerial lives in transition: Advancing age and changing times.* New York: Guilford Press, 1988.

Brown, J. A. C. *Techniques of persuasion.* Baltimore: Penguin Books, 1963.

Dalton, G. W., Thompson, P. H. and Price, R. L. The four stages of professional careers: A new look at performance by professionals. In R. Katz (Ed.), *Career issues in human resource management.* Englewood Cliffs, N.J.: Prentice-Hall, 1982.

Gouldner, A. W. Cosmopolitans and socials: Towards an

analysis of latent social roles. *Administrative Science Quarterly,* 1957, 2, 446–467.

Graen, G. Role-making processes within complex organizations. In M. D. Dunnette (Ed.), *Handbook of industrial and organizational psychology.* Chicago: Rand McNally, 1976.

Graen, G. and Ginsburgh, S. Job resignation as a function of role orientation and leader acceptance. *Organizational Behavior and Human Performance,* 1977, 19, 1–17.

Henning, M. and Jardim, A. *The managerial woman.* New York: Doubleday, 1977.

Kanter, R. M. *Work and family in the United States.* New York: Russell Sage, 1977.

Katz, R. Job longevity as a situational factor in job satisfaction. *Administrative Science Quarterly,* 1978, 23, 204–223.

Katz, R. Time and work: Toward an integrative perspective. *Research in Organizational Behavior,* 1980, 2, JAI Press, 81–127.

Katz, R. The effects of group longevity on project communication and performance. *Administrative Science Quarterly,* 1982, 27, 81–104.

Katz, R. and Tushman, M. An investigation into the managerial roles and career paths of gatekeeper and project supervisors in a major R&D facility. *R&D Management,* 1981, 11, 103–110.

Katz, R. and Tushman, M. A longitudinal study of the effects of boundary spanning supervision on turnover and promotion in research and development. *Academy of Management Journal,* 1983, 26, 437–456.

Lawrence, B. At the crossroads: A multiple-level explanation of individual attainment. *Organization Science,* 1990, 1, 65–85.

Lee, D. Job challenge, work effort, and job performance of young engineers. *IEEE Transactions on Engineering Management,* 1992, 39, 214–226.

Lee, D. Social ties, task-related communication and first job performance of young engineers. *Journal of Engineering and Technology Management,* 1994, 11, 203–228.

Lindholm, J. A study of the mentoring relationship in work organizations. Unpublished MIT Doctoral Dissertation. 1983.

Louis, M. Surprise and sense making: What newcomers experience in entering unfamiliar organizational settings. *Administrative Science Quarterly,* 1980, 25, 226–251.

McCain, B. R., O'Reilly, C. and Pfeffer, J. The effects of departmental demography on turnover. The case of a university. Working Paper, March 1981.

McCall, M. W. and Lombardo, M. M. *The lessons of experience: How successful executives develop on the job.* Lexington, Mass.: D.C. Heath & Co., 1989.

Ouchi, W. G. and Jaeger, A. M. Type Z organization: Stability in the midst of mobility. *Academy of Management Review,* 1978, 3, 305–314.

Pelz, D. C. and Andrews, F. M. *Scientists in organizations.* Ann Arbor: University of Michigan, 2nd ed., 1976.

Pfeffer, J. Management as symbolic action: The creation and maintenance of organizational paradigms. *Research in Organizational Behavior,* 1981, 3, JAI Press, 1981.

Salancik, G. R. and Pfeffer, J. A social information processing approach to job attitudes and task design. *Administrative Science Quarterly,* 1978, 23, 224–253.

Schein, E. H. *Career dynamics.* Reading, Mass.: Addison-Wesley, 1978.

Sofer, C. *Men in mid-career: A study of British managers and technical specialists.* London: Cambridge University Press, 1970.

Weick, K. E. *The social psychology of organizing.* Reading, Mass.: Addison-Wesley, 1980.

2

The Management of Creativity in Organizations

4

Managing Creative Professionals

ALBERT SHAPERO

More than 30 years of research have provided answers to some of the questions technical managers ask about creativity in organizations.

The management of creative workers has become the most critical area faced by management in both the private and public sectors. Without a great deal of fanfare, creative workers, or, more strictly, professionals, have come center stage in the United States and the rest of the developed world. Quan-

titatively, professionals now surpass all other categories in the U.S. work force. Qualitatively, professionals have a disproportionate effect on all aspects of our society, as the researchers, designers, decision makers and managers who define and direct much of what is done in society. The quality and extent of what is accomplished in the foreseeable future have become functions of the ability of management to harness and channel the efforts of creative workers. The difference in success between one effort and another, one organization and another increasingly depends on whether manage-

Reprinted with permission from *Research-Technology Management,* March–April, 1985.
Albert Shapero is Professor of Management Science at Ohio State University.

ment understands the differences between the management of professionals and the management relevant to the assembly line.

In trying to evoke and develop creativity in an organization, managers are interested in such questions as: Can creative people be identified for the purpose of hiring? Are there valid and reliable tests that can predict who will be creative? Can creativity be developed or enhanced in employees? Are there creativity techniques that can be taught to employees that will increase creativity within the organization? What kinds of management actions help or retard creativity? What kinds of environments enhance or deter creativity? What differentiates the creative organization from those that aren't creative? Researchers on creativity have generated data that provide some answers to these questions.

From the beginning, much research on creativity has focused on developing ways of predicting who will demonstrate high creativity in the future. One approach, based on biographical and autobiographical studies of individuals with demonstrated high creativity, attempts to develop predictive profiles. Included among the profile methods is factor analysis. Other attempts have produced psychometric instruments to measure intellectual capabilities considered by the researcher as central to creativity. Most of the latter have measured divergent thinking. Despite several decades of research effort on creativity and highly creative individuals, there is as yet no profile or test that reliably predicts who will be highly creative in the future. Efforts to develop tests to predict later creativity in students have borne little result. Longitudinal studies of the predictive strength of divergent-thinking tests given to students have been disappointing (Howieson, 1981; Kogan, 1974). So far, the only good indication that an individual will be highly creative in the future has been demonstrated high creativity in the past.

THE ENVIRONMENT FOR CREATIVITY

Two aspects of the environment for creativity have been examined by researchers: (1) the kinds of fa-milial and educational environments in childhood that lead to creativity in adulthood, and (2) the kinds of immediate, organizational, and physical environments associated with high creativity. The effect of childhood environments in subsequent creativity is of little utility for managers, although one finding worth noting is that high creatives, unlike those with high IQs, came from families in which parents put little stress on grades (Getzels & Jackson, 1962).

The manager of professionals is concerned with organizational environments associated with high creativity and how they might be generated. Most of the organizational characteristics that appear to enhance creativity relate to the characteristics attributed to highly creative individuals (Steiner, 1965). For example, since nonconformity in both thought and action characterizes high creatives, the organization that is tolerant of a large variety of deviance from the norm is more likely to enhance creativity. It is not surprising to find that many "high tech" companies, architectural firms, advertising organizations, and academic faculties are marked by unconventional dress and little rigidity concerning hours of work.

Characteristics of creative organizations (Steiner, 1965) include the following:

- Open channels of communications are maintained.
- Contacts with outside sources are encouraged.
- Nonspecialists are assigned to problems.
- Ideas are evaluated on their merits rather than on the status of their originator.
- Management encourages experiments with new ideas rather than making "rational" prejudgments.
- Decentralization is practiced.
- Much autonomy is allowed professional employees.
- Management is tolerant of risk-taking.
- The organization is not run tightly or rigidly.
- Participative decision making is encouraged.
- Employees have fun.

THE PROCESS OF CREATING

In spite of the apparent uniqueness of the creative process in each individual and the idiosyncratic patterns followed by many creative individuals, studies of the process are in fair agreement that it follows a recognizable overall pattern. The creative process has been variously described, but most descriptions include a series of steps, varying in number, that can be subsumed within the following four steps: (1) preparation, (2) incubation, (3) illumination, and (4) verification.

Preparation. The creative process begins with a problem perceived or experienced. Whenever humans have a problem, and don't know how to solve it by direct action, they resort to thinking, problem-solving, and creativity. The problems that lead to creative responses arise from many sources. They can be thrust upon one or assigned from the outside, be perceived as a threat or opportunity, be encountered, or be sought out because humans are dreaming, restless creatures who enjoy the creative process. Once a problem is perceived, the creative process begins.

Research shows that the conscious "creative" moment comes only after intensive preparation and a period of subconscious incubation. Louis Pasteur put it succinctly: "Chance only favors the prepared mind." Helmholtz described his own creative process: "It was always necessary, first of all, that I should have turned my problem over on all sides to such an extent that I had all its angles and complexities "in my head" and could run through them freely without writing" (McKellar, 1957). In a study of highly productive inventors, Rossman (1964) found that they all started the process by "soaking themselves in the problem." Though Rossman reports that some inventors reviewed all previous efforts to solve the problem and others avoided being influenced by previous attempts, all spent time thoroughly exploring the problem to be solved.

The preparation process can include literature searches, talking to many people about aspects of the problem, experimentation, and doodling. Some-

times the preparation process can appear as unplanned, unfocused meandering through a variety of materials. McKellar (1957) considers it as almost a form of "overlearning" to the point where some of the materials become "automatic" in one's consciousness. The gathering of information is a critical part of the process in which the individual examines the materials critically, but not negatively. The creative process requires discriminating criticism that does not reject, but builds upon the materials examined.

Incubation. This is a process that goes on below the level of consciousness. It cannot be commanded. Incubation appears to be a gestation period in which the process goes on subconsciously, and it works best when the individual is inactive with regard to the problem or working on something else. A passage of time, vital to the process, varies with the problem and individual (McKellar, 1957). The philosopher Nietzsche spoke of a period of 18 months, and the poetess Amy Lowell spoke of six months. It can be a period of frustration for the individual working against a deadline, for it cannot be pushed or rushed. It is a period when apparently nothing is happening.

One soaks oneself in the problem and then waits. The passage of time is often accomplished by sleep. It is as if sleep provides the time and the opportunity to abandon consciousness of the problem and let the unconscious work. Some great creative discoveries have surfaced in sleep. Kekulé realized his discovery of the benzene ring as the result of a dream of the image of a snake that seized hold of its own tail. Many of Descartes' basic notions of analytical geometry formed in his dreams. Everyone has had the experience of "fighting" a problem to an impasse, and having the solution suddenly crystallize while visiting with friends or discussing other things. The need for a period of incubation may explain why professionals who work on more than one project at a time are more productive than others. Having more than one project permits a person to switch to another project when apparently at an impasse. Switching from one project to another permits the first project to incubate until it is ready, while one is still doing something productive.

The incubation process is recognized but not understood. One plausible explanation is that it is a period in which the mind tests different associations, matches different frames of reference and different conceptual elements to see if they make sense. This explanation fits with the most accepted view of creativity as a process of association.

Probably the most widely held psychological conception is that creativity is the ability to call up and make new and useful combinations out of divergent bits of stored information (Guilford, 1964). The more creative the individual the greater the ability to synthesize remote bits of information. The likelihood of a solution being creative is a function of the number and uncommonness of associative elements an individual brings together (Mednick, 1962). The latter notion has been incorporated into a test for creative ability, The Remote Associations test (Mednick & Mednick, 1964). The test taker is asked to "make sense" out of each of 30 sets of three, not obviously related, terms by providing a fourth term related to them (e.g., the fourth term related to "cookies," "sixteen," and "heart" would be "sweet"). Another associationist view is Koestler's "bisociation of Matrices," expressed by the metaphor of creativity as a "dumping together on the floor the contents of different drawers in one's mind" (Koestler, 1964).

Illumination. The Gestalt psychologists refer to illumination as the "aha!" phenomenon. It is that sudden insight, that flash of understanding, in which the solution appears. The mathematician Polya describes it as entering an unfamiliar room in the dark, and stumbling around, falling over pieces of furniture, looking for the light switch. When the switch is found and activated, everything falls into place. All historic examples of the incubation process end with that moment of illumination.

Verification. After the exhilaration of illumination comes the tedious, time-consuming stage of verification. The creative idea must pass the tests of validity, reality, utility, realizability, costs, time, and acceptance in the marketplace.

CREATIVE PROBLEM SOLVING

Rules for creative problem solving can be derived from the data available on the process. They include the following:

1. Soak Yourself in the Problem. Read, review, examine, and analyze any material that you can find on the problem. Talk to people who know about the problem. Look at every side of the problem. Saturate yourself in the subject. Be critical in the positive sense. Don't accept authority. Question the premises. However, insist on finding a way to solve the problem, and do not accept that it cannot be solved.

As a manager, do not easily accept the conclusion that "it can't be done." On hearing all the reasons from his group why something wouldn't work, one successful manager of professionals counters, "I agree it can't be done, but if we had to do it or be shot what could we do?" It always changes the atmosphere, and turns the group to finding ways to attack the problem rather than judge it. Push the people in your organization to soak themselves in the problem. Provide them with all the information you can. Err on the side of overload. Encourage them to contact a wide variety of sources for information.

2. Play with the Problem. Stay loose and flexible in dealing with the problem. Try different assumptions. Leave out one of the conditions that affects the problem, and see what that suggests. Approach the problem from different directions. Turn it over and inside out. Assume different environments. Shift parts of the problem around, physically and spatially. Change the time sequence. Change the order of events. Change the situation.

As a manager, you can help by encouraging your people to explore the problem from every kind of viewpoint. You can do this by discussion, by questioning, by encouraging "wild" approaches in the early stages of a project.

3. Suspend Judgment. Fight any tendency to draw early conclusions. You will only lock yourself in, and lose several degrees of freedom. Early fixation on even part of a problem definition keeps you from seeing larger parts of the problem. It foreshortens your perception. Even worse is to get an early fix on part or all of a solution. It cuts you off from a great many possibilities, since you begin to justify your early solution. Suspension of judgment keeps you open to new information and enables you to see new possibilities as the problem unfolds. If solutions keep suggesting themselves to you, write them down in a notebook, and deliberately put them aside until later in the project. Get them out of your mind.

As a manager, help your people suspend judgment. The manager is probably the biggest obstacle to suspensions of judgment. The manager represents deadlines and budgets, and they must not be forgotten. However, it is important not to push for immediate judgments. Don't pressure your people to come up with solutions in the early stages of a project. Encourage them to note and file any early conclusions as suggested above.

4. Come Up with at Least Two Solutions. When you start out with the objective of coming up with two different solutions, it keeps you from fixating on a solution and keeps you thinking about the problem. Studies have shown second solutions tend to be more creative, and trying for two solutions results in more creative solutions. It was found, experimentally, that asking people for two solutions, as compared to one, increased the number of "creative" solutions from 16 percent to 52 percent. When pushed to the limit by being asked for three different solutions, it was found not all responded, but there was an increase by 25 percent in very good, creative solutions (Hyman and Anderson, 1965).

As a manager, insist on two independent solutions to a problem. It is not necessary to work them out in detail, but make sure they are significantly different. The first solution seems to catch all the anxieties and stiffness of an individual, while the second is more free flowing.

5. When Stuck.   Try more than one way of picturing the problem and solution. Go from a word description to a drawing. Many creative scientists, mathematicians, and writers use sketches and diagrams to put the problem into a different perspective. Go from a drawing to abstraction.

Try your problem out on other people. When you discuss your problem with others, you see it differently. You have to make sense out of it.

What the other people say is less important than your own presentation, though unexpected questions can help hook different parts of your brain together.

Give your subconscious a chance to work. Take a break. When you are really up against it, do something else for a while. Remember! It is a ripening process, and you can't force it. Spending round-the-clock sessions will only exhaust you, rather than solve the problem.

As a manager, put yourself in the way of being the person on whom the problem can be tried. Ask your people to "draw you a picture" of the problem to help you understand it. When they get too intense and are not making progress, give them another short assignment to pull them away for awhile, to let their subconscious work.

CREATIVITY FROM THE VIEWPOINT OF THE MANAGER

Can anything systematic be done to increase creativity in individuals and in an organization? Does management really want creativity and the somewhat less controlled conditions necessary to foster it?

To individuals, more creativity carries an implication of special, personally gratifying experiences. To managers, more creativity means new ideas, inventions, and solutions that will do wonderful things for the organization in the marketplace. Few, however, have thought through the consequences of having more creative people and of allowing the conditions that enhance creative behavior in their organizations.

Trying to answer the converse of the question, "Can anything be done to increase creativity?" quickly illustrates how much is generally known about conditions for creativity. Pose the question

"Can anything be done to kill creativity in an individual or an organization?" and the mind immediately fills with answers:

- Discourage and penalize risk-taking.
- Discourage and ridicule new ideas.
- Reject and discourage attempts to try unusual methods.
- Make sure all communications follow formal organizational lines and all employees cover themselves.
- Discourage reading and communications with people outside the immediate organization.
- Discourage nonconformity of any kind.
- Discourage joking and humor.
- Provide no recognition.
- Provide no resources.

We easily intuit what it takes to minimize creative behavior, which suggests that it must be possible to improve creativity or, at least, to minimize barriers to creativity. The available information strongly indicates that it is possible to improve one's own creativity and the creativity of employees. It is possible to increase the creative activities and products of an organization. Increasing creativity in an organization is achievable, but it takes a lot more effort than preventing it from occurring. Continuity and stability are important attributes in society, and, of necessity, the dice are loaded against divergence and change.

Highly creative people are attracted by the work, by the problem being worked on, which is good from an organizational viewpoint, but they don't respond in satisfactory ways to the political or organizational constraints that are involved in every problem. Creative people are nonconformists. They are jokers. They have little reverence for authority or procedures. They are short on apparent "loyalty" to the organizations they work for. They don't respond to the kinds of incentives that stir others. They are not moved by status. High creatives don't seem to care about what others think, and they don't easily become part of a general consensus. (Could a preference for consensus management be

why the Japanese have recently expressed concern about a lack of creativity in Japan?) In short, creative people can make most managers very uncomfortable. (Teachers and even parents are far more comfortable with students and children with high IQs than with those who are highly creative.)

A case can be always made for creativity, but managers should carefully and honestly think about whether they truly need more creativity and can live with it. If successful at hiring and retaining high creatives, and at generating the conditions needed to keep them creative, management may be creating conditions that make it difficult for its own natural style of doing things. New methods, processes, and products can be purchased, copied, and stolen. According to one ironic maxim, it doesn't pay to be first— pioneers get killed. Some years ago, the head of a metal machining company producing thousands of metal fasteners picked through his catalog and fondly indicated product after product that had been invented by other companies. "You know," he said, "we don't know anything about managing creative people, but we're very, very good at designing around other people's designs. What we're really competent at is production and marketing, and we beat the hell out of the creative companies. I can't wait for their next products." Cynical? Perhaps, but it highlights the questions raised here. Many can benefit from the creativity of a few, and there are industries, companies, and fields where creativity is far less needed than in others.

ON THE ROAD TO MORE CREATIVITY

If desired, creativity can be consciously and systematically enhanced in an organization through hiring, motivation, organization, and management actions.

Hiring. The number of highly creative people in an organization can be increased by a hiring policy that deliberately attempts to identify, locate, and hire them. The only valid and reliable way to identify individuals with a high probability of future creative performance is through evidence of past creative performance. The more recent and continuous the past creative performance, the more likely there will be future creative performance.

Where examples of a professional's work are not as easily demonstrated as in the arts and architecture, the task of determining past creative performance is harder. It is difficult to tease out evidence of the individual creative contributions of an engineer or scientist who has worked on a project that employed scores or hundreds of professionals. One way to tackle the problem is to put the questions directly to the individual: "What are the most creative things you have done on the job in the past three years? What are the most creative things you have ever done?" Similar questions about the individual's work can be asked of others who are familiar with it. In some fields patents, in others publications, may serve the purpose, though they should be examined for their content.

Tests, profiles of traits, and checklists are neither valid nor reliable. No available test can determine who will perform creativity in the future with any reliability. (One may be tempted to follow the example of the author who tried to hire on the basis of the apparent relationship between a good sense of humor and creativity. The rationale was, "If they don't turn out to be creative, at least they'll be a barrel of laughs.")

Motivation. Creative behavior can be maintained and enhanced through incentives that reward creative output and encourage risk-taking and through the use of new methods, processes, and materials. For those who are already highly creative, incentives can maintain and encourage their creative efforts and help retain them in the organization. For other professionals, incentives and positive feedback from management can encourage them to overcome some of the natural blocks to creativity and to take more risks and be more curious. As with any other desired behavior, feedback from management, the performance evaluation system, and the example of management can help stimulate creativity. If a manager smiles on "far out" ideas when they are ventured, lets them be

tried (even when he or she is personally sure they won't work), and will even express some extravagant ideas himself, others may feel freer to think and act creatively.

Providing the Necessaries. The availability of resources for initial creative efforts is a powerful indicator of management support for creative activities. The resources required to give an idea a preliminary investigation are seldom of any magnitude. Direct provision of resources, or turning a blind and benevolent eye on the inevitable "bootlegging" of an unauthorized project, both serve the purpose of support for creative experimentation. Providing resources for preliminary explorations of ideas without requiring exhaustive justification is a form of intellectual overhead and should be treated as such, formally or informally. (Remember that time is one of the most important resources required for creative activities.)

Some boost to creativity can be obtained through educational programs, though management should be wary of "patented" techniques. All creativity-enhancing techniques have some limited value in terms of stirring up new ideas for a short time. An inherent limitation in almost all of the techniques is that they purport to provide *the* way to the generation of creative ideas or to problem solving. The overall process follows a broad general pattern, but individuals must find their own personal approach.

Managing. Managers should assign tough deadlines but stay out of the operating details of a project. There is no conflict between a deadline and creativity. Creative people resist closure because they see new possibilities as the project unfolds. For all the complaints, deadlines are necessary. Without deadlines few creative projects would ever finish.

Both productivity and creativity can be enhanced by assigning more than one project to a professional. Not all the projects have to be of equal weight, or size, or value. The ability to switch to a second project and let the first project incubate in the subconscious is important to creativity. With only one project and a tough deadline, there is a tendency to try to force the project at times when it can't be forced. Having other projects provides a legitimate (forgivable) and productive way to back off from a stymied project when a pause is needed.

New projects need fresh, unchanneled thinking. Managers might make up project groups to include people of different backgrounds, and refrain from always assigning projects to the individuals who have done that kind of work before and are apparently most suited to it.

Each professional's assignments should provide diversity for that individual. And highly productive groups of five or more years duration should be made more diverse through the addition of new people and by making certain that the individuals in the group get occasional assignments to work with other groups.

Organizations. Organizational mechanisms to assure that new ideas don't get turned down for the wrong reasons (such as middle-management cautiousness) are important. One company set up a new products committee to which any employee, and not just professionals, could submit ideas. The committee, made up of senior scientists, product development people, and a patent lawyer, investigated and discussed each idea and wrote up a decision stating why the idea was accepted, rejected, or recommended for more research. By taking a positive and encouraging stance the company developed a strong flow of ideas from throughout the organization.

There should be a legitimate (nonthreatening) means for taking an idea up the management line if it is rejected by first-line management. The means may be a new product committee, of the type described above, or a procedure for periodic review of ideas people feel strongly about. After many attempts to correlate creativity with personal characteristics, General Electric found that a key variable was the ability not to be dissuaded from their intuitions. The former director of technical systems

and materials Jerome Suran believes that high creatives are stubborn types, "because you don't get past the first level of management in a big company unless you feel strongly about your ideas" (Cullem, 1981).

A periodic review of organizational procedures and forms, with a view to identifying and removing those that cannot pass a test of necessity, is often a good idea. Too many required administrative procedures and forms sop up time and energy and impede creative activity. Procedures and forms are pervasive forces for conformity, and the more there are, the less space and time are left for nonconforming, creative thought and effort. Professional organizations should follow the rule that for every procedure or form that is added, at least one should be removed.

REFERENCES

Cullem, T., "Stimulating Creativity," *Electronic Engineering Times,* July 20, 1981.

Getzels, J. W. and P. W. Jackson, *Creativity and Intelligence,* New York, John Wiley and Sons, 1962.

Guilford, J. P., *The Nature of Human Intelligence,* New York, McGraw-Hill, 1964.

Howieson, N., "A Longitudinal Study of Creativity: 1965–1975," *Journal of Creative Behavior,* April-June, 1981.

Hyman, R. and B. Anderson, "Solving Problems," *International Science and Technology,* September 1965.

Koestler, A., *The Act of Creation,* New York, Macmillan, 1964.

Kogan, N. and E. Pankove, "Long Term Predictive Validity of Divergent Thinking Tests. Some Negative Evidence," *Journal of Educational Psychology,* 66 (6), 1974.

McKellar, P., *Imagination and Thinking,* New York, Basic Books, 1957.

Mednick, S. A., "The Associative Basis of the Creative Process," *Psychology Review,* 69 (3), 1962.

Mednick, S. A. and M. T. Mednick, *Remote Associates Test,* Boston, Houghton Mifflin, 1964.

Rossman, J., *Industrial Creativity,* New Hyde Park, N.Y., University Books, 1964.

Steiner, G. A., *The Creative Organization,* Chicago, University of Chicago Press, 1965.

5

How to Manage Geeks

RUSS MITCHELL

Eric Schmidt, CEO of Novell, believes that "geek" is a badge of honor. (After all, he is one!) But how do you manage these geek gods? Just follow his nine-point techie tutorial.

There's a saying in Silicon Valley: "The geeks shall inherit the earth."

That's a sign, if you needed one, that we have permanently entered a new economy. Once a term of derision, the label "geek" has become a badge of honor, a mark of distinction. Anyone in any business in any industry with any hope of thriving knows that he or she is utterly dependent on geeks—those technical wizards who create great software and the powerful hardware that runs it. The geeks know it too—a fact that is reflected in the rich salaries and hefty stock options that they now command.

But how do you manage these geek gods? Perhaps no one knows better than Eric Schmidt, CEO of Novell Inc. Schmidt, 44, is a card-carrying geek himself: His resume boasts a computer-science PhD and a stint at Sun Microsystems, where he was the chief technology officer and a key developer of the Java software language. And, as if his technical skills weren't enough to prove the point, Schmidt even looks the part, with his boy-genius face, his wire-rim spectacles, and his coder's pallid complexion.

Two years ago, Schmidt left Sun and took charge at Novell, where he has engineered an im-

pressive turnaround. After years of gross mismanagement, the $1 billion networking-software company, headquartered in Provo, Utah, had been written off by competitors and industry observers alike. Since Schmidt's arrival, however, the company has become steadily profitable, its stock price has more than doubled, and, within its field, Novell has again come to be seen as a worthy competitor to Microsoft.

A good deal of the credit for Novell's turnaround must go to Schmidt, who excels at getting the best out of his geeks. He has used his tech savvy to bring focus to Novell's product line and his geek-cred to reenergize a workforce of highly skilled but (until recently) deeply dispirited technologists. In general, Schmidt speaks of his geeks in complimentary terms, while acknowledging their vulnerabilities and shortcomings. "One of the main characteristics of geeks is that they are very truthful," says Schmidt (who, in fact, uses the term "geek" only occasionally). "They are taught to think logically. If you ask engineers a precise question, they will give you a precisely truthful answer. That also tends to mean that they'll only answer the question that you asked them. If you don't ask them exactly the right question, sometimes they'll evade you—not because they're lying but because they're being so scrupulously truthful."

With that rule of geek behavior in mind, Fast Company went to Novell headquarters to ask Schmidt a series of precise, carefully worded questions. His answers add up to a short course in how to bring out the best in your geeks.

Reprinted from *Fast Company Magazine*, June 1999, pp. 174–178. Used with permission.
Russ Mitchell is a senior writer for *U.S. News and World report.*

YOU'VE GOT TO HAVE YOUR OWN GEEKS

Today innovation drives any business. And since you don't want to outsource your innovation, you need to have your own geeks. Look at trends in e-commerce: Who would have thought that all of these "old" companies would have to face huge new distribution-channel issues, all of which are driven by technology? The truth is, you need to have a stable of technologists around—not just to run your systems but also to help you figure out which strategies to pursue, which innovations to invest in, and which partnerships to form.

The geeks control the limits of your business. It's a fact of life: If the technologists in your company invent something ahead of everybody else, then all of a sudden your business will get bigger. Otherwise, it will get smaller. You simply have to recognize and accept the critical role that technologists play. All new-economy businesses share that property.

GET TO KNOW YOUR GEEK COMMUNITY

According to the traditional stereotype, geeks are people who are primarily fascinated by technology and its uses. The negative part of that stereotype is the assumption that they have poor social skills. Like most stereotypes, it's true in general—but false at the level of specifics. By society's definition, they are antisocial. But within their own community, they are actually quite social. You'll find that they break themselves into tribes: mainframe-era graybeards, Unix people who started out 20 years ago, the new PC-plus-Web generation. They're tribal in the way that they subdivide their own community, but the tribes don't fight each other. In fact, those tribes get along very well—because all of them fight management.

Perhaps the least-becoming aspect of the geek community is its institutional arrogance. Remember, just because geeks have underdeveloped social skills doesn't mean that they don't have egos. Tech people are uppity by definition: A lot of them would like to have been astronauts. They enjoy the limelight. In a power relationship with management, they have more in common with pro basketball players than they do with average workers. Think of your techies as free agents in a highly specialized sports draft. And the more specialized they are, the more you need to be concerned about what each of them needs as an individual.

LEARN WHAT YOUR GEEKS ARE LOOKING FOR

This is a golden era for geeks—it doesn't get any better than this. In the early 1970s, an engineering recession hit, and we reached a low point in engineering and technical salaries. Ever since then, salaries have been going way up. Geeks have figured out that increasing their compensation through stock options is only fair: They expect to share in the wealth that they help to create through technology. Today technology salaries are at least twice the national average. In fact, tech salaries are going through the roof, and non-tech salaries are not—which presents a serious problem for many companies.

But, as important as money is to tech people, it's not the most important thing. Fundamentally, geeks are interested in having an impact. They believe in their ideas, and they like to win. They care about getting credit for their accomplishments. In that sense, they're no different from a scientist who wants credit for work that leads to a Nobel Prize. They may not be operating at that exalted level, but the same principle applies.

CREATE NEW WAYS TO PROMOTE YOUR GEEKS

If you don't want to lose your geeks, you have to find a way to give them promotions without turning them into managers. Most of them are not going to make very good executives—and, in fact, most of them would probably turn out to be terrible managers. But you need to give them a forward career path, you need to give them recognition, and you need to give them more money.

Twenty years ago, we developed the notion of a dual career ladder, with an executive career track on one side and a technical career track on the other. Creating a technical ladder is a big first step. But it's also important to have other kinds of incentives, such as awards, pools of stock, and nonfinancial kinds of compensation. At Novell, we just added a new title: distinguished engineer. To become a distinguished engineer, you have to get elected by your peers. That requirement is a much tougher standard than being chosen by a group of executives. It's also a standard that encourages tech people to be good members of the tech community. It acts to reinforce good behavior on everyone's part.

EITHER GEEKS ARE PART OF THE SOLUTION—OR THEY'RE THE PROBLEM

Here's another thing you need to know about the geek mind-set: Because tech people are scientists or engineers by training, they love to solve really hard problems. They love to tackle a challenge. The more you can get them to feel that they're helping to come up with a solution to a tough problem, the more likely they are to perform in a way that works for you.

When you talk with them, your real goal should be to engage them in a dialogue about what you and they are trying to do. If you can get your engineering team to agree with what you're trying to accomplish, then you'll see them self-organize to achieve that outcome. You'll also need to figure out what they're trying to accomplish—because, no matter what you want, that's probably what they're going to do.

The next thing you need to remember is that you can tell them what to do, but you can't tell them how to do it. You might as well say to a great artist, "I'll describe to you what a beautiful painting is. Then I'll give you an idea for a particular painting. I'll tell you which colors to use. I'll tell you which angle to use. Now you just paint that painting." You'd never get a great painting out of any artist that way—and you'll never get great work out of your geeks if you try to talk to them

like that. You need to give them a problem or a set of objectives, provide them with a large amount of hardware, and then ask them to solve the problem.

THE BEST JUDGES OF GEEKS ARE OTHER GEEKS

Make sure that there is always peer-group pressure within your project teams. For example, if you want to motivate your project leaders, just require them to make presentations to each other. They care a great deal about how they are perceived within their own web of friends and by the professional community that they belong to. They're very good at judging their own. And they're often very harsh: They end up marginalizing the people who are terrible—for reasons that you as a manager may not quite understand.

It sounds like I'm touting tech people as gods, but there are plenty of bad projects, and there is plenty of bad engineering and bad technology. You're always going to encounter techies who are arrogant and who aren't as good as they think they are. A team approach is the best way to deal with that problem. Tech people know how to deal with the wild ducks in their group—on their own and with the right kind of peer pressure.

LOOK FOR THE NATURAL LEADERS AMONG YOUR GEEKS

In a high-tech company that is run by engineers, what matters most is being right. And what's "right" is determined by outcomes. You can listen to lots of exceptionally bright people talk about their brilliant vision. I've done it for the past 25 years. But what matters is, Which ones deliver on their vision? When a project is on the line, who actually gets the job done?

Every team has a natural leader—and often that leader is not a team's official manager. Your job is to get the team motivated. Once you do that, the natural leaders will emerge very quickly. If you keep an eye on the team, you can figure out who those natural leaders are—and then make sure that they're happy and that they have everything they need to do their job. For instance, natural leaders need to feel that they have access to the company's senior managers. Don't forget: They feel like they're changing the world—so you need to make them feel like you're helping them do that.

There are easy ways that you can help them out. For example, encourage them to bypass layers of middle management and to send you email directly. Sure, that will piss off the people in middle management, but it's better to piss off those people than to piss off your key project leaders.

BE PREPARED FOR WHEN THE GEEKS HIT THE FAN

You can divide project teams into two categories. First, there is the preferred variety: You get an engineering team that's absolutely brilliant, that executes well, and that's exactly right in its assumptions. Second, there is the more usual variety: You get an engineering team that has a very strong opinion about what it's trying to do—but that's on the wrong track, because some of its assumptions are wrong. That second kind of team is what you have to focus your attention on. But often you can't intervene from the top down. You have to find a way to come at the problem from the side.

At Novell, we have a series of checkpoints at which our teams get lateral feedback—feedback that comes from outside of the management hierarchy. Every six weeks, we have three days of product reviews. But it's not just management conducting those reviews. We also bring in smart technologists with good memories: They remind us of what everybody committed to.

In most technology companies, there are always a few people who, everyone agrees, have better taste than anyone else. Those are the people whom everyone goes to; they serve as reviewers or advisers. At Sun Microsystems, for instance, it's Bill Joy. At Novell, it's Drew Major, the founder and chief scientist. Everyone knows that when Drew gets involved in a project, he'll size up

quickly what needs to get done, and people will listen to him.

In general, as long as you consider everyone's ideas, most teams react well to management decisions. If you have to make a business decision that conflicts with what your engineers want to do, they'll accept it—as long as it is truly a business decision. On the other hand, if the decision is based on a technology analysis by someone whom the engineers do not respect professionally, then they'll never agree to it. So, if you're facing a decision that you know will affect a team negatively, you must vet that decision through a technologist who has that team's respect.

TOO MANY GEEKS SPOIL THE SOUP

If you want your geeks to be productive, keep your teams small. The productivity of any project is inversely proportional to the size of the project team. In the software business, most problems draw on the efforts of large numbers of people. Typically, companies deal with a problem by putting together a large team and then giving that team a mission. But in this industry, that approach almost never works. The results are almost invariably disappointing. Still, people keep doing it that way—presumably because that's the way they did it last year. The question is, How do you break out of that mode? It seems to be a cancer on the industry.

On a large team, the contributions of the best people are always smaller, and overall productivity is always lower. As a general rule, you can count on each new software project doubling in team size and in the amount of code involved—and taking twice as long—as the preceding project. In other words, the average duration of your projects will go from 2 years to 4 years to 8 years to 16 years, and so on. You can see that cycle with almost any technology. Two or three people invent a brilliant piece of software, and then, five years later, 1,000 people do a bad job of following up on their idea. History is littered with projects that follow this pattern: Windows, Unix, Java, Netscape Navigator.

The smaller the team, the faster the team members work. When you make the team smaller, you make the schedule shorter. That may sound counterintuitive, but it's been true for the past 20 years in this industry, and it will be true for another 20 years. The only method that I've found that works is to restrict the size of teams arbitrarily and painfully. Here's a simple rule of thumb for techie teams: No team should ever be larger than the largest conference room that's available for them to meet in.

At Novell, that means a limit of about 50 people. We separate extremely large projects into what we call "Virtual CDs." Think of each project as creating a CD-ROM of software that you can ship. It's an easy concept: Each team has to ship a CD of software in final form to someone else—perhaps to another team, perhaps to an end user. When you treat each project as a CD, you enable one group to say to another, "Show me the schedule for your CD. When is this deliverable coming?" It's the kind of down-to-earth approach that everyone can understand, that techies can respect and respond to, and that makes almost any kind of project manageable.

EDITOR'S ADDENDUM

In a book entitled *Managing Einsteins: Leading High Tech Workers in the Digital Age* (McGraw-Hill, 2002), Tom Duening and Jack Ivancevich discuss the management of different types of geeks, whom they have labeled "Einsteins." The authors contend that all managers have to deal with difficult employees, which in this case are those annoying, irritating, and brilliant employees—the best and the brightest who also present special managerial challenges. The research reported in that book identifies six of the more difficult Einstein types that technical managers may encounter and provides some advice for dealing with them.

Arrogant Einsteins. These individual contributors often receive praise and accolades for their intelligence and their ability to solve problems. They are so good at this that some begin to believe their smartness gives them a special place in the world, and therefore, they are often perceived as arrogant prima donnas. Arrogant Einsteins typically flout authority and like to challenge managerial decisions. One suggested way for dealing with this type is to be "matter of fact" about their arrogance so that it cannot become an intimidating factor. The authors contend that managers may have to accept the fact that arrogant Einsteins will often gripe publicly about them. Tolerating a small amount of this will allow them to blow off steam as long as this griping doesn't affect the performance of others, in which case immediate steps must be taken to stop it. If one decides to meet with these individuals in private and listen to their complaints, which can sometimes help defuse potentially damaging behaviors, one must be serious about trying to take actions or make improvements since they expect that their ideas and suggestions will be used and lead to change. It can be counterproductive to listen to arrogant Einstein gripes if one has no intention of doing something.

Know-It-All Einsteins. In similarity with their arrogant brethren, these know-it-all Einsteins like to test authority but do it in a quieter and more subtle manner. As a result, they are harder to spot but therefore potentially more troublesome. These know-it-all Einsteins, according to the authors, will work on their own priorities and will generally have a group of followers. The authors' study also suggests that it is best to deal with them through indirect methods rather than confronting or sanctioning them since such direct attempts will only serve to reinforce their opinion that they know better. Indirect methods include keeping track of where they are and how they are applying their skills, perhaps by asking another person who works with the Einstein to quietly keep you informed. In this manner, the manager is not monitoring or micromanaging the know-it-all directly but can help keep him or her focused on the projects and tasks that the manager thinks are high priority. The authors also claim that astute managers can actually wear down know-it-all Einsteins and reduce the large amount of maintenance typically required if these individuals realize that their skills are greatly appreciated but also that they cannot hide and work on their own priorities.

Impatient Einsteins. One of the more annoying traits uncovered in the research and displayed by these Einsteins is impatience. In its extreme form, Einstein impatience can be very challenging to deal with since many of these individuals are also the superstars of their teams. They understand things quickly and don't need to be told twice. The authors also find that although they can be very deferential and obedient, they also get bored quickly. The authors advise managers to let them do their assigned tasks, as they will dig in with amazing intensity as long as they are working and progressing in the expected directions. Impatient Einsteins don't complain when they become bored; instead, they look around for other options so they can get back to using their knowledge and expertise. They are unlikely to tell managers when they become bored, so if they don't see other options or assignments on the horizon for using their skills, they are likely to simply leave the organization.

Eccentric Einsteins. These Einsteins are likely to enjoy behaving in some eccentric way at work or in their outside hobbies or interests. They are likely to become eccentric about some aspect of their work style, too, perhaps in the way they dress, eat, sit in meetings, come to work, and so on. The authors indicate that eccentric Einsteins revel in their eccentricity and that organizations should allow, perhaps even encourage, "healthy" expression of the eccentricity as long as it doesn't interfere with the work or professionalism that the organizations is trying to achieve. These Einsteins crave their individuality, and managers can assist in this process by permitting tasteful, nondisruptive expressions of their eccentricities.

Disorganized Einsteins. These Einsteins are ensconced in clutter, surrounded by old computers, wires, components, and other artifacts of their work. However, if you ask them to find something, they will often be able to find it amidst what appears to be disorganized junk and chaos. These individuals do not want to spend a lot of time organizing and filing. And this can be okay as long as valuable assets don't get lost and costs don't mount from huge amounts of time wasted searching for needed items. The authors make several suggestions for dealing with these disorganized Einsteins. One is to assign an individual to help keep track of the firm's more valuable assets or to develop some online software for managing important assets. However, disorganized Einsteins genuinely like being surrounded by their technical paraphernalia, collecting gadgets, technical papers, books, and so on—just be very careful about putting them in charge of assets and information that may be quickly needed.

Withdrawn Einsteins. The authors identify this group of Einsteins as neither arrogant nor eccentric, but unfortunately, they tend more toward underachievement. While there may be a whole host of reasons for this, the authors hypothesize that Einsteins are different—and they've been different since their youth. They are brighter, smarter, and prone to deeper analytic thoughts than their peers and so many adopt a withdrawing style to cope. To be effective, managers have to recognize the potential within withdrawn Einsteins and strive to help them release it, helping them overcome their resistance to change and providing support during the process. The challenge is to help withdrawn Einsteins become more visible and involved. The important tactic in helping these individuals is to get them to stick to an action plan that brings out the best in them.

Duening and Ivancevich argue that managers don't have to fret, worry, or sit idly by when confronted with these kinds of difficult workers. The authors hope that by understanding more about these six types, managers might take more responsibility for managing them rather than being passive observers.

6

Managing Creativity
A Japanese Model

MIN BASADUR

Dr. Min Basadur visited several major companies in Japan to conduct comparative research on organizational creativity. Unexpected insights emerged during interviews with Japanese managers and are the basis for this article. These managers knew a great deal about North American motivational theory and how to implement it. Employee creativity is managed through deliberate structural means, not to effect direct economic outcomes, but to develop motivation, job satisfaction, and teamwork. Contrasts to North American suggestion systems are made.

The rapidly accelerating rate of technological and environmental change demands much greater organizational adaptability than in the more stable past. Attempting behavioral change has turned out to be very difficult for many North American organizations because they have, by and large, developed along bureaucratic, non-flexible, and non-adaptive lines. Recent research has indicated that people at all organizational levels in North American business and industry can learn to think more creatively, to discover and solve important interfunctional problems, and to innovate new products and new methods faster, all of which results in greater organizational adaptability.[1] Simply put, creativity in organizations is a continuous search for and solving of problems and a creating and implementing of new solutions for the betterment of the organization, its customers, and its members.

Much has been written about the recent business success of Japanese corporations. It is often implied that superior management methods are the key. At the same time, the Japanese are viewed as not being truly creative. They are accused of being very good at copying and nothing more. For example, it is pointed out they have not produced many Nobel laureates, nor have they made many basic science discoveries. It could be argued that this is because they have not yet had the world class training needed by their scientists. Some observers believe the Japanese will soon begin producing Nobel laureates by making world-class training available. This belief is based on the fact that Japanese students are being sent to top North American institutions to learn mathematics and science from the current "masters," much like North American students went to learn from the European masters in the 19th century.

> The Japanese may already be better students of creativity than North Americans. They appear to be ahead of North Americans in implementing new ideas about management from the behavioral sciences which our own managers find difficult to accept. These new ideas include improved manufacturing and service management methods for higher quality, efficiency, and flexibility, such as "Just in Time" (J.I.T.), "Statistical Process Controls" (S.P.C.), and "Quality Circles" (Q.C.C.).

Many of these ideas originated in North America in the 1940s and 1950s but have never really caught on and were left in the classroom. Attempts to ap-

From Basadur, Min, "Managing Creativity: A Japanese Model," *Academy of Management Executive,* Vol. 6, 1992, pp. 29–40. Reprinted with permission.
Min Basadur is a Professor in the College of Business at McMaster University.

ply them in the workplace have often failed. Rather than admit we just don't want to change, North American managers have found it easier to assume that there is something mysterious about Japanese culture that permits new approaches to management to work over there but not here. This article examines the ways in which management ideas that originated in North America are being applied in Japan.

FINDING OUT ABOUT JAPANESE CREATIVITY

A bilingual Japanese colleague of mine set up open-ended interviews with five major Japanese companies including second and third visits in cases when it was necessary to probe more deeply. Comparisons were made with North American firms on emerging themes. To facilitate comparisons, data were gathered during the same time period from eleven leading North American companies. These data were obtained by a combination of questionnaire, in-depth interviews, and shop floor visits. The data from the Japanese and North American companies were organized along emerging themes, similarities, and contrasts. For example, would Japanese styles of creativity favor problem finding activity more than their North American counterparts? Another purpose was to see if Japanese organizations understood creativity as the process pictured in Figure 6.1 and do they try to implement the model.

The model in Figure 6.1 provides a framework for speculation about Japanese management practices. Creativity in organizations is a continuous finding and solving of problems and a creating and implementing of new solutions. *Problem finding* activity means continuously identifying new and useful problems to be solved. This may include finding new product or service opportunities by anticipating new customer needs, discovering ways to improve existing products, services, procedures, and processes or finding opportunities to improve the satisfaction and well-being of organizational members. Finally, problem finding includes defining such new problems and opportunities accu-

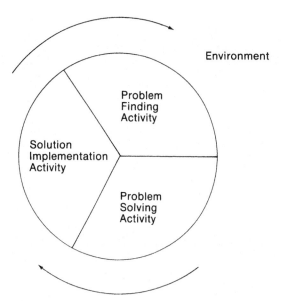

Figure 6.1. Creative activity in an organization.

rately and creatively. *Problem solving* activity means developing new, useful, imaginative solutions to found problems. *Solution implementing* activity means successfully installing such new solutions into the ongoing life of the organization.[2]

Problem finding may be the key to Japanese management success. The more emphasis placed on problem finding, the less is needed for solving and implementation. Solutions are more on target and successful implementation is facilitated. This is especially true when hierarchically lower level employees are invited to participate in the problem finding phase. Ownership and commitment are increased by early inclusion in change making. It takes less time to implement solutions when those affected have been permitted to participate from the beginning in finding and defining the problem and developing the solution.

Trained by traditional business schools in the "scientific management" approach originally identified by Frederick Taylor, most North American managers do not understand the importance of involving employees in early problem finding activities. They assume they are the only ones who know what needs to be done or that they can solve problems faster or better on their own. When these managers attempt to impose their solutions on their subordinates, there can be resentment and subordinates are often left uncommitted. Solutions fail either due to inadequate problem definition or lack of ownership. The same managers are likely to repeat the cycle over and over again hoping to find one solution that will finally be accepted and do some good. Such haphazard problem solving activity wastes human resources and detracts from managerial productivity. Training in creative problem solving is designed to improve upon these inadequate attitudes and behaviors but it is difficult to get many North American organizations to provide such training and to get it used on the job.

Organizations often view training as a luxury to be initiated only when business is good and the pressure is off. It is regarded as an educational experience serving as a reward for people having done a good job, rather than to change attitudes and behaviors. The training is not seen as something that can genuinely change and improve the way work is done. As a result, many management methods are rejected after a short trial period even when they have worked well elsewhere, notably in Japan. The real reason for rejection is often a lack of desire or willpower to make significant changes. Since change is the essence of creativity, the ability to foster change is a major indicator worth observing when comparing management practices in Japan and North America.

PROBLEM FINDING IS EMPHASIZED

The first Japanese company visited was a large international consumer electronics firm. Discussions were held with several senior R&D managers about managing this function. While there were many similarities to North American R&D management, one major difference that emerged was that newly hired R&D scientists and engineers always start their careers with six months in the sales department. The company wants them to learn first hand at the beginning of their careers about the needs and problems of their customers. In the long run,

their jobs will focus on meeting those needs and solving those problems. For the next eighteen months, the new hires gradually work their way back to R&D through stints in various other functions including manufacturing and engineering. This suggests an interesting organizational emphasis on inducing problem finding behavior (anticipating and sensing customer needs) through structural means (job placement and rotation).

The remaining four companies visited were world-class manufacturers of car parts and scale measurement instruments. This time the interviews were with manufacturing and personnel managers and centered on the nature of their Employee Suggestion Systems (E.S.S.). It is not uncommon for employees of top Japanese companies to conceive and implement between forty and one hundred new suggestions per person, per year on average. This figure might amaze most North American managers since leading U.S. companies consider themselves lucky to obtain an average of about two suggestions per person, per year (see Table 6.1). The rest of this article explains not only how it is possible to achieve the Japanese levels, but also the theoretical rationale and comprehensive organizational benefits.

The interviews were in-depth, open-ended question and answer sessions; and shop floor visits. The collected data reveal that the primary objectives of the E.S.S. are motivation, job satisfaction, and group interaction. There is an infrastructure which guarantees that all three phases of creativity are completed. Individuals are encouraged to find problems with their work and improve their own jobs. Suggestions are submitted only after the solution has been demonstrated to work successfully. All suggestions are accepted and given credit. Monetary awards for most ideas are small.

Quality circle activity provides a reservoir of problems to aid individual problem finding activity, and smart managers learn how to get individuals to select problems of strategic importance to solve. Employees are trained that suggestions desired include new and improved products as well as methods. Individuals are encouraged to ask co-

TABLE 6.1
Leading Japanese Companies

Company	# of suggestions	# of employees	per/employee
MATSUSHITA	6,446,935	81,000	79.6
HITACHI	3,618,014	57,051	63.4
MAZDA	3,025,853	23,929	126.5
TOYOTA	2,648,710	55,578	47.6
NISSAN	1,393,745	48,849	38.5
NIPPON DENSO	1,393,745	33,192	41.6
CANON	1,076,356	13,788	78.1
FUJI ELECTRIC	1,022,340	10,226	99.6
TOHOKU OKI	734,044	881	833.2
JVC	738,529	15,000	48.6
TYPICAL LEADING U.S. COMPANY	21,000	9,000	2.3

Reference: Japan Human Relations Association, April, 1988: "The Power of Suggestion"

workers for help in problem solving. If individuals or informal teams cannot solve certain problems, they are referred to a quality circle team or the engineering department for help.

Group-oriented quality circles work supportively with the individually oriented Employee Suggestion System in other ways as well. The team gets credit every time one of its members submits a suggestion. Major celebrations are held by top management each year-end honoring teams and individual members of teams who have performed well in their suggestion work. All new employees are trained the first day on the job about the importance of the E.S.S. and how it works. Managers and supervisors are trained to work closely with subordinates to help them find and solve problems, implement their solution, and provide plenty of positive feedback throughout.

R&D IS EVERYBODY'S BUSINESS

In all four companies, suggestions for improving both procedures and products are encouraged. Employees are trained from the first day on the job that "R&D is everybody's business." For example, in one company of 9,000 employees, 660,000 employee suggestions were received in one year. Of these, 6,000 were suggestions for new products or product improvements and the remainder were suggestions for new methods. New methods are improvements to the work itself—simplifying jobs, accelerating procedures and work flow, and so on.

PROBLEMS ARE GOLDEN EGGS

In the companies studied, creative activity is deliberately induced on the job in a manner that is consistent with Figure 6.1. On the first day on the job, new employees are trained that problems (discontents) are really "golden eggs." In other words, it is good to identify problems. One should be constructively "discontented" with one's job and with company products and seek ways to improve them. In some of the companies, the "golden eggs" are posted on large sheets of paper in the work area. Employees are then encouraged to interact with their co-workers to solve such problems and demonstrate that their solutions can be implemented.

In North America there is a real reluctance to identify problems. Employees, especially managers, often don't want anybody to know they've got problems because they are seen as a sign of weakness and poor performance. Subordinates soon pick up this attitude and adopt a problem avoidance approach to their work ("it didn't happen on my shift" and "that's not our problem").

This leads to neglect of important interfunctional opportunities for improvement and customer needs.

> In these Japanese firms not only are people taught, but there is also a structured mechanism for causing problem finding activity. Workers are provided with problem finding cards. If dissatisfied with something about one's job, the worker writes the discontent on the card and posts it up on a wall poster in the column marked "problems." Workers post their problems, their "golden eggs," their discontents, so other people can see them. If others notice a problem posted which is of interest to them, they will join forces to help solve it.

Group interaction is stimulated and people work together on the problems they select. Later they can write their solutions in the second column beside the problem on the wall poster. There is a third column for implementation documentation. When all three columns are complete, and the individual or small team has done the problem finding and the problem solving and has shown that the solution works, then it can be said that a suggestion has been completed, but not until then. This suggestion can now be submitted.

IMPLEMENTATION BEFORE SUBMISSION AND ALL SUGGESTIONS ACCEPTED

Although not all suggestions are actually implemented, all of them are accepted. In other words, when all three phases of the creative process are completed (problem found, problem solved, solution shown to be implementable) by the employees themselves, a suggestion has been created and is accepted. About ninety-six percent of the suggestions end up being put into practice.

An "idea" is not a "suggestion" until it has gone through all three stages of the creative process modelled in Figure 6.1. Every suggestion receives a monetary award. The vast majority of the suggestions are small $5 (500 yen) ideas. These are accepted and assigned the award by the supervisor on the spot. The suggestions that are more creative and significant are evaluated by a committee against multiple criteria including creativity and contribution to goals; they receive bigger awards of up to $10,000 and more.

The main objective is to accept all ideas and encourage the little ones as well as the big ones. It is the *process* of getting involved in one's work that counts, not the quality of any single idea. The goal is to have thinking workers and a spirit of never-ending improvement. Of the small ratio (about four percent) of accepted suggestions that do not get implemented right away, most are the kind that require skills beyond the scope of the suggestors. The team leader or the supervisor can get additional help from other departments for these ideas. Also, it may be found that the implementation of a suggestion is not timely or is inappropriate in the bigger picture. In this case, the idea is not implemented, but is given credit anyway. This is the way the system is supposed to operate and works very well in actual practice.

Employees are told they are expected to create new ideas. Some companies even establish informal goals per person per month. Each formal work group has a team leader who ensures that daily production is met and new ideas keep flowing at the same time. The team leader communicates, coordinates, and gets help across the organization as needed. This prevents the work group from worrying unduly about maintaining daily production and saying "we don't have time to work on new ideas." Workers are given overtime as needed to complete their suggestions. The overtime is usually aimed at implementation work. Much of the problem finding and problem solving work is done continuously in people's minds off the job as well as on the job. When people are creatively involved in their work, ideas about new problems and solutions can occur to them at any time.

COACHING, POSITIVE FEEDBACK, AND FACILITATOR SKILLS EMPHASIZED FOR MANAGERS

The secret to making this process work begins with getting people to take ownership of problem finding as well as evaluation and implementation. Employees learn to accept evaluation and implementation of their ideas as part of their jobs. Their supervisors and managers support them and help them to be successful throughout the process. This includes helping the employee evaluate a potential suggestion's worthiness and how to make it work.

The boss is trained to be an encourager and coach, providing positive feedback at every opportunity. The system is structured to make sure such coaching and feedback occurs. A supervisor will help a new employee find a "golden egg" and develop a suggestion as part of the orientation process. Employees are given coaching on the appropriateness of "golden eggs" to be posted and positive feedback on all contributions.

On larger projects, the team leader and supervisor make sure that additional time (including overtime) and other resources are made available to workers as needed. Also, teams routinely make presentations to the rest of the organization during working hours, typically in the company cafeteria. The plant manager acts as a master of ceremonies, giving praise, recognition, and expert commentary as each project is presented. Suggestions which require higher level consideration for awards or implementation enter a formal system of evaluation and feedback. The suggestors are given feedback and positive recognition by design at several stages of this formal process.

Managers are not permitted to submit suggestions—that is, to get directly involved in the Employee Suggestion System; however, they are trained to get indirectly involved. For example, if a manager happens to think up an idea, rather than submit it, he or she is trained to figure out what problem that idea is trying to solve. The manager then goes down into the ranks and seeks out someone willing to post that problem. The group, or anybody in the group, can solve it themselves, probably with a different solution. This is how problem ownership is built.

> Managers learn how to "dump problems into the fray" and let the ownership grow. This contrasts with the old-fashioned scientific management approach which designates management as "thinkers" and labor as "doers," which is not very scientific at all because thinking is done from the top down and wastes the minds of the workers. Worse yet, changes are usually sprung on the work force suddenly and are resisted.

First line supervisors find themselves stuck in the middle—expected to support the change but facing an unwilling, untrusting, and unaccepting group of subordinates who feel no ownership for the change. According to this research, Japanese managers are trained to facilitate change, not impose it. The Employee Suggestion System provides an excellent tool to accomplish this facilitative approach.

MOTIVATION IS THE OUTCOME

When the top managers of these leading companies were asked what the primary objective of their Employee Suggestion System was, none of them said new products or new methods. Furthermore, none of them said lower costs, or higher profits. In fact, none of them mentioned any final economic outcomes. In contrast, all of them said *motivated people*.

> These Japanese organizations believe that workers get motivated when they get a chance to be creative on the job. Employees enjoy coming to work. This is what the Japanese call "cheerfulness" and we call "job satisfaction." This creative activity also stimulates group interaction. People help each other solve problems which provides the opportunity for genuine team building.

People find real reasons to work together and feel good about their accomplishments, monetary awards, and the fact that their work team gets credit for their individual suggestions. Individual awards, especially larger ones, are shared with the team. The team decides how much the individual gets and how much they keep for their "activity fund." The activity fund is accumulated by the work teams to fund personal development, recreation, physical education, and other growth activities. The fund grows from quality circle awards and employee suggestion awards. Individuals get recognition and the team gets recognition.

All of the companies said they have found that when people are given the opportunity to engage in creative activity (as it has been described here), they become very motivated. This causes them to want to participate even more in creative activity. It also causes them to work harder on performing their normal routine jobs better—more quality, more quantity, and lower cost. This is consistent with increasing organizational efficiency and short-term organizational effectiveness. Figure 6.2[3] models this simple management process.

CONSISTENCY WITH MOTIVATION RESEARCH

Organizational research conducted by P.E. Mott showed that effective organizations have three major simultaneous characteristics: efficiency, adaptability, and flexibility.[3] Efficiency is the ability to organize for routine production. Every organization is turning out some kind of product (a needed good or service).

Efficient organizations are customer focused; they know their customer and product. Over the years they have developed good routines for making their product the best they can with current technology. They produce a high quantity, quality product, and maintain a high output over input ratio (low cost) during production.

Effective organizations are also able to respond and react to sudden temporary changes or interruptions. They can deal with unexpected disruptions and get back quickly to their normal routine without getting stuck in red tape. Flexibility is a way of preserving efficiency. Flexibility and efficiency are both necessary in the short run.

Adaptability is a longer range characteristic and refers to an organization's capacity to continually and intentionally change its routines and find new and better ways to do the work. Adaptable organizations anticipate problems and develop timely solutions. They stay abreast of new methods and technologies that may be applicable to the organization. The organization's members accept good, new ideas and make sure new solutions and techniques get installed and maintained. Acceptance of new ideas is widespread across all organizational departments.

The creative process of problem finding, problem solving and solution implementation becomes more vital as the amount of change confronting the oganization increases. Up until recently, many or-

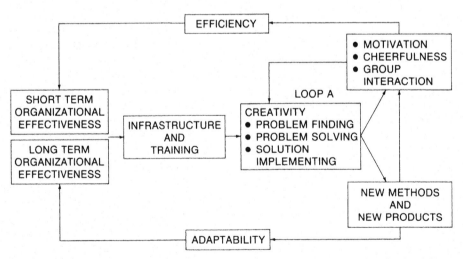

Figure 6.2. Japanese model for fostering problem finding and creativity to increase motivation, job satifaction, and teamwork.

ganizations could be effective by concentrating only on efficiency and flexibility. Today, adaptability is equally important because of the rapidly accelerating rate of change. Adaptability is crucial for long-term effectiveness.

Motivating people by providing the opportunity for creative activity is consistent with the motivation literature in industrial and organizational psychology. One major category of motivational theories are the *need* theories. Two important motivational need sets are the need for competence and the need for curiosity and activity. These two needs and related motives provide the most direct explanation of how creativity is a means for motivating people.[4]

> People have a desire to master their environment. Such mastery is intrinsically pleasurable and independent of outside rewards. This need for competence is aroused when people are faced with new challenging situations and dissipates after repeated mastery of the task. The concept of intrinsic motivation is also consistent with the notion that curiosity, activity, and exploration are enjoyed for their own sake. This was discovered in early animal research and in later studies in humans.

People develop negative attitudes toward repetitive tasks and report experiencing fatigue and boredom. Berlyne suggests that people adapt to certain levels of stimulation and take action to reduce discrepancies from these levels. The implication is similar to Herzberg's notion that challenging jobs are motivating in themselves.

Other motivation theories are also consistent with what is being practiced in the companies in this study. Herzberg proposes that the way to motivate most people is by redesigning their jobs so the work itself provides opportunity for growth, challenge, stimulation, learning, and recognition.[5] McClelland has advanced the need for achievement as the primary driving force motivating organizational members.[6] By giving employees the opportunity to find challenging problems, solve them, and implement solutions, the Employee Suggestion System taps into both the forces of intrinsically re- warding work and the need for achievement. According to Maslow, offering employees the opportunity to satisfy their higher level needs for self-esteem and for self-actualization through work accomplishment is the best way to motivate them.[7]

The Japanese Employee Suggestion System is a straightforward example of how these two highest level needs can be met. People are provided with the opportunity to use their creativity. They seek out work-related challenges of interest to themselves, then find success and recognition in developing implementable solutions that are welcomed and celebrated by the organization.

Motivation theory has not remained static since the 1950s and 1960s. Deci and Ryan provided a comprehensive review of intrinsic motivation.[8] Locke and Latham showed that when people are given a chance to set their own goals (the problem-anticipating aspect of problem finding) and the more specifically they state those goals (the problem definition aspect of problem finding), the more motivated they are to achieve those goals.[9]

The Japanese may not yet be at the forefront of initiating theoretical research, but their ability to apply motivation theory is impressive and far more than simple copying. As Japanese students continue to learn from the "masters," the time will come when their research informs our practice.

The vast majority of North American business and industry is still organized and managed on the scientific management concept made popular by Frederick Taylor in 1911.[10] One of the main premises of scientific management is that people at work are motivated by one dominant factor—money. This is the concept of "economic man." In spite of research showing that most people at work are multi-motivated (money does play a role but in a complex way), most managers continue to manage by simplistic, economic formulae.

The motivating factor in most North American employee suggestion systems is extrinsic, usually money. A few employees suggest a few big ideas that save the company large sums of money and win major cash awards for themselves. Most people don't participate.

In contrast, Japanese employee suggestion

TABLE 6.2
Contrasting Elements Summary for Employee Suggestion Systems

	New (Japanese)	Traditional (North American)
Culture	• Group & individual synchronized	• Individual
Core objectives	• Thinking workers • Never ending improvement • Individual growth • Communications • Decision making	• Breakthrough • Produce savings
Management	• Primary responsibility	• Secondary at best
Area of suggestion	• Within your job & your workplace	• Outside of your job or your workplace
Evaluation	• Simple • Quick answers • Supervisor responsibility • Lots of suggester involvement • Most accepted	• Very structured • Slow answers • Evaluator responsibility • Little suggester involvement • Most rejected
Communication	• Employee to supervisor • Employee to employee	• Employee to evaluator to supervision to management
Awards	• Intrinsic	• Extrinsic

systems emphasize a large number of small ideas and everyone participates. There are small monetary awards for each implementable suggestion shared by the participating members. Larger awards are given for ideas of greater scope, but the vast majority of suggestions win small awards. The real awards, as far as employees are concerned, are the feelings of accomplishment, recognition, and growth.

TOP DOWN IMPETUS AND STRATEGIC ALIGNMENT

In the Japanese companies interviewed, training and a well-developed infrastructure are used to make creative activity important. They are also used to align creative activity strategically with important organizational goals as an every day routine. Managers are trained to help their employees find, solve, and implement problems and solutions.

As clear company goals are articulated by top management and specific departmental objectives

and subgoals are developed, these are communicated downward to guide individuals and teams in their selection of problems. This results in a close alignment of E.S.S. activity with strategic corporate needs. Managers who are skillful in the E.S.S. learn how to influence their subordinates toward including problems which are related to specific goals and objectives for their departments.

The reward system reinforces the importance of creative activity to the company. Not only does the E.S.S. provide extrinsic and intrinsic rewards for employees, but their managers' performance appraisals are also based in part on their ability to get their subordinates to perform well in the E.S.S.

In the Japanese companies I visited, Management by Objectives (M.B.O.) is integrated with the Employee Suggestion System. Typically, the manager's objectives will include helping people create and implement suggestions. This emphasis on getting subordinates involved in creating new ideas is part of the long-range process of management. The belief is that if people are encouraged to use their thinking power on a habitual daily basis, major tangible benefits will accrue to the organization in the long run.

Quality circle (Q.C.C.) group activity also serves to help align E.S.S. activity with strategic goals. Q.C.C. work is a concentrated attack on major "theme" problems identified by upper management. These themes are assigned about every six months. A bonus of Q.C.C. activity is that it also provides a regular forum for spontaneous discussion of spinoff problems during Q.C.C. team meetings. The Q.C.C. infrastructure serves as a deliberate reservoir for problem finding to fuel the Employee Suggestion System program. Both the group-oriented Quality Circles and the individually oriented Employee Suggestion System are sparked by top management involvement. Not only is top management instrumental in setting direction and relevant goals, it also works hard insuring that such goals are followed up. Celebrations are hosted at the end of the year by presidential level management for teams which have performed well in Q.C.C. and E.S.S. activity.

JOB REDESIGN, ENRICHMENT, AND ADAPTABILITY

Proactive creative activity leads to a continuous supply of new methods and new products. This is synonymous with Mott's definition of organizational adaptability. Not only are new problems deliberately anticipated and solved, but also acceptance of the new solutions by employees is virtually assured because the employees are finding and solving their own problems and implementing their own changes. They have high ownership of the solutions and are redesigning their own jobs. This is consistent with a well-documented axiom of organizational psychology: "People don't resist change; they do resist being changed."

Herzberg's research on job satisfaction suggests that motivation can be achieved best by factors intrinsic to the work itself, such as responsibility and opportunity for growth and achievement. The validity of job enrichment, which is based on Herzberg's dual factor theory is supported by the findings reported in this article. Many companies have tried to redesign employee jobs to make them more intrinsically rewarding, however, evaluation of research results have been inconsistent. This may be because employees do not participate in it. The Japanese model goes one step further by letting employees be creative and allowing them to enrich their own jobs. Perhaps this is the missing link for North American companies who have tried other approaches to job enrichment and failed.

TEAMWORK AND INDIVIDUAL WORK HARMONIZED

When individuals start working together on problems of common interest, solve them, and implement solutions together, group cohesiveness develops. Cohesiveness is an important factor in group productivity. The E.S.S. encourages small, informal teams to develop. People who want to work together on problems of common interest join up. The attraction contributes to cohesiveness. Group cohesiveness is also built more formally through the Quality Circle approach. Even though Q.C.C. activity is, in theory, voluntary, in actual practice everybody is a member of a quality circle team because it is the same as their functional work unit team.

One of the firms stressed that in their experience Q.C.C. didn't work well alone and neither did E.S.S., but together they worked very well. The firm recommended that both be used for best results.

The Quality Circle (Q.C.C.) system provides opportunities for group performance, recognition, and initiative. The Employee Suggestion System (E.S.S.) adds opportunities for individual performance, recognition, and initiative. Q.C.C. activity is highly structured and uses analytical problem solving tools such as fishboning and root cause analysis. The team must stick to the theme and not pursue other problems or ideas. Prior to the introduction of E.S.S., this restriction bothered many people. If one were sitting in a Quality Circle working on the assigned theme and suddenly thought up an idea to solve a totally unrelated problem, it would be frustrating to not be permitted to voice the idea.

The Employee Suggestion System provides an outlet for finding, solving and implementing solutions to off-theme problems. The team gets credit for every suggestion that one of their team members submits individually and there is little conflict between the E.S.S. and Q.C.C. systems.

In contrast, attempts by some North American companies to install group-based Q.C.C. systems have run into conflict with long-established individual-based suggestion sytems. These companies have not yet figured out how to integrate the two systems.

HOW DO NORTH AMERICAN SYSTEMS COMPARE?

One key to the success of the Japanese Employee Suggestion System is the emphasis on problem finding. In North America, promotions and rewards go more often to people who appear not to have many problems. Managers don't feel they have enough time for problem finding. They feel they are too busy doing their "regular work" which often means fire-fighting activity and meeting short-term cost and profit goals. They want their people feeling the same way, and put focus on solutions, not problems. While the term "constructive discontent" is something that is often given lip service in North American organizations, the Japanese companies studied in this research are promoting and implementing it through simple structural methods.

Most North American suggestion systems use the suggestion box approach. Employees dump ideas in the suggestion box without the responsibility of evaluating them first or explaining just what the problem is that they are trying to solve. Managers evaluate the ideas and the employee waits to hear the judgment. Usually, the wait is long and most ideas are rejected. Managers find it onerous to judge so many suggestions and worry about the amount of change they represent. Many sug-

gestions are difficult to understand since they have neither been discussed, nor shared with other employees. There is no incentive to share an idea with anybody for reasons such as the boss may not want to hear about new changes, other employees will want to share in the award, or someone may claim it as their own idea. The main incentive is to make lots of money for the individual submitting the suggestion. Small ideas are not worth the effort.

> Teamwork, job satisfaction, and motivation are all secondary. In addition, many employees of North American companies do not receive awards for suggestions to improve their own job. They are rewarded only for ideas that are outside their own job. This goes against all the rules of motivation theory.

Finally, in many traditional North American companies, new product ideas are considered the job of R&D departments exclusively. Suggestion systems are concerned only with methods and procedures to save money or increase efficiency. New product ideas are not encouraged from employees of other departments and usually there are no organizational mechanisms to facilitate their emergence or development.

DISCOVERING HOW AND WHY JAPANESE ORGANIZATIONS INDUCE CREATIVITY

The major discovery of this research is that Japanese organizations demonstrate a great deal of knowledge about inducing employee creativity through deliberate structural means. They believe they derive important benefits in doing so. This study indicates that top Japanese organizations recognize, emphasize, support, and induce problem finding which is elevated to at least equal priority as problem solving and solution implementation. They recognize all three as separate important activities which is consistent with research that suggest that all three activities need to be nurtured and managed to achieve organizational creativity. They have devised structural means through the way they place R&D hires and their Employee Suggestion Systems to induce creativity throughout the organization.

Through managing the Employee Suggestion System, the Japanese companies in our study implement what theory and literature suggests needs to be done to induce creative behavior, to get creative output in the organization, and to motivate members of the organization. By doing so, they get tangible creative output like short-term costs savings and new products and procedures. They also reap other important benefits, the most important being motivated, committed people who enjoy their jobs, participate in teamwork, and get fully involved in advancing the company goals.

NOTES

The author would like to acknowledge Professor Mitsuru Wakabayashi, associate professor, Dept. of Educational Psychology, Nagoya University, Dr. Bruce Paton, vice president, Internal Consulting, Frito-Lay, Inc., and Jim O'Neal, president of Northern European Operations. Pepsi-Co Foods International for their help in laying the groundwork for this research.

1. For discussion and supporting data on organizational creativity see the author's following research. M.S. Basadur, G.B. Graen, and S.G. Green, "Training in Creative Problem Solving: Effects on Ideation and Problem Finding and Solving in an Industrial Research Organization," in *Organizational Behavior and Human Performance, 30,* 1982, 41–70; M.S. Basadur, "Needed Research in Creativity for Business and Industrial Applications," in S.G. Isaksen (ed.) *Frontiers of Creativity Research: Beyond the Basics* (Buffalo, N.Y.: Bearly, 1987); M.S. Basadur, G.B. Graen, and T.A. Scandura, "Training Effects on Attitudes Toward Divergent Thinking Among Manufacturing Engineers," in *Journal of Applied Psychology,* Vol. 71, No. 4, 1986, 612–617.

2. For more information concerning the creative process in organizations, see M.S. Basadur, G.B. Graen, and M. Wakabayashi, "Identifying Individual Differences in Creative Problem Solving Style" in *Journal of Creative Behavior,* Vol. 24, No. 2, 1990, 111–131; M.S. Basadur, "Managing the Creative Process in Organizations," in M.J. Runco (ed.) *Problem Finding, Problem Solving and Creativity* (New York: Ablex, 1991, in press). The latter is also available from the author as McMaster University Faculty of Business Research and Working Paper Series, No. 357, April 1991.

3. See P.E. Mott, *The Characteristics of Effective Organizations* (New York, NY: Harper and Row, 1972); M.S. Basadur, "Impacts and Outcomes of Creativity in Organizational Settings," in S.G. Isaksen, M.C. Murdock, R.L. Firestein, and D.J. Treffinger (ed.) *The Emergence of a Discipline: Nurturing and Developing Creativity,* Volume II (New York: Ablex, 1991; in press). The latter is also available as McMaster University Faculty of Business Research and Working Paper Series. No. 358, April 1991.

4. For more discussion on human needs and related motives see D.E. Berlyne, "Arousal and Reinforcement" in Nebraska Symposium on Motivation, D. Levine, ed. (Lincoln, NE: University of Nebraska Press, 1967) and R.W. White, "Motivation reconsidered: The concept of competence," *Psychological Review,* 66(5), 297–333.

5. For further discussion on motivation see F. Herzberg, B. Mausner, and B. Snyderman, *The Motivation to Work* (2nd ed.) (New York, NY: Wiley, 1959).

6. See D.C. McClelland, *Personality* (New York, NY: Dryden Press, 1951).

7. See A.H. Maslow, *Motivation and Personality* (New York, NY: Harper and Row, 1954).

8. See E.L. Deci and R.M. Ryan, *Intrinsic Motivation and Self-determination in Human Behavior* (New York, NY: Plenum Press, 1985).

9. See E.A. Locke and G.P. Latham, "Work Motivation and Satisfaction: Light at the End of the Tunnel," *Psychological Service,* Vol. 1, No. 4, July 1990, 240–246.

10. For review of scientific management see F.W. Taylor, *Principles of Scientific Management* (New York, NY: Norton, reprinted 1967, originally published in 1911).

7

Managing Innovation

When Less Is More

CHARLAN JEANNE NEMETH

While a "good company" requires unsurpassed management, product quality, and financial soundness, the "most admired" companies are presumed to also have "a spark that ignites the work force and allows the enterprise to respond readily to change. That ingredient is innovation and all the top companies embrace it passionately."[1]

Is this really true? Do our most admired companies emphasize innovation as much as execution? I think not. Most companies, even those considered "visionary," emphasize mechanisms of social control rather than innovation. They recognize the power of clear goals, worker participation, consistent feedback, a cohesive work force, and a reward system that underscores desired behaviors and values. In fact, the "spark" that many companies are likely to ignite is not innovation or risk taking, but rather loyalty and commitment to the company. They attempt to create a cult-like culture involving passion and excitement. Through this path, they may achieve productivity and high morale, but at the same time can thwart creativity, innovation, and an ability "to respond readily to change."

Creativity and innovation may require a "culture" that is very different and, in a sense, diametrically opposed to that which encourages cohesion, loyalty, and clear norms of appropriate attitudes and behavior. It is wrong to assume that the mechanisms of social control that heighten adherence to company rules and expectations can also be easily used to enhance innovation. Desiring and expecting creativity—and even rewarding it—do not necessarily increase its appearance. Motivation and in-creased effort may permit new variations on a theme, but they are unlikely to stimulate major changes in perspective or reformulation. Quite the contrary. In fact, there is evidence that one must be removed from this social control. One must feel free to "deviate" from expectations, to question shared ways of viewing things, in order to evidence creativity. To use a metaphor from an actual creativity task, one must be able to look "outside the box" to find new insights, e.g., to see a new use for an old product or to recognize a new market.

Minority viewpoints have importance and power, not just for the value of the ideas themselves, but for their ability to stimulate creative thought. Thus, one must learn not only to respect and tolerate dissent, but to "welcome" it. The "trick" is to balance coordinated group activity with an openness to differing views—to create unity in the organization without uniformity.

THE "GOOD" AND THE "VISIONARY"

In a recent book, Collins and Porras attempted to analyze "visionary" companies, those so nominated by CEOs as not only "good" or successful, but enduring.[2] The criterion "endurance" was defined as a company that was founded prior to 1950 and that had multiple product or service life cycles. Examples of those nominated included Hewlett Packard, IBM, Procter and Gamble, Wal-Mart, GE, Boeing, 3M, Nordstrom, Merck, and Walt Disney.

In their analysis, Collins and Porras tried to

Reprinted from the *California Management Review*, vol. 40, no. 1, by permission of the Regents. Copyright © 1997, by the Regents of the University of California.

Charlan Jeanne Nemeth is Professor of Psychology at the University of California in Berkeley.

discern what distinguished "visionary companies" from their less successful counterparts. One particular characteristic stands out. Collins and Porras argue that the visionary companies are marked by a cult-like atmosphere which includes a fervently held ideology, indoctrination, a high degree of "fit" or uniformity, and elitism. Such comparisons between strong "corporate cultures" and "cults" have been noted by other authors as well.[3]

A fervently held ideology is assumed to be important because it provides the core values, the glue that will bind the organization. Table 7.1 illustrates the core values of the "visionary" companies and includes examples, e.g., Nordstrom's credo of "service to the customer above all else." Some companies have explicit rules as well as a general statement. Thomas Watson, Sr., founder of IBM, for example, not only had core values of "respect for the individual" and "listen to the customer," but he had rules. There was a dress code which included dark suits. Marriage was encouraged; smoking was discouraged and alcohol was forbidden. At Disneyland, there is a strict grooming code, one which does not permit facial hair or dangling jewelry.[4]

Newcomers to the organization are indoctrinated into this prevailing ideology by various socialization techniques. The ideology may be succinct. For example, Nordstrom's employee "handbook" consists of a single 5"×8" card with the rule: "Use your good judgment in all situations. There will be no additional rules." While there may not be additional rules, there is considerable "on the job" socialization, including guidance in the form of approval and disapproval by peers and continual ranking and feedback as to performance. There are also intensive orientations, universities, and training centers.

Another common technique includes a mythology of "heroic deeds," which is essentially an oral history of people who have exemplified the ideology of the company—for example, extraordinary service to a customer. And, finally, there is the usage of unique language, mottoes, and corporate songs, all of which foster cohesion and a sense of group belonging.

WHY IT WORKS: MAJORITY POWER

That such a cult-like climate would enhance morale, loyalty, and adherence to normative prescriptions is not particularly surprising. These companies have used well-established principles of social control and, in so doing, recognize the importance of normative prescriptions, of the power of the judgments and approval of peers. This power of peers is one of the most established findings in social psychology. When people are faced with a majority of others who agree on a particular attitude or judgment, they are very likely to adopt the majority judgment. Literally hundreds of studies have documented this finding.[5] Even when using objective issues (such as judging the length of lines) people will abdicate the information from their own senses and adopt an erroneous majority view. The question is: Why?

The available evidence suggests that there are two primary reasons for adopting normative or majority views, even when incorrect. One is that people assume that truth lies in numbers and are quick to infer that they themselves are incorrect when faced with a unanimous majority. The other reason is that they fear disapproval and rejection for being different.[6]

It is difficult to overestimate the power of these majority views. In the early studies with unambiguous and easy judgments of line length, fully 35% of the judgments were against the individual's own senses and in conformity with the erroneous majority judgment.[7] In later studies using color

TABLE 7.1
"Core Values" (as defined by Collins and Porras)

"Respect and concern for individual employees"	Hewlett-Packard
"Exceed customer expectations"	Wal-Mart
"Being on the leading edge of aviation: being pioneers"	Boeing
"Respect for individual initiative"	3M
"Service to the customer above all else"	Nordstrom
"We are in the business of preserving and improving human life"	Merck

judgments, conformity was as high as 70%.[8] Given that participants in these studies were relative strangers, one can only imagine the magnitude of this pressure when the majority members are valued co-workers, colleagues, and bosses.

People worry about being different, about not being accepted—and worse, about being ridiculed or rejected. The evidence suggests that this is not a baseless fear. It is a highly predictable reaction. When a person "differs," especially when the group is cohesive and is of importance to the members, that "deviate" can be certain of receiving the most communication, usually aimed at changing his or her opinion. If such persuasion is unsuccessful, the person is often disliked, made to feel unwelcome, and may ultimately be rejected.[9] Such reactions can be found in even temporary groups with little at stake in the issue. It is exacerbated when the group is important and the issue is relevant to the group itself.

Recent work shows that majorities not only shape judgments and behavior, but they also shape the ways in which individuals think. We now have numerous studies showing that, when faced with a majority view that differs from their own, people tend to view the issue from the majority perspective.[10] In an attempt to find the majority to be correct, they not only adopt the majority position, but they convince themselves of the truth of that position by how they think about the issue. They consider the issue only from the majority perspective, trying to understand why the majority takes the position it does.

Faced with a majority, people search for information in a biased manner. They consider primarily information that corroborates the majority position. They also tend to adopt the majority strategy for solving problems to the exclusion of other strategies. They are also relatively unable to detect original solutions to problems.[11] In some sense, they "brainwash" themselves by finding and focusing on information consistent with the majority view.

An important limitation to the power of the majority appears to lie in the issue of unanimity. Numerous studies have documented that a single dissenter can break the power of the majority. Further, the dissenter need not agree with anyone else. It is not a question of providing support for a given viewpoint; it is a question of the majority being fragmented and divided.[12] This element is one reason why dissent can have value.

ENHANCING THE POWER OF THE MAJORITY

Many "visionary" companies seek to enhance the power of the majority so as to increase adoption of company principles. They not only have clear values, goals, and indoctrination technique, but they tend to promote interaction with like-minded individuals (i.e., within the company) and reinforce it with an intolerance for dissent. Much like cults, many of these companies tend to isolate the individual from the "outside world." They promote socializing *within* the organization and try to inhibit the maintenance of relationships outside the corporate family. These are powerful mechanisms for promoting cohesion and consolidating opinion.

One of the best documented findings in social psychology is that discussion among like-minded individuals increases both the extremeness of the views and the confidence in them. After discussion, groups of "risky" people become more risky; groups of "cautious" people become more cautious. Groups of people who favor segregation become more extreme and sure of their position. So do groups of people who favor integration.[13] They don't just become more "like" each other; they all become more extreme as well as more alike. Furthermore, people are more likely to *act* on their beliefs after discussion with like-minded individuals. Thus, promoting interaction within the company and shielding people from dissenting views are likely to make the individuals adhere to the company ideology more strongly and to act in accordance with it.

The sense of identification with the organization is further enhanced by the usage of mottoes, slogans, and company songs such as those sung at IBM and Wal-Mart.[14] Some even invoke a special language. For example, Nordstrom employees are "Nordies." Disneyland employees use theatrical language, e.g., a job is a "part," being on duty is "onstage." Such promotions of the "ingroup"—especially when they are coupled with a definition of another corporate entity as an "outgroup"—further augment a sense of pride and even superiority relative to the other group. Many successful companies purposely imbue their employees with the belief that they are the "best." Procter and Gamble, IBM, and Nordstrom epitomize the type of company where such elitism reigns. With this sense of confidence and belonging can come a fervent adherence to company ideology and to the preservation and even enhancement of the basic company philosophy. Thus, many have argued that such cult-like atmospheres are highly productive and filled with enthusiasm.

A corresponding element of this "ingroup" promotion is a tendency to monitor and punish deviance. The "visionary" companies are particularly intolerant of dissent. They "eject like a virus" those who do not fit with the corporate culture. If you don't want to be "Procterized," you don't belong at P&G; if you are not dedicated to the clean living and service atmosphere at Marriott, then stay away. Diversity is tolerated—even welcomed—as long as it is accompanied by a belief in the company ideology.[15]

THE OTHER SIDE: CREATIVITY

One might ask how can you argue with a corporate culture, cult-like or not, that results in high morale, enthusiasm, a clear vision, a sense of belonging, and a dedication that translates into profits? How can you question the value of coordinated group activity which may account for the success of large-scale projects (e.g., Boeing's launch of the 747 or Disney's entry into Epcot Center and Disneyland)? But does this type of corporate culture also promote creativity? Or an ability to respond readily to change?

There are authors and executives who suggest that the same techniques of clear goals, indoctrination, cohesion, consistent feedback, and reward

systems can be used to foster innovation—even a "cult-like culture of change." The premise is that one can have the camaraderie, a sense of "can do," and a dedicated commitment to the company's ideology while, at the same time, having flexibility, innovation, and an ability to adjust to changing circumstances.

Many companies recognize the need to "reinvent themselves,"[16] whether they call it "self renewal" (Motorola), "individual initiative" (Philip Morris), or being a "pioneer" (Sony). However, can one create a culture that promotes creativity, openness to new ideas, and innovation while at the same time having people indoctrinated in a similar ideology? Can the ideology itself be innovation? With feedback and appropriate rewards, does it work? The available research on creativity suggests it may not be that easy. Good intentions and great effort do not necessarily result in creativity. One can work very hard while engaging in convergent (single perspective) thinking that is unlikely to produce anything original. In fact, there is evidence that the atmosphere most likely to induce creativity is one diametrically opposed to the "cult-like" corporate culture.

THE NATURE OF CREATIVITY

Much of creativity starts with the proper posing of a question. And it seems aided by an ability to break premises, by being able to look "outside the box." Using that metaphor, the task of joining all the dots in Figure 7.1A in 4 straight lines without lifting pen from paper (and without retracing any lines) is made solvable by adding two hypothetical dots (■) outside the box (Figure 7.1B). Who said that you can't connect more dots than required?

Though this task serves as a metaphor, there is ample evidence that the unreflective adoption of premises and old patterns of solving problems is detrimental to creativity. If individuals are given a series of problems for which a given solution "works," they will tend to use that solution even when it is no longer the best one. An obvious better solution will not be detected.[17]

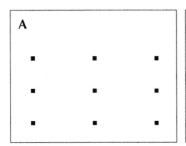 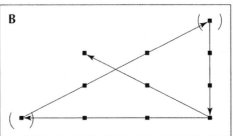

Figure 7.1.

In light of this, a firm might want to recruit relatively unconventional people and be careful about the extent to which it tames that unconventionality. Thomas Watson of IBM recognized this in his continual reference to "wild ducks." Borrowing from Kierkegaard, Watson was fond of telling the story of a man on the coast of Zealand who liked to watch the ducks fly south each fall. Well intentioned, he began to feed the ducks in a nearby pond only to find that, with time, they stopped flying south. They wintered in Denmark, feeding on what he provided. After several years, they grew so fat and lazy that they hardly flew at all. The message: You can make wild ducks tame, but you can never make tame ducks wild again.[18]

Many companies have recognized how easy it is to thwart creativity. As expressed by Richard Kinder, a good idea is fragile "like a lighted match, easily blown out by the cold winds of rigid management."[19] In a good deal of literature, however, it isn't rigid management that thwarts creativity. It is much more basic and much more insidious. People in general are loathe to suggest an original idea for fear of the ridicule and rejection. In fact, most attempts to raise the level of creativity explicitly understand that one must try to do two things: ensure that the individual's creativity is not stifled by conformity pressures or fears of social embarrassment; and take advantage of creativity enhancing forces (e.g., reinforcement for contributing or cross-stimulation).

Brainstorming, for example, explicitly asks people to generate as many ideas as possible, to refrain from evaluating or criticizing their own or others' views, and to build on ideas that are proposed (piggybacking).[20] While there appears to be modest success with such a technique, the fear of disapproval and the reluctance to voice ideas remain. Even with these admonitions, groups are still less creative than individuals alone.[21] When people are in group interaction, it is not easy to lessen the concerns about embarrassment and ridicule. Status tends to make it worse. People are particularly unlikely to challenge persons of higher status. Prof. Bennis of the University of Southern California estimates that at least 7 out of 10 in American business hush up when their opinions are at odds with their superiors. Even when they know better, they allow their bosses to make mistakes.[22]

Such problems are apparent in many different kinds of decision-making groups. In studies of Cabinet level "fiascoes," such as the Bay of Pigs, Janis found that groups marked by cohesiveness and a strong directive leader are especially likely to seek "uniformity."[23] As a result, they suffer under the illusion that there is unanimity when, in fact, there is not. Silence is often assumed to indicate assent. Then, out of a fear of ridicule and a sense of futility, each individual is reluctant to voice a differing view. Dean Rusk, for example, had strong doubts about the Bay of Pigs invasion but did not express them at the Cabinet level meetings. Arthur Schlesinger later rebuked himself, writing: "In the months after the Bay of Pigs I bitterly reproached myself for having kept so silent during those crucial discussions though my feelings of guilt were tempered by the knowledge that a course of objection would have accomplished little save to gain me a name as a nuisance."[24]

Thus, managers need to be aware of the fact that most of us are quite fragile when it comes to expressing "differing" views. People usually test the waters to see what is acceptable and, in the process, may unwittingly censor their own thoughts as well as their verbal expressions. Contrary to what strong leaders with strong cultures may think, the directiveness and strength of the leader can actually thwart creative thought and the expression of divergent views. It is even more likely when the group is highly cohesive. These are precisely the elements that are believed to cause "groupthink," the strain to uniformity that overrides a realistic appraisal of alternatives.[25]

CREATIVITY AT THE TOP: IMPLEMENTATION VS. INNOVATION

Some of the admired companies—those reputed to have not only good management and financial success, but innovation—appear to be those whose *leader* had the creative idea. Under these conditions, a strong corporate culture emphasizing uniformity, loyalty, and adherence to company expectations would be advantageous. It is advantageous precisely because it can operate in a relatively monolithic way—full of energy, morale, and a tendency not to consider alternatives or problems. This is not the same as promoting creativity from within the organization. Cohesion convergent thought, and loyalty help to *implement* an idea but tend not to enhance the *production* of a creative idea. Rather, flexibility, openness, and the welcoming of dissent are especially useful for stimulating creative thought.

Several companies showing creativity made the "most admired" list and illustrate creativity by top executives. One such example is Enron, whose CEO, Kenneth Lay, found himself with a large national network of gas lines. Trained as an economist, he favored free markets and disliked federal regulation. Yet, in viewing his national map differently than others, he saw it as a network to buy gas where it was cheap and sell it where it was needed. Thus, he found himself promoting deregulation. He created spot markets and, in the process, changed the industry. It paid off. Enron was named one of the "most admired" companies by *Fortune* magazine and ranked No. 1 in innovation.

Another example is Mirage Resorts, ranked No. 2 in innovation. CEO Steve Wynn had the idea: Giving customers of the gambling industry outstanding service, food, and entertainment. His Treasure Island resort was the first designed to generate more income from non-gambling than gambling sources. And he found a way to make his employees enthusiastic about service. He spent as much on the employees' cafeteria and corridors as on those used by "guests." He essentially created a culture of cohesion and high morale, all adhering to his core ideology of service. Again, the idea appears to be an insight of an individual—and, importantly, of the CEO himself. In that context, the cohesive culture that he created is an advantage, primarily in terms of implementation.

Honda is still another example of creativity at the top. Two owners, Honda and Fujisawa, the former the motorcycle enthusiast, the latter the businessman, merged their talents and developed the Honda 50cc Supercub in 1958. It was a marketing success, strongly appealing to small commercial enterprises in Japan. It was lightweight, affordable, and yet had strong horsepower.

When they tried to enter the U.S. marker, however, there were several problems. Motorcycle enthusiasts in the U.S. were a special breed and the emphasis was on size and power. Thus, Honda feared introducing their small bikes for fear of hurting their image in the macho U.S. market and, instead, introduced their larger motorcycles, the 250cc and 305cc bikes. However, it was the failure of these bikes that inadvertently led to the success of the smaller 50cc Supercub. Not content with using an intermediary, the Honda executives themselves drove the smaller 50cc bikes around Los Angeles on errands and, in the process, attracted attention and enquiries, including one from Sears. It was the failure of their larger bikes that led to the willingness to permit distribution of the Supercub. And the rest is history.

Honda is said to have "redefined" the U.S. motorcycle industry. After their "You meet the nicest people on a Honda" campaign, nearly one of our every two motorcycles sold was a Honda.[26] The history of this invention is not dissimilar from most creative efforts. It is a combination of talent and drive, but also a series of miscalculations, serendipity, and learning. Again, however, it was the leadership that manifested the creativity.

The problem with creative ideas generated by the CEO and implemented by a cohesive, strong culture that promotes uniformity is that the beneficial outcomes are dependent on the CEO. He had better have the best ideas because his ideas are likely to be implemented—right or wrong—in such a culture. This is part of the reason why some worry a bit about Microsoft. While tied for third place in Financial Executive's survey for "best boss,"[27] Bill Gates is seen as the quintessential "modern-day robber baron" who can create a monopoly without it appearing to be one, and who can make a product like Windows as addictive as the products of RJ Reynolds. Yet, Cusumano and Selby point out that Microsoft's weakness is precisely its dependence on Bill Gates.[28] He is the undisputed leader of Microsoft, but he is also one human being. With increasingly complex scenarios and potentials, Gates, brilliant though he may be, has limitations—be it time, attention span, motivation, or wisdom.

CREATIVITY WITHIN THE ORGANIZATION

For companies that attempt to foster innovation from within the ranks, a number recognize the importance of dissent or of being a "maverick." They often try to limit the fear of failure and promote risk raking. Anecdotal evidence abounds, such as when Robert Johnson (Johnson and Johnson) is reputed to have congratulated a manager who lost money on a failed new product by saying, "If you are making mistakes, that means you are making decisions and taking risks." Coca-Cola actually celebrates the failure of its sweet "New Coke," the venture in 1985 that proved to be ill-advised—except for the fact that Coca-Cola learned something important: Coke's strength was its image, nor necessarily its flavor.[29]

Some companies even tout an ideology that permits employees to buck top management. Marriott says that "if managers can't explain why they're asking employees to do something, they don't have to do it."[30] DuPont grants employees money to pursue projects turned down by management. Hewlett-Packard instituted a "medal of defiance," presumably based on an employee (Chuck House) who defied David Packard in continuing to work on a computer graphics project, one that eventually turned out to be a big money maker.[31]

Even with such statements and medals, it is unlikely to reduce employees' concerns about defying management. While most of us would like to believe that we are tolerant of dissenting views, the initial inclination of anyone—whether they are management or workers—is to underestimate and resist viewpoints that differ from their own. The tendency is to consider them false, unworthy of reflection, and often indicative of poor intelligence or an undeveloped moral character.[32] With such a reaction from management, no matter how subtle, most would not "defy." One has only to wonder whether Chuck House would have been given a medal had his project turned out to be a failure.

Perhaps more effective are concrete mechanisms instituted by some companies that actually limit the control of upper management or that actively remove some of the inhibitions to creativity. For example, GE has work-out groups where employees voice their gripes—though it appears that the number who actually do so are in a minority. Pfizer under Edward Steer sent its R&D center overseas. Presumably this frees the researchers since they are removed from authorities in New York; it also permits entry into overseas markets more readily. Other companies actively recruit heterogeneity in teams, hoping for diversity of viewpoints. Chrysler was one of the first to put together heterogeneous teams (manufacturing, marketing, and engineering) to work on the same automobile design.[33] Motorola and 3M regularly use "venture" teams of people from various disciplines,[34] a practice that research suggests will enhance the quality of decisions.[35]

Other companies have attempted to institute classic brainstorming techniques, the elements of which include refraining from criticism of an idea and attempting to build on the ideas of others. Raychem, for example, promotes the stealing of ideas, all for a good cause. Recognizing that employees often suffer from a "not invented here" syndrome—that is, if we didn't invent it, it's not very valuable—they give an award for adopting others' ideas from within the organization. The trophy and certificate say, "I stole somebody else's idea and I'm using it." And the originator of the idea? He or she gets a certificate saying, "I had a great idea and 'so and so' is using it."[36] This is a clever concrete mechanism for inducing the basic brainstorming tenet of building on others' ideas.

CREATIVITY WITHIN: THE CORPORATE CULTURE

Motorola is described as having an especially contentious corporate culture. Gary Tooker, CEO and Vice Chairman, is seen as "a master of the company's usually contentious, sometimes profane discussions that pit manager against manager and business unit against business unit, often in pursuit of conflicting technologies."[37] An example is the development by the Land Mobile sector of a combination cellular phone/pager/two way radio (MIRS), which utilizes specialized mobile radio frequencies (SMR). This competes directly with the alternative cellular duopolies and could undermine its other sector, General Systems. Do they worry about internal conflict? Motorola mainly believes in the free market and as Chris Galvin, President and COO explains, a "willingness to obsolete ourselves."[38]

3M is almost everyone's choice for a prototype of innovation. Well known is their "15 percent rule" where employees devote up to 15% of their time pursuing ideas of interest to them that

may have potential value for the company. And equally known are their demands for performance—25% of sales coming from products introduced in the last 5 years plus 10% growth in sales and earnings, 20% return on equity, and 27% return on capital employed. With these demands, however, are support systems ranging from seed money, mechanisms for linking employees (e.g., through e-mail) and, apparently, a deeply held conviction by management that one needs to respect ideas coming from below and that people may need to fail in order to learn.[39]

One of 3M's classic stories is the development of Post-it Notes—which, in fact, was aided by a failure. The glue eventually used for Post-it Notes was a failed attempt by Spencer Silver to develop a super strong glue. Art Fry, singing in his church choir, had trouble marking his hymnal with pieces of paper; they would slip out. Then came the idea: "What I need is a bookmark with Spence's adhesive along the edge."[40] And thus Post-it Notes were conceived. It should also be mentioned that marketing was initially skeptical, but the case was made when they found that the notes were already being used extensively among 3M's internal staff.

DuPont actually tries to teach creativity skills. Using their own in-house training center, there is a recognition that creative ideas need to be implemented. Thus, a senior manager or department controls the funds and resources that will permit the translation of the idea into concrete action. Workshops not only include managers and employees who are knowledgeable about the technology under discussion, but they make sure that there are also individuals who, while competent, are not particularly knowledgeable about the issue. They are the hoped-for "wild cards" who will bring fresh perspectives. Finally, the individuals are taught to avoid the "inhibitors" of creativity. They are vigilant about indications of convergent thought; for example, they are "on guard" about tendencies to make an idea fit with a preconceived notion. Further, people are encouraged to suspend their own knowledge, experience, and expertise and to play with each idea.[41]

THE FIT WITH "CREATIVITY" RESEARCH

Many of the devices utilized by Motorola, 3M, and DuPont are consistent with available knowledge about good problem solving and creativity. There are attempts to make groups heterogenous in the hopes of gaining diverse views. There are cautions about convergent thought, about premature criticism of an idea. People are given time and resources to "play" with ideas and yet are encouraged by clear goals. Yet it is the recognition of a "contentious atmosphere," the "bickering, tension, and dissent" described at Motorola that may provide better insight into creativity than most executives and researchers realize.[42]

While the literature on creativity is somewhat mixed, the profile used to describe the "highly creative" individual includes personality traits such as confidence and independence, a preference for complexity over simplicity, for some disorder rather than for everything neat and tidy, a tendency for being "childlike though not childish," and some indications of nonconformity and even rebellion in childhood. Creative people are characterized as high on personal dominance and forcefulness of opinion and, in addition, they tend to have a distant or detached attitude in interpersonal relations, though they are not without sensitivity or insight.[43] Play appears important, even random variations of thought.[44] Such profiles have also appeared in my own recent series of interviews with Nobel laureates. Most were aware of being "different" even as children. Most had to buck conventional wisdom and collegial skepticism in their continued pursuits. And they were childlike, even at the age of 70+, playing with metaphors from completely different disciplines.[45]

Such a portrait of the "highly creative" would suggest that they would not be drawn to nor nurtured by a highly cohesive corporate culture, one that demanded strict adherence to company norms, attitudes, and values or one that demanded "belonging" and a high degree of in-house socializing. Creative people and the creative process need independence—at least independence in thought. They need to be able to break premises, to pose questions differently. They need to be an "outsider," a person who can effectively interact with others but who remains marginalized from the group. The research often notes such outsider status on the part of the highly creative and it is a persistent theme among the Nobel laureates that I interviewed.

CREATIVITY IN GROUPS: THE ROLE OF DISSENT

While we often tend to think of creativity as an individual phenomenon, it is important to recognize the importance of groups. First of all, most discussion and decision making occurs in groups and, thus, the influence processes and interactions that occur become important in their own right. Secondly, and more importantly, groups can strongly hinder or promote individual creativity and the quality of individual judgments and solutions to problems.

As noted, most of the research literature emphasizes the negative aspects of groups—the fact that groups can hinder creativity. In groups, individuals have an overriding fear of social embarrassment and ridicule and they are reluctant to voice differing views. Thus, most attempts to raise the quality of group decision making and creativity have concentrated on how one might diminish these fears and thus give "voice" to differing viewpoints. Encouraging risk, asking people to refrain from criticism, and promoting the idea that people should build on the ideas of other are all attempts at "freeing" the individual from the fears that silence him.

Recent research, however, shows that groups are not just inhibitors of good decision making and creativity. Rather, groups can actually stimulate creativity and better problem solving on the part of each of the individuals. The key is the presence of dissent, of exposure to minority viewpoints that are assumed to be incorrect, that are likely to be ridiculed, and that are likely to invoke rejection. Such minority views stimulate more complex thinking, better problem solving, and more creativity.[46]

One aspect of good problem solving is a willingness to search the available information and, importantly, to search it in a relatively unbiased way. When we are faced with a majority dissent, we tend to look for information that corroborates the majority view. However, when the dissent comes from the minority, it stimulates us to reassess the entire issue and, in the process, search for information on all sides of the issue.[47]

In addition to information search, minority views have also been found to stimulate what we call "divergent thought," where people are more likely to consider the issue from various perspectives. For example, when faced with a minority view, people utilize all strategies in the service of problem solving. They come at the problem from all possible directions and, in the process, find more solutions. In fact, stimulated by minority views, they perform much better than they would alone.[48]

Still other studies demonstrate that minority viewpoints stimulate original thought. Thoughts are more "unique"—that is, less conventional. People exposed to minority views are also more likely to come up with original solutions or judgments.[49]

What is important about these findings is that the benefits of minority viewpoints do not depend on the "truth" of the minority position. The findings mentioned above hold whether or not the minority is correct. It is not because the minority holds a correct position that decision making and problem solving are enhanced. It is because dissenting views by a minority of individuals—right or wrong—stimulates the kinds of thought processes that, on balance, lead to better decisions, better problem solving, and more originality.

CONCLUDING THOUGHTS

Many people are wary of dissent because it poses conflict and can strain the cohesion and camaraderie of a well-functioning group. Further, it can impede organizational goals by questioning the established routes to success. This concern is given further credence by the examples of companies who have thrived on strong corporate cultures that emphasized uniformity, loyalty, and cohesion as efficient mechanisms for achieving its goals. The problem with such efficiency is that it results in a lack of reflection. If the path to success changes, there is no provision for flexibility, adaptation to new circumstances, or innovation.

It is tempting to conclude that innovation can be achieved simply by using these same cultural norms—uniformity, loyalty, and cohesion—along with a system of rewards or encouragement from the top. Yet, creativity in individuals and innovation at the organizational level are not so easily produced. Rather, the ability to think "outside the box," to find truly original solutions to old problems, requires the freedom to break the rules and to consider different options without fear of reprisals or rejection. As the research findings show, dissent actually stimulates originality and better decision making procedures. When challenged by minority views, people reappraise the situation. Without such stimulation, people tend to be complacent, relying on the agreement of their group. With exposure to minority views, people come to recognize that their own views may be incorrect, or at least partially incorrect. As a result, they search anew, they think anew, and, in the process, they consider new options and achieve greater clarity.

There needs to be a "welcoming" and not just a tolerating of dissent. Dissent is a very economical mechanism for producing innovation. By harnessing the power of conflict, one can limit complacency and even substitute robust thought. Complacency is the real danger. Even with the best of intentions, people are loathe to consider alternatives when convinced of the truth of their own position. They tend to search for confirming information, augment their own views, and punish dissenters. Dissent breaks up that complacency and sets in motion thought processes that ultimately result in better and more original solutions. If an existing idea is correct, it is likely to remain with even greater clarity as a result of the challenge of dissent; if it is flawed, most likely it will be replaced by a better one. Either way, the group and the organization will profit.

ENDNOTES

1. B. O'Reilly, "The Secrets of America's Most Admired Corporations: New Ideas, New Products," *Fortune*, 135/4 (March 1997): 60–64.

2. J.C. Collins and J.I. Porras, *Built to Last: Successful Habits of Visionary Companies* (New York, NY: Harper Collins, 1994).

3. C.A. O'Reilly and J.A. Chatman, "Culture as Social Control: Corporations, cults and commitment," *Research in Organizational Behavior*, 18 (1996): 157, 200.

4. T. Ehrenfeld, "Out of the Blue," *Inc.*, 117 (1995): 68–72.

5. V.L. Allen, "Situational Factors in Conformity," in L. Berkowitz, ed., *Advances in Experimental Social Psychology* v. 2 (New York, NY: Academic Press, 1965), pp. 133–175: J.M. Levine, "Reaction to Opinion Deviance in Small Groups," in P. Paulus, ed., *Psychology of Group Influence: New Perspectives*, 2nd edition (Hillsdale, NJ: Erlbaum, 1989).

6. M. Deutsch and H.B. Gerard. "A Study of Normative and Informational Social Influence upon Individual Judgment," *Journal of Abnormal and Social Psychology*, 195/51 (1955): 629–636.

7. S.F. Asch, "Effects of Group Pressure upon the Modification and Distortion of Judgment," in H. Guetzkow, ed., *Groups, Leadership and Men* (Pittsburgh, PA: Carnegie Press, 1951).

8. C. Nemeth and C. Chiles, "Modeling Courage: The Role of Dissent in Fostering Independence," *European Journal of Social Psychology*, 18 (1988): 275–280.

9. S. Schachter, "Deviation, Rejection and Communication," *Journal of Abnormal and Social Psychology*, 46(1951): 190–207.

10. C.J. Nemeth, "Dissent as Driving Cognition, Attitudes and Judgments," *Social Cognition* (1995), pp. 273–291; C. Nemeth and P. Owens, "Making Work Groups More Effective: The Value of Minority Dissent," in M.A. West, ed., *Handbook of Work Group Psychology* (London: John Wiley & Sons, 1996), pp. 125–141.

11. C. Nemeth and J. Wachrler, "Creative Problem Solving as a Result of Majority vs. Minority Influence," *European Journal of Social Psychology*, 13 (1983): 45–55.

12. Levine, op. cit.

13. D.G. Myers and H. Lamm, "The Group Polarization Phenomenon," *Psychological Bulletin*, 83 (1976): 602–627.

14. J. Pfeffer, "Management as Symbolic Action: The Creation and Maintenance of Organizational Paradigms," in L.L. Cummings and B.M. Staw, eds., *Research in Organizational Behavior*, Vol. 3 (Greenwich, CT: JAI Press, 1981), pp. 1–52.

15. Collins and Porras, op. cit.

16. T. Peters, "Crazy Times Call for Crazy Organizations," *Working Woman*, 19 (1994): 1–8.

17. A.S. Luchins, "Classroom Experiments on Mental Set," *American Journal of Psychology*, 59 (1946):295–298.

18. T.J. Watson, *A Business and Its Beliefs: The Ideas That Helped Build IBM* (New York, NY: McGraw-Hill, 1963).

19. B. O'Reilly, op. cit.

20. A.F. Osborne, *Applied Imagination* (New York, NY: Scribner, 1957).

21. J.E. McGrath, *Groups: Interaction and Performance* (Englewood Cliffs, NJ: Prentice-Hall Publ. Co., 1984). p. 131.

22. F. Summerfield, "Paying the Troops to Buck the System," *Business Month* (May 1990), pp. 77–79.

23. I.L. Janis, *Groupthink: Psychological Studies of Policy Decisions and Fiascoes* (Boston, MA: Houghton-Mifflin, 1972).

24. Ibid., p. 39.

25. Ibid.

26. See, generally, R.T. Pascale, "The Honda Effect," *California Management Review*, 38/4 (Summer 1996): 80–91.

27. "Who's the Best Boss?" *Financial Executive*, 12 (1996):22–24.

28. M.A. Cusumano and R.W. Selby, "What? Microsoft Weak?" *Computerworld*, 29/40 (October 1995): 105–106.

29. G. Dutton, "Enhancing Creativity," *Management Review*, 85 (1996): 44–46.

30. Collins and Porras, op. cit.

31. Summerfield, op. cit.

32. Asch, op. cit.; Nemeth and Wachtler, op. cit.

33. B. O'Reilly, op. cit.

34. A.K. Gupta and A. Singhal, "Managing Human Resources for Innovation and Creativity," *Research-Technology Management*, 36 (1993): 41–48.

35. S.E. Jackson, K.E. May, and K. Whitney, "Dynamics of Diversity in Decision Making Teams," in R.A. Guzzo and E. Salas, eds., *Team Decision Making Effectiveness in Organizations* (San Francisco, CA: Jossey-Bass, 1995).

36. Gupta and Senghal, op. cit.

37. R. Henkoff, "Keeping Motorola on a Roll," *Fortune*, 12 (1994): 67–78.

38. S. Cabana and J. Fiero, "Motorola, Strategic Planning and the Search Conference," *Journal for Quality and Participation*, 18 (1995): 22–31.

39. "Deeply Embedded Management Values at 3M," *Sloan Management Review*, 37 (1995): 20.

40. "Post-it: How a Maverick Got His Way," *Marketing*, October 28, 1993, p. 31.

41. L.K. Gundry, C.W. Prather, and J.R. Kickul,

"Building the Creative Organization," *Organizational Dynamics*, 22 (1994): 22–37.

42. G.N. Smith, "Dissent in to the Future," *Financial World*, 162 (1993): 8.

43. F. Barton, *Creative Person and Creative Process* (New York, NY: Holt, Rinehart and Winston, 1969).

44. D.T. Campbell, "Variation and Selective Retention in Sociocultural Evolution," *General Systems: Yearbook of the Society for General Systems Research*, 16 (1969): 69–85.

45. C.J. Nemeth, *Those Noble Laureates* (1997, in progress).

46. C.J. Nemeth, "Differential Contributions of Majority and Minority Influence," *Psychological Review*, 93 (1986): 23–32; C.J. Nemeth, "Dissent as Driving Cognition, Attitudes and Judgments," *Journal of Social Cognition*, 13 (1995): 273–291.

47. C.J. Nemeth and J. Rogers, "Dissent and the Search for Information," *British Journal of Social Psychology*, 35 (1996): 67–76.

48. C. Nemeth and J. Kwan, "Originality of Word Associations as a Function of Majority vs. Minority Influence Processes," *Social Psychology Quarterly*, 48 (1985): 277–282.

49. Ibid.; C.K.W. DeDreu and N.K. DeVries, "Differential Processing and Attitude Change Following Majority and Minority Arguments," *British Journal of Social Psychology*, 35 (1996): 77–90; R. Martin, "Minority Influence and Argument Generation," *British Journal of Social Psychology*, 35 (1996): 91–113.

3

The Transition from Technical Specialist to Managerial Leader

8

Why Managers Fail

MICHAEL K. BADAWY

Generally ill-equipped for a management career, engineers and scientists will fail as managers unless they understand the reasons for such failure and take steps to prevent it.

Many engineers and scientists have made, or will make, the transition to management smoothly and successfully. However, the record is less than promising. While there is no law of nature that says good technical practitioners cannot be good managers, it is unlikely that they will be. Although they are well qualified for management by virtue of their analytical skills and backgrounds, many technologists switch to management for the wrong reasons and to satisfy the wrong needs. Hence, they do not make competent managers. There is substantial evidence derived from my own research studies and those of others that the transition to management has been troublesome for many technologists, and that many of them have failed because they were generally ill-equipped for such a career.

From Michael K. Badawy, *Developing Managerial Skills in Engineers and Scientists: Succeeding as a Technical Manager*, copyright © 1982 by Van Nostrand Reinhold Co., Inc. Reprinted with permission.

Michael K. Badawy is Professor of Management at Virginia Polytechnic Institute and State University.

In order to understand why managers fail, one must first recognize that managerial competency has three interrelated components; knowledge, skills and attitudes. Although sophisticated knowledge in the principles and elements of administration is a prerequisite for managerial success, such knowledge by itself is not enough for managerial competency. While management theory is a science, management practice is an art. Therefore, to be effective, the manager must develop a set of professional skills. These skills are:

• *Technical:* Technical skills include the ability of the manager to develop and apply certain methods and techniques related to his tasks. The manager's technical skills also encompass a general familiarity with, and understanding of, the technical activities undertaken in his department and their relation to other company divisions. The manager's technical specialization, formal education, experience, and background form a strong foundation for the development of technical skills.

• *Administrative:* Administrative skills relate primarily to the manager's ability to manage. Effective management, of course, reflects the ability to organize, plan, direct, and control. It is the capacity to build a workable group or unit, to plan, to make decisions, to control and evaluate performance, and finally to direct subordinates by motivating, communicating, and leading them into a certain direction that would help the organization achieve its objectives most effectively. The core elements of administrative skills are the ability to search out concepts and catalog events; the capacity to collect, evaluate, and process pertinent information; the ability to distinguish alternatives and make a decision; and resourcefulness in directing others and communicating to them the reasons behind the decisions and actions. Superior administrative skill is, of course, related to and based on other skills such as cognitive and conceptual skills.

• *Interpersonal:* Interpersonal skills are probably the most important of all. Since managing is a group effort, managerial competency requires a superior ability to work with people. The manager, to be effective, must interact with, motivate, influence, and communicate with people. People make an organization, and through their activities, organizations either prosper or fail. Managing people effectively is the most critical and most intricate problem for the manager of today.

ATTITUDES AND MANAGERIAL COMPETENCY

Attitudes, the third ingredient of managerial competency, are essentially the manager's value system and beliefs toward self, task, and others in the organization. Attitudes include those patterns of thought that enable one to characterize the manager and predict how well he will handle a problem. Attitudes are partly emotional in origin, but they are necessary because they determine two things. First, the acquisition of knowledge and skills is, in part, a function of attitudes, and second, attitudes determine how the manager applies her knowledge and techniques.

Attitudes are also important in determining managerial competency for another reason: They tell us what needs are dominant in an individual at a certain time, and thus we can predict and identify the individual's managerial potential. This identification is crucial for enhancing future managerial effectiveness.

Modern psychological research tells us that effective managers share at least three major attitudinal characteristics: a high need to manage, a high need for power, and a high capacity for empathy.

The need or will to manage has to do with the fact that no individual is likely to learn how to manage unless he really wants to take responsibility for the productivity of others, and enjoys stimulating them to achieve better results. The "way to manage" can usually be found if there is the "will to manage." Many individuals who aspire to high-level managerial positions—including engineers and scientists—are not motivated to manage. They are motivated to earn high salaries and to attain high status, but they are not motivated to get effective results through others. Thus, they will not make competent managers.

The need to manage is a crucial factor, therefore, in determining whether a person will learn and apply what is necessary to get effective results on the job. They key point here is that an outstanding record as an individual performer does not indicate the ability or willingness to get other people to excel at the same tasks. This partly explains, for example, why outstanding scholars often make poor teachers, excellent engineers are often unable to supervise the work of other engineers, and successful salesmen are often ineffective sales managers.

Second, effective managers are characterized by a strong need for the power derived from such sources as job titles, status symbols, and high income. The point is that power seekers can be counted on to strive to reach positions where they can exercise authority over large numbers of people. Modern behavioral science research suggests that individuals who lack this drive are not likely to act in ways that will enable them to advance far up the managerial ladder.[1] Instead, they usually scorn company politics and devote their energies to other types of activities that are more satisfying to them. For many engineers and scientists, power emanates from "professional" sources other than sources of managerial power. While managerial power is based on politics, titles, and organizational status, professional power is based on knowledge and excellence in one's discipline and profession. In short, the power game is part of management, and it is played best by those who enjoy it most.

The third characteristic of effective managers is the capacity for empathy—being able to cope with the emotional reactions that inevitably occur when people work together in an organization. Effective managers cannot be mired in the code of rationality, which explains, in part, the troublesome transition of some engineers and scientists to management. Individuals who are reluctant to accept emotions as part of being human will not make "human" managers, and, in turn, they will not be managerially competent.

THE MANAGERIAL SKILL MIX

The technical, administrative, and interpersonal skills are all closely interrelated and can be significant in

determining your success in management. However, experience shows that the relative importance of these skills varies with the management level you are on and the type of responsibility you have.

As shown in Figure 8.1, technical skills are inversely related to your management level. They are most important at lower management levels but that importance tends to decrease as you advance to higher levels in the organization.

Managerial success on upper management levels is determined by your vision and ability to understand how the entire system works (the conceptual skill), as well as your capacity for organization and coordination between various divisions (the administrative skill). How much you know about the technical details of the operation becomes considerably less important. In fact, beyond middle management, "special knowledge" can actually be a detriment to the individual.

Handling people effectively is the most important skill at all levels of management. Knowing how to handle people effectively, I believe, is the art of the arts. If a manager has considerable technical and administrative skills, but his interpersonal skills are wanting, he is a likely candidate for managerial failure. Conversely, problems that occur because technical or administrative skills are not up to par will be more easily surmounted.

It is important to remember that success in management is largely determined by the manager's ability to understand, interact with, communicate with, coach, and direct subordinates. This statement should not be taken to undermine the importance of technical and administrative skills in

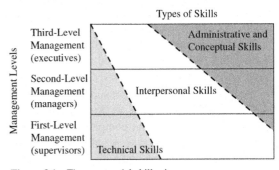

Figure 8.1 The managerial skill mix.

managerial effectiveness, but rather to underscore the ability to get along with people as a prerequisite for success in management.

While interpersonal skills are presumably important at all levels of management, they are perhaps most important at the lower and middle levels, where managers interact with subordinates, supervisors, and associates. At upper management levels, where the frequency of these interactions tends to decrease, the importance of interpersonal skills decreases as conceptual and administrative skills become more important. It is also possible that with the increased degree of power and influence upper-level managers typically have, they can afford to pay less attention to interpersonal relations, and hence they become less sensitive to human needs and individual satisfaction.

Some technical managers and supervisors find it difficult to get away from "the bench" and do what they are paid for—namely managing. But everything in life has a price. Perhaps this is one of the prices managers must pay—the price of management! I have seen quite a few technologists in management who were not willing to pay this price, so they tried to do technical and managerial tasks simultaneously—with the incompatible requirements of two different roles. I have seen others enter management without realizing what they were getting into. The paths taken by both groups turned out to be very costly—they failed! The lesson is quite clear: The "right" mix of the three skill types at different management levels must be properly maintained.

MANAGERIAL FAILURE

Organizations fail because managers fail. Managers fail because they perform poorly. Performance is one of the prime criteria that an organization uses in evaluating employee contributions. While many managers deny it in public, performance is hardly the sole basis for appraising employees. Managerial jobs are no exception. Given the difficulty of measuring managerial performance accurately and objectively, organizations consider the manager's contributions to the accomplishment of the company's goal to be the most concrete basis for distributing rewards and making salary adjustments and promotion decisions. Thus, managerial failure can be defined as the inability of the manager to meet certain performance standards imposed by his superiors. It can also be defined in terms of organizational policies. A manager has failed if his performance is considered unsatisfactory or unacceptable by virtue of some preestablished criteria. Managerial success, on the other hand, is reflected in added responsibility, promotion, title change, and increased salary. Managerial "failure" is not intended to imply that the manager is demoted because he totally failed to get the job done but rather that he failed to accomplish results that he is capable of achieving.

Why do managers fail? Causes of managerial failure can be numerous. Since managerial performance is a product of the interaction of many factors inside the individual and in the surrounding environment, managerial success or failure will be determined by factors in both areas. Bear in mind that failures in management are rarely due to a single cause. Typically, aspects of the individual interact with aspects of the environment to create the problem.

Research has shown as many as 35 types of factors as potential causes of unsatisfactory managerial performance.[2] Any one of these factors can contribute to the failure of a manager. Some of these factors are more likely to operate than others because of the nature of managerial work. As shown in Table 8.1, there are three groups of factors containing nine categories representing possible causes of performance failures. The first four categories refer to aspects of the individual. The remaining five categories refer to different aspects of the environment: categories five to eight refer to the various groups of which the individual is a member, while the ninth category is contextual, referring to nonhuman aspects of the environment and the work itself which may be strategic to performance failure.

Note the relative importance of the different factors. As shown in the table, the most dominant causes of performance failure are deficiencies of an intellectual nature, frustrated motivations, mem-

TABLE 8.1
Causes of Managerial Failure

Category	High	Low	Rare
A. Individual Factors			
I. *Problems of Intelligence and Job Knowledge*	X		
1. Insufficient verbal ability			
2. Insufficient special ability			
3. Insufficient job knowledge			
4. Defect judgment or memory			
II. *Emotional Problems*		X	
5. Frequent disruptive emotion (anxiety, depression, guilt, etc.)			
6. Neurosis (anger, jealousy, and so on, predominating)			
7. Psychosis			
8. Alcohol and drug problems			
III. *Motivational Problems*			
9. Strong motives frustrated at work (fear of failure, dominance, need for attention, and so on)	X		
10. Unintegrated means used to satisfy strong motives	X		
11. Excessively low personal work standards			X
12. Generalized low work motivation			X
IV. *Physical Problems*		X	
13. Physical illness or handicap including brain disorders			
14. Physical disorders of emotional origin			
15. Inappropriate physical characteristics			
16. Insufficient muscular or sensory ability or skill			
B. Group Factors			
V. *Family-Related Factors*	X		
17. Family crises			
18. Separation from the family and isolation			
19. Predominance of family considerations over work demands			
VI. *Problems Caused in the Work Group*	X		
20. Negative consequences associated with group cohesion			
21. Ineffective management			
22. Inappropriate managerial standards or criteria			
VII. *Problems Originating in Company Policies*	X		
23. Insufficient organizational action			
24. Placement error			
25. Organizational overpermissiveness			
26. Excessive spans of control			
27. Inappropriate organizational standards and criteria			
VIII. *Problems Stemming from Society and Its Values*			X
28. Application of legal sanctions			
29. Enforcement of societal values by means other than the law			
30. Conflict between job demands and cultural values (equity, freedom, religious values, and so on)			
IX. *Contextual Factors*			
C. Factors in the Work Context and the Job Itself			
31. Negative consequences of economic forces	X		
32. Negative consequences of geographic location	X		
33. Detrimental conditions in the work setting			
34. Excessive danger			
35. Problems in the work itself			

From Miner, John B., "The Challenge of Managing." Philadelphia: Saunders, 1975, pp. 330–331.

bership in and relations with different groups, and economic and geographic factors.

Discussing all possible causes of managerial failure in detail would be impractical. Thus, for our purposes here, I shall focus on the forces relating to the individual manager or to his or her job.

The development of management skills and the ability to convert them into effective actions are crucial to managerial competency. The lack of ingredients necessary for effective managerial performance will lead to poor results. Lack of knowledge of management principles and concepts, for example, will hamper a manager's performance. In addition, developing managerial skills is difficult when management knowledge is insufficient or lacking.

Many technical managers get fired, not because they lack technical competence, but because they lack managerial competence (another common practice, sadly, is to transfer incompetent managers back into a heavy technical role!). Managerial failure can thus result from inadequate management and administrative skills. Establishing a well-functioning unit or division with a sound structure and clear authority and responsibility relationships is crucial for achieving divisional objectives. In addition, developing policies and procedures consistent with objectives and goals, allocating resources, and monitoring progress toward goal achievement are key managerial tasks. These activities call for considerable administrative skill on the manager's part.

The major cause of managerial failure among engineers and scientists is poor interpersonal skills. Many technologists are more comfortable dealing with matters in the laboratory than they are dealing with people. Because many of them are loners, they are used to doing things for themselves. Once promoted to management, however, they have to delegate responsibility to others. They often find this extremely difficult, especially if they have less than complete confidence in their subordinates' abilities. As a result, many technologists find that their advancement—and their managerial careers—are limited more by human factors than by technical ability.

Managerial failure also occurs when an individual becomes a manager for the wrong reasons. This happens when a person seeks the attractive rewards (e.g., economic) associated with a managerial position, yet has no strong will to manage. Such managers do so poorly that they become "retired on the job."

JOB-RELATED FACTORS

A brief account of some of the job-related factors causing managerial failure follows.

Some managers, especially new managers, are unwilling to pay the price of being a manager, i.e., loneliness. They seem to be unaware that the higher one climbs on the managerial ladder, the more lonesome it becomes, the fewer peers one has to talk to, and the more restraint one must exercise over what one says. As a result, while they may enjoy the new challenges and the greater opportunity to make unilateral decisions, they find themselves nostalgic for the "good old days" when it was not so lonely. They therefore try to act like "one of the boys."[3] In doing so, they often lose the respect of the people they manage. Employees want to look up to the boss. They want to feel that they can turn to him or her for necessary decisions. The genesis of this managerial failure is the desire to be liked rather than respected! It is interesting that managers who are found to be most effective are usually categorized by their subordinates as "fair but firm."

In addition, new managers sometimes fail to adjust to the demands of their role. Managerial positions are usually characterized by a high degree of power over other people, some standards for controlling human behavior in work settings, large amounts of visibility and influence, and a keen understanding of how corporate power and politics work. These characteristics require that managers play multiple roles. If managers are uncomfortable with these factors, they could develop a strong fear of failure. This situation might very well lead to a poor managerial performance.

Other causes of managerial failure include the following:

Bias toward Objective Measurement. Having been trained in "hard" sciences, where exact measurement is one of the natural beauties of the scientific method, engineers and scientists are more comfortable working with things that they can objectively control and measure. Managers, on the other hand, must rely on intuition and judgment in dealing with attitudes, biases, perceptions, emotions and feelings. The fact that these intangible variables are hardly measurable—let alone controllable—makes the technical manager's job thoroughly frustrating. To be sure, one of the things that technologists must learn in order to succeed in management is to stop insisting on using a yardstick to measure everything. The nature of management—contrary to engineering and science—defies objective and tangible measurement.

Paralysis by Analysis. Engineers and scientists, more than others, suffer from this disease: the tendency to wait for all information to be in before they make a decision. I can think of no worse cause of managerial failure—it is a clear case of how the professional's technical training can hamper rather than enhance one's chances for success. In management, you will never have all the facts, nor will there ever be riskless decisions. All decision-making involves risk taking. An adaptation of Pareto's principle would be that 20 percent of the facts are critical to 80 percent of the outcome. Being slow to decide, waiting for more facts, is known as "paralysis by analysis."

The inability of engineering or R&D supervisors to adjust to making managerial decisions on the basis of incomplete data and in areas where they lack first-hand experience (since the information is usually provided by other people and divisions) results in managerial anxiety.[4] The fact that they must function within a highly ambiguous and unpredictable environment makes them unsure about the data available, thereby reinforcing a neurotic demand for more data in an attempt to make riskless decisions. If this cycle is not somehow broken and appropriate adjustments made, it can be a deadly time waster and a complicating factor in the transition to management.

Fear of Loss of Intimate Contact with Their Fields. Effective managers always focus on what needs to be done, when it should be done, and how much it should cost, rather than on how to do it. Since managers must get things done through other people, the question of "how" should always be left to them. Technologists usually find this difficult to understand, and in their zeal to stay professionally competent, they try to keep intimate contact with their specialties. As a result, they fail to delegate and they tend to handle the technical details as well—they try to do two jobs in the time of one! The manager, to be sure, is paid to get things done—not to do them himself; this is the job of his subordinates. Sacrificing some of their technical competence—in a relative sense—is the price technical managers must pay for staying managerially competent.

Technologists as Introverts. Many engineers and scientists are "introverts" rather than "extroverts." Research shows that introversion is usually associated with creativity. The problem is that while creating is an individual (introvert) activity, managing is a team (extrovert) activity. The ability to work with others and to be a good team player is one of the distinctive skills of successful and competent managers. The "lone wolf" nature of many technologists could, therefore, make it doubly difficult for them to function effectively as technical managers.

Poor Delegators. One of the most valuable skills a manager can possess is the ability to delegate. You should never undertake what you can delegate. You cannot grow as a technical manager unless you delegate, and your subordinates expect you to. However, technical managers have been found to be very poor in learning to achieve things through others. They are poor delegators. Technologists are doers rather than delegators because they believe, rightly or wrongly, that they perform a task better than anyone on their staff can. Developing the will to delegate requires a change in the technologist's attitudes, behavior, and assumptions about people working for him. She might even have to force herself to delegate tasks to her subordinates. At any rate, whatever it takes, delegation is one of the prime skills technical supervisors must acquire to enhance their managerial competence.

Farming out Responsibilities. Some managers do not recognize their responsibility for on-the-job training and coaching. They are all too eager to send their subordinates to courses conducted by staff agencies and outside trainers and consultants. While there are many occasions when it is wise to use these outside resources, the manager must make certain that such training is utilized by the subordinates when they return to the job. Without a standing policy of what training should cover and who should be trained and by whom, the manager runs the risk of abdicating coaching and training responsibilities. Poor development of subordinates hinders their professional growth, and the manager becomes indispensable. When a manager's responsibilities and salary remain the same over a long period of time, this stagnation is sometimes an indication of managerial failure. Upward movement on the managerial ladder, if based on one's managerial capability, is the ultimate reward of success.

In the absence of concrete figures on managerial "malpractice," it is difficult to estimate the number of ineffective or unsuccessful managers in or-ganizations. However, experience shows that there are a lot of incompetent managers around. The best way to deal with performance failure, I believe, is to prevent it. Unfortunately, managers are willing enough to put out the fires but they take little interest in fire prevention. Dealing with ineffective performance requires diagnosing the causes and then coming up with appropriate remedies. The possible causes of managerial failure have been analyzed in this article. However, the scrutiny of your own performance and the development of a personal plan of action must remain your responsibility.

NOTES

1. McClelland, David C., and Burnham, David H. "Power-Driven Managers: Good Guys Make Bum Bosses." *Psychology Today* (December, 1975): pp. 69–70.

2. Miner, John. "The Challenge of Managing." Philadelphia: Saunders, 1975, pp. 215–216.

3. This discussion is partly based on McCarthy, John F. "Why Managers Fail." Second Edition. New York: McGraw-Hill, 1978, pp. 3–34.

4. For more on this point see Steele, Lowell W. "Innovation in Big Business." New York: Elsevier, 1975, p. 186; and Thompson, Paul and Dalton, Gene. "Are R&D Organizations Obsolete?" *Harvard Business Review* (November–December 1976).

9

How Do You Feel?

TONY SCHWARTZ

"Emotional intelligence" is starting to find its way into companies, offering employees a way to come to terms with their feelings—and to perform better. But as the field starts to grow, some worry that it could become just another fad.

Appreciation, apprehension, defensiveness, inadequacy, intimidation, resentment. Twenty midlevel executives at American Express Financial Advisors are gathered in a room at a conference center outside Minneapolis. Each has been asked to try to convey a specific emotion—by reading a particular statement aloud. The challenge for listeners is to figure out which emotion each speaker is trying to evoke. It seems like a relatively straightforward exercise but only a fraction of the group comes anywhere close to correctly identifying speakers' emotions.

"I sometimes wish I had a corporate decoder for each relationship," one woman laments. "It's very hard to know what people are feeling in my office and how I should respond." Her comment prompts a discussion about the difficulty in the workplace of finding a balance between reasonable openness and respectful discretion.

"When one of my direct reports starts talking to me about her medical problems, I don't want to be unsympathetic, but it makes me very uncomfortable," says a male department head. "I find myself joking by saying to her, 'Too much information.' But I'm not really sure how to get the message across."

Conversations like that one, focusing on the importance of emotions in the workplace, are oc-curring with greater frequency in all kinds of American companies. Inside American Express, training sessions on emotional competence take place at the Minneapolis facility several dozen times a year. An unlikely pioneer in the field of emotional competence, AmEx launched its first experimental program in 1992. An eight-hour version of the course is now required of all of its new financial advisers, who help clients with money management. During a four-day workshop, 20 participants are introduced to a range of topics that comprise an emotional-competence curriculum, including such fundamental skills as self-awareness, self-control, reframing, and self-talk.

Much of that material represents new territory for these businesspeople. "The majority of those we work with are very cognitive and not very experienced with emotions," explains Darryl Grigg, a psychologist who practices in Vancouver, British Columbia and conducts about 20 workshops each year for AmEx and other organizations. "We're introducing people to a whole new language."

Most attendees of these emotional-competence workshops are compelled to learn a new language for one simple reason: They're visiting a foreign land. Over the past 50 years, large companies have embraced a business dictum that told workers to check their emotions at the door. A legacy from the days of "The Organization Man" and "The Man in the Gray Flannel Suit," this never-spoken but widely shared policy reflected the sensibility that frowned on employees who brought messy emotions and troubling personal issues to work.

Employees, for their part, complied with that prevailing mind-set. Until recently, the workplace was dominated by male employees—and most of

Reprinted from *Fast Company Magazine*, June 2000, pp. 296–308. Used with permission.

Tony Schwartz is a *Fast Company* contributing editor and author.

them were just as eager as their employers were to avoid the ambiguous complications and unexplored terrain of personal feelings.

One notable exception to that tacit pact occurred in the 1970s and early 1980s, when the influence of the human-potential movement prompted a brief corporate romance with such experiential techniques as sensitivity training and encounter groups. But those approaches lacked the rigor to endure. Before long, business got back to business. A backlash set in, and the focus returned to no-nonsense training methods that were highly quantifiable, happily free of emotions, and demonstrably able to produce results that would show up on the bottom line.

Today, more than 20 years later, companies in a variety of industries are once again exploring the role of emotions in business. This renewed interest in self-awareness is, in part, the result of the rising corporate power of baby boomers. The increasing presence of women in the workplace and the higher comfort level they bring to the territory of emotions have also nudged companies in this direction. And the arrival of the new economy has made companies realize that what they need from their workers goes beyond hands, bodies, and eight-hour days.

While the field of emotional competence appears to have emerged overnight, it has, in fact, been 15 years in the making. In 1985, Reuven Bar-On, 56, a psychologist who practices in Israel, first coined the term "emotional quotient," or EQ. Bar-On had moved to Israel at age 20 and became interested in the field while studying for his PhD in South Africa. "My simple—almost simplistic—question in the beginning was, 'Are there factors that determine one's ability to be effective in life?'" he explains. "Very quickly, I saw that people can have very high IQs, but not succeed. I became interested in the basic differences between people who are more or less emotionally and socially effective in various parts of their lives—in their families, with their partners, in the workplace—and those who aren't." For his thesis, Bar-On identified a series of factors that seemed to influence such success. He then developed a tool that assessed strengths or deficits, based on those factors.

A diminutive, bearded man with a genial style, Bar-On, now a research fellow at Haifa University, is a meticulous researcher who has gathered more scientifically validated data worldwide about emotional intelligence than anyone in his field has. His work has recently focused on developing EQ "profiles," which reveal the specific competencies that characterize high performers in a range of professions. In 1996, he launched the EQI, a self-administered test designed to assess specific emotional competencies. Companies and organizations, ranging from the Bank of Montreal to Fannie Mae to the Toronto Maple Leafs, have used the EQi for employee development. "To measure emotional intelligence is to measure one's ability to cope with daily situations and to get along in the world," argues Bar-On, whose test is marketed through Toronto-based Multi-Health Systems. "I've conceptualized emotional intelligence as another way of getting at human effectiveness.

But if Bar-On pioneered the field, Daniel Goleman, 54, formerly a behavioral- and brain-sciences writer for the "New York Times," brought it to popular attention. Drawing on the work of two academic psychologists, John D. Mayer and Peter Salovey, Goleman published "Emotional Intelligence" (Bantam, 1995). The book became an instant best-seller—with more than 5 million copies in print worldwide—and sparked inevitable criticism from Mayer and Salovey, who believed that Goleman distorted their work and made sweeping claims about the benefits of emotional intelligence.

Goleman has gone on to advance the case for emotional competence in the workplace. He published a second book, "Working with Emotional Intelligence" (Bantam, 1998), aimed specifically at businesspeople. He then authored two articles for the "Harvard Business Review." He also coedited a forthcoming book of essays written by leaders in the field; cofounded the Consortium for Research on Emotional Intelligence, a group of academics and businesspeople with interest in the field. Later, he began working with the Hay Group, a Philadelphia-based consulting firm that specializes in human-resource issues, to deliver emotional-intelligence training.

All of this should come as welcome news to residents of the new economy. Companies can continue to give top priority to financial performance—but many now also realize that technical and intellectual skills are only part of the equation for success. A growing number of organizations are now convinced that people's ability to understand and to manage their emotions improves their performance, their collaboration with colleagues, and their interaction with customers. After decades of businesses seeing "hard stuff" and "soft stuff" as separate domains, emotional competence may now be a way to close that breach and to produce a unified view of workplace performance.

But like other good ideas that started in psychology and later found new applications in business, emotional competence is confronting the challenge of its own sudden popularity. Increasingly, emotional competence is being sold as a solution to each of the categories for which companies have training budgets, from leadership to motivation to leveraging diversity—competencies that are emotional only by the most ambitious of stretches. The emerging field has sparked the almost inevitable scramble to cash in on the spreading claims of its potential applications.

As emotional competence grows in application, so do the questions: Are these new dimensions of emotional competence genuine and verifiable categories? Can they be effectively taught and measurably improved? And what is the risk that emotional competence will veer badly off course and end as the next short-lived fad?

AMEX'S EMOTIONAL-INSURANCE POLICY

It's difficult to imagine a less likely setting for a training program on emotions than American Express—that buttoned-up, by-the-numbers, financial-services giant, which last year had more than $21 billion in sales. In fact, the company launched its program in 1991 as a possible solution to a simple business problem that defied a logical solution. More than two-thirds of American Express clients were declining to buy life insurance, even though their financial profiles suggested a need for it. Jim Mitchell, then president of IDS, American Express's Minneapolis-based insurance division, commissioned a skunk-works team to analyze the problem and to develop a way to make life insurance more compelling to clients.

The team's findings took the company in an unexpected direction. The problem, the team discovered, wasn't with AmEx's product—or even with its cost. Put simply, the problem was emotional.

Using a technique called "emotional resonance," the team identified the underlying feelings that were driving client decisions. "Negative emotions were barriers," explains Kate Cannon, 51, formerly an AmEx executive who eventually headed the team and whose interest in the role of emotions in the workplace was in part sparked by her background in mental-health administration. "People reported all kinds of emotional issues—fear, suspicion, powerlessness, and distrust—involved in buying life insurance."

But the team's second finding proved the clincher: The company's financial advisers were experiencing their own emotional issues. "All kinds of on-going stuff was holding them back—feelings of incompetence, dread, untruthfulness, shame, and even humiliation," explains Cannon. The result was a vicious cycle. When clients expressed negative feelings, advisers had been trained to press harder. But this hard-sell approach only exacerbated clients' emotional conflicts, increasing their discomfort and distrust. In turn, advisers experienced more distress, stemming from their mandate to apply high-pressure tactics, which made them feel unethical. Ultimately, they became reluctant to try to sell life insurance at all.

At the same time, interviews with AmEx's most-successful advisers revealed that they took a very different approach to their jobs. They tended to take the perspective of their clients, which enabled them to forge trusting relationships. They were also more connected to their own core values and motivations for selling insurance in the first place. Perhaps most important, they were more aware of their own feelings, better able to manage those feelings, and more resilient in the face of disappointment.

"We were sitting around a conference table one day," Cannon recalls, "when it dawned on us that someone with all of these positive qualities is emotionally competent." Eager to test the hypothesis that specific nontechnical skills can influence performance, Cannon's team devised a study: One group of financial advisers receives 12 hours of training to help them understand their emotions better, while the other group received no training and served as the control.

The training, called Focus on Coping Under Stress, was relatively brief—only 12 hours—and relatively narrow in design. It used techniques to increase AmEx salespeople's awareness of their emotions, gave them tools to change negative emotions into positive ones, offered ways to rehearse mentally before stressful events, and provided a way to identify deeper personal values that motivated them at work.

At the end of the study, Cannon's team compared the sales results of the two groups: Nearly 90% of those who took the training reported significant improvements in their sales performance. In addition, the trained group, in contrast to the control group, showed significant improvement in coping capacity, as measured on standardized psychological tests. Advisers, in short, had become more emotionally competent.

After Cannon's team made adjustments to the program, including recasting it as emotional-competence training, a second, more-detailed project was launched to assess sales results. The group that participated in the more-detailed study improved its sales by 18.1%, compared to 16.1% for the control group. That 2% difference may seem insignificant—but it apparently added tens of millions of dollars in revenues. While Cannon's group quickly acknowledged that the sample was too small to be statistically significant, the results did suggest that even a modest, short-term program aimed at teaching "soft skills" could have a noticeable impact on the bottom line.

AmEx disbanded its skunk-works team in 1994, but Cannon, convinced that the group was

on to something important, found a new source of support in Doug Lennick, 47, now an executive vice president of American Express Financial Advisors. With the clean-cut, boyish looks of a high-school class president, Lennick had built his reputation at AmEx as a superstar salesman. Long before he learned about Cannon's program, Lennick had become something of a Stephen Covey-type figure in his own organization, spreading the word about self-improvement techniques and eventually writing up his ideas in two, short, folksy books published locally in Minneapolis: "The Simple Genius (You)" and "How to Get What You Want and Remain True to Yourself."

For all of his salesman's pithy aphorisms and upbeat exhortations, Lennick was also interested in people's interior lives—and specifically in the role that emotions play. "Emotional competence is the single most important personal quality that each of us must develop and access to experience a breakthrough," he says emphatically. "Only through managing our emotions can we access our intellect and our technical competence. An emotionally competent person performs better under pressure."

With Lennick's support, Cannon gathered several colleagues and six outside psychologists to develop longer versions of the initial training. The focus broadened from improving people's coping capacity to training people in the skills of emotional self-awareness, emotional self-management, and emotional connection with others. Lennick, in turn, mandated that all newly hired financial advisers receive an eight-hour version of the program as part of their job training. Since 1993, more than 5,500 new advisers have had the training, and an additional 850 "high potential" managers from other parts of AmEx have voluntarily enrolled in the full five-day course.

Cannon, who left American Express a year ago and now licenses emotional-intelligence training to corporations like Motorola, as well as to individuals, is modest but firm in her claims about the program that she helped to create. "It's a basic introduction, but what it gives people is permission, a language, and a structure for bringing their emotional lives into the workplace," she says. "It also prompts a shift in perspective. They come out seeing the world differently. For men, who are often talking about emotions for the first time, it opens a window. They finally understand what their mothers and sisters and wives have been talking about all these years when they say, 'You don't communicate with me,' and 'You never tell me what you're feeling.' For women, it's often their first confirmation that qualities like self-awareness and empathy can really make a positive difference in the workplace."

THE AIR FORCE FLIES ON EMOTIONS

If emotions and life insurance seem an unlikely match, consider instilling emotional competence within the ranks of the U.S. Air Force! That experience, according to Rich Handley, 43, an organizational-development specialist and chief of human-resources development for the Air Force Recruiting Service, has been a very valuable one.

Like AmEx, the Air Force found itself stumped by a problem that seemed to defy conventional solutions. Each year, it would hire about 400 new recruiters and charge them with finding a fresh group of recruits. And each year, within just seven months, the Air Force would dismiss as many as a quarter of those recruiters for failing to meet their quotas. The cost of that turnover was catastrophic. The Air Force spends an average of $30,000 to train a recruiter. The direct cost of replacing 100 a year was nearly $3 million. The indirect costs—which, for starters, included the missed recruiting targets—were even greater.

For Handley, the challenge was to figure out a way to assess each recruiter applicant more accurately—to predict a candidate's likelihood of success before hiring that person. After looking over a series of sales-aptitude screening instruments, Handley was most impressed by Reuven Bar-On's EQI. "It just seemed to go to the heart of it," he says. The 133-question self-administered test evaluates 15 qualities, such as empathy, self-awareness, and self-control, but also includes categories that seem less obviously a measure of emotional competence—among them

assertiveness, independence, social responsibility, and even happiness.

In early 1997, eager to learn more about the predictive capabilities of the EQI, Handley administered the test to 1,200 staff Air Force recruiters. They were divided into three groups: high performers who met 100% of their quotas, average performers who met at least 70%, and failures who met less than 30% of their quotas. The highest performers outscored the lowest in 14 of the 15 EQI competencies.

Handley found the results intriguing but not fully satisfying. "They were equivalent to telling you, 'Here are 14 ingredients that will make a good-tasting cake,' but then not giving you the exact amounts of the ingredients," he says. Taking his analysis one level deeper, Handley used a statistical-modeling technique to determine the top-five qualities that were associated with the highest-performing recruiters. They were (in order of importance) assertiveness, empathy, happiness, self-awareness, and problem solving. Disparate as these qualities may seem, they made sense to Handley. "Assertiveness is obviously important," he says. "If you're happier, you're more positive, and that's infectious. Someone with strong empathy skills can read a cold sale very quickly and won't waste time if it isn't going to work out.

And recruiters with strong problem-solving skills think on their feet more efficiently, waste less time, and feel less stressed—which makes them more effective in the long run." Indeed, the highest-performing recruiters put in the fewest number of hours. "The best ones work smarter, not harder," Handley says.

Recruiters who matched this high-performance profile turned out to have been nearly three times more likely to have met their quotas than their less-successful counterparts. The model yielded five categories for rating the probable success of new recruiters based on their EQI scores—excellent fit, good fit, fair fit, poor fit, and bad fit. This assessment turned out to be remarkably accurate. All recruiters who were considered "excellent" fits have met 100% of their recruiting quotas during the past year. More than 90% of the "good" fits met their quotas, compared to 80% of the "fairs" and less than 50% of the "poors."

The real value of that data was its ability to predict the performance of job applicants. Theoretically, the model suggests a 95% chance of success of a potential recruiter with a "good" or an "excellent" EQI profile. So Handley required every new recruiter to meet that threshold. One year later, the turnover among new recruiters had dropped from 100 to just 8. Based on an investment of less than $10,000 for EQI testing, the Air Force saved $2.76 million. "I come from an aeronautical orientation, and drag is what slows a plane down and impedes performance," says Handley. "To me, the EQI is a way to profile individual and organizational drag."

Handley went on to administer the EQI to two other groups in the Air Force—chronic substance abusers and spousal abusers. His goal was to identify their EQI deficits. Substance abusers' key deficits turned out to be problem solving, social responsibility, and stress tolerance. Spousal abusers primarily lacked empathy and had poor impulse control and an inflated self-regard. Again, the results made sense to Handley—and suggested a better approach to those problems. "We typically give people standard treatments," he says. "For spousal abusers, it might be anger management. The implication of these findings is that you need to individualize training to enhance the specific competencies that a person is lacking."

Handley has also begun to experiment with delivering such training through a Web site called EQ University.com. For $99, visitors to that site can take the EQI online, receive a seven-to-eight-page assessment, and participate in a 30-minute confidential telephone consultation with Handley or another trained professional. Based on that feedback, people can then select the competencies that they want to improve on and sign up for Web-based courses on 9 of Bar-On's 15 competencies. (The other 6 will be available by summer 2000.) Each course costs $49, and personal coaching is also available. So far, these courses are very basic and minimally interactive. It will be interesting to see whether deeply habitual behavior patterns can be transformed through Web-based training programs.

Bar-On believes that progress will occur in increments. "We've got a good start in assessment," he says. "Successful training is what we really have to tune up over the next several years."

DANIEL GOLEMAN'S EMOTIONAL JOURNEY

No one has done more than Daniel Goleman to spark corporate interest in emotional intelligence, to launch training programs that carry the message to business audiences, and to convey the ideas of emotional intelligence in an accessible manner. Much like Peter Senge with learning organizations and Michael Hammer with business-process reengineering, Goleman has become the most visible proponent of this new field. Now he must also contend with critics who argue that in broadening the appeal of emotional intelligence and tailoring it to the needs of a corporate audience, Goleman has sacrificed and diluted its original meaning.

As in any burgeoning field, a number of questions hand in the air: Will emotional competence become a better-understood concept, with clear definitions and well-defined boundaries—or will it slide into ambiguity as it becomes commercialized consulting property? Will companies use the demonstrated value of emotional competence to help improve performance and humanize employees' experiences—or will they come to view it with suspicion? In short, will emotional competence emerge as a useful and valuable tool to help businesses evolve—or will the lure of big consulting and big money distort its value, turning it into the next fad? Answers to many of those questions ultimately involve Goleman and the direction he chooses to take.

Until 1995, when Goleman published his book on the subject, emotional intelligence had languished as a relatively obscure theory. Psychologists Mayer at the University of New Hampshire and Salovey at Yale first introduced the term in 1990. Their first paper, published that same year, discussed emotional intelligence in highly technical terms. Reduced to its simplest description, emotional intelligence was defined as a group of mental abilities that help you recognize and understand your own feelings and those of others.

When Goleman tackled the subject, he expanded the definition to include the ability to motivate oneself. Where Mayer and Salovey appeared content to identify the concept of emotional intelligence. Goleman staked out a more aggressive claim for its value and benefits. Witness the subtitle of Emotional Intelligence: "Why it can matter more than IQ."

Published in the wake of Richard Hernstein and Charles Murray's controversial book, "The Bell Curve" (Free Press, 1994), which argued that IQ is the critical variable in achievement, "Emotional Intelligence" offered readers a new set of metrics. "We have gone too far in emphasizing the value and import of the purely rational," Goleman wrote. "For better or for worse, intelligence can come to nothing when the emotions hold sway."

Among Goleman's skills is his ability to move easily between different audiences and disciplines without any sense of contradiction or ambivalence. After getting his doctorate in psychology at Harvard in 1973, he went on to write popular pieces for "Psychology Today" and later became a reporter for the New York Times." He published his first book, "The Meditative Mind" (Tarcher, 1988), and for years, he lectured and held workshops on meditation, mindfulness, and Buddhist psychology.

Although "Emotional Intelligence" focused largely on applying emotional competencies to education, the book resonated with a business audience, and Goleman began getting flooded with corporate speaking invitations. He also established a consulting firm in 1996 to deliver emotional-intelligence training. But he quickly realized that he was more interested in speaking and writing than in designing training programs and running a business.

Goleman solved that dilemma a year later when he began working with the Hay Group. David McLelland, founder of one of the firm's divisions, had been Goleman's dissertation adviser at Harvard. The Hay Group largely took over the design, marketing, and delivery of a Goleman-branded emotional-intelligence training program.

The program model was based on the work of Richard Boyatzis, a classmate and fellow doctoral candidate of Goleman. Boyatzis, now 53 and a professor and department chair at Weatherhead School of Management at Case Western Reserve University, had focused his research on identifying the competencies that predict high performance in a particular job. Goleman found the data compelling. "If you want to know what will make an outstanding performer, don't look at IQ scores or specific technical skills," he says. "Look at the people who are the stars and see the abilities they exhibit that aren't found in people who are mediocre."

From there, it wasn't a huge leap to adapt that model to emotional intelligence—which is precisely what Goleman did in his third book, "Working with Emotional Intelligence," which he aimed specifically at corporate audiences. "The competency model uncovers hidden ingredients for success," he explains. "The correlations between performance and these emotional competencies have been well-established, but no overarching framework or theory could make sense of the foundation of these abilities." Goleman's competency model suggested an accessible way of understanding the connection between "soft" skills, such as adaptability, interpersonal effectiveness, leadership, and teamwork—competencies that previously seemed to stand alone.

Goleman set out to demonstrate this model's predictive power. He began by looking at research results, from studies conducted by several hundred organizations, on a range of competencies as predictors of performance. "When I sorted out those results, EQ abilities were twice as important as anything else in distinguishing stars from average performers," Goleman says. "And the higher you go in an organization, the more they matter."

In one Hay study, outstanding sales agents at a financial-services company were compared with average agents. The competencies that distinguished the two groups were ranked and then correlated with the degree to which a strength in a particular competency contributed to agents' annual revenues. High "networking-empathy skills," for example, correlated with a $50,000-a-year difference in revenues. High "team-leadership skills" increased salaries by $39,000 a year. Those agents who demonstrated a high "drive to achieve" earned $31,000 more a year. According to Goleman's and Hay's interpretation of the findings, emotional competencies accounted for a difference of 58% between earnings of high and low performers, whereas technical skills accounted for a far lower percentage.

The study seemed initially to provide resounding evidence that emotional intelligence helps predict business success. But, as is often the case when psychological measures are used in business, the data were open to more than one interpretation: In developing a model for business, Goleman had widened his definition of emotional intelligence far beyond Mayer and Salovey's original model—and even beyond the one that he had offered in his first book. The business version includes 25 separate "emotional" competencies—among them, achievement drive, commitment, conscientiousness, influence, initiative, political awareness, self-confidence, service orientation, trustworthiness, and even something called "leveraging diversity." Goleman's new model of emotional intelligence comes dangerously close to including nearly any competency that isn't explicitly cognitive or technical. He and Boyatzis have pinpointed qualities that correlate with success but not necessarily with one another. "They have included many competencies that are not really part of emotional intelligence but probably are important in determining effectiveness," argues Bar-On.

Take, for example, the "drive to achieve," which Hay has found to be the single most important predictor of success. "What makes you smarter is understanding your own feelings better," argues John Mayer. "Goleman has broadened the definition of emotional intelligence to such an extent that it no longer has any scientific meaning or utility and is no longer a clear predictor of outcome."

And what about the different forms that a drive to succeed can take? Does, say, an executive exhibit the core qualities of emotional intelligence, such as self-awareness, self-control, and empathy? Goleman acknowledges that many top executives

currently lack those competencies, which will, he argues, be increasingly more critical to success in the decade ahead, as the competition for talent escalates and hierarchical structures continue to break down. "A coercive style of leadership is a negative driver on every measure of climate in a company," he says. "Bosses who lead by coercion are the kinds of bosses whom people hate."

As a practical matter, the Goleman-Boyatzis-Hay approach has focused less on training emotional intelligence than on addressing specific deficiencies in those competencies. Boyatzis's work has been influential: At Case Western, he developed an elegant, comprehensive, highly successful approach to training competencies in graduate students, as well as in executives. His model blends work on deepening self-awareness ("the real self"), defining one's values ("the ideal self"), and implementing one's goals (changing specific behaviors to do that).

But adopting this competency model as the basis for emotional-intelligence training has proved complicated. Goleman has suggested that, broadly speaking, emotional intelligence is similar to character and virtue—that positive values go hand-in-hand with such qualities as self-awareness and self-control. In practice, however, it's simply not that neat. Used-car salesmen can just as easily use emotional-intelligence skills to sell defective cars as social activists can to inspire positive action. Intelligence—emotional or intellectual—is a value-free capacity that can be marshalled as effectively for good as for ill. "My statement might have been too unconditional," Goleman acknowledges. "This model is still evolving."

Finally, just how enthusiastically has the business community embraced offerings from consultants in emotional intelligence? Here, the evidence is mixed. Goleman has found companies that are willing to pay him as much as $40,000 for a one-hour lecture. But far fewer have been have been prepared to invest in a weeklong training program for top executives. "We're hearing a great deal of talk about emotional intelligence, but mostly it's at the inquiry stage," says Mark Van Buren, 35, director of research for the American Society for

Training & Development, which monitors trends in corporate training. "It's not yet clear whether many companies have figured out how to put the concept to work so that it produces meaningful results." Annie McKee, 45, director of management-development services at the Hay Group, who heads the company's efforts in emotional intelligence, insists that she is pleased with how things are going. "We have thousands of inquiries, many of which are from senior executives who want to introduce those concepts and practices into their organizations, and we're working steadily with dozens of clients," McKee says. Clients range from the Bank of Montreal to the Department of Defense's accounting office.

Hay's five-day emotional-competency training programs are typically customized to each of its clients. Hay begins every program by administering an Emotional Competence Inventory (ECI), that Boyatzis and Goleman developed. Unlike Bar-On's test, which is self-administered, the ECI is a 360-degree instrument given to bosses, colleagues, and direct reports. "People tend to be very poor judges of what they aren't good at, and that is particularly true of those who are having performance problems," says Goleman. Based on a dozen responses, participants receive detailed feedback on their perceived strengths and weaknesses. Next, they design an "action plan," typically in consultation with a Hay coach who regularly monitors progress during the months ahead.

Goleman argues that even with a broad defi-nition of emotional competency, Hay is offering something substantially different from ordinary technical-competency training. "When it comes to people's emotions, you're dealing with a different part of the brain," he says. "That part, the limbic system, is where you learn not just by processing information cognitively but also by repeating it. You need to practice a new habit to change your neural circuitry. It's not like sitting in a classroom. If you want a habit to stick, you have to repeat it over several months."

Whatever model finally prevails, the challenge ahead is to demonstrate to executives who are curious about emotional intelligence that their employees can actually be trained in such competencies, and that doing so will have a direct and significant impact on employee performance. Successfully demonstrating that will depend partly on developing a simpler definition of emotional intelligence; it will also depend on the ability to identify which competencies have the greatest effect on high performance and to make those the focus of corporate training.

Goleman is carefully optimistic about the future prospects of emotional-competency training in business. "I believe that this is a new paradigm for business and that paradigms shift slowly," he explains. "The first job is to raise awareness that this sort of training might actually make sense. It's less of a commitment for a company to have someone like me come in and talk about emotional intelligence for an hour than it is to do something about it."

10

Beyond the Charismatic Leader
Leadership and Organizational Change

DAVID A. NADLER AND MICHAEL L. TUSHMAN

Like never before, discontinuous organization change is an important determinant of organization adaptation. Responding to regulatory, economic, competitive and/or technological shifts through more efficiently pushing the same organization systems and processes just does not work.[1] Rather, organizations may need to manage through periods of both incremental as well as revolutionary change.[2] Further, given the intensity of global competition in more and more industries, these organizational transformations need to be initiated and implemented rapidly. Speed seems to count.[3] These trends put a premium on executive leadership and the management of system-wide organization change.

There is a growing knowledge base about large-scale organization change.[4] This literature is quite consistent on at least one aspect of effective system-wide change—namely, executive leadership matters. The executive is a critical actor in the drama of organization change.[5] Consider the following examples:

• At Fuji-Xerox, Yotaro Kobayashi's response to declining market share, lack of new products, and increasing customer complaints was to initiate widespread organization change. Most fundamentally, Kobayashi's vision was to change the way Fuji-Xerox conducted its business. Kobayashi and his team initiated the "New Xerox Movement" through Total Quality Control. The core values of quality, problem solving, teamwork, and customer emphasis were espoused and acted upon by Kobayashi and his team. Further, the executive team at Fuji instituted a dense infrastructure of objectives, measures, rewards, tools, education and slogans all in service of TQC and the "New Xerox." New heroes were created. Individuals and teams were publicly celebrated to reinforce to the system those behaviors that reflected the best of the new Fuji-Xerox. Kobayashi continually reinforced, celebrated, and communicated his TQC vision. Between 1976–1980, Fuji-Xerox gained back its market share, developed an impressive set of new products, and won the Deming prize.[6]

• Much of this Fuji-Xerox learning was transferred to corporate Xerox and further enhanced by Dave Kearns and his executive team. Beginning in 1983, Kearns clearly expressed his "Leadership Through Quality" vision for the corporation. Kearns established a Quality Task Force and Quality Office with respected Xerox executives. This broad executive base developed the architecture of Leadership Through Quality. This effort included quality principles, tools, education, required leadership actions, rewards, and feedback mechanisms. This attempt to transform the entire corporation was initiated at the top and diffused throughout the firm through overlapping teams. These teams were pushed by Kearns and his team to achieve extraordinary gains. While not completed, this transformation has helped Xerox regain lost market share and improve product development efforts.[7]

Copyright © 1990 by The Regents of the University of California. Reprinted from the *California Management Review*, Vol. 32, No. 2. By permission of The Regents.

David A. Nadler is President and Founder of the Delta Consulting Group, Inc. ***Michael L. Tushman*** is a Chaired Professor of Business Administration at the Harvard Business School.

• At General Electric, Jack Welch's vision of a lean, aggressive organization with all the benefits of size but the agility of small firms is being driven by a set of interrelated actions. For example, the "work-out" effort is a corporate-wide endeavor, spearheaded by Welch, to get the bureaucracy out of a large-old organization and, in turn, to liberate GE employees to be their best. This effort is more than Welch. Welch's vision is being implemented by a senior task force which has initiated workout efforts in Welch's own top team as well as in each GE business area. These efforts consist of training, problem solving, measures, rewards, feedback procedures, and outside expertise. Similarly, sweeping changes at SAS under Carlzon, at ICI under Harvey-Jones, by Anderson at NCR, and at Honda each emphasize the importance of visionary leadership along with executive teams, systems, structures and processes to transfer an individual's vision of the future into organizational reality.[8]

On the other hand, there are many examples of visionary executives who are unable to translate their vision into organization action. For example, Don Burr's vision at People Express not only to "make a better world" but also to grow rapidly and expand to capture the business traveller was not coupled with requisite changes in organization infrastructure, procedures, and/or roles. Further, Burr was unable to build a cohesive senior team to help execute his compelling vision. This switch in vision, without a committed senior team and associated structure and systems, led to the rapid demise of People Express.

Vision and/or charisma is not enough to sustain large-system change. While a necessary condition in the management of discontinuous change, we must build a model of leadership that goes beyond the inspired individual; a model that takes into account the complexities of system-wide change in large, diverse, geographically complex organizations. We attempt to develop a framework for the extension of charismatic leadership by building on the growing leadership literature,[9] the literature on organization evolution,[10] and our intensive consulting work with executives attempting major organization change.[11]

ORGANIZATIONAL CHANGE AND RE-ORGANIZATION

Organizations go through change all the time. However, the nature, scope, and intensity of organizational changes vary considerably. Different kinds of organizational changes will require very different kinds of leadership behavior in initiating, energizing, and implementing the change. Organization changes vary along the following dimensions:

• *Strategic and Incremental Changes.* Some changes in organizations, while significant, only affect selected components of the organization. The fundamental aim of such change is to enhance the effectiveness of the organization, but within the general framework of the strategy, mode of organizing, and values that already are in place. Such changes are called *incremental changes*. Incremental changes happen all the time in organizations, and they need not be small. Such things as changes in organization structure, the introduction of new technology, and significant modifications of personnel practices are all large and significant changes, but ones which usually occur within the existing definition and frame of reference of the organization. Other changes have an impact on the whole system of the organization and fundamentally redefine what the organization is or change its basic framework, including strategy, structure, people, processes, and (in some cases) core values. These changes are called *strategic organizational changes*. The Fuji-Xerox, People Express, ICI, and SAS cases are examples of system-wide organization change.

• *Reactive and Anticipatory Changes.* Many organizational changes are made in direct response to some external event. These changes, which are forced upon the organization, are called *reactive*. The Xerox, SAS and ICI transformations were all initiated in response to organization performance crisis. At other times, strategic organizational change is initiated not because of the need to re-spond to a contemporaneous event, but rather be-cause senior management believes that change in anticipation of events still to come will provide competitive advantage. These changes are called *anticipatory*. The GE and People Express cases as well as more recent system-wide changes at ALCOA and Cray Research are examples of system-wide change initiated in anticipation of en-vironmental change.

If these two dimensions are combined, a basic ty-pology of different changes can be described (see Figure 10.1).

Change which is incremental and anticipatory is called *tuning*. These changes are not system-wide redefinitions, but rather modifications of specific components, and they are initiated in anticipation of future events. Incremental change which is ini-tiated reactively is called *adaptation*. Strategic change initiated in anticipation of future events is called *reorientation*, and change which is prompted by immediate demands is called *re-creation*.[12]

Research on patterns of organizational life and death across several industries has provided insight into the patterns of strategic organizational change.[13] Some of the key findings are as follows:

• *Strategic organization changes are neces-sary.* These changes appear to be environmentally driven. Various factors—be they competitive, tech-nological, or regulatory—drive the organization (either reactively or in anticipation) to make system-wide changes. While strategic organization change does not guarantee success, those organizations that fail to change, generally fail to survive. Discontin-uous environmental change seems to require dis-continuous organization change.

• *Re-creations are riskier.* Re-creations are riskier endeavors than reorientations if only be-cause they are initiated under crisis conditions and under sharp time constraints. Further, re-creations almost always involve a change in core values. As core values are most resistant to change, re-creations always trigger substantial individual resistance to change and heightened political behavior. Re-creations that do succeed usually in-volve changes in the senior leadership of the firm, frequently involving replacement from the outside. For example, the reactive system-wide changes at U.S. Steel, Chrysler, and Singer were all initiated by new senior teams.

• *Re-orientations are associated more with success.* Re-orientations have the luxury of time to shape the change, build coalitions, and empower individuals to be effective in the new organization. Further, re-orientations give senior managers time to prune and shape core values in service of the re-vised strategy, structure, and processes. For exam-ple, the proactive strategic changes at Cray Re-search, ALCOA, and GE each involved system-wide change as well as the shaping of core values ahead of the competition and from a position of strength.

Re-orientations are, however, risky. When sweeping changes are initiated in advance of pre-cipitating external events, success is contingent on making appropriate strategic bets. As re-orientations are initiated ahead of the competition and in advance of environmental shifts, they require visionary ex-ecutives. Unfortunately, in real time, it is unclear who will be known as visionary executives (e.g., Welch, Iacocca, Rollwagen at Cray Research) and who will be known as failures (e.g., Don Burr at People Express, or Larry Goshorn at General Au-tomation). In turbulent environments, not to make strategic bets is associated with failure. Not all bets will pay off, however. The advantages of re-orientations derive from the extra implementation time and from the opportunity to learn from and adapt to mistakes.[14]

	Incremental	Strategic
Anticipatory	Tuning	Re-orientation
Reactive	Adaptation	Re-creation

Figure 10.1. Types of organizational changes.

As with re-creations, executive leadership is crucial in initiating and implementing strategic re-orientations. The majority of successful re-orientations involve change in the CEO and substantial executive team change. Those most successful firms, however, have executive teams that are relatively stable yet are still capable of initiating several re-orientations (e.g., Ken Olsen at DEC and An Wang at Wang).

There are, then, quite fundamentally different kinds of organizational changes. The role of executive leadership varies considerably for these different types of organization changes. Incremental change typically can be managed by the existing management structures and processes of the organization, sometimes in conjunction with special transition structures.[15] In these situations, a variety of leadership styles may be appropriate, depending upon how the organization is normally managed and led. In strategic changes, however, the management process and structure itself is the subject of change; therefore, it cannot be relied upon to manage the change. In addition, the organization's definition of effective leadership may also be changing as a consequence of the re-orientation or re-creation. In these situations, leadership becomes a very critical element of change management.

This article focuses on the role of executive leadership in strategic organization change, and in particular, the role of leadership in re-orientations. Given organization and individual inertia, re-orientations can not be initiated or implemented without sustained action by the organization's leadership. Indeed, re-orientations are frequently driven by new leadership, often brought in from outside the organization.[16] A key challenge for executives facing turbulent environments, then, is to learn how to effectively initiate, lead, and manage re-orientations. Leadership of strategic re-orientations requires not only charisma, but also substantial instrumental skills in building executive teams, roles, and systems in support of the change, as well as institutional skills in diffusing leadership throughout the organization.

THE CHARISMATIC LEADER

While the subject of leadership has received much attention over the years, the more specific issue of leadership during periods of change has only recently attracted serious attention.[17] What emerges from various discussions of leadership and organizational change is a picture of the special kind of leadership that appears to be critical during times of strategic organizational change. While various words have been used to portray this type of leadership, we prefer the label "charismatic" leader. It refers to a special quality that enables the leader to mobilize and sustain activity within an organization through specific personal actions combined with perceived personal characteristics.

The concept of the charismatic leader is not the popular version of the great speech maker or television personality. Rather, a model has emerged from recent work aimed at identifying the nature and determinants of a particular type of leadership that successfully brings about changes in an individual's values, goals, needs, or aspirations. Research on charismatic leadership has identified this type of leadership as observable, definable, and having clear behavioral characteristics.[18] We have attempted to develop a first cut description of the leader in terms of patterns of behavior that he/she seems to exhibit. The resulting approach is outlined in Figure 10.2, which lists three major types of behavior that characterize these leaders and some illustrative kinds of actions.

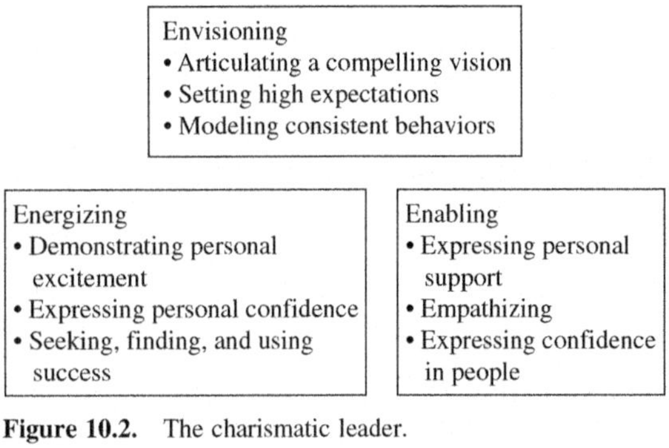

Envisioning
• Articulating a compelling vision
• Setting high expectations
• Modeling consistent behaviors

Energizing
• Demonstrating personal excitement
• Expressing personal confidence
• Seeking, finding, and using success

Enabling
• Expressing personal support
• Empathizing
• Expressing confidence in people

Figure 10.2. The charismatic leader.

The first component of charismatic leadership is *envisioning*. This involves the creation of a picture of the future, or of a desired future state with which people can identify and which can generate excitement. By creating vision, the leader provides a vehicle for people to develop commitment, a common goal around which people can rally, and a way for people to feel successful. Envisioning is accomplished through a range of different actions. Clearly, the simplest form is through articulation of a compelling vision in clear and dramatic terms. The vision needs to be challenging, meaningful, and worthy of pursuit, but it also needs to be credible. People must believe that it is possible to succeed in the pursuit of the vision. Vision is also communicated in other ways, such as through expectations that the leader expresses and through the leader personally demonstrating behaviors and activities that symbolize and further that vision.

The second component is *energizing*. Here the role of the leader is the direct generation of energy—motivation to act—among members of the organization. How is this done? Different leaders engage in energizing in different ways, but some of the most common include demonstration of their own personal excitement and energy, combined with leveraging that excitement through direct personal contact with large numbers of people in the organization. They express confidence in their own ability to succeed. They find, and use, successes to celebrate progress towards the vision.

The third component is *enabling*. The leader psychologically helps people act or perform in the face of challenging goals. Assuming that individuals are directed through a vision and motivated by the creation of energy, they then may need emotional assistance in accomplishing their tasks. This enabling is achieved in several ways. Charismatic leaders demonstrate empathy—the ability to listen, understand, and share the feelings of those in the organization. They express support for individuals. Perhaps most importantly, the charismatic leader tends to express his/her confidence in people's ability to perform effectively and to meet challenges.

Yotaro Kobayashi at Fuji-Xerox and Paul O'Neill at ALCOA each exhibit the characteristics of charismatic leaders. In Kobayashi's transformation at Fuji, he was constantly espousing his New Xerox Movement vision for Fuji. Kobayashi set high standards for his firm (e.g., the 3500 model and the Deming Prize), for himself, and for his team. Beyond espousing this vision for Fuji, Kobayashi provided resources, training, and personal coaching to support his colleagues' efforts in the transformation at Fuji. Similarly, Paul O'Neill has espoused a clear vision for ALCOA anchored on quality, safety, and innovation. O'Neill has made his vision compelling and central to the firm, has set high expectations for this top team and for individuals throughout ALCOA and provides continuous support and energy for his vision through meetings, task forces, video tapes, and extensive personal contact.

Assuming that leaders act in these ways, what functions are they performing that help bring about change? First, they provide a psychological focal point for the energies, hopes, and aspirations of people in the organization. Second, they serve as powerful role models whose behaviors, actions and personal energy demonstrate the desired behaviors expected throughout the firm. The behaviors of charismatic leaders provide a standard to which others can aspire. Through their personal effectiveness and attractiveness they build a very personal and intimate bond between themselves and the organization. Thus, they can become a source of sustained energy; a figure whose high standards others can identify with and emulate.

Limitations of the Charismatic Leader

Even if one were able to do all of the things involved in being a charismatic leader, it might still not be enough. In fact, our observations suggest that there are a number of inherent limitations to the effectiveness of charismatic leaders, many stemming from risks associated with leadership which revolves around a single individual. Some of the key potential problems are:

• *Unrealistic Expectations.* In creating a vision and getting people energized, the leader may create expectations that are unrealistic or unattainable. These can backfire if the leader cannot live up to the expectations that are created.

• *Dependency and Counterdependency.* A strong, visible, and energetic leader may spur different psychological response. Some individuals may become overly dependent upon the leader, and in some cases whole organizations become dependent. Everyone else stops initiating actions and waits for the leader to provide direction; individuals may become passive or reactive. On the other extreme, others may be uncomfortable with strong personal presence and spend time and energy demonstrating how the leader is wrong—how the emperor has no clothes.

• *Reluctance to Disagree with the Leader.* The charismatic leader's approval or disapproval becomes an important commodity. In the presence of a strong leader, people may become hesitant to disagree or come into conflict with the leader. This may, in turn, lead to stifling conformity.

• *Need for Continuing Magic.* The charismatic leader may become trapped by the expectation that the magic often associated with charisma will continue unabated. This may cause the leader to act in ways that are not functional, or (if the magic is not produced) it may cause a crisis of leadership credibility.

• *Potential Feelings of Betrayal.* When and if things do not work out as the leader has envisioned, the potential exists for individuals to feel betrayed by their leader. They may become frustrated and angry, with some of that anger directed at the individual who created the expectations that have been betrayed.

• *Disenfranchisement of Next Levels of Management.* A consequence of the strong charismatic leader is that the next levels of management can easily become disenfranchised. They lose their ability to lead because no direction, vision, exhortation, reward, or punishment is meaningful unless it comes directly from the leader. The charismatic leader thus may end up underleveraging his or her management and/or creating passive/dependent direct reports.

• *Limitations of Range of the Individual Leader.* When the leadership process is built around

an individual, management's ability to deal with various issues is limited by the time, energy, expertise, and interest of that individual. This is particularly problematic during periods of change when different types of issues demand different types of competencies (e.g., markets, technologies, products, finance) which a single individual may not possess. Different types of strategic changes make different managerial demands and call for different personal characteristics. There may be limits to the number of strategic changes that one individual can lead over the life of an organization.

In light of these risks, it appears that the charismatic leader is a necessary component—but not a sufficient component—of the organizational leadership required for effective organizational re-organization. There is a need to move beyond the charismatic leader.

INSTRUMENTAL LEADERSHIP

Effective leaders of change need to be more than just charismatic. Effective re-orientations seem to be characterized by the presence of another type of leadership behavior which focuses not on the excitement of individuals and changing their goals, needs or aspirations, but on making sure that individuals in the senior team and throughout the organization behave in ways needed for change to occur. An important leadership role is to build competent teams, clarify required behaviors, built in measurement, and administer rewards and punishments so that individuals perceive that behavior consistent with the change is central for them in achieving their own goals.[19] We will call this type of leadership *instrumental leadership*, since it focuses on the management of teams, structures, and managerial processes to create individual instrumentalities. The basis of this approach is in expectancy theories of motivation, which propose that individuals will perform those behaviors that they perceive as instrumental for acquiring valued outcomes.[20] Leadership, in this context, involves managing environments to create conditions that motivate desired behavior.[21]

In practice, instrumental leadership of change involves three elements of behavior (see Figure 10.3). The first is *structuring*. The leader invests time in building teams that have the required competence to execute and implement the re-orientation[22] and in creating structures that make it clear what types of behavior are required throughout the organization. This may involve setting goals, establishing standards, and defining roles and responsibilities. Re-orientations seem to require detailed planning about what people will need to do and how they will be required to act during different phases of the change. The second element of instrumental leadership is *controlling*. This involves the creation of systems and processes to measure, monitor, and assess both behavior and results and to administer corrective action.[23] The third element is *rewarding*, which includes the administration of both rewards and punishments contingent upon the degree to which behavior is consistent with the requirements of the change.

Instrumental leadership focuses on the challenge of shaping consistent behaviors in support of the re-orientation. The charismatic leader excites individuals, shapes their aspirations, and directs their energy. In practice, however, this is not enough to sustain patterns of desired behavior. Subordinates and colleagues may be committed to the vision, but over time other forces may influence their behavior, particularly when they are not in direct personal contact with the leader. This is particularly relevant during periods of change when the formal organization and the informal social system may lag behind the leader and communicate outdated messages or reward traditional behavior. Instrumental leadership is needed to ensure compliance over time consistent with the commitment generated by charismatic leadership.

At Xerox, for example, David Kearns used instrumental leadership to further enliven his Leadership Through Quality efforts.[24] Beyond his own sustained behaviors in support of the Leadership Through Quality effort, Kearns and his Quality Office developed comprehensive set of roles, processes, teams, and feedback and audit mechanisms for getting customer input and continuous improvement into everyday problem solving throughout Xerox. Individuals and teams across the corporation were evaluated on their ability to continuously meet customer requirements. These data were used in making pay, promotion, and career decisions.

The Role of Mundane Behaviors

Typical descriptions of both charismatic and instrumental leaders tend to focus on significant events, critical incidents, and grand gestures. Our vision of the change manager is frequently exemplified by the key speech or public event that is a potential watershed event. While these are important arenas for leadership, leading large-system change also requires sustained attention to the myriad of details that make up organizational life. The accumulation of less dramatic, day-to-day activities and mundane behaviors serves as a powerful determinant of behavior.[25] Through relatively unobtrusive acts, through sustained attention to detail, managers can directly shape perceptions and culture in support of the change effort. Examples of mundane behavior that when taken together can have a great impact include:

- allocation of time; calendar management
- asking questions, following up
- shaping of physical settings
- public statements
- setting agendas of events or meetings
- use of events such as lunches, meetings, to push the change effort

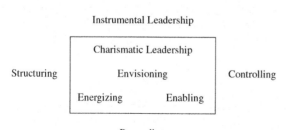

Figure 10.3. Instrumental leadership.

- summarization—post hoc interpretation of what occurred

- creating heroes

- use of humor, stories, and myths

- small symbolic actions, including rewards and punishments

In each of these ways, leaders can use daily activities to emphasize important issues, identify desirable behavior, and help create patterns and meaning out of the various transactions that make up organizational life.

The Complementarity of Leadership Approaches

It appears that effective organizational re-orientation requires both charismatic and instrumental leadership. Charismatic leadership is needed to generate energy, create commitment, and direct individuals towards new objectives, values or aspirations. Instrumental leadership is required to ensure that people really do act in a manner consistent with their new goals. Either one alone is insufficient for the achievement of change.

The complementarity of leadership approaches and the necessity for both creates a dilemma.[26] Success in implementing these dual approaches is associated with the personal style, characteristics, needs, and skills of the executive. An individual who is adept at one approach may have difficulty executing the other. For example, charismatic leaders may have problems with tasks involved in achieving control. Many charismatic leaders are motivated by a strong desire to receive positive feedback from those around them.[27] They may therefore have problems delivering unpleasant messages, dealing with performance problems, or creating situations that could attract negative feelings.[28]

Only exceptional individuals can handle the behavioral requirements of both charismatic and instrumental leadership styles. While such individuals exist, an alternative may be to involve others in leadership roles, thus complementing the strengths and weaknesses of one individual leader.[29] For example, in the early days at Honda, it took the steadying,

systems-oriented hand of Takeo Fujisawa to balance the fanatic, impatient, visionary energy of Soichiro Honda. Similarly, at Data General, it took Alsing and Rasala's social, team, and organization skills to balance and make more humane Tom West's vision and standards for the Eclipse team.[30] Without these complementary organization and systems skills, Don Burr was unable to execute his proactive system-wide changes at People Express.

The limitations of the individual leader pose a significant challenge. Charismatic leadership has a broad reach. It can influence many people, but is limited by the frequency and intensity of contact with the individual leader. Instrumental leadership is also limited by the degree to which the individual leader can structure, observe, measure and reward behavior. These limitations present significant problems for achieving re-orientations. One implication is that structural extensions of leadership should be created in the process of managing re-orientations.[31] A second implication is that human extensions of leadership need to be created to broaden the scope and impact of leader actions. This leads to a third aspect of leadership and change—the extension of leadership beyond the individual leader, or the creation of institutionalized leadership throughout the organization.

INSTITUTIONALIZING THE LEADERSHIP OF CHANGE

Given the limitations of the individual charismatic leader, the challenge is to broaden the range of individuals who can perform the critical leadership functions during periods of significant organizational change. There are three potential leverage points for the extension of leadership—the senior team, broader senior management, and the development of leadership throughout the organization (see Figure 10.4).

Leveraging the Senior Team

The group of individuals who report directly to the individual leader—the executive or senior team—is the first logical place to look for opportunities to

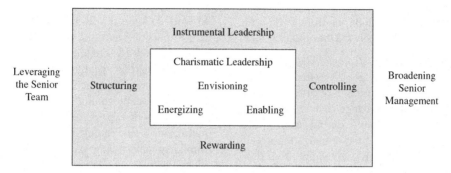

Figure 10.4. Institutionalized leadership.

extend and institutionalize leadership. Development of an effective, visible, and dynamic senior team can be a major step in getting around the problems and limitations of the individual leader.[32] Examples of such executive teams include the Management Committee established at Corning by Jamie Houghton or Bob Allen's Executive Committee at AT&T. Several actions appear to be important in enhancing the effectiveness of the senior team.

• *Visible Empowerment of the Team.* A first step is the visible empowerment of the team, or "anointing" the team as extensions of the individual leader. There are two different aspects to this empowerment: objective and symbolic. Objective empowerment involves providing team members with the autonomy and resources to serve effectively. Symbolic empowerment involves communicating messages (through information, symbols, and mundane behaviors) to show the organization that these individuals are indeed extensions of the leader, and ultimately key components of the leadership. Symbolic empowerment can be done through the use of titles, the designation of organizational structures, and the visible presence of individuals in ceremonial roles.

• *Individual Development of Team Members.* Empowerment will fail if the individuals on the team are not capable of executing their revised leadership roles. A major problem in re-orientations is that the members of the senior team frequently are the product of the very systems, structures, and values that the re-orientation seeks to change. Participating in the change, and more importantly, leading it, may require a significant switching of cognitive gears.[33] Re-orientations demand that senior team members think very differently about the business and about managing. This need for personal change at the most senior level has implications for the selection of senior team members (see below). It also may mean that part of the individual leader's role is to help coach, guide, and support individuals in developing their own leadership capabilities. Each individual need not (and should not) be a "clone" of the individual leader; but each should be able to initiate credible leadership actions in a manner consistent with their own personal styles. Ultimately, it also puts a demand on the leader to deal with those who will not or can not make the personal changes required for helping lead the re-orientation.

• *Composition of the Senior Team.* The need for the senior team to implement change may mean that the composition of that team may have to be altered. Different skills, capabilities, styles, and value orientations may be needed to both lead the changes as well as to manage in the reconfigured organization.[34] In fact, most successful re-orientations seem to involve some significant changes in the make-up of the senior team. This may require outplacement of people as well as importing new people, either from outside the organization, or from outside the coalition that has traditionally led the organization.[35]

• *The Inducement of Strategic Anticipation.* A critical issue in executing re-orientations is strategic anticipation. By definition, a re-orientation is a strategic organizational change that is initiated in anticipation of significant external events. Re-orientation occurs because the organization's leadership perceives competitive advantage from initiating change earlier rather than later. The question is, who is responsible for thinking about and anticipating external events, and ultimately deciding that re-orientation is necessary? In some cases, the individual leader does this, but the task is enormous. This is where the senior team can be helpful, because as a group it can scan a larger number of events and potentially be more creative in analyzing the environment and the process of anticipation.

Companies that are successful anticipators create conditions in which anticipation is more likely to occur. They invest in activities that foster anticipation, such as environmental scanning, experiments or probes inside the organization (frequently on the periphery), and frequent contacts with the outside. The senior team has a major role in initiating, sponsoring, and leveraging these activities.[36]

• *The Senior Team as a Learning System.* For a senior team to benefit from its involvement in leading change, it must become an effective system for learning about the business, the nature of change, and the task of managing change. The challenge is to both bond the team together, while avoiding insularity. One of the costs of such team structures is that they become isolated from the rest of the organization, they develop patterns of dysfunctional conformity, avoid conflict, and over time develop patterns of learned incompetence. These group processes diminish the team's capacity for effective strategic anticipation, and decreases the team's ability to provide effective leadership of the re-orientation.[37]

There are several ways to enhance a senior team's ability to learn over time. One approach is to work to keep the team an open system, receptive to outside ideas and information. This can be accomplished by creating a constant stream of events that expose people to new ideas and/or situations. For example, creating simulations, using critical incident techniques, creating near histories, are all ways of exposing senior teams to novel situations and sharpening problem-solving skills.[38] Similarly, senior teams can open themselves to new ideas via speakers or visitors brought in to meet with the team, visits by the team to other organizations, frequent contact with customers, and planned informal data collection through personal contact (breakfasts, focus groups, etc.) throughout the organization. A second approach involves the shaping and management of the internal group process of the team itself. This involves working on effective group leadership, building effective team member skills, creating meeting management discipline, acquiring group problem-solving and information-processing skills, and ultimately creating norms that promote effective learning, innovation, and problem solving.[39]

David Kearns at Xerox and Paul O'Neill at ALCOA made substantial use of senior teams in implementing their quality-oriented organization transformations. Both executives appointed senior quality task forces composed of highly respected senior executives. These task forces were charged with developing the corporate-wide architecture of the change effort. To sharpen their change and quality skills these executives made trips to Japan and to other experienced organizations, and were involved in extensive education and problem-solving efforts in their task forces and within their own divisions. These task forces put substance and enhanced energy into the CEO's broad vision. These executives were, in turn, role models and champions of the change efforts in their own sectors.

As a final note, it is important to remember that frequently there are significant obstacles in developing effective senior teams to lead re-orientations.

The issues of skills and selection have been mentioned. Equally important is the question of power and succession. A team is most successful when there is a perception of common fate. *Individuals have to believe that the success of the team will, in the long run, be more salient to them than their individual short-run success.* In many situations, this can be accomplished through appropriate structures, objectives, and incentives. But these actions may fail when there are pending (or anticipated) decisions to be made concerning senior management succession. In these situations, the quality of collaboration tends to deteriorate significantly, and effective team leadership of change becomes problematic. The individual leader must manage the timing and process of succession in relation to the requirements for team leadership, so that conflicting (and mutually exclusive) incentives are not created by the situation.[40]

Broadening Senior Management

A second step in moving beyond individual leadership of change is the further extension of the leadership beyond the executive or senior team to include a broader set of individuals who make up the senior management of the organization. This would include individuals one or two levels down from the executive team. At Corning, the establishment of two groups—the Corporate Policy Group (approximately the top 35) and the Corporate Management Group (about the top 120)—are examples of mechanisms used by Houghton to broaden the definition of senior management. This set of individuals is in fact the senior operating management of most sizeable organizations and is looked upon as senior management by the majority of employees. In many cases (and particularly during times of change) they do not feel like senior management, and thus they are not positioned to lead the change. They feel like participants (at best) and victims (at worst). This group can be particularly problematic since they may be more embedded in the current system of organizing and managing than some of the senior team. They may be less prepared to change, they frequently have molded themselves to

fit the current organizational style, and they may feel disenfranchised by the very act of developing a strong executive team, particularly if that team has been assembled by bringing in people from outside of the organization.

The task is to make this group feel like senior management, to get them signed up for the change, and to motivate and enable them to work as an extension of the senior team. Many of the implications are similar to those mentioned above in relation to the top team; however, there are special problems of size and lack of proximity to the individual charismatic leader. Part of the answer is to get the senior team to take responsibility for developing their own teams as leaders of change. Other specific actions may include:

- *Rites of Passage.* Creating symbolic events that help these individuals to feel more a part of senior management.

- *Senior Groups.* Creating structures (councils, boards, committees, conferences) to maintain contact with this group and reinforce their sense of participation as members of senior management.

- *Participation in Planning Change.* Involving these people in the early diagnosing of the need to change and the planning of change strategies associated with the re-orientation. This is particularly useful in getting them to feel more like owners, rather than victims of the change.

- *Intensive Communication.* Maintaining a constant stream of open communication to and from this group. It is the lack of information and perspective that psychologically disenfranchises these individuals.

Developing Leadership in the Organization

A third arena for enhancing the leadership of re-organizations is through organizational structures, systems, and process for leadership development consistent with the re-orientation. Frequently leadership development efforts lag behind the re-orientation. The management development system of many organizations often works effectively to create managers who will fit well with the organizational environment that the leadership seeks to abandon. There needs to be a strategic and anticipatory thinking about the leadership development process, including the following:

• *Definition of Managerial Competence.* A first step is determining the skills, capabilities, and capacities needed to manage and lead effectively in the re-orientation and post re-orientation period. Factors that have contributed to managerial success in the past may be the seeds of failure in the future.

• *Sourcing Managerial Talent.* Re-orientations may require that the organization identify significantly different sources for acquiring leaders or potential leaders. Senior managers should be involved in recruiting the hiring. Because of the lead time involved, managerial sourcing has to be approached as a long-term (five to ten years) task.

• *Socialization.* As individuals move into the organization and into positions of leadership, deliberate actions must be taken to teach them how the organization's social system works. During periods of re-orientation, the socialization process ought to lead rather than lag behind the change.[41]

• *Management Education.* Re-orientation may require managers and leaders to use or develop new skills, competencies, or knowledge. This creates a demand for effective management education. Research indicates that the impact of passive internal management education on the development of effective leaders may be minimal when compared with more action-oriented educational experiences. The use of educational events to expose people to external settings or ideas (through out-of-company education) and to socialize individuals through action-oriented executive education may be more useful than attempts to teach people to be effective leaders and managers.[42]

• *Career Management.* Research and experience indicate that the most potent factor in the development of effective leaders is the nature of their job experiences.[43] The challenge is to ensure that middle and lower level managers get a wide range of experiences over time. Preparing people to lead re-orientations may require a greater emphasis on the development of generalists through cross-functional, divisional, and/or multinational career experiences.[44] Diverse career experiences help individuals develop a broad communication network and a range of experiences and competences all of which are vital in managing large-system change. This approach to careers implies the sharing of the burden of career management between both the organization and the employee as well as the deliberate strategy of balancing current contribution with investment for the future when placing people in job assignments.[45]

• *Seeding Talent.* Developing leadership for change may also require deliberate leveraging of available talent. This implies thoughtful placement of individuals leaders in different situations and parts of the organization, the use of transfers, and the strategic placement of high-potential leaders.[46]

Perhaps the most ambitious and most well-documented effort at developing leadership throughout the organization is Welch's actions at GE. Welch has used GE's Management Development Institute at Crotonville as an important lever in the transformation of GE. Based on Welch's vision of a lean, competitive, agile organization with businesses leading in their respective markets, Crotonville has been used as a staging area for the revolution at GE. With Welch's active involvement, Crotonville's curriculum has moved from a short-term cognitive orientation towards longer-term problem solving and organization change. The curriculum has been developed to shape experiences and sharpen skills over the course of an individual's career in service of developing leaders to fit into the new GE.[47]

SUMMARY

In a world characterized by global competition, deregulation, sharp technological change, and political turmoil, discontinuous organization change seems to be a determinant of organization adaptation. Those firms that can initiate and implement discontinuous organization change more rapidly and/or prior to the competition have a competitive advantage. While not all change will be successful, inertia or incremental change in the face of altered competitive arenas is a recipe for failure.

Executive leadership is the critical factor in the initiation and implementation of large-system organization change. This article has developed an approach to the leadership of discontinuous organization change with particular reference to re-orientations—discontinuous change initiated in advance of competitive threat and/or performance crisis. Where incremental change can be delegated, strategic change must be driven by senior management. Charismatic leadership is a vital aspect of managing large-system change. Charismatic leaders provide vision, direction, and energy. Thus the successes of O'Neill at ALCOA, Welch at GE, Kearns at Xerox, and Rollwagen and Cray are partly a function of committed, enthusiastic, and passionate individual executives.

Charisma is not, however, enough to effect large-system change. Charismatic leadership must be bolstered by instrumental leadership through attention to detail on roles, responsibilities, structures, and rewards. Further, as many organizations are too large and complex for any one executive and/or senior team to directly manage, responsibility for large-system change must be institutionalized throughout the management system. The leadership of strategic organization change must be pushed throughout the organization to maximize the probability that managers at all levels own and are involved in executing the change efforts and see the concrete benefits of making the change effort work. O'Neill, Welch, Kearns, and Rollwagen are important catalysts in their organizations. Their successes to date are, however, not based simply on strong personalities. Each of these executives has been able to build teams, systems, and managerial processes to leverage and add substance to his vision and energy. It is this interaction of charisma, attention to systems and processes, and widespread involvement at multiple levels that seems to drive large-system change.

Even with inspired leadership, though, no re-orientation can emerge fully developed and

planned. Re-orientations take time to implement. During this transition period, mistakes are made, environments change and key people leave. Given the turbulence of competitive conditions, the complexity of large-system change and individual cognitive limitations, the executive team must develop its ability to adapt to new conditions and, as importantly, learn from both its successes and failures. As organizations can not remain stable in the face of environmental change, so too must the management of large-system change be flexible. This ability of executive teams to build-in learning and to build-in flexibility into the process of managing large-system organizational change is a touchstone for proactively managing re-orientations.

REFERENCES

1. R. Solow, M. Dertouzos, and R. Lester, *Made in America* (Cambridge, MA: MIT Press, 1989).
2. See M.L. Tushman, W. Newman, and E. Romanelli, "Convergence and Upheaval: Managing the Unsteady Pace of Organizational Evolution," *California Management Review*, 29/1 (Fall 1986):29–44.
3. E.g., K. Imai, I. Nonaka, and H. Takeuchi, "Managing the New Product Development Process: How Japanese Companies Learn and Unlearn," in K. Clark and R. Hayes, *The Uneasy Alliance* (Cambridge, MA: Harvard University Press, 1985).
4. E.g., A. Pettigrew, *The Awakening Giant: Continuity and Change at ICI* (London: Blackwell, 1985); J.R. Kimberly and R.E. Quinn, *New Futures: The Challenge of Managing Corporate Transitions* (Homewood, IL: Dow Jones-Irwin, 1984): Y. Allaire and M. Firsirotu, "How to Implement Radical Strategies in Large Organizations," *Sloan Management Review* (Winter 1985).
5. E.g., J. Gabbaro, *The Dynamics of Taking Charge* (Cambridge, MA: Harvard Business School Press, 1987); L. Greiner and A. Bhambri, "New CEO Intervention and Dynamics of Deliberate Strategic Change," *Strategic Management Journal*, 10 (1989):67–86; N.M. Tichy and M.A. Devanna, *The Transformational Leader* (New York, NY: John Wiley & Sons, 1986); D. Hambrick, "The Top Management Team: Key to Strategic Success," *California Management Review*, 30/1 (Fall 1987):88–108.
6. Y. Kobayashi, "Quality Control in Japan: The Case of Fuji Xerox," *Japanese Economic Studies* (Spring 1983).

7. G. Jacobson and J. Hillkirk, *Xerox: American Samurai* (New York, NY: Macmillan, 1986).

8. For SAS, see J. Carlzon, *Moments of Truth* (Cambridge, MA: Ballinger, 1987); for ICI, see Pettigrew, op. cit.; for NCR, see R. Rosenbloom, *From Gears to Chips: The Transformation of NCR in the Digital Era* (Cambridge, MA: Harvard University Press, 1988); for Honda, see I. Nonaka, "Creating Organizational Order Out of Chaos: Self-Renewal in Japanese Firms," *California Management Review*, 30/3 (Spring 1988):57–73.

9. Gabbaro, op. cit.; H. Levinson and S. Rosenthal, *CEO: Corporate Leadership in Action* (New York, NY: Basic Books, 1984); Greiner and Bhambri, op. cit.

10. Tushman et al., op. cit.; R. Greenwood and C. Hinings, "Organization Design Types, Tracks, and the Dynamics of Strategic Change," *Organization Studies*, 9/3 (1988):293–316; D. Miller and P. Friesen, *Organizations: A Quantum View* (Englewood Cliffs, NJ: Prentice-Hall, 1984).

11. D.A. Nadler and M.L. Tushman, "Organizational Framebending: Principles for Managing Re-orientation," *Academy of Management Executive*, 3 (1989):194–202.

12. For a more detailed discussion of this framework, see Nadler and Tushman, ibid.

13. Tushman et al., op. cit.; Greiner and Bhambri, op. cit.; Greenwood and Hinings, op. cit.; B. Virany and M.L. Tushman, "Changing Characteristics of Executive Teams in and Emerging Industry," *Journal of Business Venturing*, 1 (1986):261–274; M.L. Tushman and E. Romanelli, "Organizational Evolution: A Metamorphosis Model of Convergence and Reorientation," in B.M. Staw and L.L. Cummings, eds., *Research in Organizational Behavior*, 5 (Greenwich, CT: JAI Press, 1985), pp. 171–222.

14. J. March, L. Sproull, and M. Tamuz, "Learning from Fragments of Experience," *Organization Science* (in press).

15. R. Beckhard and R. Harris, *Organizational Transitions* (Reading, MA: Addison-Wesley, 1977).

16. See R. Vancil, *Passing the Baton* (Cambridge, MA: Harvard Business School Press, 1987).

17. J.M. Burns, *Leadership* (New York, NY: Harper & Row, 1978); W. Bennis and B. Nanus, *Leaders: The Strategies for Taking Charge* (New York, NY: Harper & Row, 1985); N.M. Tichy and D. Ulrich, "The Leadership Challenge: A Call for the Transformational Leader." *Sloan Management Review* (Fall 1984); Tichy and Devanna, op. cit.

18. D.E. Berlew, "Leadership and Organizational Excitement," in D.A. Kolb, I.M. Rubin, and J.M. McIntyre, eds., *Organizational Psychology* (Englewood Cliffs, NJ: Prentice-Hall, 1974); R.J. House, "A 1976 Theory of Charismatic Leadership," in J.G. Hunt and L.L. Larson, eds. *Leadership: The Cutting Edge* (Carbondale, IL: Southern Illinois University Press, 1977); Levinson and Rosenthal, op. cit.; B.M. Bass, *Performance Beyond Expectations* (New York, NY: Free Press, 1985); R. House et al., "Personality and Charisma in the U.S. Presidency," Wharton Working Paper, 1989.

19. Hambrick, op. cit.; D. Ancona and D. Nadler, "Teamwork at the Top: Creating High Performing Executive Teams," *Sloan Management Review* (in press).

20. V.H. Vroom, *Work and Motivation* (New York, NY: John Wiley & Sons, 1964); J.P. Campbell, M.D. Dunnette, E.E. Lawler, and K. Weick, *Managerial Behavior, Performances, and Effectiveness* (New York, NY: McGraw-Hill, 1970).

21. R.J. House, "Path-Goal Theory of Leader Effectiveness," *Administrative Science Quarterly*, 16 (1971):321–338; G.R. Oldham, "The Motivational Strategies Used by Supervisors: Relationships to Effectiveness Indicators," *Organizational Behavior and Human Performance*, 15 (1976):66–86.

22. See Hambrick, op. cit.

23. E.E. Lawler, and J.G. Rhode, *Information and Control in Organizations* (Pacific Palisades, CA: Goodyear, 1976).

24. Jacobson and Hillkirk, op. cit.

25. Gabbaro, op. cit.; T.J. Peters, "Symbols, Patterns, and Settings: An Optimistic Case for Getting Things Done," *Organizational Dynamics* (Autumn 1978).

26. R.J. House, "Exchange and Charismatic Theories of Leadership," in G. Reber, ed., *Encyclopedia of Leadership* (Stuttgart: C.E. Poeschel-Verlag, 1987).

27. M. Kets de Vries and D. Miller, "Neurotic Style and Organization Pathology," *Strategic Management Journal* (1984).

28. Levinson and Rosenthal, op. cit.

29. Hambrick, op. cit.

30. T. Kidder, *Soul of the New Machine* (Boston, MA: Little, Brown, 1981).

31. These are discussed in Nadler and Tushman, op. cit.

32. Hambrick, op. cit.

33. M. Louis and R. Sutton, *Switching Cognitive Gears* (Stanford, CA: Stanford University Press, 1987).

34. C. O'Reilly, D. Caldwell, and W. Barnett, "Work Group Demography, Social Integration, and Turnover," *Administrative Science Quarterly*, 34 (1989):21–37.

35. Hambrick, op. cit.; Virany and Tushman, op. cit.

36. See D. Ancona, "Top Management Teams: Preparing for the Revolution," in J. Carroll, ed., *Social Psychology in Business Organizations* (New York, NY: Erlbaum Associates, in press).

37. Louis and Sutton, op. cit.

38. March et al., op. cit.

39. See also C. Gersick, "Time and Transition in Work Teams," *Academy of Management Journal*, 31 (1988):9–41; Ancona and Nadler, op. cit.

40. See Vancil, op. cit.

41. R. Katz, "Organizational Socialization and the Reduction of Uncertainty," in R. Katz, *The Human Side of Managing Technological Innovation* (New York, NY: Oxford University Press, 1997, Chapter 4).

42. N. Tichy, "GE's Crotonville: A Staging Ground for Corporate Revolution," *Academy of Management Executive*, 3 (1989):99–106.

43. E.g., Gabbaro, op. cit.; V. Pucik, "International Management of Human Resources," in C. Fombrun et al., *Strategic Human Resource Management* (New York, NY: John Wiley & Sons, 1984).

44. Pucik, op. cit.

45. M. Devanna, C. Fombrun, and N. Tichy, "A Framework for Strategic Human Resource Management," in C. Fombrun et al., *Strategic Human Resource Management* (New York, NY: John Wiley & Sons, 1984).

46. Hambrick, op. cit.

47. Tichy, op. cit.

II

THE MANAGEMENT OF INNOVATIVE GROUPS AND PROJECT TEAMS

4

The Management of High-Performing Technical Teams

11

How a Team at Digital Equipment Designed the 'Alpha' Chip

RALPH KATZ

If you had attended the 1992 International Solid-State Circuit Conference in San Francisco, you would have known that something special had happened. Dan Dobberpuhl, the technical leader of Digital Equipment's Alpha Chip design team, was simply being mobbed after his presentation. To quote a German reporter, "Dan was under siege!" Technologists from many of the world's most respected organizations, including Intel, Sun, HP, IBM, Hitachi, Motorola, Siemens, and Apple, were pressed around him, anxious to catch his responses to their follow-up questions. As one of the chip's lead designers described the scene to his colleagues back at Digital, "You'd have thought they'd found Elvis!" Clearly, the ALPHA chip was being hailed as one of the more significant technical developments in the microprocessor industry in recent years.[1]

How did this team of designers accomplish such a noteworthy advancement? Why were they successful? Clearly, there were many other excellent microprocessor design teams throughout the

Reprinted with permission from the author.

Ralph Katz is Professor of Management at Northeastern University's College of Business and Research Associate at MIT's Sloan School of Management.

world; there were even other chip design teams within Digital. What were the organizational and managerial levers that permitted this group of designers to make this technological advancement and get it embedded into commercialized products? Was this technical and product development achievement a deliberately planned, well-managed, well-organized team effort; or was it more accidental, mostly a matter of luck in that the right people happened to come together at the right time to work on the right project?

Over the past years, a plethora of articles and books have surfaced emphasizing the virtues of teams and teamwork within organizations. The empowered, self-directed team is being hailed as the principal means by which organizations can revitalize their overall technological and market competitiveness. Some of the most dramatic stories in the popular press have attributed success to the effective use of crossfunctional or multidiscipline team efforts. The ability of people to commit to work together to achieve a common purpose has become one of the cornerstone ingredients of extraordinary achievement within organizations. Indeed, high performance teams have become the exemplary model for higher productivity, innovation, and breakthrough accomplishments, surpassing even the status of yesterday's entrepreneurial hero.

Given such testimonials, one might think that it would be somewhat straightforward to establish and lead a team-based work environment. More often than not, however, we are frustrated by our group, team, or task force experiences rather than energized by them. Even though we readily acknowledge the potential advantages of team-based interaction, it is not always easy to get engineers and scientists to function and behave as an effective team. It is often said that trying to manage R&D professionals is very much like "herding cats," that is, they tend to behave rather independently and are not especially responsive to or tolerant of their organization's formal rules, procedures, and bureaucratic demands. Nevertheless, to be both timely and effective in today's competitive global environment, technologists have to function collaboratively. They have to be able to integrate their in-

sights, creativity, and accomplishments not only with each other but also with the demands and activities of the functioning organization and business.

One could easily argue that the ALPHA team was successful simply because it possessed all the elements typically associated with high performing teams. After all, the ALPHA team members did have a clear performance goal with a shared sense of purpose, specifically, to design a RISC chip that would be twice as fast as any comparable commercially available chip. The team leadership was strongly committed, highly respected, and very credible both technically and organizationally. The team members were also extremely motivated, dedicated, technically talented, and tenaciously committed to solving tough problems. Furthermore, the individuals functioned as a cohesive unit, characterized by a collaborative climate of mutual support and trust, team spirit, and strong feelings of involvement and freedom of expression. They had a strong *esprit de corps*, communicating effectively with one another, working hard together, and valuing each other's ideas and contributions.

While it is certainly useful to generate such a descriptive list of high-performing team characteristics, the more critical questions surround the creation and management of these attributes. How, in fact, does one establish and manage these characteristics over time? How did the ALPHA team initially achieve and ultimately maintain its strong, unified sense of purpose and commitment to such difficult and risky technical objectives? Why were these very creative engineers able to work, trust, and communicate with each other so effectively?

In addition to this focus on internal team dynamics, it may also be important to learn how the ALPHA team interfaced with other business areas within Digital. Except perhaps in the case of startups, the activities and outcomes of most teams do not take place in organizational vacuums, and many technical teams that have enjoyed these same high performing characteristics have ended up unsuccessful in transferring their advances into commercialized products. For example, although the ALPHA design team was strongly committed to delivering its technical advances within an extremely tight and aggressive development schedule, the team did not have the strong unqualified support of senior management within the business systems and software groups, many of whom were very skeptical about what the ALPHA team was trying to do. Furthermore, although semiconductor management was supportive, outside Digital management could not assign the team any additional outside resource support from other technical or system development groups. For the most part, the team was not viewed by systems management or by many other groups within Digital as a well-disciplined, high-performing team. On the contrary, these individuals were often perceived as a somewhat arrogant, albeit respected, 'band of high-powered technical renegades.' The ALPHA team members may have trusted each other—they did not, however, trust management outside the semiconductor area. Yet, despite these apparent obstacles, the team was incredibly successful. How come?

PRISM: THE PREDECESSOR OF ALPHA

To fully appreciate the success of the ALPHA design team, it is necessary to understand more completely the historical context in which this technical effort took place. Many key members of the ALPHA team had originally worked together for several years under the leadership of Dan Dobberpuhl and Rich Witek, the chip's chief designer and chief architect, respectively, to design a new RISC/UNIX architecture that would be the fastest in the industry. They had hoped that the design of this new microprocessor would provide Digital the basis for developing and commercializing a new, more powerful, and less expensive line of computer products. Having been the major technical designer behind the first PDP-11 and the first VAX on a single chip, two of Digital's most successful product platforms, Dobberpuhl was able to attract some of the best technical minds to work with him and Witek on the design of this new RISC chip, including Jim Montanaro, one of the chip's lead designers. Given Dobberpuhl's history of successful

accomplishment within Digital, he was extremely credible within both the technical and managerial communities. His informal, sensitive, but pacesetting technical leadership also made working for him very appealing to other highly competent technologists.

This developmental effort had not been the only attempt at getting Digital into RISC technology. There were, in fact, at least three other major efforts, the most critical being led in Seattle, Washington by David Cutler. After several years of ongoing internal competitive activity, corporate research management finally decided to merge these separate efforts into a single program architecture, code-named PRISM. On the west coast, Cutler assumed responsibility for the overall PRISM program architecture and became the technical leader behind the development of a new operating system. On the east coast, Dobberpuhl and his team assumed responsibility for the design and development of PRISM's new RISC chip. Unfortunately, this convergence around PRISM did not end the program's or the chip design team's difficulties. Since Digital's existing VAX technology was still selling incredibly well, the company's need for this new RISC technology was not very great. As long as Digital's computer systems were 'flying high,' that is, selling well enough to meet or exceed projected revenues, the case for developing a whole new RISC-based line of computers to complement Digital's already successful VAX-based machines could not be made convincingly to Digital's senior executives. Furthermore, one of the major assumptions within systems management was that CISC (Complex Instruction Set Computer) technology would always be better and therefore more preferred by users over RISC technology. As a result, it was hard to marshal much outside managerial interest in or support for PRISM, especially since the management of the systems groups kept changing. The design team felt that it was consistently being 'battered around,' having been stopped and redirected many times over the course of several years.

By late 1987, however, as RISC technology became more proven and accepted within the marketplace, especially in workstations, Digital's management became increasingly interested in having its own RISC machines. Senior managers were also being heavily lobbied by representatives from MIPS Computer Systems, Inc. to license its existing and future family of RISC chips. Although PRISM was a cleaner, better-engineered design that had significantly higher performance potential than the MIPS chip, the Digital team had not completely finished its design work, nor had the chip been fabricated. More importantly, the associated computer and operating systems software were still in the early stages of development. MIPS, on the other hand, was much further along both in terms of hardware and software. Digital's senior executive committee decided it was more important for them to get a quick foothold in the open systems market than to wait for PRISM's more elegant but late design. Based on this rationale, Digital's senior leadership agreed to adopt the RISC architecture from MIPS—at the same time, however, they also decided to cancel PRISM.

ALPHA: THE OFFSPRING OF PRISM

The decision to cancel PRISM shocked everyone on the program, especially Cutler, Dobberpuhl, and Witek, none of whom had been asked to attend the MIPS presentation or had even been involved in the committee's subsequent deliberations. As a result, they all refused to accept the committee's explanation and became somewhat hostile towards them. After several years of hard work and effort, the team simply felt betrayed by senior management outside the semiconductor area. To give up their own technology and architecture and rely completely on a small outside company made no sense to them whatsoever. In the words of one designer, 'MIPS must have sold management a real "bill of goods"—it was all politics!'

This reaction is understandable in the light of what we know about such kinds of win/lose situations. Given the strong feelings and commitments that are reinforced within a team as its work progresses, the so-called 'losing team' refuses to believe that it really lost at least legitimately. Instead, it searches for outside explanations, excuses, and scapegoats upon which to blame the decision. Interestingly enough, however, if the losing group remains intact and is given a new opportunity to 'win,' the team will often become even more motivated, ready to work harder to succeed in this next round. In such a situation, the losing group tends to learn more about itself and is likely to become even more cohesive and effective as the loss is put to rest and the members begin to plan and organize for their next attempt. They can even become more savvy about what it takes to succeed if they work together to identify previous obstacles and mistakes that they either have to overcome or don't want to see recur.

This scenario is very similar to what seems to have happened—at least to the PRISM chip design team. Because of his high credibility and history of success, Dobberpuhl was able to negotiate an agreement with his management to move his hardware team to the Advanced Development area where they could conduct advanced 'clocking' studies but where they could also discreetly finish the RISC design and test it even though the project had been officially cancelled. Within a few months, an initial design was completed and a fabricated PRISM chip was produced. It worked even better than the PRISM team had expected, almost three times faster than the comparable MIPS chip being licensed by Digital. Armed with this new evidence, Dobberpuhl circulated throughout Digital a scathing memo claiming that management had just decided to throw away an in-house technology that was two to three times better than what they were purchasing from MIPS. He had hoped to start a dialogue that might get management in the systems areas to reconsider their cancellation decision but this did not occur. Although a great many engineers responded positively to his memo, management was not impressed, most likely seeing it as 'sour grapes.'

Dobberpuhl's memo did succeed, however, in arousing the interest of Nancy Kronenberg, a senior VMS expert (VMS is the software operating system for Digital's VAX machines). Kronenberg called Dobberpuhl to discuss the status of PRISM which was working but had no computer operating system that could utilize it.[2] She suggested that if PRISM's RISC architecture could be modified to 'port' (i.e., 'work with') VMS, then systems management might be convinced to use the new RISC chip as the basis for developing new machines. The limits of VAX technology were being reached faster than expected and new products would be needed as quickly as possible to increase sales revenue. Kronenberg was part of a high-level committee that had recently been chartered to examine alternative strategies for prolonging or rejuvenating Digital's VAX technology. The possibility of using a modified RISC chip that was VMS compatible was not one of the committee's current considerations, especially since Dobberpuhl and Witek were sufficiently upset with the sudden cancellation of PRISM that they had declined to be members of this task force.

The more Dobberpuhl thought about Kronenberg's option, the more appealing it became. He convinced Witek that they could bring RISC speed to the VAX customer base and once they were back in business, they might even be able to supplant the MIPS deal. After several weeks of discussion, Dobberpuhl, Witek, and Kronenberg were sufficiently excited that they were ready to present a proposal to the VAX strategy task force. Dick Sites, one of the key task force members, was very skeptical about the feasibility of the proposed option. He just didn't think it could work—it would be much too difficult to get all of the existing customer software applications to run on this RISC architecture. When challenged, however, Sites could not clearly demonstrate why it wouldn't work, and after several weeks of working on the problem, he slowly realized that it really was feasible. Sites had become a 'convert.'

With this new basis of support, the VAX task force was now more inclined to endorse the proposal to design a new RISC chip that could port all current VMS customer applications. This new RISC architecture could then be used to extend the life of VAX technology; hence, the original code name for the ALPHA chip was EVAX (i.e., Ex-

tended VAX). In 1988, the Budgeting Review Committee formally approved a proposal to 'flesh out' an overall technical and business program but left the objectives and specific details to the vice presidents and the program's leadership. In addition, since Sites' own project on cooled chips had recently been cancelled, he and his small design group were very anxious to join the Dobberpuhl and Witek team to design the new RISC chip that would be the foundation for some of Digital's new computer systems. As a result, the ALPHA team was essentially formed through the merging of two small groups of extremely talented individuals whose projects had been recently cancelled and who were determined not to get cancelled again.

STRATEGIZING FOR SUCCESS

Even though they had been blessed by senior management, the ALPHA team remained skeptical. From their prior experiences with PRISM, however, Dobberpuhl and the team had learned some very valuable lessons. Technical advances and achievements *per se* would not be sufficiently convincing to shift Digital's base architecture from VAX to RISC. The ALPHA team would need a real demonstration vehicle to truly win support for their architecture. As pointed out by Montanaro, 'You could talk technical, but you needed to put on a true test.' Management in the systems areas had never seen PRISM 'boot' (i.e., 'bring to life') a real computer system.

The team also realized they could not rely on the official systems groups within Digital to build a test computer for the chip within the project's tight schedule. These groups were just too busy working to meet their own currently scheduled product development targets and commitments. The ALPHA project was a relatively small semiconductor effort that did not as yet have sufficient clout or sponsorship to capture the immediate attention or interest of these large system-development groups. Historically, the semiconductor area within Digital had been a component organization that had not played a strong leadership role in leading the corporation's computer systems strategy. Although they were an extremely critical component, Digital's semiconductors had not been a driving, influential force. As a consequence, the design team realized that it would have to learn how to control and shape its own destiny.

To accomplish this, Dobberpuhl and Witek agreed that they would have to build their own computer system, that is, their own test vehicle. Using their informal networks of technical contacts, they successfully enrolled three very gifted technologists, namely, Dave Conroy, Chuck Thacker, and Larry Stewert, to develop on their own an ALPHA Development Unit (ADU) that would be based on the new RISC architecture. Since no computer system had ever been designed to the kind of speeds projected for ALPHA, management was incredulous as to whether a computer system could be built easily and inexpensively that could keep up with such a fast processor. The real challenge to the ADU group, therefore, was to build such a computer system using only off-the-shelf components.

The ALPHA team also learned that it would be very risky to go too long without showing management tangible results. As indicated by one designer, 'You had to show them something exciting—something that would capture their imagination.' As a result, the team would need to develop as quickly as possible a prototype version of the chip that could be demonstrated on the ADU test vehicle. This led to a two-tier approach in which the team would first design and fabricate an early version, that is, EVAX-3, that would not include ALPHA's full functionality but that could be convincingly validated. A fully functioning ALPHA chip, EVAX-4, would then be designed and fabricated using Digital's more advanced CMOS-4 process technology.

Organizing for Success

In addition to paying close attention to these outside management and organizational issues, the ALPHA team members functioned extremely effectively as a unit. The design group was comprised of individuals whose values, motivations, and work interests were of a very similar nature. Because the specific requirements of the ALPHA chip were left somewhat ambiguous by senior management, it was up to the design team to decide just how ambitious they were going to be. Extrapolating from the significant technical advances they had made in designing the PRISM chip to run at 75 MHz, the team was confident that they could 'push the technology' way beyond what others in the industry were expecting. They had learned enough from PRISM to know that a 200 MHz chip was feasible even though they really didn't know how they would accomplish it. Nonetheless, they were driven to build a chip that would be at least twice as fast as anything that might be available within the industry by the end of the project's 3-year development schedule.

They were also individuals who did not complain about hard work, long hours, or midnight E-mail as long as they were doing 'neat things.' No one had been assigned; they had all voluntarily agreed to work on ALPHA because they'd be 'testing the fringes' and 'pushing frontiers,' which were fun and neat things to do! The ALPHA design team, as was customary within Digital, recruited and interviewed its own members as they scaled up in size or replaced individuals who had to leave. In the words of Conroy, 'We looked for people with fire in the belly, people who did not try to "snow" you but who knew what they were talking about, and people who would not panic or get discouraged when they found themselves in over their heads.'

Members of ALPHA were experienced individuals who could function independently and who did not need a lot of direction, hand-holding, or cheerleading. They were not preoccupied with their individual careers; they were more interested in having their peers within the engineering community see them as being one of the world's best design teams. Ambition, promotion, and monetary rewards were not the principal driving forces. Recognition and acceptance of their accomplishments by their technical peers and by society was, for them, the true test of their creative abilities.

Although team members had very different backgrounds, experiences, and technical strengths, they were stimulated and motivated by common criteria. In the words of one ALPHA member, 'We see eye-to-eye on so many things.' This diversity of talent but singular mindset materialized within ALPHA not through any formalized staffing process but as a consequence of Digital's fluid boundaries and self-selection to projects. It is an organic process that may look messy and may lead to many unproductive outcomes; but it can also result in synergistic groups where the individual talents become greatly amplified through mutual stimulation and challenge.

'We'll Show Them!'

The intensity of the ALPHA team's motivation not only stemmed from its desire to do 'firsts,' that is, to make the world's fastest chip, but also from fear that management could cancel them at any time, as they had done with PRISM. They stuck to aggressive goals and schedules not only because it was the right thing to do but also because they perceived the possibility of cancellation as real. This 'creative tension' between the team and senior management kept the group working 'on the edge,' making them even more close-knit with a 'we'll show them attitude.' The group became resentful of management *per se* and normal management practices, stereotyping them and seeing them all as one big bureaucratic obstacle. No one on the team claimed to be a manager or wanted to be one. The group even kidded among themselves that 'Dobberpuhl is a great manager to have as long as you don't need one.' Just as the team members had a kindred spirit about technical work, they developed a uniform perspective about managerial practices and philosophies. This 'antimanagement' viewpoint allowed the ALPHA group to rely less on formal management structures and procedures for carrying out their task activities, and concentrate more on creating the team environment in which self-directed professionals could interact and problem solve together quickly and effectively.

Without the normal managerial roles, plans and reviews, the ALPHA team knew its success depended greatly on its ability to communicate openly and honestly among themselves. Because the team was relatively small and physically co-located, there was constant passing of information and decisions in the hallways. People didn't 'squirrel' away; instead, each member was cognizant of what the other team members were doing. Since the team also had to discover new circuit techniques, all kinds of wild and creative ideas had to be tried. To aid in this effort, the group instituted a series of weekly 'circuit chats'. In these hour long meetings, an individual would be given 10 minutes of preparation to present his or her work in progress. This was not meant to be a formal status report or rehearsed presentation with slides, rather it was an opportunity to see each others' problems, solution approaches, and mistakes. As explained by Montanaro, 'The intent was to establish an atmosphere where half-baked options could be freely presented and critiqued in a friendly manner and allow all team members to steal clever ideas from each other.'

The ALPHA team also realized that to get the performance speed they had targeted, the circuit design would have to be optimized across levels and boundaries rather than the suboptimizations that typically characterized previous designs. From their experience in designing PRISM, the team already knew where a lot of compromises had been made that may not have been necessary. Members would have to understand more fully the consequences of their individual design efforts on each other's work to overcome such compromises. The initial ALPHA documentation, therefore, was not so much one of detailed descriptions, specifications, and design rules, but one of intentions, guidelines, and assumptions. By knowing more about the intentions and what was trying to be accomplished, it was hoped that designers would have a broader exposure to both the microarchitecture of the chip and the circuit implementation from which the best tradeoffs could then be made. In the absence of formal management, members would have to be trusted to resolve their design conflicts by themselves. On the chance that it might be needed, a 'Critical Path Appeals Board' was created within the team to resolve intractable conflicts. The ALPHA team even discussed the advantages of holding 'Circuit Design Confessionals' during which members could admit to some error or design compromise and then solicit clever suggestions for fixing or repairing the problem.

No Room for Status Seekers

The team's effectiveness was also facilitated by certain normative and egalitarian behaviors. The group was conscious not to allow status or hierarchical differences to interfere with their joint problem solving activities. Even though there were some very high-powered senior people on the team, they were all senior people involved in technical work. They did not just tell people what to do, they were also doing it, from designing fancy circuitry to performing power and resistance calculations, to simple layout design. When team members had to stay late during particular crunch times, it was the custom that good dinners (not pizza) be brought in and served to them by the senior people. This degree of involvement and support helped solidify the group's confidence and trust in each other and prevented technical intimidation or status from becoming a problem.

A number of important behavioral norms were also established and reinforced by the ALPHA team. It was expected, for example, that one would inform other members as soon as one realized that one could not make a given deadline or milestone. It was okay to be in trouble; it was not okay to surprise people. Individuals were not expected to 'grind away' but to go for help. There was zero tolerance for trying to 'bull' your way through a problem or discussion. It was important to be tenacious and not to give up easily, but it was also essential to realize when you were no longer being productive. Pushing and working hard were okay, but it was also important to have fun. Humor and good-natured teasing were commonplace occurrences.

Dobberpuhl was also instrumental in keeping the group strongly committed to its aggressive goals. Generally speaking, when a group gets in trouble, there is a tendency for the members to want to change the nature of their commitments either by extending the schedule, enlarging the team, reducing the specified functionality, reducing features, accepting higher costs, etc. People would prefer to play it safe by saying 'I'll do my best' or 'I'll try harder' rather than voluntarily recommitting to achieve the difficult result. By not allowing these kinds of slack alternatives, Dobberpuhl challenged the team to search for creative solutions to very difficult problems. When management suddenly discovered, for example, that the ALPHA chip and its associated new products would be needed as much as a year earlier than originally scheduled, Dobberpuhl knew that by simply working harder, the team could not possibly reduce its schedule by that much. He also realized that if ALPHA was going to remain a viable option in Digital's strategic plans for filling this 'revenue gap,' then the schedule would have to be speeded up; otherwise, their efforts would once again run the risk of being cancelled in favor of some other alternative product strategy. By committing to the speeded up schedule, the team was forced to find and incorporate what turned out to be some very creative breakthrough modifications in the design of the microprocessor and in the way it was being developed. This steadfast commitment not only ensured that the project would not be cancelled; it energized the technologists to stretch their creative abilities for extraordinary results. This supports Scherr's (1989) contention that breakthrough advances are achieved only when members commit unequivocally to overcome apparent obstacles or 'breakdowns.' Scherr, an IBM Fellow who studied several high performing breakthrough teams at IBM, argues that extraordinary results cannot be attained if the organization and its teams continue to play it safe by building in lots of comfortable slack and contingency strategies.

Other Groups Were Supportive

Finally, it is important to realize that a design project like ALPHA must have considerable support from other groups if it is to be successfully completed in such an aggressive timeframe. Following the persistent lead of Montanaro, the team found ways to work around the management bureaucracy to get the help it needed from other areas. The computer-aided design (CAD) groups were especially critical for providing very advanced design and verification tools that were not generally available. They were also willing to modify and extend the software to the requirements of the new circuit design. The CAD people were responsive because they assumed that the ALPHA team was probably working on something very important and because they were an interesting and unusual group of 'techies' who delivered on their promises.

There were also a number of other design groups within Digital with good-natured rivalries taking place among them. Montanaro was able to build enough contacts within these groups that he could temporarily borrow (not raid or steal) additional personnel, resources, and even design documents to help the ALPHA team complete its design.

Even though it was customary within Digital to beg, borrow, and scrounge for resources, Montanaro soon realized that it was important for him and the team to find a way to say 'thank you.' Using adult-type toys he purchased from a mail-order catalogue, Montanaro started to give people phosphorescent insects and fishes, or floating eyeballs, and so forth, as a means of thanking them for their assistance. For example, if someone found a bug in the design, Montanaro would give them a phosphorescent roach or fish; if they found a bigger bug, then he would give them a bigger bug or fish, a phosphorescent squid for example. For the ALPHA team, these toys became a fantastic way for getting round the 'us vs. them' turf problem or stepping on people's toes. They became a great ice-breaking vehicle and as people collected them, they took on strong symbolic value. So many of these phosphorescent fishes were distributed that when the facility had a sudden blackout, many of the employees discovered that they did not have to leave the darkened building as they could easily continue their work by the glows of light from their fish toys!

CONCLUDING OBSERVATIONS

Digital Equipment is banking very heavily on ALPHA technology to help lead its resurgence in the industry. Not only is Digital designing and marketing its own line of ALPHA computers, it is also working with other vendors to use ALPHA in their product offerings. Microsoft, for example, is using ALPHA machines to demonstrate and market the capabilities of its new Windows operating system called Windows NT. Cray Computer's next generation of massive parallel processing supercomputers are designed to use large arrays of ALPHA chips. And Mitsubishi has become a second manufacturing source, adding substantial credibility to the ALPHA product's viability. Digital is even hoping that ALPHA can eventually be used and adopted as an open market PC standard.

What, then, enabled the success of the ALPHA chip development effort? Clearly, the fact that a very high-powered group of individual technologists had come together through self-selection to work towards a single-minded objective is a strong contributing factor. *These were not team-playing individuals—they were a collection of talented individual contributors willing to play together as a team!* For previously explained reasons, they were all eager to commit to a very aggressive set of goals and very willing to accept the risks that such a commitment entailed. There were no artificial barriers in the design process and the creative juices that flowed from the group's communication and problem solving interactions were exceedingly critical. At times they worked like maniacs, totally immersed in their project activities and buffered successfully from the normal managerial and bureaucratic demands and disruptions by Dobberpuhl. Because of their singular purpose and common motivational interests around technology, there was little of the in-fighting and turf-related issues that often characterize other teams. There was a genuine and unified team feeling of ownership in having contributed to the technical achievements as can be seen by the large number of names that have appeared on the technical publications and patent applications. Managers need to recognize that there is a world of difference between having team-playing people and having people playing together as a team.

All of the afore-mentioned characteristics focus primarily on the group's internal dynamics and were probably very instrumental in allowing the group to achieve its strong 'technical' success. It is less likely, however, that these dynamics contributed to the 'organizational' success of the product. PRISM was also a major technical achievement and the PRISM team was very similar to ALPHA in terms of group membership and process. Yet, the ALPHA chip and not the PRISM chip is being commercialized.

Only when the ALPHA team shifted its emphasis from concentrating on its internal group process to worrying about how it should relate to other critical areas within the company did the seeds for organizational success get planted. Only when the team learned to integrate their technical goals with the company's strategic business interests were they successful at shifting senior management's attention from relying on the company's 'core' technology to relying on promised, but unproven, advancements in a much less familiar technology.

Unlike the PRISM episode, ALPHA was able to gain and strengthen over time its sponsorship within the organization. By sending out an irate memo, Dobberpuhl had taken a risk, albeit he had a strong basis from which to take this risk, but he managed to capture the attention of a few senior sponsors who helped link his technical interests to the strategic interests of the company. By making sure they had an ADU machine ready to demonstrate the new RISC architecture, the ALPHA team was able to increase its sponsorship even further. Neither Ken Olsen, Digital's CEO at that time, nor his senior staff had seen PRISM in action. They did, however, see a true demonstration of the EVAX-3 version of ALPHA and became so excited that they soon wanted ALPHA to become an open systems platform product that could run both VMS and UNIX operating systems. The demonstration had been a galvanizing moment for the project, shifting senior management's language and dialogue from 'Will we be doing this?' to 'Of course we're going to do all this!' Dobberpuhl, Witek, and the ALPHA team had learned not only the importance of developing new technology but the importance of protecting and marketing it within a large organization so that the technology becomes effectively coupled with the strategic interests and decisions of the established businesses.

For R&D groups to be successful, especially under conditions of uncertainty, the teams must learn how to become politically effective within their organizational contexts. Either they can work in isolation and behave as spectators watching organizational events unfold; or they can actively insert themselves into the organization's decision-making and problem-solving processes. Either they can build and shape the networks, relationships, and strategies that 'make things happen' or, as in the case of PRISM, they can wake up one morning and say 'What the Hell happened?' As further evidence, Kelley and Caplan (1993) recently reported from their research on R&D productivity at Bell Labs that the true 'star performers' were not necessarily brighter than their technical colleagues, they were in fact more adept at nine particular work strategies. Interestingly enough, many of these work strategies were political in nature such as networking, navigating among competing organizational interests, promoting new ideas, coordinating efforts, taking initiative, and so on. Even in the more team-based cultures like Japan, it is readily acknowledged that important decisions are shaped and influenced through a political process, called *nemawashi*. Creating excellent intragroup processes are not enough. Technical teams have to learn to strengthen constructively their organizational sponsorship by becoming more effective players in the political processes that surround the making of key strategic decisions.

But what exactly crystalized the political energies and teamwork behaviors of the ALPHA team? Clearly, it was the unfathomable cancellation of PRISM that sparked the motivational responses and behaviors of the design team. *Without PRISM there would have been no ALPHA!* The cancellation of PRISM is what I call a 'marshaling' event, that is, an event that significantly arouses people so that they are finally willing to do something to redirect their efforts and attention.

If one examines many of the purported high performing team situations, one often finds that marshaling events have similarly influenced the motivational behaviors. Furthermore, it is not the collection of data or the presentation of analyses, logical arguments, or forecasts *per se* that typically stirs people to action. It is, instead, the emotional repercussions of seeing or feeling the reality of the data, information, or situation that awakens or persuades them. Collecting one's own disturbing benchmark data or experiencing first-hand the embarrassment of one's own product offering at an industry show relative to the products of what one had previously regarded as mediocre competitors can powerfully affect the efforts and motivations of product development teams when they return to work.

It is said, for example, that IBM's Thomas Watson Jr kept hundreds of transistorized radios on his shelf in his office. And whenever engineers came in to complain about the risks associated with designing IBM's first completely transistorized computer, called the '360,' Watson simply turned on all the radios interjecting that only when the radios failed, would he be willing to listen to their arguments. It is also alleged that Estridge, the project manager for the original IBM PC, incited the members of his core design team by buying Apple II computers and placing them on the members' desks with a note essentially affirming that this was the product they'd be going after.[3] Toshi Doi, the product development manager of Sony's most successful computer product called News, sent his senior manager a 'rotting fish' to illustrate somewhat vividly what would happen to his product development effort if certain resources and decisions were not forthcoming. Even the project to land a person on the moon by the end of the 1960s, perhaps the most quoted example of a vision, fails to recognize the motivational importance of Sputnik. Had the Russians not embarrassed and scared Americans with the surprise launching of Sputnik, and later with the first person in space, there would have been no NASA and no lunar mission in the 1960s. Like Watson's transistorized radios, Estridge's Apple II computers, and Doi's rotting fish, Sputnik was the marshaling event that enabled Kennedy's man-on-the-moon speech to be so captivating. It is not only the marshaling event *per se* that

is important; just as important is the focused leadership that arises to take advantage of the event. The combination is critical. For as in the case of ALPHA, it is the cancellation of PRISM coupled with the proactive leadership of Dobberpuhl and others that eventually gets all the design individuals to strategize, organize, and work together as a team so effectively.

SOME ADDITIONAL LESSONS

What additional managerial lessons might one deduce from R&D episodes such as ALPHA? First, projects like ALPHA do not seem to originate in a 'top-down' fashion; instead, they appear to evolve as pockets of individuals are able to come together and excite each other about relevant technological developments, problems, and possibilities. Every professional I interviewed indicated that senior management is just too caught up in the pressures of their present businesses to be receptive to the many uncertainties and risks associated with 'leapfrog-type' efforts. After all, did Intel's senior management really want to get out of the DRAM business and into microprocessors; did Hewlett-Packard executives really anticipate the strength of their laser and inkjet printer lines; did Motorola's sector managers really predict the growth of their current paging and cellular businesses; did Microsoft's Bill Gates really know how successful DOS and WINDOWS would be; and did Apple's corporate leaders really plan their transition to the MacIntosh product line? These and many similar examples from every industry suggest that if management truly wants to foster these kinds of accomplishments, then it must do more than simply encourage all technologists 'to take risks' and 'not fear failure'. It must provide the active sponsorship through which key talented technologists can come together to work on those 'far out' ideas and problems in which they have become strong believers. In some sense, the trick is to be more like the successful venture capitalist firms who realize that when they are dealing with a great deal of uncertainty, they are primarily justifying their investment in the track records, energies, and talents of the individuals behind the idea rather than in the idea itself.

In thinking about the ALPHA team's experiences, there are several important areas in which the team could have profited from stronger management sponsorship. To speed up the commercialization of technical advances from projects like ALPHA, it is essential that management be ready to strengthen and push the downstream integration of these projects. There is no question in the minds of those with whom I talked that ALPHA could have benefitted greatly from the earlier involvements of key business functions, particularly marketing and advertising. Given the strong footholds that the Pentium and Power PC chips would probably have in the marketplace, the faster ALPHA could be commercialized ahead of them, the greater its chances would be of becoming a market success.

Business managers often strategize by projecting changes in their business environments and then planning only for those technical projects that are needed to meet these expected changes. It is just as important that management strategize by working jointly with the technical part of the organization to plan and sponsor those project developments that could help create or shape environmental changes. This kind of sponsorship is not achieved by simply giving or 'empowering' technical teams with freedom and autonomy. On the contrary, it is achieved by strengthening the linkages between technical development activities and business strategies. And as in the case of ALPHA, such connections are often established as the technical leadership discovers how to present their 'neat ideas,' not in technical terms *per se*, but in terms that are meaningful to those who are managing and running the businesses.

Over time, a strong partnership feeling must develop between the technical and management parts of the business; otherwise, the organization runs the risk of amplifying their differences rather than being able to ameliorate them. Even if a project becomes a success, if the rifts and lack of supportive trust between the two camps are allowed to endure, then the images of 'good guys' versus 'bad guys'—'us versus them'—become the stories that

key players remember and reinforce. Such negative dispositions are often exacerbated when the technical people feel that they have not been equitably recognized and rewarded, as has happened in so many of these cases. What typically happens is that the technical people see the noncontributing managers, that is, the 'bad guys,' ending up with more exciting work and challenging positions as a result of the project's success while they, that is, the 'good guys,' receive relatively little for their persevering efforts. Having focused all of their attention on completing the current project, the team members soon realize that there is also no new exciting assignment or project to capture their time and energies. What unfortunately happens is that neither the team nor the organization's management has planned the members' next projects or career assignments. Not only is this misplanning not a reward, but it can be rather demoralizing after having worked so hard. Under such conditions, the long-term continuity of technical developments can be seriously impaired through decreased morale and increased levels of turnover, particularly among the key technical contributors. An organization must not only sponsor the research projects and high performing teams that *get* it into the game; it must also learn to build the bridges and longer-term relationships, career plans and assignments, and reward alternatives that *keep* it in the game.

NOTES

The material for this study was derived by interviewing representative ALPHA team members and Digital managers and engineering staff. With the cooperation of Digital, in-depth interviews were conducted individually at DEC during normal work hours. An early version of this article was read by those interviewed to ensure the accuracy of the information and the flow of events.

1. As Digital Equipment's first announced microprocessor in RISC (Reduced Instruction Set Computer) technology, the ALPHA chip runs at more than two to three times the speed of its nearest competition. As of 1993, it had been included in the Guiness Book of World Records as the world's fastest chip with a clock speed of more than 200 MHz (megahertz).

2. When PRISM was cancelled, the software group working to develop the new advanced operating system that could be used by this chip also became upset. As David

Cutler, the software development leader, told his team, "We really got screwed—years of development work just went down the drain." As they were located in Seattle, Cutler and many of the software team soon left Digital to work for Microsoft, where they used their ideas to produce the new operating system known as Windows NT.

3. The norm at IBM at this time was not to buy competitive products.

REFERENCES

Kelley, R. and Caplan, J. (1993) How Bell Labs Creates Star Performers. *Harvard Business Review,* **89,** 128–139.

Scherr, A.L. (1989) Managing for Breakthroughs in Productivity. *Human Resource Management,* **28,** 403–424.

12

The Java Saga

DAVID BANK

When *Wired Magazine* published the inside story of bringing Sun's Java to market in December of 1995, Java was the hottest thing on the Web since Netscape. Maybe hotter.

Since then, Java's history has been a tangle of lawsuits, strategy reversals, and surprising persistence. A brief tour would include Microsoft's 1996 decision to license Java and include it in Windows; Sun's federal lawsuit against Microsoft in 1998; Microsoft's decision to drop Java from Windows; Java's starring role in Microsoft's federal antitrust trial; Sun's schizophrenic attitude toward making Java an open industry standard; the contentious Java partnership between Sun and IBM; the eclipse of Java as a factor in client-side PC software applications; and its emergence as a major force in server-side programming.

But all that came later. Before Java was even released, it nearly became a business-school case study in how a good product fails.

With three minutes to go before the midnight deadline in August 1995, Sun Microsystems engineer Arthur van Hoff took one last look at Java and HotJava, the company's new software for the World Wide Web, and pondered what his colleagues call Arthur's Law: Do it right, or don't do it. Satisfied, the Dutch programming wizard encrypted the files containing the software's source code, moved them to an Internet site, and e-mailed the key to Netscape Communications Corporation,

Java's first commercial customer. Five years after the project was launched, Java was done—with a minute to spare.

As he sat at his workstation ready to push the button, van Hoff had good reason to hesitate. Since early versions of the software were released in December 1994, Java has unleashed stratospheric expectations. While today's Web is mostly a static brew—a grand collection of electronically linked brochures—Java holds the promise of caffeinating the Web, supercharging it with interactive games and animation and thousands of application programs nobody's even thought of. At the same time, Java offers Sun and other Microsoft foes renewed hope that Bill Gates's iron grip on the software business can be pried loose. Microsoft rules the desktop, but as networking expands its role, says van Hoff, Java could turn out to be "the DOS of the Internet." Indeed, Sun is rushing to make Java a de facto standard on the burgeoning Web. If Sun succeeds, even Microsoft will have a hard time muscling in.

Software developers are busy shaping Java into applications that will add new life to Web browsers like Netscape and Mosaic, producing programs that combine real-time interactivity with multimedia features that have been available only on CD-ROM. (Java is a programming language; the Java Virtual Machine, labelled HotJava, is an "interpreter" installed onto a browser, enabling Java programs delivered over the Web to run on the desktop.) What's a Java application? Point to the Ford Motor website, for instance, and all you'll get are words and pictures of the latest cars and trucks. Using Java, however, Ford's server could relay a

---

Reprinted with permission of the author.

David Bank is a staff reporter of the *Wall Street Journal* and the author of several books, including *Breaking Windows: How Bill Gates Fumbled the Future of Microsoft* (Free Press, 2001). An earlier version of this case was published in 1995 by *Wired Magazine*.

small application (called an applet) to a customer's computer.

From there, the client could customize options on an F-series pickup while calculating the monthly tab on various loan rates offered by a finance company or local bank.

Add animation to these applications and the possibilities are endless. Hollywood and Madison Avenue are salivating. "Java allows us to do the things that advertisers and studios are asking us to do," says Karl Jacob, CEO and chief technologist at Dimension X Inc., a San Francisco company creating 3-D websites using Java.[1] "Until now, everything on the Web was fizzling, not sizzling."

Even if Java turns piping hot, how might it lift profits at Sun, which turns out Unix-based workstations and servers for its bread and butter? It's rumored that Netscape paid a paltry US$750,000 to license HotJava (escaping any per-copy charges), a figure that Sun, whose annual revenues will top $6 billion this year, does not dispute. Sun is giving away Java and HotJava free for noncommercial use, in a fast-track attempt to make them the standard before Microsoft begins shipping a similar product, codenamed Blackbird, in early 1996.[2]

Java is unlikely ever to become a major profit center at Sun, though any increase in Web traffic is bound to increase sales of Sun's workstations and servers. But in this case, emotion may be at least as important as profit. Sun chief Scott McNealy is a fierce competitor, and his blood lust for Bill Gates has fueled the Java project from the beginning. McNealy is especially excited about Java's ability to run on any computer, using Windows, Mac OS, Unix, or any other operating system—posing a threat to Microsoft hegemony. Spinning into the future, McNealy even sees the day when disposable word processors and spreadsheets will be delivered over the Web via Java, priced per use. "This blows up Gates's lock and destroys his model of a shrink-wrapped software that runs only on his platform," effuses McNealy.

Maybe he's dreaming. But Java's progression thus far is a lesson in what can happen when a major company loosens the reins on some of its most precocious talent. The story of Java also highlights the sometimes serendipitous nature of technological development in the face of vague and fast-changing markets.

The origins of Java go back to 1990, when the World Wide Web was barely a glimmer in a British programmer's eye. The personal computer was in its ascendancy, and many inside and outside Sun thought the company had missed major opportunities in the desktop market. Its high-end workstation and server markets were rolling along fine, but as PC use spread across the landscape, the company faced being stranded in a narrowing slice of the computer market. Sun machines had a reputation for being too complicated, too ugly, and too nerdy for mass consumption.

Thus, McNealy was more than ready to listen when a well-regarded 25-year-old programmer with only three years at the company told him he was quitting. Patrick Naughton played on McNealy's ice hockey team. Over beers, Naughton told McNealy that he was quitting to join NeXT Computer Inc., where, he said, "they're doing it right." McNealy paused for a second then shrewdly asked Naughton a favor. "Before you go, write up what you think Sun is doing wrong. Don't just lay out the problem. Give me a solution. Tell me what you would do if you were God."

The following morning, Naughton threw his heart and soul into the challenge.

He typed out a list of Sun's shortcomings along with his own glowing appraisal of NeXT's critically acclaimed NeXTstep operating system. Twelve screens later, he e-mailed his report to McNealy, who forwarded it to the entire management chain.

A firestorm was ignited. Among Naughton's suggestions: hire an artist to pretty up Sun's uninspired interfaces; pick a single programming tool kit; focus on a single windows technology, not several; and, finally, lay off just about everybody in the existing windows group. (Naughton figured they wouldn't be needed if the previous suggestions were taken.)

Naughton held off NeXT while he awaited the response. The following morning, his e-mail box was bursting. Hundreds of CC'd readers had read his recipe for what ailed Sun and had agreed in a resounding chorus. A typical reaction: "Patrick wrote down everything I say to myself in the morning but have been afraid to admit." Another voice was that of James Gosling, a remarkable programmer whose opinions carried great weight higher up. Naughton was "brutally right," Gosling e-mailed. "Somewhere along the line, we've lost touch with what it means to produce a quality product."

Naughton joined what one participant called a "bitchfest" attended by a number of high-level engineers. It was John Gage, Sun's science office director, says Naughton, who really dug in, asking, "What is it you really want to do?" The group blue-sky'd until 4:30 the next morning. During those wee hours, they came up with some core principles for a new project: consumers are where it's at; build a small environment created by a small team—small enough to fit around a table at a Chinese restaurant; and make the environment, whatever it may become, include a new generation of machines that are personal and simple to use—computers for normal people.

If still vague, these principles were enough to get Gage's executive juices flowing. With his support, Naughton pitched the high concept to Wayne Rosing, then president of Sun Laboratories Inc. and onetime vice president of engineering at Apple. Naughton laid down key demands he'd scribbled on the back of a restaurant place mat: the project would be located offsite, away from corporate "antibodies" well known for attacking innovative ideas; the project's mission would be kept a secret from all but the top executives at Sun; the software and hardware designs would not have to be compatible with Sun's existing products; and for the first year, the team would be given a million bucks to spend.

Rosing took the idea to McNealy over dinner, and afterward called his assistant by car phone. Via e-mail, the assistant relayed Rosing's pledge to Naughton: "I expect to make 100 percent delivery on what we discussed." Within a day, two of Sun's top dogs, Bill Joy and Andy Bechtolsheim, were throwing in their approvals. That same day,

Naughton got his wish. Naughton, Gosling, and Mike Sheridan, who had come to Sun after it bought out his start-up, would be given carte blanche to pursue new projects.

But what projects? At this point the team, co-denamed Green, had only a vague notion of what it would do. Competing head-on with Microsoft was out: the Goliath had already won the battle for the mass-market desktop. Instead, the team resolved to bypass Microsoft and the PC market altogether by designing a software system that could run anywhere, even on devices that people did not yet think of as computers. That meant the system had to be compact and simple—the complete opposite of Sun's existing offerings. "We wanted computers to go away, to instead become an everyday thing," Naughton said. "We thought the third wave of computing would be driven by consumer electronics. The hardware would come from Circuit City, and the software would come from Tower Records."

But Green was still a solution in search of a problem. The shared epiphany that set the team's early direction came in the spring of '91 in a hot tub near Lake Tahoe, where Sun's high-level staff had gathered for an annual retreat. Gosling, Sheridan, and Naughton, now joined by Ed Frank, one of Sun's top hardware engineers, soaked and drank beer. Gosling made the observation that computer chips were appearing in toasters, VCRs, and many other household appliances, even in the doorknobs of their Squaw Valley ski-lodge rooms. "That's getting pretty ubiquitous when it's in the bloody doorknob," he said. Yet three remote-control devices were needed just to get a television, a VCR, and a living-room sound system to work. Needless to say, most people still couldn't program any of them. The wonder wasn't that chips were everywhere but that they were being used so badly.

"With a little computer science, all of these things could be made to work together," Gosling insisted. A light switch with a liquid crystal display and a touch pad could play little movies to demonstrate what it controlled and how, he brainstormed. Any control device could do multimedia, and multimedia could help people do real, useful things.

There in the hot tub, Gosling recalls, the Green team decided to build a prototype of a device that could control everyday consumer appliances.

Thus began Green's glory days. In April 1991, the team moved from Sun's main campus to office space above a branch of the Bank of America on Menlo Park's Sand Hill Road, cutting itself off from Sun's internal computer network. The programmers disconnected culturally as well. ''We thought if we stayed over there, we would end up with just another workstation,'' Sheridan says. ''I was obnoxious about keeping it secret.''

They cleared the center of the large room for lab benches and couches, stocked the refrigerator with Dove bars and Cokes, and spent hours playing Nintendo games—the better to understand hypnotically engaging user interfaces. Business issues were put on the back burner to give the technical ideas room to roam. Their mission statement was laid down in a business plan they called Behind the Green Door: "To develop and license an operating environment for consumer devices that enables services and information to be persuasively presented via the emerging digital infrastructure."

The key insights into the software that would run such devices came to Gosling at a Doobie Brothers concert at the Shoreline Amphitheater in Mountain View, California. As he sat slouched in front-row seats letting the music wash over him, Gosling looked up at wiring and speakers and semi-robotic lights that seemed to dance to the music. "I kept seeing imaginary packets flowing down the wires making everything happen," he recalls. "I'd been thinking a lot about making behavior flow through networks in a fairly narrow way. During the concert, I broke through on a pile of technical issues. I got a deep feeling about how far this could all go: weaving networks and computers into even fine details of everyday life."

Gosling quickly concluded that existing languages weren't up to the job. C++ had become a near-standard for programmers building specialized applications where speed is everything—computer-aided design for instance, where success is measured by the number of polygons generated per second. But C++ wasn't reliable enough for

what Gosling had in mind. It was fast, but its interfaces were inconsistent, and programs kept on breaking. In consumer electronics, reliability is more important than speed. Software interfaces had to be as dependable as a two-pronged plug fitting into an electrical wall socket. "I came to the conclusion that I needed a new programming language," Gosling says.

As it happened, Gosling, who wrote his first computer language at 14, had been working on a C++ replacement at home. "From the initial 'Oh, fuck' to getting it to a reasonable state" took only a few months, he says. Naughton, meanwhile, had been working on graphic animations, which would serve as the device's interface. By August 1991, Gosling had the graphics running in his new language, which he called Oak (named for the tree outside his office window); this was the progenitor of Java.

By now, the Green team's ambition was to build a device that would work as an interface to cyberspace. Its aim was to create a visual interface to a virtual world. If you wanted to record a TV program while you were away from home, you could control your video recorder by working a virtual video recorder in the virtual world Naughton was designing. The virtual world—in color and in 3-D—was written in Oak language and spruced up by graphic artists.

All the new programs needed now was a device to run them on. The team wanted a working box, small enough to hold, with batteries included. To build one, the members trotted out what they call "hammer technology"; as Naughton describes it, this involved finding "something that has a real cool 'mumble' (a neat piece of hardware). Then you hit it with a hammer, take the mumble off, and use it. We got a consumer-grade Sharp mini-television, hit it with a hammer, and got an active-matrix color LCD. We put a resistive touch screen on the front, making sure there'd be no moving parts on the system, no buttons, no power switches, nothing," Naughton explains. The team then wanted to add stereo speakers inside, but couldn't find any to fit the case. "We went to Fry's and bought a dozen Game Boys, played like mad for about three hours, then broke them open—that's where the speakers came from."

The guts of the device were even more remarkable: one of Sun's high-end SPARC workstations stuffed into a dark green aluminum case barely bigger than a softball. Frank's hardware team built three custom chips and designed a motherboard that folded in on itself to save space. The team hacked furiously all through the summer of 1992. "It was a blood bath," Naughton says. "We bit off way more than any seven people should chew. We were arrogant sons of bitches to think we could pull it off. We had so many free variables that we had nothing to come back to. We didn't have anything we knew would work."

The demo was shown to McNealy in August 1992. McNealy saw a hand-held contraption with a small screen and no buttons. When you touched the screen, it turned on. Cool! It opened to a cartoon world—no menus! A character named Duke, a molar-shaped imp with a big red nose, guided the user through the rooms of a cartoon house. You steered with your finger—no mouse! Sliding your finger across the screen, you picked up a virtual TV guide on the sofa, selected a movie, dragged the movie to the cartoon image of a VCR, and programmed the VCR to record the show. This was even more elegant than it first sounded. Everything was done without a keyboard, simply by ripping objects with your finger and dropping them with a "ka-ching" sound.

Sun's boss was ecstatic. Nothing like this smooth, natural interface existed at the time (no Bob, no Magic Cap, no eWorld). And nothing like Oak. A natural cheerleader, McNealy dashed off a hyperactive e-mail. "This is real breakthrough stuff. Don't fail me now. We need to sell this puppy hard, and there's tons of work to make it real. You deliver, it will win.

"I can sell this stuff. Charge! Kill H-P, IBM, MSFT, and Apple all at once." Rosing added his own exudations: "Bill Gates is gonna weep."

The prototype was not the only thing the Green team brought to the table. Oak was more than cartoons. Oak was to be an industrial-strength object-oriented language that would work over networks in a very distributed manner. Little packets of code (objects) would scoot around the Net, functioning independently of the devices they were in (computer, telephone, or toaster). Entire applications, like an e-mailer, say, could be built by stitching together objects in a modular way—and the objects wouldn't all have to live in the same place. Plus, Oak dealt with a chief concern of distributed computing: it encased security, encryption, and authentication procedures into its core so security was essentially invisible to users.

At demos, Naughton would go to the white board to show the scope of Oak, filling the blankness with lines crisscrossing from home computers, to cars, to TVs, to phones, to banks, to—well, to everything. Oak was to be the mother tongue of the network of all digital things.

The Green team had been talking to Mitsubishi Electric about using Oak-based interfaces in cellular phones, televisions, and home and industrial automation systems. France Telecom, which was looking to upgrade its Minitel system, was also interested, Sheridan says. He followed up with a detailed business plan dubbed Beyond the Green Door, proposing that Sun establish a separate subsidiary to push the Oak technology into the consumer marketplace.

Two months after the demo, Sun set up the team as FirstPerson Inc., a wholly owned subsidiary. Rosing moved over from Sun Labs to head the project. Meanwhile, Sheridan, passed over and feeling disrespected, cleaned out his desk and went home. "I thought that having brought it so far, I should be running it," says Sheridan, who now heads a technology consulting group in Virginia.[3] Gosling and Naughton stayed. "There was a changing of the guard, and it wasn't exactly smooth," Gosling remembers.

As FirstPerson grew from 14 employees to more than 60 and moved from its one-room hideout to palatial offices in downtown Palo Alto, things began to drift.

Mitsubishi and France Telecom weren't interested after all, nor was anyone else. Sun figured the cost of the chip, memory, and display device needed for Oak could be squeezed to about $50, but consumer electronics makers were used to paying nominal bucks for chips that made their products easier to use.

In the meantime, the notion of the information superhighway had taken over rational minds and declared itself to be arriving Real Soon Now. Most people understood the I-way to be interactive TV. So when Time Warner circulated proposals in March 1993 to begin interactive television trials in Orlando, FirstPerson jumped at the chance to provide the set-top boxes. As far as Sun was concerned, interactive TV was the technology of the moment, and the company was desperate to use Oak as its entry. Naughton, Gosling, and Joy, who had taken a sporadic interest in Oak, trekked many times to Time Warner's technology center in Denver. There they labored over prototype designs and hashed out cost estimates for a set-top box that would link a television to the information superhighway, using Oak to coordinate a vast complex of images, data, and money securely over a distributed network.

But the deal went to Silicon Graphics Inc., Sun's cross-town rival in the high-end workstation market. Around Sun, McNealy took heat for not pushing as hard with Time Warner as Jim Clark, SGI's founder and then-chair. "All that mattered to Time Warner was who was going to commit to delivering a $300 box on top of the TV," McNealy grouses. "Nobody knows how to do a $300 set-top box that does what they want it spec'd to do." SGI then delivered a box that cost almost 10 times that amount in the summer of '93.

In hindsight, losing the deal was a stroke of good luck, given the overruns, glitches, and weak consumer interest that made the Orlando project a disaster for both Time Warner and SGI. (Clark later left SGI, abandoning interactive TV in favor of the emerging World Wide Web, and founded Netscape—whose browser would later incorporate Java, the successor to Oak. But more of that in a moment.)

A few months later, FirstPerson got tantalizingly close to a deal with 3DO, a company having trouble selling an expensive CD-ROM game machine and thus trying to double the product as a set-top box. It took just 10 days to get Oak running on one of 3DO's game boxes but three months to negotiate a commercial agreement. Finally, the paperwork was ready, and Trip Hawkins, the founder of 3DO, weighed in demanding exclusive rights to the technology. McNealy refused. Given 3DO's precarious state—its expensive game machines had languished on store shelves for two years—that loss turned into a blessing, too.

There were few new avenues left for FirstPerson to try. "What was really dumb for us was focusing on set-top boxes and putting on blinders," Gosling said. "Interactive TV was a mistake. There was so much enthusiasm about it we didn't understand the unreality of that universe."

Those blinders kept Sun—along with many others—from grasping the significance of another emerging phenomenon. In June 1993, Marc Andreessen and Eric Bina at the National Center for Supercomputing Applications at the University of Illinois had released the first version of the Mosaic browser, and the formerly obscure World Wide Web began to take off. But it was at least three more months before Eric Schmidt, Sun's chief technology officer, saw the software for the first time. How had Sun, so long a supplier of well over 50 percent of all host computers to the Internet, missed the Web's mass appeal? "We just took our eye off the ball," Schmidt admits.

FirstPerson was in disarray. Nerves were frayed. Marching orders came down from Sun management: Find something to produce profits. Now! A new business plan was drawn up early in 1994, which unceremoniously dumped the speculative markets FirstPerson had pursued and began focusing on personal computers—the technology the project was supposed to leapfrog in the first place. The new plan was to create a corps of CD-ROM developers who would write in Oak and, ideally, stick with it as their platform language while moving applications to the commercial online services. Eventually, by the turn of the century, the team thought, broadband networks for interactive TV would finally be ready, and Oak would be established as the programming language of choice. The plan, remarkably, contained no mention of Mosaic or the Web.

For different reasons, the plan met with little enthusiasm among Sun's top executives. They wanted a strategy that would drive demand for Sun's core hardware products, something a program for CD-ROMs would not do. With no profitable strategy in sight, FirstPerson was scrapped in the spring of 1994, the set-top box project shelved, and the interactive TV crew, now renamed Sun Interactive, collapsed into an alliance with Thomson Consumer Electronics to develop scaled-down box and video servers—without Oak for now. It took Bill Joy, who again turned his attention to the project, to rescue Oak.

Joy, a Sun co-founder, is a rare hybrid: a legendary programmer who understands every line of code and carries enormous clout with execs. He is described lovingly by colleagues as a brilliant wild man whose ideas are at the lunatic fringe. In 1991, annoyed by Silicon Valley area traffic after the Loma Prieta earthquake and disillusioned about the prospect of doing anything interesting until the Microsoft juggernaut had run its course, Joy established a small Sun research lab at the foot of the ski mountain in Aspen, Colorado. (The town also had a good bookstore, he says, explaining his choice of locale.) Tired of disputes within Sun, he had mostly disengaged, though he dropped in occasionally on the FirstPerson project.

The Web's sudden emergence changed all that. For Joy, the prospect of bringing Oak to the Internet recalled his days at the University of California, Berkeley, nearly two decades earlier, when he'd developed the Berkeley flavor of the Unix operating system out of the original code from Bell Labs and pushed it into widespread use through the Net. That had laid the foundation for Sun. Because of Joy, Sun finally saw that the Internet could become Oak's redemption. Joy's support was critical in what became known as the Internet Play, the "profitless" approach to building market share—a ploy Netscape had made famous by giving away its browser. "There was a point at which I said, 'Just screw it, let's give it away. Let's create a franchise,'" Joy says.

Joy and Schmidt wrote yet another plan for Oak and sent Gosling and Naughton back to work

adapting Oak for the Internet. Gosling, whom Joy calls "the world's greatest programmer," worked on the Oak code, while Naughton set out to develop a true "killer app."

In January 1995, Gosling's version of Oak was renamed the more marketable Java. Naughton's killer app was an interpreter (or Java virtual machine) for a Web browser, later named HotJava. He wrote the bones of it in a single weekend. Following Joy's dictum, they intended to make it available free on the Web.

But Naughton and Gosling didn't entirely trust Sun to turn Java over to the Internet. Sun has always been a proponent of open standards for software interfaces that allow anyone to build his or her own compatible applications. But this strategy was going further to include the free release of a software implementation. Kim Polese, Java's senior product manager, wrote in big letters on her office white board, "Open means. . . ." and kept adding items to the list. It was one thing for college hackers to release university-grade software like Mosaic for free, or even for a start-up like Netscape to offer its browser for free noncommercial use. It was something else for a major company such as Sun to give away its technological crown jewels, the source code for some of its most valuable technology.

Even Schmidt had doubts about whether he'd be able to live up to his promise to protect the team from the pressure at Sun. "The conversation that never took place, but that I could feel all around me, was, 'Eric, you are violating every principle in the company,'" Schmidt says. "'You are taking our technology and giving it away to Microsoft and every one of our competitors. How are you going to make money?' At the time, I didn't have an answer. I would make something up. I would lie. What I really believed was that Java could create an architectural franchise. The quickest way was through volume and the quickest way to volume was through the Internet."

In December 1994, Java and HotJava (it was actually still called Oak at this stage) were posted in a secret file deep in the Net; only a select few were given pointers and invited to check it out.

Three months later, Marc Andreessen, who had gone on to start Netscape with Jim Clark, was given a copy. Andreessen gushed to the San Jose Mercury News: "What these guys are doing is undeniably, absolutely new. It's great stuff." That was how the Java team knew it was going to finally make it. "That quote was a blessing from the god of the Internet," Polese says.

Now that HotJava is being given away, Sun has to make sure it cements Java as a standard—and then figure out how Java can make money. The Netscape deal—incorporating Java into its browser—helps establish a population of Java users. But Sun has to go much further to make it easy enough for anyone other than hard-core nerds to populate the Web with applets.

Sun has promised, but not delivered, tool kits that will allow artists, writers, and other would-be Web authors to speak Java fluently. That's what's needed to increase the supply of enticing applications and to spur users to demand that software suppliers include the technology in their offerings. And Sun has to accelerate that circle into a dizzying spin before competing technologies come along to challenge Java's position. "Sun's window is six to twelve months," says Sheridan. "They need to move quickly because Microsoft will respond in a way that freezes development."

The Java applets are the key. Here's why: for a program to run on a computer, it must first be translated from a language like Basic or C into the machine's native tongue. Because this translation process is incredibly time-consuming, most software comes already translated. But that means different versions have to be created for different computers. Java gets around this problem by using an intermediate language—a sort of Esperanto that is not machine specific but that can quickly be interpreted by any computer.

The result is that small programs—applets—can fly around the Net without regard to what kind of hardware they end up on. If you need to watch an animation that requires a particular fancy doodad to run it, but you don't have that doodad, your machine will pick up the Java-coded applet along with the animation file and run both. Who cares where the software lives? Who cares what kind of machine you have? Who cares about Microsoft?

Microsoft has stated that its response will be Blackbird, a package due for initial release in January 1996. It is suppose to contain a Web application programming language and an interpreter of its own, at first based on C++ and later on Microsoft's own Visual Basic. Unlike Java, Blackbird's language will work only on the Windows platform at first, but that may not be such a problem given Windows's 80 percent share of the PC market. Anyway, Microsoft is planning a Mac version. And it also claims that Blackbird will be easier to use than professional-programmer-oriented Java.

Sun is pursuing licensing deals as fast as it can to set Java as the standard before Blackbird flutters in. It recently penned an agreement with Toshiba to use Java on a wireless Internet device and claims to have more than 25 potential Java agreements in the pipeline. Though the company says the low-price licensing deal with Netscape is a onetime exception, Sun is explicit about making its technology cheap. The published rates for licensing Java's source code for commercial use include a $125,000 up-front fee plus $2 a copy. "It's priced below our cost," Schmidt says. "This loses money in the licensing business for the foreseeable future. It's a strategic investment in market share."

Another way to avoid that fate would be to license Java to Microsoft and thus complete the penetration of the entire market.[4] Schmidt says he's willing, but Bill Gates hasn't called. Others doubt whether McNealy could bring himself to consort with the enemy even if Gates showed up at the door. They believe that Sun's desire to beat Microsoft may be even stronger than its desire to see Java succeed. "There are cheaper ways to say 'Fuck you' to Bill Gates," says Naughton, who left Sun last year after he became, as he puts it, "damaged goods" during FirstPerson's nadir. After Sun showed him the indignity of a 2 percent raise, he says, Naughton joined Starwave Corp. in Seattle, where he is using Java to create online services.

Sun is racing to stay ahead of the accelerating wave. The day after that midnight deadline for sending the finished code to Netscape, Joy was already at work pushing the limits of what Java could do. His team was aiming a videocamera at a computer-controlled water fountain in Aspen and putting its image on the Web. The idea was to let anybody with a HotJava interpreter anywhere on the Internet control the spray and interact with kids playing in the water. "I've got 15 patents I could file as soon as I type them," Joy says. "I figure I've got five years. It's like we've got a blank sheet and it says 'Internet.' Normally, the best products don't win. The Internet is an opportunity for the best products to win. Java is great technically and people want it. I'm happy to get that once in my life—or maybe twice."

Often, great technologies are born into the world without one of three essential factors for success: a committed champion, a willing marketplace, and a workable business model. Clearly, Java had its champions—believers like James Gosling, Arthur van Hoff, Bill Joy, Patrick Naughton, and the many who fell by the wayside during Java's long, twisting history. Tweaked, renamed, and repositioned, this time the idea has found a willing marketplace; in time, a distributed object-oriented language like Java will probably establish itself as the foundation of the Net. But whether the standard will be Java depends on whether Sun finds a business model to keep it alive.

NOTES

1. Dimension X was later acquired by Microsoft.

2. Blackbird was dropped in 1996. Another Microsoft initiative, dubbed ActiveX, was started several years later.

3. Sheridan later joined Sun.

4. Blackbird was soon canceled and Sun did indeed reach a licensing agreement with Microsoft in March 1996. Microsoft, however, has since introduced its own Java competitive initiative.

13

Hot Groups

HAROLD J. LEAVITT AND JEAN LIPMAN-BLUMEN

A hot group is just what the name implies: a lively, high-achieving, dedicated group, usually small, whose members are turned on to an exciting and challenging task. Hot groups, while they last, completely captivate their members, occupying their hearts and minds to the exclusion of almost everything else. They do great things fast.

At one time or another, every successful executive has seen or been part of a group that was really hot. Whether it was called a team, a committee, or even a task force, its characteristics were the same: vital, absorbing, full of debate, laughter, and very hard work.

Although hot groups are almost never consciously planned, they can turn up in just about any setting: social, organizational, academic, or political. When the conditions are right, hot groups happen, inspired by the dedication of their members to solve an impossible problem or beat an unbeatable foe. When hot groups are allowed to grow unfettered by the usual organizational constraints, their inventiveness and energy can benefit organizations enormously.

Consider predivestiture Bell Telephone Laboratories. A mature company with more than 20,000 employees located in tradition-bound suburban New Jersey, Bell Labs was a tightly controlled, conservative organization. One senior manager often stationed himself at the entrance to the main lab in the morning, noting which employees arrived late. Back then, we counted nine levels in the organizational hierarchy. No place for hot groups, one would think. Yet this was the same company that invented modern communication theory, the

transistor, and a host of other advances thanks in large part to its hot groups.

In the case of Bell Labs, we believe that hot groups thrived for two primary reasons: a strong commitment to scientific values and an equally strong commitment to maintaining independence from AT&T. First, the core scientific values underlay everything at Bell Labs and were ingrained in everyone who worked there. The highest status in the organization went to the people in basic research, the ones doing the most far out and, in the short run, the most impractical work. In many other companies, those people would have been pilloried as nerds and longhairs, irrelevant to the real power structure of the organization. At Bell Labs, they were highly valued and encouraged to pursue whatever they believed to be the most interesting avenues of scientific inquiry.

Second, Bell Labs had been designed from the start to be independent from the rest of AT&T. While AT&T paid its bills, Bell Labs was sheltered from the usual business pressures. It was given an extended period of time before being required to demonstrate practical results. Not surprisingly, Bell Labs' senior managers were not managers in the traditional sense: They were outstanding engineers and scientists who demanded discipline and responsibility on the one hand, while encouraging creativity and communication on the other.

Is something like the Bell Labs experience with hot groups possible in other organizations? Unquestionably. Must the organization be unencumbered by a need to make money, fueled only by a thirst for scientific knowledge? Not at all. To

Reprinted by permission of *Harvard Business Review*. From "Hot Groups" by Harold J. Leavitt and Jean Lipman-Blumen, Vol. 73, 1995, pp. 109–116. Copyright © 1995 by the President and Fellows of Harvard College, all rights reserved.

Harold J. Leavitt is Professor Emeritus at Stanford University's Graduate School of Management. *Jean Lipman-Blumen* is Professor of Organizational Behavior at the Peter F. Drucker Management Center at the Claremont Graduate School in California.

compete successfully, most companies today are striving to develop breakthrough products and services. To do this, managers must understand that encouraging some behaviors at the edge of accepted organizational propriety can help their companies become hot.

No one really knows enough about hot groups to draw a blueprint for building them. After years of observing and participating in hot groups, however, we can describe the conditions under which such groups flourish, the behaviors they exhibit, the types of leadership they require, and the benefits they bring. To the question, How does one build hot groups? the answer is clear. One doesn't. Like plants, they grow naturally. We can instead tackle different—but equally important—questions: Under what environmental conditions are these groups most likely to flourish? How much moisture and light do they need?

GETTING TO KNOW HOT GROUPS

Hot groups labor intensely at their task—living, eating, and sleeping their work. Members believe that their group is on to something significant, something full of meaning. As individuals, they may feel that they have been more creative, capable, and productive while in their hot group than at most other times in their lives.

For many people, membership in such a group is a peak experience, something to be remembered wistfully and in considerable detail. Despite the intensity of the experience, members usually find it impossible to specify exactly what made it so hot. They are likely to feel that their group was unique, the product of a rare conjunction of the planets, impossible to reproduce.

The excitement, chaos, and joy generated in hot groups make all the participants feel young and optimistic regardless of their chronological age. In hot groups, the usual intellectual and social inhibitions are relaxed. These qualities almost re-create the sense of exuberant confidence people feel as children. In fact, people are more likely to participate in hot groups when they are young than when

they are old because the young feel omnipotent and immortal. They are eager to take on the exciting challenges that characterize hot groups. As people age, those challenges are usually redefined as dangerous risks.

Many people may have felt the excitement of a hot group when they were at school, putting together a show or a school magazine. It may have been in the military in a squad fighting its way up an impossible hill. Perhaps it was in a research group on the trail of an elusive gene or in a cross-functional new product team building the next generation of pasta makers. We have even received unsubstantiated reports of hot groups taking root in board rooms. Overall, however, hot groups are rare, especially within traditional organizations.

Total Preoccupation

The most distinguishing characteristic of any hot group is its total preoccupation with its task. Hot group members think about their task constantly. They talk about it anywhere, anytime. It is their top priority to the exclusion of almost everything else. Closing time often slips by without anyone noticing. Sometimes, hot group members bring cots to the office so that they can work most of the night. The more challenging, unusual, or "impossible" the task, the more dedicated they are to it.

Participants in hot groups achieve this level of preoccupation because they always feel that their task is immensely significant both in terms of the challenges it represents and in terms of its intrinsic meaning. The challenge may be one of design or of implementation. From a design perspective, the task must be a puzzle, a conundrum that is difficult to solve. In terms of implementation, it must be a feat that tests the group's mettle.

Above all, the task must be uplifting, one that is worth doing because it will make some kind of positive difference. Hot groups almost always believe that they are embarking on a journey that will make the world a better place. Sometimes the goal has broad societal impact: A group has the mission of developing a vaccine for AIDS or isolating the gene for Alzheimer's disease. Sometimes the goal is more pragmatic and local but absolutely central to members nonetheless: A group has the task of instituting 24-hour customer service for a department store. Whether or not outsiders see it that way, hot groups feel that what they are doing is relevant and important.

Not surprisingly, a hot group's preoccupation with its task is accompanied by extremely high performance standards. Without exception, hot groups shoot for the stars. Their members feel that they are stretching themselves, surpassing themselves, moving beyond their own prior performance limits. Hot group members are seldom motivated by the promise of bonuses or of other material rewards. The challenge of the task is its own pot of gold.

Intellectual Intensity, Integrity, and Exchange

All members of hot groups use their heads, intensely and continuously. This intellectual energy stems partly from the absence of inhibition that we discussed earlier. Members pump out ideas and possibilities at an astonishing rate. From the outside, many of their ideas may look wildly absurd and impossible to achieve. Although hot groups indeed push the limits, many of their extreme ideas are ultimately refined into practical actions.

Members often debate loudly and passionately about issues. They are not given to easy consensus. Because they are turned on by one another's ideas, and because their primary concern is finding the best possible solution, numerous noisy and seemingly disorganized discussions are more the rule than the exception.

Emotional Intensity

Hot group members behave like people in love. They are infatuated with the challenge of their task and often with the talent around them. They frequently sacrifice themselves, including their own resources and their outside relationships, to the cause. In contrast to traditional committees and task forces whose members may try hard to avoid extra duties, hot group members tend to volunteer for extra work and even to create it for themselves.

When hot group members bring the fruits of their solo efforts back to the group and its leader, they treat it as a gift, an offering of sorts. "I was

thinking about that problem in bed last night, and I had an idea. So I got up and tried a few things on my PC, and here's what I got. What do you all think?" At first glance, one might think that such a desire to please could lead to "groupthink," but that is hardly a real danger. Hot groups are too dynamic, too open, too full of debate, challenge, and creativity for unquestioning conformity to set in.

Members of hot groups know they're hot, and they show it. They feel that their team and each individual member is something special. Even in hot groups composed of very dissimilar personalities, members respect and trust one another because they see themselves as highly capable people dedicated to an important task. Communication is typically wide open-up, down, and across the group. Members treat one another with casual respect and focus on colleagues' contributions to the task at hand, not on title, rank, or status. One aerospace executive, recounting his stint in an advanced design group, put it this way: "We even walked differently than anybody else. We felt we were way out there, ahead of the whole world. And," he added, "everybody else in the company knew it."

The emotional intensity that binds the members of a hot group together may come at a cost. First, it may isolate members from the rest of the organization. Hot groups are not pleased when their concentration is disturbed, and they dislike interruptions from outside. They do not readily welcome newcomers who might disrupt the dynamics of the group. And they certainly do not appreciate bureaucratic distractions, such as expense forms and formal progress reports. Their attitude and their freedom from organizational restrictions are apparent to outsiders who frequently resent them as both exclusive and arrogant. This resentment may ultimately harm the hot group and its mission, when the surrounding organization later fails to support its initiatives.

Second, the emotional intensity of hot groups may occasionally lead to burnout. Both leaders and members of hot groups must support one another, buoy up those who falter, and take breaks for short periods of relaxation and recharging. Although leaders need to stay alert to these critical human issues, they must also be careful not to overplan and overprogram. Hot groups usually handle these issues quite well on their own. The emotional needs of the group become especially important as the task nears completion. The sense of camaraderie is apt to dissipate as the group approaches its end, and members begin to feel a sense of sadness and loss. They begin to think again about their individual needs and interests as they prepare to re-enter the organization.

Fluid Structure/Small Size

It is difficult to define a hot group's structure. Roles and duties can change swiftly and subtly as the requirements of the task change. As priorities are identified, dealt with, and reordered, leadership may also shift from one member to another as the situation dictates. In sum, a hot group will organize as it sees fits. That said, however, we can generalize about two common structural characteristics.

First, hot groups are almost always small enough to permit close interpersonal relationships among their members. Thus they usually range in size from around 3 to perhaps 30 members, although groups as large as 30 are rare. Size varies with the organizational context, as well as the complexity and time frame of the task.

Occasionally, a large group of hundreds or even thousands of people may look hot for a period. More aptly classified as networks of small hot groups, these large hot organizations are usually held together by a small central core. Some are hot small companies, which seem to maintain their heat even as they grow. Others are protected organizations such as the National Institutes of Health (NIH) in its early days. Congress insulated the NIH, granting it freedom from oversight for several years. During that period, the NIH grew into a hot organization, innovative, risk-taking, exciting, and productive. The same was true of Bell Labs in its scientific glory years, when it was similarly insulated from the day-to-day pressures experienced by other units of the Bell System. Even as it grew in size, Bell Labs maintained its heat for a long time.

The second characteristic of hot groups is that they are almost always temporary and relatively short-lived. They share the happy attribute of dissolving when they finish their work. Unlike so many units in traditional organizations, hot groups do not try to guarantee their longevity. They are dedicated to excellence, speed, and flexibility; and when hot groups end, they end. On rare occasions, a few members remain bonded together, teaming up on other projects, like some pairs from the original Macintosh design team.

WHERE DO HOT GROUPS GROW?

One reason that hot groups are rare is because they grow only under the most special conditions. The environment inside the parent corporation must be hospitable, and the external environment must be challenging.

Internal Conditions

Like truffles, hot groups are not easily domesticated. Neatly organized institutions usually stifle them, whereas companies of the sort that Professor James March of Stanford University once called "organized anarchies" seem to provide quite fertile soil for growth. Hot groups need to feel that they can somehow be on their own, not fitting too neatly within the pre-ordained objectives of an organization.

Openness and Flexibility. While an intriguing task can be the magnet that pulls people together into a potentially hot group, often small sets of people with overlapping interests and shared values generate their own tasks and develop into hot groups. This, of course, requires easy, informal access across hierarchical levels and across departmental, divisional, and organizational boundaries. Indeed, we believe that such spontaneous conception is a very common form of genesis for hot groups. They frequently pop up in new, still-pliable start-up organizations and disappear as those organizations grow and calcify.

Consider, for example, the adolescent years of Apple Computer. In the early 1970s, Apple people, from top to bottom, averaged 20-something years old. Apple's early culture was exciting, urgent, flamboyant, defiant, ready to take on Big Blue (IBM) and anyone else in its path. Moreover, Apple's open culture was consistent with that of its Northern California location.

It is not surprising, therefore, that Apple's flagship product, the Macintosh, was developed by a small hot group consisting of people from all over the company. Led primarily by the aggressive, charismatic, fast-talking Steve Jobs, the group was spurred on by the ennobling challenge of building small computers for the masses. IBM, in contrast, made monstrously huge machines for the corporate few. Apple would occupy the moral mountain top, promising an agile little computer in every pot. Like many other hot groups, that dedicated hot Macintosh group devised and used its own emotional symbols: Members flew a skull and crossbones from a flagpole, with an Apple logo covering one eye socket. What could better express their youthful, free-spirited defiance of the stodgy, old establishment?

Independence and Autonomy. The Macintosh success story illustrates something repeatedly seen in organizations that successfully grow hot groups: To help keep them hot, it is wise to leave them alone for reasonable periods of time. Giving hot groups elbow room is difficult for many managers who understandably want to stay in close touch with what goes on in their organizations. Keeping hands off may also be bureaucratically difficult: Controllers and administrators exist to make sure that everyone in the organization abides by the rules. Nevertheless, as we have already seen, to help keep hot groups hot, senior managers must allow them substantial chunks of time before payoffs are demanded.

People First. The notion that hot people create hot groups has an organization corollary: People-first organizations of the type we saw in the Bell Labs example develop task-obsessed hot groups. Paradoxical though it seems, organizations that place more emphasis on people than on tasks spawn hot groups that focus tirelessly on tasks.

Why? The answer is straightforward: Organizations that first devote a lot of effort to selecting their people and then allow them plenty of elbow room and opportunities to interact are likely to generate groups that will build challenging tasks for themselves. The logic of traditional organizational design is quite the opposite: First, define the task with great care; then, break it down into individual sized pieces; and, finally, select people with skills and aptitudes appropriate for each piece. Such organizations constrain and place limits on their people's behavior and activities. Hot groups do not prosper in such settings.

Great universities remain largely people-first organizations. They select their faculties very carefully and then grant them enormous freedom to interact informally. Good schools usually search for the best candidates in each broad field instead of seeking people to fill narrowly defined niches. Despite all the internal politics that plague many universities, the relative openness of debate within their cultures has always provided a rich seedbed for both hot individuals and hot groups.

Consider also that segment of the business world in which temporary organizations are the rule rather than the exception. Independent-film production companies are an example. For any project, producers must assemble a collection of writers, directors, actors, camera people, financial backers, and more. Most often, those temporary groupings do not become hot groups. They are simply a collection of specialists doing their own things. But, if the producer selects the very best people and leads the project with belief and passion, and if the project itself takes on great significance for the team, then there is a reasonable chance that such a diverse assemblage can become a hot group.

The Search for Truth. Hot groups seem to prosper in organizations that are deeply dedicated to seeking truth. Many research organizations in industry and government are quite traditional and hierarchical in their management styles, so at first glance they would seem unlikely places to look for hot groups. Yet at most of these institutions, like at Bell Labs, hot groups exist because the traditions and mores of science, which place a high value on the search for the truth, prevail. Those traditional scientific values, coupled with the realization that frequent failures are an inescapable feature of the research process, combine to make even rather authoritarian research institutions quite supportive of both hot groups and hot individuals.

External Conditions

Two of the most powerful and fast-acting sources of group heat have always been crises and competitors. Under the novel conditions and tight deadlines that major crises generate, the pressing task at hand overwhelms "normal" concerns about power and control. Formal hierarchies and status systems are often suspended in the desperate search for anything that might restore equilibrium. During crises, new voices may be heard and previously ignored alternatives may be considered. Consider the events during World War II. An obscure lieutenant colonel, Dwight D. Eisenhower, was promoted to a position of enormous power; women were welcomed into jobs previously reserved for men; and hot groups of scientists were funded to conduct their atomic witchcraft under the stands of the stadium at the University of Chicago.

Or consider the Tylenol crisis of over a decade ago, when cyanide was injected into a number of bottles of Tylenol capsules, which led to the death of seven people. In that case, all of Johnson & Johnson seemed to get hot. A top management committee debated options for long hours. Almost overnight, the company's PR people produced videos for network television use. The engineering and design people quickly repackaged the product to make it more tamper proof. Thousands of employees made over a million personal visits to physicians, hospitals, and pharmacists around the nation to restore confidence in the Tylenol brand name.

Crises, of course, do not always spawn hot groups. Some organizations revert to extreme authoritarianism in crises: Leaders (or would-be leaders) take command and bark out orders. Readers may recall that when President Ronald Reagan was shot and Vice President George Bush was out of town, Secretary of State Alexander Haig got himself into trouble with his career-limiting statement,

"I am in control." The usual explanation for such behavior is that, in crises, time is of the essence, and people think, We don't have time for damn fool meetings. In many crises, that's true. In many others, however, that claim is invoked to justify precipitous action by panicky or egocentric leaders.

Competition, like crises, may generate high energy and dedication. But, while strong and visible competitors can indeed turn up group heat, it is dangerous to count too heavily on competition as a long-term motivator. For example, while high school or college teams can find the challenge of competition enormously motivating, professional teams show just a bit less of that competitive heat. Pro players, after all, are seasoned veterans. The enthusiasm of their younger days has been sandpapered away by injuries, salary disputes, and the other elements of the big business of professional sports. The same is true of mature companies. They may have been competitively hot a decade or two ago, but over the years they get a touch of arthritis. Good competitors are great for helping hot groups get started, but they're not to be depended on over the long haul.

Emerging Conditions

Modern information technology and the proliferation of alliances among organizations may facilitate the self-generating process that launches many hot groups. A critical mass of people who share interests, values, and thinking styles may be hard to find within a single organization. In our new, soft-boundaried world of networks and instantaneous communication, the birthrate of cross-organizational and even cross-national hot groups should rise. We suspect that the Internet will spawn a large number of hot groups.

WHO STARTS HOT GROUPS? AND WHO KEEPS THEM GOING?

Hot groups are always formed and carried by individuals, and mostly by individualists who by their very nature love to pursue markedly independent routes in life. A subset of the population of people who start most things, these are people with an intrinsic love of challenge. It is not ambition that drives them so much as a spirit of inquiry. They also like to prove that they can do what others insist cannot be done.

Still, individualism alone is far from sufficient. Certainly, many entrepreneurs are individualists, and, certainly, hot groups often take root in entrepreneurial soil. Yet, in our experience, most individualistic entrepreneurs are *not* great cultivators of hot groups. They prefer to run their own shows. They haven't the patience to share their goals and visions with others—nor do they feel the need. Only a subset of those strong, intrinsically driven individualists has both the interest and the ability to grow hot groups. That subset, which we call *connective individualists,* is worth a closer look.

Connective individualists are team players with strong egos; they are confident and stable enough to feel comfortable bringing other people into the act. They willingly share plans, goals, and glory, incorporating multiple approaches and multiple ideas. Unlike many other entrepreneurial individualists, connective types don't have to do it all by themselves. Much as parents can vicariously identify with their children's accomplishments, connective individualists can identify with the accomplishments of their groups, whether as leaders, members, or coaches.

Connective individualists can in turn be divided into three subgroups: *conductors,* who lead the orchestra, *patrons,* who support it, and *keepers of the flame,* who sustain it through time. When conductors find themselves challenged by an idea, they are likely to begin putting a group together immediately and to act as the group's inspirational leader. The charismatic power of conductors is contagious, attracting others to join in, help out, and identify with the group. That charismatic power may emanate from the flamboyance of many conductors, but it also flows from their extraordinary ability, which attracts the best and brightest young minds to their projects.

Conductors do best in face-to-face settings, where their personal styles inspire their people. Often, they show up as leaders of small, innovative

companies or of autonomous units within large companies. They are also to be found building volunteer organizations, leading special service units in the military, or initiating all sorts of student activities in colleges. Conductors loom large in organizations. Their strong voices, although not always loved, are always heard. They become prominent figures, much talked and gossiped about. Everybody knows their names.

Patrons behave quite differently. They are catalysts in the formation of hot groups without themselves becoming active members: the high school teacher who inspires a group of students to take on difficult and exciting challenges, the soft-spoken boss who somehow always seems to get an important group going. Patrons protect and nourish their groups. They usually operate unobtrusively, often almost invisibly—coaching, listening, offering suggestions. In large organizations, many people don't even know their names, but those who do appreciate them. Patrons are particularly valuable to organizations. While the personal energy of conductors may spark successful small outfits that last for one generation, patrons are more likely to build enduring cultures that routinely support the growth of new hot groups.

Keepers of the flame, dedicated to solving a certain basic problem, nourish hot groups sequentially throughout their careers. In the pursuit of the solution to that problem, they realize that one hot group's completion of a task usually generates new, intriguing possibilities. Keepers of the flame end up nourishing new ideas, new solutions, and new partners in a long chain of hot groups.

One critical role of leaders, whatever the type, is to provide route markers for the hot group. These indicators of progress are usually of two kinds: hard markers, indicating real, measurable movement toward completing the task; and soft markers, in the form of approval and encouragement from the leader and from one another.

Hard markers are visible to all: "We have now finished segment four of our project. We have six more to go." Or, "Once we get this problem solved, we'll have a clear road ahead." Such markers recharge batteries, signaling that the group is on a positive, progressive path; however, they are often difficult to create. In many of the trailblazing tasks typically undertaken by hot groups, hard markers simply don't exist. There are no road signs in unexplored wilderness. Soft markers then become critical in keeping a hot group's heart pumping. These markers, however, require the management of meaning. Soft markers are voices of reassurance, encouragement, and support from people whose competence and good sense are trusted by the group: "Great idea! Let's keep pushing it." Or, "OK. So those two alternatives didn't work. Let's try the third alternative."

GROWING HOT GROUPS

Twenty-first century organizations will require the capacity to keep up with an intense pace of change as well as the capacity to reshape themselves continually. No longer can we build our organizational houses on the obsolete assumption that they will last for 100 years. Rigid, old corporate styles, like the inflexible steel and stone headquarters that symbolized them, are fast becoming quaint vestiges of things past.

A hot group is one form of small group that can be especially effective at performing relatively short, intensive bursts of highly innovative work. If we wish to reap the benefits of hot groups, however, we need to recognize and accept some of their potential costs. Like any other powerful tool, hot groups can be dangerous if mishandled. Hot groups work unconventionally—a fact that is often disturbing to the larger organization. In their wholehearted dedication to a task, hot groups may ignore or challenge many conventional rules, making

waves in other parts of the organization. Isolated in a hot group, members become temporarily unavailable to other parts of the organization. They may become blind to their own shortcomings and impervious to criticism. They may appear arrogant and contemptuous to employees from other parts of the organization. They may burn out and refuse to enlist in the next hot group.

Some senior managers are likely to interpret such organizational turbulence as disruptive to smooth and orderly operation. As a result, they will often try to eliminate the hot groups that cause it. But other, more connective managers will perceive that same turbulence as a stimulant to their organization, speeding up its metabolism and helping to inculcate a sense of urgency.

For those executives who feel that more hot groups might help stir the hearts and minds of their people, there remains the question of how to make them happen. There are a few suggestions managers can follow to create an environment fertile enough for hot groups to grow: Make room for spontaneity; encourage intellectual intensity, integrity, and exchange; value truth and the speaking of it; help break down barriers; select talented people and respect their self-motivation and ability; and use information technology to help build relationships, not just manage information.

At some point in the future, we may understand hot groups well enough to be able to manufacture them, but, until that time, an agricultural metaphor is more appropriate than a manufacturing one. For the time being, hot groups must be allowed to grow; they must be nurtured rather than engineered. Like plants, they are best raised from carefully selected seed, cultivated, and given plenty of room to mature.

The Management of Cross-Functional Groups and Project Teams

14

The Discipline of Teams

JON R. KATZENBACH AND DOUGLAS K. SMITH

Early in the 1980s, Bill Greenwood and a small band of rebel railroaders took on most of the top management of Burlington Northern and created a multibillion-dollar business in "piggybacking" rail services despite widespread resistance, even resentment, within the company. The Medical Products Group at Hewlett-Packard owes most of its leading performance to the remarkable efforts of Dean Morton, Lew Platt, Ben Holmes, Dick Alberting, and a handful of their colleagues who revitalized a health care business that most others had written off. At Knight-Ridder, Jim Batten's "customer obsession" vision took root at the *Tallahassee Democrat* when 14 frontline enthusiasts turned a charter to eliminate errors into a mission of major change and took the entire paper along with them.

Such are the stories and the work of teams— real teams that perform, not amorphous groups that we call teams because we think that the label is motivating and energizing. The difference between teams that perform and other groups that don't is

Reprinted by permission of *Harvard Business Review*. From "The Discipline of Teams" by Jon R. Katzenbach and Douglas K. Smith, Vol. 17, 1993, pp. 111–120. Copyright © 1993 by the President and Fellows of Harvard College, all rights reserved.
Jon R. Katzenbach and *Douglas K. Smith* are partners at the management consulting firm of McKinsey.

a subject to which most of us pay far too little attention. Part of the problem is that *team* is a word and concept so familiar to everyone.

Or at least that's what we thought when we set out to do research for our book *The Wisdom of Teams*. We wanted to discover what differentiates various levels of team performance, where and how teams work best, and what top management can do to enhance their effectiveness. We talked with hundreds of people on more than 50 different teams in 30 companies and beyond, from Motorola and Hewlett-Packard to Operation Desert Storm and the Girl Scouts.

We found that there is a basic discipline that makes teams work. We also found that teams and good performance are inseparable; you cannot have one without the other. But people use the word *team* so loosely that it gets in the way of learning and applying the discipline that leads to good performance. For managers to make better decisions about whether, when, or how to encourage and use teams, it is important to be more precise about what a team is and what it isn't.

Most executives advocate teamwork. And they should. Teamwork represents a set of values that encourage listening and responding constructively to views expressed by others, giving others the benefit of the doubt, providing support, and recognizing the interests and achievements of others. Such values help teams perform, and they also promote individual performance as well as the performance of an entire organization. But teamwork values by themselves are not exclusive to teams, nor are they enough to ensure team performance.

Nor is a team just any group working together. Committees, councils, and task forces are not necessarily teams. Groups do not become teams simply because that is what someone calls them. The entire work force of any large and complex organization is *never* a team, but think about how often that platitude is offered up.

To understand how teams deliver extra performance, we must distinguish between teams and other forms of working groups. That distinction turns on performance results. A working group's performance is a function of what its members do as individuals. A team's performance includes both individual results and what we call "collective work-products." A collective work-product is what two or more members must work on together, such as interviews, surveys, or experiments. Whatever it is, a collective work-product reflects the joint, real contribution of team members.

Working groups are both prevalent and effective in large organizations where individual accountability is most important. The best working groups come together to share information, perspectives, and insights; to make decisions that help each person do his or her job better; and to reinforce individual performance standards. But the focus is always on individual goals and accountabilities. Working group members don't take responsibility for results other than their own. Nor do they try to develop incremental performance contributions requiring the combined work of two or more members.

Teams differ fundamentally from working groups because they require both individual and mutual accountability. Teams rely on more than group discussion, debate, and decision; on more than sharing information and best practice performance standards. Teams produce discrete work-products through the joint contributions of their members. This is what makes possible performance levels greater than the sum of all the individual bests of team members. Simply stated, a team is more than the sum of its parts.

The first step in developing a disciplined approach to team management is to think about teams as discrete units of performance and not just as positive sets of values. Having observed and worked with scores of teams in action, both successes and failures, we offer the following. Think of it as a working definition or, better still, an essential discipline that real teams share.

> A team is a small number of people with complementary skills who are committed to a common purpose, set of performance goals, and approach for which they hold themselves mutually accountable.

The essence of a team is common commitment. Without it, groups perform as individuals; with it, they become a powerful unit of collective performance. This kind of commitment requires a purpose in which team members can believe. Whether the purpose is to "transform the contributions of suppliers into the satisfaction of customers," to "make our company one we can be proud of again," or to "prove that all children can learn," credible team purposes have an element related to winning, being first, revolutionizing, or being on the cutting edge.

Teams develop direction, momentum, and commitment by working to shape a meaningful purpose. Building ownership and commitment to team purpose, however, is not incompatible with taking initial direction from outside the team. The often-asserted assumption that a team cannot "own" its purpose unless management leaves it alone actually confuses more potential teams than it helps. In fact, it is the exceptional case—for example, entrepreneurial situations—when a team creates a purpose entirely on its own.

Most successful teams shape their purposes in response to a demand or opportunity put in their path, usually by higher management. This helps teams get started by broadly framing the company's performance expectation. Management is responsible for clarifying the charter, rationale, and performance challenge for the team, but management must also leave enough flexibility for the team to

develop commitment around its own spin on that purpose, set of specific goals, timing, and approach.

The best teams invest a tremendous amount of time and effort exploring, shaping, and agreeing on a purpose that belongs to them both collectively and individually. This "purposing" activity continues throughout the life of the team. In contrast, failed teams rarely develop a common purpose. For whatever reason—an insufficient focus on performance, lack of effort, poor leadership—they do not coalesce around a challenging aspiration.

The best teams also translate their common purpose into specific performance goals, such as reducing the reject rate from suppliers by 50% or increasing the math scores of graduates from 40% to 95%. Indeed, if a team fails to establish specific performance goals or if those goals do not relate directly to the team's overall purpose, team members become confused, pull apart, and revert to mediocre performance. By contrast, when purposes and goals build on one another and are combined with team commitment, they become a powerful engine of performance.

Transforming broad directives into specific and measurable performance goals is the surest first step for a team trying to shape a purpose meaningful to its members. Specific goals, such as getting a new product to market in less than half the normal time, responding to all customers within 24

Not All Groups Are Teams: How to Tell the Difference

Working Group	Team
• Strong, clearly focused leader	• Shared leadership roles
• Individual accountability	• Individual and mutual accountability
• The group's purpose is the same as the broader organizational mission	• Specific team purpose that the team itself delivers
• Individual work-products	• Collective work-products
• Runs efficient meetings	• Encourages open-ended discussion and active problem-solving meetings
• Measures its effectiveness indirectly by its influence on others (e.g., financial performance of the business)	• Measures performance directly by assessing collective work-products
• Discusses, decides, and delegates	• Discusses, decides, and does real work together

hours, or achieving a zero-defect rate while simultaneously cutting costs by 40%, all provide firm footholds for teams. There are several reasons:

Specific team performance goals help to define a set of work-products that are different both from an organizationwide mission and from individual job objectives. As a result, such work-products require the collective effort of team members to make something specific happen that, in and of itself, adds real value to results. By contrast, simply gathering from time to time to make decisions will not sustain team performance.

The specificity of performance objectives facilitates clear communication and constructive conflict within the team. When a plant-level team, for example, sets a goal of reducing average machine changeover time to two hours, the clarity of the goal forces the team to concentrate on what it would take either to achieve or to reconsider the goal. When such goals are clear, discussions can focus on how to pursue them or whether to change them; when goals are ambiguous or nonexistent, such discussions are much less productive.

The attainability of specific goals helps teams maintain their focus on getting results. A product development team at Eli Lilly's Peripheral Systems Division set definite yardsticks for the market introduction of an ultrasonic probe to help doctors locate deep veins and arteries. The probe had to have an audible signal through a specified depth of tissue, be capable of being manufactured at a rate of 100 per day, and have a unit cost less than a preestablished amount. Because the team could measure its progress against each of these specific objectives, the team knew throughout the development process where it stood. Either it had achieved its goals or not.

As Outward Bound and other team-building programs illustrate, specific objectives have a leveling effect conducive to team behavior. When a small group of people challenge themselves to get over a wall or to reduce cycle time by 50%, their respective titles, perks, and other stripes fade into the background. The teams that succeed evaluate what and how each individual can best contribute to the team's goal and, more important, do so in terms of the performance objective itself rather than a person's status or personality.

Specific goals allow a team to achieve small wins as it pursues its broader purpose. These small wins are invaluable to building commitment and overcoming the inevitable obstacles that get in the way of a long-term purpose. For example, the Knight-Ridder team mentioned at the outset turned a narrow goal to eliminate errors into a compelling customer-service purpose.

Performance goals are compelling. They are symbols of accomplishment that motivate and energize. They challenge the people on a team to commit themselves, as a team, to make a difference. Drama, urgency, and a healthy fear of failure combine to drive teams who have their collective eye on an attainable, but challenging, goal. Nobody but the team can make it happen. It is their challenge.

The combination of purpose and specific goals is essential to performance. Each depends on the other to remain relevant and vital. Clear performance goals help a team keep track of progress and hold itself accountable; the broader, even nobler, aspirations in a team's purpose supply both meaning and emotional energy.

Virtually all effective teams we have met, read or heard about, or been members of have ranged between 2 and 25 people. For example, the Burlington Northern "piggybacking" team had 7 members, the Knight-Ridder newspaper team, 14. The majority of them have numbered less than 10. Small size is admittedly more of a pragmatic guide than an absolute necessity for success. A large number of people, say 50 or more, can theoretically become a team. But groups of such size are more likely to break into subteams rather than function as a single unit.

Why? Large numbers of people have trouble interacting constructively as a group, much less doing real work together. Ten people are far more likely than fifty are to work through their individual, functional, and hierarchical differences toward a common plan and to hold themselves jointly accountable for the results.

Large groups also face logistical issues, such as finding enough physical space and time to meet.

And they confront more complex constraints, like crowd or herd behaviors, which prevent the intense sharing of viewpoints needed to build a team. As a result, when they try to develop a common purpose, they usually produce only superficial "missions" and well-meaning intentions that cannot be translated into concrete objectives. They tend fairly quickly to reach a point when meetings become a chore, a clear sign that most of the people in the group are uncertain why they have gathered, beyond some notion of getting along better. Anyone who has been through one of these exercises knows how frustrating it can be. This kind of failure tends to foster cynicism, which gets in the way of future team efforts.

In addition to finding the right size, teams must develop the right mix of skills, that is, each of the complementary skills necessary to do the team's job. As obvious as it sounds, it is a common failing in potential teams. Skill requirements fall into three fairly self-evident categories:

Technical or Functional Expertise. It would make little sense for a group of doctors to litigate an employment discrimination case in a court of law. Yet teams of doctors and lawyers often try medical malpractice or personal injury cases. Similarly, product-development groups that include only marketers or engineers are less likely to succeed than those with the complementary skills of both.

Problem-Solving and Decision-Making Skills. Teams must be able to identify the problems and opportunities they face, evaluate the options they have for moving forward, and then make necessary trade-offs and decisions about how to proceed. Most teams need some members with these skills to begin with, although many will develop them best on the job.

Interpersonal Skills. Common understanding and purpose cannot arise without effective communication and constructive conflict, which in turn depend on interpersonal skills. These include risk taking, helpful criticism, objectivity, active listening, giving the benefit of the doubt, and recognizing the interests and achievements of others.

Obviously, a team cannot get started without some minimum complement of skills, especially technical and functional ones. Still, think about how often you've been part of a team whose members were chosen primarily on the basis of personal compatibility or formal position in the organization, and in which the skill mix of its members wasn't given much thought.

It is equally common to overemphasize skills in team selection. Yet in all the successful teams we've encountered, not one had all the needed skills at the outset. The Burlington Northern team, for example, initially had no members who were skilled marketers despite the fact that their performance challenge was a marketing one. In fact, we discovered that teams are powerful vehicles for developing the skills needed to meet the team's performance challenge. Accordingly, the team member selection ought to ride as much on skill potential as on skills already proven.

Effective teams develop strong commitment to a common approach, that is, to how they will work together to accomplish their purpose. Team members must agree on who will do particular jobs, how schedules will be set and adhered to, what skills need to be developed, how continuing membership in the team is to be earned, and how the group will make and modify decisions. This element of commitment is as important to team performance as is the team's commitment to its purpose and goals.

Agreeing on the specifics of work and how they fit together to integrate individual skills and advance team performance lies at the heart of shaping a common approach. It is perhaps self-evident that an approach that delegates all the real work to a few members (or staff outsiders), and thus relies on reviews and meetings for its only "work together" aspects, cannot sustain a real team. Every member of a successful team does equivalent amounts of real work; all members, including the team leader, contribute in concrete ways to the team's work-product. This is a very important element of the emotional logic that drives team performance.

When individuals approach a team situation,

especially in a business setting, each has preexisting job assignments as well as strengths and weaknesses reflecting a variety of backgrounds, talents, personalities, and prejudices. Only through the mutual discovery and understanding of how to apply all its human resources to a common purpose can a team develop and agree on the best approach to achieve its goals. At the heart of such long and, at times, difficult interactions lies a commitment-building process in which the team candidly explores who is best suited to each task as well as how individual roles will come together. In effect, the team establishes a social contract among members that relates to their purpose and guides and obligates how they must work together.

No group ever becomes a team until it can hold itself accountable as a team. Like common purpose and approach, mutual accountability is a stiff test. Think, for example, about the subtle but critical difference between "the boss holds me accountable" and "we hold ourselves accountable." The first case can lead to the second; but without the second, there can be no team.

Companies like Hewlett-Packard and Motorola have an ingrained performance ethic that enables teams to form "organically" whenever there is a clear performance challenge requiring collective rather than individual effort. In these companies, the factor of mutual accountability is commonplace. "Being in the boat together" is how their performance game is played.

At its core, team accountability is about the sincere promises we make to ourselves and others, promises that underpin two critical aspects of effective teams: commitment and trust. Most of us enter a potential team situation cautiously because ingrained individualism and experience discourage us from putting our fates in the hands of others or accepting responsibility for others. Teams do not succeed by ignoring or wishing away such behavior.

Mutual accountability cannot be coerced any more than people can be made to trust one another. But when a team shares a common purpose, goals, and approach, mutual accountability grows as a natural counterpart. Accountability arises from and reinforces the time, energy, and action invested in figuring out what the team is trying to accomplish and how best to get it done.

When people work together toward a common objective, trust and commitment follow. Consequently, teams enjoying a strong common purpose and approach inevitably hold themselves responsible, both as individuals and as a team, for the team's performance. This sense of mutual accountability also produces the rich rewards of mutual achievement in which all members share. What we heard over and over from members of effective teams is that they found the experience energizing and motivating in ways that their "normal" jobs never could match.

On the other hand, groups established primarily for the sake of becoming a team or for job enhancement, communication, organizational effectiveness, or excellence rarely become effective teams, as demonstrated by the bad feelings left in many companies after experimenting with quality circles that never translated "quality" into specific goals. Only when appropriate performance goals are set does the process of discussing the goals and the approaches to them give team members a clearer and clearer choice: they can disagree with a goal and the path that the team selects and, in effect, opt out, or they can pitch in and become accountable with and to their teammates.

The discipline of teams we've outlined is critical to the success of all teams. Yet it is also useful to go one step further. Most teams can be classified in one of three ways: teams that recommend things, teams that make or do things, and teams that run things. In our experience, each type faces a characteristic set of challenges.

Teams That Recommend Things. These teams include task forces, project groups, and audit, quality, or safety groups asked to study and solve particular problems. Teams that recommend things almost always have predetermined completion dates. Two critical issues are unique to such teams: getting off to a fast and constructive start and dealing with the ultimate handoff required to get recommendations implemented.

The key to the first issue lies in the clarity of the team's charter and the composition of its membership. In addition to wanting to know why and how their efforts are important, task forces need a clear definition of whom management expects to participate and the time commitment required. Management can help by ensuring that the team includes people with the skills and influence necessary for crafting practical recommendations that will carry weight throughout the organization. Moreover, management can help the team get the necessary cooperation by opening doors and dealing with political obstacles.

Missing the handoff is almost always the problem that stymies teams that recommend things. To avoid this, the transfer of responsibility for recommendations to those who must implement them demands top management's time and attention. The more top managers assume that recommendations will "just happen," the less likely it is that they will. The more involvement task force members have in implementing their recommendations, the more likely they are to get implemented.

To the extent that people outside the task force will have to carry the ball, it is critical to involve them in the process early and often, certainly well before recommendations are finalized. Such involvement may take many forms, including participating in interviews, helping with analyses, contributing and critiquing ideas, and conducting experiments and trials. At a minimum, anyone responsible for implementation should receive a briefing on the task force's purpose, approach, and objectives at the beginning of the effort as well as regular reviews of progress.

Teams That Make or Do Things. These teams include people at or near the front lines who are responsible for doing the basic manufacturing, development, operations, marketing, sales, service, and other value-adding activities of a business. With some exceptions, like new-product development or process design teams, teams that make or do things tend to have no set completion dates because their activities are ongoing.

In deciding where team performance might

have the greatest impact, top management should concentrate on what we call the company's "critical delivery points," that is, places in the organization where the cost and value of the company's products and services are most directly determined. Such critical delivery points might include where accounts get managed, customer service performed, products designed, and productivity determined. If performance at critical delivery points depends on combining multiple skills, perspectives, and judgments in real time, then the team option is the smartest one.

When an organization does require a significant number of teams at these points, the sheer challenge of maximizing the performance of so many groups will demand a carefully constructed and performance-focused set of management processes. The issue here for top management is how to build the necessary systems and process supports without falling into the trap of appearing to promote teams for their own sake.

The imperative here, returning to our earlier discussion of the basic discipline of teams, is a relentless focus on performance. If management fails to pay persistent attention to the link between teams and performance, the organization becomes convinced that "this year we are doing 'teams.' " Top management can help by instituting processes like pay schemes and training for teams responsive to their real time needs, but more than anything else, top management must make clear and compelling demands on the teams themselves and then pay constant attention to their progress with respect to both team basics and performance results. This means focusing on specific teams and specific performance challenges. Otherwise "performance," like "team" will become a cliché.

Teams That Run Things. Despite the fact that many leaders refer to the group reporting to them as a team, few groups really are. And groups that become real teams seldom think of themselves as a team because they are so focused on performance results. Yet the opportunity for such teams includes groups from the top of the enterprise down through the divisional or functional level. Whether it is in charge of thousands of people or a handful, as long as the group oversees some business, on-going program, or significant functional activity, it is a team that runs things.

The main issue these teams face is determin-ing whether a real team approach is the right one. Many groups that run things can be more effective as working groups than as teams. The key judg-ment is whether the sum of individual bests will suffice for the performance challenge at hand or whether the group must deliver substantial incre-mental performance requiring real, joint work-products. Although the team option promises greater performance, it also brings more risk, and managers must be brutally honest in assessing the trade-offs.

Members may have to overcome a natural re-luctance to trust their fate to others. The price of faking the team approach is high: at best, members get diverted from their individual goals, costs out-weigh benefits, and people resent the imposition on their time and priorities; at worst, serious animosi-ties develop that undercut even the potential per-sonal bests of the working-group approach.

Working groups present fewer risks. Effective working groups need little time to shape their pur-pose since the leader usually establishes it. Meet-ings are run against well-prioritized agendas. And decisions are implemented through specific indi-vidual assignments and accountabilities. Most of the time, therefore, if performance aspirations can be met through individuals doing their respective jobs well, the working-group approach is more comfortable, less risky, and less disruptive than trying for more elusive team performance levels. Indeed, if there is no performance need for the team approach, efforts spent to improve the effec-tiveness of the working group make much more sense than floundering around trying to become a team.

Having said that, we believe the extra level of performance teams can achieve is becoming criti-cal for a growing number of companies, especially as they move through major changes during which company performance depends on broad-based be-havioral change. When top management uses teams to run things, it should make sure the team suc-ceeds in identifying specific purposes and goals.

This is a second major issue for teams that run things. Too often, such teams confuse the broad mission of the total organization with the specific purpose of their small group at the top. The disci-pline of teams tells us that for a real team to form there must be a *team* purpose that is distinctive and specific to the small group and that requires its members to roll up their sleeves and accomplish something beyond individual end-products. If a group of managers looks only at the economic per-formance of the part of the organization it runs to assess overall effectiveness, the group will not have any team performance goals of its own.

While the basic discipline of teams does not differ for them, teams at the top are certainly the most difficult. The complexities of long-term chal-lenges, heavy demands on executive time, and the deep-seated individualism of senior people con-spire against teams at the top. At the same time, teams at the top are the most powerful. At first we thought such teams were nearly impossible. That is because we were looking at the teams as defined by the formal organizational structure, that is, the leader and all his or her direct reports equals the team. Then we discovered that real teams at the top were often smaller and less formalized—White-head and Weinberg at Goldman, Sachs; Hewlett and Packard at HP; Krasnoff, Pall, and Hardy at Pall Corp; Kendall, Pearson, and Calloway at Pepsi; Haas and Haas at Levi Strauss; Batten and Ridder at Knight-Ridder. They were mostly twos and threes, with an occasional fourth.

Nonetheless, real teams at the top of large, complex organizations are still few and far be-tween. Far too many groups at the top of large cor-porations needlessly constrain themselves from achieving real team levels of performance because they assume that all direct reports must be on the team; that team goals must be identical to corpo-rate goals; that the team members' positions rather than skills determine their respective roles; that a team must be a team all the time; and that the team leader is above doing real work.

As understandable as these assumptions may be, most of them are unwarranted. They do not ap-ply to the teams at the top we have observed, and

Building Team Performance

Although there is no guaranteed how-to recipe for building team performance, we observed a number of approaches shared by many successful teams.

Establish urgency, demanding performance standards, and direction. All team members need to believe the team has urgent and worthwhile purposes, and they want to know what the expectations are. Indeed, the more urgent and meaningful the rationale, the more likely it is that the team will live up to its performance potential, as was the case for a customer-service team that was told that further growth for the entire company would be impossible without major improvements in that area. Teams work best in a compelling context. That is why companies with strong performance ethics usually form teams readily.

Select members for skill and skill potential, not personality. No team succeeds without all the skills needed to meet its purpose and performance goals. Yet most teams figure out the skills they will need after they are formed. The wise manager will choose people both for their existing skills and their potential to improve existing skills and learn new ones.

Pay particular attention to first meetings and actions. Initial impressions always mean a great deal. When potential teams first gather, everyone monitors the signals given by others to confirm, suspend, or dispel assumptions and concerns. They pay particular attention to those in authority: the team leader and any executives who set up, oversee, or otherwise influence the team. And, as always, what such leaders do is more important than what they say. If a senior executive leaves the team kickoff to take a phone call ten minutes after the session has begun and he never returns, people get the message.

Set some clear rules of behavior. All effective teams develop rules of conduct at the outset to help them achieve their purpose and performance goals. The most critical initial rules pertain to attendance (for example, "no interruptions to take phone calls"), discussion ("no sacred cows"), confidentiality ("the only things to leave this room are what we agree on"), analytic approach ("facts are friendly"), end-product orientation ("everyone gets assignments and does them"), constructive confrontation ("no finger pointing"), and, often the most important, contributions ("everyone does real work").

Set and seize upon a few immediate perfor-mance-oriented tasks and goals. Most effective teams trace their advancement to key performance-oriented events. Such events can be set in motion by immediately establishing a few challenging goals that can be reached early on. There is no such thing as a real team without performance results, so the sooner such results occur, the sooner the team congeals.

Challenge the group regularly with fresh facts and information. New information causes a team to redefine and enrich its understanding of the performance challenge, thereby helping the team shape a common purpose, set clearer goals, and improve its common approach. A plant quality improvement team knew the cost of poor quality was high, but it wasn't until they researched the different types of defects and put a price tag on each one that they knew where to go next. Conversely, teams err when they assume that all the information needed exists in the collective experience and knowledge of their members.

Spend lots of time together. Common sense tells us that team members must spend a lot of time together, scheduled and unscheduled, especially in the beginning. Indeed, creative insights as well as personal bonding require impromptu and casual interactions just as much as analyzing spreadsheets and interviewing customers. Busy executives and managers too often intentionally minimize the time they spend together. The successful teams we've observed all gave themselves the time to learn to be a team. This time need not always be spent together physically; electronic, fax, and phone time can also count as time spent together.

Exploit the power of positive feedback, recognition, and reward. Positive reinforcement works as well in a team context as elsewhere. "Giving out gold stars" helps to shape new behaviors critical to team performance. If people in the group, for example, are alert to a shy person's initial efforts to speak up and contribute, they can give the honest positive reinforcement that encourages continued contributions. There are many ways to recognize and reward team performance beyond direct compensation, from having a senior executive speak directly to the team about the urgency of its mission to using awards to recognize contributions. Ultimately, however, the satisfaction shared by a team in its own performance becomes the most cherished reward.

when replaced with more realistic and flexible assumptions that permit the team discipline to be applied, real team performance at the top can and does occur. Moreover, as more and more companies are confronted with the need to manage major change across their organizations, we will see more real teams at the top.

We believe that teams will become the primary unit of performance in high-performance organizations. But that does not mean that teams will crowd out individual opportunity or formal hierarchy and process. Rather, teams will enhance existing structures without replacing them. A team opportunity exists anywhere hierarchy or organizational boundaries inhibit the skills and perspectives needed for optimal results. Thus, new-product innovation requires preserving functional excellence through structure while eradicating functional bias through teams. And frontline productivity requires preserving direction and guidance through hierarchy while drawing on energy and flexibility through self-managing teams.

We are convinced that every company faces specific performance challenges for which teams are the most practical and powerful vehicle at top management's disposal. The critical role for senior managers, therefore, is to worry about company performance and the kinds of teams that can deliver it. This means that top management must recognize a team's unique potential to deliver results, deploy teams strategically when they are the best tool for the job, and foster the basic discipline of teams that will make them effective. By doing so, top management creates the kind of environment that enables team as well as individual and organizational performance.

15

Managing Creative Performance in R&D Teams

RALPH KATZ

The general neglect of a temporal perspective—the fact that group activities and reactions can change significantly over the course of a long project—has been one of the major problems in the study of project groups and teams. Yet until it is addressed, questions about how well a group is doing will receive answers that are, at best, incomplete. Engineers and scientists have long recognized the problems facing a technical group should its membership remain constant too long. As individuals are born, grow up, and grow older—first feeling their way uncertainly, then seeking out new challenges and experiences as they gain confidence, and finally, becoming a bit self-satisfied about their own knowledge and achievements—so the same process seems to occur within groups whose members have worked together for a long time. Research and Development groups seem to have performance curves analogous to the human life cycle—tentative youth, productive energy, and decline with maturity.

The analogy is a convenient one, though subject in both cases to variation: age *need not* mean stagnation in either an individual or a group. Still, a field study of research and development project teams, which Professor Tom Allen and I have been engaged in for some years, does tend to support a general finding of less intense involvement in job demands and challenges with increasing stability in project membership.

It is, of course, natural for both individuals and groups to attempt to structure their work activities to reduce stress and ensure a level of certainty. People do not deal well with uncertainty; they like to know, as much as possible, what will happen next,

how they will be affected, etc. Given this, group members interacting over a long time are likely to develop standard work patterns that are both familiar and comfortable, patterns in which routine and precedent play a relatively large part—perhaps at the expense of unbiased thought and new ideas. On the other hand, an environment devoid of structure and definition, one wholly unfamiliar and enigmatic, is equally undesirable. Without some sort of established pattern or perspective to serve as a basis for action, nothing at all would be accomplished. The task of management, then, is to create and maintain an atmosphere in which employees are both familiar with their job requirements and challenged by them.

THE REQUISITE FAMILIARITY

How long it takes to acquire the requisite familiarity with one's job to function efficiently depends on the length of time it takes an employee to feel accepted and competent in his or her new environment. This feeling is influenced both by the nature of the individual and the socialization process that unfolds for that person. In general, the time varies according to the level of complexity involved in the job requirements, ranging from as little as a few months to as much as a year or more in non-routine kinds of professions such as R&D.

In engineering, for example, strategies and solutions are usually peculiar to specific settings. Research and development teams in different organizations may face similar problems, yet approach their solutions with widely divergent methods.

Reprinted with permission from the author.

Ralph Katz is Professor of Management at Northeastern University's College of Business and Research Associate at MIT's Sloan School of Management.

Thus, even though one may have received an excellent education in, say, mechanical engineering principles, one must still figure out how to be an effective mechanical engineer at Westinghouse, Alcoa, or General Electric.

In the course of long-term job tenure, an individual may be said to pass through three broad stages: *socialization, innovation,* and *stabilization.* A graphic representation of the model is shown in Figure 15.1.

During the *socialization* period, employees are primarily concerned with understanding and coming to terms with their new and unknown social and task environments. Newcomers must learn the customary norms of behavior within their groups, how reward systems operate, the expectations of supervisors, and a host of other considerations that are necessary for them to function meaningfully. These considerations may vary to a surprisingly large extent even within a single organization. This is important for, while the necessity of such a "breaking-in" period has long been recognized in the case of recently hired members of an organization, it should also be understood that veteran employees assigned to new groups must also "resocialize" themselves since they, too, must now deal with unfamiliar tasks and colleagues. It is in this period that employees learn not only the

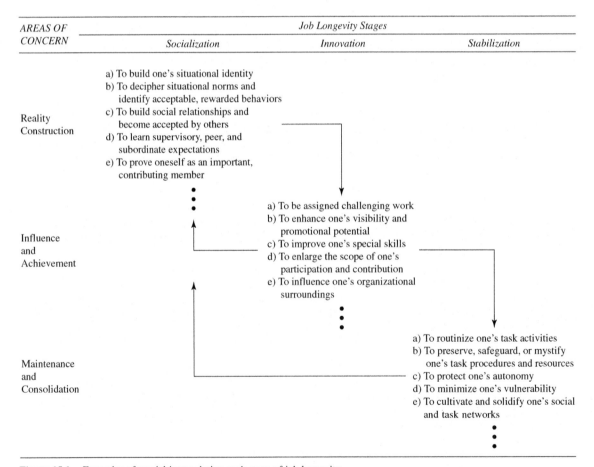

Figure 15.1.   Examples of special issues during each stage of job longevity.

technical requirements of their new job assignments, but also the behaviors and attitudes that are acceptable and necessary for becoming a true contributing member of the group.

As individuals gain familiarity with their work settings, they are freer to devote their energies and concerns less toward socialization and more toward performance and accomplishment. In the *innovation* stage of a job, employees become capable, to a greater extent, of acting in a responsive and undistracted manner. The movement from socialization to innovation implies that employees no longer require much assistance in deciphering their new job and organizational surroundings. Instead, they can divert their attention from an initial emphasis on psychological "safety and acceptance" to concerns for achievement and influence. Opportunities to change the status quo and to respond to new challenging demands within job settings become progressively more pertinent to employees in this stage. As the length of time spent in the same job environment stretches out, however, employees may gradually enter the *stabilization* phase, in which there is a slow shift away from a high level of involvement and receptivity to the challenges in their jobs and toward a greater degree of unresponsiveness to these challenges.

In time, even the most engaging job assignments and responsibilities can appear less exciting, little more than habit, to people who have successfully mastered and become accustomed to their everyday task requirements. It makes sense, then, that with prolonged job stability, employees' perceptions of their conditions at present and possibilities for the future will become increasingly impoverished. If employees cannot maintain, redefine, or expand their jobs for continued change and growth, then their work enthusiasm will deteriorate. If possibilities for development *are* continued, however, then the stabilization period may be held off indefinitely.

The irony of the situation is that employees with the greatest initial responsiveness to job challenges seem to retain that responsiveness for a *shorter* length of time than those expressing less of a need for high job challenge. The greater initial enthusiasm of high-need employees appears to drive them more swiftly through their socialization and innovation periods and into stabilization. They become bored more quickly with tasks that are now too familiar and routine to their growth-oriented natures.

Of course, job longevity does not exist in a vacuum. Many other factors may influence the level of job interest. New technological developments, rapid growth and expansion, or strong competitive pressures, or new excited co-workers could all help sustain or even enhance one's involvement in his or her job-related activities. On the other hand, working closely with a group of unresponsive peers in a relatively unchanging situation might shorten an individual's responsive period on that particular job rather dramatically.

Despite these other influences, though, the general trend does hold. In moving from innovation to stabilization, employees who continue to work in the same overall job situation for long periods gradually adapt to such steadfast employment by becoming increasingly indifferent to the challenging aspects of their assignments. And as employees come to care less about the intrinsic nature of the work they do, their absorption in contextual features such as salary, benefits, vacations, friendly co-workers, and compatible superiors tends to increase.

Interestingly, entry into the stabilization period does not necessarily imply a reduced level of job satisfaction. On the contrary, in fact. As employees enter the stabilization stage, they have typically adapted by becoming very satisfied with the comfortableness and predictability of their work environments; for when the chances of future growth and change become limited, existing situations become accepted as the desired. Only when a reasonable gap remains between what individuals desire and what they are presently able to achieve will there be energy for change and accomplishment.

With stability comes a greater loyalty to precedent, to the established patterns of behavior. In adapting to high job longevity, employees become increasingly content with customary ways of doing things, comfortable routines, and familiar sets of task demands that promote a feeling of security and confidence while requiring little exceptional effort or vigilance. The preservation of such patterns is likely to be a prime consideration, with the result that contact with information and ideas that threaten change may be curtailed. Moreover, strong biases may develop in the selection and interpretation of information, in abilities to generate new options and strategies creatively, and in the level of willingness to innovate or implement alternative courses of action. In a sense, the differences between the innovation and stabilization stages are indicative of the distinctions between creative performance and routine performance; between job excitement and work satisfaction. What they are asked to do, they do well—but that spark is missing. The willingness to go beyond what is requested, to experiment, to try something new, or to seek out new responsibilities is not there.

What is also important to note from the model portrayed in Figure 15.1 is that individuals can easily cycle between socialization and innovation (with on-going job changes and promotions) or they can slowly proceed from innovation to stabilization over time. Direct movement from stabilization back to innovation, however, is very unlikely without the individual first going through a new socialization (or resocialization) experience in order to unfreeze previously defined and reinforced habits and perspectives. Thus, rotation per se is not the solution to rejuvenation; instead, it is rotation coupled with a new socialization experience that provides the individual with a new opportunity to regain responsiveness to new task challenges and environmental demands. The intensity of the resocialization experience, moreover, must match the strength of the prior stabilization period. Organizations often spend much time and effort planning the movements and rotations of personnel. They often, however, spend very little time managing the socialization process that occurs *after* rotation. This is unfortunate, for it is the experiences and interactions that take place after rotation that are so important for influencing and framing an individual's attitudes and eventual responsiveness.

So far, we have described what happens to individual professionals as if they work independently and autonomously. Most of the time, technical professionals function interdependently either as members of specific project teams or specific technology-based groups. It may be more important, therefore, to know not only what happens to an individual over time but also what happens to the performances of teams or groups of individuals who have been working together over time. In any group, there is a changing mix of individuals, some of whom may be in socialization, some in the innovation stage, and still others in stabilization. It is not how this particular mix of individuals act that is important, but how they interact both amongst themselves and outside their group. Successful innovation, after all, is not a function of how well individuals act, but how they interact. While "invention" can result from individualistic actions, effective "innovation" is a function of collective activities and teamwork.

To investigate these issues, Professor Tom Allen and I conducted a study to examine how group longevity affects project performance and communication behavior, where group longevity (i.e., group age) measures the average length of time that project members have worked and shared experiences with one another. Group longevity or group age was determined simply by averaging the individual project tenures of all group members. The measure, therefore, is *not* the length of time the project has been in existence. Rather it represents the length of time group members have been working together in a particular project or technical area.

Data collection for the study took place at the R&D facility of a large American chemical company, employing 345 engineering and scientific professionals in 61 distinct project groups or work areas. Project groups were organized around specific, long-term kinds of problem areas such as fiber-forming development and urethane development, and ranged across three broad categories of R&D activity: (1) applied research, (2) product & process development, and (3) technical service and support.

The purpose of the study was twofold: first, to examine the level of communication activity by project groups at various stages in the group's "life" (that is, its group longevity) and, second, to discover any possible correspondence between a lessening of communication and a possible drop in performance. The focus was on interpersonal communication, which, as many previous studies have demonstrated, is the primary means by which engineering and scientific professionals transfer and process technical ideas and information.

To measure communication activity, participants kept track of all other professionals with whom they had work-related interaction on a randomly chosen day each week for fifteen weeks. Contacts both inside and outside the R&D facility were measured. Based on this frequency data, three independent measures of communication were calculated for each project to each of three separate areas of important information:

1. Intraproject Communication: The amount of communication reported among all project members.

2. Organizational Communication: The level or amount of contact reported by project members with individuals outside the R&D facility but within other corporate divisions, principally marketing and manufacturing.

3. Professional Communication: The amount of communication reported by project members with professionals outside the parent organization, including professionals in universities, consulting firms, and professional societies.

For all three areas or sources of information, project groups whose longevity index was greater than four years reported much lower levels of actual contact than project groups whose longevity index fell between one and a half and four years. Intraproject, organizational, and outside professional interaction were considerably lower for the longer-tenured groups. Members of these groups, therefore, were significantly more isolated from external sources of new ideas and technological advances and from information within other organizational divisions, especially marketing and manufacturing. Project members in these long-tenured groups even communicated less often amongst themselves about technically related matters.

In addition to these measures of actual communication behavior, a direct evaluation of the current technical performance of the project groups was obtained. All department managers and laboratory directors assessed the overall performance of all projects with which they were technically familiar, based on their knowledge of and experience with the various projects. The managers, in making their evaluations, considered such elements as schedule and cost performance; innovativeness; adaptability; and the ability to coordinate with other parts of the organization. Each project was independently rated by an average of five higher-level managers; consensus among the ratings was extremely high.

On average, the association between project performance and group longevity closely paralleled the association of longevity and communication trends. This is to say that for these 61 project teams, there was a strong curvilinear relationship between group age and project performance—the best performing groups being those with longevities between one and a half and four years. Performance was significantly lower for the relatively new teams and for teams that had been together for more than four years. In fact, *none* of the ten project groups with the highest levels of group age (i.e., five or more years) were among the facility's higher performing project teams, all being rated by the facility's management as either average or below average. It is also interesting to note that none of the managerial evaluators knew which project teams were the long-term ones or whether their organization even had any, since rotations and movements were always ongoing. In reality, over 20 percent of the R&D effort within this organization was being conducted by these ten lower-performing, long-term technical teams.

Almost by definition, projects with higher mean group tenure were staffed by older engineers. This raises the possibility that performance may be lower as a result of the increasing obsolescence of individuals' skills as they aged, rather than because of anything to do with the group's tenure composition. The data, however, do not bear this out. For both the communication and the performance data, it was found that group longevity and not the chronological age of individuals was more likely to have influenced the results.

Another possibility is that long-tenured project teams had simply come to be staffed by less technically competent or perhaps less motivated engineers and scientists. Follow-up visits to this facility over the next five years, however, reveal that about the same proportion of professionals from both the long- and medium-tenured teams were awarded promotions to higher level managerial positions above the project leadership level. Over this five-year period, 15 percent of the engineers who had worked in the medium-tenured groups attained managerial positions of either laboratory supervisor or laboratory manager, while the comparable proportion from the longer-tenured groups was 13 percent. In addition, the percentage of technical professionals promoted to the "technical" side of the facility's dual ladder promotional system was slightly greater for members in the longer-tenured project groups than the medium-tenured ones, 19 percent compared to 12 percent. This seems to indicate a relative parity in the area of individual competence and capability between the respective group memberships of the medium and long-term categories of group longevity.

Despite the parallel declines in both project communication and performance with increasingly high levels of group longevity, one must be careful not to jump to the conclusion that decays in all areas of communication contributed equally to the lower levels of project performance. Different categories of project tasks require different patterns of communication for more effective performance. Research project groups, for example, have been found to be higher performing when project members maintain high levels of technical communication with outside professionals. Performance in development projects, on the other hand, is related more to contact within the organization, primarily with divisions such as marketing and manufactur-

ing. Finally, for technical-service projects, communication within the team appears most crucial.

Significantly, for each project type, the deterioration in interaction was particularly strong in the area *most* important for high technical performance. This suggests that it is not a reduction in project communication per se that leads to less effective project performance; but rather it is an isolation from sources that can provide the most critical kinds of evaluation, information, and new ideas. Thus, overall effectiveness suffers when research project members fail to pay attention to events and information within the larger technical community outside the organization; or when development project members lose contact with client groups from marketing and manufacturing; or when technical-service project members do not interact among themselves.

Clearly—at least in the case of the groups studied here—there are strong relationships between longevity within a group and decreased levels of communication activity and project performance. In order to develop strategies that circumvent these unfortunate outcomes, the processes through which they occur must be understood in greater detail. What happens in long-term groups that leads to their being relatively cut off from sources of new ideas and information?

Essentially, project newcomers in the midst of socialization are trying to navigate their way through new and unfamiliar territories without the aid of adequate or even accurate perceptual maps. During this initial period, they are relatively more malleable and more susceptible to change, dependent as they are on other project members to help them define and interpret the numerous activities taking place around them. As they become more familiar with their project settings, however, they also become more capable of relying on their own perceptions and knowledge for interpreting events and executing their everyday project requirements. Having established their own social and task supports, their own outlooks and work identities, they become less easily changed and influenced.

If this process is allowed to continue among project members, healthy levels of self-reliance can easily degenerate into problematic levels of closed-mindedness. Rigidity in problem-solving activities—a kind of functional fixedness—may result from this, reducing the group's ability to react flexibly to changing conditions. Novel situations are either ignored or forced into established categories; new or changing circumstances either trigger old responses or none at all. New ideas, opportunities, or creative suggestions are greeted with resistance and easily disposed of with remarks like "it won't work," "it's infeasible," "it's too difficult," "we've never done that before," or "that's not our business."

Furthermore, the longer group members are called upon to follow and justify their problem-solving strategies and decisions, the more ingrained these approaches are likely to become. As as result, alternative ideas that were probably considered and discarded during previous discussions may never be reconsidered even though they may have become more appropriate or feasible. In fact, members may end up devoting much of their efforts to the preservation of their particular approaches against the encroachment of competing methods and negative evaluations. Essentially, they become overly committed to the continuation of their existing ideas and solutions, often without sufficient regard to their "true" applicability.

With this perspective, as one might suspect, the extent to which group members are willing or even feel the need to expose themselves to alternative ideas, solution strategies, or constructive criticism is likely to be diminished. A pattern of *increasing isolation* from external changes and new technological developments, coupled with a growing complacence about work-related challenges may be the result. Project groups with high levels of group longevity, then, appear to behave as if they possess so much expertise in their specialized technical areas that it is unlikely that outsiders might be producing important new ideas or information relevant to the performance of their project tasks. Rather than face the anxiety and discomfort inherent in learning or change, they tacitly assume that their abilities and experienced know-how are far better than those ideas or suggestions coming from outside their group. They become increasingly reliant on their own technology and knowledge base, creating an appearance to the outside world of decreased relevance, which leads to a decrease in the team's motivation to communicate with and respond to the outside. It is this isolation and more narrow focus which in turn leads to poorer performance.

Another explanation contributing to the reduced levels of project member interaction is the principle of *selective exposure,* the tendency for group members to communicate only with those whose ideas and outlooks are in accord with their own current interests, needs, and existing attitudes. And group members tend to become more alike over time. Just as it is sometimes said that close friends or husbands and wives seem to grow closer in appearance, so groups may take on a kind of collective viewpoint after interacting for an extended period. As members stabilize their work settings and patterns of communication, a greater degree of similarity is likely to emerge. This, in turn, leads to further stability in communication, and, therefore, even greater isolation from different-thinking others.

There is at least one advantage to this. People who think alike are able to communicate more effectively and economically. This advantage is more than outweighed, however, by the fact that such communication is likely to yield less creative and innovative outcomes than communications containing a variety of differing perspectives.

It should also be recognized that under these kinds of circumstances even the outside information that is processed by long-tenured groups may not be viewed in the most open or unbiased fashion. Many kinds of cognitive defenses and distortions are commonly used by members in *selectively perceiving* outside information in order to support and maintain their decisional policies and strategies. Such defenses can easily be used to argue against any disquieting information and evidence in order to maintain their present courses of action. Such selectivity can also result in a more restricted perspective of one's situation, which can be very detrimental to the group's overall effectiveness, for it often screens out vitally important information cues.

These trends of *increasing isolation, selective exposure,* and *selective perception* can all feed off each other in a kind of vicious circle, leaving group members in a state of greater and greater distance from new advances and ideas, and greater and greater reliance on an increasingly narrow and homogeneous set of alternatives. The prior curvilinear relationship between group age and project performance, therefore, can really be thought of as the composite result of two component forces, as shown in Figure 15.2. One component term rises rapidly as group age begins to increase, showing the positive effects of "team-building." Group members develop a common language and frame of reference. They have more concrete understandings of each others' capabilities, contributions, and working styles. And such improvements in communication and working relationships translate into higher levels of group performance.

At the same time, however, a decay component term sets in, resulting in part from the previously described problem-solving, communication, and cognitive processes that become more estab-lished, reinforced, and habitual as group members reduce uncertainty together within their group setting. This decay component, as shown in Figure 15.2, describes processes underneath the well-recognized "Not-Invented Here" or "NIH" syndrome in which groups gradually define themselves into a narrow field of specialization and convince themselves that they have a monopoly on knowledge and capability in their area of specialty. Between these two component curves lies the area for potentially influencing a project's innovative performance.

MANAGING FOR INNOVATION

Are these processes inevitable? Or can management alter the composition of current R&D groupings in order to minimize the effects of extended group longevity on project performance and still ensure an adequate level of stability for relatively smooth operation? What follows are a few suggestions toward the goal of managing for a continuously high level of innovation.

Employee perspectives and behaviors, and their subsequent effects on performance, can be significantly affected through the systematic and creative use of staffing and career decisions. For example, regular placement of new members into project groups may perform an energizing and destabilizing function—keeping the group longevity index from rising, thereby preventing the group from developing some of the tendencies described here (particularly isolation from critical information areas). New members have the advantage of fresh ideas and approaches, and of a fresh eye for old ones. With their active participation, established members might be kept responsive to the generation of new methods and behaviors as well as the reconsideration of alternatives that might otherwise be ignored. In short, project newcomers create a novelty-enhancing situation, challenging and improving the scope of existing methods and accumulated knowledge, provided of course that the newcomer(s) socialization process is appropriately managed (see Selection 3 in this volume, entitled

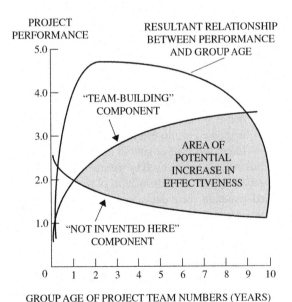

Figure 15.2. Relationship between group age and project performance analyzed by component forces.

"Organizational Socialization and the Reduction of Uncertainty").

Clearly, the longevity framework suggests that periodic additions or rotations can help *prevent* the onset of the stabilization processes associated with high longevity. Provided the socialization period is not overly conforming, project groups can simply remain in an innovation cycle. While prevention is clearly easier, it is also suggested that the replacement or reassignment of certain long-tenured professionals to different project groups may be necessary for improving the performance of high longevity teams as well as for keeping such groups stimulated, flexible, and vigilant with respect to their project environments. Continued growth and development comes from adaptations to new challenges, often requiring the abandonment of familiar and stable work patterns in favor of new ones.

Interestingly, managers are usually not aware of the tenure demographics of their project groups. In our studies, managers are usually unable to identify which of their projects have high levels of group longevity. Many are even surprised that any of their project teams have a mean group age of five or more years. Once the behavioral patterns characterizing the NIH decay component are described to them, however, they can easily spot the appropriate groups.

Of course, rotations and promotions are not always possible, especially when there is little organizational growth. As important as job mobility is, it is no doubt equally crucial to determine whether project groups can circumvent the effects of high longevity without new assignments or rejuvenation from new project members. To do this, we must learn considerably more about the effects of increasing job and group longevities. For example, in the study presented here, none of the long-tenured project groups was above average in project performance. Yet different trends might have emerged with different kinds of organizational climates, different personnel and promotional policies, different economic and marketing conditions, or even different types of organizational structures. Can project groups keep *themselves* energized and innovative over long periods, or are certain kinds

of structures and managerial practices needed to maintain effectiveness and high performance as a team ages? Just how deterministic are these curves and relationships?

In more recent and extensive project data collected from twelve different technology-based organizations involving more than 300 R&D project groups, of which approximately fifty have group longevity scores of five or more years, it turns out that a large number of these long-tenured groups were judged to have a high level of performance. The data are still being processed, but preliminary analyses seem to indicate that the nature of the project's supervision may be the most important factor differentiating the more effective long-tenured teams from those less effective. In particular, engineers belonging to the high-performing, long-tenured groups perceived their project supervisors to be superior in dealing with conflicts between groups and individuals, in obtaining necessary resources for project members, in setting project goals, and in monitoring the activities and progress of project members toward these goals. Furthermore, in performing these supervisory functions, project managers of the more effective long-tenured groups were *not* very participative in their approaches, instead, they were extremely demanding of their teams, challenging them to perform in new ways and directions. In fact, the most participative managers (as viewed by project members) were significantly less effective in managing teams with high group longevity. Our study also revealed that not all managers may be able to gain the creative performances out of long-term technical groups. Typically, the managers of the higher performing long-term groups had been with their teams less than 3 years and had come to this assignment with a strong history of prior managerial success. It was not their first managerial experience! To the contrary, most were well-respected technical managers who had "made things happen" and who had developed strong power bases and strong levels of senior managerial support within their R&D units or divisions. It was this combination of technical credibility and managerial respect and power that enabled these managers to be effective with their long-term stabilized R&D project teams.

These and other preliminary findings suggest the following strategies for managing project groups with high levels of long-term stability and group age:

1. More emphasis should be placed on the particular skills and abilities of the project manager. Members of long-tenured groups are more responsive to the nature of their supervision than to the intrinsic nature of their work content.

2. In terms of managerial styles, project managers should place less emphasis on participative management and more emphasis on direction and control. As long as members of long-tenured groups are unresponsive to the challenges in their tasks, participative management will only be related to job satisfaction—not project performance.

3. Project managers, on the other hand, should be very responsive to the challenging nature of their project's work. Consequently, they should be given considerable authority and freedom to execute their project responsibilities, but they, in turn, should be "tight-fisted" with respect to their subordinates.

In a sense, then, traditional managers may be effective for managing high group longevity teams. In a broader context, however, we need to learn how to manage workers, professionals, and project teams as they proceed through different stages of longevity. Clearly; different kinds of managerial styles and practices may be more appropriate at different stages of the process. Delegative or participative management, for example, may be very effective when individuals are highly responsive to their work, but much less successful when employees are not, as in the stabilization phase. As perspectives and responsiveness shift over time, the actions required of managers will vary as well. Managers may be effective to the extent that they can recognize and react to such developments. As in so many areas, it is the ability to manage change that seems most important in providing careers that keep employees responsive and organizations effective.

6

Managing Organizational Roles and Structures in Project Groups

16

Organizing and Leading "Heavyweight" Development Teams

KIM B. CLARK AND STEVEN C. WHEELWRIGHT

Effective product and process development requires the integration of specialized capabilities. Integrating is difficult in most circumstances, but is particularly challenging in large, mature firms with strong functional groups, extensive specialization, large numbers of people, and multiple, ongoing operating pressures. In such firms, development projects are the exception rather than the primary focus of attention. Even for people working on development projects, years of experience and the established systems—covering everything from career paths to performance evaluation, and from reporting relationships to breadth of job definitions—create both physical and organizational distance from other people in the organization. The functions themselves are organized in a way that creates further complications: the marketing organization is based on product families and market segments; engineering around functional disciplines and technical focus; and manufacturing on a

Kim B. Clark is Dean of the Harvard Business School and Professor of Business Administration. *Steven C. Wheelwright* is Professor of Business Administration at the Harvard Business School.

mix between functional and product market structures. The result is that in large, mature firms, organizing and leading an effective development effort is a major undertaking. This is especially true for organizations whose traditionally stable markets and competitive environments are threatened by new entrants, new technologies, and rapidly changing customer demands.

This article zeros in on one type of team structure—"heavyweight" project teams—that seems particularly promising in today's fast-paced world yet is strikingly absent in many mature companies. Our research shows that when managed effectively, heavyweight teams offer improved communication, stronger identification with and commitment to a project, and a focus on cross-functional problem solving. Our research also reveals, however, that these teams are not so easily managed and contain unique issues and challenges.

Heavyweight project teams are one of four types of team structures. We begin by describing each of them briefly. We then explore heavyweight teams in detail, compare them with the alternative forms, and point out specific challenges and their solutions in managing the heavyweight team organization. We conclude with an example of the changes necessary in individual behavior for heavyweight teams to be effective. Although heavyweight teams are a different way of organizing, they are more than a new structure; they represent a fundamentally different way of working. To the extent that both the team members and the surrounding organization recognize that phenomenon, the heavyweight team begins to realize its full potential.

TYPES OF DEVELOPMENT PROJECT TEAMS

Figure 16.1 illustrates the four dominant team structures we have observed in our studies of development projects: functional, lightweight, heavyweight, and autonomous (or tiger). These forms are described below, along with their associated project leadership roles, strengths, and weaknesses. Heavyweight teams are examined in detail in the subsequent section.

Functional Team Structure

In the traditional functional organization found in larger, more mature firms, people are grouped principally by discipline, each working under the direction of a specialized subfunction manager and a senior functional manager. The different subfunctions and functions coordinate ideas through detailed specifications all parties agree to at the outset, and through occasional meetings where issues that cut across groups are discussed. Over time, primary responsibility for the project passes sequentially—although often not smoothly—from one function to the next, a transfer frequently termed "throwing it over the wall."

The functional team structure has several advantages, and associated disadvantages. One strength is that those managers who control the project's resources also control task performance in their functional area; thus, responsibility and authority are usually aligned. However, tasks must be subdivided at the project's outset, i.e., the entire development process is decomposed into separable, somewhat independent activities. But on most development efforts, not all required tasks are known at the outset, nor can they all be easily and realistically subdivided. Coordination and integration can suffer as a result.

Another major strength of this approach is that, because most career paths are functional in nature until a general management level is reached, the work done on a project is judged, evaluated, and rewarded by the same subfunction and functional managers who make the decisions about career paths. The associated disadvantage is that individual contributions to a development project tend to be judged largely independently of overall project success. The traditional tenet cited is that individuals cannot be evaluated fairly on outcomes over which they have little or no control. But as a practical matter, that often means that no one directly involved in the details of the project is responsible for the results finally achieved.

Finally, the functional project organization brings specialized expertise to bear on the key technical issues. The same person or small group of people may be responsible for the design of a par-

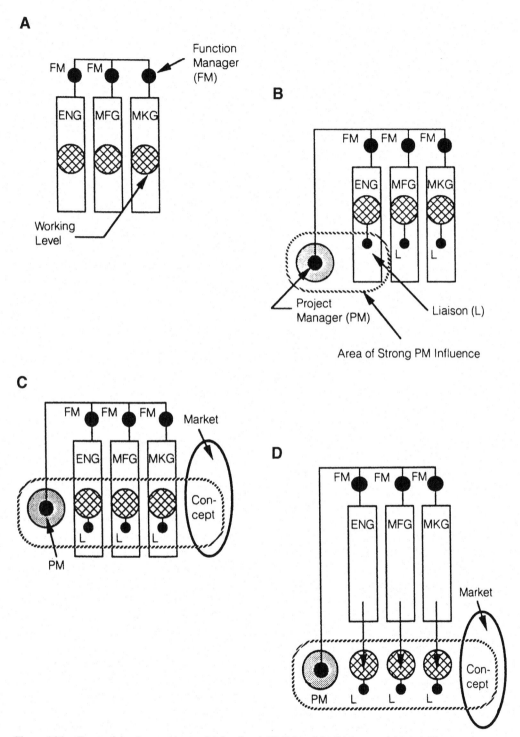

Figure 16.1. Types of development teams: (a) functional, (b) lightweight, (c) heavyweight, and (d) autonomous.

ticular component or subsystem over a wide range of development efforts. Thus the functions and subfunctions capture the benefits of prior experience and become the keepers of the organization's depth of knowledge while ensuring that it is systematically applied over time and across projects. The disadvantage is that every development project differs in its objectives and performance requirements, and it is unlikely that specialists developing a single component will do so very differently on one project than on another. The "best" component or subsystem is defined by technical parameters in the areas of their expertise rather than by overall system characteristics or specific customer requirements dictated by the unique market the development effort aims for.

Lightweight Team Structure

Like the functional structure, those assigned to the lightweight team reside physically in their functional areas, but each functional organization designates a liaison person to "represent" it on a project coordinating committee. These liaison representatives work with a "lightweight project manager," usually a design engineer or product marketing manager, who coordinates different functions' activities. This approach usually figures as an add-on to a traditional functional organization, with the functional liaison person having that role added to his or her other duties. The overall coordination assignment of lightweight project manager, however, tends not to be present in the traditional functional team structure.

The project manager is a "lightweight" in two important respects. First, he or she is generally a middle- or junior-level person who, despite considerable expertise, usually has little status or influence in the organization. Such people have spent a handful of years in a function, and this assignment is seen as a "broadening experience," a chance for them to move out of that function. Second, although they are responsible for informing and coordinating the activities of the functional organizations, the key resources (including engineers on the project) remain under the control of their respective functional managers. The lightweight project manager does not have power to reassign people or reallocate resources, and instead confirms schedules, updates time lines, and expedites across groups. Typically, such project leaders spend no more than 25% of their time on a single project.

The primary strengths and weaknesses of the lightweight project team are those of the functional project structure. But now at least one person over the course of the project looks across functions and seeks to ensure that individual tasks—especially those on the critical path—get done in a timely fashion, and that everyone is kept aware of potential cross-functional issues and what is going on elsewhere on this particular project.

Thus, improved communication and coordination are what an organization expects when moving from a functional to a lightweight team structure. Yet, because power still resides with the subfunction and functional managers, hopes for improved efficiency, speed, and project quality are seldom realized. Moreover, lightweight project leaders find themselves tolerated at best, and often ignored and even preempted. This can easily become a "no-win" situation for the individual thus assigned.

Heavyweight Team Structure

In contrast to the lightweight set-up, the heavyweight project manager has direct access to and responsibility for the work of all those involved in the project. Such leaders are "heavyweights" in two respects. First, they are senior managers within the organization; they may even outrank the functional managers. Hence, in addition to having expertise and experience, they also wield significant organizational clout. Second, heavyweight leaders have primary influence over the people working on the development effort and supervise their work directly through key functional people on the core teams. Often, the core group of people are dedicated and physically co-located with the heavyweight project leader. However, the longer-term career development of individual contributors continues to rest not with the project leader—although that heavyweight leader makes significant input to individual performance evaluations—but with the functional manager, because members are not assigned to a project team on a permanent basis.

The heavyweight team structure has a number of advantages and strengths, along with associated weaknesses. Because this team structure is observed much less frequently in practice and yet seems to have tremendous potential for a wide range of organizations, it will be discussed in detail in the next section.

Autonomous Team Structure

With the autonomous team structure, often called the "tiger team," individuals from the different functional areas are formally assigned, dedicated, and co-located to the project team. The project leader, a "heavyweight" in the organization, is given full control over the resources contributed by the different functional groups. Furthermore, that project leader becomes the sole evaluator of the contribution made by individual team members.

In essence, the autonomous team is given a "clean sheet of paper"; it is not required to follow existing organizational practices and procedures, but allowed to create its own. This includes establishing incentives and rewards as well as norms for behavior. However, the team will be held fully accountable for the final results of the project: success or failure is its responsibility and no one else's.

The fundamental strength of the autonomous team structure is focus. Everything the individual team members and the team leader do is concentrated on making the project successful. Thus, tiger teams can excel at rapid, efficient new product and new process development. They handle cross-functional integration in a particularly effective manner, possibly because they attract and select team participants much more freely than the other project structures.

Tiger teams, however, take little or nothing as "given"; they are likely to expand the bounds of their project definition and tackle redesign of the entire product, its components, and subassemblies, rather than looking for opportunities to utilize existing materials, designs, and organizational relationships. Their solution may be unique, making it more difficult to fold the resulting product and

process—and, in many cases, the team members themselves—back into the traditional organization upon project completion. As a consequence, tiger teams often become the birthplace of new business units or they experience unusually high turnover following project completion.

Senior managers often become nervous at the prospects of a tiger team because they are asked to delegate much more responsibility and control to the team and its project leader than under any of the other organization structures. Unless clear guidelines have been established in advance, it is extremely difficult during the project for senior managers to make midcourse corrections or exercise substantial influence without destroying the team. More than one team has "gotten away" from senior management and created major problems.

THE HEAVYWEIGHT TEAM STRUCTURE

The best way to begin understanding the potential of heavyweight teams is to consider an example of their success, in this case, Motorola's experience in developing its Bandit line of pagers.

The Bandit Pager Heavyweight Team

This development team within the Motorola Communications Sector was given a project charter to develop an automated, on-shore, profitable production operation for its high-volume Bravo pager line. (This is the belt-worn pager that Motorola sold from the mid-1980s into the early 1990s.) The core team consisted of a heavyweight project leader and a handful of dedicated and co-located individuals, who represented industrial engineering, robotics, process engineering, procurement, and product design/CIM. The need for these functions was dictated by the Bandit platform automation project and its focus on manufacturing technology with a minimal change in product technology. In addition, human resource and accounting/finance representatives were part of the core team. The human resource person was particularly active early on as subteam positions were defined and jobs posted throughout Motorola's Communications Sector, and played an important subsequent role in training and development of operating support people. The accounting/finance person was invaluable in "costing out" different options and performing detailed analyses of options and choices identified during the course of the project.

An eighth member of the core team was a Hewlett Packard employee. Hewlett Packard was chosen as the vendor for the "software backplane," providing an HP 3000 computer and the integrated software communication network that linked individual automated workstations, downloaded controls and instructions during production operations, and captured quality and other operating performance data. Because HP support was vital to the project's success, it was felt essential they be represented on the core team.

The core team was housed in a corner of the Motorola Telecommunications engineering/manufacturing facility. The team chose to enclose in glass the area where the automated production line was to be set up so that others in the factory could track the progress, offer suggestions, and adopt the lessons learned from it in their own production and engineering environments. The team called their project Bandit to indicate a willingness to "take" ideas from literally anywhere.

The heavyweight project leader, Scott Shamlin, who was described by team members as "a crusader," "a renegade," and "a workaholic," became the champion for the Bandit effort. A hands-on manager who played a major role in stimulating and facilitating communication across functions, he helped to articulate a vision of the Bandit line, and to infuse it into the detailed work of the project team. His goal was to make sure the new manufacturing process worked for the pager line, but would provide real insight for many other production lines in Motorola's Communications Sector.

The Bandit core team started by creating a contract book that established the blueprint and work plan for the team's efforts and its performance expectations; all core team members and senior management signed on to this document. Initially, the team's executive sponsor—although not formally identified as such—was George Fisher, the

Sector Executive. He made the original investment proposal to the Board of Directors and was an early champion and supporter, as well as direct supervisor in selecting the project leader and helping get the team underway. Subsequently, the vice president and general manager of the Paging Products division filled the role of executive sponsor.

Throughout the project, the heavyweight team took responsibility for the substance of its work, the means by which it was accomplished, and its results. The project was completed in 18 months as per the contract book, which represented about half the time of a normal project of such magnitude. Further, the automated production operation was up and running with process tolerances of five sigma (i.e., the degree of precision achieved by the manufacturing processes) at the end of 18 months. Ongoing production verified that the cost objectives (substantially reduced direct costs and improved profit margins) had indeed been met, and product reliability was even higher than the standards already achieved on the off-shore versions of the Bravo product. Finally, a variety of lessons were successfully transferred to other parts of the Sector's operations, and additional heavyweight teams have proven the viability and robustness of the approach in Motorola's business and further refined its effectiveness throughout the corporation.

The Challenge of Heavyweight Teams

Motorola's experience underscores heavyweight teams' potential power, but it also makes clear that creating an effective heavyweight team capability is more than merely selecting a leader and forming a team. By their very nature—being product (or process) focused, and needing strong, independent leadership, broad skills and cross-functional perspective, and clear missions—heavyweight teams may conflict with the functional organization and raise questions about senior management's influence and control. And even the advantages of the team approach bring with them potential disadvantages that may hurt development performance if not recognized and averted.

Take, for example, the advantages of ownership and commitment, one of the most striking advantages of the heavyweight team. Identifying with the product and creating a sense of esprit de corps motivate core team members to extend themselves and do what needs to be done to help the team succeed. But such teams sometimes expand the definition of their role and the scope of the project, and they get carried away with themselves and their abilities. We have seen heavyweight teams turn into autonomous tiger teams and go off on a tangent because senior executives gave insufficient direction and the bounds of the team were only vaguely specified at the outset. And even if the team stays focused, the rest of the organization may see themselves as "second class." Although the core team may not make that distinction explicit, it happens because the team has responsibilities and authority beyond those commonly given to functional team members. Thus, such projects inadvertently can become the "haves" and other, smaller projects the "have-nots" with regard to key resources and management attention.

Support activities are particularly vulnerable to an excess of ownership and commitment. Often the heavyweight team will want the same control over secondary support activities as it has over the primary tasks performed by dedicated team members. When waiting for prototypes to be constructed, analytical tests to be performed, or quality assurance procedures to be conducted, the team's natural response is to "demand" top priority from the support organization or to be allowed to go outside and subcontract to independent groups. While these may sometimes be the appropriate choices, senior management should establish make-buy guidelines and clear priorities applicable to all projects—perhaps changing service levels provided by support groups (rather than maintaining the traditional emphasis on resource utilization)—or have support groups provide capacity and advisory technical services but let team members do more of the actual task work in those support areas. Whatever actions the organization takes, the challenge is to achieve a balance between the needs of the individual project and the needs of the broader organization.

Another advantage the heavyweight team brings is the integration and integrity it provides through a system solution to a set of customer needs. Getting all of the components and subsystems to complement one another and to address effectively the fundamental requirements of the core customer segment can result in a winning platform product and/or process. The team achieves an effective system design by using generalist skills applied by broadly trained team members, with fewer specialists and, on occasion, less depth in individual component solutions and technical problem solving.

The extent of these implications is aptly illustrated by the nature of the teams Clark and Fujimoto studied in the auto industry.[1] They found that for U.S. auto firms in the mid-1980s, typical platform projects—organized under a traditional functional or lightweight team structure—entailed full-time work for several months by approximately 1500 engineers. In contrast, a handful of Japanese platform projects—carried out by heavyweight teams—utilized only 250 engineers working full-time for several months. The implications of 250 versus 1500 full-time equivalents (FTEs) with regard to breadth of tasks, degree of specialization, and need for coordination are significant and help explain the differences in project results as measured by product integrity, development cycle time, and engineering resource utilization.

But that lack of depth may disclose a disadvantage. Some individual components or subassemblies may not attain the same level of technical excellence they would under a more traditional functional team structure. For instance, generalists may develop a windshield wiper system that is complementary with and integrated into the total car system and its core concept. But they also may embed in their design some potential weaknesses or flaws that might have been caught by a functional team of specialists who had designed a long series of windshield wipers. To counter this potential disadvantage, many organizations order more testing of completed units to discover such possible flaws and have components and subassemblies reviewed by expert specialists. In some cases, the quality assurance function has expanded its role to make sure sufficient technical specialists review designs at appropriate points so that such weaknesses can be minimized.

Managing the Challenges of Heavyweight Teams

Problems with depth in technical solutions and allocations of support resources suggest the tension that exists between heavyweight teams and the functional groups where much of the work gets done. The problem with the teams exceeding their bounds reflects in part how teams manage themselves, in part, how boundaries are set, and in part the ongoing relationship between the team and senior management. Dealing with these issues requires mechanisms and practices that reinforce the team's basic thrust—ownership, focus, system architecture, integrity—and yet improve its ability to take advantage of the strengths of the supporting functional organization—technical depth, consistency across projects, senior management direction. We have grouped the mechanisms and problems into six categories of management action: the project charter, the contract, staffing, leadership, team responsibility, and the executive sponsor.

The Project Charter. A heavyweight project team needs a clear mission. A way to capture that mission concisely is in an explicit, measurable project charter that sets broad performance objectives and usually is articulated even before the core team is selected. Thus, joining the core team includes accepting the charter established by senior management. A typical charter for a heavyweight project would be the following:

> The resulting product will be selected and ramped by Company X during Quarter 4 of calendar year 1991, at a minimum of a 20% gross margin.

This charter is representative of an industrial products firm whose product goes into a system sold by its customers. Company X is the leading customer for a certain family of products, and this project is dedicated to developing the next generation platform offering in that family. If the heavyweight program results in that platform product being chosen by the leading customer in the segment by a certain date and at a certain gross margin, it will have demonstrated that the next generation platform is not only viable, but likely to be very successful over the next three to five years. Industries and settings where such a charter might be found would include a microprocessor being developed for a new computer system, a diesel engine for the heavy equipment industry, or a certain type of slitting and folding piece of equipment for the newspaper printing press industry. Even in a medical diagnostics business with hundreds of customers, a goal of "capturing 30% of market purchases in the second 12 months during which the product is offered" sets a clear charter for the team.

The Contract Book. Whereas a charter lays out the mission in broad terms, the contract book defines, in detail, the basic plan to achieve the stated goal. A contract book is created as soon as the core team and heavyweight project leader have been designated and given the charter by senior management. Basically, the team develops its own detailed work plan for conducting the project, estimates the resources required, and outlines the results to be achieved and against which it is willing to be evaluated. (The table of contents of a typical

---

Heavyweight Team, Contract Book— Major Sections

- Executive Summary
- Business Plan and Purposes
- Development Plan
   —Schedule
   —Materials
   —Resources
- Product Design Plan
- Quality Plan
- Manufacturing Plan
- Project Deliverables
- Performance Measurement and Incentives

heavyweight team contract book are shown in Heavyweight Team, Contract Book—Major Sections.) Such documents range from 25 to 100 pages, depending on the complexity of the project and level of detail desired by the team and senior management before proceeding. A common practice following negotiation and acceptance of this contract is for the individuals from the team and senior management to sign the contract book as an indication of their commitment to honor the plan and achieve those results.

The core team may take anywhere from a long week to a few months to create and complete the contract book; Motorola, for example, after several years of experience, has decided that a maximum of seven days should be allowed for this activity. Having watched other heavyweight teams—particularly in organizations with no prior experience in using such a structure—take up to several months, we can appreciate why Motorola has nicknamed this the "blitz phase" and decided that the time allowed should be kept to a minimum.

Staffing. As suggested in Figure 16.1, a heavyweight team includes a group of core cross-functional team members who are dedicated (and usually physically co-located) for the duration of the development effort. Typically there is one core team member from each primary function of the organization; for instance, in several electronics firms we have observed core teams consisting of six functional participants—design engineering, marketing, quality assurance, manufacturing, finance, and human resources. (Occasionally, design will be represented by two core team members, one each for hardware and software engineering.) Individually, core team members represent their functions and provide leadership for their function's inputs to the project. Collectively, they constitute a management team that works under the direction of the heavyweight project manager and takes responsibility for managing the overall development effort.

While other participants—especially from design engineering early on and manufacturing later on—may frequently be dedicated to a heavyweight

team for several months, they usually are not made part of the core team though they may well be co-located and, over time, develop the same level of ownership and commitment to the project as core team members. The primary difference is that the core team manages the total project and the coordination and integration of individual functional efforts, whereas other dedicated team members work primarily within a single function or subfunction.

Whether these temporarily dedicated team members are actually part of the core team is an issue firms handle in different ways, but those with considerable experience tend to distinguish between core and other dedicated (and often co-located) team members. The difference is one of management responsibility for the core group that is not shared equally by the others. Also, it is primarily the half a dozen members of the core group who will be dedicated throughout the project, with other contributors having a portion of their time reassigned before this heavyweight project is completed.

Whether physical colocation is essential is likewise questioned in such teams. We have seen it work both ways. Given the complexity of development projects, and especially the uncertainty and ambiguity often associated with those assigned to heavyweight teams, physical colocation is preferable to even the best of on-line communication approaches. Problems that arise in real time are much more likely to be addressed effectively with all of the functions represented and present than when they are separate and must either wait for a periodic meeting or use remote communication links to open up cross-functional discussions.

A final issue is whether an individual can be a core team member on more than one heavyweight team simultaneously. If the rule for a core team member is that 70% or more of their time must be spent on the heavyweight project, then the answer to this question is no. Frequently, however, a choice must be made between someone being on two core teams—for example, from the finance or human resource function—or putting a different individual on one of those teams who has neither the experience nor stature to be a full peer with the other core team members. Most experienced organizations we have seen opt to put the same person on two teams to ensure the peer relationship and level of contribution required, even though it means having one person on two teams and with two desks. They then work diligently to develop other people in the function so that multiple team assignments will not be necessary in the future.

Sometimes multiple assignments will also be justified on the basis that a function such as finance does not need a full-time person on a project. In most instances, however, a variety of potential value-adding tasks exist that are broader than finance's traditional contribution. A person largely dedicated to the core team will search for those opportunities and the project will be better because of it. The risk of allowing core team members to be assigned to multiple projects is that they are neither available when their inputs are most needed nor as committed to project success as their peers. They become secondary core team members, and the full potential of the heavyweight team structure fails to be realized.

Project Leadership. Heavyweight teams require a distinctive style of leadership. A number of differences between lightweight and heavyweight project managers are highlighted in Project Manager Profile. Three of those are particularly distinctive. First, a heavyweight leader manages, leads, and evaluates other members of the core team, and is also the person to whom the core team reports throughout the project's duration. Another characteristic is that rather than being either neutral or a facilitator with regard to problem solving and conflict resolution, these leaders see themselves as championing the basic concept around which the platform product and/or process is being shaped. They make sure that those who work on subtasks of the project understand that concept. Thus they play a central role in ensuring the system integrity of the final product and/or process.

Finally, the heavyweight project manager carries out his or her role in a very different fashion than the lightweight project manager. Most lightweights spend the bulk of their time working at a desk, with paper. They revise schedules, get frequent updates, and encourage people to meet previously agreed upon deadlines. The heavyweight project manager spends little time at a desk, is out talking to project contributors, and makes sure that decisions are made and implemented whenever and wherever needed. Some of the ways in which the heavyweight project manager achieves project results are highlighted by the five roles illustrated in

---

Project Manager Profile

	Lightweight (limited)	Heavyweight (extensive)
Span of coordination responsibilities	⊢———————————————————⊣	
Duration of responsibilities	⊢———————————————————⊣	
Responsible for specs, cost, layout, components	⊢———————————————————⊣	
Working level contact with engineers	⊢———————————————————⊣	
Direct contact with customers	⊢———————————————————⊣	
Multilingual/multi-disciplined skills	⊢———————————————————⊣	
Role in conflict resolution	⊢———————————————————⊣	
Marketing imagination/concept champion	⊢———————————————————⊣	
Influence in: engineering	⊢———————————————————⊣	
marketing	⊢———————————————————⊣	
manufacturing	⊢———————————————————⊣	

The Heavyweight Project Manager

Role	Description
Direct Market Interpreter	First hand information, dealer visits, auto shows, has own marketing budget, market study team, direct contact and discussions with customers
Multilingual Translator	Fluency in language of customers, engineers, marketers, stylists; translator between customer experience/requirements and engineering specifications
"Direct" Engineering Manager	Direct contact, orchestra conductor, evangelist of conceptual integrity and coordinator of component development; direct eye-to-eye discussions with working level engineers; shows up in drafting room, looks over engineers' shoulders
Program Manager "in motion"	Out of the office, not too many meetings, not too much paperwork, face-to-face communication, conflict resolution manager
Concept Infuser	Concept guardian, confronts conflicts, not only reacts but implements own philosophy, ultimate decision maker, coordination of details and creation of harmony

a Heavyweight Project Manager on a platform development project in the auto industry.

The *first role* of the heavyweight project manager is to provide for the team a direct interpretation of the market and customer needs. This involves gathering market data directly from customers, dealers, and industry shows, as well as through systematic study and contact with the firm's marketing organization. A *second role* is to become a multilingual translator, not just taking marketing information to the various functions involved in the project, but being fluent in the language of each of those functions and making sure the translation and communication going on among the functions—particularly between customer needs and product specifications—are done effectively.

A *third role* is the direct engineering manager, orchestrating, directing, and coordinating the various engineering subfunctions. Given the size of many development programs and the number of types of engineering disciplines involved, the project manager must be able to work directly with each engineering subfunction on a day-to-day basis and ensure that their work will indeed integrate

and support that of others, so the chosen product concept can be effectively executed.

A *fourth role* is best described as staying in motion: out of the office conducting face-to-face sessions, and highlighting and resolving potential conflicts as soon as possible. Part of this role entails energizing and pacing the overall effort and its key subparts. A *final role* is that of concept champion. Here the heavyweight project manager becomes the guardian of the concept and not only reacts and responds to the interests of others, but also sees that the choices made are consistent and in harmony with the basic concept. This requires a careful blend of communication and teaching skills so that individual contributors and their groups understand the core concept, and sufficient conflict resolution skills to ensure that any tough issues are addressed in a timely fashion.

It should be apparent from this description that heavyweight project managers earn the respect and right to carry out these roles based on prior experience, carefully developed skills, and status earned over time, rather than simply being designated "leader" by senior management. A qualified heavyweight project manager is a prerequisite to an effective heavyweight team structure.

Team Member Responsibilities. Heavyweight team members have responsibilities beyond their usual functional assignment. As illustrated in Responsibilities of Heavyweight Core Team Members, these are of two primary types. Functional hat responsibilities are those accepted by the individual core team member as a representative of his or her function. For example, the core team member from marketing is responsible for ensuring that appropriate marketing expertise is brought to the project, that a marketing perspective is provided on all key issues, that project sub-objectives dependent on the marketing function are met in a timely fashion, and that marketing issues that impact other functions are raised proactively within the team.

But each core team member also wears a team hat. In addition to representing a function, each member shares responsibility with the heavyweight project manager for the procedures followed by the team, and for the overall results that those procedures deliver. The core team is accountable for the success of the project, and it can blame no one but itself if it fails to manage the project, execute the tasks, and deliver the performance agreed upon at the outset.

Finally, beyond being accountable for tasks in their own function, core team members are responsible for how those tasks are subdivided, organized, and accomplished. Unlike the traditional functional development structure, which takes as given the subdivision of tasks and the means by which those tasks will be conducted and completed, the core heavyweight team is given the power and responsibility to change the substance of those tasks to improve the performance of the project. Since this is a role that core team members do not lay under a lightweight or functional team structure, it is often the most difficult for them to accept fully and learn to apply. It is essential, however, if the heavyweight team is to realize its full potential.

The Executive Sponsor. With so much more accountability delegated to the project team,

---

Responsibilities of Heavyweight Core Team Members

Functional Hat Accountabilities

- Ensuring functional expertise on the project
- Representing the functional perspective on the project
- Ensuring that subobjectives are met that depend on their function
- Ensuring that functional issues impacting the team are raised pro-actively within the team

Team Hat Accountabilities

- Sharing responsibility for team results
- Reconstituting tasks and content
- Establishing reporting and other organizational relationships
- Participating in monitoring and improving team performance
- Sharing responsibility for ensuring effective team processes
- Examining issues from an executive point of view (Answering the question, "Is this the appropriate business response for the company?")
- Understanding, recognizing, and responsibly challenging the boundaries of the project and team process

---

establishing effective relationships with senior management requires special mechanisms. Senior management needs to retain the ability to guide the project and its leader while empowering the team to lead and act, a responsibility usually taken by an executive sponsor—typically the vice president of engineering, marketing, or manufacturing for the business unit. This sponsor becomes the coach and mentor for the heavyweight project leader and core team, and seeks to maintain close, ongoing contact with the team's efforts. In addition, the executive sponsor serves as a liaison. If other members of senior management—including the functional heads—have concerns or inputs to voice, or need current information on project status, these are communicated through the executive sponsor. This reduces the number of mixed signals received by the team and clarifies for the organization the reporting and evaluation relationship between the team and senior management. It also encourages the executive sponsor to set appropriate limits and bounds on the team so that organizational surprises are avoided.

Often the executive sponsor and core team identify those areas where the team clearly has decision-making power and control, and they distinguish them from areas requiring review. An electronics firm that has used heavyweight teams for some time dedicates one meeting early on between the executive sponsor and the core team to generating a list of areas where the executive sponsor expects to provide oversight and be consulted; these areas are of great concern to the entire executive staff and team actions may well raise policy issues for the larger organization. In this firm, the executive staff wants to maintain some control over:

- resource commitment—head count, fixed costs, and major expenses outside the approved contract book plan;

- pricing for major customers and major accounts;

- potential slips in major milestone dates (the executive sponsor wants early warning and recovery plans);

- plans for transitioning from development project to operating status,

- thorough reviews at major milestones or every three months, whichever occurs sooner;

- review of incentive rewards that have company-wide implications for consistency and equity; and

- cross-project issues such as resource optimization, prioritization, and balance.

Identifying such areas at the outset can help the executive sponsor and the core team better carry out their assigned responsibilities. It also helps other executives feel more comfortable working through the executive sponsor, since they know these "boundary issues" have been articulated and are jointly understood.

THE NECESSITY OF FUNDAMENTAL CHANGE

Compared to a traditional functional organization, creating a team that is "heavy"—one with effective leadership, strong problem-solving skills and the ability to integrate across functions—requires basic changes in the way development works. But it also requires change in the fundamental behavior of engineers, designers, manufacturers, and marketers in their day-to-day work. An episode in a computer company with no previous experience with heavyweight teams illustrates the depth of change required to realize fully these teams' power.[2]

Two teams, A and B, were charged with development of a small computer system and had market introduction targets within the next twelve months. While each core team was co-located and held regular meetings, there was one overlapping core team member (from finance/accounting). Each team was charged with developing a new computer system for their individual target markets but by chance, both products were to use an identical, custom-designed microprocessor chip in addition to other unique and standard chips.

The challenge of changing behavior in creating an effective heavyweight team structure was highlighted when each team sent this identical, custom-designed chip—the "supercontroller"—to the vendor for pilot production. The vendor quoted a 20-week turnaround to both teams. At that time, the supercontroller chip was already on the critical path for Team B, with a planned turnaround of 11 weeks. Thus, every week saved on that chip would save one week in the overall project schedule, and Team B already suspected that it would be late in meeting its initial market introduction target date. When the 20-week vendor lead time issue first came up in a Team B meeting, Jim, the core team member from engineering, responded very much as he had on prior, functionally structured development efforts: because initial prototypes were engineering's responsibility, he reported that they were working on accelerating the delivery date, but that the vendor was a large company, with whom the computer manufacturer did substantial business, and known for its slowness. Suggestions from other core team members on how to accelerate the delivery were politely rebuffed, including one to have a senior executive contact their counterpart at the vendor. Jim knew the traditional approach to such issues and did not perceive a need, responsibility, or authority to alter it significantly.

For Team A, the original quote of 20-week turnaround still left a little slack, and thus initially the supercontroller chip was not on the critical path. Within a couple of weeks, however, it was, given other changes in the activities and schedule, and the issue was immediately raised at the team's weekly meeting. Fred, the core team member from manufacturing (who historically would not have been involved in an early engineering prototype), stated that he thought the turnaround time quoted was too long and that he would try to reduce it. At the next meeting, Fred brought some good news: through discussions with the vendor, he had been able to get a commitment that pulled in the delivery of the supercontroller chip by 11 weeks! Furthermore, Fred thought that the quote might be reduced even further by a phone call from one of the computer manufacturer's senior executives to a contact of his at the vendor.

Two days later, at a regular Team B meeting, the supercontroller chip again came up during the status review, and no change from the original schedule was identified. Since the finance person, Ann, served on both teams and had been present at Team A's meeting, she described Team A's success in reducing the cycle time. Jim responded that he was aware that Team A had made such efforts, but that the information was not correct, and the original 20-week delivery date still held. Furthermore, Jim indicated that Fred's efforts (from Team A) had caused some uncertainty and disruption internally, and in the future it was important that Team A not take such initiatives before coordinating with Team B. Jim stated that this was particularly true when an outside vendor was involved, and he closed the topic by saying that a meeting to clear up the situation would be held that afternoon with Fred from Team A and Team B's engineering and purchasing people.

The next afternoon, at his Team A meeting, Fred confirmed the accelerated delivery schedule for the supercontroller chip. Eleven weeks had indeed been clipped out of the schedule to the benefit of both Teams A and B. Subsequently, Jim confirmed the revised schedule would apply to his team as well, although he was displeased that Fred had abrogated "standard operating procedure" to achieve it. Curious about the differences in perspective, Ann decided to learn more about why Team A had identified an obstacle and removed it from its path, yet Team B had identified an identical obstacle and failed to move it at all.

As Fred pointed out, Jim was the engineering manager responsible for development of the supercontroller chip; he knew the chip's technical requirements, but had little experience dealing with chip vendors and their production processes. (He had long been a specialist.) Without that experience, he had a hard time pushing back against the vendor's "standard line." But Fred's manufacturing experience with several chip vendors enabled him to calibrate the vendor's dates against his best-case experience and understand what the vendor needed to do to meet a substantially earlier commitment.

Moreover, because Fred had bought into a clear team charter, whose path the delayed chip would block, and because he had relevant experience, it did not make sense to live with the vendor's initial commitment, and thus he sought to change it. In contrast, Jim—who had worked in the traditional functional organization for many years—saw vendor relations on a pilot build as part of his functional job, but did not believe that contravening standard practices to get the vendor to shorten the cycle time was his responsibility, within the range of his authority, or even in the best long-term interest of his function. He was more concerned with avoiding conflict and not roiling the water than with achieving the overarching goal of the team.

It is interesting to note that in Team B, engineering raised the issue, and, while unwilling to take aggressive steps to resolve it, also blocked others' attempts. In Team A, however, while the issue came up initially through engineering, Fred in manufacturing proactively went after it. In the case of Team B, getting a prototype chip returned from a vendor was still being treated as an "engineering responsibility," whereas in the case of Team A, it was treated as a "team responsibility." Since Fred was the person best qualified to attack that issue, he did so.

Both Team A and Team B had a charter, a contract, a co-located core team staffed with generalists, a project leader, articulated responsibilities, and an executive sponsor. Yet Jim's and Fred's understanding of what these things meant for them personally and for the team at the detailed, working level was quite different. While the teams had been through similar training and team startup processes, Jim apparently saw the new approach as a different organizational framework within which work would get done as before. In contrast, Fred seemed to see it as an opportunity to work in a different way—to take responsibility for reconfiguring tasks, drawing on new skills, and reallocating resources, where required, for getting the job done in the best way possible.

Although both teams were "heavyweight" in theory, Fred's team was much "heavier" in its operation and impact. Our research suggests that heaviness is not just a matter of structure and mechanism, but of attitudes and behavior. Firms that try to create heavyweight teams without making the deep changes needed to realize the power in the team's structure will find this team approach problematic. Those intent on using teams for platform projects and willing to make the basic changes we have discussed here, can enjoy substantial advantages of focus, integration, and effectiveness.

NOTES

1. See Kim B. Clark and Takahiro Fujimoto, *Product Development Performance* (Boston, MA: Harvard Business School Press, 1991).

2. Adapted from a description provided by Dr. Christopher Meyer, Strategic Alignment Group, Los Altos, CA.

17

Lessons for an Accidental Profession

JEFFREY K. PINTO AND OM P. KHARBANDA

Projects and project management are the wave of the future in global business. Increasingly technically complex products and processes, vastly shortened time-to-market windows, and the need for cross-functional expertise make project management an important and powerful tool in the hands of organizations that understand its use. But the expanded use of such techniques is not always being met by a concomitant increase in the pool of competent project managers. Unfortunately, and perhaps ironically, it is the very popularity of project management that presents many organizations with their most severe challenges. They often belatedly discover that they simply do not have sufficient numbers of the sorts of competent project managers who are often the key driving force behind successful product or service development. Senior managers in many companies readily acknowledge the ad hoc manner in which most project managers acquire their skills, but they are unsure how to better develop and provide for a supply of well-trained project leaders for the future.

In this article, we seek to offer a unique perspective on this neglected species. Though much has been written on how to improve the process of project management, less is known about the sorts of skills and challenges that specifically characterize project managers. What we do know tends to offer a portrait of successful project managers as strong leaders, possessing a variety of problem-solving, communication, motivational, visionary, and team-building skills. Authors such as Posner (1987), Einsiedel (1987), and Petterson (1991) are correct: Project managers are a special breed. Man-

aging projects is a unique challenge that requires a strategy and methodology all its own. Perhaps most important, it requires people willing to function as leaders in every sense of the term. They must not only chart the appropriate course, but provide the means, the support, and the confidence for their teams to attain these goals. Effective project managers often operate less as directive and autocratic decision makers than as facilitators, team members, and cheerleaders. In effect, the characteristics we look for in project managers are varied and difficult to pin down. Our goal is to offer some guidelines for an accidental profession, based on our own experiences and interviews with a number of senior project managers—most of whom had to learn their own lessons the hard way.

"Accidental" Project Managers

Project managers occupy a unique and often precarious position within many firms. Possessing little formal authority and forced to operate outside the traditional organizational hierarchy, they quickly and often belatedly learn the real limits of their power. It has been said that an effective project manager is the kingpin, but not the king. They are the bosses, it is true, but often in a loosely defined way. Indeed, in most firms they may lack the authority to conduct performance appraisals and offer incentives and rewards to their subordinates. As a result, their management styles must be those of persuasion and influence, rather than coercion and command.

Because of these and other limitations on the flexibility and power of project managers, project

Reprinted from *Business Horizons*, March–April, 1995, pp. 41–50. Copyright 1995 by the Foundation for the School of Business at Indiana University. Used with permission.

Jeffrey K. Pinto is Professor of Management at Penn State University in Erie, PA. *Om P. Kharbanda* is an independent management consultant in Bombay, India.

management has rightly been termed the "accidental profession" by more than one writer. There are two primary reasons for this sobriquet. First, few formal or systematic programs exist for selecting and training project managers, even within firms that specialize in project management work. This results at best in ad hoc training that may or may not teach these people the skills they need to succeed. Most project managers fall into their responsibilities by happenstance rather than by calculation. Second, as Frame (1987) cogently observed, few individuals grow up with the dream of one day becoming a project manager. It is neither a well-defined nor a well-understood career path within most modern organizations. Generally, the role is thrust upon people, rather than being sought.

Consider the typical experiences of project managers within many corporations. Novice managers, new to the company and its culture, are given a project to complete with the directive to operate within a set of narrowly defined constraints. These constraints most commonly include a specified time frame for completion, a budget, and a set of performance characteristics. Those who are able to quickly master the nature of their myriad duties succeed; those who do not generally fail. This "fly or die" mentality goes far toward creating an attitude of fear among potential project managers. Generation after generation of them learn their duties the hard way, often after having either failed completely or stumbled along from one crisis to another. The predictable result is wasteful: failed projects; managers battling entrenched bureaucracy and powerful factions; money, market opportunities, and other resources irretrievably lost to the company.

The amazing part of this scenario is that it is repeated again and again in company after company. Rather than treating project management as the unique and valuable discipline it is, necessitating formal training and selection policies, many companies continue to repeat their past mistakes. This almost leads one to believe they implicitly view experience and failure as the best teacher.

We need to shed light on the wide range of demands, opportunities, travails, challenges, and vexations that are part of becoming a better project manager. Many of the problems these individuals struggle with every day are far more managerial or behavioral in nature than technical. Such behavioral challenges are frequently vexing, and though they can sometimes seem inconsequential, they have a tremendous impact on the successful implementation of projects. For example, it does not take long for many project managers to discover exactly how far their personal power and status will take them in interacting with the rest of the organization. Hence, an understanding of influence tactics and political behavior is absolutely essential. Unfortunately, notice project managers are rarely clued into this important bit of information until it is too late—until, perhaps, they have appealed through formal channels for extra resources and been denied.

Consider the following examples:

• A long-distance telephone company whose CEO became so enamored of the concept of high-profile project teams—or "skunkworks," as they have come to be called—that he assigned that title to the few most highly visible, strategically important projects. Quickly, both senior and middle managers in departments across the organization came to realize that the only way to get their pet projects the resources necessary to succeed was to redesignate all new projects as "skunkworks." At last report, there were more than 75 high-profile skunkworks projects whose managers report directly to the CEO. The company now has severe difficulties in making research allocation decisions among its projects and routinely underfunds some vital projects while overfunding other, less important ones.

• A large computer hardware manufacturer has been dominated by the members of the hardware engineering department to such an extent that practically all new product ideas originate internally, within the department. By the time marketing personnel (sneeringly called "order takers" by the engineering department) are brought on board, they are presented with a fait accompli: a finished product they are instructed to sell. Marketing managers are now so cynical about new projects that they usually do not even bother sending a representative to new product development team meetings.

• A medium-sized manufacturing firm made it a policy to reward and punish project managers on the basis of their ability to bring projects in on time and under budget. These project managers were never held to any requirement that the project be accepted by its clients or become commercially successful. They quickly learned that their rewards were simply tied to satisfying the cost accountants, so they began to cut corners and make decisions that seriously undermined product quality.

• Projects in one division of a large, multinational corporation are routinely assigned to new managers who often have less than one year of experience with the company. Given a project scheduling software package and the telephone number of a senior project manager to be used "only in emergencies," they are instructed to form their project teams and begin the development process without any formal training or channels of communication to important clients and functional groups. Not surprisingly, senior managers at this company estimate that fewer than 30 percent of new product development efforts are profitable. Most take so long to develop, or incur such high cost overruns, that they are either abandoned before scheduled introduction or never live up to their potential in the marketplace.

This ad hoc approach to project management—coupled, as it frequently is, with an on-the-job training philosophy—is pervasive. It is also pernicious. Under the best of circumstances, project managers are called upon to lead, coordinate, plan, and control a diverse and complex set of processes and people in the pursuit of achieving project objectives. To hamper them with inadequate training and unrealistic expectations is to unnecessarily penalize them before they can begin to operate with any degree of confidence or effectiveness. The successful management of projects is simultaneously a human and technical challenge, requiring a farsighted strategic outlook coupled with the flexibility to react to conflicts and trouble areas as they arise on a daily basis. The project managers who are ultimately successful at their profession must learn to deal with and anticipate the constraints on their project team and personal freedom of action while consistently keeping their eyes on the ultimate prize.

From Whence Comes the Challenge?

One of the most intriguing and challenging aspects of project management lies in the relationship of project teams to the rest of the parent organization. With the exception of companies that are set up with matrix or project structures, most firms using project management techniques employ some form of standard functional structure. When project teams are added to an organization, the structural rules change dramatically. The vast majority of personnel who serve on project teams do so while maintaining links back to their functional departments. In fact, they typically split their time between the project and their functional duties.

The temporary nature of projects, combined with the very real limitations on power and discretion most project managers face, constitutes the core challenge of managing projects effectively. Clearly the very issues that characterize projects as distinct from functional work also illustrate the added complexity and difficulties they create for project managers. For example, within a functional department it is common to find people with more homogenous backgrounds. This means that the finance department is staffed with finance people, the marketing department is made up of marketers, and so on. On the other hand, most projects are constructed from special, cross-functional teams composed of representatives from each of the relevant functional departments, who bring their own attitudes, time frames, learning, past experiences, and biases to the team. Creating a cohesive and potent team out of this level of heterogeneity presents a challenge for even the most seasoned and skilled of project managers.

But what is the ultimate objective? What determines a successful project and how does it differ from projects we may rightfully consider to have failed? Any seasoned project manager will usually tell you that a successful project is one that has come in on time, has remained under budget, and performs as expected (that is, it conforms to specifications). Recently, though, there has been a reassessment of this traditional model for project success. The old triple constraint is rapidly being replaced by a new model, invoking a fourth hurdle for project success: client satisfaction. This means that a project is only successful if it satisfies the needs of its intended user. As a result, client satisfaction places a new and important constraint on project managers. No wonder, then, that there is a growing interest in the project manager's role within the corporation.

THE VITAL DOZEN FOR PROJECT MANAGERS

Over the last several years, we have conducted interviews with dozens of senior project managers in which we asked them a simple question: "What information were you never given as a novice project manager that, in retrospect, could have made your job easier?" From the data gathered in these interviews, we have synthesized some of the more salient issues, outlined in Twelve Points to Remember and detailed below, that managers need to keep in mind when undertaking a project implementation effort. While not intended to appear in any particular order, these 12 rules offer a useful way to understand the challenge project managers face and some ways to address these concerns.

1. Understand the Context of Project Management

Much of the difficulty in becoming an effective project manager lies in understanding the particular challenges project management presents in most corporations. Projects are a unique form of organizational work, playing an important role within many public and private organizations today. They act as mechanisms for the effective introduction of new products and services. They offer a level of intraorganizational efficiency that all companies

---

Twelve Points to Remember

1. **Understand** the context of project management.
2. **Recognize** project team conflict as progress.
3. **Understand** who the stakeholders are and what they want.
4. **Accept** and use the political nature of organizations.
5. **Lead** from the front.
6. **Understand** what "success" means.
7. **Build** and maintain a cohesive team.
8. **Enthusiasm** and despair are both infectious.
9. **One look** forward is worth two looks back.
10. **Remember** what you are trying to do.
11. **Use time** carefully or it will use you.
12. **Above** all, plan, plan, plan.

---

seek but few find. But they also force managers to operate in a temporary environment outside the traditional functional lines of authority, relying upon influence and other informal methods of power. In essence, it is not simply the management of a project per se that presents such a unique challenge; it is also the atmosphere within which the manager operates that adds an extra dimension of difficulty. Projects exist outside the established hierarchy. They threaten, rather than support, the status quo because they represent change. So it is important for project managers to walk into their assigned role with their eyes wide open to the monumental nature of the tasks they are likely to face.

2. Recognize Project Team Conflict as Progress

One of the common responses of project managers to team conflict is panic. This reaction is understandable in that project managers perceive—usually correctly—that their reputation and careers are on the line if the project fails. Consequently, any evidence they interpret as damaging to the prospects of project success, such as team conflict, represents a very real source of anxiety. In reality, however, these interpersonal tensions are a natural result of putting individuals from diverse backgrounds together and requiring them to coordinate their activities. Conflict, as evidenced by the stages of group development, is more often a sign of healthy maturation in the group.

The result of differentiation among functional departments demonstrates that conflict under these circumstances is not only possible but unavoidable. One of the worst mistakes a project manager can make when conflicts emerge is to immediately force them below the surface without first analyzing the nature of the conflict. Although many interpersonal conflicts are based on personality differences, others are of a professional nature and should be addressed head-on.

Once a project manager has analyzed the nature of the conflict among team members, a variety of conflict handling approaches may be warranted, including avoidance, defusion, or problem-solving. On the other hand, whatever approach is selected should not be the result of a knee-jerk reaction to suppress conflict. In our experience, we have found many examples that show that even though a conflict is pushed below the surface, it will continue to fester if left unaddressed. The resulting eruption, which will inevitably occur later in the project development cycle, will have a far stronger effect than would the original conflict if it had been handled initially.

3. Understand Who the Stakeholders Are and What They Want

Project management is a balancing act. It requires managers to juggle the various and often conflicting demands of a number of powerful project stakeholders. One of the best tools a project manager can use is to develop a realistic assessment early in the project identifying the principal stakeholders and their agendas. In some projects, particularly those with important external clients or constituent groups, the number of stakeholders may be quite large, particularly when "intervenor" groups are included. Intervenors, according to Cleland (1983), may include any external group that can drastically affect the potential for project success, such as environmental activists in a nuclear plant construction project. Project managers who acknowledge the impact of stakeholders and work to minimize their effect by fostering good relations with them are often more successful than those who operate in a reactive mode, continually surprised by unexpected demands from groups that were not initially considered.

As a final point about stakeholders, it is important for a project manager's morale to remember that it is essentially impossible to please all the stakeholders all the time. The conflicting nature of their demands suggests that when one group is happy, another is probably upset. Project managers need to forget the idea of maximizing everyone's happiness and concentrate instead on maintaining satisfactory relations that allow them to do their job with a minimum of external interference.

4. Accept the Political Nature of Organizations and Use It to Your Advantage

Like it or not, we exist in a politicized world. Unfortunately, our corporations are no different. Important decisions involving resources are made through bargaining and deal-making. So project managers who wish to succeed must learn to use the political system to their advantage. This involves becoming adept at negotiation as well as using influence tactics to further the goals of the project.

At the same time, it is important to remember that any project representing possible organizational change is threatening, often because of its potential to reshuffle the power relationships among the key units and actors. Playing the political system simply acknowledges this reality. Successful project managers are those who can use their personal reputations, power, and influence to ensure cordial relations with important stakeholders and secure the resources necessary to smooth the client's adoption of the project.

Pursuing a middle ground of political sensibility is the key to project implementation success. There are two alternative and equally inappropriate approaches to navigating a firm's political waters: becoming overly political and predatory—we call these people "sharks"—and refusing to engage in politics to any degree—the politically "naive." Political sharks and the politically naive are at equal disadvantage in managing their projects: sharks because they pursue predatory and self-interested tactics that arouse distrust, and the naive because they insist on remaining above the fray, even at the cost of failing to attain and keep necessary resources for their projects.

Table 17.1 illustrates some of the philosophical differences among the three types of political actors. The process of developing and applying appropriate political tactics means using politics as it can most effectively be used: as a basis for negotiation and bargaining. "Politically sensible" implies being politically sensitive to the concerns (real or imagined) of powerful stakeholder groups. Legitimate or not, their concerns over a new project are real and must be addressed. Politically sensible managers understand that initiating any sort of organizational disruption or change by developing a new project is bound to reshuffle the distribution of power within the firm. That effect is likely to make many departments and managers very nervous as they begin to wonder how the future power relationships will be rearranged.

Appropriate political tactics and behavior include making alliances with powerful members of other stakeholder departments, networking, nego-tiating mutually acceptable solutions to seemingly insoluble problems, and recognizing that most organizational activities are predicated on the give-and-take of negotiation and compromise. It is through these uses of political behavior that managers of project implementation efforts put themselves in the position to most effectively influence the successful introduction of their systems.

5. Lead from the Front; the View Is Better

One message that comes through loud and clear is that project management is a "leader intensive" undertaking. Strong, effective leaders can go a long way toward helping a project succeed even in the face of a number of external or unforeseen problems. Conversely, a poor, inflexible leader can often ruin the chances of many important projects ever succeeding. Leaders are the focal point of their projects. They serve as a rallying point for the team and are usually the major source of information and communication for external stakeholders. Because their role is so central and so vital, it is important to recognize and cultivate the attributes project "leaders" must work to develop.

The essence of leadership lies in our ability to use it flexibly. This means that not all subordinates or situations merit the same response. Under some circumstances an autocratic approach is appropriate; other situations will be far better served by

TABLE 17.1
Characteristics of Political Behaviors

Characteristics	Naive	Sensible	Sharks
Underlying Attitude	Politics is unpleasant	Politics is necessary	Politics is an opportunity
Intent	Avoid at all costs	Further departmental goals	Self-serving and predatory
Techniques	Tell it like it is	Network; expand connections; use system to give and receive favors	Manipulate; use fraud and deceit when necessary
Favorite Tactics	None—the truth will win out	Negotiate, bargain	Bully; misuse information; cultivate and use "friends" and other contacts

adopting a consensual style. Effective project leaders seem to understand this idea intuitively. Their approach must be tailored to the situation; it is self-defeating to attempt to tailor the situation to a preferred approach. The worst leaders are those who are unaware of or indifferent to the freedom they have to vary their leadership styles. And they see any situation in which they must involve subordinates as inherently threatening to their authority. As a result, they usually operate under what is called the "Mushroom Principle of Management." That is, they treat their subordinates the same way they would raise a crop of mushrooms—by keeping them in the dark and feeding them a steady diet of manure.

Flexible leadership behavior consists of a realistic assessment of personal strengths and weaknesses. It goes without saying that no one person, including the project manager, possesses all necessary information, knowledge, or expertise to perform the project tasks on his own. Rather, successful project managers usually acknowledge their limitations and work through subordinates' strengths. In serving as a facilitator, one of the essential abilities of an exceptional project manager is knowing where to go to seek the right help and how to ask the right questions. Obviously, the act of effective questioning is easier said than done. However, bear in mind that questioning is not interrogation. Good questions challenge subordinates without putting them on the spot; they encourage definite answers rather than vague responses, and they discourage guessing. The leader's job is to probe, to require subordinates to consider all angles and options, and to support them in making reasoned decisions. Direct involvement is a key component of a leader's ability to perform these tasks.

6. Understand What "Success" Means

Successful project implementation is no longer subject to the traditional "triple constraint." That is, the days when projects were evaluated solely on adherence to budget, schedule, and performance criteria are past. In modern business, with its increased emphasis on customer satisfaction, we have to re-train project managers to expand their criteria for project success to include a fourth item: client use and satisfaction. What this suggests is that project "success" is a far more comprehensive word than some managers may have initially thought. The implication for rewards is also important. Within some organizations that regularly implement projects, it is common practice to reward the implementation manager when, in reality, only half the job has been accomplished. In other words, giving managers promotions and commendations before the project has been successfully transferred to clients, is being used, and is affecting organizational effectiveness is seriously jumping the gun.

Any project is only as good as it is used. In the final analysis, nothing else matters if a system is not productively employed. Consequently, every effort must be bent toward ensuring that the system fits in with client needs, that their concerns and opinions are solicited and listened to, and that they have final sign-off approval on the transferred project. In other words, the intended user of the project is the major determinant of its success. Traditionally, the bulk of the team's efforts are centered internally, mainly on their own concerns: budgets, timetables, and so forth. Certainly, these aspects of the project implementation process are necessary, but they should not be confused with the ultimate determinant of success: the client.

7. Build and Maintain a Cohesive Team

Many projects are implemented through the use of cross-functional teams. Developing and maintaining cordial team relations and fostering a healthy intergroup atmosphere often seems like a full-time job for most project managers. However, the resultant payoff from a cohesive project team cannot be overestimated. When a team is charged to work toward project development and implementation, the healthier the atmosphere within that team, the greater the likelihood the team will perform effectively. The project manager's job is to do whatever is necessary to build and maintain the health (cohesion) of the team. Sometimes that support can be accomplished by periodically checking with team members to determine their attitudes and satisfaction with the process. Other times the project manager may have to resort to less conventional methods, such as throwing parties or organizing field trips. To effectively intervene and support a team, project managers play a variety of roles—movitator, coach, cheerleader, peacemaker, conflict resolver. All these duties are appropriate for creating and maintaining an effective team.

8. Enthusiasm and Despair Are Both Infectious

One of the more interesting aspects of project leaders is that they often function like miniaturized billboards, projecting an image and attitude that signals the current status of the project and its likelihood for success. The team takes its cue from the attitudes and emotions the manager exhibits. So one of the most important roles of the leader is that of motivator and encourager. The worst project managers are those who play their cards close to their chests, revealing little or nothing about the status of the project (again, the "Mushroom Manager"). Team members want and deserve to be kept abreast of what is happening. It is important to remember that the success or failure of the project affects the team as well as the manager. Rather than allowing the rumor mill to churn out disinformation, team leaders need to function as honest sources of information. When team members come to the project manager for advice or project updates, it is important to be honest. If the manager does not know the answer to their questions, he should tell them that. Truth in all forms is recognizable, and most project team members are much more appreciative of honesty than of eyewash.

9. One Look Forward Is Worth Two Looks Back

A recent series of commercials from a large computer manufacturer had as their slogan the dictum that the company never stop asking "What if?" Asking "What if?" questions is another way of saying we should never become comfortable with the status of the project under development. One large-scale study found that the leading determinant of project failure was the absence of any trou-

bleshooting mechanisms—that is, no one was asking the "What if?" questions. Projecting a skeptical eye toward the future may seem gloomy to some managers. But in our opinion, it makes good sense. We cannot control the future but we can actively control our response to it.

A good example of the failure to apply this philosophy is evidenced by the progress of the "Chunnel" intended to link Great Britain with France. Although now in full operation, it was not ready for substantial traffic until some 15 months later than originally scheduled. As a result, chunnel traffic missed the major summer vacation season with a concomitant loss in revenue. At the same time, the final cost (£15 billion) is likely to be six times the original estimate of £2.3 billion (O'Connor 1993). It is instructive to take note of a recent statement by one of the project's somewhat harassed directors who, when pressed to state when the Chunnel would be ready, replied, "Now it will be ready when it's ready and not before!" Clearly, the failure to apply adequate contingency planning has led to the predictable result: a belief that the project will simply end when it ends.

10. Remember What You Are Trying to Do

Do not lose sight of the purpose behind the project. Sometimes it is easy to get bogged down in the minutiae of the development process, fighting fires on a daily basis and dealing with thousands of immediate concerns. The danger is that in doing so, project managers may fail to maintain a view of what the end product is supposed to be. This point reemphasizes the need to keep the mission in the forefront—and not just the project manager, but the team as well. The goal of the implementation serves as a large banner the leader can wave as needed to keep attitudes and motives focused in the right direction. Sometimes a superordinate goal can serve as a rallying point. Whatever technique project managers use, it is important that they understand the importance of keeping the mission in focus for all team members. A simple way to discover whether team members understand the project is to intermittently ask for their assessment of its status. They should know how their contributions fit into the overall installation plan. Are they aware of the specific contributions of other team members? If no, more attention needs to be paid to reestablishing a community sense of mission.

11. Use Time Carefully or It Will Use You

Time is a precious commodity. Yet when we talk to project managers, it seems that no matter how hard they work to budget it, they never have enough. They need to make a realistic assessment of the "time killers" in their daily schedule: How are they spending their time and what are they doing profitably or unprofitably? We have found that the simple practice of keeping a daily time log for a short time can be an eye-opening experience. Many project managers discover that they spend far too much of their time in unproductive ways: project team meetings without agendas that grind on and on, unexpected telephone calls in the middle of planning sessions, quick "chats" with other managers that end up taking hours, and so forth. Efficient time management—one of the keys to successful project development—starts with project managers. When they actively plan their days and stick to a time budget, they usually find they are operating efficiently. On the other hand, when they take each problem as it comes and function in an ad hoc, reactive mode, they are likely to remain prisoners of their own schedules.

A sure recipe for finding the time and resources needed to get everything done without spending an inordinate amount of time on the job or construction site is provided by Gosselin (1993). The author lists six practical suggestions to help project managers control their tasks and projects without feeling constantly behind schedule:

- Create a realistic time estimate without overextending yourself.

- Be absolutely clear about what the boss or client requires.

- Provide for contingencies (schedule slippage, loss of key team member).

- Revise the original time estimate and provide a set of options as required.

- Be clear about factors that are fixed (specifications, resources, and so on).

- Learn to say "Yes, and . . ." rather than "No, but. . . ." Negotiation is the key.

12. Above All, Plan, Plan, Plan

The essence of efficient project management is to take the time to get it as right as possible the first time. "It" includes the schedule, the team composition, the project specifications, and the budget. There is a truism that those who fail to plan are planning to fail. One of the practical difficulties with planning is that so many of us distinguish it from other aspects of the project development, such as doing the work. Top managers are often particularly guilty of this offense as they wait impatiently for the project manager to begin doing the work.

Of course, too much planning is guaranteed to elicit repeated and pointed questions from top management and other stakeholders as they seek to discover the reason why "nothing is being done." Experienced project managers, though, know that it is vital not to rush this stage by reacting too quickly to top management inquiries. The planning stage must be managed carefully to allow the project manager and team the time necessary to formulate appropriate and workable plans that will form the basis for the development process. Dividing up the tasks and starting the "work" of the project too quickly is often ultimately wasteful. Steps that were poorly done are often steps that must be redone.

A complete and full investigation of any proposed project does take significant time and effort. However, bear in mind that overly elaborate or intricate planning can be detrimental to a project; by the time an opportunity is fully investigated, it may no longer exist. Time and again we have emphasized the importance of planning, but it is also apparent that there comes a limit, both to the extent and the time frame of the planning cycle. A survey among entrepreneurs, for example, revealed that only 28 percent of them drew up a full-scale plan (Sweet 1994). A lesson here for project managers is that, like entrepreneurs, they must plan, but they must also be smart enough to recognize mistakes and change their strategy accordingly. As is noted in an old military slogan, "No plan ever survives its first contact with the enemy."

PROJECT MANAGERS IN THE TWENTY-FIRST CENTURY

In our research and consulting experiences, we constantly interact with project managers, some with many years of experience, who express their frustration with their organizations because of the lack of detailed explication of their assigned tasks and responsibilities. Year after year, manager after manager, companies continue to make the same mistakes in "training" their project managers, usually through an almost ritualized baptism of fire. Project managers deserve better. According to Rodney Turner (1993), editor of the *International Journal of Project Management*:

> Through the 90's and into the 21st century, project-based management will sweep aside traditional functional line management and (almost) all organizations will adopt flat, flexible organizational structures in place of the old bureaucratic hierarchies. . . . [N]ew organizational structures are replacing the old. . . . [M]anagers will use project-based management as a vehicle for introducing strategic planning and for winning and maintaining competitive advantage.

Turner presents quite a rosy future, one that is predicated on organizations recognizing the changes they are currently undergoing and are likely to continue to see in the years ahead. In this challenging environment, project management is emerging as a technique that can provide the competitive edge necessary to succeed, given the right manager.

At the same time, there seems to have been a sea of change in recent years regarding the image of project managers. The old view of the project manager as essentially that of a decision maker, expert, boss, and director seems to be giving way to a newer ideal: that of a leader, coach, and facilitator. Lest the reader assume these duties are any easier, we would assert that anyone who has attempted to perform these roles knows from personal expe-

rience just how difficult they can be. As part of this metamorphosis, says Clarke (1993), the new breed of project manager must be a natural salesperson who can establish harmonious customer (client) relations and develop trusting relationships with stakeholders. In addition to some of the obvious keys to project managers' success—personal commitment, energy, and enthusiasm—it appears that, most of all, successful project managers must manifest an obvious desire to see others succeed.

For successful project managers, there will always be a dynamic tension between the twin demands of technical training and an understanding of human resource needs. It must be clearly understood, however, that in assessing the relative importance of each challenge, the focus must clearly be on managing the human side of the process. As research and practice consistently demonstrate, project management is primarily a challenge in managing people. This point was recently brought to light in an excellent review of a book on managing the "human side" of projects (Horner 1993):

> There must be many project managers like me who come from a technological background, and who suffered an education which left them singularly ill-prepared to manage people.

Leading researchers and scholars perceive the twenty-first century as the upcoming age of project management. The globalization of markets, the merging of many European economies, the enhanced expenditures of money on capital improvement both in the United States and abroad, the rapidly opening borders of Eastern European and Pacific Rim countries, with their goals of rapid infrastructure expansion—all of this offers an eloquent argument for the enhanced popularity of project management as a technique for improving the efficiency and effectiveness of organizational operations. With so much at stake, it is vital that we immediately begin to address some of the deficiencies in our project management theory and practice.

Project management techniques are well known. But until we are able to take further steps toward formalizing training by teaching the necessary skill set, the problems with efficiently developing, implementing, and gaining client acceptance for these projects are likely to continue growing. There is currently a true window of opportunity in the field of project management. Too often in the past, project managers have been forced to learn their skills the hard way, through practical experience coupled with all the problems of trial and error. Certainly, experience is a valuable component of learning to become an effective project manager, but it is by no means the best.

What conclusions are to be drawn here? If nothing else, it is certain that we have painted a portrait of project management as a complex, time-consuming, often exasperating process. At the same time, it is equally clear that successful project managers are a breed apart. To answer the various calls they continually receive, balance the conflicting demands of a diverse set of stakeholders, navigate tricky corporate political waters, understand the fundamental process of subordinate motivation, develop and constantly refine their leadership skills, and engage in the thousands of pieces of detailed minutiae while keeping their eyes fixed firmly on project goals requires individuals with special skills and personalities. Given the nature of their duties, is it any wonder successful project managers are in such short supply and, once identified, so valued by their organizations?

There is good news, however. Many of these skills, though difficult to master, can be learned. Project management is a challenge, not a mystery. Indeed, it is our special purpose to demystify much of the human side of project management, starting with the role played by the linchpin in the process: the project manager. The problem in the past has been too few sources for either seasoned or novice project managers to turn to in attempting to better understand the nature of their unique challenge and methods for performing more effectively. Too many organizations pay far too little attention to the process of selecting, training, and encouraging those people charged to run project teams. The predictable result is to continually compound the mistake of creating wave after wave of accidental project managers, forcing them to learn through trial and error with minimal guidance in how to perform their roles.

Managing projects is a challenge that requires a strategy and methodology all its own. Perhaps most important, it requires a project manager willing to function as a leader in every sense of the term. We have addressed a wide range of challenges, both contextual and personal, that form the basis under which projects are managed in today's organizations. It is hoped that readers will find something of themselves as well as something of use contained in these pages.

REFERENCES

B.N. Baker, P.C. Murphy, and D. Fisher, "Factors Affecting Project Success," in D.I. Cleland and W.R. King, eds., *Project Management Handbook* (New York: Van Nostrand Reinhold, 1983): 778–801.

K. Clarke, "Survival Skills for a New Breed," *Management Today,* December 1993, p. 5.

D.I. Cleland, "Project Stakeholder Management," in D.I. Cleland and W.R. King, eds., *Project Management Handbook* (New York: Van Nostrand Reinhold, 1983): 275–301.

J.C. Davis, "The Accidental Profession," *Project Management Journal, 15,* 3 (1984): 6.

A.A. Einsiedel, "Profile of Effective Project Managers," *Project Management Journal, 18,* 5 (1987): 51–56.

J. Davidson Frame, *Managing Projects in Organizations* (San Francisco: Jossey-Bass, 1987).

T. Gosselin, "What to Do With Last-Minute Jobs," *World Executive Digest,* December 1993, p. 70.

R.J. Graham, "A Survival Guide for the Accidental Project Manager," *Proceedings of the Annual Project Management Institute Symposium* (Drexel Hill, PA: Project Management Institute, 1992), pp. 355–361.

M. Horner, "Review of 'Managing People for Project Success,'" *International Journal of Project Management, 11* (1993): 125–126.

P.R. Lawrence and J.W. Lorsch, "Differentiation and Integration in Complex Organizations," *Administrative Science Quarterly, 11,* (1967): 1–47.

M. Nichols, "Does New Age Business Have a Message for Managers?" *Harvard Business Review,* March-April 1994, pp. 52–60.

L. O'Connor, "Tunneling Under the Channel," *Mechanical Engineering,* December 1993, pp. 60–66.

N. Pettersen, "What Do We Know About the Effective

Project Manager?" *International Journal of Project Management, 9* (1991): 99–104.

J.K. Pinto and O.P. Kharbanda, *Successful Project Managers: Leading Your Team to Success* (New York: Van Nostrand Reinhold, 1995).

J.K. Pinto and D.P. Slevin, "Critical Factors in Successful Project Implementation," *IEEE Transactions on Engineering Management,* EM-34, 1987, p. 22–27.

B.Z. Posner, "What it Takes to be a Good Project Manager," *Project Management Journal, 18,* 1 (1987): 51–54.

W.A. Randolph and B.Z. Posner, "What Every Manager Needs to Know About Project Management," *Sloan Management Review, 29,* 4 (1988): 65–73.

P. Sweet, "A Planner's Best Friend," *Accountancy, 113* (1994): 56–58.

H.J. Thamhain, "Developing Project Management Skills," *Project Management Journal, 22,* 3 (1991): 39–53.

R. Turner, "Editorial," *International Journal of Project Management, 11* (1993): 195.

18

How Project Performance Is Influenced by the Locus of Power in the R&D Matrix

RALPH KATZ AND THOMAS J. ALLEN

ABSTRACT: This study examines the relationship between project performance and the relative influence of project and functional managers in 86 R&D teams in nine technology-based organizations. Analyses show higher project performance when influence over salaries and promotions is perceived as balanced between project and functional managers. Performance reaches its highest level, however, when organizational influence is centered in the project manager and influence over technical details of the work is centered in the functional manager.

The matrix structure was first developed in research and development organizations in an attempt to capture the benefits and minimize the liabilities of two earlier forms of organization, the functional structure and the project form of organization (Crawford, 1986; Allen, 1986).

The functional alternative, in which departments are organized around disciplines or technologies, enables engineers to stay in touch more easily with new developments in those disciplines or technologies than does the project form. It has, however, the disadvantage of creating separations between technologies, thereby making interdisciplinary projects more difficult to coordinate. The functional approach is generally considered incapable of dealing with the added complexity and information requirements associated with a critical program or project-development effort whose tasks are significantly interdependent. Organizing activities primarily around functional expertise contributes to miscommunications and bottlenecks not only because there is no formal coordination mechanism but also because functional specialists tend to adopt a more restricted and parochial view of the overall project.

The project form of organization overcomes the coordination problem by grouping engineers together on the basis of the project or program on which they are working, regardless of their disciplines. Speed and more focused alignment around common project objectives tend to be the attributes of project teams, fostering a greater sense of collective identity, commitment, and ownership. Although it eases the integration of multidisciplinary efforts, the project structure does not allow as much sharing of resources across projects and it removes technical specialists from their disciplinary departments. This resulting detachment makes it more difficult for professionals to keep pace with the most recent developments in their areas of expertise. This can be very detrimental in long-term programs, especially if there is a fast rate of change in the underlying technologies. It can also be problematic for individual members' careers once the program is over if they have become so wrapped up and specialized in their project's activities and demands that they are now viewed as or have become somewhat obsolete in their basic disciplines.

Reprinted with permission from the authors. An earlier version of this paper was published in 1985 titled Project Performance and the Locus of Influence in the R&D Matrix. *Academy of Management Journal*, 28(1):67–87.

Ralph Katz is Professor of Management at Northeastern University's College of Business and Research Associate at MIT's Sloan School of Management. *Thomas J. Allen* is the Howard Johnson Professor of Management at MIT's Sloan School of Management.

FORCES INHERENT IN THE MATRIX

The matrix, by creating an integrating force in a program or project office, attempts to overcome the divisions that are inherent in the basic functional structure. In the matrix, project or program managers and their staffs are charged with the responsibility of integrating the efforts of engineers who draw upon a variety of different disciplines and technical specialities in the development of new products or processes. The managers of functional departments, on the other hand, are responsible for making sure that the organization is aware of the most recent developments in its relevant technologies, thereby insuring the technical integrity of products and processes that the program or project office is attempting to develop. As emphasized by Larson and Gobeli (1988), the project manager is responsible for defining what needs to be done, while the functional managers are concerned with how it will be accomplished.

While both parties are supposed to work closely together and jointly communicate and approve work-flow decisions, these disparate responsibilities often lead to conflict between the two arms of the matrix. Project managers are often forced by market needs and schedule pressures to assume a shorter range view of the research and development activities than functional managers need to have. Since they are responsible for developing a product that can be successfully produced and marketed, project managers take on a perspective that is sometimes more closely aligned to that of persons in marketing or manufacturing than to the perspective held in the research and development organization. Functional department heads, with their closer attachment to underlying technologies, are inclined to take a longer term view and consequently may be more concerned with the organization's capability to deliver the highest quality solution with the most relevant up-to-date technologies than with meeting immediate customer needs.

Both of these perspectives are necessary to the survival of the organization. Someone has to be concerned with getting new products out into the market, and someone has to be concerned with maintaining the organization's long-term capability to develop and incorporate technical advancements into future products. Research and development organizations, no matter how they are organized, always have both of these concerns. The matrix structure merely makes them explicit by vesting the two sets of concerns in separate managers.

In formalizing these two distinct lines of managerial influence, the R&D organization is generating "deliberate conflict" between two essential managerial perspectives as a means of balancing these two organizational needs (Cleland, 1968). Project managers whose prime directive is to get the product "out the door" are matched against functional managers who tend to hold back because they can always make the product "a little bit better," given more time and effort (Allen, 1984). When these two opposing forces are properly balanced, the organization should achieve a more nearly optimum balance, both in terms of product completion and technical excellence. Unfortunately, a balanced situation is not easy to achieve. Often one or the other arm of the matrix will dominate, and then, what appears to be a matrix on paper becomes either a project or a functional organization in operation.

These two conflicting forces of a matrix affect R&D project performance principally through their respective influences on the behaviors and attitudes of individual engineers. It is the engineers who perform the actual problem-solving activities that result in new products or processes. How they view the relative power of project and functional managers over their work lives will strongly influence how they respond to the different sets of pressures and priorities confronting them in the performance of their everyday tasks.

In any matrix organization, there are at least three broad areas of decision making in which both project and functional managers are supposed to be involved: (1) technical decisions regarding project work activities and solution strategies; (2) determination of salaries and promotional opportunities; and (3) staffing and organizational assignments of engineers to particular project activities. These are critical areas in which project and functional managers contend for influence, for it is through these supervisory activities that each side of the matrix attempts to motivate and direct each engineer's efforts and performance (Kingdon, 1973). The degree to which each side of the matrix is successful in building its power and influence within the R&D organization will have a strong bearing on the outcomes that emerge from the many interdependent engineering activities (Wilemon & Gemmill, 1971).

Although a great deal has been written about matrix organizations, there is very little empirical evidence about the effectiveness of these structures. What are the relationships between project performance and the distribution of power and influence within the organization? Will a balance in power between project and functionally oriented forces result in higher project performance? To answer such questions, the present study examines the relationships between project performance and the relative dominance of project and functional managers for 86 matrix project teams from nine technology-based organizations.

HYPOTHESES

Details of Project Work

This is the arena in which project and functional interests are most likely to come into direct conflict. The project manager has ultimate responsibility for bringing the new product into being and is, therefore, intimately concerned with the technical approaches used in accomplishing that outcome. However, if the project side of the matrix is allowed to dominate development work, two quite different problems can develop. At one extreme, there is the possibility that sacrifices in technical quality and long-term reliability will be made in order to meet budget, schedule, and immediate market demands (Knight, 1977). At the other extreme, the potential of products is often oversold by making claims that are beyond the organization's current technological capability to deliver.

To guard against these shortcomings, functional managers can be held accountable for the overall integrity of the product's technical content.

If the functional side of the matrix becomes overly dominant, however, the danger is that the product will include not only more sophisticated, but also perhaps less proven and riskier technology. The functional manager's desire to be technologically aggressive—to develop and use the most attractive, most advanced technology—must be countered by forces that are more sensitive to the operational environment and more concerned with moving developmental efforts into final physical reality (Mansfield & Wagner, 1975; Utterback, 1974).

To balance the influence of both project and functional managers over technical details is often a difficult task. While an engineer may report to both managers in a formal sense, the degree to which these managers both actively influence the direction or clarification of technical details and solution strategies will vary considerably from project to project, depending on the ability and willingness of the two managers to understand and become involved in the relevant technology and its applications. Nonetheless, project performance should be higher when team members can take both perspectives into account. Accordingly, the following is proposed:

Hypothesis 1: Project performance will be higher in a matrix structure when both project and functional managers are seen to exert equal influence over the detailed technical work of engineers.

Salaries and Promotions

Advocates of matrix organizations (e.g., Davis & Lawrence, 1977) have long agreed on the importance of achieving balanced influence over salary and promotion decisions. Both Roberts (1988) and Pinto, Pinto, and Prescott (1993) emphasize that matrix organizations require matching control systems to support their multidimensional structures; otherwise, they would be undermined by reward systems that are based on assumptions of unitary authority. The underlying argument is that when engineers view either their project or their functional managers as having more control over chances for salary increases and promotions, the engineers' behaviors and priorities are more likely to be influenced and directed solely by the side with that control.

This is one of the key issues in what are often described as "paper matrix" situations: management assumes that by drawing overlapping structures and by prescribing areas of mutual responsibility, balance will be achieved among appropriate supporting management systems. In practice, however, one of the two components of the matrix comes to dominate or appears to dominate in key areas such as determination of salaries or of promotions. It is important to stress here that it is the engineer's perception that counts. Unless engineers see both managers as affecting their progress in terms of income and status, there will be a natural tendency for them, particularly in conflict situations, to heed the desires of one manager to the neglect of the other. The matrix then ceases to function, resulting in a structure that is more likely to resemble either the pure project or the pure functional form of organization despite any "paper" claims to the contrary. We therefore expect that:

Hypothesis 2: Project performance will be higher in a matrix structure when both project and functional managers are seen to exert equal influence over the promotions and rewards of engineers than it will be when one or the other manager is seen as dominating.

Personnel Assignments

Personnel assignments often provide the focus for the priority battles that frequently afflict matrix organizations. With the pressure on them from both management and customers to produce, project managers often find themselves in tight competition for the resources necessary to provide results. One of the most critical of these resources is technical talent. Each functional department employs engineers of varying technical backgrounds, experiences, and capabilities. Every project manager learns quickly which engineers are the top performers and naturally wants them assigned to his project. As a result, an intense rivalry develops

among project managers, with each attempting to secure the most appropriate and most talented engineers for his project (Cleland & King, 1968). Functional managers, on the other hand, have a different motivation. They have no difficulty finding resources to support their top performers, but they also have to keep the rest of their engineering staff employed. They must therefore allocate or market the services of their less talented engineers to all project groupings.

At this point we must make a distinction between performance at the project level and performance at the level of the entire R&D organization. Organizational performance might be higher when project and functional managers have equal influence over personnel assignments. The performance of a single project, however, will probably be higher when that project's manager has greater influence over personnel assignments, since presumably that project will then obtain the best talent. Since our study is at the project level, and although we realize that high individual project performance may be suboptimal for the entire R&D organization, we expect that:

Hypothesis 3: Project performance will, on the average, be higher when project managers are seen to exert greater influence over personnel assignments to their projects than functional managers.

Organizational Influence

We must consider more than the bases of supervisory influence that exist within a project group. Considerable research has shown that managers of high performing projects are also influential outside their project teams (e.g., Pfeffer, 1993; Katz et al., 1995). According to such studies, managers affect the behaviors and motivations of subordinates not only through leadership directed *within* the project group but also through their organizational influence *outside* the project. The critical importance of organizational influence on project outcomes has also been confirmed by many studies of technological innovation (e.g., Ancona & Caldwell,

1990; Howell & Higgins, 1990). In almost every instance, successful innovation required the strong support and sponsorship of organizationally powerful managers who could provide essential resources, mediate intergroup conflicts, and were positioned to protect the developmental effort from outside sources of interference.

Based on these findings, if engineers see either arm of the matrix as having greater power in the organization at large, their behavior should be affected, particularly in situations of conflict. Engineers want to be on the "winning team" (Kidder, 1981). Perceptions of organizational influence, therefore, will be an important determinant of what actually occurs in a project, for an imbalance would probably result in engineers' paying greater attention and attributing greater importance to the more powerful side of the matrix.

This does not mean that the locus of organizational influence necessarily determines the loci of influence over work, rewards, and assignments. There may be, for example, many instances in which the less organizationally powerful manager exerts greater influence over one of the other dimensions. Such incongruences place engineers in uncomfortable positions, particularly if there is strong disagreement between their two managers. Discomfort over technical direction can often lead to the postponement of critical technical decisions and the failure to narrow the scope of technical alternatives, resulting in lower project performance. From an exploratory standpoint, our research examines two important questions involving organizational and project influence. First, is there a strong association between perceived organizational influence and the relative dominance of project and functional managers over the rewards, personnel assignments, and technical work of project engineers? And second, to what extent do these dimensions of organizational and internal dominance interact to affect project performance? Do they independently relate to project performance or do they interact in determining performance?

The basic model (Figure 18.1) underlying our study, then, is that the loci of power between project and functional managers relate to project performance through their respective effects on the behaviors and efforts of engineers in matrix situations. More specifically, how professionals perceive the distribution of influence between their functional and project managers over the technical details of their project work, over their chances for organizational rewards, and over their assignments to particular project activities will significantly affect their performances on their project teams, as hypothesized. These perceived loci of influence, moreover, may be strongly related to how engineers in a matrix organization see the relative power of the two managers within the larger organization.

On the other hand, the locus of organizational influence may not be associated with any of these three measures of internal influence; instead, it may be an additional factor that interacts with these measures to affect project performance. Most likely, the locus of organizational influence will interact with a particular measure of internal project influence only when the two influence measures are not strongly interconnected; otherwise, it is more likely that they will covary with project performance.

RESEARCH METHODS

Setting

The data presented in this paper derive from a study of R&D project teams in nine major U.S. organizations. Although the selection of participating organizations could not be made random, they were chosen to represent several distinct work sectors and markets. Two of the sites are government laboratories; three are not-for-profit firms receiving most of their funding from government agencies. The four remaining companies are in private industry, two in aerospace, one in electronics, and one in consumer goods and products.

In each organization, we met with senior managers and administrators to ascertain the project assignments and reporting relationships of all R&D staff professionals. The questionnaires were then tailored to the specific membership and reporting structure of each project group using the appropriate terminology of that project. To ensure higher-

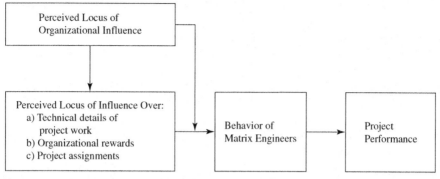

Figure 18.1. A model of the relationships between project performance and loci of influence.

quality data responses through voluntary participation, individuals were asked to complete their questionnaires on their own time and to mail them back to us directly using pre-stamped, return envelopes. Questionnaires were distributed through a number of brief explanatory meetings followed up by several reminder notes and "thank you" letters that were mailed over the course of the next few weeks. Response rates across the organizations were extremely high, ranging from as low as 82 percent in one organization to a high of 96 percent in another.

Although these procedures yielded over 2,000 respondents from 201 project teams, only 86 projects involved engineers and scientists in matrix dual-reporting relationships. A total of 486 engineers worked in matrix relationships in the 86 projects, an average of almost 6 engineers per project. Responses of these engineers, averaged to provide project measures, are the basic data analyzed. The proportion of the project team working in dual-reporting relationships varied from 20 percent (1 project) to 100 percent (18 projects); 19 projects had 70–99 percent, 21 projects had 60–69 percent, 20 projects had 50–59 percent, and 7 projects had 21–49 percent in matrix dual-reporting relationships. Results reported here are based only upon responses of engineers in matrix dual-reporting structures. Since the percentage of matrix engineers within projects varies considerably across the sample, significant findings will be reexamined as a function of these variations.

Matrix Relationships

We asked respondents in matrix structures to indicate on 7-point Likert-type scales the degree to which their project and functional managers influenced: (1) the technical details of their project work; (2) their salary increases and promotions; (3) their having been selected to work on the project; and (4) the overall conduct of the organization. For each of these dimensions of influence, scale responses ranged from 1 for "my project manager dominates" to 7 for "my functional manager dominates"; the middle point, 4, indicated that influence was balanced between the two. For each question, we averaged individual member responses to calculate overall project scores for the four influence areas.

For each of the four influence measures, appropriate statistical tests (cf. Katz and Tushman, 1979) were used to make sure it was permissible to combine individual perceptions to derive aggregated project scores. Except for a few isolated cases, there was sufficient consensus and reliability of individual responses within projects for each measure of influence. All of our analyses, therefore, are conducted at the project level; responses of individual engineers are averaged to provide project measures. Also, the influence scores for each project are based *only* on the aggregated perceptual responses of those engineers or scientists who were actually in matrix-reporting relationships. For each dimension of influence, lower scores indicate project manager dominance, while higher scores represent functional manager dominance. Because some questions were not included in the initial stages of the research (it took 21 months to conduct the study across all nine organizations), the number of project teams from which complete data are obtained ranges from 63 to 86.

Project Performance

Since measures of objective performance that are comparable across different technologies have yet to be developed, we used a subjective measure similar to that of many previous studies. In each organization, we measured project performance by interviewing managers who were at least one hierarchical level above the project and functional managers, asking them to indicate on a 5-point Likert-type scale whether a project team was performing above, below, or at the level expected of them, given the particular technical activities on which they were working. Managers evaluated only those projects that they were personally familiar with and knowledgeable about. Evaluations were made independently and submitted confidentially to the investigators. On the average, between four and five managers evaluated each project. The evaluations showed very strong internal consensus within each organization (Spearman-Brown reliabilities range from a low of .74 to a high of .93). It was therefore safe to average the ratings of individual managers to yield reliable project performance scores. However, performance data were missing from two projects. Right after we collected

our data, an expert panel of independent, outside R&D professionals exhaustively evaluated a small subset of our project base ($N = 8$). The ordering of their project performance evaluations agreed perfectly with the ordering of our own aggregated measures of performance. Such agreement between two separate sources provides considerable support for the validity of our project performance measures. Finally, our measures of relative project performance were not significantly related to the overall number of project members, to the number of matrix project members, or to the proportion of project members in matrix reporting relationships. To clarify the distinction between high and low project performance, performance measures were converted to normalized scores, with a mean of 0 (the original sample mean was 3.32).

RESULTS

As previously explained, we averaged responses to classify projects according to the degree to which project or functional managers exerted influence over each of four activity areas. Project scores of 1 through 3 were coded as signifying dominant influence by the project manager, while scores of 5 through 7 were taken to indicate functional manager dominance. Intermediate values, greater than 3 and less than 5, were considered as signifying balanced influence.

The locus of influence, as shown in Table 18.1, varies considerably both among projects and

TABLE 18.1
Distribution of Managerial Influence by Area as Perceived by Project Members

Area of influence	Locus of influence			
	Functional manager	Balanced	Project manager	N^a
Influence within the project				
Technical content of project work	14.0%	50.0%	36.1%	86
Salaries and promotions	58.1	34.9	7.0	86
Personnel assignments	28.6	54.0	17.5	63
Influence within the organization	30.2	38.3	31.4	86

[a]As previously explained, N varies by area of influence.

across dimensions of influence. Influence over technical details of work and over personnel assignments is balanced in the majority of cases. On the other hand, over half of the functional managers are seen as having greater influence over salaries and promotions: functional managers are viewed as controlling these rewards in almost 60 percent of the projects, project managers in only 7 percent. It is important to remember that it is the perceptions of engineers in matrix-reporting relationships that was measured, for it is perceived reality—not the reality itself—that influences engineers' behavior. Project managers may in fact have equal influence over salaries and promotions, but unless this equality is clearly apparent to engineers, it cannot affect their behavior.

Organizational influence, in contrast, is almost equally distributed across the three influence categories, with 30 percent of the projects having a more dominant functional side, 31 percent a more dominant project side, and 38 percent a reasonably balanced situation.

Because the projects under investigation come from government, not-for-profit, and industrial organizations, it is also important to see if there are major differences among these sectors. Generally speaking, there are no significant differences in the distributions of managerial influence for the dimensions of technical content and personnel assignments. In each sector, the distributions are consistent with the percentages reported in Table 18.1. For the other two loci of influence, however, there are significant variations from the distributions of Table 18.1 by the type of organization. In the not-for-profit sector, functional managers are seen as having considerably more influence within their organizations than their project management counterparts and are perceived as dominating rewards in over 80 percent of the projects. In sharp contrast, project managers are viewed as having stronger organizational influence than functional managers in over half of the projects in the industrial and government sectors. These differences are not surprising, since not-for-profit organizations are somewhat more oriented to academic research and probably place greater emphasis on the disciplines

than either industry or government organizations; industry and government organizations, in turn, probably put more emphasis on project management and the clear-cut product or system that must be brought into being. We present these descriptive distributions not to test any specific hypothesis, but simply to give the reader a better view of our data base, especially since we could not undertake randomized sampling of organizations.

Project Performance

As the above distributions show, it is very clear that the degree to which project or functional managers exert influence over dual-reporting engineers differs considerably among projects. The locus of influence also differs for each dimension of influence. The next step, therefore, is to test our hypotheses by seeing how project performance varies with these loci of influence. To examine the proposed relationships, we performed an analysis of variance on each dimension of internal project influence. In each analysis, project performance was the dependent variable, and the three categories of managerial dominance and balance represent the comparative levels within the independent variable.

Technical Details of Project Work. Table 18.2 presents results on the relationship between project performance and the locus of influence over the technical details of project work. Performance does not vary significantly with the locus of influence over technical content. Although there is a slight tendency toward higher performance when the project manager is perceived to have moder-

TABLE 18.2

Project Performance as a Function of the Locus of Influence over Technical Content of Project Work

Locus of influence	Number of projects	Project performance[a]
Project manager	31	0.07
Balanced	42	−0.08
Functional manager	11	0.10

[a]Normalized means; a one-way analysis of variance indicated that mean performance did not differ significantly ($F = 1.43$)

ately high influence or the functional manager to have strong influence, neither of these tendencies is significant. Also, the latter result stems from only 11 development projects. In any event, balanced involvement in technical matters of both sides of the matrix is not related to higher project performance; the data do not support hypothesis 1.

Salaries and Promotions. In the area of salaries and promotions, the ANOVA results of Table 18.3 show that project performance varies significantly across the loci of managerial influence. The mean performance levels in Table 18.3 indicate that project performance is highest when influence is either balanced or when project managers are viewed as controlling organizational rewards, although there are only six project cases in this latter category. Nevertheless, mean performance is significantly lower when functional managers are seen by project members as having more influence over their salaries and promotion opportunities.

Because the distribution of projects along the influence continuum is so extremely skewed towards functional control, we used Tukey's (1977) smoothing procedures as most appropriate for obtaining a more complete descriptive picture of the association between project performance and the locus of managerial influence over salaries and promotions. An examination of the resulting plot of smoothed performances (Figure 18.2) reveals a fairly regular pattern of decreasing performance with increasing functional control over monetary and career rewards. Although Tukey's smoothing procedures do not yield specific statistical tests, the

pattern that emerges from Figure 18.2, together with the results from Table 18.3, supports the hypothesis that project performance is directly associated with the degree to which project managers are seen as influential over the salaries and promotions of their subordinates.

Personnel Assignments. Project performance does not vary significantly with the locus of influence over personnel assignments (Table 18.4). Although we hypothesized that project performance would be higher when the project manager was seen to have greater influence than the functional manager over the staffing of project work, this does not turn out to be the case—at least not to the extent that its effects are evident in the different project groupings. It is interesting to note, however, that the lowest-performing set of projects are those in which the functional managers are seen as controlling the allocation of project personnel.

Organizational Influence. To what extent is organizational influence associated with the three measures of internal influence within the project? The correlations in Table 18.5 show that the locus of organizational influence is closely related to the locus of influence over salaries and promotions and to the locus of influence over personnel assignments. The way in which engineers in a matrix structure view the relative power of project and functional managers within the organization is not independent of how they view their managers' relative power over organizational rewards and staffing decisions. The locus of organizational influence is, however, independent of how they see their project and functional managers' influencing the detailed technical content of their work. The correlation between these two areas of influence is close to zero.

Since there is not a strong connection between organizational influence and influence over technical content of the work, the final question is whether the loci of influence in these two areas operate separately on performance or whether they interact to affect project performance. A two-way analysis of variance (Table 18.6) reveals once again that influence over the technical details of project

TABLE 18.3
Project Perfomance as a Function of the Locus of Influence over Salaries and Promotions

Locus of influence	Number of projects	Project performance[a]
Project manager	6	0.40
Balanced	29	0.37
Functional manager	49	−0.27

[a]Normalized means; one-way analysis of variance indicated that mean performance differed significantly at the .02 level ($F = 4.69$).

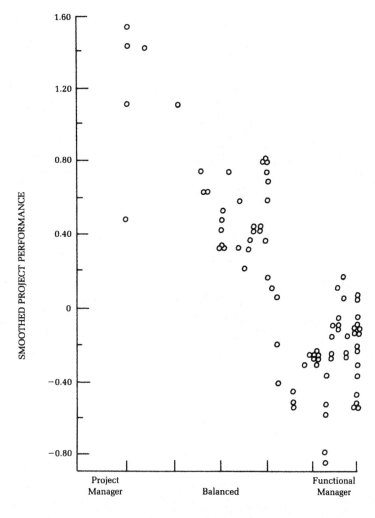

Figure 18.2. Smoothed project performance as a function of the locus of influence over salaries and promotions.

work is not related, at least as a main effect, to project performance. The locus of organizational influence, on the other hand, is significantly associated with project performance in that projects with relatively more powerful project managers are somewhat higher performing than are other projects.

More important, however, the ANOVA results also reveal an interaction effect on project performance between these two modes of influence. As shown by the performance means at the top of

Table 18.6, project performance is higher when project managers are seen as having relatively more influence within the organization and functional managers are seen as having relatively more influence over the technical content of what goes into the project. Performance is lowest when functional managers are seen as dominant in both of these areas. Additional analyses did not uncover any interference with these findings by project size or organization sector; nor did they uncover any other significant interaction effects on project performance among the other influence combinations.

As previously discussed, it is important to investigate the robustness of the results from Table 18.6 since projects varied widely in their percentages of matrix engineers. It is possible, for example, that the relationships between project performance and the distributions of power will be significantly different for projects with high proportions of matrix engineers than for projects with relatively low proportions of matrix engineers. To test for this possibility, the 86 projects were split into high and low subsamples based on the proportion of matrix engineers, and separate 2-way ANOVA tests were run on the split samples. In each case, the pattern of performance means strongly paralleled the results of Table 18.6. Project performance, for both subsamples, was highest when project managers were perceived as having relatively more organizational influence and functional managers were perceived as having relatively more influence over the technical details of the project work; in both instances, project performance was lowest when functional managers were seen as dominant along both of these influence dimensions.

TABLE 18.4
Project Performance as a Function of the Locus of Influence over Personnel Assignments

Locus of influence	Number of projects	Project performance[a]
Project manager	10	0.04
Balanced	33	0.08
Functional manager	18	−0.41

[a]Normalized means; one-way analysis of variance indicated that mean performance did not differ significantly ($F = 1.07$).

TABLE 18.5
Correlation of Locus of Organizational Influence and Loci of Internal Influence

Internal influence over:	r
Technical content of work	−.02
Salaries and promotions	.65*
Personnel assignments	.49*

*$p < 0.001$

DISCUSSION

Our findings suggest an appropriate separation of roles between the managers of R&D professionals in matrix structures. The project manager should be concerned with external relations and activities. He or she should have sufficient power within the

TABLE 18.6
Project Performance[a] as a Function of the Loci of Influence over Technical Content of Project Work and Influence in the Organization

Locus of influence within the organization	Locus of influence over technical content	
	Project manager	Functional manager
Project Manager	0.10 ($N = 30$)	0.80 ($N = 12$)
Functional manager	−0.05 ($N = 30$)	−0.59 ($N = 12$)

Sources of Variation for Two-Way ANOVA:	df	F	p
Influence over technical content	1	0.36	N.S.[b]
Influence in organization	1	4.88	.03
Interaction	1	6.45	.01

[a]Normalized means
[b]N.S. = not significant

organization to gain the backing and continued support of higher management, to obtain critical resources, and to coordinate and couple project efforts with marketing and manufacturing. The concern of functional managers, on the other hand, should be more inward-directed, focusing chiefly on the technology that goes into the project. They are usually more closely associated with the necessary technologies, and consequently, should be better able than project managers to make informed decisions concerning technical content.

But these roles can never be completely separate since, for example, relations with marketing and manufacturing have critical implications for technical content, and vice-versa. A strong working relationship must therefore exist between project and functional managers. However, the results of this study suggest that clearer distinction of managerial roles leads to more effective project performance than does managers' sharing responsibilities and involvement. Performance appears to be highest when project managers focus principally on external relations and the output side of the project work, leaving the technological input side to be managed primarily by the functional side of the matrix. These results are also very consistent with the recent study of Larson and Gobeli (1988). In their mail survey of some 500 managers, they reported that the most successful projects were organized as a project matrix in which project managers had primary authority and responsibility for overseeing and completing the project, while functional managers had primary authority and responsibility for providing the requisite technical expertise. This was also the most preferred structure, even among the subsample of functional managers.

Despite these findings, most of the projects in our sample do not have this role separation pattern, at least as judged by the project members themselves. In more than half of our project groups, for example, members report their project managers have substantially more influence over the technical content of project work than their functional counterparts. Perhaps this is not too surprising since it is the project manager who manages the output and who is ultimately responsible for the project's success. It is, moreover, the project manager's reputation and career that are most intimately tied to project outcomes. Nevertheless, according to our study, overall performance might be improved if functional managers, who know more than project managers about the technologies involved, had greater influence over the technical activities of personnel assigned to project managers.

An enhanced role for the functional manager might also provide some additional benefit in mitigating one of the problems characteristic of matrix organizations. Functional managers have often felt threatened by the introduction of the matrix. Where they formerly had power and visibility in their functional structures, they see, under the matrix, a drift of all of this "glamour" to the project side of the organization. As a result, the matrix has often been undermined by recalcitrant or rebellious department heads, who saw the technical content of their responsibilities diminishing, and their careers sinking into an abyss of personnel decisions and human relations concerns. A clearer delineation of technical responsibilities and an explicitly defined contributive role for functional managers in the technical content of project work may well alleviate this problem.

Over the years, there has been considerable discussion concerning the need to maintain a balance of power in matrix organizations. Very little has been done, however, to investigate the elements or components of power and influence that should be balanced. Using project performance as our criterion, the present study's results provide very little support for the theories of balanced responsibility. Except for joint influence over the areas of salary and promotion, higher project performance is not associated with a balanced state of influence within any of the other three areas of supervisory activity.

Where does this leave all of the theories and propositions regarding matrix balance? Our findings imply that it is *not* through mutual balance or joint responsibilities along single dimensions of influence that the matrix should be made to work, but rather that the matrix should be designed and organized around more explicit role differentiation among dimensions of influence. The project manager's role is distinctly different from that of a functional manager. The two have very different concerns and should relate to both project team members and the larger organization in distinctly different ways. It therefore makes sense that the influence which each should exert over the behaviors of matrix project members to bring about effective project performance will be along different dimensions.

Project performance appears to be higher when project managers are seen as having greater organizational influence. They have an outward orientation. As a result, they should be concerned with gaining resources and recognition for the project and with linking it to other parts of the business to insure that the project's direction fits the overall business plan of the organization. Functional managers, on the other hand, should be concerned with technical excellence and integrity, seeing that the project's inputs include reliable state-of-the-art technology. Their orientation is inward, focusing on the technical content of the project. Detailed technical decisions should be made by those who are closest to the technology. The localization of technical decision-making in functional departments, however, implies an important integrating role for project managers, who are responsible for making sure that the technical decisions overseen by several different functional managers all fit together to yield the best possible end result. Clearly, the greater project managers' organizational influence, the easier it will be for them to integrate and negotiate with functional managers whose technical goals are often in conflict.

Balanced authority need not exist along each dimension of managerial influence. Instead, the distribution of influence seems better accomplished through differentiation of input- and output-oriented roles to functional and project managers, respectively; although the joint involvement of both managers in the area of organizational rewards is important for fostering higher project performance. Since the data reported in this study are cross-sectional, it is important to realize that we cannot really be sure of what happens to a project team as

its members continue to interact throughout the different innovative phases of a project (Roberts & Fusfeld, 1988). For example, the locus of influence that is most effective in the "upstream" or early phases of an innovation process may be very different from what is required as R&D efforts move further "downstream" into the engineering and manufacturing stages. Furthermore, while our discussion has emphasized the direct role that project and functional managers can play in influencing the overall performance of matrix project groups, the reverse situation is just as possible. With higher project performance, for example, project managers may come to be seen as more powerful and influential within an organization. Clearly, it remains for future research to look even more closely at these kinds of relationships so that we can learn how to alleviate the conflicts and frustrations that have become so pervasive as organization's try to structure and manage their portfolio of research and development projects.

REFERENCES

Allen, T. J. 1984. *Managing the Flow of Technology.* Cambridge, MA: MIT Press.

Allen, T. J. 1986. Organizational structure, information, technology, and R&D productivity. *IEEE Transactions on Engineering Management,* 33 (4): 212–17.

Ancona, D., & Caldwell, D. 1990. Improving the performance of new product development teams. *Research-Technology Management,* 33 (2): 25–29.

Cleland, D. I. 1968. The deliberte conflict. *Business Horizons,* 11 (1): 78–80.

Cleland, D. I., & King, W. R. 1968. *Systems Analysis and Project Management.* New York: McGraw-Hill.

Crawford, C. M. 1986. *New Products Management.* New York: Irwin.

Davis, S., & Lawrence, P. 1977. *Matrix.* Reading, MA: Addison-Wesley.

Howell, J. M., & Higgins, C. A. 1990. Champions of technological innovation. *Administrative Science Quarterly,* 35 (2): 317–41.

Katz, R., & Allen, T. J. 1982. Investigating the "not invented here" syndrome. *R&D Management,* 12 (1): 7–19.

Katz, R., & Allen, T. J. 1985. Project performance and the locus of influence in the R&D matrix. *Academy of Management Journal,* 28 (1): 67–87.

Katz, R., & Tushman, M. 1979. Communication patterns, project performance, and task characteristics: An empirical investigation in an R&D setting. *Organizational Behavior and Human Performance,* 23: 139–62.

Katz, R., Tushman, M., & Allen, T. J. 1995. The influence of supervisory promotion and network location on subordinate careers in a dual ladder RD&E setting. *Management Science,* 41 (5): 848–63.

Kidder, T. 1981. *The Soul of a New Machine.* New York: Little Brown.

Kingdon, O.R. 1973. *Matrix Organization: Managing Information Technologies.* London: Tavistock.

Knight, K. 1977. *Matrix Management.* London: Gower.

Larson, E. W., & Gobeli, D. H. 1988. Organizing for product development projects. *Journal of Product Innovation Management,* 5: 180–90.

Mansfield, E., & Wagner, S. 1975. Organizational and strategic factors associated with probability of success in industrial research. *Journal of Business,* 48: 179–98.

Pfeffer, J. 1993. *Managing with Power.* Cambridge, MA: Harvard Business School Press.

Pinto, M. B., Pinto, J. K., & Prescott, J. E. 1993. Antecedents and consequences of project team cross-functional cooperation. *Management Science,* 39 (10): 1281–90.

Roberts, E. 1988. Managing invention and innovation: What we've learned. *Research-Technology Management,* 31 (1): 11–29.

Roberts, E. B., & Fusfeld, A. R. 1988. Critical functions: Needed roles in the innovation process. In R. Katz (Ed.) *Managing Professionals in Innovative Organizations: A Collection of Readings:* 101–20, New York: Harper Business.

Tukey, J. 1977. *Exploratory Data Analysis.* Reading, MA: Addison-Wesley.

Utterback, J. M. 1974. Innovation in industry and the diffusion of technology. *Science,* 183: 620–26.

Wilemon, D. L., & Gemmill, G. R. 1971. Interpersonal power in temporary management systems. *Journal of Management Studies,* 8: 315–28.

III

LEADERSHIP ROLES IN THE INNOVATION PROCESS

7

Formal Problem-Solving Roles in Leading Innovation

19

Enlightened Experimentation

The New Imperative for Innovation

STEFAN THOMKE

The high cost of experimentation has long put a damper on companies' attempts to create great new products. But new technologies are making it easier than ever to conduct complex experiments quickly and cheaply. Companies now have an opportunity to take innovation to a whole new level—if they're willing to rethink their R&D from the ground up.

Experimentation lies at the heart of every company's ability to innovate. In other words, the systematic testing of ideas is what enables companies to create and refine their products. In fact, no product can be a product without having first been an idea that was shaped, to one degree or another, through the process of experimentation. Today, a major development project can require literally thousands of experiments, all with the same objective: to learn whether the product concept or proposed technical solution holds promise for addressing a new need or problem, then incorporating that information in the next round of tests so that the best product ultimately results.

In the past, testing was relatively expensive, so companies had to be parsimonious with the number of experimental iterations. Today, however, new technologies such as computer simulation, rapid prototyping, and combinatorial chemistry allow companies to create more learning more rapidly, and that knowledge, in turn, can be incorporated in more experiments at less expense. Indeed, new information-based technologies have driven down the marginal costs of experimentation, just as they have decreased the marginal costs in some production and distribution systems. Moreover, an experimental system that integrates new information-based technologies does more than lower costs; it also increases the opportunities for innovation. That is, some technologies can make existing experimental activities more efficient, while others introduce entirely new ways of discovering novel concepts and solutions.

Millennium Pharmaceuticals in Cambridge, Massachusetts, for instance, incorporates new technologies such as genomics, bioinformatics, and combinatorial chemistry in its technology platform for conducting experiments. The platform enables factory-like automation that can generate and test drug candidates in minutes or seconds, compared with the days or more that traditional methods require. Gaining information early on about, say, the toxicological profile of a drug candidate significantly improves Millennium's ability to predict the drug's success in clinical testing and, ultimately, in the marketplace. Unpromising candidates are eliminated before hundreds of millions of dollars are invested in their development. In addition to reducing the cost and time of traditional drug development, the new technologies also enhance Millennium's ability to innovate, according to Chief Technology Officer Michael Pavia. Specifically, the company has greater opportunities to experiment with more diverse potential drugs, including those that may initially seem improbable but might eventually lead to breakthrough discoveries.

This era of "enlightened experimentation" has thus far affected business with high costs of product development, such as the pharmaceutical, automotive, and software industries. By studying them, I have learned several valuable lessons that I believe have broad applicability to other industries. As the cost of computing continues to fall, making all sorts of complex calculations faster and cheaper, and as new technologies like combinatorial chemistry emerge, virtually all companies will discover that they have a greater capacity for rapid experimentation to investigate diverse concepts. Financial institutions, for example, now use computer simulations to test new financial instruments. In fact, the development of spreadsheet software has forever changed financial modeling; even novices can perform many sophisticated what-if-experiments that were once prohibitively expensive.

A SYSTEM FOR EXPERIMENTATION

Understanding enlightened experimentation requires an appreciation of the process of innovation. Namely, product and technology innovations don't drop from the sky; they are nurtured in laboratories and development organizations, passing through a *system* for experimentation. All development organizations have such a system in place to help them narrow the number of ideas to pursue and then refine that group into what can become viable products. A critical stage of the process occurs when an idea or concept becomes a working artifact, or prototype, which can then be tested, discussed, shown to customers, and learned from.

Perhaps the most famous example of the experimental system at work comes from the laboratories of Thomas Alva Edison. When Edison noted that inventive genius is "99% perspiration and 1% inspiration," he was well aware of the importance of an organization's capability and capacity to experiment. That's why he designed his operations in Menlo Park, New Jersey, to allow for efficient and rapid experimental iterations.

Edison knew that the various components of a system for experimentation—including personnel, equipment, libraries, and so on—all function interdependently. As such, they need to be jointly optimized, for together they define the system's performance: its speed (the time needed to design, build, test, and analyze an experiment), cost, fidelity (the accuracy of the experiment and the conditions under which it is conducted), capacity (the number of experiments that can be performed in a given time period), and the learning gained (the amount of new information generated by the experiment and an organization's ability to benefit from it). Thus, for example, highly skilled machinists worked in close proximity to lab personnel at Menlo Park so they could quickly make improvements when researchers had new ideas or learned something new from previous experiments. This system led to landmark inventions, including the electric lightbulb, which required more than 1,000 complex experiments with filament materials and shapes, electromechanical regulators, and vacuum technologies.

Edison's objective of achieving great innovation through rapid and frequent experimentation is especially pertinent today as the costs (both financial and time) of experimentation plunge. Yet many companies mistakenly view new technologies solely in terms of cost cutting, overlooking their vast potential for innovation. Worse, companies with that limited view get bogged down in the confusion that occurs when they try to incorporate new technologies. For instance, computer simulation doesn't simply replace physical prototypes as a cost-saving measure; it introduces an entirely different way of experimenting that invites innovation. Just as the Internet offers enormous opportunities for innovation—far surpassing its use as a low-cost substitute for phone or catalog transactions—so does state-of-the-art experimentation. But realizing that potential requires companies to adopt a different mind-set.

Indeed, new technologies affect everything, from the development process itself, including the way an R&D organization is structured, to how new knowledge—and hence learning—is created. Thus,

for companies to be more innovative, the challenges are managerial as well as technical, as these four rules for enlightened experimentation suggest:

1. Organize for Rapid Experimentation

The ability to experiment quickly is integral to innovation: as developers conceive of a multitude of diverse ideas, experiments can provide the rapid feedback necessary to shape those ideas by reinforcing, modifying, or complementing existing knowledge. Rapid experimentation, however, often requires the complete revamping of entrenched routines. When, for example, certain classes of experiments become an order of magnitude cheaper or faster, organizational incentives may suddenly become misaligned, and the activities and routines that were once successful might become hindrances. (See the sidebar "The Potential Pitfalls of New Technologies.")

Consider the major changes that BMW recently underwent. Only a few years ago, experimenting with novel design concepts—to make cars withstand crashes better, for instance—required expensive physical prototypes to be built. Because that process took months, it acted as a barrier to innovation because engineers could not get timely feedback on their ideas. Furthermore, data from crash tests arrived too late to significantly influence decisions in the early stages of product development. So BMW had to incorporate the information far downstream, incurring greater costs. Nevertheless, BMW's R&D organization, structured around this traditional system, developed award-winning automobiles, cementing the company's reputation as an industry leader. But its success also made change difficult.

Today, thanks to virtual experiments—crashes simulated by a high-performance computer rather than through physical prototypes— some of the information arrives very early, before BMW has made major resource decisions. The costs of experimentation (both financial and time) are therefore lower because BMW eliminates the creation of physical prototypes as well as the expense of potentially reworking bad designs after the company has committed itself to them. (Physical prototypes

The Essentials for Enlightened Experimentation

New technologies such as computer simulations not only make experimentation faster and cheaper, they also enable companies to be more innovative. But achieving that requires a thorough understanding of the link between experimentation and learning. Briefly stated, innovation requires the right R&D systems for performing experiments that will generate the information needed to develop and refine products quickly. The challenges are managerial as well as technical:

1. Organize for Rapid Experimentation

- Examine and, if necessary, revamp entrenched routines, organizational boundaries, and incentives to encourage rapid experimentation.

- Consider using small development groups that contain key people (designers, test engineers, manufacturing engineers) with all the knowledge required to iterate rapidly.

- Determine what experiments can be performed in parallel instead of sequentially. Parallel experiments are most effective when time matters most, cost is not an overriding factor, and developers expect to learn little that would guide them in planning the next round of experiments.

2. Fail Early and Often, But Avoid Mistakes

- Embrace failures that occur early in the development process and advance knowledge significantly.

- Don't forget the basics of experimentation. Well-designed tests have clear objectives (what do you anticipate learning?) and hypotheses (what do you expect to happen?). Also, mistakes often occur when you don't control variables that could diminish your ability to learn from the experiments. When variability can't be controlled, allow for multiple, repeated trials.

3. Anticipate and Exploit Early Information

- Recognize the full value of front-loading: identifying problems upstream, where they are easier and cheaper to solve.

- Acknowledge the trade-off between cost and fidelity. Experiments of lower fidelity (generally costing less) are best suited in the early exploratory stages of developing a product. High-fidelity experiments (typically more expensive) are best suited later to verify the product.

4. Combine New and Traditional Technologies

- Do not assume that a new technology will necessarily replace an established one. Usually, new and traditional technologies are best used in concert.

- Remember that new technologies emerge and evolve continually. Today's new technology might eventually replace its traditional counterpart, but it could then be challenged by tomorrow's new technology.

---

are still required much further downstream to verify the final designs and meet safety regulations.) In addition, the rapid feedback and the ability to see and manipulate high-quality computer images spur greater innovation: many design possibilities can be explored in "real time" yet virtually, in rapid iterations.

To study this new technology's impact on innovation, BMW performed the following experiment. Several designers, a simulation engineer, and a test engineer formed a team to improve the side-impact safety of cars. Primarily using computer

simulations, the team developed and tested new ideas that resulted from their frequent brainstorming meetings.

Because all the knowledge required about safety, design, simulation, and testing resided within a small group, the team was able to iterate experiments and develop solutions rapidly. After each round of simulated crashes, the team analyzed the results and developed new ideas for the next round of experiments. As expected, the team benefited greatly from the rapid feedback: it took them only a few days to accept, refine, or reject new de-

The Potential Pitfalls of New Technologies

New technologies can slash the costs (both financial and time) of experimentation and dramatically increase a company's ability to develop innovative products. To reap those benefits, though, organizations must prepare themselves for the full effects of such technologies.

Computer simulations and rapid prototyping, for example, increase not only a company's capacity to conduct experiments but also the wealth of information generated by those tests. That, however, can easily overload an organization that lacks the capability to process information from each round of experiments quickly enough to incorporate it into the next round. In such cases, the result is waste, confusion, and frustration. In other words, without careful and thorough planning, a new technology might not only fail to deliver on its promise of lower cost, increased speed, and greater innovation, it could actually decrease the overall performance of an R&D organization, or at a minimum disrupt its operations.

Misaligned objectives are another common problem. Specifically, some managers do not fully appreciate the trade-off between response time and resource utilization. Consider what happens when companies establish central departments to oversee computing resources for performing simulations. Clearly, testing ideas and concepts virtually can provide developers

with the rapid feedback they need to shape new products. At the same time, computers are costly, so people managing them as cost centers are evaluated by how much those resources are being used.

The busier a central computer is, however, the longer it takes for developers to get the feedback they need. In fact, the relationship between waiting time and utilization is not linear—queuing theory has shown that the waiting time typically increases gradually until a resource is utilized around 70%, and then the length of the delays surge. (See Figure 19.1.) An organization trying to shave costs may become a victim of its own myopic objective. That is, an annual savings of perhaps a few hundred thousand dollars achieved through increasing utilization from 70% to 90% may lead to very long delays for dozens of development engineers waiting for critical feedback from their tests.

A huge negative consequence is that the excessive delays not only affect development schedules but also discourage people from experimenting, thus squelching their ability to innovate. So in the long term, running additional computer equipment at a lower utilization level might well be worth the investment. An alternative solution is to move those resources away from cost centers and under the control of developers, who have strong incentives for fast feedback.

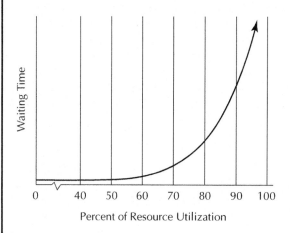

Figure 19.1. Waiting for a resource. According to queuing theory, the waiting time for a resource such as a central mainframe computer increases gradually as more of the resource is used. But when the utilization passes 70%, delays increase dramatically.

sign solutions: something that had once taken months.

As the trials accrued, the group members greatly increased their knowledge of the underlying mechanics, which enabled them to design previously unimaginable experiments. In fact, one test completely changed their knowledge about the complex relationship between material strength and safety. Specifically, BMW's engineers had assumed that the stronger the area next to the bottom of a car's pillars (the structures that connect the roof of an auto to its chassis), the better the vehicle would be able to withstand crashes. But one member of the development team insisted on verifying this assumption through an inexpensive computer simulation.

The results shocked the team: strengthening a particular area below one of the pillars substantially *decreased* the vehicle's crashworthiness. After more experiments and careful analysis, the engineers discovered that strengthening the lower part of the center pillar would make the pillar prone to folding higher up, above the strengthened area. Thus, the passenger compartment would be more penetrable at the part of the car closer to the midsection, chest, and head of passengers. The solution was to weaken, not strengthen, the lower area. This counterintuitive knowledge—that purposely weakening a part of a car's structure could increase the vehicle's safety—has led BMW to reevaluate all the reinforced areas of its vehicles.

In summary, this small team increased the side-impact crash safety by about 30%. It is worth noting that two crash tests of physical prototypes at the end of the project confirmed the simulation results. It should also be noted that the physical prototypes cost a total of about $300,000, which was more than the cost of all 91 virtual crashes combined. Furthermore, the physical prototypes took longer to build, prepare, and test than the entire series of virtual crashes.

But to obtain the full benefits of simulation technologies, BMW had to undertake sweeping changes in process, organization, and attitude—changes that took several years to accomplish. Not only did the company have to reorganize the way

different groups worked together; it also had to change habits that had worked so well in the old sequential development process.

Previously, for example, engineers were often loath to release less-than-perfect data. To some extent, it was in each group's interest to hold back and monitor the output from other groups. After all, the group that submitted its information to a central database first would quite likely have to make the most changes because it would have gotten the least feedback from other areas. So, for instance, the door development team at BMW was accustomed to—and rewarded for—releasing nearly flawless data (details about the material strength of a proposed door, for example), which could take many months to generate. The idea of releasing rough information very early, an integral part of a rapid and parallel experimentation process, was unthinkable—and not built into the incentive system. Yet a six-month delay while data were being perfected could derail a development program predicated on rapid iterations.

Thus, to encourage the early sharing of information, BMW's managers had to ensure that each group understood and appreciated the needs of other teams. The crash simulation group, for example, needed to make the door designers aware of the information it required in order to build rough models for early-stage crash simulations. That transfer of knowledge had a ripple effect, changing how the door designers worked because some of the requested information demanded that they pay close attention to the needs of other groups as well. They started to understand that withholding information as long as possible was counterproductive. By making these kinds of organizational changes, BMW in Germany significantly slashed development time and costs and boosted innovation.

2. Fail Early and Often, But Avoid Mistakes

Experimenting with many diverse—and sometimes seemingly absurd—ideas is crucial to innovation. When a novel concept fails in an experiment, the failure can expose important gaps in knowledge. Such experiments are particularly desirable when they are performed early on so that unfavorable options can be eliminated quickly and people can refocus their efforts on more promising alternatives. Building the capacity for rapid experimentation in early development means rethinking the role of failure in organizations. Positive failure requires having a thick skin, says David Kelley, founder of IDEO, a leading design firm in Palo Alto, California.

IDEO encourages its designers "to fail often to succeed sooner," and the company understands that more radical experiments frequently lead to more spectacular failures. Indeed, IDEO has developed numerous prototypes that have bordered on the ridiculous (and were later rejected), such as shoes with toy figurines on the shoelaces. At the same time, IDEO's approach has led to a host of bestsellers, such as the Palm V handheld computer, which has made the company the subject of intense media interest, including a *Nightline* segment with Ted Koppel and coverage in *Serious Play*, a book by Michael Schrage, a co-director of the e-markets initiative at the MIT Media Lab, that describes the crucial importance of allowing innovators to play with prototypes.

Removing the stigma of failure, though, usually requires overcoming ingrained attitudes. People who fail in experiments are often viewed as incompetent, and that attitude can lead to counterproductive behavior. As Kelley points out, developers who are afraid of failing and looking bad to management will sometimes build expensive, sleek prototypes that they become committed to before they know any of the answers. In other words, the sleek prototype might look impressive, but it presents the false impression that the product is farther along than it really is, and that perception subtly discourages people from changing the design even though better alternatives might exist. That's why IDEO advocates the development of cheap, rough prototypes that people are invited to criticize—a process that eventually leads to better products. "You have to have the guts to create a straw man," asserts Kelley.

To foster a culture in which people aren't afraid of failing, IDEO has created a playroomlike atmosphere. On Mondays, the different branches hold show-and-tells in which employees display and talk about their latest ideas and products. IDEO also maintains a giant "tech box" of hundreds of gadgets and curiosities that designers routinely rummage through, seeking inspiration among the switches, buttons, and various odd materials and objects. And brainstorming sessions, in which wild ideas are encouraged and participants defer judgment to avoid damping the discussion, are a staple of the different project groups.

3M is another company with a healthy attitude toward failure. 3M's product groups often have skunkworks teams that investigate the opportunities (or difficulties) that a potential product might pose. The teams, consisting primarily of technical people, including manufacturing engineers, face little repercussion if an idea flops—indeed, sometimes a failure is cause for celebration. When a team discovers that a potential product doesn't work, the group quickly disbands and its members move on to other projects.

Failures, however, should not be confused with mistakes. Mistakes produce little new or useful information and are therefore without value. A poorly planned or badly conducted experiment, for instance, might result in ambiguous data, forcing researchers to repeat the experiment. Another common mistake is repeating a prior failure or being unable to learn from that experience. Unfortunately, even the best organizations often lack the management systems necessary to carefully distinguish between failures and mistakes.

3. Anticipate and Exploit Early Information

When important projects fail late in the game, the consequences can be devastating. In the pharmaceutical industry, for example, more than 80% of drug candidates are discontinued during the clinical development phases, where more than half of total project expenses can be incurred. Yet although companies are often forced to spend millions of dollars to correct problems in the later stages of product development, they generally underestimate the cost savings of early problem solving. Studies of software development, for instance, have shown that late-stage problems are more than 100 times as costly as early-stage ones. For other environments that involve large capital investments in production equipment, the increase in cost can be orders of magnitude higher.

In addition to financial costs, companies need to consider the value of time when those late-stage problems are on a project's critical path—as they often are. In pharmaceuticals, shaving six months off drug development means effectively extending patent protection when the product hits the market. Similarly, an electronics company might easily find that six months account for a quarter of a product's life cycle and a third of all profits.

New technologies, then, can provide some of their greatest leverage by identifying and solving problems upstream—best described as *front-loaded development*. In the automotive industry, for example, "quick-and-dirty" crash simulations on a computer can help companies avoid potential safety problems downstream. Such simulations may not be as complete or as perfect as late-stage prototypes will be, but they can force organizational problem solving and communication at a time when many downstream groups are not participating directly in development. (See the sidebar "The Benefits of Front-Loaded Development.")

Several years ago, Chrysler (now Daimler-Chrysler) discovered the power of three-dimensional computer models, known internally as digital mock-ups, for identifying certain problems in early development stages. When Chrysler developed the 1993 Concorde and Dodge Intrepid models, the process of decking—placing the power train and related components like the exhaust and suspension in the prototype automobile—took more than three weeks and required many attempts before the powertrain could be inserted successfully. By contrast, the early use of digital mock-ups in the 1998 Concorde and Intrepid models allowed the company to simulate decking to identify (and solve) numerous interference problems before the physical decking took place. Instead of taking weeks, decking was completed in 15 minutes because all obstruction problems had been resolved earlier—when it was relatively inexpensive and fast to do so.

The Benefits of Front-Loaded Development

In the 1990s, Toyota made a major push to accelerate its product development cycle. The objective was to shorten the time from the approval of a body style to the first retail sales, thereby increasing the likelihood that Toyota kept up with the rapidly changing tastes of consumers.

Toyota made a concerted effort to identify and solve design-related problems earlier in product development—a concept known as front-loading. To accomplish that, the company implemented a number of initiatives, such as involving more manufacturing engineers during the product-engineering stage, increasing the transfer of knowledge between projects, investing substantially in computer-aided design and engineering tools, and developing rapid-prototyping capabilities.

To measure the benefits of these initiatives—and to monitor the company's evolving capabilities for early problem solving—Toyota tracked problems over multiple development projects. (See Figure 19.2.) The knowledge that a higher percentage of problems were being solved at earlier stages reassured Toyota's managers that they could aggressively reduce both development time and cost without risking product quality. In particular, between the first and third front-loading initiatives, Toyota slashed the cost (including the number of full physical prototypes needed) and time of development by between 30% and 40%.

It should be noted that in the early 1990s Toyota substantially reorganized its development activities, resulting in more effective communication and coordination between the different groups. This change most likely accounted for some of the performance improvements observed, particularly during the first front-loading initiatives.

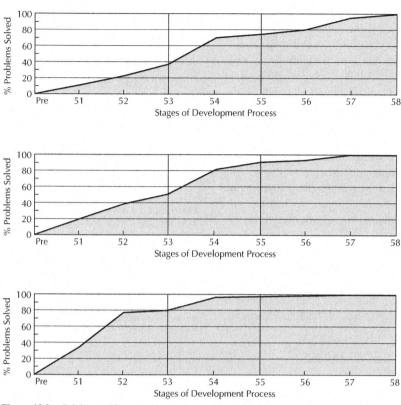

Figure 19.2. Solving problems earlier. As Toyota intensified its front-loading efforts, it was able to identify and solve problems much earlier in the development process. In the early 1990s (see top graph), the first initiatives for front-loading began. Formal, systematic efforts to improve face-to-face communication and joint problem solving between the prototype shops and production engineers resulted in a higher relative percentage of problems found with the aid of first prototypes. Communication between different engineering sections (for instance, between body, engine, and electrical) also improved. In the mid-1990s (see middle graph), the second front-loading initiatives called for three-dimensional computer-aided design, resulting in a significant increase of problem identification and solving prior to stage 3 (first prototypes). In the ongoing third front-loading initiatives (see bottom graph), Toyota is using computer-aided engineering to identify functional problems earlier in the development process, and the company is transferring problem and solution information from previous projects to the front end of new projects. As a result, Toyota expects to solve at least 80% of all problems by stage 2—that is, before the first prototypes are made. And because the second-generation prototypes (stage 5) are now less important to overall problem solving, Toyota will be able to eliminate parts of that process, thereby further reducing time and cost without affecting product quality. Source: Stefan Thomke and Takahiro Fujimoto, "The Effect of 'Front-Loading' Problem-Solving on Product Development Performance," *The Journal of Product Innovation Management*, Vol. 17, No. 2, March 2000.

Of course, it is neither pragmatic nor economically feasible for companies to obtain all the early information they would like. So IDEO follows the principle of three R's: rough, rapid, and

right. The final R recognizes that early prototypes may be incomplete but can still get specific aspects of a product right. For example, to design a telephone receiver, an IDEO team carved dozens of pieces of foam and cradled them between their heads and shoulders to find the best possible shape for a handset. While incomplete as a telephone, the model focused on getting 100% of the shape right. Perhaps the main advantage of this approach is that it forces people to decide judiciously which factors can initially be rough and which must be right. With its three R's, IDEO has established a process that generates important information when it is most valuable: the early stages of development.

In addition to saving time and money, exploiting early information helps product developers keep up with customer preferences that might evolve over the course of a project. As many companies can attest, customers will often say about a finished product: "This is exactly what I asked you to develop, but it is not what I want." Leading software businesses typically show incomplete prototypes to customers in so-called beta tests, and through that process they often discover changes and problems when they are still fairly inexpensive to handle.

4. Combine New and Traditional Technologies

New technologies that are used in the innovation process itself are designed to help solve problems as part of an experimentation *system.* A company must therefore understand how to use and manage new and traditional technologies together so that they complement each other. In fact, research by Marco Iansiti of Harvard Business School has found that, in many industries, the ability to integrate technologies is crucial to developing superior products.

A new technology often reaches the same general performance of its traditional counterpart much more quickly and at a lower cost. But the new technology usually performs at only 70% to 80% of the established technology. For example, a new chemical synthesis process might be able to obtain a purity level that is just three-quarters that of a mature technique. Thus, by combining new and established technologies, organizations can avoid the performance gap while also enjoying the benefits of cheaper and faster experimentation. (See Figure 19.3.)

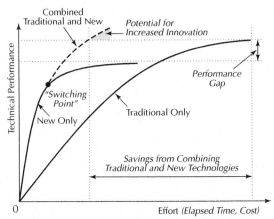

Figure 19.3. Combining the new with the traditional. A new technology will reach perhaps just 70% to 80% of the performance of a traditional technology. A new computer model, for instance, might be able to represent real-world functionality that is just three-quarters that of an advanced prototype model. To avoid this performance gap—and potentially create new opportunities for innovation—companies can use the new and traditional technologies in concert. The optimal time for switching between the two occurs when the rates of improvement between the new and mature technologies are about the same—that is, when the slopes of the two curves are equal.

Indeed, the true potential of new technologies lies in a company's ability to reconfigure its processes and organization to use them in concert with traditional technologies. Eventually, a new technology can replace its traditional counterpart, but it then might be challenged by a newer technology that must be integrated. To understand this complex evolution, consider what has happened in the pharmaceutical industry.

In the late nineteenth century and for much of the twentieth century, drug development occurred through a process of systematic trial-and-error experiments. Scientists would start with little or no knowledge about a particular disease and try out numerous molecules, many from their company's chemical libraries, until they found one that happened to work. Drugs can be likened to keys that need to fit the locks of targets, such as the specific nerve cell receptors associated with central nervous diseases. Metaphorically, then, chemists were once blind, or at least semiblind, locksmiths who have had to make up thousands of different keys to find the one that matched. Doing so entailed synthesizing compounds, one at a time, each of which usually required several days at a cost from $5,000 to $10,000.

Typically, for each successful drug that makes it to market, a company investigates roughly 10,000 starting candidates. Of those, only 1,000 compounds make it to more extensive trials in vitro (that is, outside living organisms in settings such as test tubes), 20 of which are tested even more extensively in vivo (that is, in the body of a living organism such as a mouse), and ten of which make it to clinical trials with humans. The entire process represents a long and costly commitment.

But in the last ten years, new technologies have significantly increased the efficiency and speed at which companies can generate and screen chemical compounds. Researchers no longer need to painstakingly create one compound at a time. Instead, they can use combinatorial chemistry, quickly generating numerous variations simultaneously around a few building blocks, just as today's locksmiths can make thousands of keys from a dozen basic shapes, thereby reducing the cost of a compound from thousands of dollars to a few dollars or less.

In practice, however, combinatorial chemistry has disrupted well-established routines in laboratories. For one thing, the rapid synthesis of drugs has led to a new problem: how to screen those compounds quickly. Traditionally, potential drugs were tested in live animals—an activity fraught with logistical difficulties, high expense, and considerable statistical variation.

So laboratories developed test-tube-based screening methodologies that could be automated. Called high-throughput screening, this technology requires significant innovations in equipment (such as high-speed precision robotics) and in the screening process itself to let researchers conduct a series of biological tests, or assays, on members of a chemical library virtually simultaneously.

The large pharmaceutical corporations and academic chemistry departments initially greeted such "combichem" technologies (combinatorial chemistry and high-throughput screening) with skepticism. Among the reasons cited was that the purity of compounds generated via combichem was relatively poor compared to traditional synthetic chemistry. As a result, many advances in the technology were made by small biotechnology companies.

But as the technology matured, it caught the interest of large corporations like Eli Lilly, which in 1994 acquired Sphinx Pharmaceuticals, one of the start-ups developing combichem. Eli Lilly took a few years to transfer the new technologies to its drug discovery division, which used traditional synthesis. To overcome the internal resistance, senior management implemented various mechanisms to control how the new technologies were being adopted. For example, it temporarily limited the in-house screening available to chemists, leaving them no choice but to use some of the high-throughput screening capabilities at the Sphinx subsidiary and interact with the staff there.

Until now, pharmaceutical giants like Eli Lilly have used combinatorial chemistry primarily to optimize promising new drug candidates that resulted from an exhaustive search through chemical libraries and other traditional sources. But as combinatorial chemistry itself advances and achieves levels of purity and diversity comparable to the compounds in a library, companies will increasingly use it at the earlier phases of drug discovery. In fact, all major pharmaceutical companies have had to use combichem and traditional synthesis in concert, and the companies that are best able to manage the new and mature technologies together so that they fully complement each other will have the greatest opportunity to achieve the highest gains in productivity and innovation.

ENLIGHTENED IMPLICATIONS

New technologies reduce the cost and time of experimentation, allowing companies to be more innovative. Automotive companies, for example, are currently advancing the performance of sophisticated safety systems that measure a passenger's position, weight, and height to adjust the force and speed at which airbags deploy. The availability of fast and inexpensive simulation enables the massive and rapid experimentation necessary to develop such complex safety devices.

But it is important to note that the increased automation of routine experiments will not remove the human element in innovation. On the contrary, it will allow people to focus on areas where their value is greatest: generating novel ideas and concepts, learning from experiments, and ultimately making decisions that require judgment. For example, although Millennium's R&D facilities look more and more like factories, the value of knowledge workers has actually increased. Instead of carrying out routine laboratory experiments, they now focus on the early stages (determining which experiments to conduct, for instance) and making sense of the information generated by the experimentation.

The implications for industries are enormous. The electronic spreadsheet has already revolutionized financial problem solving by driving down the marginal cost of financial experimentation to nearly zero; even a small startup can perform complex cash-flow analyses on an inexpensive PC. Similarly, computer simulation and other technologies have enabled small businesses and individuals to rapidly experiment with novel designs of customized integrated circuits. The result has been a massive wave of innovation, ranging from smart toys to electronic devices. Previously, the high cost of integrated-circuit customization made such experimentation economical to only the largest companies.

Perhaps, though, this era of enlightened experimentation is still in its bare infancy. Indeed, the ultimate technology for rapid experimentation might turn out to be the Internet, which is already turning countless users into fervent innovators.

20

Product-Development Practices That Work
How Internet Companies Build Software

ALAN MACCORMACK

Software is an increasingly pervasive part of the New Economy. As a result, today's general managers need to be aware of the most effective methods for developing and deploying software products and services within their organizations. Delegating such decisions to a technical staff, however skilled, can be a risky strategy. A study completed last year contains a surprising insight for managers: Dealing with the software revolution requires a process that is not revolutionary but evolutionary.

Evidence of the increasing importance of software abounds. In the United States alone, sales of software products and services exceeded $140 billion during 1998, a gain of more than 17% from the previous year.[1] In 2000, the software industry's contribution to the U.S. economy was expected to surpass that of the auto industry and overtake all other manufacturing industry groups for the first time.[2] Employment in software-related positions is growing, too. In 1998, the U.S. software industry directly employed more than 800,000 people, with an average salary twice the national figure.[3] More than 2 million people are now employed as software programmers, showing that software is not developed at a Microsoft or an Oracle but within the information-technology departments of large, traditional organizations.[4]

Software also is playing a larger role in the content delivered to customers in many industries. Nowadays, the average family sedan or high-end coffee maker may contain more software than the first Apollo spacecraft. What's more, the software features in those products may be the most critical differentiating factors. And even in industries in which software is not yet part of the products, it is playing a greater role in the products' development. As companies adopt new computer-aided design technologies, the development processes for many products increasingly resemble those found in the software industry.

DEVELOPING PRODUCTS ON INTERNET TIME

Given the importance of software, the lack of research on the best ways to manage its development is surprising. Many different models have been proposed since the much-cited waterfall model emerged more than 30 years ago. Unfortunately, few studies have confirmed empirically the benefits of the newer models. The most widely quoted references report lessons from only a few successful projects.[5]

Now a two-year empirical study, which the author and colleagues Marco Iansiti and Roberto Verganti completed last year, reveals thought-provoking information from the Internet-software industry—an industry in which the need for a responsive development process has never been greater.[6] The researchers analyzed data from 29 completed projects and identified the characteristics most associated with the best outcomes. (See the box "Four Software-Development Practices That Spell Success.") Successful development was evolutionary in nature. Companies first would release a low-functionality version of a product to se-

Reprinted from *MIT Sloan Management Review*, vol 42, no. 1, Winter, 2001, pp. 75–84. Reprinted with permission.
Alan MacCormack is a Professor of Technology and Operations Management at Harvard Business School.

Four Software-Development Practices That Spell Success

Analysis of Internet-software-development projects in a recent study uncovered successful practices:

- An early release of the evolving product design to customers
- Daily incorporation of new software code and rapid feedback on design changes
- A team with broad-based experience of shipping multiple projects
- Major investments in the design of the product architecture

---

lected customers at a very early stage of development. Thereafter work would proceed in an iterative fashion, with the design allowed to evolve in response to the customers' feedback. The approach contrasts with traditional models of software development and their more sequential processes. Although the evolutionary model has been around for several years, this is the first time the connection has been demonstrated between the practices that support the model and the quality of the resulting product.

MICROSOFT MEETS THE CHALLENGE— INTERNET EXPLORER 3.0

Consider Microsoft and its development of Internet Explorer. In the Internet's early years, small, nimble competitors such as Netscape and Yahoo! established leading positions—in part, through highly flexible development techniques.[7] In late 1995, many analysts thought Microsoft would be another incumbent that stumbled when faced with a disruptive innovation in its core business. Microsoft had been slow to recognize the potential of the Internet and was considered at least a generation behind Netscape in browser technology. Yet in the course of one project, Microsoft succeeded

in making up the ground and introducing a product—Internet Explorer 3.0—that many considered the equal of Netscape's offering. To a great extent, the achievement relied on the Explorer team's development process. (See Figure 20.1.)

Internet Explorer 3.0 (IE3) was Microsoft's first browser release with a major internal-development component.[8] The project started on Nov. 1, 1995, with the white paper "How We Get 30% Market Share in One Year." A small team started putting together the initial specifications, which were released to Microsoft's development partners on Dec. 7. The project was designated a "companywide emergency." As one IE3 manager explained it, the designation meant that "if you were smart and had time on your hands, you should help out the IE3 team. Given that we have a bunch of people here who are incredibly smart, we got a lot of great help. People realized this was a group that was going to determine what their stock was worth."

During December, detailed coding of the individual modules started. But the IE3 team was still making decisions about the overall product architecture—decisions that would not only affect the features in the final product but also the development process itself. A team member explained, "We had a large number of people who would have to work in parallel to meet the target ship date. We therefore had to develop an architecture where we could have separate component teams feed into the product. Not all of these teams were necessarily inside the company. The investment in architectural design was therefore critical. In fact, if someone asked what the most successful aspect of IE3 was, I would say it was the job we did in 'componentizing' the product."

The first integration of the new component modules into a working system occurred in the first week of March 1996. Although only about 30% of the final functionality was included in IE3 at that point, it was enough to get meaningful feedback on how the product worked. It also provided a baseline product, or alpha version, that could be handed to Microsoft's development partners. From that point on, the team instituted a process of "daily builds," which integrated new code into a complete

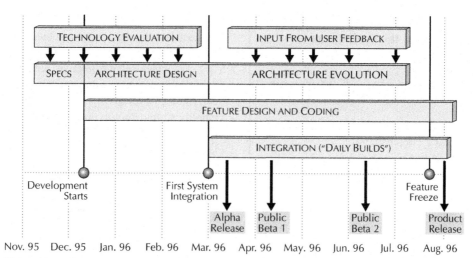

Figure 20.1. The development of Internet Explorer 3.0.

product every day. Once new code was "checked in" (integrated into the master version), getting performance feedback through a series of automated tests typically took less than three hours. With the rapid feedback cycle, the team could add new functionality to the product, test the impact of each feature and make suitable adjustments to the design.

In mid-April, Microsoft distributed the first beta version of IE3 to the general public. That version included about 50% to 70% of the final functionality in the product. A second beta version followed in June and included 70% to 90% of IE's final functionality. The team used the beta versions (as well as the alpha version) to gather feedback on bugs and on possible new features. Customers had a chance to influence the design at a time that the development team had the flexibility to respond. A significant proportion of the design changes made after the first beta release resulted from direct customer feedback. Some of the changes introduced features that were not even present in the initial design specification.

The cycle of new-feature development and daily integration continued frenetically through the final weeks of the project. As one program manager said, "We tried to freeze the external components of the design three weeks before we shipped.

In the end, it wasn't frozen until a week before. There were just too many things going on that we had to respond to . . . but, critically, we had a process that allowed us to do it."

MODELS OF THE SOFTWARE-DEVELOPMENT PROCESS

The Explorer team's process, increasingly common in Internet-software development, differs from past software-engineering approaches. (See the box "The Evolution of the Evolutionary-Delivery Model.") The waterfall model emerged 30 years ago from efforts to gain control over the management of large custom-software-development projects such as those for the U.S. military.[9] (See Figure 20.2.) The model features a highly structured, sequential process geared to maintaining a document trail of the significant design decisions made during development. A project proceeds through the stages of requirements analysis, specification, design, coding, and integration and testing—with sign-off points at the end of each stage. In theory, a project does not move to the next stage until all activities associated with the previous one have been completed.

The Evolution of the Evolutionary-Delivery Model

Companies that develop software are constantly improving the development models:

- The Waterfall Model (a sequential process maintains a document trail)
- The Rapid-Prototyping Model (a disposable prototype helps establish customer preferences)
- The Spiral Model (a series of prototypes identifies major risks)
- The Incremental, or Staged-Delivery, Model (a system is delivered to customers in chunks)
- The Evolutionary-Delivery Model (iterative approach in which customers test an actual version of the software)

The waterfall model, which has been compared to ordering a mail-order suit based upon a five-page text specification, is best for environments in which user requirements (and the technologies required to meet those requirements) are well understood. Its application in more uncertain environments, such as Internet-software engineer-ing, is problematic. Uncertain environments call for interactivity that lets customers evaluate the design before the specification has been cast in stone.

To achieve that objective, several alternative models use prototypes that are shown to customers early in the development process. Some companies employ a rapid-prototyping model that emphasizes the construction of an early prototype to help establish customer requirements.[10] Similarly, the spiral model moves through a series of prototype builds to help developers identify and reduce the major risks associated with a project.[11]

In both those models, however, the prototypes are not part of the design itself but merely representations that are thrown away after fulfilling their function. The bulk of the design work carried out thereafter is performed in a similar manner to the waterfall model.[12] In contrast, the process used to develop Microsoft's IE3 browser had at its heart the notion that a product can be developed in an iterative fashion. Critical parts of the functionality were delivered to customers early in the process; subsequent work added to the core design and responded to customers' feedback. Although the core functionality was continually improved, the early design that customers tested was an actual working version of the product.[13]

One development model with similarities to the IE3 process is the incremental, or staged-

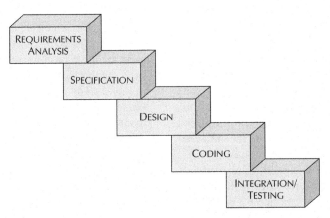

Figure 20.2. The waterfall model of software development is the traditional approach.

delivery, model.[14] In its basic form, it involves a system that is delivered to the customer in discrete chunks. However, it is unlike IE3's iterative process in that it assumes that the entire product design is specified in the early stages of development. Staged delivery is used only as a means of partitioning work so that some functionality can be delivered to customers early. By contrast, an iterative process is founded upon the belief that not everything can be known upfront—the staged delivery of the product actually helps determine the priorities for work to be done in subsequent stages.

The iterative process is best captured in the evolutionary-delivery model proposed by Tom Gilb.[15] In Gilb's model, a project is broken down into many microprojects, each of which is designed to deliver a subset of the functionality in the overall product. (See Figure 20.3.) The microprojects give a team early feedback on how well the evolving design meets customer requirements. At the same time, they build in flexibility: The team can make changes in direction during development by altering the focus of subsequent microprojects. Furthermore, the number and length of the microprojects can be tailored to match the context of a project. In its most extreme form, each individual feature within a product could be developed in a separate microproject. To a large extent, the model mirrors the way IE3 was built.

RESEARCH ON THE INTERNET-SOFTWARE INDUSTRY

Our study of projects in the Internet-software industry asked the question, Does a more evolutionary development process result in better performance? The study was undertaken in stages. First, the researchers conducted face-to-face interviews with project managers in the industry to understand the types of practices being used. Next, they developed metrics to characterize the type of process adopted in each project. Finally, the metrics were incorporated into a survey that went to a sample of Internet-software companies identified through a review of industry journals. The final sample contained data on 29 projects from 17 companies.[16]

To assess the performance of projects in the industry, we examined two outcome measures—one related to the performance of the final product and the other to the productivity achieved in terms of resource consumption (resource productivity). To assess the former, the researchers asked a panel of 14 independent industry experts to rate the comparative quality of each product relative to other products that targeted similar customer needs at the time the product was launched.[17] Product quality was defined as a combination of reliability, technical performance (such as speed) and breadth of functionality. Experts' ratings were gathered using

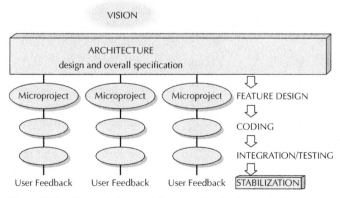

Figure 20.3. The evolutionary-delivery model of software development.

a two-round Delphi process (in which information from the first round is given to all experts to help them make their final assessment).[18] To assess the resource productivity of each project, the researchers calculated a measure of the lines of new code developed per person-day and adjusted for differing levels of product complexity.[19] Analysis of the data uncovered four practices critical to success.

Early Release of the Evolving Product Design to Customers

The most striking result to emerge from the research concerned the importance of getting a low-functionality version of the product into customers' hands at the earliest opportunity. (See Figure 20.4.) The research provided data on the percentage of the final product functionality that was contained in the first beta version (the first working version distributed to external customers).[20] Plotting the functionality against the quality of the final product demonstrated that projects in which most of the functionality was developed and tested *prior* to releasing a beta version performed uniformly poorly. In contrast, the projects that performed best were those in which a low-functionality version of the product was distributed to customers at an early stage.

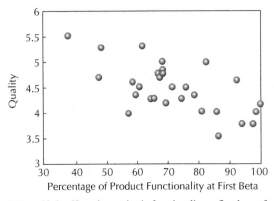

Figure 20.4. How the product's functionality at first beta affects quality. If customers test products early in development, when the products have low functionality, the final products are likely to have higher quality.

The differences in performance are dramatic. That one parameter explains more than one-third of the variation in product quality across the sample—a remarkable result, given that there are hundreds of variables that can influence the effectiveness of a development project, many of which are out of the project team's control.[21]

Consider the development of a simple Web browser. Its core functionality—the ability to input a Web-site address, have the software locate the server, receive data from the server and then display it on a monitor—could be developed relatively rapidly and delivered to customers. Although that early version might not possess features such as the ability to print a page or to e-mail a page to other users, it would still represent the essence of what a browser is supposed to do.

Of course, getting a low-functionality version to the customer early has profound implications for task partitioning. For example, let's say the aim of a project called BigBrain is to develop a new software application encompassing 10 major features. The traditional approach would involve dividing the team in such a way that all the features were worked on in parallel. Although progress would be made on each, the first opportunity to integrate a working version of the system would not occur until late in the project.

In an evolutionary process, however, the team might work first on only the three most important features—the essence of the system. Once those features were complete, the team would integrate them into a working version that could provide early feedback on how well the core modules interact. More important, the team would be able to distribute that early version to customers. As successive sets of features were completed and added to the product offering, their development would be guided by the customers' feedback.

The team might find that, of the seven remaining features planned for BigBrain, customers value only five (something customers may not have realized prior to testing a working version). In addition, customers might identify several features that had not previously been part of the design, giving designers the opportunity to make midcourse corrections—and thereby deliver a superior product.

By allowing the team to react to unforeseen circumstances, an evolutionary approach also reduces risk. Suppose that during the first part of project BigBrain, problems emerge in getting the core technical components to work together. With the evolutionary approach, the team can reschedule later-stage work—perhaps by eliminating one or more features of the original design. If development had proceeded in a more traditional fashion, feedback on such problems would not have been received until all the various component modules were integrated—much later in the process. The flexibility to react to new information would have been lost, and BigBrain would have shipped late.

Given the marked benefits of early beta testing, we considered whether the number of separate beta versions released to customers contributed to a product's performance. The Netscape Navigator 3.0 development team, for example, released six beta versions to external customers, each one following two to three weeks after the previous one.[22] The process of distributing an early release, gathering feedback, updating the design and redistributing the product to customers would seem an ideal way to ensure that the evolving functionality meshes with emerging customer needs. Surprisingly, however, the data showed no relationship between the performance of the final product and the number of beta releases. (See Figure 20.5.)

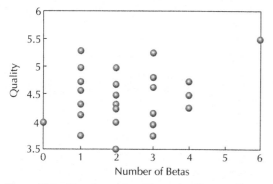

Figure 20.5. How the number of beta tests affects quality. The number of beta tests an Internet-software-development group uses does not affect the quality of the end product.

Our interviews revealed that the benefits obtained from the evolving product's early release to customers depended not upon the number of releases but on the intensity with which companies worked with customers after the first release. In general, the number of releases was not a good proxy for how well a company chose its beta customers or how well they subsequently worked with those customers.

Indeed, although the project that produced the highest-quality product in the sample—Netscape Navigator 3.0—released the largest number of beta versions, a member of its development team noted that multiple versions can create version-control problems: "The majority of beta testers who give us feedback don't necessarily tell us which beta version they have been working with. So the problem is they might be pointing out a bug that has already been fixed or one that manifests itself in a different way in a later release of the product. The result is we can spend as much time tracking down the problem as we do fixing it."

Daily Incorporation of New Software Code and Rapid Feedback on Design Changes

The need to respond to feedback generated through the release of early product versions to customers requires a process that allows teams to interpret new information quickly, then make appropriate design changes. In more than half the projects in the study, such changes were made through a daily build of the software code. In the same way that one checks books out of a library, developers working on the project would check out parts of the main code base to work on during the day. At the end of each day, they would take any newly modified code they had finished working on and check it back into the system. At check-in or overnight, a set of automated tests would run on the main code base to ensure that the new additions did not cause problems. At the start of the next day, the first task for each developer would be to fix any problems that had been found in his or her latest submissions.

Because daily builds have become an accepted approach to Internet-software development, they did not differentiate successful projects in the study. However, a measure of rapid feedback produced an intriguing result. (See Figure 20.6.) We looked at final product quality and plotted it against the time it took to get feedback on the most comprehensive set of automated tests performed on the design. None of the projects with extremely long feedback times (more than 40 hours) had a quality level above the mean. The conclusion, supported by interviewees' comments, is that rapid feedback on new design choices, is a necessary component of an evolutionary process. However, rapid feedback alone is not sufficient to guarantee that evolutionary software development will result in success. Indeed, projects with short feedback times were just as likely to perform poorly as to perform well.[23]

A Team with Broad-Based Experience of Shipping Multiple Projects

One might be forgiven for thinking that the value of experience is limited in a revolutionary environment such as the Internet-software industry. Much academic research has pointed out that in dynamic environments, experience may cause trouble, given that the knowledge of specific technologies and design solutions atrophy fast.[24] Indeed, the development ranks of many leading Internet-software companies often are filled with programmers barely out of college. Yet the view that a less experienced team is somehow better at developing products in such environments defies common sense—after all, a team with no experience at all would not know where to start. Even in a development process with the capacity to run thousands of design experiments, there is still a need to decide *which* experiments to run—and then to interpret the results. The question that must be asked therefore is What *types* of experience have value in revolutionary environments?

To answer that question, we studied two different measures of experience. The first was associated with the more traditional view of experience—the average tenure of the development team in years. The second measure reflected a different form of experience—namely, the number of project generations a team member had completed (generational experience).[25]

Our thinking was that the experience of completing many different projects would give developers a more abstract approach—one that evolves overtime as lessons are learned from successive projects. As a result, they would be better equipped to adapt to novel contexts and applications. In addition, the completion of each project would help developers see how their work fit into the system. The more projects completed, the greater their knowledge of how to design effectively at the module level while keeping the system level in view.

The results showed that the traditional measure of experience had no association with either product quality or resource productivity. The measure of generational experience had no association with product quality either, but it turned out to be a powerful predictor of resource productivity.[26] This suggests that the value of completing multiple projects in an evolutionary-development environment does not derive from an ability to predict specific customer requirements. Rather, such experience—by providing knowledge that helps developers analyze and respond to the data during development—allows greater efficiency in ongoing design activities.

The findings provide some insight into why a youthful development team is not necessarily one that lacks relevant experience. Given that many

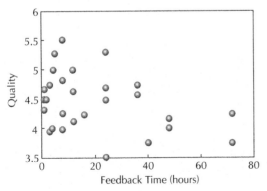

Figure 20.6. How feedback time affects quality. Rapid feedback on changes made to software facilitates better product performance.

software-development projects have short lead times, it is possible for a developer to complete quickly a large number of projects, thereby gaining substantial generational experience. That experience benefits future projects by helping the developer frame and direct an experimentation strategy that can resolve design problems quickly, even when the problems faced are novel.

Major Investments in the Design of the Product Architecture

In most development projects, the main design criterion for the product architecture is that it provide the highest possible level of performance. Often, the way that occurs is through an architecture that tightly links the various component modules. In an evolutionary process, however, there is another important criterion—flexibility in the development process. Designing the architecture so that a version of the product can be assembled at an early stage and distributed to customers requires explicit architectural choices. Building in the ability to accept additional functionality during late project stages adds further demands.

The key to an evolutionary process is to develop an architecture that is both modular and scaleable.[27] A more modular system is better at accommodating changes to the design of individual modules without requiring corresponding changes to other modules in the system. The loosely coupled nature of a modular system buffers the effect of changes. It is therefore suited to uncertain environments—at least, to the degree that the design isolates uncertainties within modules and does not allow them to affect the interfaces between modules. A more scaleable system allows initially unanticipated functions and features to be added at a late stage without disrupting the existing design. That requires a solid underlying infrastructure, such as that of the Linux operating system, one of the best examples of a modular and scaleable architecture. (See the box "A Modular and Scaleable Architecture: The Linux Operating System.")

In lieu of being able to examine the details of each product's architecture for our study, we focused on the relative investments in architectural

design that companies made.[28] Our assumption was that those investments reflected the degree to which companies were trying to resolve potential conflicts between a highly optimized architecture (one that is tightly coupled) and an architecture that facilitates process flexibility (one that is both modular and scaleable). Our analysis confirmed that a high level of investment in architectural design did indeed have a strong association with higher-quality products.[29]

PUTTING IT ALL TOGETHER

Although the study demonstrated that early customer involvement in an evolutionary process is vital, companies must take care to select suitable beta partners. We learned from the fieldwork that a valuable avenue for identifying beta partners is through exploring a company's customer-support database to identify customers who stretch the performance envelope. Customers who initiate numerous calls for support are good candidates for beta programs; however, it is the *nature* of those calls that is critical. The most effective beta groups include distinct customers for each performance dimension (say, reliability, speed or functionality) rather than customers who make demands on many fronts. Asking support employees which customers experience the strangest problems is one way of identifying those who are using the product in novel ways and who therefore might provide useful insights on performance.

With regard to the number of beta customers involved, we noticed that some companies emphasized a broad release of early versions to the entire customer base, whereas others employed a narrower distribution to a select group. The former strategy seemed most useful for product segments in which the software was meant to operate on a variety of different hardware platforms and alongside other software products. In such situations, mass distribution helped identify bugs related to the interactions among systems and products. In contrast, enterprises with sophisticated products that placed greater demands upon users preferred to

A Modular and Scaleable Architecture: The Linux Operating System

The initial version of the Linux kernel, the core of the Linux system, was developed in 1991 as part of an open-source project to develop a freely available Unix-like operating system. (In open-source projects, the underlying code that makes the software work is distributed to users so they can improve upon or customize it.) At the time, the kernel comprised only 10,000 lines of code and ran on only one hardware platform—the In-tel 386. However, as developers around the world began using the system and contributing to the project, its functionality expanded dramatically. By 1998, the kernel had grown to more than 1.5 million lines of code and was used on hardware platforms from supercomputers to robotic dogs (See Table 20.1.) Yet estimates of the amount of code added by the system's originator, Linus Tor-valds, were typically less than 5%.

TABLE 20.1
Evolution of the Linux Kernel

Year	Version	Lines of Code	Users
1991	0.01	10,000	1
1992	0.96	40,000	1,000
1993	0.99	100,000	20,000
1994	Linux 1.0	170,000	100,000
1995	Linux 1.2	250,000	500,000
1996	Linux 2.0	400,000	1,500,000
1997	Linux 2.1	800,000	3,500,000
1998	Linux 2.1.110	1,500,000	7,500,000

Source: Forbes, Aug. 10, 1998.

Note: When Torvalds released the original version of the code, he could not have predicted the functionality that would be added in subsequent years. So he based the design upon a modular architecture even though the inner core was monolithic. Developers around the world would be able to contribute to Linux without having to worry about the effect their code would have on other modules. Torvalds also made Linux scaleable—able to accept new functionality in a way that minimized changes to the existing core. Although the Linux architecture owes much to the long heritage of Unix, it is also a reflection of the true genius of its author.

work with a smaller group, given the extra support that such users required.

The benefits of an evolutionary approach to software development have been evangelized in the software-engineering literature for many years. However, the precise form of an evolutionary model and the empirical validation of its supposed advantages have eluded researchers. The model has now been proved successful in the Internet-software industry. When combined with the insights gained in fieldwork, our research suggests a clear agenda for managers: Get a low-functionality version of the product into customers' hands at the earliest possible stage and thereafter adopt an iterative approach to adding functionality. The results also underscore the importance of having a development team with experience on multiple projects and creating a product architecture that facilitates flexibility.[30]

The usefulness of the evolutionary model extends beyond developing software in environments with rapidly changing markets and technologies. By dividing tasks into microprojects, a company can tailor the process to reflect any particular context. Uncertainty in the Internet-software industry dictates short microprojects—down to the level of individual features. Traditional market research has limited value here, so companies need an early working version to gain feedback on the product concept.[31] In more-mature environments, however, companies can specify more of the product design upfront, use longer microprojects and develop greater functionality before needing feedback. In a world where customer needs and the underlying technologies in a product are known with certainty, only one large microproject is necessary, and the waterfall model suffices. An evolutionary-delivery model represents a transcendent process for managing the development of *all* types of software, with the details tailored to reflect each project's unique challenges.

NOTES

1. "Forecasting a Robust Future," www.bsa.org/statistics/index.html?/statistics/global_economic_studies_c.html.

2. Measured in terms of value added. Ibid.

3. Ibid.

4. "A Survey of the Software Industry," The Economist, May 25, 1996, 14.

5. See, for example: M.A. Cusumano and R. Selby, "Microsoft Secrets" (New York: Free Press, 1995); and F.P. Brooks, "The Mythical Man-Month" (Reading, Massachusetts: Addison-Wesley, 1995).

6. A. MacCormack, R. Verganti and M. Iansiti, "Developing Products on Internet Time: The Anatomy of a Flexible Development Process," Management Science 47, no. 1 (January 2001).

7. M. Iansiti and A. MacCormack, "Developing Products on Internet Time," Harvard Business Review 75 (September-October 1997): 108–117.

8. The first two versions of Internet Explorer relied extensively on licensed technology.

9. W.W. Royce, "Managing the Development of Large Software Systems: Concepts and Techniques" (Procedures of WESCON [Western Electric Show and Convention], Los Angeles, August 1970).

10. J.L. Connell and L. Shafer, "Structured Rapid Prototyping: An Evolutionary Approach to Software Development" (Englewood Cliffs, New Jersey: Yourdon Press, 1989).

11. B. Boehm, "A Spiral Model of Software Development and Enhancement," IEEE Computer 21 (May 1988): 61–72.

12. For example, in Boehm's spiral model, the outer layer of the spiral contains the activities of detailed design, coding, unit testing, integration testing, acceptance testing and implementation. Those activities are carried out sequentially.

13. That does not preclude the fact that "throwaway" prototypes are used in such a process. Indeed, they are likely to be extremely important in establishing a direction for the initial design work.

14. See, for example: C. Wong, "A Successful Software Development," IEEE Transactions on Software Engineering (November 1984): 714–727.

15. T. Gilb, "Principles of Software Engineering Management" (Reading, Massachusetts: Addison-Wesley, 1988), 84–114.

16. The survey was distributed to 39 firms, of which 17 responded with data on completed projects. The resulting sample of products is quite diverse and includes products and services targeted at both commercial and consumer users.

17. Quality was assessed on a seven-point scale, with level four indicating the product was at parity with competitive offerings.

18. H.A. Linstone and M. Turoff, eds., "The Delphi Method: Techniques and Applications" (Reading, Massachusetts: Addison-Wesley, 1975).

19. Projects in our sample differed significantly with regard to the number of lines of code developed. We therefore normalized the resources consumed in each project to reflect the development of an application of standard size. We adjusted the resulting measure for scale effects (larger projects were found to consume relatively fewer resources) and complexity effects (projects to develop Web-based services were found to consume relatively fewer resources, because of the specifics of the programming language used).

20. A beta version, as defined, is not a throwaway prototype. It is a working version of the system. The measure of the percentage of product functionality contained in the first beta was adjusted for scale effects.

21. We also examined the relationship an early beta release has with resource productivity. One might have imagined that in an evolutionary process there is a penalty to pay in terms of productivity, given the possibility that some early design work will be thrown away as customer requirements become clearer. However, our results showed no association between an early release to customers and lower productivity. The benefits from an early release appear to overcome the potential drawbacks of multiple iterations.

22. M. Iansiti and A. MacCormack, "Developing Products on Internet Time," Harvard Business Review 75 (September-October 1997): 108–117.

23. As a result, the correlation between feedback time and product quality is not statistically significant.

24. See, for example, R. Katz and T.J. Allen, "Investigating the Not-Invented-Here (NIH) Syndrome: A Look at the Performance, Tenure and Communication Patterns of 50 R&D Project Groups," in "Readings in the Management of Innovation," eds. M. Tushman and W. Moore (New York: HarperBusiness, 1982), 293–309.

25. We used the term "generations" to distinguish between major "platform" projects (that is, those in which major changes were made to the previous version of a product) and minor derivative/incremental projects.

26. The measure of generational experience we used in our analysis was the percentage of the development team that had previously completed more than two generations of software projects. Note that the variation in generational experience explains more than 24% of the variation in resource productivity.

27. Note that there is a relationship between those two characteristics. Namely, a scaleable architecture is likely to be modular. A modular architecture, however, is not necessarily scaleable.

28. The measure we used, adjusted to control for scale effects, was a ratio of the resources dedicated to architectural design relative to the resources dedicated to development and testing.

29. Those investments explain more than 15% of the variation in product quality. We found no significant association between them and differences in resource productivity.

30. In our sample, measures of the parameters in combination explain almost half the variation in product quality and a quarter of the variation in resource productivity.

31. For example, consider attempting back in early 1996 to conduct market research into the features that a browser should contain. Most people would have had no clue what a browser was meant to do. Hence traditional market research techniques (focus groups, surveys and the like) would have had less value.

ADDITIONAL RESOURCES

Readers interested in the general topic of managing product development are directed to a popular textbook, "Revolutionizing Product Development," by Steven Wheelwright and Kim Clark, published in 1992 by Free Press. The most practical publication specifically on software development may be the 1996 Microsoft Press book "Rapid Development," by Steve McConnell.

A deeper discussion of the open-source approach can be found at www.opensource.org/. To read more about Linux and one of the companies involved in its distribution, see the Harvard Business School case "Red Hat and the Linux Revolution," by Alan MacCormack, no. 9-600-009.

For a discussion of Microsoft's approach to developing software, see "Microsoft Secrets," by Michael Cusumano and Richard Selby, a 1995 Free Press book. Harvard Business School's multimedia case "Microsoft Office 2000," by Alan MacCormack, illustrates that approach in detail (case no. 9-600-023), and the accompanying CD-ROM contains interviews with team members and a demonstration of Microsoft's Web-based project-management system.

A new model of software development with similarities to the evolutionary model is "extreme programming." Details can be found at www.ExtremeProgramming.org/. The Software Engineering Institute at Carnegie-Mellon University is a useful source of research on software-engineering management. See www.sei.cmu.edu/.

21

Meeting the Challenge of Global Team Management

EDWARD F. MCDONOUGH III AND DAVID CEDRONE

Assembling product developers into global teams is an important mechanism for developing products that meet globally-consistent needs; however, these global teams differ from traditional project teams in several respects, including:

- They are composed of individuals who are globally dispersed.
- They meet face-to-face rather infrequently, if at all, during the course of a project.
- Members are from different cultures and speak different languages.

As a consequence, managing global teams presents new, and in many respects, more difficult challenges. In an effort to manage their global teams effectively, companies have used a variety of information technologies, such as videoconferencing, audioconferencing, and e-mail.[1]

Although many of the shortcomings associated with these technologies have been overcome, most companies have found that technology alone has not been the answer to achieving satisfactory global team performance. As recent research indicates, the principal problems facing these companies involve the failure to effectively manage their people.[2]

Digital Equipment Corporation was one of the many companies that attempted to rely on an array of telecommunications systems to overcome the problems created by the geographic dispersion of team members. Because there has been little research on the *management* of global teams, Digital had little choice but to try a variety of approaches in an effort to increase its speed of response, enhance its flexibility, and improve the quality of its products. Despite their best efforts, management was not satisfied with the results. In this article, we discuss Digital's experience using what we refer to as a Globally Dispersed Team (GDT) approach and lessons learned about managing in an environment where face-to-face meetings are the exception rather than the rule.

DEVELOPING A STRATEGIC PLAN

Digital's Telecommunications Group was asked to develop a transmission systems strategic plan. Traditionally, this task was performed by a central staff, which handed the completed plan to the line organization for implementation. Although this approach had worked well in the past, the telecommunications environment facing Digital had now become bewilderingly complex, with the globalization of many of Digital's business initiatives demanding intergrated, international telecommunications systems. These, in turn, required the development and adoption of worldwide technical standards within the Telecom organization. At the same time, the organizations that Telecom supported were expecting Telecom to implement global systems that were sensitive to local markets worldwide.

Consequently, it was no longer reasonable to expect a traditional cross-functional team co-located in the headquarters building to be aware of critical information that resided with people and

Reprinted with permission from *Research-Technology Management*, June–July, 2000.

Edward F. McDonough III is Professor of Organizational Behavior at Northeastern University's College of Business. **David Cedrone** is Director of Planning and Business Development for telecommunications industry solutions at Compaq Computer in Littleton, MA.

markets spread over the globe. The deliberate pace that typified the development of a strategic plan had suddenly become a liability. Clearly, there was a need for a new approach, but just what it should be was not entirely clear. Digital had been an early adopter of cross-functional teams to develop new products, but Telecom's global needs presented challenges that traditional cross-functional teams were not forced to confront.

TRADITIONAL CROSS-FUNCTIONAL TEAMS

The success of a cross-functional team is largely dependent on the creation of a shared team identity, the development of mutual respect and trust among team members, and the formation of supportive and collaborative personal relationships between team members.[3] These characteristics foster the free exchange of information, the sharing of risks, and a problem-solving climate.

Creating such an environment is easier when the team is co-located.[4] One approach used by Digital and other companies has been to relocate members of the team to a single location for a few days or even several months. After individuals return to their original home office location to work on the project, they meet periodically to rekindle the bonds that had begun to form at their initial meeting.

The dispersion issue has also been dealt with by using the "pony express" manager approach: team members remain in their respective locations while the leader travels to visit with each of the members. By doing this, the company seeds ideas from one location to another and facilitates the development of a team culture without having to disrupt the lives and work of each team member.

The appeal of both of these approaches is that they permit face-to-face meetings between the team members and the team leader, which helps team members to establish trusting, supportive, and collaborative personal relationships. The disadvantages of each, however, quickly become apparent at Digital.

Relocating individuals inflicted a terrible toll on their psychic energy. Many individuals found

that being away from home for an extended period caused major disruptions in their personal and family lives. While such burdensome disruptions may have been acceptable in the past, today's workforce has come to see such disruptions as being less so. In addition, from the company's point of view, taking people away and relocating them meant that they were then unavailable for consultation on other current projects. It also meant that the person could only be used on a single project as opposed to working on several simultaneously.

The pony express manager facilitated face-to-face interactions with the team—when he was there! The problem was that often he wasn't. As a result, members of the team often had to wait for the arrival of their manager to secure an answer or to obtain agreement on proceeding with a request. In addition, while there was no question that the pony express manager was the formal leader of the team and the central node in the team network, because he was away so much a great deal of time was taken up getting "reacquainted" at the beginning of each meeting.

These drawbacks become major deficiencies when new product needs shift from a national setting to a global scale. Relocating individuals means relocating them not just three states away, but three countries away. When the costs of relocating individuals, flying pony express managers around the globe, and not having people available when other situations call for their services are added to these other major deficiencies, it's not difficult to see why Digital continued to search for a better way to develop its strategic plan.

GLOBALLY DISPERSED TEAMS

The Globally Dispersed Team approach used to develop the transmission systems strategic plan was dubbed the "Columbus Team." Telecom's environment demanded diverse inputs from individuals located in different parts of the world, greater responsiveness to local situations, and sensitivity to market issues. To meet these requirements, a diverse team was assembled from Switzerland,

---

Managing a Globally Dispersed Team

- Recognize the importance of each individual's network as a supporting resource for risk taking early in the project and an ongoing source of motivation throughout the project.
- Identify and link individual work assignments to the priorities and values of the team members' networks, thereby ensuring early commitment to the team and support of the members by their networks.
- Distinguish between "safe" work and "risky" work, and create environments that support the team members' needs to accomplish both, either within the team or with their networks.
- Employ technology that is appropriate to the needs and abilities of the team members and is consistent with the work, for example, audio conferencing for status reporting and computer conferencing for workgroup collaboration.

Pay careful attention to meeting management techniques, such as moderating audio conference calls to ensure equal and full participation, creating and maintaining distribution lists and structuring computer conference topics to provide a "roadmap" for discussions.

Level "the electronic playing field" so that all team members have equal access to, and proficiency with, the communications technologies, and ensure that team members are not disadvantaged due to a lack of familiarity with technology.

France, Japan, and within the United States from New York and four New England locations. Importantly, each team member remained in his or her own country while participating as a member of this team.

Although the organization of the GDT met many of the needs of the Telecom Group, we found that using a GDT also raised three particularly vexing issues, discussed below.

1. Motivating GDT Members

While motivating project team members can often be challenging, motivating GDT members who are globally dispersed provides its own unique set of challenges. First, each member was a member of several other projects besides Columbus, thus creating the potential for divided loyalties as members came under pressure from their local managers to devote the majority of their time to local priorities at the expense of the global project. Second, the Columbus project had a number of sub-goals, each of which was seen as most important by a different member of the team, thus making it difficult to motivate individuals to act in concert to achieve the project's common goals. Another problem was that because team members rarely met face-to-face,

they had little opportunity to get to know one another and to develop trusting relationships.

One means of fostering motivation among team members is to leverage the individual's network. A network is similar to the "invisible college"; it is composed of individuals who possess similar functional expertise and who speak the same "language" (i.e., jargon) and who can, therefore, interact at the same intellectual level. As we use the term, a network is a set of relationships that an individual develops informally over time with other individuals who may be in another part of the company or in another company altogether. Other members of this network are, in all likelihood, working on their own projects for their own organization.

In the case of the Columbus project, each member's network was widely dispersed, in some cases, across several countries. However, continual interaction around common issues and the use of common jargon helped to establish a common bond and trust between individuals and their networks. This, in turn, played an important role in fostering an environment in which members of the network could talk openly, thus making high-quality decisions much more likely. The strong network bonds so necessary to building trust were forged from the regular contact that individuals in the network maintained via telephone conversations, electronic mail, computer conferences, and even video conferences.

As a result of the wide diversity of cultures and positions in the organizational hierarchy, the goals of the Columbus GDT members varied significantly. The top priority of the Europeans was the first-time implementation of a telecommunications inventory management system, while the group representing corporate headquarters' interests was focused on the conversion of existing databases. At the same time, the applications engineers were concerned with systems integration and development standards, while the product managers wanted to shorten development schedules.

The combination of the geographic dispersion of GDT members, their networks, and their different goals—functional vs. local organization—made it impossible to rely on developing a team identity as the sole means of motivating individual team members. The individual's strong affiliation with the network meant that common project goals alone could not be relied upon to motivate the Columbus team. Instead, it was necessary to use the individual's connection with the network as a motivator, in large part because members placed greater importance on being valued by their network for their knowledge, experience and results, and less importance on belonging to a Telecom GDT. In sum, the motivation for members of GDTs was based more on their professional influences (their network), than on any shared allegiance with team members.

While some traditional motivators do exist even in global team situations, our experience suggests that they are significantly less effective. Thus, it is important to recognize and leverage this additional source of motivation—the network. Indeed, we found that it was the individual's "global network of peers" that provided the strongest source of individual motivation. It was the desire to "perform well" in the eyes of other members of the network that had the strongest impact on each person's behavior and actions. Thus, to leverage the network and motivate the individual team member, it was necessary to first recognize the importance of the

network to the individual and then to allow the individual to have at least some say in selecting tasks that would both contribute to the project and to his or her stature within the network.

In selecting members for the Columbus team the manager sought individuals with specific skills in a range of telecommunications disciplines, as well as geographic and organizational representation, to contribute to the strategic plan. The expectation was that this diverse body of specialists would find synergies in their collaboration across disciplines, yielding a more integrated and effective plan.

Before this collaboration could happen within the GDT, however, the individuals needed some source of motivation to prioritize the Columbus effort in the context of other, pressing demands on their time from their local organizations. Consequently, the GDT manager encouraged the team members to identify elements of work that would "feed" the strategic planning effort but that were also relevant to outstanding issues or current pressing concerns in their individual disciplines. This allowed them to select tasks that were of high priority to their network and also relevant to the Columbus effort.

For example, the strategy for telecommunications transmission and routing systems and services was an essential element of the Columbus effort but also affected current contractual commitments with existing vendors. The ability of transmission engineers on the team to direct the longer-range strategy while simultaneously influencing near-term vendor selection or term and volume purchase decisions, was of great importance to both the Columbus effort and to the interests of the members' networks.

Recognizing this need to link the early work assignments of members of the GDT to the priorities and values of the members' networks is key to achieving the early commitment of the members to the team and support of the members by their networks. How well the GDT performed, then, was dependent on how well the influence of each individual's *network* could be leveraged to the benefit of the project.

2. Creating a "Safe" Environment

A safe work environment is a *psychological* environment in which team members feel free to brainstorm, present untested ideas, and react quickly to each others' ideas. This environment usually comes about as a result of interpersonal, face-to-face interaction among team members, but globally dispersed teams have little opportunity to forge bonds of trust and mutual respect. Also, with team members having less access to each other's work and ideas, there is no "common" repository for their work in process and no way for individuals to access the work of other team members, comment on it, or use it.

In order for the strategic plan that the Columbus team was developing to be coherent, members had to be able to see what others had done and understand how what they were doing fit in with what others were doing. Because a physical "place" for the location of the work was not possible, an "electronic workplace" had to be created.

Having an electronic workplace helped meet the need for a common repository for work but it did not immediately mean that individuals made use of it. While ideas were being tested and approaches kicked around in each person's network, team members were reluctant to leave work in process in an "electronic place" that was open to all members of the team. To do so would require that they "expose" themselves by testing ideas or admitting a lack of knowledge in "public."

While personal exposure is routine in development work, it nevertheless requires a certain level of security and safety. In globally dispersed teams, creating a safe environment takes more time and requires a different approach. The first step involves distinguishing between "safe" work and "risky" work. Safe work involves more objective tasks such as helping to establish the project's parameters, milestones, schedules, budgets, and overall goals. Risky work is work that is usually undertaken on an individual basis and includes such tasks as developing innovative solutions to problems and trying out new approaches and techniques.

Initially, Columbus team members were not asked to expose their risky work to the team. Instead, the exposed work involved decisions about schedules, budgets, milestone development, and task prioritization. These tasks required a lower level of trust among team members because they involved issues that were more objective and thus "safer." Risky work—i.e., the development work on the project—was still being done. However, instead of sharing it, the outcomes of risky tasks were initially presented to members of each person's network because it was here that team members felt safe and secure that their ideas could be shared even when they were only partially formed. Over time, as the level of trust among the project team members increased, they became increasingly comfortable making riskier decisions, and distinguishing between safe and risky work became less important.

3. Managing Communications

Although many researchers have discussed the use of telecommunications technologies to improve communication among employees in dispersed locations,[1] it was the ability to manage those technologies in the absence of face-to-face meetings that contributed to the success of the Columbus GDT. Understanding how to use these technologies at the appropriate time and in different situations, however, was not obvious and required considerable trial and error by the GDT manager.

We found, for example, that using an audioconference in an initial meeting allowed people to be introduced on a more personal level than if the first meeting had been conducted via e-mail or computer conferencing, e.g., VAXNotes, UseNet Newsgroups or Lotus Notes. Because audioconferencing permits real-time dialogue between team members, it helped to establish a baseline safety zone and to gradually increase the comfort and trust within the GDT, which is imperative to achieving effective communication.

We also discovered that computer conferencing at the start of the project was a poor choice, even with experienced users. It is one thing to meet new teammates, be introduced, and establish a rapport on an audioconference call, but something altogether different to have these "conversations" recorded and retained in a computer conference.

Because an audioconference call does not become a permanent document, conversation can be much more casual and open, with less concern for precision of expression. A computer conference tends to inhibit the development of relationships. The GDT manager found that later in the project, however, computer conferencing and e-mail became important tools because they allowed the team to work on the basis of individual schedules rather than on a slower team schedule.

Audioconference meeting techniques were found to be equally crucial, as was the selection of appropriate technologies.[5] For example, conducting a successful audioconference, where the subtleties of eye contact and body language are non-existent, makes it especially important to e-mail a structured meeting agenda to participants ahead of time and to use a moderator. Unstructured audioconferences can easily run astray, resulting in reduced confidence in a project's success and, in turn, a loss of commitment to the project by the team.

Other skills often need to be subtly taught or reinforced by the GDT manager. For example, it was important that each team member be introduced at the beginning of the meeting and that individuals identify themselves before speaking. While calling on silent members of the team for opinions is something that team leaders always need to do, when team members are not physically present quiet members can totally disappear. In this circumstance, it is crucial for the leader to make note of who is and who is not participating so that they too do not overlook silent members. Finally, encouraging vivid verbal descriptions of ideas was a very useful means of offsetting the lack of visual cues.

Similar techniques are also critical to successful e-mail and computer conferencing. Creating and maintaining distribution lists and structuring computer conference topics, for example, provide roadmaps for individuals to follow. It is the role of the GDT manager to create and manage this electronic environment in order to ensure a successful launch of the project and to maintain the commitment of team members.

In addition, in an environment where team members do not meet face-to-face, ensuring that members feel knowledgeable about and comfortable with the use of various electronic technologies has an impact on their ability to perform their work effectively. A team member's lack of facility in using computer conferencing, for example, can exacerbate the tensions that already exist between individuals who are unfamiliar with one another to begin with. Thus, an important role for the GDT manager to play was to create and maintain a "level electronic playing field."

Elements of this electronic playing field may include e-mail, audioconferencing, computer and videoconferencing, and common databases (for storage and reference). On the Columbus project, these allowed the team to collaborate on written, graphical and image products of their work. Although most GDT members were familiar with a variety of electronic technologies, it became apparent that not all members of the team were equally familiar with and skilled at using electronic tools, nor did they have a similar understanding of the norms of behavior required to work in a purely electronic medium.

Leveling the electronic playing field was important to fostering a baseline of trust within the GDT and to ensure that less knowledgeable team members were not further disadvantaged in an environment that had already been stripped of traditional visual cues. The GDT manager helped to foster trust among the team by conducting the first team meeting using a familiar and not technically threatening technology—audioconferencing. In addition, he had everyone call in from their own offices rather than having groups assemble around a speakerphone in a conference room. This prevented the isolation of one member on a single phone from those using a speakerphone.

During the course of the project, it was also important to make deliberate decisions about which technologies to use. The benefits of a particular technology had to be traded off with each member's familiarity with it. Videoconferencing, for example, permits participants to see the body language of other team members and other visual cues, white boards, flip charts, and the like. But, while this technology contributes several additional communication dimensions, it also introduces technological and behavioral complexities that must be managed to prevent pockets of advantage and disadvantage within the GDT.

Where there was a large gap between one member's knowledge and familiarity with a technology and another's, it was necessary to provide training to members who were unfamiliar with the ins and outs of a particular technology. While it is safe to assume everyone is familiar with the technical operation of the telephone, for example, it is not equally safe to expect the same level of comfort and experience with behavioral practices on a conference call. Thus, it was necessary to make sure that all team members announced themselves, that all members had an opportunity to speak, and that concurrence on a topic was confirmed by a roll call vote.

ONGOING RESEARCH

The role of the GDT manager is far more complex than the role of the traditional project manager. GDT managers need to interact with the individual's network in such a way that the manager can leverage the motivational influence of the network on the team member. At the same time, the GDT manager must establish and manage an electronic workplace that is based on a variety of telecommunications systems and tools that support the needs of the GDT, and reflect the readiness of each member to use these tools. While we are only just beginning to understand how to manage global teams effectively, our research on effectively managing global new product development is ongoing. Readers interested in reading more about our findings on this subject are directed to our other articles referenced below. Companies wishing to participate in our ongoing research are invited to contact E. F. McDonough at e.mcdonough@neu.edu.

REFERENCES AND NOTES

1. Johnson, L. "Advances in Telecommunications Technologies That May Affect the Location of Business Activities." Rand Note, Santa Monica, 1992; Charan, R. "How Networks Reshape Organizations—For Results." *Harvard Business Review* 69:104–115 (1991); Ciborra, C. U. *Teams, Markets and Systems*, New York: Cambridge University Press, 1993; Dubini, P. and Aldrich, H. "Personal and Extended Networks Are Central to the Entrepreneurial Process." *Journal of Business Venturing* 6:305–312 (1991).

2. Edward F. McDonough III, Kenneth B. Kahn. "Using 'Hard' and 'Soft' Technologies for Global New Product Development." *R&D Management* 26:241–253 (1996); Edward F. McDonough III, Kenneth B. Kahn, and Abbie Griffin, "Managing Communication in Global Product Development Teams." *IEEE Transactions on Engineering Management* 46:375–386 (Nov. 1999); Edward F. McDonough III, Kenneth B. Kahn, and Gloria Barczak, "Effectively Managing Global New Product Teams." *Proceedings of the International Product Development & Management Association Conference*, Atlanta, GA, 1998.

3. Edward F. McDonough III. "An Investigation of Factors Contributing to the Success of Cross-Functional Teams." *Journal of Product Innovation Management*, forthcoming; Hershock, R. J., Cowman, C. D., and Peters, D. "From Experience: Action Teams That Work." *Journal of Product Innovation Management*: 1195–104 (1994).

4. Nancy Ross-Flanigan. "The Virtues and Vices of Virtual Colleagues." *Technology Review*, March/April: 53–59 (1998).

5. Initially it was thought that videoconferencing would be the preferred means of meeting. However, we quickly discovered its limitations. Many members of the team were required to spend considerable time traveling to the videoconference site and, equally important, it increased the formality of the meetings—just the opposite from what was desired.

8

Informal Critical Roles in
Leading Innovation

22

Critical Functions

Needed Roles in the Innovation Process

EDWARD B. ROBERTS AND ALAN R. FUSFELD

This article examines the main elements of the technology-based innovation process in terms of certain usually informal but critical "people" functions that can be the key to an effective organizational base for innovation. This approach to the innovation process is similar to that taken by early industrial theorists who focused on the production process. Led by such individuals as Frederick W.

Taylor, their efforts resulted in basic principles for increasing the efficiency of producing goods and services. These principles of specialization, chain of command, division of labor, and span of control continue to govern the operation of the modern organization (despite their shift from popularity in many modern business schools). Hence, routine tasks in most organizations are arranged to facili-

From E. Roberts and A. Fusfeld, "Critical Functions: Needed Roles in the Innovation Process," in *Career Issues in Human Resource Management*, R. Katz (ed.), © 1982, pp. 182–207. Reprinted by permission of Prentice-Hall, Inc., Englewood Cliffs, NJ.

Edward B. Roberts is Sarnoff Professor of Management at MIT's Sloan School of Management. ***Alan R. Fusfeld*** is President of The Fusfeld Group, a management of technology consulting firm.

tate work standardization with expectations that efficient production will result. However, examination of how industry has organized its innovation tasks—that is, those tasks needed for product/process development and for responses to nonroutine demands—indicates an absence of comparable theory. And many corporations' attempts to innovate consequently suffer from ineffective management and inadequately staffed organizations. Yet, through tens of studies about the innovation process, conducted largely in the last fifteen years, we now know much about the activities that are requisite to innovation as well as the characteristics of the people who perform these activities most effectively.

The following section characterizes the technology-based innovation process via a detailed description of a typical research and development project life cycle. The types of work activities arising in each project phase are enumerated. These lead in the third section to the identification of the five basic critical roles that are needed for effective execution of an innovative effort. Problems associated with gaps in the fulfillment of the needed roles are discussed. Detailed characteristics and specific activities that are associated with each role filler are elaborated upon in the fourth section. The multiple roles that are sometimes performed by certain individuals are observed, as are the dynamics of role changes that tend to take place over the life span of a productive career. The fifth section presents several areas of managerial implications of the critical functions concepts, beginning first with issues of manpower planning, then moving to considerations of job design and objective setting and to the determination of appropriate performance measures and rewards. How an organizational assessment can be carried out in terms of these critical functions dimensions is discussed in the last section.

THE INNOVATION PROCESS

The major steps involved in the technology-based process are shown in Figure 22.1. Although the project activities do not necessarily follow each other in a linear fashion, there is more or less clear demarcation between them. Moreover, each stage, and its activities, require a different mix of people skills and behaviors to be carried out effectively.

This figure portrays six stages as occurring in the typical technical innovation project, and sixteen representative activities that are associated with innovative efforts. The six stages are here identified as:

1. Pre-project
2. Project possibilities
3. Project initiation
4. Project execution
5. Project outcome evaluation
6. Project transfer

These stages often overlap and frequently recycle.[1] For example, problems or findings that are generated during project execution may cause a return to project initiation activities. Outcome evaluation can restart additional project execution efforts. And, of course, project cancellation can occur during any of these stages, redirecting technical endeavors back into the pre-project phase.

A variety of different activities are undertaken during each of the six stages. Some of the activities, such as generating new technical ideas, arise in all innovation project stages from pre-project to project transfer. But our research studies and consulting efforts in dozens of companies and government labs have shown other activities to be concentrated mainly in specific stages, as discussed below.

1. *Pre-Project.* Prior to formal project activities being undertaken in a technical organization, considerable technical work is done that provides a basis for later innovation efforts. Scientists, engineers, and marketing people find themselves involved in discussions internal and external to the organization. Ideas get discussed in rough-cut ways and broad parameters of innovative interests get established. Technical personnel work on problem-

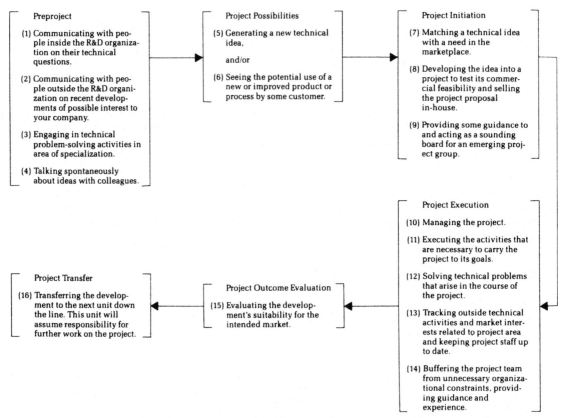

Figure 22.1. A multi-stage view of a technical innovation project.

solving efforts to advance their own areas of specialization. Discussions with numerous industrial firms in the United States and Europe suggest that from 30 to 60% of all technical effort is devoted to work outside of or prior to formal project initiation.

2. *Project Possibilities*. Arising from the preproject activities, specific ideas are generated for possible projects. They may be technical concepts for assumed-to-be-feasible developments. Or they may be perceptions of possible customer interest in product or process changes. Customer-oriented perspectives may be originated by technical or marketing or managerial personnel out of their imagination or from direct contact with customers or competitors. Recent evidence indicates that many of these ideas enter as "proven" possibili-

ties, having already been developed by the customers themselves.[2]

3. *Project Initiation*. As ideas evolve and get massaged through technical and marketing discussions and exploratory technical efforts, the innovation process moves into a more formal project initiation stage. Activities occurring during this phase include attempts to match the directions of technical work with perceived customer needs. (Of course, such customer needs may exist either in the production organization or in the product's marketplace.) Inevitably, a specific project proposal has to be written up, proposed budgets and schedules have to get produced, and informal pushing as well as formal presentations have to be undertaken in order to sell the project. A key input during this stage is the counseling and encouragement that senior technical professionals or laboratory and marketing management may provide to the emerging project team.

4. Project Execution. With formal approval of a project aimed at an innovative output, activities increase in intensity and focus. In parallel, someone usually undertakes planning, leadership, and coordination efforts related to the many continuing technical idea-generating and problem-solving activities being done by the engineers and scientists assigned to the project. Technical people often make special attempts to monitor (and transfer in) what had been done previously as well as what is then going on outside the project that is relevant to the project's goals. Management or marketing people frequently take a closer look at competitors and customers to be sure the project is appropriately targeted.[3] Senior people try to protect the project from overly tight control or from getting cut off prematurely, and the project manager and other enthusiasts keep fighting to defend their project's virtues (and budget). Unless canceled, the project work continues toward completion of its objectives.

5. Project Outcome Evaluation. When the technical effort seems complete, most projects undergo another often intense evaluation to see how the results stack up against prior expectations and current market perceptions. If a successful innovation is to occur, some further implementation must take place, either by transfer of the interim results to manufacturing for embodiment in its process or for volume production activities, or by transfer to later stages of further development. All such later stages involve heavier expenditures and the post-project evaluation can be viewed as a pre-transfer screening activity.

6. Project Transfer. If the project results survive this evaluation, transfer efforts take place (e.g., from central research to product department R&D, or development to manufacturing engineering).[4] The project's details may require further technical documentation to facilitate the transfer. Key technical people may be shifted to the downstream unit to transfer their expertise and enthusiasm, since downstream staff members, technical or marketing, often need instruction to assure effective continuity. Within the downstream organizational unit, the cycle of stages may begin again, perhaps bypassing the earliest two stages and starting with project initiation or even project execution. This "pass-down" continues until successful innovation is achieved, unless project termination occurs first.

NEEDED ROLES

Assessment of activities involved in the several-stage innovation process, as just described, points out that the repeated direct inputs of five different work roles are critical to innovation. The five arise in differing degrees in each of the several steps. Furthermore, different innovation projects obviously call for variations in the required role mix at each stage. Nevertheless, all five work roles must be carried out by one or more individuals if the innovation is to pass effectively through all six steps.

The five critical work functions are:

- *Idea Generating.* Analyzing and/or synthesizing (implicit and explicit) information (formal and informal) about markets, technologies, approaches, and procedures, from which an idea is generated for a new or improved product or service, a new technical approach or procedure, or a solution to a challenging technical problem.[5]

- *Entrepreneuring or Championing.* Recognizing, proposing, pushing, and demonstrating a new (his or her own or someone else's) technical idea, approach or procedure for formal management approval.[6]

- *Project Leading.* Planning and coordinating the diverse sets of activities and people involved in moving a demonstrated idea into practice.[7]

- *Gatekeeping.* Collecting and channeling information about important changes in the internal and external environments; information gatekeeping can be focused on developments in the market, in manufacturing, or in the world of technology.[8]

- *Sponsoring or Coaching.* "Behind-the-scene" support-generating function of the protector and advocate, and sometimes of the "bootlegger" of funds; the guiding and developing of less-experienced personnel in their critical roles (a "Big Brother" role).[9]

Lest the reader confuse these roles as mapping one-for-one with different people, three points need emphasis: (1) some roles (e.g., idea generating) frequently need to be fulfilled by more than one person in a project team in order for the project to be successful; (2) some individuals occasionally fulfill more than one of the critical functions; and (3) the roles that people play periodically change over a person's career with an organization.

Critical Functions

These five critical functions represent the various roles in an organization that must be carried out for successful innovation to occur. They are critical from two points of view. First, each role is different or unique, demanding different skills. A deficiency in any one of the roles contributes to serious problems in the innovation effort, as we illustrate below. Second, each role tends to be carried out primarily by relatively few individuals, thereby making even more unique the critical role players. If any one critical function-filler leaves, the problem of recruiting a replacement is very difficult—the specifics of exactly who is needed is dependent on usually unstated role requirements.

We must add at this point that another role clearly exists in all innovative organizations, but it is not an *innovative* role! "Routine" technical problem-solving must be carried out in the process of advancing innovative efforts. Indeed the vast bulk of technical work is probably routine, requiring professional training and competence to be sure, but nonetheless routine in character for an appropriately prepared individual. A large number of people in innovative organizations do very little "critical functions" work; others who are important performers of the critical functions also spend a good part of their time in routine problem-solving activity. Our estimate, supported now by data from

numerous organizations, is that 70 to 80% of technical effort falls into this routine problem-solving category. But the 20 to 30% that is unique and critical is the part we emphasize.

Generally, the critical functions are not specified within job descriptions since they tend to fit neither administrative nor technical hierarchies; but they do represent necessary activities for R&D, such as problem definition, idea nurturing, information transfer, information integration, and program pushing. Consequently, these role behaviors are the underlying informal functions that an organization carries out as part of the innovation process. Beyond the five above, different business environments may also demand that additional roles be performed to assure innovation.[10]

It is desirable for every organization to have a balanced set of abilities for carrying out these roles as needed, but unfortunately few organizations do. Some organizations overemphasize one role (e.g., idea generating) and underplay another role (e.g., entrepreneuring). Another organization might do just the reverse. Nonetheless, technical organizations tend to assume that the necessary set of activities will somehow be performed. As a consequence, R&D labs often lack sensitivity to the existence and importance of these roles, which, for the most part, are not defined within the formal job structure. How the critical functions are encouraged and made a conscious part of technology management is probably an organization's single most important area of leverage for maintaining and improving effective innovation. The managerial capabilities required for describing, planning, diagnosing problems, and developing the necessary teamwork in terms of the people functions demanded by an innovative program are almost entirely distinct from the skills needed for managing the technical requirements of the tasks.

Impact of Role Deficiencies

Such an analytic approach to developing an innovative team has been lacking in the past and, consequently, many organizations suffer because one or more of the critical functions is not being performed adequately. Certain characteristic signs can provide evidence that a critical function is missing.

Idea generating is deficient if the organization is not thinking of new and different ways of doing things. However, more often than not when a manager complains of insufficient ideas, we find the real deficiency to be that people are not aggressively entrepreneuring or championing ideas, either their own or others'. Evidences of entrepreneuring shortages are pools of unexploited ideas that seldom come to a manager's attention.[11]

Project leading is suspect if schedules are not met, activities fall through cracks (e.g., coordinating with a supplier), people do not have a sense for the overall goal of their work, or units that are needed to support the work back out of their commitments. This is the role most commonly recognized formally by the appointment of a project manager. In research, as distinct from development, the formal role is often omitted.

Gatekeeping is inadequate if news of changes in the market, technology, or government legislation comes without warning, or if people within the organization are not getting the information that they need because it has not been passed on to them. When, six months after the project is completed, you suddenly realize that you have just succeeded in reinventing a competitor's wheel, your organization is deficient in needed gatekeeping! Gatekeeping is further lacking when the wheel is invented just as a regulatory agency outlaws its use.

Inadequate or inappropriate sponsoring or coaching often explains projects that get pushed into application too soon, or project managers who have to spend too much time defending their work, or personnel who complain that they do not know how to "navigate the bureaucracy" of their organizations.

The importance of each critical function varies with the development stage of the project. Initially, idea generation is crucial; later, entrepreneurial skill and commitment are needed to develop the concept into a viable activity. Once the project is established, good project leading/managing is needed to guide its progress. Of course, the need for each critical function does not abruptly appear and disappear. Instead, the need grows and diminishes, being a focus at some point, but of lesser importance at others. Thus, the absence of a function when it is potentially very important is a serious weakness regardless of whether or not the role had been filled at an earlier, less crucial time. As a corollary, assignment of an individual to a project, at a time when the critical role that he or she provides is not needed, leads to frustration for the individual and to a less effective project team.

Frequently, we have observed that personnel changes that occur because of career development programs often remove critical functions from a project at a crucial time. Since these roles are usually performed informally, job descriptions are made in terms of technical specialties, and personnel replacements are chosen to fill those job vacancies, rather than on their ability to fill the needs of the vacated critical roles. Consequently, the project team's innovative effectiveness is reduced, sometimes to the point of affecting the project's success. Awareness of which roles are likely to be required at what time will help to avoid this problem, as well as to allow people performing functions no longer needed to be moved to other projects where their talents can be better utilized.

CHARACTERISTICS OF THE ROLE PLAYERS

Compilation of several thousand individual profiles of staff in R&D and engineering organizations has demonstrated patterns in the characteristics of the people who perform each innovation function.[12] These patterns are shown in Table 22.1, indicating which persons are predisposed to be interested in one type of activity more than another and to perform certain types of activity well. For example, a person who is theoretically inclined and comfortable with abstractions feels better suited to the idea-generating function than does someone who is very practical and uncomfortable with seemingly discrepant data. In any unit of an organization, people with different characteristics can work to complement each other. Someone good at idea generating might be teamed with a colleague good at gatekeeping and another colleague good at en-

TABLE 22.1
Critical Functions in the Innovation Process

Personal characteristics	Organizational activities
Idea Generating	
Is expert in one or two fields	Generates new ideas and tests their feasibility
Enjoys conceptualization, comfortable with abstractions	Good at problem solving
	Sees new and different ways of viewing things
Enjoys doing innovative work	Searches for the break-throughs
Usually is an individual contributor, often will work alone	
Entrepreneuring or Championing	
Strong application interests	Sells new ideas to others in the organization
Possesses a wide range of interests	Gets resources
Less propensity to contribute to the basic knowledge of a field	Aggressive in championing his or her "cause"
	Takes risks
Energetic and determined; puts himself or herself on the line	
Project Leading	
Focus for the decision making, information, and questions	Provides the team leadership and motivation
	Plans and organizes the project
Sensitive to accommodating to the needs of others	Ensures that administrative requirements are met
Recognizes how to use the organizational structure to get things done	Provides necessary coordination among team members
Interested in a broad range of disciplines and how they fit together (e.g., marketing, finance)	Sees that the project moves forward effectively
	Balances the project goals with organizational needs
Gatekeeping[a]	
Possesses a high level of technical competence	Keeps informed of related developments that occur outside the organization through journals, conferences, colleagues, other companies
Is approachable and personable	
Enjoys the face-to-face contact of helping others	Passes information on to others; finds it easy to talk to colleagues
	Serves as an information resource for others in the organization (i.e., authority on whom to see and/or what has been done)
	Provides informal coordination among personnel
Sponsoring or Coaching	
Possesses experience in developing new ideas	Helps develop people's talents
Is a good listener and helper	Provides encouragement and guidance and acts as a sounding board to the project leader and others
Can be relatively more objective	
Often a more senior person who knows the organizational ropes	Provides access to a power base within the organization—a senior person
	Buffers the project team from unnecessary organizational constraints
	Helps the project team to get what it needs from the other parts of the organization
	Provides legitimacy and organizational confidence in the project

[a]Our empirical studies have pointed out three different types of gatekeepers: (1) technical, who relates well to advancing world of science and technology; (2) market, who senses and communicates information relating to customers, competitors, and environmental and regulatory changes affecting the marketplace; and (3) manufacturing, who bridges the technical work to the special needs and conditions of the production organization.

trepreneuring to provide necessary supporting roles. Of course, each person must understand his or her own expected role in a project and appreciate the roles of others for the teaming process to be successful. Obviously, as will be discussed later, some people have sufficient breadth to perform well in multiple roles.

Table 22.1 underlies our conclusion that each

of the several roles required for effective technical innovation presents unique challenges and must be filled with essentially different types of people, each type to be recruited, managed, and supported differently, offered different sets of incentives, and supervised with different types of measures and controls. Most technical organizations seem not to have grasped this concept, with the result that all technical people tend to be recruited, hired, supervised, monitored, evaluated, and encouraged as if their principal roles were those of creative scientists or, worse yet, routine technical problem-solvers. But only a few of these people in fact have the personal and technical qualifications for scientific inventiveness and prolific idea generating. A creative idea-generating scientist or engineer is a special kind of professional who needs to be singled out, cultivated, and managed in a special way. He or she is probably an innovative technically well-educated individual who enjoys working on advanced problems, often as a "loner."

The technical champion or entrepreneur is a special person, too—creative in his own way, but his is an aggressive form of creativity appropriate for selling an idea or product. The entrepreneur's drives may be less rational, more emotional than those of the creative scientist; he is committed to achieve, and less concerned about how to do so. He is as likely to pick up and successfully champion someone else's original idea as to push something of his own creation. Such an entrepreneur may well have a broad range of interests and activities; and he must be recruited, hired, managed, and stimulated very differently from the way an idea-generating scientist is treated in the organization.

The person who effectively performs project leading or project managing activities is a still different kind of person—an organized individual, sensitive to the needs of the several different people she is trying to coordinate, and an effective planner; the latter is especially important if long lead time, expensive materials, and major support are involved in the development of the ideas that she is moving forward in the organization.

The information gatekeeper is the commu-

nicative individual who in fact, is the exception to the truism that engineers do not read—especially that they do not read technical journals. Gatekeepers link to the sources of flow of technical information into and within a research and development organization that might enhance new product development or process improvement. But those who do research and development need market information as well as technical information. What do customers seem to want? What are competitors providing? How might regulatory shifts affect the firm's present or contemplated products or processes? For answers to questions such as these, research and development organizations need people we call the "market gatekeepers"—engineers or scientists, or possibly marketing people with technical background, who focus on market-related information sources and communicate effectively with their technical colleagues. Such a person reads journals, talks to vendors, goes to trade shows, and is sensitive to competitive information. Without him, many research and development projects and laboratories become misdirected with respect to market trends and needs.

Finally, the sponsor or coach may in fact be a more experienced, older project leader or former entrepreneur who now has matured to have a softer touch than when he was first in the organization. As a senior person, she can coach and help subordinates in the organization and speak on their behalf to top management, enabling ideas or programs to move forward in an effective, organized fashion. Many organizations totally ignore the sponsor role, yet our studies of industrial research and development suggest that many projects would not have been successful were it not for the subtle and often unrecognized assistance of such senior people acting in the role of sponsors. Indeed, organizations are more successful when chief engineers or laboratory directors naturally behave in a manner consistent with this sponsor role.

The significant point here is that the staffing needed to cause effective innovation in a technical organization is far broader than the typical research and development director has usually assumed. Our studies indicate that many ineffective technical organizations have failed to be innovative solely because one or more of these five quite different critical functions has been absent.

Multiple Roles

As indicated earlier, some individuals have the skills, breadth, inclination, and job opportunity to fulfill more than one critical function in an organization. Our data collection efforts with R&D staffs show that a few clusters explain most of these cases of multiple role-playing. One common combination of roles is the pairing of gatekeeping and idea generating. Idea-generating activity correlates in general with the frequency of person-to-person communication, especially external to the organization.[13] The gatekeeper, moreover, in contact with many sources of information, can often synergistically connect these bits into a new idea. This seems especially true of market gatekeepers, who can relate market relevance to technical opportunities.

Another role couplet is between entrepreneuring and idea generating. In studies of formation of new technical companies the entrepreneur who pushed company formation and growth was found in half the cases also to have been the source of the new technical idea underlying the company.[14] Furthermore, in studies of M.I.T. faculty, 38% of those who had ideas they perceived to have commercial value also took strong entrepreneurial steps to exploit their ideas.[15] The idea-generating entrepreneuring pair accounts for less than one half of the entrepreneurs, but not the other half.

Entrepreneuring individuals often become project leaders, usually in what is thought to be a logical organizational extension of the effective selling of the idea for the project. And some people strong at entrepreneuring indeed also have the interpersonal and plan-oriented qualities needed for project leading. But on numerous occasions the responsibility for managing a project is mistakenly seen as a necessary reward for successful idea championing. This arises from lack of focus upon the functional differences. What evidence indicates that a good salesperson is going to be a good manager? If an entrepreneur can be rewarded appropriately and more directly for his or her own function, many project failures caused by ineffective project managers might be avoided. Perhaps giving the entrepreneur a prominent project role, together with a clearly designated but different project manager, might be an acceptable compromise.

Finally, sponsoring, although it should be a unique role, occasionally gives way to a takeover of any or all of the other roles. Senior coaching can degenerate into idea domination, project ownership, and direction from the top. This confusion of roles can become extremely harmful to the entire organization. Who will bring another idea to the boss, once he steals some junior's earlier concept? Even worse, who can intervene to stop the project once the boss is running amok with his new pet?

All of the critical innovative roles, whether played singly or in multiples, can be fulfilled by people from multiple disciplines and departments. Obviously, technical people—scientists and engineers—might carry out any of the roles. But marketing people also generate ideas for new and improved products, "gatekeep" information of key importance to a project—especially about use, competition and regulatory activities—champion the idea, sometimes sponsor projects, and in some organizations even manage innovation projects. Manufacturing people periodically fill similar critical roles, as do general management personnel.

The fact of multiple role filling can affect the minimum-size group needed for attaining "critical mass" in an innovative effort. To achieve continuity of a project, from initial idea all the way through to successful commercialization, a project group must have all five critical roles effectively filled while satisfying the specific technical skills required for project problem-solving. In a new high-technology company this critical mass may sometimes be ensured by as few as one or two co-founders. Similarly, an "elite" team, such as Cray's famed Control Data computer design group, or Kelly Johnson's "Skunk Works" at Lockheed, or McLean's Sidewinder missile organization in the Navy's China Lake R&D center, may concentrate in a small number of select multiple-role players the staff needed to accomplish major objectives.

But the more typical medium-to-large company organization had better not plan on finding "Renaissance persons" or superstars to fill its job requirements. Staffing assumptions should more likely rest on estimates that 70% of scientists and engineers will turn out to be routine problem-solvers only, and that most critical role-players will be single dimensional in their unique contributions.

Career-Spanning Role Changes

We showed above how some individuals fulfill multiple critical roles concurrently or in different stages of the same project. But even more people are likely to contribute critically but differently at different stages of their careers. This does not reflect change of personality, although such changes do seem partly due to the dynamics of personal growth and development. But the phenomenon also clearly reflects individual responses to differing organizational needs, constraints, and incentives.

For example, let us consider the hypothetical case of a bright, aggressive, potentially multiple-role contributor, newly joining a company fresh from engineering school. What roles can he play? Certainly, he can quickly become an effective routine technical problem-solver and, hopefully, a productive novel idea generator. But even though he may know many university contacts and also be familiar with the outside literature, he cannot be an effective information gatekeeper, for he does not yet know the people inside the company with whom he might communicate. He also cannot lead project activities. No one would trust him in that role. He cannot effectively act as an entrepreneur, as he has no credibility as champion for change. And, of course, sponsoring is out of the question. During this stage of his career, the limited legitimate role options may channel the young engineer's productive energies and reinforce his tendencies toward creative idea output. Alternatively, wanting to offer more and do more than the organization will "allow," this high-potential young performer may feel rebuffed and frustrated. His perception of what he can expect from the job and, perhaps more important, what the job will expect from him, may become set in these first few months on the job.

Some disappointeds may remain in the company, but "turn off" their previously enthusiastic desire for multidimensional contributions. More likely, the frustrated high-potential will "spin-off" and leave the company in search of a more rewarding job, perhaps destined to find continuing frustration in his next one or two encounters. For many young professionals the job environment moves too slowly from encouraging idea generating to even permitting entrepreneurial activities.

With two or three years on the job, however, the engineer's role options may broaden. Of course, routine problem solving and idea generating are still appropriate. But some information gatekeeping may now also be possible, as communication ties increase within the organization. Project leading may start to be seen as legitimate behavior, particularly on small efforts.[16] And the young engineer's work behavior may begin to reflect these new possibilities. But perhaps his attempts at entrepreneurial behavior would still be seen as premature. And sponsoring is not yet a relevant consideration.

With another few years at work, the role options are still wider. Routine problem-solving, continued idea generating, broad-based gatekeeping (even bridging to the market or to manufacturing), responsible project managing, as well as project championing may become reasonable alternatives. Even coaching a new employee becomes a possibility. The next several years can strengthen all these role options, a given individual tending usually to focus on one of these roles (or on a specific multiple-role combination) for this midcareer period.

Getting out of touch with a rapidly changing technology may later narrow the role alternatives available as the person continues to age on the job. Technical problem-solving effectiveness may diminish in some cases, idea generating may slow down or stop, technical information gatekeeping may be reduced. But market and/or manufacturing gatekeeping may continue to improve with increased experience and outside contacts, project managing capabilities may continue to grow as more past projects are tucked under the belt, entrepreneuring may be more important and for higher stakes, and sponsoring of juniors in the company may be more generally sought and practiced. This career phase is too often seen as characterized by the problem of technical obsolescence, especially if the organization has a fixation on assessing engineer performance in terms of the narrow but traditional stereotypes of technical problem solving and idea generating. "Retooling" the engineer for an earlier role, usually of little current interest and satisfaction to the more mature, broader, and differently directed person, becomes a source of mutual grief and anxiety to the organization and the individual. An aware organization, thinking in terms of critical role differences, can instead recognize the self-selected branching in career paths that has occurred for the individual. Productive technically trained people can be carrying out critical functions for their employers up to retirement, if employers encourage the full diversity of vital roles.

At each stage of his evolving career, an individual can encounter severe conflicts between his organization's expectations and his personal work preferences. This is especially true if the organization is inflexible in its perception of appropriate technical roles. In contrast, with both organizational and individual adaptability in seeking mutually satisfying job roles, the scientist or engineer can contribute continuously and importantly to accomplishing innovation. As suggested in this illustrative case, during his productive career in industry the technical professional may begin as a technical problem solver, spend several years primarily as a creative idea generator, add technical gatekeeping to his performance while maintaining his earlier roles, shift toward entrepreneuring projects and leading them forward, gradually grow in his market-linking and project managing behavior, and eventually accrue a senior sponsoring role while maintaining senior project-program-organizational leadership until retirement. But this productive full life is not possible if the engineer is pushed to the side early as a technically obsolete contributor. The perspective taken here can lead to a very different approach to career development for professionals than is usually taken by industry or government.

MANAGING THE CRITICAL FUNCTIONS FOR ENHANCED INNOVATION

To increase organizational innovation, a number of steps can be taken that will facilitate implementation of a balance among the critical functions. These steps must be addressed explicitly or organizational focus will remain on the traditionally visible functions that produce primarily near-term incremental results, such as problem solving. Indeed, the "results-oriented" reward systems of most organizations reinforce this short-run focus, causing other activities to go unrecognized and unrewarded.

We are not suggesting that employees should ignore the problem-solving function for the sake of the other functions. Rather, we are emphasizing the need for a balance of time and energy distributed among all functions. As indicated earlier, our impressions and data suggest that 70 to 80% of the work of most organizations is routine problem-solving. However, the other 20 to 30% and the degree of teamwork among the critical functions make the difference between an innovative and a noninnovative organization.

Implementing of the results, language, and concepts of a critical functions perspective is described below for the selected organizational tasks of manpower planning, job design, measurement, and rewards. If critical functions awareness dominated managerial thinking, other tasks, not dealt with here, would also be done differently, including R&D strategy, organizational development, and program management.

Manpower Planning
The critical functions concept can be applied usefully to the recruiting, job assignment, and development or training activities within an organization. In recruiting, an organization needs to identify not only the specific technical or managerial requirements of a job, but also the critical function activities that the job requires. That is, does the job require consulting with colleagues as an important part of facilitating teamwork? Or does it require the coaching and development of less experienced personnel to ensure the longer-run productivity of that

area? To match a candidate with the job, recruiting should also include identification of the innovation skills of the applicant. If the job requires championing, the applicant who is more aggressive and has shown evidence of championing new ideas in the past should be preferred over the less-aggressive applicant who has shown much more technically oriented interests in the past.

As indicated above, there is room for growth from one function to another, as people are exposed to different managers, different environments, and jobs that require different activities. Although this growth occurs naturally in most organizations, it can be explicitly planned and managed. In this way, the individual has the opportunity to influence his growth along the lines that are of most interest to him, and the organization has the opportunity to oversee the development of personnel and to ensure that effective people are performing the essential critical functions.

Industry has at best taken a narrow view of manpower development alternatives for technical professionals. The "dual ladder" concept envisions an individual as rising along either "scientific" or "managerial" steps. Attempted by many but with only limited success ever attained, the dual ladder reflects an oversimplification and distortion of the key roles needed in an R&D organization.[17] As a minimum, the critical function concept presents "multiladders" of possible organizational contribution; individuals can grow in any and all of the critical roles while benefiting the organization. And depending on an organization's strategy and manpower needs, manpower development along each of the paths can and should be encouraged. Most job descriptions and statements of objectives emphasize problem-solving, and sometimes project leading. Rarely do job descriptions and objectives take into account the dimensions of a job that are essential for the performance of the other critical functions. Yet availability of unstructured time in a job can influence the performance of several of the innovation functions. For example, to stimulate idea generating, some slack time is necessary so that employees can pursue their own ideas and explore new and interesting ways of doing things. For gatekeeping to occur, slack time also needs to be available for employees to communicate with colleagues and pass along information learned, both internal to and external to the organization. The coaching role also requires slack time, during which the "coach" can guide less experienced personnel. Table 22.2 elaborates our views on the different emphasis on deadlines (i.e., the alternative to slack time) for each of the critical functions and the degree of specificity of task assignments (i.e., another alternative to slack) for each function.

These essential activities also need to be in-

TABLE 22.2
Job Design Dimensions

Dimension of job	Critical function				
	Idea generating	Entrepreneuring or championing	Project leading	Gatekeeping	Sponsoring or coaching
Emphasis on deadlines	Little emphasis; exploring encouraged	Jointly set deadlines emphasized by management	Management identifies; needs strong emphasis	Set by the job (i.e., the person needing the information)	Little emphasis
Emphasis on specifically assigned tasks	Low; freedom to pursue new ideas	High; assignments mutually planned and agreed by management and champion	High with respect to overall project goals	Medium; freedom to consult with others	Low

TABLE 22.3
Measuring and Rewarding Critical Function Performance

Dimension of management	Critical function				
	Idea generating	Entrepreneuring or championing	Project leading	Gatekeeping	Sponsoring or coaching
Primary contribution of each function for appraisal of performance	Quantity and quality of ideas generated	Ideas picked up; percent carried through	Project technical milestones accomplished; cost/schedule constraints met	People helped; degree of help	Staff developed; extent of assistance provided
Rewards appropriate	Opportunities to publish; recognition from professional peers through symposia, etc.	Visibility; publicity; further resources for project	Bigger projects; material signs of organizational status	Travel budget; key "assists" acknowledged; increased freedom and use for advice	Increased freedom; discretionary resources for support of others

cluded explicitly in the objective of a job. A gatekeeper would, for example, see his or her goals as including provision of useful information to colleagues. A person who has the attitudes and skill to be an effective champion or entrepreneur could also be made responsible for recognizing good new ideas. This person might have the charter to roam around the organization, talk with people about their ideas, encourage their pursuit, or pursue the ideas himself.

This raises a very sticky question in most organizations: Who gets the credit? If the champion gets the credit for recognizing the idea, not very many idea generators will be eager to let the champion carry out his job. This brings us to the next item, measures and rewards.

Performance Measures and Rewards

We all tend to do those activities that get rewarded. If personnel perceive that idea generating does not get recognized but that idea exploitation does, they may not pass their ideas on to somebody who can exploit them. They may try to exploit them themselves, no matter how unequipped or uninterested they are in carrying out the exploitation activity.

For this reason, it is important to recognize the distinct contributions of each of the separate critical functions. Table 22.3 identifies some measures relevant to each function. Each measure has both a quantity and quality dimension. For example, the objective for a person who has the skills and information to be effective at gatekeeping could be to help a number of people during the next twelve months. At the end of that time, his manager could survey the people whom the gatekeeper feels he helped to assess the gatekeeper's effectiveness in communicating key information. In each organization specific measures chosen will necessarily be different.

Rewarding an individual for the performance of a critical function makes the function both more discussable and manageable. However, what is seen as rewarding for one function may be seen as less rewarding, neutral, or even negative for another function because of the different personalities and needs of the role fillers. Table 22.3 presents some rewards seen as appropriate for each function. Again, organizational and individual differences will generate variations in rewards selected. Of course, the informal positive feedback of managers to their day-to-day contacts is a major source of motivation and recognition for any individual performing a critical innovation function, or any job for that matter.

Salary and bonus compensation are not included here, but not because they are unimportant to any of these people. Of course, financial rewards should also be employed as appropriate, but they do not seem to be explicitly linked to any one innovative function more than another. Table 22.3 identifies the rewards that are related to critical roles.

The preceding sections demonstrate that the critical functions concept provides an important way of describing an organization's or a project team's resources for effective innovation activity. One technical organization, for example, in seeing the contributions of all five critical roles, made several important changes in their recruiting and staffing procedures. To begin with, the characteristic strengths behind each critical function were explicitly employed for identifying the skills necessary to do a particular job. This led to a framework useful for interviewing and evaluating candidates to determine how they might contribute over time to projects or organizational efforts. Upper management also became conscious of the unintended bias in their recruiting and staffing activities, as much of their focus had been on creative idea-generation needs and capabilities. As a result, upper management was more careful to have the mix of critical functions represented within the interviewing and staffing considerations.

Other values resulted from the analyses which were less tangible than those listed above but equally important. Jobs were no longer defined solely in technical terms (i.e., the educational background and/or work experience necessary). For example, whether or not a job involved idea generation or exploitation was defined, and these typical activities were included in the description of the job and the skills needed to perform it well. The objectives of the job, in the company's management-by-objectives (MBO) procedure, were then expanded to include the critical functions. However, since all five functions are essential to innovation and it is the very rare person who can do all five equally well, the clear need for a new kind of teamwork was also developed. Finally, the critical functions concept provided the framework for the selection of people and the division of labor on the "innovation team" that became the nucleus for all new R&D programs. In summary, to the extent that innovative outcomes, rather than routine production, are the outputs sought, we have confidence that the critical functions approach will afford useful insights for organizational analyses and management.

NOTES

1. For a different and more intensive quantitative view of project life cycles, see Edward B. Roberts, *The Dynamics of Research and Development* (New York: Harper & Row, 1964).

2. Eric von Hippel, "Users as Innovators," *Technology Review, 80*, No. 3 (January 1978), 30–39.

3. See Alan R. Fusfeld, "How to Put Technology into Corporate Planning," *Technology Review, 80*, No. 6, for issues that need to be highlighted in a comparative technical review.

4. For further perspectives on project transfer, see Edward B. Roberts, "Stimulating Technological Innovation: Organizational Approaches," *Research Management, 22*, No. 6 (November 1979), 26–30.

5. D.C. Pelz and F.M. Andrews, *Scientists in Organizations* (New York: Wiley, 1966).

6. E.B. Roberts, "Entrepreneurship and Technology," *Research Management, 11*, No. 4 (July 1968), 249–66.

7. D.G. Marquis and I.M. Rubin, "Management Factors in Project Performance," M.I.T. Sloan School of Management, Working Paper, Cambridge, Mass., 1966.

8. T.J. Allen, *Managing the Flow of Technology* (Cambridge, Mass.: MIT Press, 1977); and R.G. Rhoades, et al., "A Correlation of R&D Laboratory Performance with Critical Functions Analysis," *R&D Management, 9*, No. 1 (October 1978), 13–17.

9. Roberts, "Entrepreneurship and Technology," p. 252.

10. One role we have frequently observed is the "quality controller," who stresses high work standards in projects. Other critical roles relate more to organizational growth than to innovation. The "effective trainer" who could absorb new engineers productively into the company was seen as critical to one firm that was growing 30% per year. The "technical statesman" was a role label developed by an electronic components manufacturer which valued the ability of some engineers to generate a leadership technical reputation through authorship and presentation of advanced concepts.

11. One study that demonstrated this phenomenon is N.R. Baker, et al., "The Effects of Perceived Needs and Means on the Generation of Ideas for Industrial Research and Development Projects," *IEEE Transactions on Engineering Management, EM-14* (1967), 156–65.

12. For their research and consulting activities, the authors have developed a questionnaire methodology for collecting these kind of data.

13. Allen, *Managing the Flow of Technology.*

14. Roberts, "Entrepreneurship and Technology."

15. E.B. Roberts and D.H. Peters, "Commercial Innovations from University Faculty," *Research Policy, 10,* No. 2 (April 1981), 108–26.

16. One study showed that engineers who eventually became managers of large projects began supervisory experiences within an average of 4.5 years after receiving their B.S. degrees. I.M. Rubin and W. Seelig, "Experience as a Factor in the Selection and Performance of Project Managers," *IEEE Transactions on Engineering Management, EM-14,* No. 3 (September 1967), 131–35.

17. For a variety of industrial approaches to the dual ladder, see the special July 1977 issue of *Research Management* or, more recently, *Research Management,* November 1979, 8–11.

23

Innovation Through Intrapreneuring

GIFFORD PINCHOT III

In-house entrepreneurs—those "dreamers who do"—can increase the speed and cost-effectiveness of technology transfer from R&D to the marketplace.

The economy of the United States is on an innovation treadmill. Our competitors enjoy cheaper labor, cheaper capital, and more government support than we. To maintain our competitive position, we need superior technology, more proprietary products and services, and better processes. As our competitors become more scientifically and managerially sophisticated, it takes them less and less time to understand and copy our innovations. We have to increase our speed and cost-effectiveness of innovation in our country to match our competitors' increasing sophistication in copying and capitalizing on our technology.

Most large companies operate stable businesses well. However, they are not as adept at starting new ones. Most are good at developing a new business from the idea stage on through research and prototype development. But they falter at the start-up stage—the stage of commercialization. Inefficient commercialization by big business has created opportunity for venture capitalists. The venture capital industry is producing 35 percent return on investment by taking frustrated R&D people and their rejected ideas out of large companies, and financing the commercialization of those ideas. That the venture capital community can make 35 percent ROI on rejected ideas and people should be a constant rebuke to everyone in the R&D community. Venture capitalists have found a different

way of managing innovation that gets returns which few of us can equal inside large organizations.

A MISSING FACTOR IN CORPORATE INNOVATION

The primary secret of the venture capitalists' success is revealed in the way they select ventures for investment. They say: "I would rather have a class A entrepreneur with a class B idea than a class A idea with a class B entrepreneur." They put their faith in choosing the right people and then sticking with them, while many corporate managers would feel uncomfortable with a strategy dependent on trusting the talent, experience, and commitment of those implementing it. I believe the primary cause for the lower returns of corporate managers of innovation is their failure to understand the importance of backing the right people—this is their failure to identify, support, and exploit the "intrapreneurs" who drive innovation to successful conclusions.

Imagine the organization as a cell, with R&D producing new genes. In the cell, there are also the productive capacity of the ribosomes, which are like factories ready to use the information in those new genes to produce new products. What's missing in most large organizations is linkage from idea to operation—by analogy the RNA. In most large organizations there are exciting new genes—new technologies but no broadly effective system of technology transfer. What is absent are large numbers of intrapreneurs devoted to turning new technologies into profitable new businesses, cost re-

Reprinted with permission from *Research-Technology Management,* Vol. 30, March–April, 1987.
Gifford Pinchot III is the founding Chairman of the consulting firm Pinchot & Company.

ductions, new features, and competitive advantages. Because we have tended to have scientific standards of excellence in R&D, we have tended to honor the inventor more than the implementor, more than the intrapreneur. The result is that we not only reward inventing more than intrapreneuring, but our management systems are far more supportive of invention than of commercialization.

The future role of R&D, the size of its budgets and its degree of autonomy all depend on efficient technology transfer. Older "hand-off systems" of development which ignore the role of the intrapreneur don't work, or at best are so slow and expensive they make R&D appear ineffective. Cost-effective innovation happens when someone becomes the passionate champion of a new idea and acts with great courage to push it through the system despite the "Not Invented Here" syndrome, and all the other forms of resistance which large organizations supply. It is therefore important for R&D managers to understand and recognize intrapreneurs who can, when properly managed, greatly increase the speed and cost effectiveness of technology transfer.

DREAMERS WHO DO

Intrapreneurs are the "dreamers who do." In most organizations people are thought to be either dreamers or doers. Both talents are not generally required in one job. But the trouble with telling the doers not to bother about their dreams is that they dream anyway. When they are blocked from implementing dreams of how to help your company, they're dreaming dreams of revenge. A mind is meant to imagine and then act. It is a terrible thing to split apart the dreamer and the doer.

What we need, then, is to restore the place for vision in everyone's job. One of my favorite stories is the story of Nikola Tesla who invented the three-phase electric motor and a host of other things. It is said that he would build a model in his mind of a machine, such as a new generator, and then push it into the background of his consciousness, set it running, and leave it going for weeks

while he went about his other business. At the end of that time he'd pull it back into the foreground of his mind, tear it down and check the bearings for wear. With such detailed imagination, what need is there for computer-aided design and finite element analysis?

While few of us can match Tesla's talent, imagination is the most concrete mental skill that people have. It is more concrete than all the tools we have for analyzing businesses and all the formulas we have for analyzing stresses. Imagination is simply the ability to see something that doesn't yet exist as it might be. Unless we have Tesla's clarity of imagination, what we see may not be as precise as the results we can reach from doing calculations, but our vision is more concrete and more whole than any formula describing some aspect of a new design. And without this concrete skill, we do not have innovation.

An intrapreneur's imagination is very different from an inventor's. Inventors look five or ten years ahead and say, "wouldn't it be wonderful if such and such." They imagine how a customer would respond to their new product, what the technology would be, how the technology could produce desired features, and all those sorts of things. Good inventors have the customer in mind, but their vision is usually incomplete unless they are also intrapreneurs. They don't imagine in detail how to get from the here and now to that desired future. An intrapreneur, on the other hand, having seen the Promised Land, moves back to the present and takes on the rather mundane and practical task of turning the prototype into a marketplace success. This too requires enormous imagination.

Intrapreneurs ask questions such as, "Who would I need to help me with this? How much would it cost? What things have to happen first?" and so forth. They may ask, "Could we release this technology onto the marketplace in product form aimed at such-and-such a customer need? No. If we did that it would immediately bite into a very important market of one of our competitors who has the ability to respond, and before we produced our second generation products there would be a tremendous competitive response. Let's back up a little bit. What if we put it out in this way instead? Well it wouldn't do quite as well on the first round, but I begin to see it would give us a little more time to develop unbeatable second generation products."

Intrapreneurs have to constantly juggle potential implementation plans. They do this in their imaginations initially. Of course, intrapreneurs also juggle implementation plans on paper as business plans and drawings, but much of the initial work is done in the shower, or when driving the car, or any situation in which one neither feels guilty about not doing something useful nor can one get to pencil and paper. At such times, we are forced to use our imaginations, and thus often do our most creative work.

DISTINGUISH INTRAPRENEURS FROM PROMOTERS

One of the keys to managing innovation cost-effectively is to choose the right people to trust. Too often when managers look for intrapreneurs they choose promoters instead. Promoters are very good at convincing people to back their ideas, but they lack the ability to follow through. Thus, one of the keys to managing innovation is to be able to distinguish between intrapreneurs and promoters.

One of the best ways to separate the intrapreneurs from the promoters is to see how they handle, and even how they think about, barriers to their ideas. When analyzing a potential intrapreneur, think of some of the ways their project might go wrong. Ask them how they might handle such a problem. Real intrapreneurs will have explored these problems in their imagination. They will have considered them while driving to work or taking a shower. The real intrapreneur has thought of three, five, or even ten possible solutions. They may pause for a moment trying to figure out which of those answers would appeal most to you because intrapreneurs do have a certain ability to sell, but they are not hearing the question for the first time. It will be very hard for you to think of a problem which they haven't considered.

Promoters, on the other hand, respond by saying the problem you bring up will never occur.

They remind you again of how wonderful things will be ten years from now, of the hundreds of millions of dollars their product will be making. They will not even talk about the problem because they have no interest in the barriers along the way to implementation. They are counting on you to solve all problems by giving them enough funding. They just want to tell you why their idea is so much better than anyone else's. They are, in fact, so focused on getting approvals and funding that they haven't planned how to get the job done. If you give them money in the name of intrapreneurship, you will not only give intrapreneurship a bad name, but you will waste everything you invested. The most important thing a manager can do when managing innovation is to separate out the promoters, and invest only in intrapreneurs.

Many people doubt that they want entrepreneurial people in their organizations. Entrepreneurs, they believe, are driven by greed. They are high risk-takers, they shoot from the hip, and furthermore, they are dishonest. Fortunately every one of these myths is false. In fact, entrepreneurs seem to be driven by a vision which they believe is so important that they are willing to dedicate their lives to it even when it starts to have trouble. Every new idea runs into terrible obstacles. People who are driven only by a desire for money, or promotion, or status, simply do not have the persistence to move a new idea forward. It is the person with the commitment to carry through who will move an idea into a practical reality.

Intrapreneurs and entrepreneurs are not high risk-takers, as many studies have shown. They like a 50-50 set of odds—not too easy, not too hard. Having chosen a challenging objective, they do everything they can to reduce the risk.

Intrapreneurs seem to be equally right brain and left brain, equally intuitive and analytic. They make decisions based on intuition when data or time don't permit analytical solutions. When analysis will work, they use it.

Intrapreneurs may operate a little differently than other people. They often have personalities which make them difficult to live with, but their difficulties stem less from dishonesty than excessive directness. They often get themselves in trouble by saying exactly what they think because they don't seem to be good at compromising—strong politics are inherent in the cultures of very large organizations.

A NEW MONITOR FOR THE FAA

Vision and imagination make up half of "the dreamers that do." Action is the other half.

Intrapreneurs are often in trouble because they act when they are supposed to wait. They tend to act beyond the territory of their own job description and function. This boundary crossing is important. Charles House at Hewlett-Packard is a perfect example. House developed a new monitor for the Federal Aviation Administration that turned out to not quite meet the specs. (Failure is a typical way for stories of innovation to begin.) He responded to the disappointment by observing that despite not meeting the spot size criteria for this particular application, the fact that he had a monitor which was half as heavy, used half the power, and cost half as much meant he should find out what else it could be used for. He took the idea to the marketing people who asked the division's traditional customers if they would like a monitor that was cheaper, but which had a slightly blurry display.

Nobody seemed to want it. Being an intrapreneur, as opposed to just a researcher, House wasn't satisfied with talk. He took out the front seat of his Volkswagen Bug, put the monitor in its place, and visited 40 customers in three weeks. At each stop he moved the monitor into the prospective customer's shop, hooked it up to their equipment, and asked whether this thing would do anything that's useful. By the end of the trip, he had found several new markets. House succeeded because he took the actions which were necessary for his prototype to go from technology to business reality.

There are two important points in this story. One is that intrapreneurs perform their own market research. If your scientists and engineers are not allowed to do their own market research, then you have a major barrier to innovation.

The second point is that generally a new idea is so ugly only its mother could love it. Consequently, it is unrealistic to think that people in marketing will understand a research idea in its early stages well enough to do valid marketing research. In general, they ask the wrong questions. They are trying to find out if it is a good idea, which in the early stages is the wrong question. The right question is: "I know this is a good idea; how am I going to present it in a way that some class of customers will agree? What are the ways in which this is a good idea? Who really needs it? How do I have to say this so that they will understand?"

The early stage of market research is searching for the market, not testing whether or not it is there. It is only after we have found a group of customers and learned how to talk to them, redesigned the product to meet their needs, and figured out how to position the product, that we can do the traditional form of market research which asks, "Will they buy it—is this a good idea?"

The idea of technically-driven research is drifting into disrepute. We are told that we must first carefully identify market needs and then invent what customers already know they want. This is rarely the way fundamental innovation works because we are not smart enough to invent to order. We are lucky to invent anything with fundamentally new and protectable properties, and when we do so, we must then hunt for the most applicable markets.

To be sure, researchers do pursue what they perceive to be marketplace needs, but the final applications often turn out to be in some entirely different market. Scotch Tape was invented to better insulate refrigerated railroad cars. Radio was invented for point-to-point communication—missing the broadcast market entirely. Riston circuit board systems began with a failure to produce a new photopolymer-based photographic film.

It is important for researchers to know about the marketplace, but important also to realize that for all of the thousands of unfilled or poorly filled marketplace needs each of us wishes to invent a proprietary solution for, we have the ability to invent a few. We know an anti-gravity device would be useful and probably well received by customers. We don't work on it because we don't know how to begin.

We know that television sets with better reception are desirable. Most of us don't work on them because we believe others have a competitive advantage in making them inexpensively.

We left Hewlett-Packard's Charles House doing his own market research and thus doing somebody else's job, as intrapreneurs often do. He came home enthusiastic and his boss's boss, Dar Howard, believed in him and told him to go ahead for another year. Unfortunately, a few months later the chairman visited the laboratory in Colorado. David Packard listened to the marketing people say that the idea was no good, even after House's research. He also heard a negative vote from the corporate chief of technology, who was backing a different technology.

At the time, Tektronix was giving Hewlett-Packard a hard time in the division's core business, and Packard said that when he came back to this laboratory next year, he did not want to see this product in the lab. Dar Howard went back to House and told him he just didn't know what excuse he could give for going on now. With that remark he left the door open just wide enough for Chuck to get his foot in. He showed that he felt for Chuck, but . . .

House said, "What exactly did Packard say?" "When I come back to this laboratory next year, I don't want to see this product in the lab." "Good," said Chuck, "we'll have it out of the lab and into manufacturing." And so it was. The monitor was used in the first manned moon landing and turned out to be a great success.

A few years later, Packard awarded House the Hewlett-Packard Award for Meritorious Defiance. "For contempt and defiance above and beyond the call of engineering duty," the certificate read. He made it clear that at Hewlett-Packard, courage counts more than obedience. Innovation requires this attitude.

SUCCEEDING AT INTRAPRENEURSHIP

Every new idea will have more than its share of detractors. There is no doubt that being an intrapreneur is difficult, even in the most tolerant of companies. So how can people succeed at it?

1. *Do anything needed to move your idea forward.* If you're supposed to be in research but the problem is in a manufacturing process, sneak into the pilot plant and build a new process. If it is a marketing problem, do your own marketing research. If it means sweeping the floor, sweep the floor. Do whatever has to be done to move the idea forward. Needless to say, this isn't always appreciated, and so you have to remember that:

2. *It is easier to ask for forgiveness than for permission.* If you go around asking, you are going to get answers you don't want, so just do the things that need to be done and ask later. Managers have to encourage their people to do this. It may be necessary to remove some layers of management that complicate and slow down the approval process.

3. *Come to work each day willing to be fired.* I began to understand this more from talking to an old sergeant who had seen a lot of battle duty. He said, "You know, there is a simple secret to surviving in battle; you have to go into battle each day knowing you're already dead. If you are already dead, then you can think clearly and you have a good chance of surviving the battle."

Intrapreneurs, like soldiers, have to have the courage to do what's right instead of doing what they know will please the myriad of people in the hierarchy who are trying to stop them. If they are too cautious, they are lost. If they are fearful, the smell of fear is a chemical signal to the corporate immune system, which will move in quickly to smother the "different" idea.

I find that necessary courage comes from a sure knowledge that intrapreneurs have—that if their employer were ever foolish enough to fire them, they could rapidly get a better job. There is no way to have innovation without courage, and no real courage without self-esteem.

4. *Work underground as long as you can.* Every organization has a corporate immune system. As soon as a new idea comes up the white blood cells come in to smother it. I'm not blaming the organization for this. If it did not have an immune system it would die. But we have to find ways to hide the right new ideas in order to keep them alive. It is part of every manager's job to recognize which new ideas should be hidden and which new ideas should be exposed to the corporate immune system and allowed to die a natural death. Too often it is the best ideas that are prematurely exposed.

THE INTRAPRENEURIAL SHORTAGE

I've made an interesting discovery since I wrote *Intrapreneuring*. I used to think potential intrapreneurs were commonplace, that they were hard to find because they were in hiding. But I have found they are more rare in most large organizations than the 10 percent who are entrepreneurial in the population at large. There is a scarcity of people who are brave enough to take on the intrapreneurial role; therefore, we have to lower the barriers and increase the rewards.

If there are not enough intrapreneurs in your company, you can hire more. There are two ways to go about it: raiding successful intrapreneurs from other companies, and hiring more intrapreneurial people in entry positions.

Were I running an R&D organization, I would even take ads saying, "Wanted: Intrapreneurs." One could capitalize on widespread intrapreneurial frustration and selectively hire a fair number of courageous people who would move innovation forward. Second, I would focus on hiring potential intrapreneurs out of school. Here are two hints: One is that candidates' transcripts should contain both A's and D's. When intrapreneurial people are interested they get A's. When they are not interested, they don't pretend. They are self-driven.

The second hint is that any history of self-employment predicts intrapreneurial success. The strongest demographic predictor of intrapreneurial success is having one or more self-employed parents. It is more important than birth order or any of the other commonly cited predictors. I guess it is a matter of having an entrepreneurial role model.

It is a particularly good idea to hire farm kids. They seem to make good intrapreneurs. I guess farm kids grow up with a kind of a can-do attitude and it never occurs to them that there is anything they aren't supposed to do. If the hay is on the ground, the bailer is broken and it is going to rain in six hours, you don't worry that you don't have a degree in bailer mechanics. Somehow farmers learn to get the job done.

TRAINING INTRAPRENEURS

Training your people in acquiring intrapreneuring skills is as important as knowing whom to hire. Though most people imagine that intrapreneurs are born and not made, we have had good results training intrapreneurs. In our Intrapreneur Schools we ask for volunteers. This way we are training a select group of people who are courageous enough to volunteer for an intrapreneurial role. Training succeeds partly because it gives people permission to use a part of themselves that their supervisors have been trying to beat out of them for quite some time. They look around the room and say, "My goodness, there are other people like me in this world and it seems that the corporation is really serious now about wanting this aspect of me employed." They get a tremendous rejuvenation and rebirth of vision and drive.

In addition, most intrapreneurs are missing skills for which training can help. They have some functional abilities which are often technical, and they've been convinced that they really cannot understand some things like accounting or marketing. They believe that those blind spots keep them from being the general manager of a new idea. They do not have to become excellent at all functions; they just have to understand enough to work easily with others in those fields. In fact, if the idea is good, success does not require great sophistication in many disciplines, just a journeyman-like job that doesn't overlook the obvious. Training should be structured to build teams and so the whole team should work together while training.

MANAGING INTRAPRENEURS

Managers must choose intrapreneurs who are persistent, impatient, who laugh, and who face the barriers. Then they have to be willing to trust that the intrapreneurs know how to do their jobs and must give them what they are asking for—resources and people to help carry forward their ideas. Since resources are not infinite, they may have to take these things away from other people who are not intrapreneurs.

I know we are living in an age of head-count restrictions. Too often this means that everything stays the same. Whoever has three people gets three people next year. Anything new and growing will have too few people resources, and any thing old and over the hill is going to have too many. We have to be courageous in sweeping out the old and giving the right people the resources they need to get the job done. The most effective use of a manager's time is in choosing whom to trust.

One very effective approach is to create heroes so intrapreneurs have role models within the company. Select a few of the most courageous intrapreneurs and publish their stories for everyone in the company to read. These stories should be written honestly, so that all the difficulties and problems faced by the intrapreneurs are presented so that people can see how barriers were overcome.

KEEP R&D CLOSE TO THE ACTION

It is important to bring your researchers close to model shops and pilot plants that allow dirty-finger research. R&D people need to be able to test their ideas themselves—if they can't, they will fall back on more intellectual forms of research. Obvi-

---

Rewards Are the Litmus Test

For many intrapreneurs who have given up and are hiding in the woodwork, rewards for innovation are the litmus test of a company's sincerity. If a company isn't willing to reward intrapreneurship, it does not really want it. I underestimated the importance of rewards when I wrote *Intrapreneuring* four years ago. It requires care since you can make a lot of mistakes designing a reward system. If you reward just the leader and not the whole team, for instance, you will have a disaster on your hands. But if you don't reward, then people will say you don't really want innovation.

Several kinds of rewards are useful. First is recognition programs which, though obvious, are generally underused. We advise our clients to create many award programs—awards for process innovation and various awards for different kinds of new product innovation. Each recipient will be one of a few who received that award even though in total large numbers are recognized. But recognition, no matter how well done, is not enough.

Financial rewards are also important and must be arranged so as not to arouse excessive jealousy in the managers of stable and mature businesses. One technique that works well in designing predetermined (prospective) rewards is to

ask those who are signing up for a program that promises unusual rewards in the event of success to take personal risk. We often advise putting 10–20 percent of salary at risk or freezing salary until the rewards are due.

Compensation alone or even combined with recognition still does not make an adequate reward. In fact, if it is not combined with increasing freedom to try new things, bonuses may simply provide seed money for successful intrapreneurs to start their own businesses. The essential reward is freedom.

Entrepreneurs, in fact, find freedom to try their new ideas their most important reward. Their wealth is not mainly used for personal consumption: the bulk is used to find the next idea. The most tangible form of freedom in a large organization is a budget. We have developed a reward system called intracapital—a one-time earned discretionary budget to be used on behalf of the corporation to try out new ideas.

We know we must give intrapreneurs, inventors, and their collaborators an unusual degree of freedom. We all do this instinctively. As organizations grow and develop levels of bureaucracy, we must do it systematically.

ously, we'll hear more about discretionary time, the so-called 15 percent rules that many companies have. Other useful reward tools are seed money programs, the creation of cross-functional teams, and other ways to reduce the bureaucracy.

In conclusion, I issue a challenge to get your people to display courage, to display integrity and honesty, to have a sense of proprietorship—as if the business belonged to them. Help them to make the kind of decisions that would have to be made if that were true, rather than the kinds they have to make in order to negotiate the turfs of a hostile bureaucracy. Encourage them to go into action and not wait for permission. Talking about these ideas is not enough. Between the words of top management and the intrapreneurs who can carry them out there are layers of management which punish independent thought, courage, impatience, and blunt honesty. This is not something that you can devote a few hours to and fix. It is probably the most important aspect of your job, more important than getting the strategy right, because enough attention is being paid to strategy already.

You cannot have cost-effective innovation unless you hire, train and encourage intrapreneurs. The future legitimacy of R&D, the success of America's companies and of her economy depends on you, the R&D community, to do it right.

24

Virtual Teams

Technology and the Workplace of the Future

ANTHONY TOWNSEND, SAMUEL DEMARIE, AND ANTHONY HENDRICKSON

You have no choice but to operate in a world shaped by globalization and the information revolution. There are two options: Adapt or die. . . . You need to plan the way a fire department plans. It cannot anticipate fires, so it has to shape a flexible organization that is capable of responding to unpredictable events.

—**Andrew S. Grove**
Intel Corporation

Just as the personal computer revolutionized the workplace throughout the 1980s and 1990s, recent developments in information and communication technology are on the verge of creating a new revolution in the coming decade. A group of technologies, including desktop video conferencing, collaborative software, and Internet/Intranet systems, converge to forge the foundation of a new workplace. This new workplace will be unrestrained by geography, time, and organizational boundaries; it will be a virtual workplace, where productivity, flexibility, and collaboration will reach unprecedented new levels.

This exciting new potential comes at a time when increasing global competition and recent advancements in information technologies have forced organizations to reevaluate their structure and work processes. Many organizations have downsized and there are continuing pressures to implement increasingly flat (or horizontal) organizational structures. While these new organizational structures may achieve gains in efficiency, flat organizational structures, of necessity, disperse employees both geographically and organizationally, which makes it more difficult for those members to collaborate in an effective manner.

One popular response to this challenging new environment has been to outsource a number of organizational functions, replacing traditional structure with an interorganizational network or virtual organization. Virtual organizations have received substantial attention in both popular and academic literature.[2] While the interorganizational challenges presented by virtual organizations are important, this leaner new competitive landscape presents important intraorganizational challenges as well.

During the past several years, one of the most dominant intraorganizational initiatives has been the development of team-based work systems. Many organizations have recognized that team-based structures have the potential to create a more productive, creative, and individually fulfilling working environment. A majority of U.S. corporations use some form of team structures in their organizations, and many report that teams enhance their ability to meet organizational goals.[3] In general, teams have provided firms with significant gains in productivity, and as such, have become a fixture among contemporary organizations. But what happens to the team advantage when fundamental organizational structures begin to change? Can teams survive amidst radical transitions in the

Reprinted with permission of *Academy of Management Executive*, "Virtual Teams: Technology and the Workplace," Anthony Townsend, Samuel DeMarie, and Anthony Hendrickson, vol. 12, no. 3, 1998.

Anthony Townsend and ***Samuel DeMarie*** are Professors of Management at the University of Nevada's College of Business in Las Vegas. ***Anthony Hendrickson*** is Professor of Management at Iowa State University.

greater organization? Perhaps more importantly, can radically transformed organizations recapture the productive potential of team-based work?

Recapturing the benefits of team systems will require flat organizations to create teams whose members may no longer be located together, or may even include members from outside the organization. Fortunately, this period of radical organizational change has been accompanied by an equally radical change in telecommunications and computer technology. Thanks to these new technologies, teams can now be effectively reconstituted from formerly dispersed members. Thus, a key component of successful, 21st century organization will be the effective use of virtual teams.

Virtual teams are groups of geographically and/or organizationally dispersed coworkers that are assembled using a combination of telecommunications and information technologies to accomplish an organizational task. Virtual teams rarely, if ever, meet in a face-to-face setting. They may be set up as temporary structures, existing only to accomplish a specific task, or may be more permanent structures, used to address ongoing issues, such as strategic planning. Further, membership is often fluid, evolving according to changing task requirements.[4]

Virtual teams provide additional benefits in that they also can be used to address evolving interorganizational challenges that occur when organizations outsource some of their key processes to more specialized firms. By creating virtual teams, both within virtual organizations and within organizations undergoing other forms of transformation, firms can ultimately realize the competitive synergy of teamwork and exploit the revolution in telecommunications and information technology.

WHY VIRTUAL TEAMS?

Although the modern organization faces a number of challenges in its competitive environment,[5] the imperative for moving from traditional face-to-face teams to virtual teams derives primarily from five specific factors:

- The increasing prevalence of flat or horizontal organizational structures.
- The emergence of environments that require interorganizational cooperation as well as competition.
- Changes in workers' expectations of organizational participation.
- A continued shift from production to service/knowledge work environments.
- The increasing globalization of trade and corporate activity.

The emergence of the flat or horizontal organization is largely a response to intensifying competitive operating environments brought about by increased global competition and recent advancements in both information and transportation technologies.[6] Organizational flattening pushes decision authority to lower levels in the organization, reducing the need for several layers of management. With fewer layers of centralized, hierarchical management structure, organizations become increasingly characterized by structurally and geographically distributed human resources. While the organization may retain the collective talent it requires, there is a reduction in the opportunity for linkages between remaining employees (e.g., personnel and offices close enough to facilitate traditional interaction). This kind of environment occasions the need to reconstitute the benefits of the large, resource rich organization within the context of the new flattened organization.

A second trend is a shift from traditional competitive business environments toward strategic cooperation among a synergistic group of firms that may not only coexist, but also actually nurture each other.[7] In the past, firms vertically integrated to maintain more control of processes from the acquisition of raw materials to the manufacture of the final product. However, diversification and specialization have made direct management of far-flung processes unwieldy. Thus, firms have responded to this problem by eliminating their superfluous processes to concentrate on their core, value-added processes. Strategic partnering and/or outsourcing allows efficient span of control while maintaining larger economies of scale for the cooperative organizational group.

Although this segmentation enables more efficient management of each individual process, it often fails to provide an overarching structure by which these specialized organizations can compete within a large global market. These cooperative groups of organizations become increasingly interdependent, with the success of each individual organization enhancing the success of the cooperative organizational system.

A prominent example of this synergistic cooperation is the collaboration among a number of computer hardware and software developers. Unlike IBM in previous decades, firms such as Intel and Microsoft have avoided vertical integration and achieved unprecedented growth and dominance in the distributed computing environment. This success is largely due to their concentration on their respective core disciplines, thus avoiding the lack of focus inherent in vertically integrated organizations. While they have created and nurtured an environment in which both organizations flourish, the ultimate value of each is dramatically dependent upon the other. Without significant advances in chip technology, demand for personal computing software tools and distributed computing systems is limited. Conversely, advancements in computing software have led to an insatiable demand for faster, more powerful microprocessors.

Group success is dependent on effective communications and knowledge sharing among members. Microsoft's success in a variety of industries, including personal computers, corporate computing, telecommunications, and consumer electronics, is directly attributable to the firm's networking with software developers within these supplier organizations. By providing developmental versions of new software, Microsoft facilitates communication with its customers and acquires invaluable feedback prior to releasing final versions of its products. Product development is no longer an isolated task within the organization, but a collaborative effort in which product identity and loyalty are created via close customer involvement in the de-velopment process. Virtual teams provide an effective platform for these groups by using advanced technologies to facilitate their complex communication processes.

The third major trend in the business environment centers on changes in employee expectations of how they will participate in the workplace. Future employees, who have grown up in an environment of personal computers, cellular phones, and electronic classrooms, will be more likely to expect organizational flexibility. The new generation of workers will be technologically sophisticated and will expect technological sophistication from their employing organization.[8] An example of how changing employee expectations are already affecting the workplace can be seen in the increasing number of employees who are opting for telework alternatives. Teleworkers operate from their homes or some other remote location, connected to a home office primarily through telephones, fax machines, computer modems, and electronic mail. Telework provides cost savings to employees by eliminating time-consuming commutes to central offices and offers employees more flexibility to coordinate their work and family responsibilities. Teleworkers currently make up the fastest growing segment of the workforce.[9]

Virtual teams provide a platform for organizations to actually exceed these new employee expectations. For example, telework is usually limited to relatively independent job categories that involve low levels of collaboration. A virtual team format can expand telework's potential range by allowing employees involved in highly collaborative teamwork to participate from remote locations.

A fourth factor encouraging the development of virtual teams is the continued shift from manufacturing and production jobs to service and knowledge work. Production processes, by their very nature, are often more structured and defined. Service activities often require cooperation of team members in dynamic work situations that evolve according to customer requirements. The hallmark of successful service firms has been their ability to flexibly respond to the customer's needs as quickly as possible. This requisite flexibility fuels the movement from highly structured organizational forms to more ad hoc forms. Virtual teams enable this organizational flexibility because they integrate the effectiveness of traditional teamwork with the power of advanced communication and information technologies, allowing them to accommodate increased dynamism in both team membership and task structure.

Finally, the increasing importance of global trade and corporate activity has radically altered the working environment of many organizations. Recent trade agreements, such as GATT and NAFTA, coupled with economic reforms in China and eastern Europe, have created increased opportunities for international trade. Whereas in the past, multinational operations were solely the domain of the world's largest corporations, technological advances in both communications and logistics have enabled smaller firms to compete in the global marketplace. Regardless of firm size, multinational operations require high levels of cooperation and collaboration across broad geographical boundaries.[10] Turning these networks of collaborators into fully connected virtual teams has the potential to increase both the efficiency and quality of communications in this challenging environment.

THE TECHNOLOGY OF VIRTUAL TEAMS

Virtual teams are possible only because of recent advances in computer and telecommunications technology. Because these technologies define the operational environment of the virtual team, it is critical to examine how these technologies come together to form the infrastructure of virtual teamwork. Although all of the systems are somewhat interdependent, it is helpful to consider them as belonging to one of three broad categories of technology: desktop videoconferencing systems (DVCS); collaborative software systems; and Internet/Intranet systems. These three technologies provide an infrastructure across which the virtual team will interact and provide technological empowerment to the virtual teams' operation.[11]

Desktop Videoconferencing Systems (DVCS)

DVCS are the core system around which the rest of virtual team technologies are built. Although virtual teams would be possible with simple e-mail systems and telephones, DVCS recreate the face-to-face interactions of conventional teams, making possible more complex levels of communication among team members. While the technology of videoconferencing is not new, traditional videoconferencing systems typically involved dedicated meeting rooms that were very costly to set up and maintain. These videoconference rooms were also cumbersome and inconvenient to use, requiring specially trained technicians to facilitate even the simplest of meetings. The most sophisticated DVCS currently cost less than $1000 per station, can be added to most any new generation of personal computer, and can be used with no outside facilitation. This combination of affordability and operational simplicity makes DVCS an affordable organizational communications solution.[12]

Although technologically sophisticated, the DVCS is a relatively simple system for users to operate. A small camera mounted atop a computer monitor provides the video feed to the system; voice transmissions operate through an earpiece/microphone combination or speakerphone. Connection to other team members is managed through software on the user's computer; to ensure user familiarity, the software uses an on-screen version of a traditional telephone to control the system. The final component of the system is a high-speed data connection, which may be accomplished through local area network connections, or specialized digital phone lines. DVCS create the potential for two primary types of group communication:

- All team members are actively connected in a session. With current technology, groups of up to 16 team members can simultaneously videoconference, meaning that each user can see and hear up to 15 other team members on his or her computer monitor. Functioning in this mode, the entire team or subunits of the team can conference as needed.

• A face-to-face group can interact with a nonpresent team member or outside resource. The same DVCS used for individual interaction also permits a conference table of team members to have a traditional teleconference with one or more outside parties. Because the DVCS allows for multiple conference connections, a local group can connect with up to 15 different individuals or groups.[13]

In addition to providing video and audio connections, most DVCS provide users with the ability to share information and even applications while they are interconnected. For example, users can simultaneously work on documents, analyze data, or sketch out ideas on shared whiteboards. In many respects, the DVCS creates a work environment where users have more options available to help them collaborate and share data than would be possible working around a conference table or huddled around an office computer.

Collaborative Software Systems

Collaborative software systems are the second component of the virtual team technical infrastructure. Effective collaboration requires team members to work both interactively and independently; collaborative software is designed to augment both types of group work activity and to empower teamwork processes.[14]

The simplest collaborative software application involves sharing traditional software products through the DVCS. As noted above, most DVCS allow users to share any application running on any one of their individual computers. Used in this manner, a variety of existing software applications become powerful collaborative tools, allowing multiple team members to create, revise, and/or review important information.

A second category of collaborative software systems is designed to empower real time group decision making and other creative activities. These systems, called group support systems (GSS), are specifically designed to create an enhanced environment for brainstorming, focus group work, and group decision making. These systems provide

their users with a variety of support tools to poll participants and assemble statistical information relevant to the decision activity. Finally, these systems also allow users to "turn off" their individual identities during a brainstorming session and interact with relative anonymity, which can be very helpful in certain contexts.[15]

As with traditional teams, a substantial portion of the work of virtual team members may be conducted independently, and then passed along to the rest of the team at appropriate stages of the team's project. For this noninteractive aspect of the virtual team's work, there is also a developing body of software. This family of software provides specific support for collaborative accomplishment (e.g., project management, product design, document creation, and information analysis) when team members are working independently on team projects. The major focus of these collaborative software applications is to facilitate multiple authorship of documents and presentations, and joint development of databases, spreadsheets, and other information resources.

Collaborative software systems also may provide a comprehensive environment for group work. Lotus Notes, a dominant collaborative software product, is designed specifically for asynchronous teamwork (e.g., communication and data sharing where parties are working either at different times or independently) and combines scheduling, electronic messaging, and document and data sharing into one common product. By combining a number of collaborative applications and communications systems into an integrated framework, products like Lotus Notes facilitate both the production and communication necessary to effective teamwork. Although most of these types of software systems have been designed to facilitate teamwork in traditional work environments, they provide an equally powerful foundation for the collaborative empowerment of virtual teams.

The Internet and Intranets

The enormous popularity of the Internet is a significant indicator that a friendly medium can overcome the technophobia of a vast number of people, and this lesson has not been lost on business organizations. Recognizing that the explosion of the Internet is a microcosmic glimpse of the potential for employee interest and use of this new interconnective technology, a number of firms have adapted state of the art Internet technologies into internal Internets, or Intranets. The Federal Express Corporation provides a good example of this adaptive process. After finding that its Internet website was a cost-saving solution for customer service, the company decided to try out the technology on an internal basis. In 1995, the firm operated over 60 Intranet websites among 30,000 worldwide office employees.[16]

Intranets provide organizations the advantage of using Internet technology to disseminate organizational information and enhance interemployee communication, while still maintaining system security. With the Internet and Intranets, organizational users realize the benefits of the familiarity of the same connective interface, whether working with internal or external information. For the virtual team, the Internet and Intranets provide an important communicative and informational resource. They allow virtual teams to archive text, visual, audio, and numerical data in a user-friendly format. The Internet and Intranets also allow virtual teams to keep other organizational members and important outside constituents such as suppliers and customers up-to-date on the team's progress, and enable the team to monitor other ongoing organizational projects that might affect the task at hand.

The Internet and Intranets make a significant contribution to the collaborative environment because of the way that information is managed on both systems. They have proven to be a rich source of qualitative information, and new methods of information search and retrieval have been developed to effectively sort through their enormous volume of information. Systems such as Digital Equipment Corporation's AltaVista search engine provide a means to quickly and effectively locate information—first on the Internet, and now on Intranets and individual computers. Unlike traditional database software, which requires high structured data,

advanced search engines are able to find text-based information from within a jumble of file types and formats. Most recently, these products have been enhanced to incorporate the very latest in user-friendly interfaces, further improving users effectiveness by making an information search a more intuitive process. By enabling users to locate documents and text-based information from anywhere in their workgroup, these new data management tools provide a workable way to manage the distributed information resources of virtual teams.

Taken together, DVCS, collaborative software applications, and Internet/Intranet technologies form an informational infrastructure within which virtual teams can match or even surpass the effectiveness of face-to-face teams. Unfortunately, technology provides only a foundation for virtual teamwork; the real challenge to virtual team effectiveness is learning how to work with these new technologies. Although these new technical systems provide an incredibly rich communication context for virtual team members, they do not truly replicate the face-to-face environment. As such, virtual team members are challenged to recapture the effectiveness of face-to-face interactions using the virtual tools that are available to them.

VIRTUAL TEAM BUILDING

Developing effective virtual teams goes well beyond the technical problem of linking them together. As workers increasingly interact in a virtual mode, it is imperative that they rebuild the interpersonal interaction necessary for organizational effectiveness. While the virtual team presents a number of challenges in this area, it also presents the potential to recreate the way work is done. Within the virtual connection lies an opportunity for efficiencies and team synergy unrealized in traditional work interaction. John Verity writes:

> That is the essence of virtualization: rather than simply recreating in digital form the physical thing we know as a letter, e-mail reinvents and vastly enhances letter-writing. Unbound by barriers of

time and space and endowed with new powers, the electronic letter does something new altogether. The same sort of thing happens when business, the arts, or government are reborn in digital form.[17]

Recreating teams in virtual mode requires resolution of the challenges and opportunities inherent in virtual team technology, as well as the development of a new team sociology.

New Challenges in Structure, Technology, and Function

As discussed earlier, changes in organizational structure and advances in informational technology define the environment in which the virtual team operates. While many of these challenges are present in traditional work settings, they become more pronounced in the virtual environment. Consider the following:

- More so than a traditional workgroup, the virtual team will probably have membership representing a number of different geographic locations within the organization, and may also include contingent workers from outside the organization.
- Virtual team members will be challenged to adapt to the telecommunication and informational technologies that link its members. Virtual team members will have to learn to use effectively new telecommunications systems in an environment where an important client or coworker is frequently never physically present.
- The virtual team's role transcends traditional fixed functional roles, requiring virtual team members to be prepared to adapt to a changing variety of assignments and tasks during the life of any particular team.

All of these factors affect the environment in which the individual members of virtual teams must learn to operate.

Virtual teams, because they have the potential to significantly decrease the amount of travel required of team members, can significantly increase the productive capacity of individual members. For this reason, virtual team members may be asked to participate in a higher number of separate team situations than was practical in traditional face-to-face teamwork. Thus, virtual team members may have multiple (and even competing) alliances outside their specific virtual team. This same challenge has been observed in traditional work settings, both in situations where contingent workers interact with permanent workers and when members of teams or workgroups are also members of other groups competing for their time and attention. Although problems associated with these factors are not new, outsourcing, organizational partnering, and the efficiencies afforded by advanced information technologies have increased the potential for conflict caused by multiple organizational roles.

Furthermore, the virtual team environment represents a pronounced structural difference from traditional workgroup participation because of its ability to transform quickly according to changing task requirements and responsibilities. Virtual team membership will be substantially more dynamic than traditional teams and virtual teams will be more likely to include members from locations that would not traditionally have worked together. This dynamism requires virtual team members to be particularly adaptable to working with a wide variety of potential coworkers.

Differences in the functional role of the virtual team within the broader organization also create a different environment for the virtual team and its members. Virtual teams provide the capability for more flexible organizational responses, which means that the role of the virtual team, as well as the roles attributed to its members, will be substantially more dynamic than in traditional settings. The Danish hearing aid manufacturer, Oticon, exemplifies this concept. After several years of attempting to turn the company around using traditional cost-cutting and strategic marketing techniques, the president decided to restructure the organization into what is essentially a giant virtual team. Conceptualizing the entire staff as one large 150-member team, the firm now draws the necessary skills for specific projects from a pool of workers whose diverse skills most appropriately fit the project and task requirements.[18]

Each employee's physical location is no longer a barrier to effective team structure. What remains critical is how individual skill sets meet project requirements driven by an ever-evolving business environment. Virtual teams like these are more capable of addressing an evolutionary mission because their technological infrastructure is designed to facilitate transformations in response to changing organizational requirements.

By far the greatest difference in the working environment of virtual team members is the process of virtual interaction. Although electronic mail and various document-sharing capabilities have been in use in traditional work settings for some time, these systems have generally been supported by face-to-face meetings and geographic proximity to other workgroup members. In the virtual work environment, traditional social mechanisms that facilitate communication and decision making are effectively lost and participants must find new ways to communicate and interact, enabling effective teamwork within the new technical context.

Changes in Work and Interaction

The challenges detailed above have the potential to create a radically different work environment for the virtual team participant, both because of the change from face-to-face to some degree of virtual interaction, and because the virtual team is expected to operate in a different form of organization and assume new organizational roles. These changes in the work setting affect the way that team members conduct their work and how they communicate and express themselves:

- Virtual team members must learn new ways to express themselves and to understand others in an environment with a diminished sense of presence.

- Virtual team members will be required to have superior team participation skills. Because team membership will be somewhat fluid, effective teams will require members who can quickly assimilate into the team.

- Virtual team members will have to become proficient with a variety of computer-based technologies.

- In many organizations, virtual team membership will cross national boundaries, and a variety of cultural backgrounds will be represented on the team. This will complicate communications and work interactions and will require additional team member development in the areas of communication and cultural diversity.

Research has indicated that when the trappings of traditional communicative patterns are absent, communication dynamics are substantially altered. For example, in workgroup systems where members' primary interactions are through some form of electronic mail, the absence of traditional communicative cues (i.e., facial expression, gesture, and vocal inflection) make subtleties in communication more difficult to convey.[19] Additionally, when participants are able to use a communication system anonymously, the group also begins to lose distinctions among members' social and expert status.[20] Thus, the loss of traditional cues creates an environment that is substantially different from face-to-face interaction, requiring participants to reconstruct a viable workgroup dynamic. Within this reconstructed environment, there is an opportunity for enhanced organizational democracy and participation in work and decision making. Although technology certainly presents an opportunity for such development, the team's sociology will ultimately be a function of technology, the larger organizational culture, and the team's task requirements.[21]

Within the larger organizational culture and the technical environment, the group dynamic of a virtual team depends on the socialization process of the individual team. Unlike many traditional teams, virtual teams will be expected to be able to repeatedly change membership without losing productivity; little time will be available for team members to learn how to work together. Thus, effective virtual team members will have to be particularly adept at fitting into a variety of team situations.

The traditional factors identified with high team performance come into play in the virtual environment as well. Effective communication skills, clarity of goals, and a performance orientation will continue to be critical attributes for virtual team members.[22] To fully exploit the advantages of the new environment, virtual team members will require basic teamwork training and development and will also need training to enhance team workers' facility with the new information and communication technologies. Effective training in such virtual function skills as how to best use telecommunicative capability and collaborative systems may ultimately result in teams that function as naturally in a technologically empowered, virtual environment as teams currently do around a conference table. Additionally, when team members represent a variety of national or cultural groups, there will also be the need to teach team members how each of their respective cultures may differ and how they can overcome these differences and use them to the team's advantage.

Advanced technologies may also be used to improve or streamline the socialization of new team members. For example, Intranets allow teams to archive a wide range of information. New team members potentially could access a complete electronic history of the team's work, including not only text-based and graphical information, but also video and audio recordings of important team meetings. The availability of this rich history may allow new members to be brought up to speed on team task, culture, and members' personalities much more quickly than in traditional face-to-face teams.

Recall too that technology presents the opportunity to enhance a team's effectiveness by empowering the teams' collaborative activity. Both research and industry experience indicate that collaborative systems can augment a group's decision quality and performance potential,[23] and will likely do so in the virtual environment as well.[24] Given a proper set of communicative protocols (e.g., telephones, DVCS, electronic mail, and Internet/Intranets), collaborative software systems will add enormous performance potential to the virtual teams' environment. While learning to use collaborative tools is no more difficult than learning many other software systems, the effective use of col-

laborative tools likely will reorient the attitude of users toward the process of work. Michael Schrage writes:

> Collaborative tools and environments will spark the same kinds of questions and concerns as other fundamental technologies, which will in turn determine the effectiveness of both individuals and enterprises. "Why won't he get out the good collaborative tools with me?" is a question not unlike "Why won't he talk with me on the phone?" . . . The technology becomes a frame of reference and a new infrastructure for the way people relate to one another.[25]

Thus, among virtual team members, collaborative tools not only enhance the productive capacity of the team; they also become a central medium of the team's work process.

CAPITALIZING ON VIRTUAL TEAMS

It is important to stress that virtual teams are not an organizational panacea and that the degree to which organizations will benefit may differ. Developing the technology and employee skills necessary for effective virtual team implementation carries a cost in time and financial investment that must be offset by the competitive advantage virtual teams afford. Digital Equipment Corporation provides an excellent example of the productive potential of virtual teams, having used them to develop computer systems for years. These teams share databases, simulation and modeling systems, and advanced communication systems to support the collaborative design of new products. This organizational structure has enabled Digital to increase the productive capacity of its technical experts and maintain its position as a leader in its field.[26]

Therefore, although virtual teams provide many exciting opportunities, organizations require clear understanding of the purpose and goals they have for virtual team implementation. The organizational challenge is first to effectively create the virtual team and, second, to overcome the inherent resistance that inevitably accompanies large scale innovation.

Creating Virtual Teams

Once an organization determines that it has a need for virtual teams, the next challenge is to actually put them in place. At this juncture, the organization must define the teams' function and organizational role, develop the technical systems to support the teams, and assemble individual teams, as well as a cadre of potential team workers.

Managerial Direction and Control

Just as in any team environment, managers will need to clearly establish expectations about the virtual team's performance and criteria for assessing the team's success. Because of the dispersion of team members, effective supervision and control of the virtual team may appear problematic. However, the virtual team's rich communicative environment, along with the system's capacity for archiving data and communications, actually empowers considerably more managerial monitoring than is possible in traditional environments. Managers could, for example, actually view archived recordings of team meetings to assess member contribution and team progress. Finally, the reporting and administrative relationship between the team and its external manager or managers must be clearly established. Again, because none of the team members will necessarily be located in the same place as external management, clear schedules must be established of when the team will provide reports, interim deliverables, and final product.

It is also critically important that managers clearly define the virtual team's role within the context of the organization's greater mission, including the limits of the team's scope and responsibility. This will help the team to focus its efforts on activities that support the strategic direction of the firm.

Defining the Team's Organizational Role and Function

Virtual teams may be implemented as a response to one or a number of conditions detailed in the preceding sections. In turn, these underlying reasons for the introduction of virtual teams should determine the configuration of individual teams, dictate their mission, and ultimately determine the type of technical system required and the requisite skills and orientation of the team and its members. The following description of two types of team roles, while certainly not exhaustive, illustrates some of the range of the role and function of the virtual team.

Teams that are created to provide strategic responses to rapidly changing market conditions will operate in the most fluid of all virtual team environments. In addition to all of the challenges associated with virtual teamwork, these teams will be required to continuously evolve to meet changing tactical conditions. The configuration of these teams will be highly dynamic and dependent on current task and planning requirements. The role of these teams (and of the range of their potential participants) will be as highly adaptive response units, whose mission is to respond to market challenges and exploit market potentials. For example, Lithonia Lighting developed virtual marketing response teams from among independent sales agents, outside distributors, and their own electrical engineers. These virtual teams, which represent both product developers and end-user suppliers, provide the company with a unique capability to respond quickly to changes in the market and customer needs. Using these virtual teams, the company has dramatically increased both sales volume and customer satisfaction, while supply chain administrative costs have remained constant or decreased.[27]

Although flexibility is one of the potential virtues of the virtual team, virtual teams may also be created to operate in environments characterized by long-term membership and long task cycles. Virtual teams involved in complex development projects, for example, will capitalize on their ability to access a broader range of expertise and to more easily link to diverse functional resources. The role of these teams will be to manage and execute traditional organizational processes, but with the advantage of resources and expertise unavailable save for their virtual construction.

Developing the Teams' Technical Systems

Once the role and mission of the teams have been clarified, the technical systems that will enable the teams' work will have to be designed and brought on line.

Teams whose task environment requires a high degree of informational integration and/or creative group participation are candidates for greater use of collaborative software applications, in addition to DVCS. This collaborative software will benefit teams whose members must produce group documents and presentations, interactively develop and analyze data, or engage in complex team-level planning and decision making. System design must also reflect an understanding that not all teams need all systems; system design must be task-oriented in order to avoid unnecessary technical overload of team members. The Xerox Corporation met this challenge when it connected two groups of scientists (one in Palo Alto, the other in Portland, Oregon) by arranging for constantly active phone, computer linking, and video conferencing between central areas in the two offices. The two sets of scientists, although 500 miles apart, could communicate with each other just as easily as if they were walking into the next room. No communication systems had to be operated, no complex protocols followed. Scientists simply walked up to the camera and started talking or shared information back and forth through their computer links. Scientists working on the project report that the richness of the communication system significantly assists them in their work; because the system provides such high-quality communication, users regard their communications with long distance colleagues casually and indicate that their geographical separation no longer inhibits their collaboration.[28]

Teams whose task environment requires a high degree of personal interaction may simply require basic DVCS systems. However, depending on the frequency of use and the number of participants, these teams may need more advanced DVCS systems that use dedicated high-speed phone lines that allow information to flow more quickly between participants and allow for near broadcast quality video transmission. Although the more advanced systems add considerable cost to the DVCS, the increased costs may be justified where video interaction must be as absolutely seamless as is possible, such as in client presentations or sensitive negotiation sessions.

Developing Teams and Team Members

In addition to developing the hardware and software infrastructure for virtual teams, it is equally critical to develop the teams themselves and to develop employees who can effectively participate in this new environment. This means that current potential team members must be trained and acclimated to the virtual team environment. Additionally, to fully exploit the virtual team's potential for optimized membership, organizations must extend their definition of human resources to include the broad range of consultants and contingent workers who may potentially participate on a team in only a virtual model.

Training and developing virtual team members is in many ways no different from training and developing good team members in general; developing skills in communication, goal setting, planning, and task proficiency are all as important for the virtual team as for the traditional team. What is different about the virtual team is the amount of technical training that is required to empower the team member to function in the virtual environment. Learning to use all of the traditional team skills in an environment where most interactions take place through a telecommunications medium is a critical challenge. This is particularly true since technology continues to evolve and reinvent itself at an ever-increasing rate. Training to maintain technical proficiency will be an important component of any virtual team member's continuing education program.

Since virtual team members' interactions may take place across a relatively alien set of telecommunications systems, the first priority in virtual team preparation is to effectively teach team members how to fluently communicate with each other within the new media. Although team members can easily be taught to operate new technologies, they must be given an opportunity, through training and team development, to establish their own slang terminology and communications protocols. Over time, the team will develop a variety of methods to ensure that their communication is both efficient and accurate.

When team flexibility is highly stressed, team members will also require a very different attitude toward the team than would traditional team workers. Traditional teams provide members with feelings of cohort and social presence; in an extremely flexible virtual team environment, employees will have to learn to join teams and accept new members into teams without the benefit of time-related socialization. Thus, teams will benefit from learning to express explicit norms and role expectations to new members, who will, in turn, be required to quickly acculturate according to the team's guidelines. It will be critical to the functionality of the virtual team that members are instilled with the same commitment to the virtual group activities as they would to any traditional team function.

In addition to training and developing team resources, human resources planners will have to identify potential team members from outside traditional organizational boundaries. As noted earlier in the paper, the virtual team provides the opportunity to build teams out of personnel who could not possibly work together under traditional circumstances. If the potential of virtual teams is fully realized, firms will have the opportunity to greatly expand access to expertise, overcoming constraints that might have been prohibitive in the past. Additionally, organizations will have to rethink how to compensate these individuals, whose contribution to a particular team may be less than full-time.

Given the diversity of potential personnel available to the virtual team and the potential fluidity of team membership, organizations may want to consider the development of team development specialists. Team development specialists would function as resources to teams, assisting them with technical problems and facilitating their interaction when necessary. Providing this level of support would allow the virtual team to focus more on its objectives, rather than on the processes associated with teamwork in the virtual environment.

CHALLENGES AND OBSTACLES

Like any organizational innovation, the introduction of virtual teams will encounter a number of challenges and obstacles. Virtual teams require organizational restructuring and the introduction of new work technologies. The potential for startup problems and deliberate resistance is substantially greater than for changes in structure or technology alone. In discussing virtual teams with professional managers, the following four areas of potential resistance were consistently identified.[29]

Technophobia

Although an increasing percentage of the workforce is computer-literature (and even computer-oriented), a significant number of valuable employees are uncomfortable with computers and other telecommunications technologies. One of the greatest challenges in the introduction of virtual teams is the successful incorporation of valuable, technophobic personnel into the virtual team environment. Part of this problem will be obviated as both computer and telecommunications technologies become more user-friendly. The introduction of graphical operating systems (such as Microsoft Windows 95) opened up computing to a number of new users, and similar introductions of simplified operating systems, intuitive programs, and speech recognition capabilities should encourage even the most technologically recalcitrant to use sophisticated computer systems. In the meantime, organizations can more easily facilitate migration to the virtual team environment by providing training and technical support specifically geared to system novices.

Trust and Cohesion Issues

In an environment where one's primary interaction with others takes place through an electronic medium, it is only natural to expect that participants will wonder whether the system is being used to monitor and evaluate them. The free flow of team members' communication, which once might have taken place away from the office, may now be inhibited by concerns about privacy and system security. To counter this problem, organizations must establish clear policy regarding communications privacy, and must then strictly adhere to that policy. Over time, participants will realize that the virtual team system is a safe medium across which to share ideas and concerns.

Burnout and Stress

One of the benefits of the virtual team environment is its ability to efficiently connect people and enable greater levels of productivity. This may result in employees' being assigned to more teams, creating a more complex and potentially stressful work environment. Organizations must be careful not to overextend virtual team members and saddle them with levels of responsibility that they cannot reasonably satisfy. One important supervisory role will be to ensure that virtual team members have enough private time to complete their individual assignments and prepare for their team participation.

Structural Resistance

The introduction of virtual teams will require significant amounts of organizational restructuring. Aside from the reasons detailed above, some resistance will occur because organizational members do not see this particular kind of change as desirable or necessary. To overcome their concerns, management must carefully design an implementation program that highlights the contribution that virtual teams will make and ties these contributions to important organizational values.[30]

LOOKING TO THE FUTURE

The world of the virtual team is far from static; continuing changes in technology and competitive environments will present new opportunities and imperatives for virtual teamwork. Nicholas Negroponte writes, "Computers are getting smaller and smaller. You can expect to have on your wrist tomorrow what you have on your desk today, what filled a room yesterday."[31] As telecommunication technologies continue to evolve, the virtual interface will provide more realistic presence, while

simultaneously costing less and becoming easier to use.

Many of these same technological advances will create more virtual interaction in workers' private lives as well. This change will increase employee expectation of working in a virtual mode; as an increasing number of people socialize and shop in cyberspace, these same virtually savvy people will be expecting a similar experience in their workplace. The economic imperative for virtual teams, combined with changing societal experience of the virtual, may well transform the virtual team from an innovative source of competitive advantage into a dominant organizational form.

ENDNOTES

1. Grove, A. S. 1995. A high-tech CEO updates his views on managing and careers. *Fortune*, September 18: 229–230.

2. See Dess, G., Rasheed, A., McLaughlin, K. and Priem, R. 1995. The new corporate architecture. *Academy of Management Executive*, 9(3): 7–20; Davidow, W. H. and Malone, M. S. 1992. *The virtual organization*. New York: Harper Collins; Byrne, J., Brandt, R. and Port, O. 1993. The virtual corporation. *Business Week*, February 8: 98–102.

3. Ranney, J. and Deck, M. 1995. Making teams work: Lessons from the leaders in new product development. *Planning Review*, 23(4): 6–13; Lawler, E. 1992. *Ultimate advantage*. San Francisco: Jossey-Bass.

4. Townsend, A. M., DeMarie, S. M. and Hendrickson, A. R. 1996. Are you ready for virtual teams? *HR Magazine*, 41(9): 122–126; Pape, W. R. 1997. Group insurance: Virtual teams can quickly gather the knowledge of even farflung staff. *Inc.* June 15: 29–30.

5. Bettis, R. and Hitt, M. The new competitive landscape, 1995. *Strategic Management Journal*, 16(S1): 7–19; Moore, J. 1996. *The death of competition: Leadership and strategy in the age of business ecosystems.* New York: Harper Collins; Schrage, M. 1995. *No more teams! Mastering the dynamics of creative collaboration.* New York: Currency/Doubleday.

6. Bettis and M. Hitt, *op. cit.*

7. Moore, *op. cit.*

8. Schrage, *op. cit.*

9. Yap, C., and Tng, H. 1990. Factors associated with attitudes towards telecommuting. *Information and Management*, 19(4): 227–235.

10. Hitt, M. A., Keats, B. W. and DeMarie, S. M. 1998. Navigating in the new competitive landscape:

Building strategic flexibility and competitive advantage in the 21st century. *Academy of Management Executive*, forthcoming.

11. Osterlund, J. 1997. Competence management by informatics in RandD: The corporate level, *IEEE Transactions on Engineering Management*, 44(2): 135–145.

12. Brookshaw, C. 1997. Virtual meeting solutions. *Infoworld*, 19(22): 96–108.

13. Powell, D. 1996. Group communication. *Communications of the ACM*, 39(4): 50–53.

14. Schrage, *op cit.*

15. Townsend, A., Whitman, M. and Hendrickson, A. 1995. Computer support system adds power to group processes. *HR Magazine*, 40(9): 87–91.

16. *Ibid.*

17. Verity, J. 1994. The information revolution, *Business Week*, Special Bonus Issue: 12–18.

18. Lucas, H. 1996. *The T-form organization: Using technology to design organizations for the 21st century.* San Francisco: Jossey-Bass.

19. Kiesler, S. and Sproull, L. 1992. Group decision making and communication technology. *Organizational Behavior and Human Decision Processes*, 52(1): 96–123; Siegel, J., Dubrovsky, V., Kiesler, S. and McGuire, T. 1986. Group processes in computer-mediated communication. *Organizational Behavior and Human Decision Processes*, 37(1): 157–187.

20. Dubrovsky, V., Kiesler, S. and Sethna, B. 1991. The equalization phenomenon: Status effects in computer-mediated and face-to-face decision-making groups. *Human-Computer Interaction*, 6(1): 119–146; Finholt, T. and Sproull, L. 1990. Electronic groups at work. *Organization Science*, 1: 41–64.

21. Mantovani, G. 1994. Is computer-mediated communication intrinsically apt to enhance democracy in organizations? *Human Relations*, 47(1): 45–62.

22. Scott, K. D. and Townsend, A. M. 1994. Teams: Why some succeed and others fail. *HR Magazine*, 39(8): 62–67.

23. Alavi, M. 1991. Group decision support systems: A key to business team productivity. *Journal of Information Systems Management*, 8(3): 36–41; Jessup, L. M. and Kukalis, S. 1990. Better planning using group support systems. *Long Range Planning*, 23(3): 100–105; McCartt, A. T. and Rohrbaugh, J. 1989. Evaluating group decision support system effectiveness: A performance study of decision conferencing. *Decision Support Systems*, 5(2): 243–253.

24. Jessup, L. M., Connolly, T., and Galegher, J. 1990. The effects of anonymity on GDSS group process with an idea-generating task. *MIS Quarterly*, 14: 313–321; Jessup, L. M., and Tansik, D. A. 1991. Decision making in an automated environment: The effects of anonymity and proximity with a group decision support system. *Decision Sciences*, 22: 266–279.

25. Schrage, *op cit.*

26. Grenier, R. and Metes, G. 1995. *Going virtual: Moving your organization into the 21st century.* Upper Saddle River, NJ: Prentice Hall.

27. Lucas, *op cit.*

28. Schrage, *op cit.*

29. For a more in-depth discussion of the challenges of implementing change, see *Academy of Management Executive* 8 (4), 1994, special issue on restructuring, reengineering, and right-sizing.

30. Reger, R. K., Mullane, J. V., Gustafson, L. T., and DeMarie, S. M. 1994. Creating earthquakes to change organizational mindsets. *Academy of Management Executive*, 8(4): 31–46.

31. Negroponte, N. 1995. *Being digital*, New York: Alfred A. Knopf.

IV

MANAGING KNOWLEDGE WORK WITHIN INNOVATIVE ORGANIZATIONS

9

Managing Technical Communications and Technology Transfer

25

Distinguishing Science from Technology

THOMAS J. ALLEN

Technology is not science—engineers are not scientists. Few would contest these statements; and yet, the failure to recognize the distinction has created untold confusion in the literature. Despite the fact that they should be the last to commit such an egregious error, social scientists studying the behavior of scientists and engineers seldom distinguish properly between the two groups. The social science literature is replete with studies of "scientists," who upon closer examination turn out to be engineers. Worse still, in many studies the populations are mixed, and no attempt is made to distinguish between the two subsets.[1] Many social scientists still view the two groups as essentially the same and feel no need to distinguish between them. This sort of error has led to an unbelieveable amount of confusion over the nature of the populations that have been studied and over the applicability of research results to specific real-life situations. A common practice is to use the term *scientist* throughout a presentation, preceded by a disclaimer to the effect that "for ease of presentation, the term *scientist* will be assumed to include both engineers and scientists." This approach to-

Reprinted with permission from the author.

Thomas J. Allen is the Howard Johnson Professor of Management at MIT's Sloan School of Management.

tally neglects the vast differences between the two professions. Managers are not immune from this problem either. Many managers of R&D fail to recognize the true differences and often assume differences that are really non-existent.

At this point, many readers will accuse the author of magnifying what they may consider a trivial issue. But it is just that failure to recognize the distinction that has resulted in so much misdirected policy. In the field of information science, it has often resulted in heavy investments in solutions to the wrong problem. Engineers differ from scientists in their professional activity, their attitudes, their orientations, and even in their typical family background. To interpret the results of research, it is essential to know whether those results were derived from the study of technical professionals working either as engineers or as scientists because the behavior of the two is so different.

One area in which distinctions are very marked is technical communication. Engineers and scientists communicate about their work in very different ways. The reasons for this are many. Not only are the two groups socialized into entirely different subcultures but their educational processes are vastly different, and there is a considerable amount of evidence to show that they differ in personality characteristics and family backgrounds as well. Krulee and Nadler (1960) contrast the values and career orientation of science and engineering undergraduates in the following ways:

> [Students] choosing science have additional objectives that distinguish them from those preparing for careers in engineering and management. The science students place a higher value on independence and on learning for its own sake, while, by way of contrast, more students in the other curricula are concerned with success and professional preparation. Many students in engineering and management expect their families to be more important than their careers as major sources of satisfactions, but the reverse pattern is more typical for science students. Moreover, there is a sense in which the science students tend to value education as an end in itself, while the others value it as a means to an end.

Note that Krulee and Nadler do not distinguish between engineering students and students in management. There is considerable evidence to show that many engineering students see the profession as a transitional phase in a career leading to higher levels of management. Krulee and Nadler go on to argue that engineering students are less concerned than those in science with what one does in one's specialty and are more concerned with the attainment of organizational rewards and promotions. They are more prepared than their fellow scientists to sacrifice some of their independence and opportunities for innovative work in order to take on particular organizational or managerial responsibilities.

In the same vein, Ritti (1971) finds a marked contrast between the work goals of scientists and engineers after graduation. Ritti found, for example, that over 60 percent of the engineers in his sample indicated that it was "very important" for them to know the company's management policies and practices and to help the company increase its profits. Less than 30 percent of the scientists indicated that these were very important work goals. On the other hand, more than 80 percent of the scientists said that it was very important for them to publish articles in technical journals and to establish a professional reputation outside the company—the corresponding percentages for engineers were less than 30 percent. From all the analyses of his data, Ritti draws the following general conclusions:

> First, the notion of a basic conflict in goals between management and the professional is misapplied to engineers. If the goals of the business require meeting schedules, developing products that will be successful in the marketplace, and helping the company expand its activities, then the goals of engineering specialists are very much in line with these ends.
>
> Second, engineers do not have the goals of scientists. And evidently they never had the goals of scientists. While publication of results and professional autonomy are clearly valued goals of Ph.D. scientists, they are just as clearly the least valued goals of the baccalaureate engineer. The

reasons for this difference can be found in the work functions of engineers as opposed to research scientists. Furthermore, both groups desire career development or advancement but for the engineer advancement is tied to activities within the company, while for the scientist advancement is dependent upon the reputation established outside the company.

The type of person who is attracted to a career in engineering is fundamentally quite different from the type who pursues a scientific career. On top of all of this lies the most important difference: level of education. Engineers are generally educated to the baccalaureate level; some go on to a Master of Science degree; some have no college degree at all. The scientist is almost always assumed to have a doctorate. The long, complex process of academic socialization that is involved in reaching this stage is bound to result in a person who differs considerably in his lifeview. These differences in values and attitudes toward work will almost certainly be reflected in the behavior of the individuals. To treat both professions as one and then to search for consistencies in behavior and outlook is almost certain to produce error and confusion of results.

THE NATURE OF TECHNOLOGY

The differences between science and technology lie not only in the kinds of people who are attracted to them; they are basic to the nature of the activities themselves. Both science and technology develop in a cumulative manner, with each new advance building on and being a product of vast quantities of work that have gone before. In science all of the work up to any point can be found permanently recorded in literature, which serves as a repository for all scientific knowledge. The cumulative nature of science can be demonstrated quite clearly (Price, 1965a, 1970) by the way in which citations among scientific journal articles cluster and form a regular pattern of development over time.

A journal system has been developed in most technologies that in many ways emulates the system originally developed by scientists; yet the literature published in the majority of these journals lacks, as Price (1965b, 1970) has shown, one of the fundamental characteristics of the scientific literature: it does not cumulate or build upon itself as does the scientific literature. Citations to previous papers or patents are fewer and are more often to the author's own work. Publication occupies a position of less importance than it does in science where it serves to document the end product and establish priority. Because published information is at best secondary to the actual utilization of the technical innovation, this archival function is not as essential to ensure the technologist that he is properly credited by future generations. The names of Wilbur and Orville Wright are not remembered because they published papers. The technologist's principal legacy to posterity is encoded in physical, not verbal, structure. Consequently, the technologist publishes less and devotes less time to reading than do scientists.

Information is transferred in technology primarily through personal contact. Even in this, however, the technologist differs markedly from the scientist. Scientists working at the frontier of a particular specialty know each other and associate together in what Derek Price has called "invisible colleges." They keep track of one another's work through visits, seminars, and small invitational conferences, supplemented by an informal exchange of written material long before it reaches archival publication. Technologists, on the other hand, keep abreast of their field by close association with co-workers in their own organization. They are limited in forming invisible colleges by the imposition of organizational barriers.

BUREAUCRATIC ORGANIZATION

Unlike scientists, the vast majority of technologists are employed by organizations with a well-defined mission (profit, national defense, space exploration, pollution abatement, and so forth). Mission-oriented organizations necessarily demand of their technologists a degree of identification unknown in most scientific circles. This organizational identification works in two ways to exclude the technologist from informal communication channels outside his or her organization. First, they are inhibited by the requirements that they work only on problems that are of interest to their employer, and second, they must refrain from early disclosure of the results of their research in order to maintain their employer's advantage over competitors. Both of these constraints violate the strong scientific norms that underlie and form the basis of the invisible college. The first of these norms demands that science be free to choose its own problems and that the community of colleagues be the only judges of the relative importance of possible areas of investigation, and the second is that the substantive findings of research are to be fully assigned and communicated to the entire research community. The industrial organization, by preventing its employers from adhering to these two norms, impedes the formation by technologists of anything resembling an invisible college.

Impact of "Localism" on Communication

What is the effect of this enforced "localism" on the communication patterns of engineers? Because proprietary information must be protected to preserve the firm's position in a highly competitive marketplace, free communication among engineers of different organizations is greatly inhibited. It is always amusing to observe engineers from different companies interacting in the hallways and cocktail lounges at conventions of professional engineering societies. Each one is trying to draw the maximum amount of information from his competitors while giving up as little as possible of his own information in return. Often the winner in this bargaining situation is the person with the strongest physical constitution.

Another result of the concern over divulging proprietary information will be observed in looking at an engineer's reading habits. A good proportion of the truly important information generated in an industrial laboratory cannot be published in the open literature because it is considered proprietary and must be protected. It is, however, published within the organization, and, for this reason, the informal documentation system of his or her parent organization is an important source of information for the engineer.

THE EFFECT OF TURNOVER

It is this author's suspicion that much of the proprietary protectionism in industry is far overplayed. Despite all of the organizational efforts to prevent it, the state of the art in a technology propagates quite rapidly. Either there are too many martinis consumed at engineering conventions or some other mechanism is at work. This other mechanism may well be the itinerant engineer, who passes through quite a number of organizations over the course of a career. Whenever engineers leave an employer, voluntarily or otherwise, they carry some knowledge of the company's operations, experience, and current technology with them. We are gradually coming to realize that human beings are the most effective carriers of information and that the best way to transfer information between organizations or social systems is to physically transfer a human carrier. Roberts's studies (Roberts and Wainer, 1967) marshal impressive evidence for the effective transfer of space technology from quasi-academic institutions to the industrial sector and eventually to commercial application in those instances in which technologists left university laboratories to establish their own businesses. This finding is especially impressive in view of the general failure to find evidence of successful transfer of space technology by any other mechanism, despite the fact that many techniques have been tried and a substantial amount of money has been invested in promoting the transfer.

This certainly makes sense. Ideas have no real existence outside of the minds of people. Ideas can be represented in verbal or graphic form, but such representation is necessarily incomplete and cannot be easily structured to fit new situations. The human brain has a capacity for flexibly restructuring information in a manner that has never been approached by even the most sophisticated computer programs. For truly effective transfer of technical information, we must make use of this human ability to recode and restructure information so that it fits into new contexts and situations. Consequently, the best way to transfer technical information is to move a human carrier. The high turnover among engineers results in a heavy migration from organization to organization and is therefore a very effective mechanism for disseminating technology throughout an industry and often to other industries. Every time an engineer changes jobs he brings with him a record of his experiences on the former job and a great amount of what his former organization considers "proprietary" information. Now, of course, the information is usually quite perishable, and its value decays rapidly with time. But a continual flow of engineers among the firms of an industry ensures that no single firm is very far behind in knowledge of what its competitors are doing. So the mere existence of high turnover among R&D personnel vitiates much of the protectionism accorded proprietary information.

As for turnover itself, it is well known that most organizations attempt to minimize it. Actually, however, a certain amount of turnover may be not only desirable but absolutely essential to the survival of a technical organization, although just what the optimum turnover level is for an organization is a question that remains to be answered. It will vary from one situation to the next and is highly dependent upon the rate at which the organization's technical staff is growing. After all, it is the influx of new engineers that is most beneficial to the organization, not the exodus of old ones. When growth rate is high, turnover can be low. An organization that is not growing should welcome or encourage turnover, despite the costs of hiring and processing new personnel. Although it is impossible to place a price tag on the new state-of-the-art information that is brought in by new employees, it may very well more than counterbalance the costs of hiring. This would be true at least to the point where turnover becomes dis-

ruptive to the morale and functioning of the organization.

COMMUNICATION PATTERNS IN SCIENCE AND TECHNOLOGY

Scientists all share a common concern and responsibility for processing information, which is the essence of scientific activity. As physical systems consume and transform *energy*, so too does the system of science consume, transform, produce, and exchange *information*. Scientists talk to one another, they read each other's papers, and most important, they publish scientific papers, their principal tangible product. Both the input and output of this system we call science are in the form of information. Each of the components, whether individual investigations or projects, consume and produce information. Furthermore, whether written or oral, this information is always in the form of human language. Scientific information is, or can be, nearly always encoded in a verbal form.

Technology is also an ardent consumer of information. The engineer must first have information in order to understand and formulate the problem confronting him. Then she must have additional information from either external sources or memory in order to develop possible solutions to her problem. Just like his counterpart in science, the technologist requires verbal information in order to perform his work. At this level, there is a very strong similarity between the information input requirements of both scientists and technologists.

It is only when we turn to the nature of the outputs of scientific and technological activity that really striking differences appear. These, as will be seen, imply very real and important second-order differences in the nature of the information input requirements.

Technology consumes information, transforms it, and produces a product in a form that can still be regarded as information bearing. The information, however, is no longer in a verbal form. Whereas science both consumes and produces information in the form of human language, engineers transform information from this verbal format to a physically encoded form. They produce physical hardware in the form of products or processes.

The scientist's principal goal is a published paper. The technologist's goal is to produce some physical change in the world. This difference in orientation, and the subsequent difference in the nature of the products of the two, has profound implications for those concerned with supplying information to either of the two activities.

The information-processing system of science has an inherent compatibility between input and output. Both are in verbal form (Figure 25.1). The output of one stage, therefore, is in the form in which it will be required for the next stage. The problem of supplying information to the scientist thus becomes one of systematically collecting and organizing these outputs and making them accessible to other scientists to employ in their work.

In technology, on the other hand, there is a fundamental and inherent incompatibility between input and output. Because outputs are in form basically different from inputs, they usually cannot serve directly as inputs to the next stage. The physically encoded format of the output makes it very difficult to retrieve the information necessary for further developments. That is not to say that this is impossible: technologists frequently analyze a competitor's product in order to retrieve information; competing nations often attempt to capture one another's weapon systems in order to analyze them for their information content. This is a difficult and uncertain process, however. It would be much simpler if the information were directly available in verbal form. As a consequence, attempts are made to decode or understand physically encoded information only when one party to the exchange is unavailable or unwilling to cooperate. Then an attempt is made to understand how the problems were approached by analyzing the physical product. In cases where the technologists responsible for the product are available and cooperative, this strategy is seldom used. It is much more effective to communicate with them directly, thereby obtaining the necessary information in a verbal form.

A question that arises concerning the documentation produced in the course of most techno-

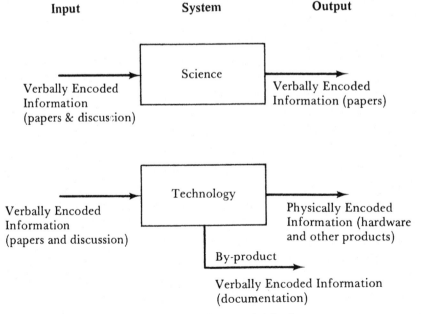

Figure 25.1. Information processing in science and technology.

TABLE 25.1
Sources of Messages Resulting in Technical Ideas Considered During the Course of Nineteen Projects*

Channel	Seventeen technological projects		Two scientific research projects	
	Number of messages produced	Percentage of total	Number of messages produced	Percentage of total
Literature	53	8	18	51
Vendors	101	14	0	0
Customer	132	19	0	0
Other sources external to the laboratory	67	9	5	14
Laboratory technical staff	44	6	1	3
Company research programs	37	5	1	3
Analysis and experimentation	216	31	3	9
Previous personal experience	56	8	7	20

*From Allen (1984).

logical projects is why it cannot serve to meet the information needs of subsequent stages in technological development. The answer is that it is not quite compatible with other input requirements although it meets the requirements of verbal structure. First, as seen in Figure 25.1, it is merely a by-product. The direct output is still physical, consequently it is incomplete. It generally assumes a considerable knowledge of what went into the physical product. Those unacquainted with the actual development therefore require some human intervention to supplement and interpret the information contained in this documentation. Thus, technological documentation is often most useful only when the author is directly available to explain and supplement its content.

Now if all of this is true, it leads to an interesting conclusion: whereas the provision of information in science involves the gathering, organizing, and distribution of publications, the situation in technology is very different. The technologist must obtain his information either through the very difficult task of decoding and translating physically encoded information or by relying upon direct personal contact and communication with other technologists. His reliance upon the written word will be much less than that of the scientist. Thus, there are very different solutions to the problems of improving the dissemination and availability of in-

formation in the two domains. If, for example, one were to develop an optimum system for communication in science, there is no reason to suspect that it would be at all appropriate for technology. It is essential that we bear these distinctions in mind while exploring the nature of the communication processes in technology. Much has been written about scientific information flow; we may even understand something about it. One must be extremely cautious, however, in extrapolating or attempting to apply this understanding of science to the situation in technology.

The difference between communication patterns in science and technology is amply illustrated by the data originally reported by Allen (1984). Among a sample of nineteen projects, seventeen were clearly developmental in their nature. The remaining two had clear-cut goals, but these were directed toward an increased understanding of a particular set of phenomena. While the information generated by these two teams would eventually be used to develop new hardware, this was not the immediate goal of the teams, who were far more interested in the phenomena than in the application. For this reason, their work can be considered to be much more scientific than technological in nature.

A comparison of the two scientific projects with the seventeen technological projects (Table 25.1) shows a marked disparity in the use of eight information channels. The scientists engaged in the phenomena-oriented project concentrated their attention heavily upon the literature and upon colleagues outside their laboratory organization. The engineers spread their attention more evenly over the channels and received ideas from two sources unused by the scientists. The customer (in this case, a government laboratory) suggested a substantial number of ideas, demonstrating the importance of the marketplace for technologists. Vendors are another important channel in technology because they are important potential suppliers of components or subsystems, and they provide information that they hope will stimulate future business. Involvement in the marketplace, either through the customer or potential vendors, exerts a significant influence upon the communication system, providing channels for the exchange of information in two directions and connecting buyers and sellers through both the procurement and marketing functions of the organization.

Scientists and engineers also differ in the way they allocate their time between oral and written channels of communication. From his studies of R&D projects, Allen (1984) shows that scientists spend substantially more of their time reading and communicating with each other about their disciplines' literatures than do engineers. In contrast, engineers spend more of their time in personal contact than in reading. The comparisons are quite revealing. Despite all the discussion of informal contact and invisible colleges among scientists (and scientists do make use of personal contacts), it is the engineer who is more dependent upon colleagues. The difference between communication behavior of scientists and engineers is not simply quantitative, however. The persons contacted by scientists are very different from those contacted by engineers, and the relationship between the engineer and those with whom she or he communicates is vastly different from the relationship that exists among scientists. In written channels, too, there are significant differences. The literature used by scientists differs qualitatively from that used by engineers. And engineers not only read different journals, but as discussed in Allen (1984), they use the literature for entirely different purposes.

THE RELATION BETWEEN SCIENCE AND TECHNOLOGY

Given the vast differences between science and technology, how do the two relate to each other? This is a question that has intrigued a number of researchers in recent years. It is generally assumed that the two are in some way related, and in fact national financing of scientific activity is normally justified on the basis of its eventual benefits to technology. Is there any basis for this, and what, if any, is the relation of science to technology? How are the results of scientific activity incorporated into technological developments? To what extent is technology dependent upon science? What are the time lags involved?

The Process of Normal Science

Kuhn (1962) describes three classes of problems that are normally undertaken in science:

1. The determination of significant facts that the research paradigm has shown to be particularly revealing of the nature of things.

2. The determination of facts, which (in contrast with problems of the first class) may, themselves, be of little interest, but which can be compared directly with predictions made by the research paradigm.

3. Empirical work undertaken to articulate the paradigm theory.

The first two of these—the precise determination and extension to other situations of facts and constants that the paradigm especially values (for example, stellar position and magnitude, specific gravities, wave lengths, boiling points) since they have been used in solving paradigmatic problems, and the test of hypotheses derived from the central body of theory—will not concern us here. These are the normally accepted concerns of science, but the third-listed function is probably the most important, and I shall address myself to this category of activity that comprises empirical work undertaken to extend and complete the central body of theory. It may, itself, be subdivided into three classes of activity (Kuhn, 1970):

1. The determination of physical constants (gravitational constants; Avogadro's Number; Joule's Coefficient; etc.).

2. The development of quantitative laws. (Boyle's, Coulomb's, and Ohm's Laws).

3. Experiments designed to choose among alternative ways to applying the paradigm to new areas of interest.

Within the third class lie problems that have resulted from difficulties encountered during the course of scientific research or during the process of technological advance. This, as we shall see, is a form of scientific activity of extreme interest and importance.

The Dependence of Technology on Science

Despite the long-held belief in a continuous progression from basic research through applied research to development, empirical investigation has found little support for such a situation. It is becoming generally accepted that technology builds upon itself and advances quite independently of any link with the scientific frontier, and often without any necessity for an understanding of the basic science which underlies it. Price (1965b), a strong advocate of this position, cites Toynbee's view that

> physical science and industrialism may be conceived as a pair of dancers, both of whom know their steps and have an ear for the rhythm of the music. If the partner who has been leading chooses to change parts and to follow instead, there is perhaps no reason to expect that he will dance less correctly than before.

Price goes on to marshal evidence refuting the idea of technology as something "growing out of" science and to make the claim that communication between the two is at best a "weak interaction." Communication between the two is restricted almost completely to that which takes place through the process of education.

Kuhn (1970) describes science as the activity or process of knowledge making. It is a stream of human activity devoted to building a store of knowledge and can be traced back to the beginning of recorded history. Science can thereby be represented as a stream of events over time cumulating in a body of knowledge. There are two other streams of human activity that operate parallel to science and that function both as contributors to scientific development and as beneficiaries of scientific accomplishment. First there is the activity we have labeled "technology." This is a stream of human activity oriented toward incorporating human knowledge into physical hardware, which will eventually meet with some human use. Then there is a much more general form of human activity in which the ideas of science and the hardware of technology are actually put to some use in the stream of human affairs. This last stream we will label *utilization* (Figure 25.2).

The activities of technology and of utilization in commerce, industry, welfare, and war, while at various times in close harness with science, have developed for the most part independently. Science builds on prior science; technology builds on prior technology; and utilization grows and spreads in response to needs and benefits.

The familiar notion of science providing the basis upon which technology is built to be later utilized in commerce or industry has been shown by the historians of science to have only a limited basis in historical fact. Civilizations have often emphasized activity in one or two of these areas to the exclusion of the others. The Greeks, for example, were very active in science, but they were relatively little concerned with the practical applications or implications of their discoveries. The Romans, in contrast, developed a highly practical civilization, which was greatly concerned with the building of artifices to aid in coping with the physical and social environment. They devoted much effort to the construction of roads and aqueducts and of improvement of armor and weapons without much concurrent increase in their understanding of the natural basis of their developments. History shows quite independent paths through the succeeding centuries to the present time. The three streams appear now in rapid parallel growth; an increased emphasis in one is usually accompanied by an increase

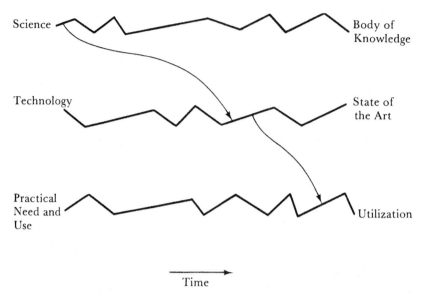

Figure 25.2. Science, technology, and the utilization of their products, showing the normal progression from one to the other.

in the other two. It is probable that the streams are more closely coupled now than they have been historically, but the delays encountered in any of the communication paths between them remain substantial.

The Flow of Information Between Science and Technology

Over the past ten years several studies have attempted to trace the flow of information from science to technology. In one of the earlier of these, Price (1965b), after investigating citation patterns in both scientific and technological journals, concluded that science and technology progress quite independently of one another. Technology, in this sense, builds upon its own prior developments and advances in a manner independent of any link with the current scientific frontier and often without any necessity for an understanding of the basic science underlying it.

Price's hypothesis certainly appears valid in light of more recent evidence. There is little support for direct communication between science and technology. The two do advance quite independently, and much of technology develops without

a complete understanding of the science upon which it is built.

Project Hindsight was the first of a series of attempts to trace technological advances back to their scientific origins. Within the twenty-year horizon of its backward search, Hindsight was able to find very little contribution from basic science (Sherwin and Isenson, 1967). In most cases, the trial ran cold before reaching any activity that could be considered basic research. In Isenson's words, "It would appear that most advances in the technological state of the art are based on no more recent advances than Ohm's Law or Maxwell's equations."

Project TRACES (IIT Research Institute, 1968), partially in response to the Hindsight results, succeeded in tracing the origins of six technological innovations back to the underlying basic sciences but only after extending the time horizon well beyond twenty years. In a follow-up, Battelle (1973) investigators found similar lags in five more innovations. In yet another study, Langrish found little support for a strong science-technology interaction. In tracing eighty-four award-winning innovations to their origins, he found that "the role of university as a source of ideas for [industrial] innovation is fairly small" and that "university science and industrial technology are two quite separate activities which occasionally come into contact with each other" (Langrish, 1971). He argued very strongly that most university basic research is totally irrelevant to societal needs and can be only partially justified for its contributions through training of students.

Gibbons and Johnston (1973) attempted to refute the Langrish hypothesis. They presented data from thirty relatively small-scale technological advances and found that approximately one-sixth of the information needed in problem solving came from scientific sources. Furthermore, they claimed greater currency in the scientific information that was used. The mean age of the scientific journals they cited was 12.2 years. This is not quite twenty, but with publication lags, it can safely be concluded that the work was fourteen or fifteen years old at the time of use. They showed considerable use of personal contact with university scientists, but nearly half of these were for the purpose of either referral to other sources of information or to determine the existence of specialized facilities or services. So, while Gibbons and Johnston may raise some doubt over the Price-Langrish hypothesis, the contrary evidence is hardly compelling.

The evidence, in fact, is very convincing that the normal path from science to technology is, at best, one that requires a great amount of time. There are certainly very long delays in the system, but it should not be assumed that the delays are always necessarily there. Occasionally, technology is forced to forfeit some of its independence. This happens when its advance is impeded by a lack of understanding of the scientific basis of the phenomena with which it is dealing. The call then goes out for help. Often a very interesting basic research problem can result, and scientists can be attracted to it. In this way, science often discovers voids in its knowledge of areas that have long since been bypassed by the research front. Science must, so to speak, backtrack a bit and increase its understanding of an area previously bypassed or neglected.

Morton (1965) described several examples in which technology has defined important problems for scientific investigation. For example, he pointed out that progress in electronic tube technology had advanced without a real understanding of the principles involved. It did this largely by "cut and dry" methods, manipulating the geometry of the elements and the composition of the cathode materials with little real understanding of the fundamental physics underlying the results. This block to the advance of a burgeoning technology forced a return to basic classical physics and a more detailed study of the interactions of free electrons and electromagnetic waves. The return allowed scientists to fill a gap in their understanding and subsequently permitted the development of such microwave amplifiers as the magnetron, klystron, and traveling wave tube. From such examples, it is important to note that there was first of all communication of a problem from technology to science, followed by a relatively easy transfer of scientific results back to the technologists. The two conditions are clearly related. When technology is the source of the problem, technologists are ready and capable of understanding the solution and putting it to work.

Additional support for this idea is provided by Project Hindsight (Sherwin and Isenson, 1967). While is most cases, Hindsight was unable to find any contribution to technology from basic science, it is the exception to this discontinuity between science and technology that chiefly concerns us at the present. Isenson reports[2] that he discovered exceptions to his general finding and that these exceptions are usually characterized by a situation in which, similar to Morton, technology has advanced to a limit at which an understanding is required of the basic physical science involved. Thus technology defines a problem for science. When this problem is attacked and resolved by scientists, its solution is passed immediately into technology. A close coupling thus exists for at least an isolated point in time, and the researchers of Project Hindsight were able to trace the record back from an improved system in what we have labeled the "utilization stream" through an advance in the technological state of the art to the closure of a gap in the body of scientific knowledge. To distinguish this latter form of research from "frontier science," I propose calling it "technology-pull" science.

Technology-pull science is by its nature directly responsive to technological need, and the advance of technology is often contingent upon the pursuit of such science. So when the connection between science and technology is of this form, little delay is encountered in the transfer of information (Figure 25.3). Communication is rapid and direct, and the long delays of the normal transfer process are circumvented. The transfer from technology-pull science can be further accelerated by including in the technological development team former scientists or individuals whose training was in science. The advantages of such a strategy were clearly demonstrated during World War II when many scientists became engineers, at least temporarily, and were very effective in implementing the results of fundamental research.

A similar phenomenon is occurring at the present time in genetic engineering. Molecular biologists have been attracted by the potential economic benefits into what is now becoming a new technology. These former basic research scientists both carry with them substantial scientific knowledge and retain information ties to the scientific community.

The point to be made is that at least a segment of basic science is not conducted at what is called the "frontier" of knowledge. Technology—and often investigation in a different scientific area—will raise problems that attract investigators to an area that has been worked on before. The investigation then proceeds, looking perhaps from a somewhat different vantage at items that had not previously been deemed important phenomena. That such investigations are searching in what had been considered secure territory makes them no less fundamental in their nature.

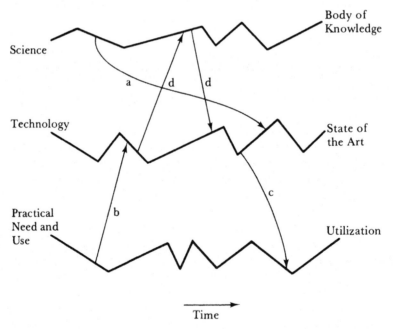

Figure 25.3.   Science, technology, and the utilization of their products, showing communication paths among the three streams ([a] The normal process of assimilation of scientific results into technology. [b] Recognized need for a device, technique, or scientific understanding. [c] The normal process of adoption of technology for use. [d] Technological need for understanding of physical phenomena and its response [from Marquis and Allen, 1966]).

While technology and science in general may progress quite independently of each other, there very probably are some technologies that are more closely connected with science than others. For example, electronics technology is more closely related to frontier work in physics than say, mechanical technology. Nuclear technology should be more closely coupled to the advance of physical knowledge than either of these two. And of course, genetic engineering still retains a very close association with its parent science. Citation studies of engineering and scientific journals, for example, have shown the ratio of technological to scientific citations to range from about six to one to less than one to one. The data clearly indicate that a wide variation exists in the degree to which technologies are coupled to their respective sciences.

NOTES

1. Even Pelz and Andrews (1966), who are careful to preserve the distinction throughout most of their book, a study of 1,300 engineers and scientists, chose *Scientists in Organizations* as its title, forgetting the majority of their sample.

2. Personal communication.

REFERENCES

Allen, T. J. 1984. *Managing the Flow of Technology.* Cambridge: MIT Press.

Battelle Memorial Institute. 1973. *Interactions of science and technology in the innovation process: Some case studies.* Final report to the National Science Foundation NSF-C667, Columbus, Ohio.

Gibbons, M., and Johnston, R. D. 1974. The roles of science in technological innovation. *Research Policy* 3:220–242.

IIT Research Institute. 1968. *Technology in retrospect and critical events in science.* Report to the National Science Foundation NSF C-235.

Krulee, G. K., and Nadler, E. B. 1960. Studies of education for science and engineering: Student values and curriculum choice. *IEEE Transactions on Engineering Management* 7:146–158.

Kuhn, T. B. 1970. *Structure of Scientific Revolutions.* Rev. ed. Chicago: University of Chicago Press.

Langrish, J. 1971. Technology transfer: Some British data. *R&D Management* 1:133–136.

Marquis, D. G., and Allen, T. J. 1966. Communication patterns in applied technologies. *American Psychologist* 21:1052–1060.

Morton, J. A. 1965. From physics to function. *IEEE Spectrum* 2:62–64.

Pelz, D. C., and Andrews, F. M. 1966. *Scientists in Organizations.* New York: Wiley.

Price, D. J. DeSolla. 1965a. Networks of scientific papers. *Science* 149:510–515.

—1965b. Is technology independent of science? *Technology and Culture* 6:553–568.

—1970. In D. K. Pollock and Nelson, C. E. (eds.) *Communication Among Scientists and Technologists.* Lexington, Mass.: Heath.

Ritti, R. R. 1971. *The Engineer in the Industrial Corporation.* New York: Columbia University Press.

Roberts, E. B., and Wainer, H. A. 1971. Some characteristics of technical entrepreneurs. *IEEE Transactions on Engineering Management*, EM-18, 3.

Sherwin, E. W., and Isenson, R. S. 1967. Project Hindsight. *Science* 156:1571–1577.

26

Communication Networks in R&D Laboratories

THOMAS J. ALLEN

Communication networks in R&D laboratories are shown to have structural characteristics, which when properly understood can be employed to more effectively keep the laboratories' personnel abreast of technological developments. Informal relations and physical location are shown to be important determinants of this structure.

INTRODUCTION

To date, attempts to automate the transmission of scientific and technological information have been most notable for their failure. The reason for this does not lie in any lack of attention or inadequate effort allocated to the problem, since very large sums of money have been expended on storage and retrieval systems for scientific and technological information. Rather, it is due to the nature and complexity of the information itself, and to the uncertainty and very personal nature of each user's needs.

For this reason, the human being is still the most effective source of information, communication with a technically competent colleague being conducted on a two-way basis, with the output of the source tracking and responding to the expressed needs of the user. In this manner, the ability of the source to adapt flexibly and respond rapidly to communicated needs enables it to cope effectively with the uncertain nature of those needs.

A large number of recent studies show that increased use of organizational colleagues for information is strongly related to scientific and technological performance. The relation to performance

is, perhaps, demonstrated most clearly in a recent study by a group at M.I.T. Some of the results of this study are presented in this paper as a basis for a discussion on strategies for properly structuring the flow of technical information in research and development organizations.

THE INTERNAL CONSULTING STUDY

Eight pairs of individuals in different organizations, but working on identical problems, were compared on the extent to which each of them consulted with organizational colleagues. Since there were always two individuals attempting to solve the same problem, their solutions could be compared for relative quality and the sample split between 'high' performers and 'low' performers. Performance evaluations were made by competent technical evaluators in the government laboratories that had sponsored the projects. Dividing the sample into high and low performers allowed a further comparison to be made, now on an aggregate basis, of behaviour leading to high or low performance.

When such a comparison was made with respect to the number of times organizational colleagues were consulted during the project, it showed that high performers made far greater use of this source of technical information (Figure 26.1). As a matter of fact, high performers not only reported a significantly greater frequency of consultation with organizational colleagues, they also spent significantly more time in their discussions with colleagues.

Furthermore, they relied on more people both

Thomas J. Allen is the Howard Johnson Professor of Management at MIT's Sloan School of Management.

within their own technical speciality and on other specialties (Figures 26.2a and 26.2b). The high performer was in closer touch than the low performer with developments in his own field. Through his wide range of contact within his specialty he is less likely to miss an important development which might have some impact on the problem to which he is assigned. He also had wider contact with people in specialties other than his own. In fact, it was only the high performers who showed any real contact outside of their specialty (Figure 26.2b). The low performers seldom ventured outside of their field. These findings agree with those of Pelz & Andrews (1966) who noted that colleague contacts both within the immediate work group and with other groups in the organization were positively related to a person's performance and that the variety of contacts and their frequency each contributed independently to performance.

One cannot of course determine very easily whether communication causes high performance or whether high performers merely communicate more. Pelz & Andrews (1966), in their study, obtained data on which of the two parties initiated the contact. They then assumed that an individual's high performance would be more likely to attract contact from others than to induce him to initiate contacts himself. They then looked only at contacts initiated by the information user and found that the relation with performance remained strong. They concluded:

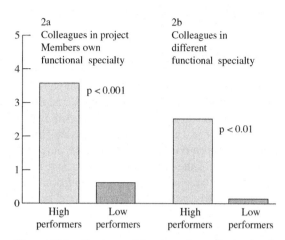

Figure 26.2. Number and location of organizational colleagues with whom project members communicate.

large amounts of colleague contact tended to go with high performance even when one looked only at scientists who themselves were the primary initiators of the contacts. Under these conditions it was difficult to believe that the contacts were primarily the *result* of previous high performance. Thus the hypothesis that contacts with colleagues stimulated performance seemed to be supported (p. 47).

SUPPORT FROM OUTSIDE THE PROJECT: A PARADOX

Given the benefits to be derived from internal consultation, one would expect project members to rely heavily upon their technical staff for information. In fact, this is not the case. During the nineteen projects studied by Allen (1966), project members actually obtained more of their ideas from outside of their firms than from their own technical staff—although in general a poor performance was shown by sources outside of the firm. In fact, when individuals inside and outside of the firm were compared as sources of ideas, there was an inverse relation between frequency of outside use and performance.

Those information sources that reward the user by contributing more to his performance are used less than those that do not. Such a situation

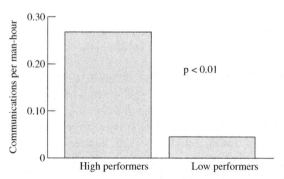

Figure 26.1. Extent of communication between R&D project members and organizational colleagues not assigned to the project.

would seem to conflict with the principles of psychological learning theory. One might expect a person to return more frequently to those channels that reward him most consistently. The data show just the opposite to be true. The paradox can, however, be resolved with the introduction of an additional parameter. It can be safely predicted that an individual will repeat a behaviour that is rewarded more frequently than one that is unrewarded, only if the cost to him of the rewarded behaviour is less than or equal to the cost of the unrewarded behaviour. In other words, both cost and benefit may be taken into consideration when deciding upon a source of information.

Gerstberger and Allen (1968) actually studied this decision process in some detail. They found no relation at all between the engineers' perception of the benefits to be gained from an informal information channel and the extent to which the channel was used. However, a very strong relation existed between extent of use and the engineers' perception of the amount of effort that it took to use the channel. Cost in that case was the overriding determinant of the decision. Working back from this finding, one might speculate that the failure to consult with organizational colleagues is attributable to a high cost associated with such consultation. In fact, there is evidence to indicate that the organizational colleague is a high cost source of information for research and development project teams (Allen *et al.*, 1968). It can, for example, be very costly for a project member to admit to a colleague that he needs his help.

TECHNOLOGICAL "GATEKEEPERS"

A number of recent studies have indicated that technologists do not read very much, and one might conclude that literature is not a very effective vehicle for bringing new information into the organization; and while it is found that outside personal contact is used very heavily by organizational technologists, further analysis suggests that this means of transfer is not much more instrumental than literature. The reason for this is that the average technologist cannot communicate effectively with outsiders. This is reflected in the results of several research studies which are consistent in their discovery of an inverse relation between outside personal contact and technical performance (Allen, 1964; Shilling & Bernard, 1964).

How then does information enter the organization? First of all, it is clear that entry does occur, because without it no R&D organization could long survive. No R&D organization, no matter how large, can be fully self-sustaining. In order for the organization to survive its members must maintain themselves abreast of current developments in those technologies which are central to the organization's mission. It must, in other words, constantly import technical information. Not only were the organizations under study surviving; they were, to all appearances, thriving. They were extremely successful, and highly regarded technically. They must, therefore have been successful, somehow, in acquiring information from outside, and disseminating it within their borders. The question remains, how?

The first important clue lies in the observation that, of all possible information sources, only one appears to satisfactorily meet the needs of R&D project members. That one source is the organizational colleague. This has been shown in the case of R&D proposal competitions (Allen, 1964) for preliminary design studies (Allen, 1966; Allen *et al.*, 1968); for 'idea generating groups' (Baker *et al.*, 1967); for engineers and scientists in a wide variety of industrial, governmental and university settings (Pelz & Andrews, 1966); and for the members of 64 laboratories in the biological sciences (Allen, 1964).

Following this clue, Allen & Cohen (1969) discovered that the process by which organizations most effectively import information is an indirect one (Figure 26.3). There existed, in the organizations that they studied, a small number of key people upon whom others relied very heavily for information. These key people, or 'technological gatekeepers,' differ from their colleagues in their orientation toward outside information sources. They read far more, particularly the 'harder' liter-

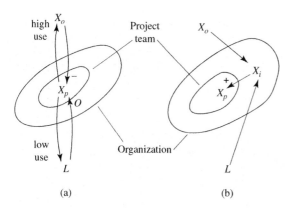

X_p = project team member, in need of information;
X_o = person outside of the organization;
X_i = organizational colleague;
L = literature.

Figure 26.3. The dilemma of importing information into the organization. Direct paths do not work (A), because literature is little used by the average technologist and because the direct contact with outside persons is ineffective. An indirect route, through the technological gatekeeper (B) has been shown to be more effective. Symbols next to incoming arrows indicate the polarity of the correlation with performance.

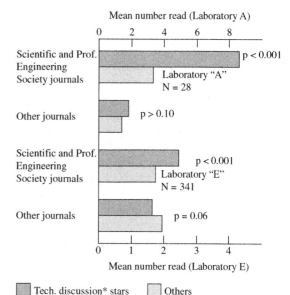

Figure 26.4. Journal readership by technical discussion stars. Laboratory 'A' is the original laboratory reported in Allen & Cohen (1969). Laboratory 'E' is the advanced technology component of a large aerospace firm.

ature. Their readership of professional engineering and scientific journals is significantly greater than that of the average technologist (Figure 26.4). They also maintain broader-ranging and longer-term relationships with technologists outside of their organizations (Figure 26.5). The technological gatekeeper mediates between his organizational colleagues and the world outside, and he effectively couples the organizational to scientific and technological activity in the world at large.

NETWORKS OF GATEKEEPERS

Using the techniques of the earlier study (Allen & Cohen, 1969), the structure of the communication network in the research and advanced technology division of a large aerospace firm was measured. The laboratory under study was organized on a functional basis around five engineering specialties and three scientific disciplines.

The gatekeepers in each specialist department were identified, as well as the structure of the com-

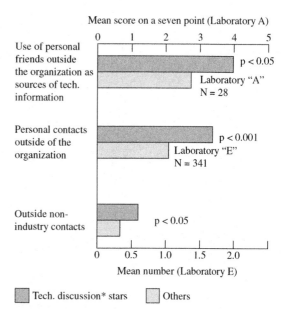

Figure 26.5. Personal contact outside of the organization by technical discussion stars. Laboratory 'A' is the original laboratory reported in Allen & Cohen (1969). Laboratory 'E' is the advanced technology component of a large aerospace firm.

*Persons receiving one standard deviation or more above the overall mean number in their laboratory (Laboratory "A") or in their department (Laboratory "B").

munication network in that department. Because of the complexity of the networks in such a large organization (Figure 26.6), an attempt was made to simplify them through graph-theoretic reduction.

A communication network (or portions thereof) can be characterized according to the degree of interconnectedness that exists among its nodes. There are several degrees of interconnectedness or 'connectivity' that can exist in a network (Flament, 1963). In the present analysis, only that degree of connectivity which Flament has called 'strong' will be considered. A strongly connected component, or strong component in a network, is one in which all nodes are mutually reachable. In a communication network, a potential exists for the transmission of information between any two members of a strong component (Flament, 1963; Harary

et al., 1965). For this reason, the laboratory's communication network was reduced into its strong components and their membership was examined.

When the departmental networks of the organization are reduced in this manner, two things become apparent. First of all, the formation of strong components is not aligned with formal organizational groups, and second, while there were in each functional department anywhere from one to six non-trivial strong components, nearly all of the gatekeepers can be found together as members of the same strong component (see, for an example, Figure 26.7). On the average, 64% of all gatekeepers can be found in eight strong components, one for each of the five technological and three scientific specialties. In each technical specialty, there is one strongly connected network in which most

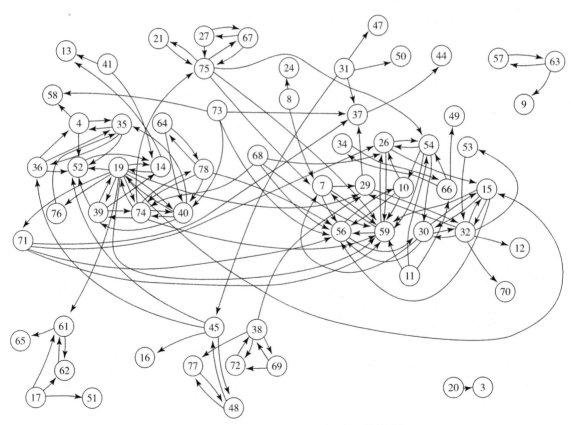

Figure 26.6. Typical communication network of a functional department in a large R&D laboratory.

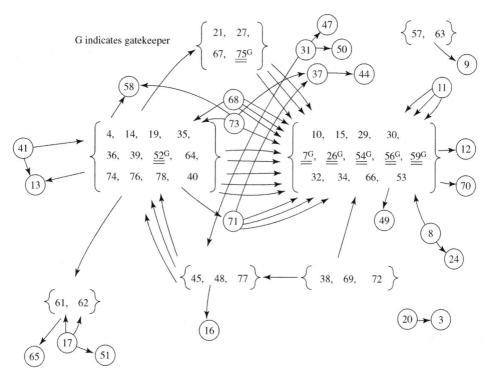

Figure 26.7. Departmental communication network after reduction into strong components. Strong components are shown in brackets, and gatekeepers are shown by underlining with 'G' superscript.

of the gatekeepers are members. The gatekeepers, therefore, maintain close communication among themselves, thus increasing substantially their effectiveness in coupling the organization to the outside world.

In fact, if one were to sit down and attempt to design an optimal system for bringing in new technical information and disseminating it within the organization, it would be difficult to produce a better one than that which exists.

New information is brought into the organization through the gatekeeper. It can then be communicated quite readily to other gatekeepers through the gatekeeper network and disseminated outward from one or more points to other members of the organization (Figure 26.8). Perhaps the most interesting aspect of this functioning of the organizational communication network is that it has de-

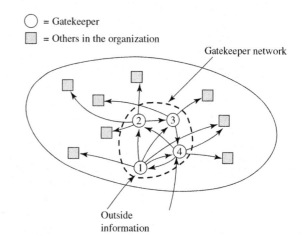

Figure 26.8. The functioning of the gatekeeper network. New information is brought into the organization by 1. It can be transmitted to 2, 3, and 4 via the gatekeeper network. It reaches its eventual users (squares) through their contacts with gatekeepers.

veloped spontaneously, with no managerial intervention. In fact, there was scarcely a suspicion on the part of management that the network operated in this way.

THE INFLUENCE OF NONORGANIZATIONAL FACTORS ON THE STRUCTURE OF COMMUNICATION NETWORKS

An organization's formal structure (that which generally appears on an organizational chart) is, as one would expect, a very important determinant of communication patterns. It is not the sole determinant, however. In addition to formal organizational structure, there are available to management at least two other factors that can be used to promote (or discourage) communication. The first of these operates through the extension of informal friendship-type relations within the organization. Allen & Cohen (1969) have shown how informal relations influence the structure of communication networks, and Allen *et al.* (1968) explore in detail how this influence comes about. Simply stated, people are more willing to ask questions of others whom they know, than of strangers. The key lies in the expected damage sustained by the ego if one's question is met with a critical response. To be told that you have asked a dumb or foolish question is the ultimate in rebuffs. Few people are willing to entertain such a risk. Now, out of all the people in the world, there are hopefully only a small percentage who would meet even a truly stupid question with such a retort. Even given that this percentage is very small, however, many people will follow the strategy of minimum regret and assume that everyone belongs to this set unless proven otherwise. This results in a situation in which, of all people who are known, only a small percentage are unapproachable, but all unknowns are unapproachable. To increase the proportion of people in the organization, who can be approached for information, management would be well advised to increase the number of acquaintanceships among its technical personnel. This it can do very easily.

People will not become acquainted until they first meet. There are, however, a number of ways through which technical people can come to meet one another. Interdepartmental projects are one such device. People who come to know one another through service on projects or other inter-functional teams retain their effectiveness as channels between departments for some time even after the project or team had been disbanded. Interdepartmental teams can, and do, provide an indirect benefit, through the persistence of the relations that they establish, over and above their direct contributions to coordination. The same thing can be said for transfers within the organization. For a period of time following a transfer, the transferred individual will provide a communication path back to his old organization. His influence extends far beyond this direct link, though. Probably the most important contribution of the transferred person lies in his ability to make referrals. The number of communication paths that potentially become available when a man is transferred is the product of the number of acquaintanceships which he developed in the two parts of the organization. For some people this can be a very large number. So with only a very few transfers, a large number of communication paths can be created and coordination thereby improved.

Of course, the effect diminishes with time, since both people and activities will change in the old group, and the transferred person will gradually lose touch. Kanno (1968) has shown that following a transfer between divisions of a large chemical firm, the transferred persons provided an effective communication link back to their old divisions for $1\frac{1}{2}$ years. The duration over which communications remain effective following a transfer is determined by many factors; principal among these are the rate of change of activities and turnover of personnel in the old organization. If projects are of short duration, with many new ones constantly being initiated and the turnover of personnel is high, one would expect that the effect of a transfer in promoting communication would be short-lived. Where the activity is more stable and turnover low, the transfer can be effective over a longer period of time. With estimates of these parameters and of the number of people (and their work) with whom the average transfer is acquainted, a systematic program of intra-organizational transfer can be developed. Such a program would contribute directly to communication, coordination and empathy among the sub-elements of the organization.

THE EFFECTS OF GEOGRAPHICAL LOCATION

In addition to formal organizational and information relations there is a third very important factor that can be used to influence the structure of organizational communication networks, i.e., the physical configuration of the facilities in which that organization is placed.

The data on the effect of spatial separation to be presented now were obtained in three very different organizations. The first organization is a 48-man department in a medium-sized aero-space firm. The 48 people were all engineers and scientists, primarily in electrical and mechanical engineering and applied physics. The second organization that was studied is a 52-man section of a medical school laboratory. The third organization comprised 57 social psychologists, economists and applied mathematicians in a management school.

To determine the influence of physical separation on the probability of two people communicating, the distance between every possible pair of people was measured. Moving outward in 5-yard intervals from each person, a measurement was made of the proportion of people within each interval with whom the focal person communicated. The measurement of distance was the actual distance that the focal person would have to walk in order to reach another person's desk. All measurements were taken on a single floor.

The proportion of people with whom an individual communicates, or the 'probability of communication' as it is labelled in the figure, decays with the square of distance outward from the focal person (Figure 26.9). The fact that the probability of communication decays with the distance sepa-

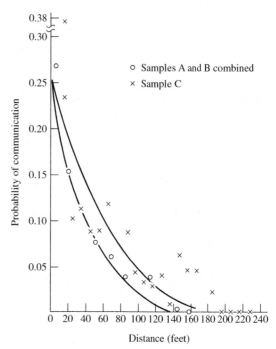

Figure 26.9. Probability of communication as a function of the distance separating pairs of people.

rating people is not too surprising. Nor is the fact that it follows an inverse square law. What is surprising is the extreme sensitivity of probability to distance. The function, naturally, must become asymptotic beyond the minimum point of the parabola. The striking thing is that it reaches this asymptote within 25 yards. This was true in all three organizations. In fact, for the first two organizations, the curves fall so close together that the data are combined in Figure 26.9. The result, therefore, appears to be general and independent of the nature of the technical work being performed.

As though, by itself, physical separation were not serious enough, there appear to be circumstances which can exacerbate its effect. The amount of difficulty, by way of corners to be turned, indirect paths to be followed, etc., encountered in traversing a path intensifies the effect of separation on communication probability. One index of this difficulty, something which might be called a 'nuisance factor' is the difference between the straight line and actual travel distances (Figure 26.10) separating two people. When communication probability is plotted as a function of the magnitude of the 'nuisance factor' (Figure 26.10) the effect is quite startling. This effect holds true whether the nuisance factor is computed on an absolute basis or as a proportion of straight line separation distance.

ORGANIZATIONAL STRUCTURE

To encourage communication between project teams and the supporting technical staff, separation distances must be kept to a minimum. To locate a project in a separate facility is essentially to cut it off from support by the rest of the laboratory staff.

There is a trade-off that must be made in locating project members. Effective coordination of all elements of project activity may require that all or most of the team members be located together in a specially assigned place. On the other hand, to maintain the specialists assigned to the project abreast of developments in their technical fields demands that they be kept in contact with the specialist colleagues. This, in turn, favours locating them with their specialist groups. Marquis (1969) has argued for the latter alternative on very large projects. All of the projects in Marquis's sample were of fairly long duration; several years. This may well hold the key to the trade-off. For long-term projects, technical personnel should remain in the same location with their specialist colleagues. Assignment to a project of long duration can force

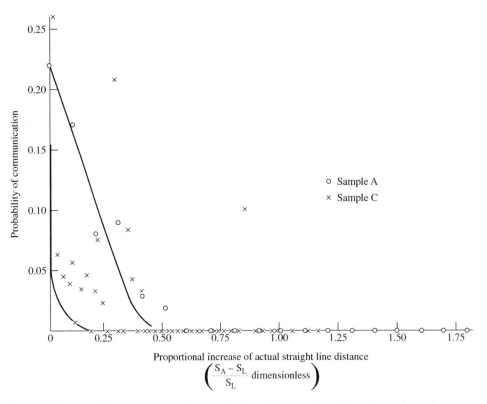

Figure 26.10. Probability of communication as a function of the magnitude of the nuisance factor (0 = Sample A; X = Sample C).

a person to lose touch with his field unless steps are taken to enable his free interaction with colleagues in the field. The result is technical obsolescence and difficulty for both the person and the organization in dealing with future assignments. In the case of a brief project, the length of separation will be too short to have these results, and the balance swings in favour of locating all project members together.

Long projects demand functional organization, while short-duration projects may be organized on a project basis with all team members located together.

Functional organization has the undesirable consequence of making intra-project coordination difficult. A possible solution to this problem lies in overlaying a coordinating team across the functional departments in what has come to be known as a matrix organization. This is not always as easy to accomplish as it first sounds, but when it functions properly it can achieve the desired goals of the functional organization without the loss in project coordination.

Project and matrix organization have the undesirable consequence of making communication between functional departments difficult. Transfers, where possible, and short-duration interdepartmental projects, assist in countering this problem. In addition, the overall configuration of the laboratory should be structured in such a way that inter-functional communication is eased. Where it is desirable to have communication between groups, they should be located near each other. Where this is impossible they can be made to share certain facilities that will force interaction. The nature of the facility is secondary. It may be as humble as a coffee pot or men's room or as grandiose as a computer or an expensive instrument. What matters is that it brings people into contact who would otherwise not meet. It is quite easy, in any organization, to think of a large number of such facilities for promoting interaction. Where possible they should be located where they will promote the desired patterns of group interaction. In those cases in which it is not feasible to manipulate the position of the interaction facility, then desired patterns of interaction should be given serious consideration in allocating the use of the facility among groups and in positioning groups around it. In the latter situation it must be borne in mind that the extent to which a facility will be used is also an inverse function of distance. Frohman (1968), for example, found the principal determinant of use of a technical library in an industrial firm to be the distance separating users from it. Interaction facilities must be positioned in such a way that they promote interaction among groups that would not otherwise interact, while at the same time they are not so far removed from any of the groups that they lose their effectiveness.

SUMMARY AND CONCLUSIONS

The importance to research and development projects of technical staff support cannot be overstressed. Seldom, if ever, is management able satisfactorily to predict and obtain all of the talents that will be needed in a project and incorporate them in the project team. The project must, therefore, obtain much of its required information from sources beyond its own membership.

Research shows very clearly that the best source for this support lies in the technical staff of the laboratory itself. Attempts to bring information to the project directly from outside of the organization usually have been ineffective. The process by which an organization imports and disseminates outside information is more complex than people normally assume. The best way to maintain the project team abreast of outside developments lies in understanding and making proper use of existing information systems. This includes the use of technological gatekeepers for project support. Outside information can then be delivered to the project quite effectively, albeit by an indirect route. Research evidence indicates quite strongly that the indirect approach is far more effective than any direct approach to coupling project members to outside sources, whether personal or written.

There are available a wide variety of techniques for improving communication and coordination between projects and their supporting staff. A number of formal organizational mechanisms have been described in detail. In addition to these, use may also be made of the informal relationships that will develop when people come into contact with one another. A very effective means for increasing the level of acquaintanceships in an organization is the inter-group transfer. Physical location is also a very strong determinant of interaction patterns. People are more likely to communicate with those who are located nearest to them. Individuals and groups can therefore be positioned in ways that will either promote or inhibit communication. Architectural design thus becomes an important determinant of the structure that an organization's communication network will assume. Shared facilities or equipment can also be used to promote interaction between groups.

All of these factors must be taken into account and properly arranged in order to effectively couple the research and development project to its supporting information system.

REFERENCES

Allen, T. J. (1964) 'The Use of Information Channels in R&D Proposal Preparation,' Cambridge, Mass.: M.I.T. Sloan School of Management, Working Paper No. 97–64.

Allen, T. J. (1966) 'Managing the Flow of Scientific and Technological Information,' Ph.D. Dissertation, Cambridge, Mass.: M.I.T. Sloan School of Management.

Allen, T. J., Gerstenfeld, A. & Gerstberger, P. G. (1968) 'The Problem of Internal Consulting in the R&D Laboratory,' Cambridge, Mass.: M.I.T. Sloan School of Management, Working Paper No. 319–68.

Allen, T. J. & Cohen, S. I. (1969) 'Information flow in two R&D laboratories,' Administrative Science Quarterly, Vol. 14.

Baker, N. R., Siegmann, J. & Rubenstein, A. H. (1967) 'The effects of perceived needs and means on the generation of ideas for industrial research and development project,' I.E.E.E. Transactions on Engineering Management, Vol. 14.

Flament, C. (1963) 'Applications of Graph Theory to Group Structure,' New York: Prentice Hall.

Frohman, A. (1968) 'Communication problems in an industrial laboratory', Unpublished term paper, Cambridge, Mass.: M.I.T. Sloan School of Management.

Gerstberger, P. G. & Allen, T. J. (1968) 'Criteria used in the selection of information channels by R&D engineers,' Journal of Applied Psychology, Vol. 52.

Harary, F., Norman, R. S & Cartwright, D (1965) 'Structural Models,' New York: Wiley.

Kanno, M. (1968) 'Effect on Communication Between Labs and Plants of the Transfer of R&D Personnel,' S.M. Thesis, Cambridge, Mass.: M.I.T. Sloan School of Management.

Marquis, D. G. (1969) 'Organizational factors in project performance,' Program Management (ed. J. Galbraith), Cambridge, Mass.: M.I.T. Press.

Pelz, D. C. & Andres, F.M. (1966) 'Scientists in Organizations,' New York: Wiley.

Shilling, C. W. & Bernard, C. W. (1964) 'Informal Communication Among Bio-Scientists,' George Washington University Biological Sciences Communication Project, Report 16A–64.

27

A Study of the Influence of Technical Gatekeeping on Project Performance and Career Outcomes in an R&D Facility

RALPH KATZ AND MICHAEL L. TUSHMAN

This study investigated the role of gatekeepers in the transfer of information within a single R&D location by comparing directly the performance of project groups with and without gatekeepers. The results show that gatekeepers performed a linking role only for projects performing tasks that were 'locally-oriented' while 'universally-oriented' tasks were most effectively linked to external areas by direct project member communication. Gatekeepers also appear to facilitate the external communication of other technologists, in addition to influencing positively the career outcomes of young engineers.

R&D project teams must process information from outside sources in order to keep informed about relevant external developments and new technological innovations. Furthermore, empirical studies over the past 25 years have demonstrated that oral communications, rather than written technical reports or publications, are the primary means by which engineering professionals collect and disseminate important new ideas and information into their project groups (Allen, 1984). While such personal contacts may be essential, there are alternative communication structures by which R&D groups can effectively draw upon information outside their organizations (Katz & Tushman, 1979).

In particular, the research reported here focuses explicitly on the role played by gatekeepers in the effective transfer and utilization of external technology and information. (Gatekeepers are defined as those key individual technologists who are strongly connected to both internal colleagues and external sources of information (Allen & Cohen, 1969).) Since most gatekeepers are also project supervisors, this study also examines the broader leadership role of gatekeeping supervisors in influencing the subsequent rates of organizational turnover and promotion among technical subordinates.

COMMUNICATION AND PERFORMANCE

Generally speaking, previous research has shown that project performance is strongly associated with high levels of technical communication by all project members to information sources within the organization (i.e., high levels of internal communication). The positive findings of Allen (1984), Pelz & Andrews (1976), and others strongly argue that direct contacts between project members and other internal colleagues can enhance project effectiveness.

While direct communication by all project members may be effective for internal communications, the particular method for effectively keeping up-to-date with technical advances outside the organization are probably very different. Numerous

Earlier versions of this paper were published in Katz, R. & Tushman, M. (1981) "An investigation into the managerial roles and career paths of gatekeepers and project supervisors in a major R&D facility," *R&D Management*, Vol. 11, 103–10; and Katz, R. & Tushman, M. (1983) "A longitudinal study of the effects of boundary spanning supervision on turnover and promotion in research and development," *Academy of Management Journal*, Vol. 26, 437–56.

Ralph Katz is Professor of Management at Northeastern University's College of Business and Research Associate at MIT's Sloan School of Management. *Michael L. Tushman* is a Chaired Professor of Business Administration at the Harvard Business School.

studies, for example, have shown that project performance is not positively associated with direct project member communication to external information areas. In fact, most studies have found them to be inversely related (e.g, Allen, 1984; Katz and Tushman, 1979). It seems that most engineers are simply unable to communicate effectively with extraorganizational information sources. Widespread direct contact by all engineering project members, then, is *not* an effective method for transferring technical information into a project from external sources.

One explanation for these significant differences stems from the idea that technological activities are strongly local in nature in that their problems, strategies, and solutions are defined and operationalized in terms of particular strengths and interests of the organizational subculture in which they are being addressed (Katz & Kahn, 1978; Allen, 1984). Such localized definitions and shared language schemes gradually unfold from the constant interactions among organizational members, the tasks' overall objectives and requirements, and the common social and task related experiences of organizational members. These idiosyncratic developments are a basic determinant of attitudes and behaviours in that they strongly influence the ways in which project members think about and define their various problems and solution strategies.

Such localized perspectives eventually become a double-edged sword. As long as individuals share the same common language and awareness, communication is rather easy and efficient. Conversely, when individuals do not share a common coding scheme and technical language, their work-related communications are less efficient, often resulting in severe misperceptions and misinterpretations. Thus, the evolution of more localized languages and technological approaches enables project members to deal effectively with their more local information processing activities within the organization; yet at the same time, it hinders the acquisition and interpretation of information from areas outside the organization. This lack of commonality across organizational boundaries serves as a strong communication impedance causing considerable difficulty in the communications of most engineers with external consultants and professionals.

Given this burden in communicating across differentiated organizational boundaries, how can project groups be effectively linked to external information areas? One way is through the role of project gatekeeper, that is, certain project members who are strongly connected to outside information domains but who are also capable of translating technical developments and ideas across contrasting coding schemes (Allen & Cohen, 1969). Through these key members, external information can be transferred effectively into project groups by means of a two-step communication process. First, gatekeepers gather and understand outside information, and subsequently they channel it in more meaningful and relevant terms to their locally constrained colleagues. Gatekeepers, as a result, perform an extremely valuable function, for they may be the principal means by which external ideas and information can be effectively transferred into R&D project groups.

While substantial literature applauds this gatekeeper concept, there is virtually no direct evidence that gatekeepers enhance project performance. Support has to be inferred indirectly either from the empirical findings of Katz & Tushman (1979) and Allen, Tushman & Lee (1979) or from the case studies in project SAPPHO (Achilladeles, Jervis & Robertson, 1971). Our initial research question, then, concerns the association between project gatekeepers and technical performance. Is this relationship positive across all forms of R&D activity or are some project areas more effectively linked to external technology through direct contact by all project members rather than through a gatekeeper? Moreover, if gatekeepers are necessary for effective technology transfer, must they then be the primary source for collecting outside information, or can they also serve to facilitate the external communication of their more locally constrained colleagues?

GATEKEEPERS, PERFORMANCE, AND THE NATURE OF THE TASK

The need for a two-step process of information flow depends on a strong communication impedance between the project group and its external information areas. To the extent that different technical languages and coding schemes exist between project members and their external technical environments, communication across organizational boundaries will be difficult and inefficient. Most technological activities (unlike the sciences) are strongly local in nature. The coupling of bureaucratic interests and demands with localized technical tasks and coding schemes produces a communication boundary that differentiates these project groups from their outside areas. Product development groups in different organizations, for example, may face similar problems yet may define their solution approaches and parameters very differently. As a result, it becomes increasingly difficult for most technologists to integrate external ideas, suggestions and solutions with internal technology that has become locally defined and constrained. It is hypothesized, therefore, that locally oriented projects (i.e., development and technical service projects) will require gatekeepers to provide the necessary linkages to external information areas—without gatekeepers, direct external contacts by members of local projects will be ineffective.

In contrast, if external information sources do not have different language and coding schemes from members of the project group, then a significant communication impedance will not exist. Work that is more universally defined (scientific or research work, for example) is probably less influenced and constrained by local organizational factors, resulting in less difficulty *vis-à-vis* external communications. Under these conditions, project members are more likely to share similar norms, values, and language schemes with outside professional colleagues, thereby permitting effective communication across organizational and even national boundaries. They are simply more capable of understanding the nature of the problems and corresponding solution approaches employed by their relevant external colleagues. Hagstrom (1965), for instance, found a strong positive correlation between the productivity of scientists and their levels of contact with colleagues from other universities. For universally defined tasks, therefore, it is hypothesized that gatekeepers are not required to link projects with their relevant external

information areas; instead, direct outside interaction by all project members is more advantageous. The nature of a project's work, therefore, should be a critical factor affecting the development of localized languages and orientations and consequently will moderate significantly the relationship between project performance and the usefulness of gatekeepers.

ROLE OF GATEKEEPERS

If gatekeepers enhance the performance of project groups working on locally defined tasks, then what specific information processing activities of gatekeepers contribute to higher project performance? There are at least two alternatives. The more traditional explanation is that gatekeepers function as the primary link to external sources of information and technology—information flows through these key individuals to the more local members of the project team (Allen & Cohen, 1969). Relevant external information is transferred effectively into a project group because of the capable boundary spanning activities of the project's gatekeeper.

Another possibility is that gatekeepers also assume an active training, development, and socialization role within their work groups. From this perspective, gatekeepers not only gather, translate, and encode external information, but they also facilitate the external contacts of their project colleagues. By helping to direct, coach, and interpret the external communications of their fellow project members, gatekeepers act to reduce the communication boundary separating their projects from outside information areas.

If gatekeeping permits other project members to communicate effectively with external areas, then for localized projects with gatekeepers, there should be a positive association between a project's external communication and its performance. On the other hand, if gatekeepers do not play this more active role, then an inverse relation is more likely to exist between the external communications of locally oriented group members and project performance. Because they work and interact so closely with other project members about technically related problems, it is argued that gatekeepers fulfill this larger role of both gathering outside information and facilitating the external communications of their project colleagues. Since most technical gatekeepers are also first-level project supervisors, it is hypothesized that there will be a significant positive association between project performance and external interaction for projects *with* gatekeeping supervisors. On the other hand, for those projects *without* gatekeeping supervisors, that is, project groups led by non-gatekeeping supervisors, there will be an inverse relationship between project performance and external communication.

GATEKEEPERS AND CAREER OUTCOMES

Even though boundary spanning gatekeeping has been recognized as one of the more important elements of effective leadership in Research, Development, and Engineering (i.e., RD&E) settings, very little is known about how these communication and network activities affect other important managerial functions and organizational outcomes. If there is a significant distinction between the information processing contributions and capabilities of project supervisors who function as gatekeepers and those who do not, then to what extent will they also have different career paths within the technical organization? Are gatekeeping supervisors more likely to be promoted to particular laboratory positions than non-gatekeeping supervisors, for example?

In addition to examining the career outcomes of gatekeeping supervisors, it is also important to investigate how they affect the job experiences and work activities of their technical subordinates. If career decisions are strongly influenced by how well individuals are connected to their company's formal and informal networks, then engineers working for gatekeeping supervisors should have substantial career advantages over technologists assigned to non-gatekeeping supervisors. Through their more elaborate contacts, valuable technical insights, and close working relationships, gate-

keeping supervisors may directly impact the personal growth and development of their technical subordinates (Tushman and Katz, 1980). And to the extent that gatekeeping supervisors help project members participate more effectively within their work settings and gain clearer working relationships with and exposure to other corporate areas and higher-level managers, project members are not only less likely to leave the organization but also more likely to be recognized and subsequently promoted.

From an exploratory point of view, therefore, this research focuses on the formal and informal aspects of leadership by investigating (over a 5-year period) the influence of gatekeeping supervisors on the organizational turnover and promotional outcomes of project subordinates. Although career outcome effects will be generally explored, younger employees are more likely to benefit from the socialization and developmental role played by gatekeeping supervisors (Katz, 1980), especially since most turnover occurs within the first years of organizational employment (Schein, 1978) and because engineers usually expect to be promoted to a managerial position between the ages of 30 and 40 (Dalton et al, 1982). Finally, as previously discussed, boundary spanning gatekeeping is more critical in development work than in either research or technical service. As a result, the influence of gatekeeping supervisors on the career outcomes of engineering subordinates is most likely to be especially important in development project areas.

METHODOLOGY

This study was conducted among all project members working in a large corporate RD&E facility. At the start of the study, the facility's professionals ($N = 345$) were divided into seven separate functional departments, which in turn were subdivided into 61 projects organized around specific, long term types of discipline and product focused problems. Each professional was a member of only one project group.

Communication

To measure actual communication activity, project members reported (on specially provided lists) all those individuals with whom they had work-related oral communication on a randomly chosen day each week for 15 weeks. Professionals were asked to report all contacts both within and outside the laboratory, including how many times they may have talked to a given person that day. Social and written communications were not reported. An overall response rate of 93 percent was achieved over the 15 weeks and almost 70 percent of all pairwise communication episodes within the RD&E facility were reciprocally reported by both parties. Given these high rates of response and mutual agreement, these sociometric methods provide a rather clear and accurate picture of each project member's communication network.

For each project member, internal communications was measured by summing the number of work-related contacts reported over the 15 weeks between that member and all other professionals within the organization, including all project and departmental colleagues and supervisors. External or outside communication was measured by summing the member's reported communications to other professional individuals outside the organization, including R&D consultants, professors, vendors, customers, and the like. As discussed by Katz & Tushman (1979), these individual scores were also aggregated to obtain project measures of internal and external communication.

Conceptually, project gatekeepers are defined as those members who are high internal communicators and who also maintain a high degree of outside communication. In line with previous studies (see Allen, 1984), this study operationalized gatekeepers as those project members who were in the top fifth of both the internal and external communication distributions. Gatekeepers were identified in 20 project groups while 40 projects had no gatekeepers within their memberships.

Project Type

R&D tasks differ along several dimensions, including time span of feedback, specific vs. general problem-solving orientation, and the generation of new knowledge vs. utilization of existing knowledge and experience (Rosenbloom & Wolek, 1970). Based on these dimensions, distinct project categories were defined ranging from research to development to technical service. Such a categorization also forms a universal (research) to local (technical service) project continuum. As discussed by Katz & Tushman (1979), respondents were asked to use these specific project definitions and indicate how well each category represented the objectives of their task activities. A second question asked respondents to indicate what percentage of their project work fell into each of the project categories. A weighted average of these two answers was calculated for each respondent (Spearman-Brown reliability = 0.91).

To categorize projects, however, the homogeneity of members' perceptions of their task characteristics had to be examined to check for the appropriateness of pooling across individual project members (see Tushman, 1977 for details). As pooling was appropriate, individual responses were averaged to get final project scores, yielding 14 Research, 23 Development, and 23 Technical Service projects. Research projects carried out more universally oriented scientific work (discovering new knowledge in glass physics, for instance) while development and technical service projects were more locally oriented in that they worked on organizationally defined problems and products.

Project Performance

To get comparable measures of project performance across significantly different technologies, all departmental and laboratory managers ($N = 9$) were interviewed individually and asked to evaluate the overall technical performance of all projects with which they were sufficiently familiar. When they could not make an informed judgment, they did not rate that project. Each project was independently evaluated by an average of five managers using a 7-point Likert type scale ranging from (1) very low to (7) very high. Individual ratings were averaged to yield overall project performance scores (Spearman-Brown reliability = 0.81).

CAREER OUTCOMES

Almost five years after the collection of the preceding data, we returned to the RD&E facility to collect data on the organizational turnover and dual ladder promotions (i.e., managerial and technical) of all those project members in our original sample. Despite the facility's strong growth, 31 percent of the project members and 19 percent of the project supervisors had left the company over the course of this five year interval, and an additional 8 percent had retired. In this organization, the formal managerial and technical ladder positions and titles start within the departments above the project supervisory level.

RESULTS

Gatekeeper Presence and Project Performance

The performance means reported in the first row of Table 27.1 clearly indicate that, in general, the performances of projects with gatekeepers were not significantly different from the performances of projects without gatekeepers. As previously discussed, however, locally oriented projects (i.e., development and technical service) should display a positive association between gatekeeper presence and project performance. Universal-type or research projects, on the other hand, should show an inverse relation between gatekeeper presence and project performance.

The breakdown of performance means by project type strongly supports these differences in the appropriateness of the gatekeeping function. As shown by Table 27.1, research projects *without* gatekeepers were significantly higher performing than research projects *with* gatekeepers. It may be that research projects are more effectively linked to external information areas through direct member contacts.

In sharp contrast, development projects *with* gatekeepers were significantly more effective than development projects *without* gatekeepers. Unlike research groups, then, development projects are linked to outside information areas more effectively through the use of gatekeepers. No significant differences in project performance, however, were discovered between technical service groups with and without gatekeepers.

Role of Gatekeepers

It was suggested that on locally oriented tasks, gatekeepers may do much more than simply channel outside information into their project groups. They may also act to reduce communication impedance by facilitating the external communica-

TABLE 27.1
Project Performance as a Function of Gatekeeper Presence and Project Type

Project type	Mean performance for projects		Mean difference in performance
	With gatekeepers	Without gatekeepers	
All projects	4.70	4.53	0.17
	($N = 20$)	($N = 40$)	
Project type			
Research	4.22	4.92	−0.70**
	($N = 5$)	($N = 9$)	
Development	4.91	4.15	0.76***
	($N = 8$)	($N = 15$)	
Technical service	4.80	4.67	0.13
	($N = 7$)	($N = 16$)	

**p < 0.05 and ***p < 0.01 indicate significant mean differences in project performance.

TABLE 27.2

Correlations Between Project Performance and External Communication by Project Type and Gatekeeper Presence

	Correlation with performance for projects	
Measures of external communication for	With gatekeeping leaders	Without gatekeeping leaders
Research projects		
(a) All project members	0.53	0.46*
(b) All members excluding project leaders†	0.37	0.70**
(c) Project leaders	0.55	0.29
	(N = 5)	(N = 9)
Development projects		
(a) All project members	0.31	−0.45**
(b) All members excluding project leaders	0.55*	−0.21
(c) Project leaders	0.37	−0.51**
	(N = 8)	(N = 15)
Technical service projects		
(a) All project members	0.31	−0.19
(b) All members excluding project leaders	0.64*	−0.03
(c) Project leaders	0.77*	−0.34*
	(N = 7)	(N = 16)

†In the first column of correlations, project leader refers to the project's gatekeeper, 75% of whom were also project supervisors. In the second column, project leader simply refers to the project's supervisor.

*$p < 0.10$; **$p < 0.05$; pairwise correlations that are significantly different at the $p < 0.10$ level or less have been underlined.

tions of their fellow project colleagues. In contrast, locally oriented projects without gatekeepers will have no effective link to external areas. Results reported in Table 27.2 support this notion for projects within the development and technical service categories. For local projects without gatekeepers, there was a consistent inverse association between members' external communication and project performance. For projects with gatekeepers, however, a significantly different pattern emerged—external communication was positively associated with project performance. Furthermore, these correlation differences were strong even after the direct communication effects of gatekeepers were removed. For both development and technical service groups, gatekeepers *and* their project colleagues were able to communicate effectively with outside professionals.

The significant correlational differences between projects with and without gatekeepers strongly support the argument that gatekeepers influence the ability of local project members to com-municate effectively with external sources of technical information. Members of research projects, on the other hand, do not seem to face a communication impedance when communicating externally, for Table 27.2 shows that the level of outside interaction by all research project members was positively associated with performance independent of a gatekeeper's presence within the group. Gatekeepers as a result, may not play an important information processing role in the more universally oriented research projects, but they appear to play a vital role in the more locally defined development and technical service projects.

Gatekeepers and Project Supervisors

Can project supervisors substitute for gatekeepers in linking their projects to external information areas? The correlations reported in Table 27.2 do not support this position. For development and technical service projects, the more non-gatekeeping project supervisors communicated with information sources outside the organization, the lower their project's performance (correlations of −0.51 and −0.34, respectively). On the other hand, the associations between outside contact and project performance were very positive for supervisors who were also gatekeepers (correlations of 0.37 and 0.77, respectively). Such significant correlational differences strongly imply that these two kinds of supervisors play a very different role and contribute very differently within their project groups.

In light of these differences, were gatekeeping and non-gatekeeping supervisors likely to receive the same kinds of promotions? Our results suggest they did not. The follow-up study of the facility some five years later revealed that almost all of the gatekeeping supervisors had been promoted up the management ladder. Of the 12 gatekeeping supervisors remaining with the company, 11, or 92 percent, were in higher-level managerial positions. The one remaining gatekeeping supervisor was on the technical side of the organization's dual-ladder reward system.

Although non-gatekeeping supervisors were almost as likely to be promoted, they were not as likely to receive managerial promotions. Only 46 percent of the 35 remaining non-gatekeeping supervisors were in higher-level managerial positions some five years later. Surprisingly enough, almost as many were promoted up the technical ladder, about 34 percent. In fact, of the thirteen project supervisors who were promoted to the technical ladder, only one had been functioning as a technical gatekeeper at the be-

ginning of our research study. (Actually, this one gatekeeper had initially been promoted up the managerial ladder but was switched to a technical ladder position when it became clear that he was not functioning effectively as a manager.)

While gatekeeping supervisors were essentially promoted up the managerial ladder, could one have differentiated between non-gatekeeping supervisors promoted managerially and those promoted technically? The means reported in Table 27.3 indicate that there were significant communication differences between these two promotional categories. Project supervisors promoted up the technical ladder had only half as many internal interactions as project supervisors selected for managerial positions. Interestingly enough, there were almost identical levels of internal communications for gatekeeping and nongatekeeping supervisors promoted to managerial positions. External communications did not differentiate between the promotional ladders of non-gatekeeping supervisors. Thus, the level of interpersonal activity and skills that one has demonstrated within the organization may have been a strong factor in shaping one's promotional ladder within this dual ladder system.

TURNOVER INFLUENCE

Our first analysis here concerned the influence of boundary spanning gatekeeping supervisors on the

TABLE 27.3
Comparisons of Mean Internal and External Communications of Project Supervisors Promoted Over the Next 5 Years

Promotional positions above the project level	Mean internal communication (per person per week)	Mean external communication (per person per week)
(a) Gatekeeping supervisors promoted to managerial positions ($N = 11$)	74.3[a]	4.8[a]
(b) Non-gatekeeping supervisors promoted to managerial positions ($N = 16$)	70.6[a]	1.5[b]
(c) Supervisors promoted to technical positions[†] ($N = 12$)	39.7[b]	1.4[b]

[†]Of these project supervisors, only one had functioned as a project gatekeeper.
Note: In each column, means with superscript 'a' are significantly greater than means with superscript 'b' at the $p < 0.01$ level.

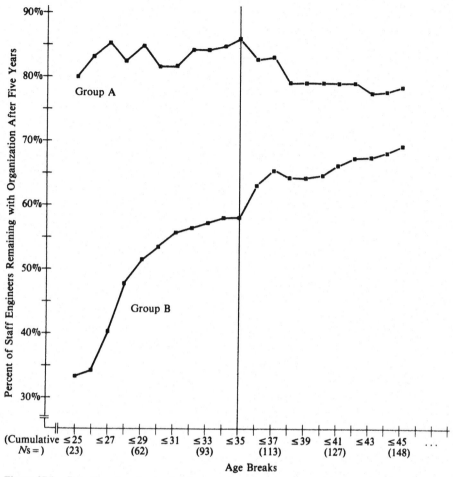

Figure 27.1. Retention of engineers after five years by prior type of reporting relationships at successive age breaks. Group A are staff engineers who reported directly to a gatekeeping supervisor, while Group B are staff engineers who reported directly to a nongatekeeping supervisor. The vertical line indicates that pairwise percentages between Groups A and B remained significantly different ($p < 0.05$) through an age break of thirty-five years. N's represent the total number of engineers within both groups at six representative age breaks.

turnover rates of project engineers. For the sample as a whole, project members who had reported to gatekeeping supervisors had a significantly lower rate of organizational turnover (17 percent turnover over the five-year interval) than did engineers assigned to non-gatekeeping supervisors, (33 percent turnover for the same five years). It appears that working for supervisors who also function as gatekeepers has a strong positive effect on employee retention rates.

Because 70 percent of the turnover occurred for project members less than 36 years of age, additional comparisons were carried out for younger versus older age groupings. In fact, to pinpoint the influence of gatekeepers on subordinate turnover, Figure 27.1 plots as a function of age the cumulative retention rates of projection members reporting to gatekeeping supervisors (Group A), versus those members not reporting to gatekeeping supervisors (Group B).

Of those project subordinates who were age 25 or less, only 33 percent remained in the organization if they had *not* reported to a gatekeeping supervisor. The comparable percentage for project members assigned to a gatekeeping supervisor was almost 80 percent. Similarly, of those project subordinates who were 35 years old or less, only 57 percent remained with the organization if they had *not* been working with a gatekeeping supervisor. The comparable percentage in Group A was 84 percent. Although this difference is statistically significant, the figure clearly shows that most of the difference in retention rates between Groups A and B occurs among members less than 30 years of age. Almost 85 percent of these young engineers were still with the organization after five years if they had been assigned to a gatekeeping type supervisor, while only 51 percent of their counterparts who had not reported to gatekeeping supervisors were still employed.

Although gatekeeping supervisors may help reduce the turnover rates among young professionals, it was previously suggested that this influence might be especially strong for members of development and technical service projects. This possibility was examined. However, the turnover differences between project members with and without gatekeeping supervisors was consistently positive for all three types of project categories. In each case, over half the young project members not reporting to a gatekeeping supervisor had left the company within the five-year interval (despite substantial growth within the RD&E facility). In sharp contrast, none of the corresponding turnover rates for project members reporting to gatekeeping supervisors exceeded 25 percent.

Clearly gatekeeping supervisors had considerable influence over the turnover rates of young professionals within this facility. What is it about gatekeeping supervisors that brings about these lower levels of turnover? As previously discussed, turnover may be a function of how well young professionals get integrated into their organization's formal and informal networks. Because gatekeepers are key individuals in these networks, reporting to a gatekeeping supervisor may well facilitate the young professional's linkages to important information sources both within and outside the organization. An investigation was therefore made as to whether young project members reporting to gatekeeping supervisors had different interaction patterns than did project members reporting to non-gatekeeping supervisors. The analyses did not reveal any significant differences in their comparative levels of collegial interaction. Young engineers who left had as much communication activity with internal colleagues as those young engineers who remained. What did differentiate young stayers from leavers was their much higher levels of contact not only with their gatekeeping supervisor but also, and perhaps more important, with their departmental manager. Thus, it may not be the assignment of young project members to gatekeeping supervisors per se that enhances long-term retention. What really makes the difference is the stronger degree of hierarchical interaction and integration—the kind of communication and connectedness that is so important and helpful for reducing stress and uncertainty during the socialization phase of any young professional's career (Katz, 2004).

Promotion Influence

During the five-year interval, 23 of the remaining project members had been promoted to managerial positions. As with turnover, project members who had reported to gatekeeping supervisors had a slightly higher rate of managerial promotion than did project members reporting to non-gatekeeping supervisors, 14.6 versus 11.2 percent, respectively. However, because almost 70 percent of the project members promoted to managerial positions were between the ages of 27 and 32 at the start of our study, the data within this more limited age range were reanalyzed for the whole subsample, as well as within each type of project category.

As shown in Table 27.4, the promotion rate for all project members reporting to gatekeeping supervisors within this restricted age subsample was 41.2 percent, whereas only 17.4 percent of the comparable engineers reporting to a non-gatekeeping supervisor were similarly promoted—a signif-

TABLE 27.4

Proportion of Engineers Promoted to High Level Managerial Positions Over the Next Five Years by Prior Reporting Relationship and Project Task Areas[a]

Prior project areas	Prior reporting relationship percent		Proportional differences (%)
	Assigned to a gatekeeping supervisor (%)	Not assigned to a gatekeeping supervisor (%)	
Across all areas	41.2	17.4	23.8*
	($n = 17$)	($n = 46$)	
By project area			
Applied research	33.3	20.0	13.3
	($n = 6$)	($n = 10$)	
Product/process development	66.7	18.5	48.2**
	($n = 6$)	($n = 27$)	
Technical service	20.0	11.1	8.9
	($n = 5$)	($n = 9$)	

[a]Table includes only those engineers in the age range (27 through 32) in which almost 70 percent of the promotions took place.
*$p < 0.10$
**$p < 0.05$.

icant difference of 23.8 percent. Although proportionately more engineers who had reported to gatekeepers were promoted to management within this general subsample, Table 27.4 also reveals that most of this difference takes place in the area of product and process development. No significant advantage was uncovered in the promotion rates of project engineers reporting to gatekeepers in either research or technical service areas. On the other hand, two-thirds of the engineers reporting to gatekeepers in development projects received management promotions, in contrast to only 18.5 percent of the engineers reporting to non-gatekeeping supervisors. Even though the subsample sizes in Table 27.4 are rather small, development work is precisely the project area in which gatekeepers were hypothesized to have the strongest influence over career outcomes.

As in the turnover analyses, the communication patterns of project members within the 27–32 year old age range were examined to see if those promoted also had differential patterns of contacts and interactions within their work settings. None of the communication measures, however, was significantly related to managerial promotions for these individuals.

DISCUSSION

In engineering and scientific environments, there are at least two distinct methods by which R&D project groups can keep abreast of technical ideas and developments outside their organizations: (1) by direct contact by all project members and (2) by contact mediated by project gatekeepers. Our findings suggest that the effectiveness of these two alternatives is strongly affected by the communication impedance separating project groups from their external information areas. Universally oriented research projects, for example, face little communication impedance when processing outside ideas and information since their work is less constrained by local organizational factors. Therefore, instead of relying on gatekeepers to keep informed about outside developments and advances, members of higher performing research groups were able to rely on their own external contacts. In fact, a significant inverse relationship between project performance and gatekeeper presence was uncovered among the facility's 14 research groups.

As project activities become more specialized and locally defined, however, language and cognitive differences between project members and external professionals increase, creating substantial communication impedance and more tendentious information flows. As a result, individual interaction across organizational boundaries becomes more difficult and ineffective. To wit, higher performing development and technical service groups had significantly less outside contact by all project members. Nevertheless, important technical information must be acquired from relevant outside sources. Gatekeeping, as a result, can be a necessary and effective process for transferring external technology into localized project groups. In particular, within our sample of development projects, those with gatekeepers were considerably more effective than those without gatekeepers. Thus, what are needed to introduce outside information effectively into development projects are specialized project individuals who keep current technically, are readily conversant across different technologies, and who are contributing to their project's work in direct and meaningful ways, i.e., technical gatekeepers.

Unlike development projects, the performances of technical service projects were not positively related to the presence of gatekeepers even though their project members could not communicate effectively with outside information areas (see Table 27.2). Perhaps the specialized gatekeeping role may not be as necessary in technical service projects because the work tends to deal with more mature technologies and existing knowledge and products than in development projects. The managerial hierarchy, instead, may be able to keep members sufficiently informed about external events and information through formal reporting channels and operating procedures.

Generally speaking, the particular method by which R&D projects can effectively connect with external technical information appears to differ significantly across the research, development, and technical service spectrum of R&D activities. The particular method is strongly contingent on both the nature of the project's work and the stability of the involved technologies. Thus, it seems that the combination of localized yet dynamic technologies necessitates the active presence and participation of gatekeepers within engineering project groups.

The Gatekeeping Role and Project Supervision

In linking local project groups to extra-organizational areas, our results indicate that gatekeepers not only bring in outside information, but just as important, they facilitate the external communication of their more locally oriented colleagues. As a result, localized engineering projects with gatekeepers are in a better position to take advantage of external technology since other members are now capable of communicating effectively across organizational boundaries. This additional capacity lessens the project's complete dependence on gatekeepers for gathering and disseminating all important outside information.

In research-type tasks, on the other hand, gatekeepers are not an effective method for obtaining external information; nor does it appear that they serve in any communication facilitating capacity. In higher performing research projects, members did not rely on gatekeepers for their external information; in a sense, they functioned as their own technical gatekeepers.

One should also note that many supervisors of locally-oriented projects could not adequately perform a gatekeeping role in linking their projects to outside technology. In contrast to gatekeeping supervisors, the external interactions of supervisors who were not gatekeepers were negatively associated with project performance. While these non-gatekeeping supervisors may have developed important internal linkages, they are unable to fulfill the same external function as their gatekeeping peers. Such findings suggest distinguishing between two types of project leaders: (1) locally oriented supervisors who may be appropriate for more administrative and technical support and development projects in which the rates of information and technology change are not high and (2) gatekeeping supervisors who may be more contributive on product- and process-development activities, especially if the rates of change in the underlying technologies are high.

These different capabilities also seem to have led to different kinds of career paths. All project gatekeepers remaining in the organization over a 5-year period were promoted along the managerial ladder. Almost all non-gatekeeping supervisors were also promoted during this interval. However, only about half were positioned on the managerial ladder—the other half being promoted along the technical ladder. While there were no strong differences between the technical performances of project groups which had supervisors promoted managerially versus those which had supervisors promoted technically, there had been very strong differences between their communication activities. Those selected for managerial positions had been high internal communicators; in fact, they were as high as project gatekeepers. In sharp contrast, supervisors promoted along the technical ladder had been extremely low internal communicators. Thus, what differentiated between these two alternative career paths for non-gatekeeping supervisors was not technical competence but interpersonal connectedness and competence. Thus, supervisors promoted along the technical ladder were probably not the most networked or influential individuals. And by promoting people to the technical ladder based on this differentiation, the organization over time runs the risk of reinforcing even more their feelings of isolation and lack of influence or power.

Perhaps it is these network and work experience differences that have caused so many companies to have substantial difficulty with the effective implementation of dual ladder reward systems (for a more recent discussion of the influence of gatekeepers on the dynamics of dual ladder promotions, see Katz, Tushman, & Allen, 1995). Supervisors who had behaviorally demonstrated their ability to interact effectively with other professionals within the organization were given higher-level managerial responsibilities and positions. Such findings clearly illustrate that technical skills were not considered sufficient for attaining these high-level managerial positions; rather, technical and interpersonal skills have to be combined. As expected, RD&E managers not only have to be technically compe-

tent, but they must also be able to communicate and interface effectively with other individuals, since most of their responsibilities have to be carried out and/or coordinated with these people. But this combination of skill requirements should not be so exclusively one-sided that it leads to perceptions that those promoted managerially are competent interpersonally while those promoted to the technical ladder are not. Under such conditions, the technical ladder will not be viewed as a reward but rather as a consolation prize for being regarded as *not* having the potential for being a "good manager."

SUBORDINATE CAREER OUTCOMES

Our research findings also support the idea that supervisory behavior has a critical influence on an engineer's organizational career. Not all supervisors, however, had comparable relationships with the career outcomes of their technical subordinates. Only boundary spanning gatekeeper supervisors were significantly associated with reduced turnover rates and higher rates of managerial promotion. These relationships, moreover, were particularly strong for young professionals; they disappeared for older, more experienced engineers.

Why were proportionately more of the young engineers who had gatekeeping supervisors still working and contributing to the organization after five years than those engineers whose supervisors were not functioning as gatekeepers? The strong communication differences that emerged suggest that gatekeepers lower turnover through the higher levels of interaction and activity that they foster for the young engineers, not only with themselves but also with more senior departmental managers. Thus, it may not be the gatekeeping role or supervisory status per se that influences turnover. What seems most beneficial are meaningful amounts of exposure and support coupled with work-related contact and involvement with relevant competent supervisors—interactions and work experiences that occur most frequently with gatekeeping supervisors.

In addition to these turnover relationships,

gatekeeping supervisors were linked to the managerial promotions of project subordinates who were between the ages of 27 and 32 at the start of the study. Within this age range, project members working for gatekeeping supervisors attained a significantly higher rate of promotion to management than did project members working for non-gatekeeping supervisors. Furthermore, the promotion rate for young project members reporting to gatekeepers in development areas was more than three times the promotion rate of development project members assigned to non-gatekeeping supervisors. In development work, gatekeepers are highly influential individuals who strongly enhance project performance by connecting engineers to more useful ideas and information outside the project. Having better access to critical information, along with working for influential supervisors, may be associated with greater work opportunities and organizational visibility which, in turn, lead to higher rates of management promotion.

From a broader perspective, the relationships between lower turnover and high interpersonal involvement with gatekeeping supervisors affirms the important role that project supervisors can and should play during the early socialization years of young professionals. As discussed by many researchers, young employees build perceptions of their work environment and establish their new organization identities through the plethora of interactions and interpersonal activities that take place during the early years of their laboratory integration. Young engineers, therefore, not only need to interact with their colleagues and peers, but they also require considerable interaction with and feedback from relevant supervisors to learn what is expected of them and to decipher how to be a high performing contributor.

Because they are well-connected professionally and organizationally, gatekeepers are particularly qualified to meet the breaking-in concerns of young professionals, directing and coupling the professional orientations of young engineers with a more appropriate organizational focus. Most likely, the high level of interpersonal contact between gatekeeping supervisors and young project engineers not only facilitates socialization but also results in more accurate expectations, perceptions, and understanding about one's role in the project and in the larger organization—all of which are important in decreasing the turnover and increasing the contribution of newcomers. Gatekeeping supervisors, therefore, fulfill an important leadership and training function for young engineers, operating effectively as important socializing agents as well as early sponsors of and network builders for these young technical professionals.

Organizations have to recognize that the problems and concerns of young engineers are real and must be dealt with before they become effective organizational members. Although specific training programs could be developed to teach project managers how to "break-in" the young engineer more effectively, the careful selection of supervisors for young professionals also would go a long way toward alleviating many of the problems that usually occur during this joining-up process. It is also important to note that gatekeeping supervisors were not related to the career outcomes of older, more established project members. The turnover and promotion rates of these more veteran technologists are probably influenced more by individual differences and by task and organizational factors than by the characteristics of their immediate supervisors.

Finally, one should realize that in a longitudinal field study of this sort, the random assignment of project members to gatekeeping and nongatekeeping supervisors was not possible. Although the thinking here emphasizes the direct role that gatekeepers might play in influencing project members' careers, it also is possible that gatekeepers either attracted or were assigned members who were more likely to stay or were of higher promotion potential. Furthermore, other uncontrolled organizational factors could have influenced the results. For example, data were not collected on how long project members worked for their respective supervisors. What is known is that project members who reported to gatekeeping supervisors early in their careers had more successful organizational outcomes. It remains for future research to look even more closely at these types of relationships.

CONCLUSIONS

In conclusion, gatekeepers perform a critical role within R&D settings that often goes unrecognized. By realizing the importance of the gatekeeping role within development tasks, R&D managers can link their product or process efforts to sources of external technology more effectively. A manager could examine, for example, the extent to which important technologies utilized within various development projects are actually 'covered' by a gatekeeping type person. However, the degree to which these communication activities can be managed may be limited. Gatekeeping is an informal role in that other project engineers must feel sufficiently secure and comfortable psychologically to approach gatekeepers with their technical problems, mistakes, and questions without fear of personal evaluation or other adverse considerations (Allen, 1984). Therefore, to the extent that the organization tries to formalize such a gatekeeping function, it runs the risk of inhibiting the very kinds of interaction it wishes to promote.

This is not meant to imply that gatekeeping cannot be managed or helped; on the contrary, it can. In fact, a number of R&D facilities have instituted formal gatekeeper programmes. What is important to recognize is that the interest and ability of individuals to link with external technology cannot be suddenly 'decreed' by management. Typically, such outside professional interests are a 'given' and are not easily influenced by the organization, although they can be made easier to pursue. What can be more easily influenced is the degree to which gatekeepers are actually present and participating in project tasks as well as their accessibility to other project members. Their work positions, for example, could be located close to other project engineers to foster easier and more frequent communication. However, the develop-

ment of sufficient internal contacts and communications to be an effective gatekeeper takes time. In the present sample, for example, all of the gatekeepers had been working in their present project groups for a period of at least two years. In short, the external side of the gatekeeping role is usually being performed by the gatekeeper anyway. It is the internal side that can be facilitated and made more effective.

REFERENCES

Achilladeles, A., Jervis, P. & Robertson, A. (1971) *Success and Failure in Innovation.* Project Sappho, Sussex: University of Sussex Press.

Allen, T. J. (1977) *Managing the Flow of Technology.* Cambridge, MA: M.I.T. Press

Allen, T. J. & Cohen, S. (1969) "Information flow in R&D laboratories," *Administrative Science Quarterly,* Vol. 14, 12–19.

Allen, T. J., Tushman, M. & Lee, D. (1979) "Technology transfer as a function of position on research, development, and technical service continuum," *Academy of Management Journal,* Vol. 22, 694–708.

Dalton, G., Thompson, P. & Price, R. (1982) "The four stages of professional careers; A new look at performance by professionals." In R. Katz (Ed.), *Career Issues in Human Resource Management,* Englewood Cliffs, NJ: Prentice-Hall, 129–53.

Hagstrom, W. (1965) *The Scientific Community.* New York: Basic Books.

Katz, R. (1980) "Time and work: Toward an integrative perspective." In B. Staw & L. L. Cummings (Eds.), *Research in Organizational Behavior* (Vol. 2), Greenwich, CT: JAI Press, 81–127.

Katz, R. (2004) "Organizational socialization and the reduction of uncertainty." In R. Katz (Ed.), *The Human Side of Managing Technological Innovation: A Collection of Readings,* New York: Oxford University Press.

Katz, D. & Kahn, R. (1966) *The Social Psychology of Organizations.* New York: Wiley Co.

Katz, R. & Tushman, M. (1979) "Communication patterns, project performance, and task characteristics," *Organizational Behavior and Human Performance,* Vol. 23, 139–62.

Katz, R., Tushman, M., & Allen, T. (1995) "The influence of supervisory promotion and network location on subordinate careers in a dual ladder RD&E setting," *Management Science,* Vol. 41, 848–63.

Pelz, D. & Andrews, F. M. (1966) *Scientists in Organizations.* New York: Wiley Co.

Rosenbloom, R. & Wolek, F. (1970) *Technology and Information Transfer.* Boston, MA: Harvard Business School.

Schein, E. H. (1978) *Career Dynamics: Matching Individual and Organizational Needs.* Reading, MA: Addison-Wesley.

Tushman, M. (1977) "Technical communication in R&D laboratories: the impact of project work characteristics," *Academy of Management Journal,* Vol. 20, 624–45.

Tushman, M. & Katz, R. (1980) "External communication and project performance: An investigation into the role of gatekeepers," *Management Science,* Vol. 26, 1071–85.

28

Why Information Technology Inspired but Cannot Deliver Knowledge Management

RICHARD MCDERMOTT

Knowledge is experience. Everything else is just information.

—Albert Einstein

A few years ago British Petroleum placed a full-page ad in the *London Times* announcing that it learned a key technology for deep-sea oil exploration from its partnership with Shell Oil Company in the Gulf of Mexico and was beginning deep-sea exploration on its own, west of the Shetland Islands. British Petroleum's ability to leverage knowledge is key to its competitive strategy. Rather than conducting its own basic research, British Petroleum learns from its partners and quickly spreads that knowledge through the company. It does this not by building a large electronic library of best practices, but by connecting people so they can think together.

Information technology has led many companies to imagine a new world of leveraged knowledge. E-mail and the Internet have made it possible for professionals to draw on the latest thinking of their peers no matter where they are located. A chemist in Minnesota can instantly tap all his company's research on a compound. A geologist can compare data on an oil field to similar fields across the globe to assess its commercial potential. An engineer can compare operational data on machine performance with data from a dozen other plants to find patterns of performance problems. As a result, many companies are rethinking how work gets

done, linking people through electronic media so they can leverage each other's knowledge. A consulting company set up a best practices database with detailed descriptions of projects so consultants around the globe could draw from each other's experience. A computer company's systems design group created an electronic library of system configurations so designers could draw from a store of pre-developed components. These companies believe that if they could get people to simply document their insights and draw on each other's work, they could create a web of global knowledge that would enable their staff to work with greater effectiveness and efficiency.

While information technology has inspired this vision, it itself cannot bring the vision into being. Most companies soon find that leveraging knowledge is very hard to achieve. Several years ago Texaco's Information Technology group installed Lotus Notes, hoping it would lead to more collaboration. They soon discovered that they only used Notes for e-mail. Not until they found an urgent need to collaborate and changed the way they worked together did they use Notes effectively. Studies show that information technology usually reinforces an organization's norms about documenting, sharing information, and using the ideas of others. People send most e-mail to those they work with daily. Computer mediated interaction is usually more polite than face-to-face, despite occasional flaming. Computer-aided decision making is no more democratic than face-to-face decision

Richard McDermott is President of McDermott Consulting in Boulder, CO.

making. Virtual teams need to build a relationship, often through face-to-face meetings, *before* they can effectively collaborate electronically.[1] The difficulty in most knowledge management effort lies in changing organizational culture and people's work habits. It lies in getting people to take the time to articulate and share the really good stuff. If a group of people don't already share knowledge, don't already have plenty of contact, don't already understand what insights and information will be useful to each other, information technology is not likely to create it. However, most knowledge management efforts treat these cultural issues as secondary, implementation issues. They typically focus on information systems—identifying what information to capture, constructing taxonomies for organizing information, determining access, and so on. *The great trap in knowledge management is using information management tools and concepts to design knowledge management systems.*

Creating Information Junkyards

A good example of how information technology alone cannot increase the leverage of professional knowledge comes from a large consumer products company. As part of reorganization, the company decided to improve professional work. Professional staff were instructed to document their key work processes in an electronic database. It was a hated task. Most staff felt their work was too varied to capture in a set of procedures. But after much berating by senior managers about being "disciplined," they completed the task. Within a year the database was populated, but little used. Most people found it too general and generic to be useful. The help they needed to improve their work processes and share learning was not contained in it. The result was an expensive and useless information junkyard. Creating an information system without understanding what knowledge professionals needed, or the form and level of detail they needed, did little to leverage knowledge.

Knowledge is different from information and sharing it requires a different set of concepts and tools. Six characteristics of knowledge distinguish it from information:

- Knowing is a human act
- Knowledge is the residue of thinking
- Knowledge is created in the present moment
- Knowledge belongs to communities
- Knowledge circulates through communities in many ways
- New knowledge is created at the boundaries of old

Leveraging knowledge involves a unique combination of human and information systems.

KNOWING, THINKING, AND COMMUNITY

Knowing is a human act. Discussions of knowledge management often begin with definitions of data, information, and knowledge. I would like to take a different starting point: an inquiry into our own experience using, discovering, and sharing knowledge. By reflecting on our own individual experience, we can gain a deeper understanding of the nature of knowledge and how we and *others* use it.[2] As Maurice Merleau-Ponty observed, "We arrive at the universal, not by abandoning our individuality, but by turning it into a way of reaching out to others."[3]

Reflecting on our experience, the first thing that comes into view is that *we* know. Knowledge always involves a person who knows. My bookcase contains a lot of information on organizational change, but we would not say that it is knowledgeable about the subject. The same is true for my computer, even though it can store, sort, and organize information much more quickly than my bookcase. Thinking of our minds as a biochemical library is little different from treating it as a bookcase or computer. To know a topic or a discipline is not just to possess information about it. It is the very human ability to *use* that information.

The art of professional practice is to turn information into solutions. To know a city is to know its streets, not as a list of street names or a map, but as a set of sights and routes useful for different *purposes*. Driving through your hometown to avoid rush-hour traffic, find an interesting restaurant, bring relatives sightseeing, or go bargain hunting, you not only draw on a vast amount of information, you *use* the information in different ways. Your purpose determines the information you focus on and remember, the routes that come to mind. Professionals do the same thing. They face a stream of problems; when to run a product promotion, how to estimate the size of an oil field, how to reduce the weight and cost of a structure. To solve these problems, professionals *piece information together, reflect* on their experience, *generate* insights, and *use* those insights to *solve* problems.

Thinking is at the heart of professional practice. If we look at our own experience, *thinking* is key to making information useful. Thinking transforms information into insights and insights into solutions. When jamming, jazz musicians get a feel for where the music is going, adjust to their partners' moves, change direction, and readjust. They take in information, make sense of it, generate new musical ideas, and apply their insights to the ongoing musical conversation. Responding to each other, they draw on tunes, chords, progressions, and musical "feels" they have known before, even though at any moment they could not predict "what's next." Jamming is a kind of musical thinking.[4] Science, architecture, engineering, marketing, and other practical professions are not that different. Professionals do not just cut and paste "best practice" from the past to the current situation. They draw from their experience to *think about* a problem. An architect looking for a design that will work on a steeply sloping site looks at the site "through the eyes" of one idea, discards it and sees it again "through the eyes" of a different idea, drawing on different information about the site in each thought experiment. In running these experiments, the architect is not just looking for pre-made solutions, but thinking about how those solutions might apply and letting ideas *seep* from one framework

to the next, so a new, creative idea can emerge.[5] Professional practice is also a kind of improvisation within a territory, whether that's a keyboard, a science, or a computer application. As knowledgeable practitioners, we move around the territory, sometimes with accuracy and efficiency, sometimes with grace and inspiration. A group of systems designers for a computer company tried to leverage their knowledge by storing their system documentation in a common database. They soon discovered that they did not need each other's system documentation. They needed to understand the logic other system designers used—why *that* software, with *that* hardware and *that* type of service plan. They needed to know the path of thinking other system designers took through the field. *To know a field or a discipline is to be able to think within its territory.*

Knowledge is the residue of thinking. Knowledge comes from experience. However, it is not just raw experience. It comes from experience that we have reflected on, made sense of, tested against other's experience. It is experience that is *informed by* theory, facts, and understanding. It is experience we make sense of in relationship to a field or discipline. Knowledge is what we retain as a result of thinking through a problem, what we remember from the route of thinking we took through the field. While developing a report on a competitor, a researcher deepens her understanding of her research question, the competitor, and the information sources she used, particularly if she used a new question, source, or approach. *From the point of view of the person who knows, knowledge is a kind of sticky residue of insight about using information and experience to think.*

Knowledge is always *recreated* in the present moment. Most of us cannot articulate what we know. It is largely invisible and often comes to mind only when we need it to answer a question or solve a problem. This isn't because knowledge is hard to find in our memory. It is because knowledge resides in our body. To find it we don't search. We engage in an act of knowing.[6] Knowledge is what a lathe operator has in his hands about the feel of the work after turning hundreds of blocks of wood. Knowledge is the insight an engineer has in the back of her mind about which analytic tools work well together and when to use them. To use our knowledge we need to make sense of our experience again, here in the present. When I think through how to champion an organizational change, I draw from the constantly evolving landscape of what I know now about change, my evolving "mental models" of change. I put that insight together in a new sense, one created just here and now. Sometimes it includes new insights freshly made. Sometimes it forgets old ones. Learning from past experience, sharing insights, or even sharing "best practices" is always rooted in the present application, the thinking we are doing now. *Insights from the past are always mediated by the present, living act of knowing.*[7]

To share knowledge we need to *think about* the present. Sharing knowledge involves guiding someone through our thinking or using our insights to help them see their own situation better. To do this we need to know something about those who will use our insights, the problems they are trying to solve, the level of detail they need, maybe even the style of thinking they use. For example, novices frequently solve problems by following step-by-step procedures, but experts solve problems in an entirely different way. They typically develop a theory of potential causes based on their experience and test to see if the theory is correct, often testing the least complex or expensive theories, rather than the logically correct ones, first.[8] The knowledge useful to novices is very different from the knowledge useful to experienced practitioners. Sharing knowledge is *an act of knowing* who will use it and for what purpose. For peers, this often involves mutually discovering which insights from the past are relevant in the present. To document for a general audience, like writing a textbook, also involves imagining a user, the novice. It is our picture of the user—their needs and competencies—that determines the level of detail, tone and focus of the insights we share.

Knowledge belongs to *communities*. The idea that knowledge is the stuff "between the ears of the individual" is a myth. We don't learn on our own.

Playing "Give Me Your Best Line"

Years ago a geoscientist at Shell Oil Company who had an uncanny knack for finding oil initiated an odd lunch-time game. Geoscientists explore prospective sites using seismic data, which give a two-dimensional picture of the earth, like the side of a slice of cake. The more lines of seismic data, the more complete and three-dimensional the picture of the prospect. His game was to gather a group of geoscientists and guess at the structure of the prospect, using the fewest number of seismic lines, and therefore the least amount of information possible. Since a prospect's geology is key to finding oil, the game had serious practical consequences. This game caused people to think together about the prospect. With very little data, it was easy to pose different theories, challenge assumptions, and reformulate their ideas. The lack of data encouraged them to consider a wider variety of models of the geography than they would have with more complete data. As they collected more data about the prospect, they continued the game and discovered which theories had been correct. The game was a powerful exercise in leveraging knowledge. It enabled this group to share their thinking and reformulate their assumptions as they expanded their understanding.

We are born into a world already *full* of knowledge, a world that already makes sense to other people—our parents, neighbors, church members, community, country. We learn by participating in these communities and come to embody the ideas, perspective, prejudices, language, and practices of that community.[9] The same is true for learning a craft or discipline. When we learn a discipline, whether at school or on the job, we learn more than facts, ideas, and techniques. We enter a territory already occupied by others and learn by participating with them in the language of that discipline and seeing the world through its distinctions. We learn a way of thinking. Marketing specialists learn market survey methods; but they also learn a marketing perspective. They learn to ask questions about product use, customer demographics, lifestyle, product life cycles, and so forth. This perspective is embedded in the discipline and handed down through generations of practitioners. It is part of the background knowledge and accumulated wisdom of the discipline. Architects from different schools approach problems in characteristically different ways. Each school's approach is embedded in the everyday practices of its faculty, shared as they see the logic of each other's thinking. *Knowledge flows through professional communities, from one generation to the next.* Even though we do most of our thinking alone, in our office or study, we are building on the thinking of others and to contribute to a discipline, we must put our ideas out into the "public"—just stewards for a moment. Even when we develop ideas that contradict the inherited wisdom of the profession, our "revolutionary" ideas are meaningful only in relation to the community's beliefs. They are still a form of participation in that discipline.[10] Despite changes in membership and dominant paradigms, the discipline itself continues often with its basic assumptions and approaches relatively intact for generations.

Knowledge circulates through communities in many ways. We typically think of a community's knowledge as the stuff in textbooks, articles, written procedures, individual file cabinets, and people's minds. However, many other "objects" contain a community's knowledge: unwritten work routines, tools, work products, machinery, the layout of a workspace or tools on a tray, stories, specialized language, and common wisdom about cause-effect relationships.[12] These unwritten artifacts circulate through the community in many ways. Stories are told at conferences and chance hallway meetings. People see each other's thinking as they solve problems together, in peer reviews, or in notes in the margins of work products. People observe and discuss informal work routines in the everyday course of work. So where does a community's knowledge reside? From the practitioner's perspective, only a small percentage is written. Most is in these informal, undocumented practices and artifacts. All contacts within the community can be vehicles for sharing knowledge, even though most are not intended to be. As Wallace Stevens wrote, "Thought is an infection, some thoughts are an epidemic."

New knowledge is created at the boundaries of old. If you reflect on how you learn new things, you probably find that most of the time you learn by comparing the new idea, fact, or tool to ones you already know. The everyday practice of professional work involves thinking that draws from experience and current information. But new

---

Thinking with Information at the Center for Molecular Genetics

Researchers at the Center for Molecular Genetics in Heidelberg use photographs extensively in their work. They take pictures of radioactively marked DNA and RNA strands using X-ray film. Their challenge is to make sense of these pictures, interpreting what the markings on the film indicate about the structure of the material and its implications for their experiments. As they pull photos from the darkroom, other people in the lab gather around to discuss what they see. These discussions frequently refer to other research, both published and current. They see the film through the eyes of one set of research findings, then another. Through these informal gatherings, the researchers think aloud together, challenge each other, try dead ends, draw metaphors from other disciplines, and use visual models and metaphors to make sense of their data and reach conclusions. Their collective know-how and knowledge of the research literature are the living backdrop for these discussions. Sometimes they talk through a procedure, looking for the meaning of a result in its minute details. Other times they focus on research findings, letting their procedures fade into the background as they compare their results to others. In these discussions, they use their knowledge of the literature and their lab know-how to think about and solve the current research problem.

knowledge typically does not come from thinking within the ordinary bounds of a discipline or craft. It comes from thinking at the edge of current practice. New, disruptive technology is often developed by small companies at the edge of a marketplace.[13] Scientists are frequently most productive a few years after they have crossed over from one specialty to another. New ideas in science frequently emerge, not from paradigm shifts at the heart of the discipline, but when scientists run out of interesting research questions—and publication opportunities—at the heart of their discipline and either shift to subspecialties on the margin of the discipline or combine the perspectives of different disciplines, forming new specialties such as psychopharmacology.[14] *New ideas emerge in the conflict of perspective, the clash of disciplines, the murky waters at the edge of a science, the technology that doesn't quite work, on the boundaries of old knowledge.*

In summary, when we look at our own experience, knowledge is much more—and much more elusive—than most definitions allow. Knowing is a *human act,* whereas information is an *object* that can be filed, stored, and moved around. Knowledge is a *product of thinking,* created in the *present moment,* whereas information is fully made and can sit in storage. To share knowledge, we need to *think about the current situation,* whereas we can simply move information from one mailbox to another. *However, knowledge is more than you think.* Knowledge settles into our body. It is a kind of "under the fingernails" wisdom, the background know-how from which we draw. Most of us find it hard or impossible to articulate what we know; whereas information can be written or built into machinery. We acquire knowledge by participating in a community—using the tools, ideas, techniques, and unwritten artifacts of that community; whereas we acquire information by reading, observing, or otherwise absorbing it. *Ironically, when we look at our experience, the heart of knowledge is not the great body of stuff we learn, not even what the individual thinks, but a community in discourse, sharing ideas.*

IMPLICATIONS FOR LEVERAGING KNOWLEDGE

What are the implications of these philosophical reflections on the movement to manage knowledge? Clearly, leveraging knowledge involves much more than it seems. It is not surprising that documenting procedures, linking people electronically, or creating web sites is often not enough to get people to think together, share insights they didn't know they had, or generate new knowledge. Using our own experience as a starting point to design knowledge management systems leads to a different set of design questions. Rather than identifying information needs and tools, we identify the community that cares about a topic and then enhance their ability to think together, stay in touch with each other, share ideas with each other, and connect with other communities. *Ironically, to leverage knowledge we need to focus on the community that owns it and the people who use it, not the knowledge itself.*

To Leverage Knowledge, Develop Existing Communities

Develop natural knowledge communities without formalizing them. Most organizations are laced with communities in which people share knowledge, help each other, and form opinions and judgments. *Increasing an organization's ability to leverage knowledge typically involves finding, nurturing, and supporting the communities that already share knowledge about key topics.* Allied Signal supports learning communities by giving staff time to attend community meetings, funding community events, creating community bulletins, and developing a directory of employee skills. If too formalized, learning communities can become bureaucratic structures, keepers of the discipline's "official story" that act as approval hurdles for operations groups. The key to nurturing communities is to tap their natural energy to share knowledge, build on the processes and systems they already use, and enhance the role of natural leaders.

Implications for Leveraging Knowledge

1. To leverage knowledge, develop communities.
2. Focus on knowledge important to both the business and the people.
3. Create forums for thinking as well as systems for sharing information.
4. Let the community decide what to share and how to share it.
5. Create a community support structure.
6. Use the communitys terms for organizing knowledge.
7. Integrate sharing knowledge into the natural flow of work.
8. Treat culture change as a community issue.

Focus on Knowledge Important to Both the Business and the People

Learning communities are organized around important topics. Developing communities takes considerable effort. The best way to insure that the effort is well spent is to identify topics where leveraging knowledge will provide value to the business as well as community members. People naturally seek help, share insights, and build knowledge in areas they care about. At Chaparral Steel, blue-collar employees meet with customers, solve problems, create new alloys, and continually redesign the steel-making process. They share insights with each other as they search for innovative, cost-competitive solutions to customer needs. Sharing knowledge helps staff solve problems directly related to their day-to-day work.[15] *Natural learning communities focus on topics that people feel passionate about.* In Shell's Deepwater Division, most learning communities are formed around disciplines, like geology, or topics that present new challenges to the business or their field. They are topics people need to think about to do their work. Most are topics people have studied, find intrinsically interesting, and have become skillful at moving around in. As one geologist said, "With so many meetings that aren't immediately relevant to your work, it's nice to go to one where we talk about rocks."

Create Forums for Thinking and Sharing Information

The ways to share knowledge should be as multidimensional as knowledge itself. Most corporate knowledge sharing efforts revolve around tools, typically electronic ones. The company finds a tool, or develops one, and then finds groups to use it. This may be good for sharing information, but since *knowing involves thinking about a field full of information*, a *knowledge management* system should include both systems for sharing information and forums for thinking. To paraphrase Henry Adams, facts without thinking are dead, and thinking without facts is pure fantasy.[16] The field of information can include statistics, maps, procedures, analyses, lessons learned, and other information with a long shelf life, but it can also include interpretations, half-formed judgments, ideas, and other perishable insights that are highly dependent on the context in which they were formed. The forums, whether face-to-face, telephone, electronic, or written, need to spark collaborative thinking, not just make static presentations of ideas. In Shell's Deepwater Division, most learning communities hold regular collaborative problem-solving meetings facilitated by a community coordinator. These sessions have two purposes. First, by solving real day-to-day problems, community members help each other and build trust. Second, by solving problems in a public forum, they create a common understanding of tools, approaches, and solutions. One learning community in Shell, composed mostly of geologists, asks people to bring in paper maps and analyses. During the meetings, people literally huddle around the documents to discuss problems and ideas. The community coordinator encourages community members to make their assumptions visible. The combination of information and thinking leads to a rich discussion. While they discuss the issues, someone types notes on a laptop, so key points are captured. This community's process for leveraging knowledge includes thinking and information, human contact and IT. In the course of problem-solving discussions such as these, most communities discover areas where they need to create common standards or guidelines, commission a small group to develop them, and incorporate their recommendations. Most have a web site where they post meeting notes and guidelines. Some have even more elaborate community libraries.

Let the Community Decide What to Share and How to Share It

Knowledge needs to have an "owner" who cares. It is tempting to create organization-wide systems for sharing knowledge so everyone can access it. This can be useful if all members of the organization truly need to work with that body of knowledge, but the further away you get from community's actual needs, the less useful the information. Communities vary greatly in the kind of knowledge they need to share. In Shell's Deepwater Division, operations groups need common standards to re-

duce redundancy and insure technology transfer between oil platforms. Their learning communities focus on developing, maintaining and sharing standards. Geologists, on the other hand, need to help each other approach technical problems from different directions to find new solutions. They need to understand the logic behind each other's interpretations. Another community found they were each collecting exactly the same information from external sources, literally replicating each other's work. They needed a common library and someone responsible for document management. Since information is meaningful only to the community that uses it, the community itself needs to determine the balance of how much they need to think together, collect and organize common information, or generate standards. Since knowledge includes both information and thinking, only the community can keep that information up-to-date, rich, alive, available to community members at just the right time, and useful. Only community members can understand what parts of it are important. When communities determine what they need to

---

Many Forums for Discussing Petrophysics

A division of a Shell Oil Company recently organized its professional engineers into permanent cross-functional teams. Team members are located together and some engineering disciplines have only one member on the team. Once organized into this new structure, a group of petrophysicists realized how much they needed other members of their discipline to get advice and think through issues. In the past they just walked down the hall to get help; they now needed to go several floors away to find a peer.

So they decided to create a process for sharing knowledge with each other that crosses the boundaries of the operations teams. Some parts of this system are organic, some informal, some explicit, and some formal. For consulting each other on interpretations of data, they hold informal, agenda-less weekly meetings where anyone can

get input on any topic. These are different from most agenda-driven meetings in that they emphasize open dialogue for exploring issues, with no pressure to come to resolution. To share knowledge that is more explicit, they have formal presentations on new technology. To ensure their data are consistent and widely available, they opened a common electronic data library that lets them compare data from many different sites. To ensure that informal help is available at any time, they established a senior coordinator who facilitates interaction among members of the discipline as well as provides his own insights and answers. Each of these forums is useful for sharing a different kind of knowledge, from fuzzy know-how to concrete data. Having all of them available ensures that each is used for sharing the knowledge most appropriate to it.

share and what forum will best enable them to share it, they can more readily own both the knowledge and the forums for sharing it.

Create a Community Support Structure

Communities are held together by people who care about the community. In most natural communities, an individual or small group takes on the job of holding the community together, keeping people informed of what others are doing and creating opportunities for people to get together to share ideas. In intentional communities, this role is also critical to the community's survival, but it typically needs to be designed. Community coordinators are usually a well-respected member of the community. Their primary role is to keep the community alive, connecting members with each other, helping the community focus on important issues, and bringing in new ideas when the community starts to lose energy. In Allen's study, project engineers used information from technical consultants and suppliers more readily when it was funneled through an internal gatekeeper than when the consultants met with them directly.[17]

Use information technology to support communities. Most companies use information technology to support individual work, leaving it to each individual to sort through the information that comes their way, decide what is important, clean, and organize it. However, if communities own knowledge, then the community can organize, maintain, and distribute it to members. This is another key role of community coordinators or core group members. They use their knowledge of the discipline to judge what is important, groundbreaking, and useful and to enrich information by summarizing, combining, contrasting, and integrating it. When IBM introduced its web-based Intellectual Competencies system, anyone could contribute to the knowledge base. However, like many other companies, IBM soon discovered that their staff did not want to hunt through redundant entries. Now a core group from each community organizes and evaluates entries, weeding out redundancies and highlighting particularly useful or groundbreaking work. Frequently, technical professionals see this as a "glorified librarian" role and many communities also have librarians or junior technical staff to do the more routine parts of organizing and distributing information.

Use the Community's Terms for Organizing Knowledge

Organize information naturally. Since knowledge is the sense we make of information, then the way information is organized is also a sense-making device. A good taxonomy should be intuitive for those who use it. To be "intuitive" it needs to tell the story of the key distinctions of the field, reflecting the natural way discipline members think about the field. Like the architecture of a building, a taxonomy enables people to move about within a bank of information, find familiar landmarks, use standard ways to get to key information, create their own "cowpaths," and browse for related items. This is a common way to spark insight. Of course, this means that if you have multiple communities in an organization, they are likely to have different taxonomies, not only in the key categories through which information is organized, but also in the way that information is presented. A group of geologists, who often work with maps, asked that their web site for organizing information be a kind of visual picture. They think in pictures. However, a group of engineers in the same organization wanted their web site to be organized like a spreadsheet. They think in tables. *The key to making information easy to find is to organize it according to a scheme that tells a story about the discipline in the language of the discipline.*

How standard should company taxonomies be? Only as wide as the community of real users. There is a great temptation to make all systems for organizing knowledge the same. Certainly formatting information so it can easily be transferred—having the same metadata so it can be searched, indexed, and used in different contexts—can be very useful. However, beyond that the systems for organizing information should be the community's. If a community of people sharing knowledge spans several disciplines, then such things as terms and structures should be common among those communities.

Integrate Learning Communities into the Natural Flow of Work

Community members need to connect in many ways. Because communities create knowledge in the present moment, they need frequent enough contact to find commonality in the problems they face, see the value of each other's ideas, build trust, and create a common etiquette or set of norms on how to interact. When people work together or sit close enough to interact daily, they naturally build this connection. It simply emerges from their regular contact. When developing intentional learning communities, it is tempting to focus on their "official tasks:" developing standards, organizing information, or solving cross-cutting technical problems. However, it is also important for them to have enough open time for "technical schmoozing," sharing immediate work problems or successes, helping each other, just as they would if they were informally networking down the hall. This informal connection is most useful if it can happen in the spontaneous flow of people's work as they encounter problems or develop ideas. So community members need many opportunities to talk one-on-one or in small groups on the telephone, through e-mail, face to face, or through an Internet site. In Shell's Deepwater Division, community coordinators "walk the halls," finding out what people are working on, where they are having problems, and making connections to other community members. This informal connecting ensures that issues don't wait until community meetings to get discussed and keeps other channels of communication open.

Treat Culture Change as a Community Issue

Communities spread cultural change. Failures in implementing knowledge management systems are often blamed on the organization's culture. It is argued that people were unwilling to share their ideas or take the time to document their insights. However, organizational culture is hard to change. It rarely yields to efforts to change it directly by manipulation of rewards, policies, or organizational structure. Often it changes more by contagion than decree. People ask trusted peers for advice, teach newcomers, listen to discussions between experts, and form judgments in conversations. In the course of that connection with community members, they adopt new practices. Despite massive efforts by public health organizations to educate physicians, most physicians abandon old drugs and adapt new ones only after a colleague has personally recommended it. They rely on the *judgment* of their peers—as well as the information they get from them—to decide. New medical practices spread through the medical community like infectious diseases, through individual physician contact. Learning communities thrive in a culture that supports sharing knowledge. However, they are also vehicles for creating a culture of sharing. While it is important to align measurement, policies, and rewards to support sharing knowledge,[18] the key driver of a change toward sharing knowledge is likely to be within communities.[19]

CONCLUSION

Today, the "knowledge revolution" is upon us, but the heart of this revolution is not the electronic links common in every office. Ironically, while the knowledge revolution is inspired by new information systems, it takes human systems to realize it. This is not because people are reluctant to use information technology. It is because knowledge involves thinking with information. If all we do is increase the circulation of information, we have only addressed one of the components of knowledge. To leverage knowledge we need to enhance both thinking and information. The most natural way to do this is to build communities that cross teams, disciplines, time, space, and business units.

There are four key challenges in building these communities. The *technical challenge* is to design human and information systems that not only make information available, but help community members think together. The *social challenge* is to develop communities that share knowledge and still maintain enough diversity of thought to encourage thinking rather than sophisticated copying. The *management challenge* is to create an environment that truly values sharing knowledge. The *personal challenge* is to be open to the ideas of others, willing to share ideas, and maintain a thirst for new knowledge.

By combining human and information systems, organizations can build a capacity for learning broader than the learning of any of the individuals within it.

NOTES

1. M. Lynne and Robert Benjamin Marcus, "The Magic Bullet Theory in IT-Enabled Transformation," *Sloan Management Review*, 38/2 (Winter 1997): 55–68; Matt Alvesson, "Organizations as Rhetoric: Knowledge-Intensive Firms and the Struggle with Ambiguity," *Journal of Management Studies*, 30/6 (November 1993): 997–1020; Ronald Rice, August Grant, Joseph Schmitz, and Jack Torbin, "Individual and Network Influences on the Adoption and Perceived Outcome of Electronic Messaging," *Social Networks*, 12 (1990): 27–55; Jolene Galegher, Robert E. Kraut, and Carmen Egido, *Technology for Intellectual Teamwork: Perspective on Research and Design* (Hillsdale, NJ: Lawrence Erlbaum Associates, 1990); Guiseppe Mantovani, "Is Computer Mediated Communication Intrinsically Apt to Enhance Democracy in Organizations?" *Human Relations*, 47/1 (1994): 45–62.

2. This is the starting point of philosophical inquiry. See Husserl and Merleau-Ponty on the importance of founding scientific theory on philosophical inquiry. Edmund Husserl, *The Crisis of European Sciences and Transcendental Phenomenology* (Evanston, IL: Northwestern University Press, 1970); Maurice Merleau-Ponty, *The Phenomenology of Perception* (London: Routledge & Kegan Paul, 1962).

3. Maurice Merleau-Ponty, *Sense and Non-Sense* (Evanston, IL: Northwestern University Press, 1964).

4. David Sudnow, *Ways of the Hand: The Organization of Improvised Conduct* (Cambridge, MA: Harvard University Press, 1978).

5. Donald Schon, *The Reflective Practitioner* (New York, NY: Basic Books, 1983).

6. Merleau-Ponty (1962), op. cit.

7. Ibid.; Schon, op. cit.; Peter Senge, *The Fifth Discipline* (New York, NY: Doubleday, 1990); Maurice Merleau-Ponty, *The Visible and the Invisible* (Evanston, IL: Northwestern University Press, 1969).

8. Udo Konradt, "Strategies of Failure Diagnosis in Computer-Controlled Manufacturing Systems," *International Journal of Human Computer Studies*, 43 (1995): 503–521.

9. Thomas Kuhn, *The Structure of Scientific Revolutions* (Chicago, IL: University of Chicago Press, 1962); Etienne Wenger, *Communities of Practice* (Cambridge: Cambridge University Press, 1998).

10. Michel Foucault, *The Order of Things* (New York, NY: Vintage Press, 1970).

11. Klaus Amann and Karin Knorr-Cetina, "Thinking through Talk: An Ethnographic Study of a Molecular Biology Laboratory," *Knowledge and Society: Studies in the Sociology of Science Past and Present*, 8 (1989): 3–36.

12. Michel Foucault, *The Birth of the Clinic* (New York, NY: Vintage Press, 1975).

13. Joseph Bower and Clayton Christensen, "Disruptive Technologies: Catching the Wave," *Harvard Business Review*, 73/1 (January/February 1995): 43–53.

14. M. F. Mulkay, "Three Models of Scientific Development," *Kolner Zeitshrift* (1974): Dorothy Leonard-Barton, *Wellsprings of Knowledge: Building and Sustaining the Sources of Innovation* (Boston, MA: Harvard Business School Press, 1995).

15. Leonard-Barton, op. cit.

16. Henry Adams, *The Education of Henry Adams* (London: Penguin, 1995).

17. Tom Allen, *Managing the Flow of Technology* (Cambridge, MA: MIT Press, 1977).

18. Q. Wang and A. Majchrzak, "Breaking the Functional Mindset in Process Organizations," *Harvard Business Review*, 74/5 (September/October 1996): 92–99.

19. Etienne Wenger, op. cit.

10

Managing Performance and Productivity in Technical Groups and Organizational Settings

29

Project Management Scorecard

DENNIS P. SLEVIN

4:32 P.M. Russ McKinley, a production supervisor, stops by Harry's office. Ralph Kingsley, his boss, has assigned him the role of project manager for the new system of quality and cost control that Harry has wanted for some time. Harry has asked for this meeting to make sure that Russ understands the project mission and objectives, and to size him up concerning how well he will do on his first project.

"Harry, Ralph wanted me to stop by to brief you on my plan of attack for our quality-cost program," says Russ.

"Great," replies Harry. "You know, Russ, this has been an important issue for me for some time, but we have just never seemed to be successful in implementing it. Ralph seems to have a lot of confidence in you and thinks that you can pull it off. Give this project a ten-plus priority rating, and you can be assured of my wholehearted support. But I am curious; how do you plan to manage this project? What is your plan for the next six months?"

"Well," says Russ enthusiastically, "I've got my team selected, and with your approval, we can

Reprinted with permission of the author.

Dennis P. Slevin is Professor of Management at the Katz Graduate School of Business, University of Pittsburgh.

get started next week. And to make sure we hit all of the important areas, I plan to use the Project Implementation Profile."

"The Project Implementation Profile?" queries Harry with a frown.

"Yes," responds Russ, "it has been developed by a couple of professors of business administration to make sure that things don't fall through the cracks during implementation of a complex project. They also have data that indicate it is useful for keeping score on how a project is going. It also helps make sure that the project manager attends to all of the dimensions for project success. Here, let me show you," says Russ as he pulls a document out of his folder. "You see, there are ten basic dimensions for project success."

"Interesting," muses Harry . . . and they review the profile together.

Harry Thorpe is getting his first briefing on the Project Implementation Profile—and by a subordinate, no less! The Project Implementation Profile was developed to assist practicing managers in the implementation of complex projects. This chapter has two functions that help you use this device in your project management:

1. It describes the ten-factor model.

2. It provides you with the specific Project Implementation Profile items.

PROJECT IMPLEMENTATION PROFILE MODEL

The Project Implementation Profile (PIP) is based on a model of successful project implementation to help the project manager understand how the process works. Figure 29.1 shows the ten key factors that need to be considered in managing the implementation of a project.[1] Seven of these factors lie on a critical time path. The sequence indicates which factors are most important in the beginning of a project and which become important as the project moves forward. For example, the project manager needs to be aware of the importance of re-

ceiving top management support and of developing the project schedule before even considering the selection and training of personnel.

Three other factors in the model are shown as occurring at all steps in the implementation process. For example, communication and feedback are vital at every stage of the project to determine whether information is being traded and if complaints or errors are being acted upon.

In order for you to improve your project manager skills, you should (1) understand the ten key factors and (2) be able to use the Project Implementation Profile.

Factor 1: Project Mission

The initial step of the implementation process is to clarify the goals of the project. Implementation of any new project is an expensive use of organizational time, money, and energy. Are the project's mission and goals clearly defined, so that you know exactly where it is going and how it can help the organization? Your considerations should be:

1. Is the mission clear?

2. Do I understand why the project is being considered?

3. Is it necessary?

4. Do I think it can succeed?

5. Are the goals specific and operational?

Factor 2: Top Management Support

After you get a clear idea of what the mission and benefits of the project are, it is crucial to gain the support of top management. Without their vocal and visible support, the project may be seen as unnecessary, pointless, or unimportant by the rest of the organization. Early in a new project's life, no single factor is as predictive of its success as the support of top management. Your considerations should be:

1. Is top management convinced the project is necessary?

2. Are they convinced it will succeed?

3. Have they made their support clear to everyone affected by the project?

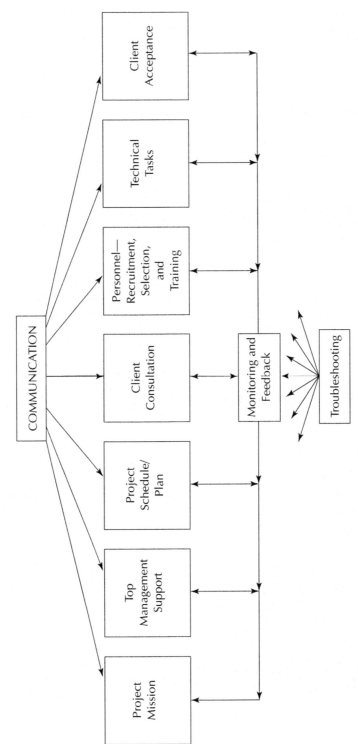

Figure 29.1.   Ten key factors of the Project Implementation Profile.

4. Do top management and I understand my role in the implementation process?

5. Do I have their confidence?

6. Will they support me, even in a crisis?

Factor 3: Project Schedule or Plan

For a project to get off the ground, it needs a well-thought-out, workable plan. All activities necessary for successful implementation need to be scheduled. Furthermore, all necessary people, money, time, and other resources to complete the project must be allocated. Finally, there must be a way of measuring the progress of the implementation against schedule projections. Your considerations should be:

1. Does the plan make sense? Is it workable?

2. Is the allocation of time, money, and people acceptable?

3. Am I satisfied that I had enough input in the planning?

4. Will the organization follow through on the plan?

5. Do I need to worry about funds being cut or schedules being altered without consultation?

6. Does the plan have enough slack to allow for cost or time overruns?

Factor 4: Client Consultation

"Client" here refers to whoever will ultimately be using the result of the project. It could be a customer or a department within the company. Because this project is for the client's benefit, close and frequent consultation with the client are imperative to make sure the effort is and stays in line with his needs.

Outside client considerations are:

1. Do I understand the client?

2. Do I know what he wants? (Or is it what I want him to want?)

3. Have I scheduled regular meetings with the client to keep him up to date on the project's progress?

Inside client considerations are:

1. Who are the key people who must support this project?

2. Is political activity needed to get client acceptance?

3. Is the client accepting or resisting?

Factor 5: Personnel—Recruitment, Selection, Training

As often occurs when implementing a new and unfamiliar project, we cannot always be sure that we have the necessary people for the project team. As a result, pay attention to selecting and training key personnel who can help make this a successful project. Neglecting this factor can force you to use personnel of convenience (simply because they are there), whether they will be helpful or destructive to the project. Your considerations should be:

1. Do I have the opportunity to pick and train my project team personnel? Or must I use personnel already in place?

2. Do I get along with them? Have I worked with them before?

3. Can I trust my key subordinates?

4. Does the organization have the kinds of personnel I need? Will I need to recruit them from outside?

5. Am I satisfied with the team's technical training and skills?

Factor 6: Technical Tasks

It is important that the implementation be well managed, by people who understand it. In addition, you must have adequate technology available to support the project (i.e., equipment, training, and the like). For a project to be successfully implemented, skilled people and proper technology are equally significant. Your considerations should be:

1. Have I assigned the correct technical problems to the right people?

2. Have I adequately documented and detailed the required technology?

3. Does the technology work well?

4. Does my team understand all aspects of the technology necessary for success?

5. Have I made provisions to update technology as minor changes in the project occur?

Factor 7: Client Acceptance

The obvious bottom line, in determining whether a project has been successfully implemented, is "Has the client bought it?" This question must be asked whether the client is internal or external to the organization. Too often managers make the mistake of believing that, if they handle all the other steps well, the client will automatically accept the resulting project. The truth is that client acceptance is a stage in project implementation that must be managed like any other. Your considerations should be:

1. Have I considered in advance a strategy to sell this project to the client?

2. Do I have leeway to negotiate with the client?

3. In the event of problems in the "teething" period of the project, do I have troubleshooters in place to help the client?

4. Will the project team be allowed to assist in follow-up? Or will it be disbanded immediately upon completion?

5. Does the organization view this as a one-shot deal? Or are organization members helping us identify other potential clients?

Factor 8: Monitoring and Feedback

It is important that, at each step of the implementation process, key personnel receive feedback on how the project is going. Monitoring and feedback mechanisms allow the project manager to be on top of any problems, to oversee any corrective measures, to prevent deficiencies from being overlooked. They ensure a quality project along the way. Your considerations should be:

1. Do I regularly ask team members for feedback on how the project is going?

2. Do I regularly assess the performance of the team members?

3. Is the project ahead, behind, or on time?

4. Are all project team members kept up to date regarding any snags in the schedule?

5. Have I established formal feedback channels? Or am I relying on informal methods?

6. Is the monitoring system working? Or are we being told what we want to hear?

Factor 9: Communication

As can be seen from the model, communication is a key component in every factor of the implementation process and must be all-pervading. Communication is essential within the project team, between the team and the rest of the organization, and with the client. Project implementation cannot take place in a vacuum; there must be constant communication. Your considerations should be:

1. Have I clearly communicated to the project team the goals and objectives of the project?

2. Do team members have formal channels for communicating with me? Or must they hope to catch me at my desk?

3. Do I regularly provide the team with written status reports?

4. Am I using the team to keep communication channels open within the organization as well as to the client?

5. Is this project viewed as open or secret by the rest of the organization?

6. Have I attempted to control rumors about this project?

Factor 10: Troubleshooting

No project operates without a few hitches. Constant fine-tuning, adjusting, and troubleshooting are required at each step of the implementation process. It is important to realize that each project team member is capable of functioning as a "lookout" for problems. Actually, each team should contain technically competent people with the specific assignment of dealing with problems when and wherever they arise. Your considerations should be:

1. Do I encourage all team members to monitor the project, to be alert to problem areas?

2. Does the team have the capabilities required to answer problems as they arise? Or must I go outside for needed expert help?

3. If problems arise, can we solve them quickly?

4. Are there any potential problems that could kill the project?

5. Do I take immediate corrective action? Or do I let problems slide?

6. Do we have sufficient troubleshooting capability?

PROJECT LIFE CYCLE

The concept of a project life cycle[2] provides a useful framework for looking at project dynamics over time. The idea is familiar to most managers; it is used to conceptualize work stages and the budgetary and organizational resource requirements of each stage.[3] As Figure 29.2 shows,[4] this frame of reference divides projects into four distinct phases of activity:

1. *Conceptualization.* The initial project stage. Top managers determine that a project is necessary. Preliminary goals and alternative project approaches are specified, as are the possible ways to accomplish these goals.

2. *Planning.* The establishment of formal plans to accomplish the project's goals. Activities include scheduling, budgeting, and allocation of other specific tasks and resources.

3. *Execution.* The actual "work" of the project. Materials and resource are procured, the project is produced, and performance capabilities are verified.

4. *Termination.* Final activities that must be performed once the project is completed. These include releasing resources, transferring the project to clients, and, if necessary, reassigning project team members to other duties.

As Figure 29.2 shows, the project life cycle is useful for project managers because it helps define the level of effort needed to perform the tasks associated with each stage. During the early stages, requirements are minimal. They increase rapidly during late planning and execution, and they diminish during termination. Project life cycles are also helpful because they provide a method for tracking the status of a project in terms of its development.

Take another look at Figure 29.1. In theory, the factors are sequenced logically rather than randomly. For example, it is important to set goals or define the mission and benefits of the problem before seeking top management support. Similarly, unless consultation with clients occurs early in the process, chances of subsequent client acceptance

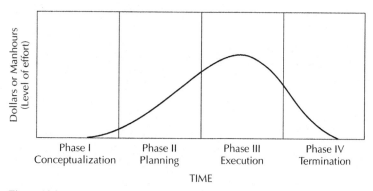

Figure 29.2. Stages in the project life cycle.

will be lowered. In actual practice, however, considerable overlap can occur among the various factors and their sequencing is not absolute. The arrows in the model represent information flows and sequences, not causal or correlational relationships.

In addition to the seven factors that can be laid out on a sequential, critical path, three factors are hypothesized to play a more overriding role in the project implementation. These factors—monitoring and feedback, communication, and troubleshooting—must all necessarily be present at each point in the implementation process. Further, a good argument could be made that these three factors are essentially different facets of the same general concern (i.e., project communication). Communication is vital for project control, for problem solving, and for maintaining beneficial contacts with both clients and the rest of the organization.

Strategy and Tactics

As one moves through the ten-factor model, it becomes clear that the general characteristics of the factors change. The first three—mission, top management support, and schedule—are related to the "planning" phase of project implementation. The other seven are concerned with the actual implementation, or "execution," of the project. These planning and execution elements, respectively, can usefully be considered strategic—the process of establishing overall goals and of planning how to achieve those goals—and tactical—using human, technical, and financial resources to achieve strategic ends. Briefly, the critical success factors of project implementation fit into a strategic-tactical breakout in the following way:

- Strategic—mission, top management support, project schedule or plans
- Tactical—client consultation, personnel, technical tasks, client acceptance, monitoring and feedback, communication, troubleshooting

Strategy and Tactics over Time

While both strategy and tactics are essential for successful project implementation, their importance shifts as the project moves through its life cycle. Strategic issues are most important at the beginning; tactical issues gain in importance toward the end. There should be continuous interaction and testing between the two: Strategy often changes in a dynamic corporation, so regular monitoring is essential. Nevertheless, a successful project manager must make the transition between strategic and tactical considerations as the project moves forward.

Our research in studying over 400 projects shows that the importance of strategy lessens over the project life cycle but *strategy never becomes unimportant*.[5] Similarly, the importance of tactics starts out low and grows as the project moves through its life cycle.

These changes have important implications. A project manager who is a brilliant strategist but an ineffective tactician has a strong potential for committing certain types of errors as the project moves forward. These errors may occur after substantial resources have been expended. In contrast, the project manager who is excellent at tactical execution but weak in strategic thinking has a potential for committing different kinds of errors. These will more likely occur early in the process, but may remain undiscovered because of the manager's effective execution.

STRATEGIC AND TACTICAL PERFORMANCE[6]

Figure 29.3 shows the four possible combinations of strategic and tactical performance, and the kinds of problems likely to occur in each scenario.[7] The values "High" and "Low" represent strategic and tactical quality—that is, effectiveness of operations performed.

A Type I error occurs when an action that should have been taken was not. Consider a situation in which strategic actions are adequate and suggest development and implementation of a project. A Type I error has occurred if tactical activities are inadequate, little action is subsequently taken, and the project is not developed.

A Type II error happens if an action is taken

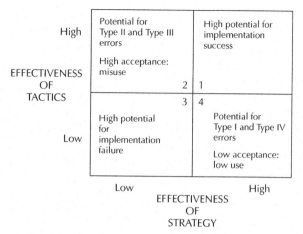

Type I error: not taking an action when one should be taken
Type II error: taking an action when none should be taken
Type III error: taking the wrong action (solving the wrong problem)
Type IV error: addressing the right problem, but solution is not used

Figure 29.3. Strategy-tactics effectiveness matrix.

when it should not have been. In practical terms, a Type II error is likely to occur if the project strategy is ineffective or inaccurate, but goals and schedules are implemented during the tactical stage of the project anyway.

A Type III error can be defined as solving the wrong problem, or "effectively" taking the wrong action. In this scenario, a problem is identified or a project is desired, but because of a badly performed strategic sequence, the wrong problem is isolated so the implemented project has little value—it does not address the intended target. Such situations often involve large expenditures of human and budgetary resources (tactics), for which there is inadequate initial planning and problem recognition (strategy).

Type IV is the final kind of error common to project implementation: The action taken does solve the right problem, but the solution is not used. That is, if project management correctly identifies a problem, proposes an effective solution, and implements the solution using appropriate tactics—but the project is not used by the client for whom it was intended—then a Type IV error has occurred.

As Figure 29.3 suggests, each of these errors is most likely to occur given a particular set of circumstances:

Cell 1: High Strategy/High Tactics. Cell 1 is the setting for projects rated effective in carrying out both strategy and tactics. Not surprisingly, most projects in this situation are successful.

Cell 3: Low Strategy/Low Tactics. The reciprocal of the first is the third cell, where both strategic and tactical functions are inadequately performed. Projects in this cell have a high likelihood of failure.

The results of projects in the first two cells are intuitively obvious. Perhaps a more intriguing question concerns the likely outcomes for projects found in the "off diagonal" of Figure 29.3, namely, high strategy/low tactics and low strategy/high tactics.

Cell 4: High Strategy/Low Tactics. In Cell 4, the project strategy is effectively developed but subsequent tactics are ineffective. We would expect projects in this cell to have a strong tendency toward "errors of inaction," such as low acceptance and low use by organization members or clients for whom the project was intended. Once a suitable strategy has been determined, little is done in the way of tactical follow-up to operationalize the goals of the project or to "sell" the project to its prospective clients.

Cell 2: Low Strategy/High Tactics. The final

cell reverses the preceding one. Here, project strategy is poorly conceived or planning is inadequate, but tactical implementation is well managed. Projects in this cell often suffer from "errors of action." Because of poor strategy, a project may be pushed into implementation even though its purpose has not been clearly defined. In fact, the project may not even be needed. However, tactical follow-up is so good that the inadequate or unnecessary project is implemented. The managerial attitude is to "go ahead and do it"; not enough time is spent early in the project's life assessing whether the project is needed and developing the strategy.

Case Studies

The four examples shown in Figure 29.4 are based on actual experiences of project managers or those involved in projects. The figure shows how the PIP can be used in real-world situations.[8]

High Strategy/High Tactics: The New-Alloy Development. One department of a large corporation was responsible for coordinating the development and production of new stainless-steel alloys for the automotive-exhaust market. This task meant overseeing the efforts of the metallurgy, research, and operations departments. The project grew out of exhaust-component manufacturers' demands for more formable alloys.

As Figure 29.4(1) demonstrates, the scores for this project, as assessed by the project team member, were uniformly high across the ten critical success factors. The project's high priority was communicated to all personnel, and this led to a strong sense of project mission and top management support. The strategy was clear and was conveyed to all concerned parties, including the project team, which was actively involved in early planning meetings. Because the project team would include personnel from the research, metallurgy, operations, production, and commercial departments, great care was taken in its selection and coordination.

In the new-alloy development project, a strong, well-conceived strategy was combined with highly competent tactical follow-up. The seeds of project success were planted during the conceptual and planning stages and were allowed to grow to their potential through rigorous project execution. Success in this project can be measured in terms of technical excellence and client use, as well as project team satisfaction and commercial profitability. In a recent follow-up interview, a member of a major competitor admitted that the project was so successful that the company still has a virtual lock on the automotive-exhaust market.

Low Strategy/Low Tactics: The Automated Office. A small, privately owned company was attempting to move from a nonautomated paper system to a fully integrated, automated office that would include purchasing, material control, sales order, and accounting systems. The owner's son, who had no previous experience with computers, was hired as MIS director. His duties consisted of selecting hardware and software, directing installation, and learning enough about the company to protect the family's interests. Figure 29.4(2) shows a breakdown of the ten critical success factors, as viewed by a project team member.

Several problems emerged immediately. Inadequate "buy-in" on the part of organization members, perceived nepotism, and lack of interaction with other top managers in purchasing decisions were seen as problems while the project was still in its strategy phase. A total lack of a formal schedule or implementation plan emphasized other strategic inadequacies destined to lead to tactical problems as well.

Tactically, the project was handled no better. Other departments that were expected to use the system were not consulted about their specific needs; the system was simply forced upon them. Little effort was made to develop project control and troubleshooting mechanisms, perhaps as a direct result of inadequate scheduling.

Project results were easy to predict. As the team members indicated and Figure 29.4(2) reinforces, the project was over budget, behind schedule, and coolly received—all in all, an expensive failure. The owner's son left the company, the manager of the computer department was demoted, the mainframe computers were found to be wholly inadequate and were sold, and upper management forfeited a considerable amount of employee goodwill.

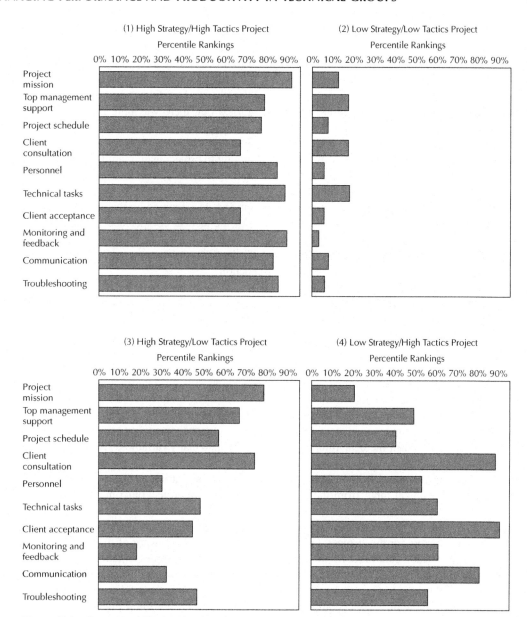

Figure 29.4. Examples of PIP uses in the real world.

High Strategy/Low Tactics: The New Bank Loan Setup. The purpose of this project was to restructure the loan procedures at a major bank. The project was intended to eliminate duplicate work done by branches and the servicing department and to streamline loan processes. These goals were developed and strongly supported by upper manage-ment, which had clearly conveyed them to all con-cerned parties. The project was kicked off with a great deal of fanfare; there was a high expectation of speedy and successful completion. Trouble started when the project was turned over to a small team that had not been privy to the initial planning, goal-setting, and scheduling meetings. In fact, the project team leader was handed the project after only three months with the company.

Project tactics were inadequate from the beginning. The team was set up without any formal feedback channels and with few communication links with either the rest of the organization or top management. The project was staffed on an ad hoc basis, often with nonessential personnel from other departments. This staffing method resulted in a diverse team with conflicting loyalties. The project leader was never able to pull the team together. As the project leader put it, "Although this project hasn't totally failed, it is in deep trouble." Figure 29.4(3) illustrates the breakdowns for the project as reported by two team members. Almost from the start of its tactical phase, the project suffered from the team's inability to operationalize the initial goals. Whether or not it achieves its final performance goals, this project will be remembered with little affection by the rest of the organization.

Low Strategy/High Tactics: The New Appliance Development. A large manufacturing company initiated the development of a new kitchen appliance to satisfy what upper management felt would be a consumer need in the near future. The project was perceived as the pet idea of a divisional president and was rushed along without adequate market research or technical input from the R&D department.

Figure 29.4(4) shows the breakdowns of the ten critical success factors for this project. Organizational and project team commitment was low. Other members of upper management felt the project was being pushed along too fast and refused to get behind it. Initial planning and scheduling developed by the divisional president and staff were totally unrealistic.

What happened next was interesting. It was turned over to an experienced, capable manager who succeeded in taking the project, which had gotten off to such a shaky start, and successfully implementing it. He reopened channels of communication within the organization, bringing R&D and marketing on board. He met his revised sched-

ule and budget, using troubleshooting and control mechanisms. Finally, he succeeded in getting the product to the market in a reasonable time frame.

In spite of the project manager's effective tactics, the product did not do well in the market. As it turned out, there was little need for it at the time, and second-generation technology would make it obsolete within a year. This project was highly frustrating to project team members who felt, quite correctly, that they had done everything possible to achieve success. Through no fault of their own, this project was doomed by the poor strategic planning. All the tactical competence in the world could not offset the fact that the project was poorly conceived and indifferently supported, resulting in an "error of action."

IMPLICATIONS FOR MANAGERS

These cases, and the strategy-tactics effectiveness matrix, suggest practical implications for managers wishing to better control project implementation.

Use a Multiple-Factor Model

Project management is a complex task requiring attention to many variables. The more specific a manager can be regarding the definition and monitoring of those variables, the greater the likelihood of a successful project outcome. It is important to use a multiple-factor model to do this, first to understand the variety of factors affecting project success, then to be aware of their relative importance across project implementation stages. This chapter offers such a model: Ten critical success factors that fit into a process framework of project implementation; within the framework, different factors become more critical to project success at different points in the project life cycle.

Additionally, both the project team and clients need to perform regular assessments to determine the health of the project. The time for accurate feedback is when the project is beginning to develop difficulties that can be corrected, not down the road when the troubles have become insurmountable. Getting the project team as well as the clients to perform status checks has the benefit of giving insights from a variety of viewpoints, not just from that of the project manager. Furthermore, it reinforces the goals the clients have in mind, as well as their perceptions of whether the project satisfies their expectations.

Think Strategically Early in the Project Life Cycle

It is important to consider strategic factors early in the project life cycle, during conceptualization and planning, As a practical suggestion, organizations implementing a project should bring the manager and his or her team on board very soon. Many managers make the mistake of not involving team members in early planning and conceptual meetings, perhaps assuming that the team members should only concern themselves with their specific jobs. In fact, it is very important that both the manager and the team members "buy in" to the goals of the project and the means to achieve those goals. The more team members are aware of the goals, the greater the likelihood of their taking an active part in monitoring and troubleshooting.

Think More Tactically

In the later project stages, strategy and tactics are of almost equal importance to project implementation success. Consequently, it is important that the project manager shift the team's emphasis from "What do we do?" to "How do we want to do it?" The specific critical success factors associated with project tactics tend to reemphasize the importance of focusing on the "how" instead of the "what." Factors such as personnel, communication, and monitoring are concerned with better managing specific action steps in the implementation process. While it is important to bring the project team on board during the initial strategy phase, it is equally important to manage their shift into a tactical, action mode in which their specific duties help move the project toward completion.

Consciously Plan for and Communicate the Transition from Strategy to Tactics

Project monitoring will include an open, thorough assessment of progress at several stages of imple-

mentation. The assessment must acknowledge that the transition from a strategic to a tactical focus introduces an additional set of critical success factors.

Project managers should regularly communicate with team members about the shifting status or focus of the project. Communication reemphasizes the importance of a joint effort, and it reinforces the status of the project relative to its life cycle. The team is kept aware of the degree of strategic versus tactical activity necessary to move the project to the next life-cycle stage. Finally, communication helps the manager track the various activities performed by the project team, making it easier to verify that strategic vision is not lost in the later phases of tactical operationalization.

Make Strategy and Tactics Work

Neither strong strategy nor strong tactics by themselves ensure project success. When strategy is strong and tactics are weak, there is a potential for creating projects that never get off the ground. Cost and schedule overruns, along with general frustration, are often the side effects of projects that encounter "errors of inaction." On the other hand, a project that starts off with a weak or poorly conceived strategy and receives strong subsequent tactical operationalization is likely to be successfully implemented but address the wrong problem. New York advertising agencies can tell horror stories of advertising campaigns that were poorly conceived but still implemented, sometimes costing millions of dollars, and that were ultimately judged disastrous and scrapped.

In addition to having project strategy and tactics working together, it is important to remember (again following Figure 29.3) that strategy should be used to drive tactics. Strategy and tactics are not independent of each other. At no point do strategic factors become unimportant to project success; instead, they must be continually assessed and reassessed over the life of the project in light of new project developments and changes in the external environment.

PROJECT IMPLEMENTATION PROFILE

Now that you understand the general factors in the model, it is time to get some experience with the

Project Implementation Profile (Form 29.1; see Box).[9] Try the profile now on a current or past project. Each time you use it, you will improve your project management skills. For current projects you may wish to complete the profile once a month or so to track the results.

Form 29.1

PROJECT IMPLEMENTATION PROFILE

Project name: _____

Project manager: _____

Profile completed by: _____

Date: _____

   Briefly describe your project, giving its title and specific goals:

_____

_____

_____

_____

   Think of the project implementation you have just named. Consider the statements on the following pages. Using the scale provided, please circle the number that indicates the *extent* to which you agree or disagree with the following statements as they relate to activities occurring in the project about which you are reporting.

FACTOR 1—PROJECT MISSION

	Strongly Disagree		Neutral			Strongly Agree	
1. The goals of the project are in line with the general goals of the organization	1	2	3	4	5	6	7
2. The basic goals of the project were made clear to the project team	1	2	3	4	5	6	7
3. The results of the project will benefit the parent organization	1	2	3	4	5	6	7
4. I am enthusiastic about the chances for success of this project	1	2	3	4	5	6	7
5. I am aware of and can identify the beneficial consequences to the organization of the success of this project	1	2	3	4	5	6	7

Factor 1—Project Mission Total	

Form 29.1—*Continued*

FACTOR 2—TOP MANAGEMENT SUPPORT

	Strongly Disagree		Neutral			Strongly Agree
1. Upper management will be responsive to our requests for additional resources, if the need arises.	1	2	3	4	5	6	7
2. Upper management shares responsibility with the project team for ensuring the project's success	1	2	3	4	5	6	7
3. I agree with upper management on the degree of my authority and responsibility for the project.	1	2	3	4	5	6	7
4. Upper management will support me in a crisis.	1	2	3	4	5	6	7
5. Upper management has granted us the necessary authority and will support our decisions concerning the project.	1	2	3	4	5	6	7

Factor 2—Top Management Support Total	

FACTOR 3—PROJECT SCHEDULE/PLAN

	Strongly Disagree		Neutral			Strongly Agree
1. We know which activities contain slack time or slack resources that can be utilized in other areas during emergencies.	1	2	3	4	5	6	7
2. There is a detailed plan (including time schedules, milestones, manpower requirements, etc.) for the completion of the project.	1	2	3	4	5	6	7
3. There is a detailed budget for the project.	1	2	3	4	5	6	7
4. Key personnel needs (who, when) are specified in the project plan	1	2	3	4	5	6	7
5. There are contingency plans in case the project is off schedule or off budget.	1	2	3	4	5	6	7

Factor 3—Project Schedule/Plan Total	

Form 29.1—*Continued*

FACTOR 4—CLIENT CONSULTATION

	Strongly Disagree			Neutral			Strongly Agree
1. The clients were given the opportunity to provide input early in the project development stage	1	2	3	4	5	6	7
2. The clients (intended users) are kept informed of the project's progress.	1	2	3	4	5	6	7
3. The value of the project has been discussed with the eventual clients	1	2	3	4	5	6	7
4. The limitations of the project have been discussed with the clients (what the project is *not* designed to do)	1	2	3	4	5	6	7
5. The clients were told whether or not their input was assimilated into the project plan	1	2	3	4	5	6	7

Factor 4—Client Consultation Total	

FACTOR 5—PERSONNEL

	Strongly Disagree			Neutral			Strongly Agree
1. Project team personnel understand their role on the project team	1	2	3	4	5	6	7
2. There is a sufficient manpower to complete the project.	1	2	3	4	5	6	7
3. The personnel on the project team understand how their performance will be evaluated.	1	2	3	4	5	6	7
4. Job descriptions for team members have been written and distributed and are understood.	1	2	3	4	5	6	7
5. Adequate technical and/or managerial training (and time for training) are available for members of the project team	1	2	3	4	5	6	7

Factor 5—Personnel Total	

Form 29.1—*Continued*

FACTOR 6—TECHNICAL TASKS

	Strongly Disagree			Neutral		Strongly Agree	
1. Specific project tasks are well managed.	1	2	3	4	5	6	7
2. The project engineers and other technical people are competent	1	2	3	4	5	6	7
3. The technology that is being used to support the project works well.	1	2	3	4	5	6	7
4. The appropriate technology (equipment, training programs, etc.) has been selected for project success	1	2	3	4	5	6	7
5. The people implementing this project understand it	1	2	3	4	5	6	7

Factor 6—Technical Tasks Total	

FACTOR 7—CLIENT ACCEPTANCE

	Strongly Disagree			Neutral		Strongly Agree	
1. There is adequate documentation of the project to permit easy use by the clients (instructions, etc.)	1	2	3	4	5	6	7
2. Potential clients have been contacted about the usefulness of the project.	1	2	3	4	5	6	7
3. An adequate presentation of the project has been developed for clients.	1	2	3	4	5	6	7
4. Clients know who to contact when problems or questions arise	1	2	3	4	5	6	7
5. Adequate advance preparation has been done to determine how best to "sell" the project to clients.	1	2	3	4	5	6	7

Factor 7—Client Acceptance Total	

Form 29.1—*Continued*

FACTOR 8—MONITORING AND FEEDBACK

	Strongly Disagree		Neutral		Strongly Agree	

1. All important aspects of the project are monitored, including measures that will provide a complete picture of the project's progress (adherence to budget and schedule, manpower and equipment utilization, team morale, etc.).   1   2   3   4   5   6   7

2. Regular meetings to monitor project progress and improve the feedback to the project team are conducted   1   2   3   4   5   6   7

3. Actual progress is regularly compared with the project schedule   1   2   3   4   5   6   7

4. The results of project reviews are regularly shared with all project personnel who have impact upon budget and schedule.   1   2   3   4   5   6   7

5. When the budget or schedule requires revision, input is solicited from the project team   1   2   3   4   5   6   7

Factor 8—Monitoring and Feedback Total	

FACTOR 9—COMMUNICATION

	Strongly Disagree		Neutral		Strongly Agree	

1. The results (decisions made, information received and needed, etc.) of planning meetings are published and distributed to applicable personnel.   1   2   3   4   5   6   7

2. Individuals/groups supplying input have received feedback on the acceptance or rejection of their input.   1   2   3   4   5   6   7

3. When the budget or schedule is revised, the changes *and* the reasons for the changes are communicated to all members of the project team   1   2   3   4   5   6   7

4. The reasons for the changes to existing policies/ procedures are explained to members of the project team, other groups affected by the changes, and upper management.   1   2   3   4   5   6   7

5. All groups affected by the project know how to make problems known to the project team.   1   2   3   4   5   6   7

Factor 9—Communication Total	

Form 29.1—*Continued*

FACTOR 10—TROUBLESHOOTING

	Strongly Disagree		Neutral		Strongly Agree
1. The project leader is not hesitant to enlist the aid of personnel not involved in the project in the event of problems	1　2	3	4	5	6　7
2. Brainstorming sessions are held to determine where problems are most likely to occur	1　2	3	4	5	6　7
3. In case of project difficulties, project team members know exactly where to go for assistance	1　2	3	4	5	6　7
4. I am confident that problems that arise can be solved completely.	1　2	3	4	5	6　7
5. Immediate action is taken when problems come to the project team's attention	1　2	3	4	5	6　7

Factor 10—Troubleshooting Total	

Project Performance

	Strongly Disagree		Neutral		Strongly Agree
1. This project has/will come in on schedule	1　2	3	4	5	6　7
2. This project has/will come in on budget.	1　2	3	4	5	6　7
3. The project that has been developed works (or, if still being developed, looks as if it will work).	1　2	3	4	5	6　7
4. The project will be/is used by its intended clients	1　2	3	4	5	6　7
5. This project has directly benefited/will directly benefit the intended users, through either increasing efficiency or employee effectiveness.	1　2	3	4	5	6　7
6. Given the problem for which it was developed, this project seems to do the best job of solving that problem—i.e., it was the best choice among the set of alternatives	1　2	3	4	5	6　7
7. Important clients, directly affected by this project, will make use of it.	1　2	3	4	5	6　7
8. I am/was satisfied with the process by which this project is being/was completed.	1　2	3	4	5	6　7

Form 29.1—*Continued*

	Strongly Disagree		Neutral			Strongly Agree	
9. We are confident that nontechnical start-up problems will be minimal, because the project will be readily accepted by its intended users	1	2	3	4	5	6	7
10. Use of this project has led/will lead directly to improved or more effective decision making or performance for the clients.	1	2	3	4	5	6	7
11. This project will have a positive impact on those who make use of it	1	2	3	4	5	6	7
12. The results of this project represent a definiteimprovement in performance over the way clients used to perform these activities	1	2	3	4	5	6	7

PROJECT PERFORMANCE TOTAL	

Percentile Scores

Now see how your project scored in comparison to a database of 409 projects. If you are below the 50th percentile on any factor, you may wish to devote extra attention to that factor.

Percentile Score	Raw Score				
% of Individuals Scoring Lower	Factor 1 Project Mission	Factor 2 Top Management Support	Factor 3 Project Schedule/ Plan	Factor 4 Client Consultation	Factor 5 Personnel— Recruitment, Selection, Training
100	35	35	35	35	35
90	34	34	33	34	32
80	33	32	31	33	30
70	32	30	30	32	28
60	31	28	28	31	27
50	30	27	27	30	24
40	29	25	26	29	22
30	28	23	24	27	20
20	26	20	21	25	18
10	25	17	16	22	14
0	7	6	5	7	5

Form 29.1—*Continued*

Percentile Score			Raw Score			
% of Individuals Scoring Lower	Factor 6 Technical Tasks	Factor 7 Client Acceptance	Factor 8 Monitoring and Feedback	Factor 9 Commu-nication	Factor 10 Trouble-shooting	Project Performance
100	35	35	35	35	35	84
90	34	34	34	34	33	79
80	32	33	33	32	31	76
70	30	32	31	30	29	73
60	29	31	30	29	28	71
50	28	30	29	28	26	69
40	27	29	27	26	24	66
30	26	27	24	24	23	63
20	24	24	21	21	21	59
10	21	20	17	16	17	53
0	8	8	5	5	5	21

Tracking Critical Success Factors Grid

After you have compared your scores, you can plot them below and mark any factors that need special effort.

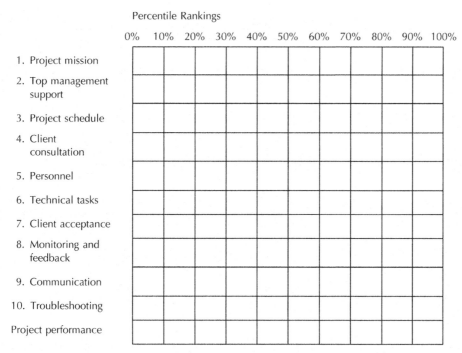

Percentile Rankings

0% 10% 20% 30% 40% 50% 60% 70% 80% 90% 100%

1. Project mission
2. Top management support
3. Project schedule
4. Client consultation
5. Personnel
6. Technical tasks
7. Client acceptance
8. Monitoring and feedback
9. Communication
10. Troubleshooting
Project performance

NOTES

1. Copyright © 1984 Randall L. Schultz and Dennis P. Slevin. Used with permission.

2. Much of the following discussion is reprinted from "Balancing Strategy and Tactics in Project Implementation," by Dennis P. Slevin and Jeffrey K. Pinto, *Sloan Management Review*, Fall 1987, pp. 33–41, by permission of the publisher. Copyright © 1987 by the Sloan Management Review Association. All rights reserved.

3. The four-stage project life cycle is based on J. Adams and S. Barndt, "Behavioral Implications of the Project Life Cycle," in David I. Cleland and William R. King, eds., *Project Management Handbook* (New York: Van Nostrand Reinhold, 1988), p. 211.

4. J. Adams and S. Barndt, "Behavioral Implications of the Project Life Cycle," in David I. Cleland and William R. King, eds., *Project Management Handbook* (New York: Van Nostrand Reinhold, 1988). Reprinted by permission.

5. Jeffrey K. Pinto, "Project Implementation: A De-termination of Its Critical Success Factors, Moderators, and Their Relative Importance Across Stages in the Project Life Cycle," unpublished doctoral dissertation, University of Pittsburgh, December 1986.

6. Much of this section is based on Randall L. Schultz, Dennis P. Slevin, and Jeffrey K. Pinto, "Strategy and Tactics in a Process Model of Product Implementation," *Interfaces*, Vol. 17, No. 3, May–June 1987, pp. 34–46. Copyright 1987, The Institute of Management Sciences, 290 Westminster Street, Providence, Rhode Island 02903.

7. Randall L. Schultz, Dennis P. Slevin, and Jeffrey K. Pinto, "Strategy and Tactics in a Process Model of Implementation," *Interfaces*, May–June 1987, p. 43.

8. Used with permission of Project Management Institute. Jeffrey K. Pinto and Dennis P. Slevin, "The Project Implementation Profile: New Tool for Project Managers," *Project Management Journal*, September 1986, Vol. 17, No. 4, pp. 57–70.

9. Copyright © 1987 Jeffrey K. Pinto and Dennis P. Slevin. Used with permission.

30
Measuring R&D Effectiveness

ROBERT SZAKONYI

Compare your department's performance of ten basic activities with the performance of an "average" R&D department.

Improving the effectiveness of R&D is the most important issue in R&D management. It sums up the major concerns of R&D managers such as: Are we selecting the right R&D projects? Are we managing our projects well? Are we contributing as much as we should to the company's businesses? Although R&D managers and researchers have attempted to measure R&D effectiveness for over 30 years, there are still no methods that are widely accepted for doing this. The crux of the problem seems to lie in the great difficulty in measuring R&D *output*. R&D managers have sensed correctly that any quantitative measures would focus, not on the quality of the R&D and its contribution to the company's businesses, but on countable items such as patents or citations in technical journals. Consequently, even though R&D managers have wanted to measure R&D effectiveness, they have not had an acceptable method for doing so.

As summarized in my review of the literature (see *Research-Technology Management,* March–April, 1994), methods for linking R&D to profits, sales, or other financial benefits, even given their shortcomings, can be helpful, provided everyone recognizes (1) that there are major limitations in any method for translating R&D output into financial payoffs, and (2) that there is a certain artificiality in isolating the contributions of R&D to profits, and so on. The methods have been useful in providing a common frame of reference for how and how much R&D influences a company's businesses. Methods for evaluating R&D output that are based on judgments about the success of individual R&D projects also can be useful. As long as the people involved are aware of (1) problems of converting qualitative judgments into rankings and (2) problems related to the relative weights of different variables, these methods can help clarify how much various R&D projects have accomplished. Measuring R&D output in terms of how many patents, publications, or citations to publications were produced is not very useful. Although such measures may at times reveal interesting patterns regarding the progress of technical work, they cannot be thought of as legitimate indicators of R&D output.

Where all the methods for measuring R&D output fall short, however, is that they really do not measure R&D effectiveness. *R&D output and R&D effectiveness are not the same thing.* As a result, the purpose of this article is to present a new approach to measuring R&D effectiveness, an approach that overcomes major shortcomings of past efforts. Such a system should require as little qualitative judgment as possible. Evaluators should be able to identify the presence or absence of particular features, rather than trying to judge their performance. It should also provide a logical set of benchmarks so that a company can compare itself against the experiences of others.

The framework of my approach to measuring R&D effectiveness consists of the following ten R&D activities: (1) selecting projects; (2) planning and managing projects; (3) generating new product ideas; (4) maintaining the quality of the R&D process and methods; (5) motivating technical people; (6) establishing cross-disciplinary teams; (7)

Reprinted with permission from *Research-Technology Management,* May–June, 1994.
Robert Szakonyi is Director of the Center on Technology Management at Chicago's IIT Center.

coordinating R&D and marketing; (8) transferring technology to manufacturing; (9) fostering collaboration between R&D and finance; and (10) linking R&D to business planning. The substance of this new approach consists of 60 examples of how an R&D department performs at each of six levels with regard to these ten R&D activities. The examples are drawn from my experience in consulting or doing research from 1978 to 1992 at over 300 companies in 27 different industries. In the tables that follow, each R&D activity level is scored from 0 to 5. These scores could be used to derive overall benchmark scores, so that R&D departments can measure themselves against my "average" as a whole rather than against each individual activity.

Although more than just the R&D department must be effective for a company to do well in these activities, the *focus* in this approach to measuring R&D effectiveness is on the R&D department's operations. Obviously, a department cannot link its R&D effectively to business planning if there are no business plans. Similarly, an R&D department's relations with the marketing, manufacturing, or finance department are not one-way. If the other department refuses to cooperate, then the R&D department is hampered. A low score on these activities, therefore, does not necessarily mean that an R&D department is operating poorly. What it definitely does mean is that the company needs improvement with regard to the effectiveness of R&D. Finally, although all 10 activities are important, they do not necessarily carry the same weight. Each company can choose the relative weight it wants to place on each activity as well as how such weights might vary over time for any particular business.

1. SELECTING R&D

Level A: Issue is not recognized. Although every R&D department recognizes that the projects it selects should serve the company's interests, some R&D departments have their own unique perception of what those interests are. For example, the R&D department at one aerospace company picked projects to build up the technical skills of its engineers instead of picking projects that the company needed.

Level B: Initial efforts are made toward addressing issue. An R&D department of an automotive company recognized that its projects should serve the company's needs, but it lacked the skills to select projects in a particular technical area. Although it had skills in mechanical engineering and selected well in this area, it lacked skills in electronics engineering. In this area, the department picked projects that were more appropriate for a university than an industrial laboratory.

Level C: Right skills are in place. An R&D department of an instrument company that had traditionally been oriented toward mechanical engineering had weaknesses in electronics engineering, but gradually built up skills in this area until it was able to select projects competently in both areas. The department had other problems, however, because it lacked methods for clarifying priorities among projects.

Level D: Appropriate methods are used. A telecommunication company's R&D department developed methods for selecting projects competently for its service businesses and for its businesses that sold products. It also had methods for clarifying priorities among these two sets of projects. Nonetheless, this R&D department still had difficulties integrating its projects into company businesses because it picked these projects on its own; i.e., without the help of the marketing department.

This is the level at which the *average R&D department* operates. The average department has the technical skills to pick projects in the required areas; it also has developed reasonably workable procedures for selecting projects. However, the average department usually picks its projects with inadequate input from the managers of the other functions in the company.

Level E: Responsibilities are clarified. An R&D department of a consumer products company overcame the difficulties of the average R&D department by working with its company's operating divisions. Over a couple years, the department worked out ways of sharing responsibility for selecting projects. A few years later, however, relations with the operating divisions returned to their previous state. After defining their responsibilities clearly, the managers of R&D and of the operating divisions had allowed their relations to stagnate.

Level F: Continuous improvement is underway. An R&D department of a pharmaceutical company and the operating divisions had exceptional working relations in selecting R&D projects. They also worked hard at maintaining those relations. For example, both parties were committed to having the company undertake a significant amount of exploratory research. Together they looked for better ways of evaluating how much exploratory research to support, such as models for determining the optimal amount of long-range research.

2. PLANNING AND MANAGING PROJECTS

Level A: Issue is not recognized. In order to plan and manage projects, they must first exist;

yet, this was not the case at the R&D department of a steel company. Technical work was done by the technical people more or less as they saw fit, except when they had to do it in response to requests from the operating divisions for technical services. R&D done in this company seldom had clear objectives; it also was not evaluated in terms of meeting technical milestones, which did not even exist.

Level B: Initial efforts are made toward addressing issue. An R&D department of a consumer products company had projects with objectives, but it still had major difficulties in executing them. This was because the members of the R&D department did not understand planning and tracking of projects. In this company, R&D project planning was done poorly, if at all. Furthermore, the project plans that did exist were not taken seriously.

Level C: Right skills are in place. In an R&D department of a pulp and paper company, the technical people had the skills to plan, but project planning was still poor. The reason for this was that the technical people did not use the methods for project planning that had been worked out. Even though there was a manual on project planning, they chose to ignore it.

TABLE 30.1
Activity 1—Selecting R&D

		Points
Level A (Not recognized)	R&D department blindly picks projects to build up technical skills instead of selecting projects that are needed (*aerospace*).	0
Level B (Initial efforts)	Has skills in mechanical engineering, but lacks skills to select in electronics engineering (*automotive*).	1
Level C (Skills)	Has skills in mechanical and electronics engineering, but lacks methods to clarify priorities of projects (*instruments*).	2
Level D (Methods)	Develops methods to select projects for service businesses as well as businesses that sell products, but picks projects on own without marketing (*telecommunications*).	3
Level E (Responsibilities)	R&D department works with operating divisions to compare priorities and jointly select projects, but after a couple of years both parties lose their sense of mission (*consumer products*).	4
Level F (Continuous improvements)	R&D department and operating divisions look at various models for evaluating payoffs from exploratory research so as to find better ways of determining how much far-out research to do (*pharmaceutical*).	5

☐ = Average R&D department.

Level D: Appropriate methods are used. In an R&D department of an instruments company, projects were planned well in terms of objectives, schedules, and costs. Difficulties still arose in the execution of these projects, however, because the company lacked "rules of the road" concerning who was responsible for a project. In this company, the role of a project manager was unclear. In addition, there were no guidelines on how conflicts between projects should be resolved.

The *average R&D department* operates at this level with regard to planning and managing R&D projects. Over the last 30 years, most R&D departments have realized how important project planning and management is and have established procedures for planning and managing projects. The average R&D department, however, has not worked out clearly what the respective responsibilities of a project manager and of a line manager in R&D are for carrying out projects.

Level E: Responsibilities are clarified. An R&D department of a metals company planned projects well and defined the responsibilities of its project managers clearly. Although successful, this R&D department still had weaknesses. One weakness was that it did not seek further improvements in how pro-

jects were managed, such as in how well the senior business managers understood project results.

Level F: Continuous improvement is underway. An R&D department of a computer company not only mastered the challenges of project planning and management, but also took initiatives to improve the capabilities of its technical people. It fostered training to sharpen their planning and technical skills, and it developed better tools for planning projects, thereby shortening the duration of its projects significantly.

3. GENERATING NEW PRODUCT IDEAS

Level A: Issue is not recognized. Although technical people are not the only ones who should be responsible for generating new product ideas, they certainly should be among the leaders in this area. In the R&D department of a food processing company, however, the technical people had to be led to develop new product concepts. These people could handle the technical work required after a new project concept was developed, but they were not aware of their responsibility for developing these concepts.

TABLE 30.2
Activity 2—Planning and Managing Projects

		Points
Level A (Not recognized)	Projects do not exist; technical work is done without clear objectives; milestones, or accountability (*steel*).	0
Level B (Initial efforts)	R&D wants to plan, but does not understand what planning and tracking of projects consists of (*consumer products*).	1
Level C (Skills)	Has skills to plan, but cannot get planning methods accepted (*pulp and paper*).	2
Level D (Methods)	Projects are planned, but responsibilities of project managers within a project and coordination between projects are lacking (*instruments*).	3
Level E (Responsibilities)	Projects are planned well, but improvements, such as improving senior business management's understanding of projects, are not sought (*metals*).	4
Level F (Continuous improvements)	Training is fostered in the R&D department to sharpen technical people's planning and technical skills in order to shorten projects (*computer*).	5

◻ = Average R&D department.

Level B: Initial efforts are made toward addressing issue. In a steel company's R&D department, people wanted to generate new product ideas but lacked experience with the company businesses, so that they were not able to generate relevant ideas. These R&D people also lacked the skills to think creatively about businesses outside the company's traditional interests.

Level C: Right skills are in place. In an R&D department of a chemical company, the technical people had the skills to develop new product concepts and developed new ideas, but most of these ideas were lost because the department did not have procedures for collecting and evaluating suggestions about new products.

This is the level at which the *average R&D department* operates. Many of the technical people have the skills to develop new product ideas and they do develop new ideas, but the average R&D department has poor methods for capturing and evaluating these ideas. As a result, many promising new product ideas are lost.

Level D: Appropriate methods are used. In an R&D department of a tire and rubber company,

mechanisms were developed to instill an innovative climate, including procedures to capture and evaluate new product ideas. However, the department still had major problems in getting new product ideas utilized because: (1) The individual suggesters did not understand their responsibilities, and (2) The staff responsible for fostering innovation did not understand its duties. The individual suggesters did not realize that they would need to take certain steps to ensure that their ideas were practical, while those responsible for fostering innovation did not know how to develop support for new ideas.

Level E: Responsibilities are clarified. An R&D department of a building materials company developed a system for effectively integrating the generation of new product ideas into mainstream activities. It did this by incorporating the process of generating ideas into the planning process. New product ideas were elicited and then evaluated within the context of defining the direction for the company. Although this process worked well at the conception of a product, this company still had problems because it was not fully committed to innovation. It reached the level of fostering new

TABLE 30.3
Activity 3—Generating New Product Ideas

		Points
Level A (Not recognized)	Technical people have to be led to develop new product ideas (*food processing*).	0
Level B (Initial efforts)	R&D people want to generate new ideas, but lack the business experience to generate new ideas applicable to new businesses or the skills to think creatively (*steel*).	1
Level C (Skills)	R&D department has new ideas, but not the methods to capture and evaluate them (*chemicals*).	2
Level D (Methods)	Has mechanisms to instill a new innovative climate, but does not understand duties of the innovation staff or responsibilities of individual suggesters (*tire and rubber*).	3
Level E (Responsibilities)	Develops a system for integrating suggestions about new product concepts into the planning process, but cannot maintain a commitment toward new ideas (*building materials*).	4
Level F (Continuous improvements)	Pushes not only to develop new product ideas, but looks at them from the perspective of the customers; studies customers' behavior (*office equipment*).	5

☐ = Average R&D department.

ideas, but it did not maintain a commitment to supporting the ideas.

Level F: Continuous improvement is underway. An R&D department at an office equipment company not only developed new product ideas but also worked on improving how it handled new ideas. This department took actions to look at its customers' needs from the perspective of the customers. It also conducted studies of the behavior of people who work in offices so that it could identify customer needs that were not even apparent to the customer. Finally, it made conscious efforts to explain all of its new ideas to key customers.

4. MAINTAINING QUALITY OF R&D PROCESS/METHODS

Level A: Issue is not recognized. Engineers in an R&D department of a petroleum equipment company were unaware of how the poor quality of their designs contributed to problems in manufacturing products. Rather than using analytical tools to test the quality of their designs, they passed the designs on as they saw fit. Design flaws were found by manufacturing people, who then sent a flood of engineering change requests back to the R&D department.

Level B: Initial efforts are made toward addressing issue. At a household products company, the technical personnel wanted to improve the quality of their R&D but lacked detailed knowledge of the company's products. Without this knowledge, they found it difficult to design experiments and tests so as to meet the required quality standards.

Level C: Right skills are in place. At a photographic equipment company, the technical personnel had the training to do quality work. The problem was that they did not use quality tools (e.g., fishbone diagrams or quality function deployment) for increasing the rigor of R&D, except when these tools were absolutely required. Typically, these technical personnel felt that because they were under so

much pressure to meet deadlines they did not have the time to use the quality tools.

This is the level at which the *average R&D department* operates. Most of the technical personnel in the average department have the skills to do quality work and have had some training in tools for improving quality. In practice though, they do not use these tools. The pressures to meet schedules usually seem so large to them that they do not believe they can afford the extra time required to use quality tools.

Level D: Appropriate methods are used. An R&D department of a chemical company overcame many of the problems in getting quality improvement methods used. Still, although these methods were used, they were not always used as effectively as they should have been. One of the reasons for this was that no one in the R&D department was designated as being responsible for seeing that such methods were used regularly.

Level E: Responsibilities are clarified. An R&D department at another chemical company made major progress in getting quality tools used. A knowledgeable and experienced technical person was assigned the tasks of: (1) Finding the best quality tools available within and outside the company, and (2) Teaching technical personnel how the tools could be used. Many of the people began using these tools regularly. However, the person who was responsible for quality was still not successful in inducing the technical people to make these tools their own. For instance, he was not able to persuade them to improve the tools by modifying them to fit the purposes of their work.

Level F: Continuous improvement is underway. The technical people in an R&D department of one semiconductor company took the lead in seeking ways to improve R&D. For example, they looked for commonalities across the designs produced in various engineering groups, so that fewer designs would need to be done from scratch. By finding common building blocks across designs, they were able to eliminate many errors, to develop more robust designs, and to save time.

TABLE 30.4
Activity 4—Maintaining Quality of R&D Processes and Methods

		Points
Level A (Not recognized)	Technical people do not understand that their own designs have flaws (*petroleum equipment*).	0
Level B (Initial efforts)	Wants to improve, but lacks detailed knowledge of the products (*household products*).	1
Level C (Skills)	Has the technical training to do quality work, but does not use the methods (e.g., fishbone diagram) for ensuring quality (*photographic equipment*).	2
Level D (Methods)	Working at improving quality within the lab, but has not designated anyone to see that it happens (*chemical*).	3
Level E (Responsibilities)	Quality tools are used, but technical people have still not made them their own (*chemical*).	4
Level F (Continuous improvements)	Technical people take the lead in looking for commonalities across designs so that fewer designs need to be done from scratch (*semiconductors*).	5

☐ = Average R&D department.

5. MOTIVATING TECHNICAL PEOPLE

Level A: Issue is not recognized. R&D managers at a chemical company lacked an understanding of how to manage technical people. Instead of recognizing that technical people need to be motivated to do good work, they managed their staff autocratically, telling people what they wanted done and how to do it. Consequently, the creative output of the staff suffered.

Level B: Initial efforts are made toward addressing issue. At an automotive company, the R&D managers recognized that they needed to motivate their staff better. Specifically, they knew that they had to get the staff to understand how they fit in the company. These R&D managers believed that when the technical people understood this, they would be more inspired to do creative work and understand better what they should contribute. Although these managers recognized what was needed, they did not find ways to accomplish this.

This is the level at which the *average R&D department* operates. The managers understand that they must motivate their staff, and, in general, they carried out technical projects earlier in their careers. However, understanding that something needs to be done is not the same as knowing how to do it. R&D managers, on the whole, are more capable in technical areas than in management. Usually their entire education and most of their working experience have involved mainly technical issues. Therefore, the challenge of motivating other people is often something with which they have had little experience and about which they are either uncomfortable or unsure of themselves.

Level C: Right skills are in place. At an appliance company, the R&D managers made progress in motivating technical people. They were able to get them to be more creative and more proactive in seeking out new opportunities. What these managers still needed to find were methods of performance evaluation to solidify the progress they had made in motivating their staff. In other words, they needed formal procedures to reinforce their own personal initiatives to motivate people.

Level D: Appropriate methods are used. At an oil company, the R&D managers put in place procedures for motivating technical people and for achieving technical excellence. Although major progress was made, further progress was hampered by the practice of middle-level R&D managers

TABLE 30.5
Activity 5—Motivating Technical People

		Points
Level A (Not recognized)	R&D managers manage technical people autocratically (*chemical*).	0
Level B (Initial efforts)	R&D managers recognize that technical people do not understand how they fit in the company, but have not found ways to correct this (*automotive*).	1
Level C (Skills)	Has made progress in encouraging technical people to be more creative and proactive, but still needs methods of performance evaluation to solidify progress (*appliances*).	2
Level D (Methods)	Has procedures in place for achieving technical excellence, but middle-level R&D managers hamper progress by making too many of the technical decisions involving projects themselves (*oil*).	3
Level E (Responsibilities)	Has succeeded in instituting a new system of rewards revolving around R&D project management, but still needs to deal with "political" problems in the lab related to authority to certain R&D groups (*pharmaceutical*).	4
Level F (Continuous improvements)	Develops a culture that makes R&D managers push responsibility downward, thus allowing technical people to expand their jobs (*instruments*).	5

☐ = Average R&D department.

making too many of the technical decisions about projects. Although these R&D managers supported the idea of giving technical people more authority to take initiatives, they found it difficult to trust their judgment about project issues.

Level E: Responsibilities are clarified. At a pharmaceutical company, the R&D managers motivated technical personnel by instituting a system of rewards revolving around project management. Project managers were given the authority to make major decisions regarding their projects, and they were rewarded for doing this. As part of this system, the responsibilities of middle-level R&D managers were also clarified so that they did not interfere too much with projects. Although this system was effective, other issues still needed to be addressed and were not. One issue was that some of the middle-level managers still did not accept giving certain R&D groups or project managers authority and thus created "political" problems in the laboratory.

Level F: Continuous improvement is underway. The R&D managers at an instrument com-

pany developed a culture that made middle-level R&D managers want to push responsibility downward, thus allowing technical people to expand their jobs and seek out opportunities on their own. In order to maintain a culture such as this in the laboratory, the R&D managers had to continually seek ways of improving everyone's understanding of what was required to get creative technical work done.

6. ESTABLISHING CROSS-DISCIPLINARY TEAMS

Level A: Issue is not recognized. A food processing company's R&D managers had great difficulty in getting cross-disciplinary—more accurately, cross-functional—teams accepted. Allegiance to the R&D department was so strong that many technical people did not appreciate the benefits of teams. The same held true for members of other functions. Therefore, even though efforts at establishing teams were continually made, these teams almost always floundered.

Level B: Initial efforts are made toward addressing issue. At a computer company, there were many cross-disciplinary teams involving technical people. However, these teams were usually only moderately successful because the technical people lacked training in what being a representative on a team involved.

This is the level of development at which the *average R&D department* operates with regard to both cross-disciplinary and cross-functional teams, although the average R&D department has somewhat more success with the former. Technical people—and people in the rest of the company—usually do not have much knowledge about how one participates on a team. Often, they do not understand the purpose and benefits of a cross-disciplinary or cross-functional team and make only half-hearted commitments when involved with team activities.

Level C: Right skills are in place. At a health care products company, cross-functional teams were manned with competent people—people who understood how to participate on a team. Nevertheless, teams still had problems because the company lacked "rules-of-the-road" regarding the managing of teams and the running of projects. There were no guidelines on how teams should be

run or on what was expected from projects, nor was there a common understanding of how more than one team could draw upon common resources. Therefore, even though the people themselves were capable of executing well, the teams often did not produce the results desired.

Level D: Appropriate methods are used. At a photographic equipment company, there were teams that functioned well. People who were knowledgeable about how to work on teams had good procedures for coordinating their actions. The limitations of teams in this company stemmed from senior managers not giving teams enough authority to take actions. Therefore, even though the teams were capable of being very successful, they were handicapped in what they could do.

Level E: Responsibilities are clarified. At a health care products company, teams were very effective when the company was small. These teams were manned by capable people and were given sufficient authority. When the company grew larger, however, many of the conditions that led to well-functioning teams disappeared. From then on, team members cooperated less and the authority of the teams diminished.

TABLE 30.6
Activity 6—Establishing Cross-Disciplinary Teams

		Points
Level A (Not recognized)	Has difficulty getting the idea of cross-disciplinary teams accepted (*food processing*).	0
Level B (Initial efforts)	Wants to have teams, but technical people need to be trained to participate on teams (*computer*).	1
Level C (Skills)	Has teams manned with competent people, but lacks "rules-of-the-road" regarding the running of projects (*health care products*).	2
Level D (Methods)	Has teams that function well, but is handicapped because teams are not given enough authority (*photographic equipment*).	3
Level E (Responsibilities)	Had effective teams in the past, but does not know how to make them effective since the company has grown larger (*health care products*).	4
Level F (Continuous improvements)	The idea of cross-disciplinary teams is inculcated in everyone; many models for teams are available in company, and each team can choose its own model (*aerospace*).	5

☐ = Average R&D department.

TABLE 30.7
Activity 7—Coordinating R&D and Marketing

		Points
Level A (Not recognized)	R&D department does not think that it needs to work with marketing in developing new products (*aerospace*).	0
Level B (Initial efforts)	Technical people want better coordination with marketing, but lack the skills to analyze the business applications of a technical idea (*petroleum equipment*).	1
Level C (Skills)	Technical people know how to develop applications of a technology, but lack methods for working backward from a customer need to selecting technical projects (*chemical*).	2
Level D (Methods)	Works closely with marketing, but has difficulties in sorting out where responsibilities lie between technical concept and product concept (*food processing*).	3
Level E (Responsibilities)	Close coordination between R&D and marketing departments, but has not figured out how to develop new products effectively (*chemical*).	4
Level F (Continuous improvements)	Close coordination, with a former technical person in charge of marketing and taking the lead in technical marketing and new market development (*industrial equipment*).	5

☐ = Average R&D department.

Level F: Continuous improvement is underway. At an aerospace company, everyone understood cross-disciplinary teams, and the teams remained effective as the company grew. In this company, there were many models of what teams could be like. Members of a team could choose a model, (e.g., "Skunk Works"), adapt this model to their circumstances, and then modify their own model as a project progressed.

7. COORDINATING R&D AND MARKETING

Level A: Issue is not recognized. The R&D department at an aerospace company did not understand that it needed to work with the marketing department in developing new products. The senior technical managers of this company believed that their businesses were strictly technology-driven. Therefore, they ignored whatever input the marketing department tried to make regarding which technical projects were needed.

Level B: Initial efforts are made toward addressing issue. At a petroleum equipment company, the technical people recognized that their

R&D had to meet market needs. However, because many of these people lacked the skills to analyze the business applications of their technical ideas, they had great difficulty in communicating with the marketing people.

Level C: Right skills are in place. At a chemical company, the technical people knew how to develop applications of a technology. They could also communicate effectively with marketing people concerning the benefits of a technology they had developed. Where these technical people had problems, however, was in working backward from a customer need to selecting technical projects. They lacked methods for choosing the right technical work to do.

The *average R&D department* operated at this level. The technical people usually have the skills to communicate to marketing what the results of R&D can lead to. Their problem is in communicating with marketing people regarding how a market need could be filled. In other words, the average R&D department has not found methods of matching technical capabilities and market needs, except for those cases in which it can demonstrate the market advantages of a technology it has developed.

Level D: Appropriate methods are used. In a food processing company, the technical people and the marketing people worked out methods for communicating effectively. Nevertheless, problems still occurred in new product development. Many of these problems involved an unclear definition of responsibilities regarding the overlap between developing a technical concept and developing a product concept. There was a gray area between the conclusion of an R&D project that was technically successful and the start of a commercial development whose business prospects were promising but uncertain. The technical people and the marketing people in this company had difficulties sorting out who should deal with development issues.

Level E: Responsibilities are clarified. At a chemical company, the R&D and marketing departments worked in close coordination. R&D groups were responsible for the technical issues related to new product development; marketing groups were responsible for the long-term marketing issues. Both sides agreed on their responsibilities; yet, they still had difficulties. Although they worked together effectively, they had not figured out how to develop new products effectively. They had excellent communications, but they needed to improve the content of what they communicated.

Level F: Continuous improvement is underway. At an industrial equipment company, the communication between R&D and marketing was excellent. The person in charge of marketing was a former technical person who was as knowledgeable about the company's technologies as the person in charge of R&D. Together they developed new products for new markets; the senior R&D manager took the lead in developing the technology, while the senior marketing manager took the lead in technical marketing and in market development.

8. TRANSFERRING TECHNOLOGY TO MANUFACTURING

Level A: Issue is not recognized. At a telecommunications company, engineers did not consider manufacturing when they were doing their technical work, except at the end. During a project, they were preoccupied with a product's technical performance. Eventually, these engineers did seek input from the manufacturing people regarding producibility of the product. By then, however, it was too late to redesign the product so that it could be produced more easily and cheaply.

Level B: Initial efforts are made toward addressing issue. At a power generation equipment company, the R&D department wanted to have a better transfer of technology to manufacturing, but it lacked process engineering skills. Traditionally, the R&D department neglected issues related to manufacturing. Almost all of the technical people were oriented toward developing new products or conducting field services. Consequently, when the department transferred technology to manufacturing, it passed on designs that could not be manufactured easily.

Level C: Right skills are in place. Although the technical people at an automotive company had the skills to transfer technology, major problems remained in transferring technology. These technical people could not get manufacturing people to agree to the procedures for managing a technology transfer; e.g., one stage for prototype development, one stage for testing, etc.

This is the level at which the *average R&D department* operates. It is committed to transferring technology and it has the necessary skills. However, the average R&D department lacks the procedures for transferring technology to the manufacturing department, and a stalemate often occurs: The R&D has not been tested enough under manufacturing conditions, and the manufacturing people are reluctant to introduce a new technology into a plant until it has been tested enough.

Level D: Appropriate methods are used. A computer company's R&D department had the skills and methods for transferring technology, but problems still existed because the R&D and manufacturing departments disagreed about their responsibilities for testing and for documentation. Their

TABLE 30.8
Activity 8—Transferring Technology to Manufacturing

		Points
Level A (Not recognized)	Engineers do not consider manufacturing when doing technical work (*telecommunications*).	0
Level B (Initial efforts)	Wants better technology transfer, but lacks process engineering skills in the R&D department (*power generation equipment*).	1
Level C (Skills)	Technical people have the skills to transfer technology, but cannot develop methods with manufacturing to manage a phased transfer (*automotive*).	2
Level D (Methods)	Has methods for transferring technology, but there are disagreements about responsibilities for testing and documentation (*computer*).	3
Level E (Responsibilities)	Has a technical services group in the plant that is responsible for handling technology transfers, but this group normally focuses on current operations (*steel*).	4
Level F (Continuous improvements)	Not only has a group between R&D and manufacturing that aids technology transfer, but also tries to find new ways of integrating designs in order to link engineering and manufacturing more effectively (*semiconductors*).	5

☐ = Average R&D department.

differences of opinion concerned what kinds of tests should be conducted, when those tests should be conducted, what kinds of data were needed, when the data should be provided, and who was responsible for conducting tests and providing data at various stages during a transfer of technology.

Level E: Responsibilities are clarified. At a steel company, these problems of responsibility for a technology transfer were solved. The R&D department worked closely with a technical services group in the manufacturing plant, which was responsible for handling technology transfers. This group conducted the tests and made sure that all of the required data were provided. Although this group functioned very competently, it normally focused on maintaining the current operations of the plant. Thus, it did not seek better ways of transferring technology.

Level F: Continuous improvement is underway. At a semiconductor company, there was a group that worked with the engineering and manufacturing departments in transferring technology. This group not only aided technology transfers, but also tried to find new ways of integrating designs

in order to link engineering and manufacturing more effectively. This group consisted of design engineers from the engineering department and test and manufacturing engineers and manufacturing workers from the manufacturing department.

9. FOSTERING COLLABORATION BETWEEN R&D AND FINANCE

Level A: Issue is not recognized. At a health care products company, the R&D department did not recognize how poor were its relations with the finance department. Although the R&D department naturally was concerned about its budget and about the financial payoffs from its R&D, it understood little about finance and communicated with the finance department only when required to. In return, the finance department had grave doubts about the value of the R&D.

This is the level at which the *average R&D department* operates. Researchers know that finance is important because there are continually questions regarding the size of the R&D budget and because questions about the return on investment from R&D are inescapable. Nonetheless, most members of the average R&D department do not know much about the financial affairs of their company and have few interactions with members of the finance or accounting department. The finance department, on the other hand, knows little about what is going on in the R&D department.

Level B: Initial efforts are made toward addressing issue. At a food processing company, the R&D managers were interested in working better with finance managers, but lacking knowledge about discounted cash flow, depreciation rates, capital projections, corporate taxes, etc., they conducted their R&D in a partial vacuum. They also had great difficulty in explaining the value of R&D to finance managers and to financially oriented business managers.

Level C: Right skills are in place. The R&D managers at an aerospace company understood financial matters, but they had difficulty communicating with finance people because they sought, but could not find, methods for determining the financial benefits of R&D. Although these R&D managers understood concepts such as discounted cash flow and depreciation rates, they could not trans-

late technical results in a way that would reveal the financial benefits of R&D. In other words, they understood the technology and they more or less understood financial matters, but they could not bridge the gap.

Level D: Appropriate methods are used. At a chemical company, there were technically knowledgeable economic analysts who worked with R&D people in bridging this gap between technology and finance. These analysts had training in engineering economics and other financial disciplines, but also had worked with R&D for many years. Nevertheless, although R&D managers found the analysts to be helpful, there still were disagreements about when they should get involved in evaluating the worth of an R&D project and about what their responsibilities were for the evaluation of R&D.

Level E: Responsibilities are clarified. At a consumer products company, a finance person was transferred to the R&D department to serve as a bridge between R&D and finance. He worked closely with R&D managers to improve the financial management of the laboratory and he explained the value of R&D to finance managers. Nonethe-

TABLE 30.9
Activity 9—Fostering Collaboration Between R&D and Finance

		Points
Level A (Not recognized)	R&D department does not recognize how poor its relations with the finance or accounting department are (*health care products*).	0
Level B (Initial efforts)	R&D managers are interested in working better with finance, but lack knowledge about the financial affairs of the company (*food processing*).	1
Level C (Skills)	Understands financial matters, but lacks methods for determining the financial benefits of R&D (*aerospace*).	2
Level D (Methods)	Economic analysts work closely with R&D people, but there are disagreements about involvement and responsibilities (*chemical*).	3
Level E (Responsibilities)	A finance person is transferred to R&D to serve as a bridge with finance, but company's accounting procedures short-change benefits of technology (*consumer products*).	4
Level F (Continuous improvements)	R&D managers have option of discussing with finance manager how economic analyses of technology are conducted if it looks like strategic benefits of technology are neglected (*defense electronics*).	5

☐ = Average R&D department.

less, even with these improvements the value of R&D was not always appreciated in this company because its accounting procedures shortchanged the benefits of technology.

Level F: Continuous improvement is underway. A defense electronics company solved this problem of shortchanging the benefits of technology by giving R&D managers the option of discussing with the finance managers how economic analyses were conducted, if they thought that these analyses neglected the strategic benefits of technology. For example, if an R&D manager thought that a benefit such as the capability for providing better field services to existing customers, or for making quicker deliveries of products, or for improving quality, would not be taken into account, he or she could discuss with finance managers how these benefits could be reckoned with when decisions about technology were made.

10. LINKING R&D TO BUSINESS PLANNING

Level A: Issue is not recognized. The R&D department of a pulp and paper company took a completely unstructured approach to managing technology. It did not plan its own R&D, and it did not try to integrate its R&D into the company's short- and medium-range planning. Instead, the R&D department pursued long-term research based on its own ideas about what the business would need in the future. Because the company did not have long-term plans, the research was almost never linked to the business goals.

Level B: Initial efforts are made toward addressing issue. At a food processing company, the R&D department did want to have technology plans and to integrate them into the business plans. It set up a group of planners that was assigned the responsibility of developing technology plans for the laboratory as a whole. Because the planners knew far more about data collection than about developing plans and getting planning accepted and implemented, they failed in their mission and the

group was eventually disbanded. This group of planners was committed to planning, but it did not know how to plan.

The *average R&D department* operates at this level. It wants better planning of overall R&D goals and use of R&D resources, but it does not know how to develop technology plans. The selection of individual projects in the average R&D department is done without reference to technology plans. The budgeting of R&D resources consists of funding the projects that are selected. Rather than having a clear overall direction, the average R&D department has merely a collection of diverse projects.

Level C: Right skills are in place. At an instruments company, the R&D department had the skills to plan but it lacked the methods for analyzing technologies. Without techniques for evaluating the company's technical strengths and weaknesses, for assessing the potential of new technologies, and for translating technical capabilities into business applications, it was not able to develop the kind of technology plans it wanted to.

Level D: Appropriate methods are used. In a natural resources company, the R&D department developed its own techniques for analyzing technology. Consequently, it was able to develop technology plans that delineated where the laboratory was going. However, disputes arose between company business planners and the R&D planners about what part technology planning should play in business planning. The business planners had their own plans and did not want technology incorporated into them. They felt that it was not the responsibility of the R&D department to get its technology plans incorporated into company plans.

Level E: Responsibilities are clarified. At a telecommunications company, the R&D department and the business units developed plans and coordinated those plans with each other. Although this company made significant progress in planning, each year the planners had to revive interest in planning. The R&D managers and business managers developed plans when required, but they did

TABLE 30.10
Activity 10—Linking R&D to Business Planning

		Points
Level A (Not recognized)	R&D department takes a completely unstructured approach to managing technology (*pulp and paper*).	0
Level B (Initial efforts)	R&D planners want better planning, but know more about data collection than about developing plans and getting planning accepted and implemented (*food processing*).	1
Level C (Skills)	R&D department has the skills to plan, but lacks methods for analyzing technologies (*instruments*).	2
Level D (Methods)	Although there is planning in both R&D and in the businesses, there are disputes about what part technology should play in business planning (*natural resources*).	3
Level E (Responsibilities)	Technology and business planning are accepted and done, but the planning process is taken as a given (*telecommunications*).	4
Level F (Continuous improvements)	An R&D planning group orchestrates technology audits, but also tracks planning decisions about technology and sponsors an audit of its own activities (*specialty materials*).	5

☐ = Average R&D department.

not embrace the ideas of planning or look for ways of improving the planning process.

Level F: Continuous improvement is underway. At a specialty materials company, planning was fully embraced within the R&D department and in the rest of the company, where it was part of the mainstream of company activities. For example, there was an R&D planning group that orchestrated audits of various technical groups' technologies. This planning group helped technical groups and business groups to make assessments of the strengths and weaknesses of the technologies in their businesses and to plan the development of new technologies. In addition, this planning group worked at improving the planning process by, for example, helping technical and business groups track their planning decisions about technology, and by sponsoring an audit of the planning group's own activities.

THE AVERAGE R&D DEPARTMENT

In reviewing all ten R&D activities, one can see the "average" R&D department operates at various levels, depending upon the activity. The average score for the average R&D department on all ten activities is 1.7, somewhat below the mid-point of 2.5. The average R&D department operates best when selecting R&D and planning and managing projects. In these two activities, the average department uses appropriate methods but has not clarified the responsibilities of various groups. It operates fairly effectively in these two areas for two reasons: First, all R&D managers continually try to find better ways of selecting R&D; second, most R&D managers now accept that in order to accomplish technical objectives on time and within budget an R&D department must have effective methods for planning and managing projects.

In four of the ten activities—generating new product ideas, maintaining quality of the R&D processes and methods, coordinating R&D and marketing, and transferring technology to manufacturing—the average R&D department does somewhat less well: It has the right skills in place, but it does not use the appropriate methods. There are three reasons for this:

1. Finding and implementing good methods of generating new product ideas require a different kind of management talent than selecting R&D or planning and managing projects does. To generate new product ideas, an innovative environment must exist; otherwise, methods for stimulating creativity will not work well. The managers of the average R&D department are much better at, say, selecting projects than in creating an innovative environment.

2. Until five or ten years ago, the average R&D department did not pay much attention to the quality of the R&D processes and methods. This is illustrated by the dearth of articles about quality in R&D before the early 1980s. Currently, the average R&D department is looking for methods of improving quality within R&D and, on the whole, has just started implementing them.

3. The average R&D department has the skills to work with the marketing and manufacturing departments but has not developed the right mechanisms for improving coordination with them. To develop such mechanisms, R&D departments need to establish common goals with marketing and manufacturing—which, of course, requires improved coordination. In other words, the average R&D department is caught in a vicious circle as it tries to improve its relations with the marketing and manufacturing departments—it cannot develop the right mechanisms until it improves coordination, and it cannot improve coordination until it develops the right mechanisms.

In three of the four remaining activities—motivating technical people, establishing cross-disciplinary teams, and linking R&D to business planning—the average R&D department has made initial efforts toward addressing the issues, but does not now have the right skills in place to be effective. With regard to motivating technical people and establishing cross-disciplinary teams, the main reason for the average R&D department's weakness is the managers' poor skills in dealing with people. R&D managers, on the whole, are technically trained and oriented much more toward technical than management issues. To motivate technical people, an R&D manager must first understand what technical people want and value. To establish cross-disciplinary teams effectively, the manager must appreciate the interests of many potentially conflicting individuals or groups. In general, managers in the average R&D department fall short in these areas.

With regard to planning, the average R&D department does not do well, partly because the average company does not do planning well. Long-range planning requires a dedication to asking tough questions about future possibilities, and to making a solid commitment to take actions in accordance with one's plans. Neither the average R&D department nor the average company knows how to do this effectively.

With regard to the remaining activity—fostering collaboration between R&D and finance—the average R&D department does very poorly. Very few technical people have an appreciation of financial matters or close contact with members of the finance department. To the average technical person, the finance department seems like a distant operation that is preoccupied with numbers. People in the average R&D department also do not understand how much their lack of understanding about financial issues hampers company support for technology.

31

Assessing the Value of Your Technology

JAMES W. TIPPING, EUGENE ZEFFREN, AND ALAN R. FUSFELD

R&D's role in the innovation process can be meaningfully represented by a hierarchy of managerial factors (The Technology Value Pyramid) that provide the foundations, links to strategy and financial outcomes for the corporation. The recognition of these TVP factors, together with an assembled menu of metrics, allows the model to be used to track the contribution to innovation performance at different levels of the TVP. The TVP model can be used to track the performance both prospectively and retrospectively, to diagnose weaknesses in the R&D organization and to plan for improvement in R&D contribution to the corporation. The various R&D stakeholders have different interests and perspectives on the innovation process, and these are accommodated by the TVP model and the menu of metrics.

Management's most serious responsibility is to account for, and to use as effectively as possible, the corporate assets with which it has been entrusted. To date, R&D management has not devised a methodology through which it can satisfactorily discharge this responsibility in any way other than in a passive fiduciary sense, and this often has little or no relationship to agreed business goals. Consequently, top corporate executives have no mechanism with which to either judge or participate in what is potentially the corporation's best competitive weapon—its technology!

Even though they possess some of the world's best-equipped R&D facilities, staffed by the most talented scientists, it is doubtful that U.S. corporations are deriving anything like the competitive advantage possible from their R&D efforts. Despite the recognized importance of technology it would appear that many corporations are finding it easier to look elsewhere for competitive advantage—for example, from marketing, acquisition, capital investment, and so on. Even those world-class corporations that do look seriously to technology for competitive advantage are frustrated in their attempts to couple their R&D effectively into their businesses, in part because of the absence of an accepted methodology to measure effectiveness (value) and continuously improve their R&D. We believe it is the responsibility of R&D management to develop this methodology.

Accounting for, and in turn improving, R&D performance and effectiveness to produce value calls for a shared understanding with the stakeholders of the role that R&D plays in that corporation, to agree to that role, and to be partners in the R&D process. This partnership may well be different for different stakeholders. It calls for R&D to clearly communicate with the stakeholders so that the stakeholders become truly involved with R&D; this involvement must be at a level which to date is rarely seen in U.S. corporations. It is only through

Acknowledgment: The authors recognize and appreciate the contributions from the members of the R&D effectiveness subcommittee of the IRI's Research-on-Research Committee.

Reprinted with permission from *Research-Technology Management*, September–October, 1995.

James W. Tipping is an independent consultant. Previously, he was responsible for corporate R&D at ICI America. *Eugene Zeffren* is President of Helene Curtis U.S.A. *Alan R. Fusfeld* is President of the Fusfeld Group, a management of technology consulting firm.

this high level of interaction that a partnership can be established that both allows leading and lagging indicators of value-producing R&D performance to be credibly established, and ultimately allows leverageable value to be created by R&D.

The true value from R&D becomes apparent only when one looks closely at the role that R&D plays throughout the creation and development of the innovations necessary to both defend and grow the corporation's businesses. One cannot judge the value of an R&D organization to a corporation simply by looking at the new products it has produced recently, just as one cannot judge the value of a house by looking at the exterior brickwork. R&D protects and supports the day-to-day operations of businesses every day of the year but rarely accounts for these activities and therefore rarely receives credit for doing so. Furthermore, would it be worthwhile to put a value on the know-how that, say, a Du Pont or a General Electric has built up over the past 100 years? Certainly one wouldn't do this by simply counting patents or measuring recent achievements; this would miss the tremendous technology platform of potential value these companies have built up, and any worthwhile measurement of R&D value must provide a way of giving credit for this value and, in particular, changes in the value from year to year. While, arguably, the value of R&D is reflected in the share price of a corporation, it is our contention that current approaches do not allow a true assessment of this value to be made.

We have developed an approach to these important considerations whereby the technology development and innovation processes are described in a pyramid of values which derive from R&D. This Technology Value Pyramid (TVP) model (Figure 31.1) allows the stakeholder to be easily involved with factors important in these processes in the most direct and relevant manner, and to be presented with simple measurement tools that can be used in both a leading and lagging manner.

CEOs, business heads, board members, the financial community, and leaders and managers of R&D will benefit most from use of the TVP model. All businesses will find the TVP model to be rel-

evant but its precise use will vary from business to business, depending on the balance of the research within individual portfolios. (Although the TVP approach can be applied to the public R&D measurement area, this article focuses on the private sector and only makes reference to the public sector for comparison purposes.)

FIVE MANAGERIAL FACTORS

As shown in Figure 31.1, the Technology Value Pyramid represents the hierarchical integration of the five managerial factors that describe the innovative capability of the firm. Metrics associated with each factor allow the TVP to be used to analyze the performance of the R&D organization, and to guide improvement efforts. Used prospectively, the TVP can be a "leading indicator" of R&D's contribution to the firm. The TVP is based on the beliefs that in order to create and sustain technological advantage for the corporation:

- The R&D effort must defend and enhance the value of the corporation.
- There must be a linkage of R&D and the strategic aims of the corporation.
- R&D must be able to sustain its capability to produce useful output over the long term.

These beliefs imply that not only will one have to assess the ability of an R&D organization to create value for the corporation, but also, one must be able to assess how well R&D is linked to the business goals, and how well it actually carries out its R&D functions. Effective measurements in these areas are both the key to understanding and the basis for improvement.

Two managerial factors, *Practice of R&D Processes To Support Innovation* and *Asset Value of Technology,* are the core operational elements of the R&D enterprise, the nuts and bolts of the operation. Effective management at these levels is characterized by high technical output and efficient technology management practices. Metrics to assess the adequacy of the foundation would be those

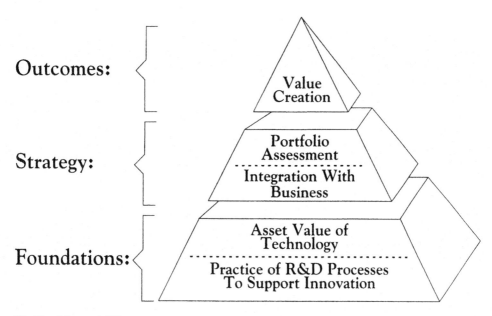

The Five Managerial Factors

1. *Value Creation (VC).—Demonstrates the value of R&D activities to the positioning, profitability and growth of the corporation and to the creation of shareholder value.*

2. *Portfolio Assessment (PA).—Communicates the total R&D program arrayed across various dimensions of interest, including time horizon, level of risk, core competency exploitation, and new/old business. This allows optimization of the total program for the corporation's benefit.*

3. *Integration With Business (IWB).—Indicates the degree of integration, the commitment of the business to the R&D processes and programs, teamwork, and ability to exploit technology across the organization.*

4. *Asset Value of Technology (AVT).—Indicates the strength and vitality of the firm's technology (e.g., proprietary assets, know-how, people, etc.) and foreshadows the potential of the R&D organization to create future value for the firm.*

5. *Practice of R&D Processes To Support Innovation (PRD).—Indicates the efficiency and effectiveness of R&D processes in producing useful output for the firm. The processes include project management practices, idea generation, communication, and other "best practices" in managing R&D.*

Figure 31.1. Technology value pyramid.

that allow measurement of parameters relevant to those R&D practices that help spur innovation, that allow an assessment of the value of the technology development capability of the organization (its asset value), and that allow the assessment of the likelihood of sustaining or enhancing that value over time. This level is the foundation of the pyramid.

The next level in the pyramid relates to business and technology strategy. The managerial factors at this level are *Integration with Business* and

Portfolio Assessment. These factors reflect on how well a firm's technology strategy is linked with its business strategy, from both a corporate and a business unit perspective. Using the appropriate metrics, one can define how the various steps in a firm's process or product creation activity link to the various levels of the model that show how the business creates value for its customers. Measurements in this area also allow an assessment of the portfolio of projects, both from the point of view of how well they will accomplish the firm's business objectives and how well they allow the developed competitive advantage to be sustained. Thus, measurements here deal with such issues as a quantitative assessment of strategic alignment as well as the extent of investment in a firm's core technical competencies.

Value Creation is the managerial factor at the top of the pyramid, and is the overarching goal of every R&D program. The TVP's foundations are linked to Value Creation through strategy. Measurements of Value Creation demonstrate how the technology development effort has contributed to the growth of the firm and the growth (if any) of its profitability. Thus, these metrics lead to an assessment of R&D's contribution to shareholder value. Used prospectively with respect to the current project portfolio, they can also indicate the potential of the R&D program to create future value for the firm.

The TVP model rests on three basic assumptions:

- *Wealth creation stems from the innovation process.* Hence, the output of R&D is directly linked to enhancing the value of the firm.

- *The various stakeholders in R&D have different interests.* Hence, measurements key to one group will be secondary to others.

- *The time scale of interest of the various stakeholders will vary.* Therefore, measurements must allow different stakeholders to address their particular period of interest for measuring the R&D activity.

Our research tells us that the primary stakeholders and the TVP factors expected to be of greatest interest to them vary as follows:

- The stakeholders most interested in value creation will be the board, financial community, and the CEO.

- Business management will be keenly interested in metrics that assess the degree of integration of R&D with the business and the balance within the project portfolio relative to business unit needs.

- R&D management will be appropriately concerned with all levels of the pyramid but will likely be most concerned with both portfolio assessment and the value of the technology assets within the firm.

- Finally, the R&D staff is likely to be most keenly interested in measurements of the practice of R&D to support innovation.

THE MENU OF METRICS

In building the model, we focused on those measurements relevant to both the building and sustaining of the firm's competitive advantage. The particular set of measures appropriate for a firm will vary by industry type, and within an industry by a firm's competitive strategy (e.g., cost, differentiation, combinations of cost and differentiation). The menu of metrics that forms the substance of the Technology Value Pyramid provides a broad selection of measures; firms can select those that make sense for them based upon their individual business strategies and their industry type. The menu of metrics is presented in Table 31.1.

The menu was derived from almost two years of discussion and debate within IRI's Research on Research Subcommittee on the Measurement of the Effectiveness and Productivity of R&D, with input from the Quality Directors Network. The metrics chosen represent those considered to be the most generally relevant and useful set of measurements considered for the various factors. Information on more than one factor can sometimes be gleaned from a single metric.

UTILITY AND APPLICATION

Previous attempts to deal with the measurement of R&D value have been dissatisfying in many ways. Those that treated only sales or profit generation often did not address the needs of the technical community. These needs include the need to protect the longer-term future of the firm (addressed

TABLE 31.1
Technology Value Pyramid Menu of Metrics

Metric/Factor	Definition	Comments
1. Financial return/VC		
a. New Sales Ratio (NSR)	The ratio of sales revenue in year i from product developments commercialized in years $i - j$ to $i - l$ to total sales revenue in year i, or: $$NSR = (1/Sales) \times \sum_{N=i-j}^{N=i-l} (NP\ Sales)$$ where: NP Sales = sales revenue in year i from product developments commercialized in year N; Sales = total sales revenue in year i.	Metrics for determining the value of the investment in R&D require a system for identifying and tracking over time the cost and commercial outcomes of R&D activities, by programs or by individual projects as well as on a total basis. In this category, three basic metrics are needed to determine the value-creation benefits of R&D: New Sales Ratio (NSR, 1a); Cost Savings Ratio (CSR, 1b), and R&D Intensity (RI, 1d). Most other useful financial metrics can be derived from these three basic metrics. NSR is essentially the fraction of sales derived from new products, with the time frame being chosen as appropriate for the company and/or industry. Since technical effort is usually the source of cost savings from process developments and for formula modifications, CSR is an attempt to capture that in a reliable manner. As seen in Metric 5, Gross Profit Margin, the gross profit number used in CSR represents the revenues minus cost of goods sold, including product cost and direct manufacturing cost. The latter are those costs that can be reduced by technology development efforts, and to the extent they are, they represent contributions from R&D.
b. Cost Savings Ratio (CSR)	The ratio of savings in the cost of goods sold in year i from process developments or product changes adopted in years $i - k$ to $i - l$ to average gross profits in year i, or: $$CSR = (1/GP) \times \sum_{N=i-k}^{N=i-l} (CG\ Savings)$$ where: CG Savings = reduction in cost of goods sold in year i from process developments or product changes adopted in year N; GP = average gross profit for business unit in year i.	
c. R&D Yield	The gross profit (GP) contribution from the sale of new and improved products (NP) and from the lower cost of goods from new and improved processes or new formulations (CR). R&D Yield = NP + CR, where: NP = GP × NSR CR = GP × CSR, or: R&D Yield = GP (NSR + CSR)	The time frame is important for proper use of these metrics. When i = current year or earlier, the metrics are retrospective; when i = next year or later, the metrics are prospective. In these definitions, j and k are constants which reflect the strategic emphasis and/or industry characteristics of the business enterprise. To utilize these metrics, appropriate values must be selected for j and k, and operational definitions must be established for product development, process development and R&D expenditures. This should be straightforward in most firms.
d. R&D Return	The ratio of R&D benefits to R&D investment. R&D Return = R&D Yield/ R&D Effort = (GP/Sales) × (NSR + CSR)/RI where: R&D Effort = Annual expenditure on R&D. RI = R&D Intensity $$= \frac{R\&D\ Effort_i}{Sales_i}$$	

Table 31.1—*Continued*

Metric/Factor	Definition	Comments
2. Projected value of the R&D pipeline/VC, PA		
a. Projected Sales Value from Pipeline	Fraction of future sales by year projected from projects in the R&D pipeline, incorporating probability of attaining objective for each project.	Most companies attempt to project future sales and income from projects in the pipeline; the cumulative value of these projects represents the projected value of the pipeline in total. Such projections must always be done with a careful understanding of the assumptions behind the probability determinations. Care must be taken not to let biases for or against individual projects influence the assignment of probabilities of sucess for the various projects
b. Projected Income Value from Pipeline	Fraction of future net income (and/or return) by year projected from projects in the R&D pipeline, incorporating NPV times probability of attaining objective for each project.	
3. Comparative manufacturing cost/VC, AVT	Benchmarked manufacturing cost data vs. competition for substantially identical manufacturing steps or for producing substantially the same product (eliminating factors unrelated to technology-based differences).	While most firms have accurate data on their own manufacturing costs, the generation of accurate manufacturing cost data for the competition is considerably more difficult. Therefore, when using this measurement, the firmness of the estimate of the competitive manufacturing cost should be identified.
4. Product quality and reliability/VC, AVT		
a. Customer or Consumer Evaluation	Relative quality (as evaluated by the customer/consumer) vs. competitive product in blinded product evaluation by techniques appropriate to the industry segment.	The two measures provided here are best used together. Each firm will generally have a preferred technique for evaluating with customers or consumers how well their products perform versus competition.
b. Reliability/Defect Rate Assessment	At the firm level, fraction of company's product output that meets or exceeds the established quality standards. At the product level, fraction of a given product's quarterly output that meets or exceeds the established quality standards.	
5. Gross profit margin/VC, AVT	Gross Profit as a percentage of sales, where gross profit equals net sales minus cost of goods sold (product costs plus direct manufacturing costs). Value assessment should be based on *change* in gross profit margin from period to period. (Periods should be appropriate to an industry and may be in excess of 1 year.)	Gross profit margin is used in assessing the value of the technology assets of the firm and R&D's contribution to value creation, since beyond direct manufacturing costs and costs of goods sold, other activities of the firm have significant influences on profit margin. In any event, users of this metric must be sure of the connection to the R&D enterprise.

Table 31.1—*Continued*

Metric/Factor	Definition	Comments
6. Market share/VC, AVT		
a. Direct Market Share	Company (or business unit) market share in various categories measured as appropriate for the industry or category. As with gross profit margin, changes should be assessed at least annually to determine rate of progress or decline.	Market share determinations need to be approached cautiously. While product advances should have a beneficial impact on market share and can be representative of value creation for the firm, there can be many confounding factors in a market share determination. These would include such things as the size and quality of the marketing effort, the competitive response, the relative state of the economy, etc. One should also be sensitive to share data in related markets, since market transformation is often created by unexpected competitors.
b. Related Market Share	Share data in related markets for indication of threats or opportunities.	
7. Strategic alignment/ PA, IWB, AVT		
a. Corporate and Business Unit	Fraction of total R&D portfolio which is consistent with corporate goals and, where applicable, for each major business unit. Should also include specific identification of the corporate or business goal with which a project is identified.	While it seems unlikely that much, if any, of the R&D portfolio would not be consistent with corporate goals, this question needs to be explicitly addressed. If the number of projects not consistent with corporate or business unit goals is substantial, explanations need to be given. For goal coverage, serious questions must be asked if major business goals are not addressed by the R&D portfolio. In particular, either the total size of the R&D effort or its distribution must be questioned.
b. Goal Coverage	Fraction of corporate or business unit goals requiring technology development that are addressed by the R&D portfolio.	
8. Distribution of technology investment/PA, AVT	Analysis of the fraction of the total R&D investment along various dimensions. Each company should describe its portfolio of technical activities along dimensions that are important to decision making and communication between stakeholders. Some of the more common dimensions are: A. Dimensions potentially of greater interest to the CEO (and to business management): • Categorization of reward vs. risk • By product line or business unit • For maintenance of current business, expansion of current business, or creation of new business • Environmentally driven vs. non-environmental	This measurement area provides much texture for the analysis of how well a firm's R&D program is protecting the technology investment and technical position of the company. Important output from assessments of technology investment distribution are the questions that are the analysis causes to be asked regarding what distribution is desired. This helps set the direction for modifying the R&D portfolio.

Table 31.1—*Continued*

Metric/Factor	Definition	Comments
	• Distribution according to time of commercialization B. Dimensions potentially of greater interest to business management (and some CEOs): • For cost reduction, applications development, or performance differentiation • For technical service, basic research, applied research, product development and process development • For current markets, markets new to the company, or markets new to the world • For current technology, technology new to the company, or technology new to the world C. Dimensions potentially of greater interest to R&D management: • Distribution by project stage • Distribution by technical discipline • External R&D vs. Internal • Base, key or pacing technology • Core vs. new competencies • U.S. vs. non-U.S.	
9. Number of ways technology is exploited/PA, AVT	Assessment of the number of different product types or business segments utilizing or planning to utilize a given technical asset (the exploitation of functional competencies).	Measurements in this area cause a firm to stretch the limits of application of its technology. By conscious examination of this, market opportunities beyond one's current markets can be created. While such opportunities may or may not be pursued, this examination allows R&D to maximize the possible output from its creative endeavors.
10. Number of project definitions having business/ marketing approval/IWB	Fraction (or percent) of projects in the total R&D portfolio with explicit business unit and/or corporate business management sign-off.	Ideally, all projects except "skunk works projects" would have either corporate or business unit sign-off. In practice, exploratory feasibility determination work is likely not to; such work would be expected, however, to be aimed at areas of known interest.
11. Use of project milestone system/ PRD, IWB	Fraction of projects in the total portfolio going through a defined project management system with defined milestones.	The ideal value for this measurement must be determined by each firm for itself, as it will depend on a number of aspects of the Distribution of the Technology Investment (Metric 8). Firms with a larger percentage of exploratory projects would be expected to have a lower fraction going through a defined project management system. Once

Table 31.1—*Continued*

Metric/Factor	Definition	Comments
		feasibility has been demonstrated, most firms would have a fairly high percentage of their projects going through a defined system.
12. Percent funding by the business/IWB	Fraction (or percent) of the R&D budget from business unit sources.	This measurement has most utility for corporate labs or for centralized R&D organizations. Again, the ideal value for this measurement will depend on the individual firm's circumstances. The number will also depend on the strategic intent of the firm; if R&D is being asked to break ground in a new business area for which no revenues currently exist, one would expect the fraction of the R&D budget from business unit sources to be lower.
13. Technology transfer to manufacturing/ IWB, PRD	Semi-quantitative assessment (e.g., an interval scale from 1 to 5) of the effectiveness of the transfer with ratings obtained from both R&D and manufacturing sides of the interface.	This measurement is important for its reflection on the closeness of the working relationship between R&D and manufacturing. As with several other measures described later, the suggested measurement technique involves an interval scale, with ratings from 1 to 5. Such a scale must be carefully created with proper anchoring descriptions for the ends of the scale, and for the middle.
14. Use of cross-functional teams/IWB, PRD	Fraction of projects in the R&D portfolio with specific cross-functional teams assigned. This analysis can be further subdivided by project types; e.g., short-term development, long-term development, process development, applied research, etc.	Current assessment of best practices of technology management includes the use of cross-functional teams to ensure alignment of research with the other functions and with the business strategy. Research has shown it to be one of the more useful leading indicators for financial success and R&D output. Metrics should evaluate not only the number of teams but also their composition and role. Internal trend analysis is more important than absolute benchmarking against competitors.
15. Rating of product technology benefits/AVT, VC		
a. Customer Rating	Numerical ranking of a firm's product by a given customer divided by that customer's ranking of the best competitive product. This ratio can be averaged across customers and/or market segments using the product to obtain an average ratio value.	Ideally, this metric will measure the customers' perceptions of how each competitor delivers value. A customer survey should be used to determine how the products from each competitor rank on a numerical scale in delivering benefits. To handle the individual opinions, a ratio should be made of the firm's numerical rating to the

Table 31.1—*Continued*

Metric/Factor	Definition	Comments
b. Economic Evaluation	{Price differential per unit obtained by virtue of quality feature(s) derived from technical effort minus the cost per unit of providing the feature(s)} times unit sales volume for products containing the feature(s).	competitor. The ratios can be averaged across many customers and different market segments to obtain an average ratio value. The Economic Evaluation measure provides an estimate of the commercial value of a technology benefit in the firm's products, based on the proposition that the benefit allows a premium price relative to the competition. The Market Share evaluation assumes that competitive pressures may prevent the firm from gaining a price premium for the technology benefit, but for the benefit to be worth the cost in this situation, market share gains should provide the economic gain that is desired.
c. Market Share Evaluation	Differential share gain(s) at a constant price for product(s) containing the quality feature(s).	
16. Response time to competitive moves/AVT, PRD	Time required for the firm to match competitors' newest product benefit(s) divided by time required for competitor to match firm's newest product benefit(s).	This metric relates to the ability of the firm to either maintain a leadership position or to match technology moves by the competition. It is a part of Customer Satisfaction.
17. Current investment in technology/AVT	Current annual expenditures for R&D staff and equipment ratioed to best competitor, to industry average, and to industry total.	This metric measures the rate of current activity in developing the technology of interest with the intent of predicting whether the firm is expected to gain or lose ground in the technology. It should be kept separately for the key and pacing technologies most critical to the strategy. It is best used as a ratio of investment by the firm compared to the best competitor. Ideally, quality of the research staff should also be factored into the metric, since output of the investment will be determined not only by the effort but by the abilities of the researchers.
18. Quality of personnel/ AVT, PRD	Several measures are possible; the ones chosen should be appropriate for the firm's technology development strategy.	Research has shown this metric to be a leading indicator for many intermediate factors, but only a weak leading indicator of new products. Quality of personnel relative to the competitors is a difficult metric to evaluate in most cases. In some cases, it is clear that one competitor considers one competitive area more strategic than another competitor and consequently puts the best resources there. In using this metric it is
a. Internal Customer Rating	Interval rating scale (e.g., a scale from 1 to 5) from R&D's internal customers, such as marketing, manufacturing, etc.	
b. External Customer Rating	Interval rating scale from the firm's major customers.	

Table 31.1—*Continued*

Metric/Factor	Definition	Comments
c. External Recognition	External awards and invited lectures by the professional staff over a relevant time period.	important to decide whether one is addressing the quality in a specific arena or the general overall quality of the research group.
d. Published Works	Publications and patents by the professional staff over a relevant time period.	
19. Development cycle time/ AVT, PRD		
a. Market Cycle Time	Elapsed time from identification of a customer product need until commercial sales commence.	The literature suggests that short cycle time will greatly influence financial return, other factors being equal. However, there is also an indication that shortening the cycle time may result in shortcuts and poor quality work which will more than overpower the positive gains. Another concern is that the metric could cause a firm to move to short-term, faster projects which inherently do not have the impact of longer, more strategic projects.
b. Project Management Cycle Time	Elapsed time from establishment of a discrete project to address an identified customer product need until commercial sales commence. For both (a) and (b) the end point can be the time when manufacturing feasibility is established for those cases where no commercialization occurs. Compare to historical values and benchmark vs. competition, if possible. Group by categories of projects (e.g., major new product, minor product variation, etc.). Can also be used to track milestone attainment rate for firms using a stage gate management process.	The metric is probably most useful as an internal standard for continual improvement, since obtaining competitor information for comparison is likely to be very difficult. The cycle time that should be measured and charted is the interval between the first indication of a customer need or opportunity to the time that a successful solution, by an appropriate measure for the industry, is available for general use by the customers.
20. Customer rating of technical capability/AVT	Average customer rating (internal or external) of overall technical capability of the firm (interval rating scale) in providing technical service and/or new product innovations. Can be ratioed to ratings for relevant competitors for benchmarking purposes.	This metric relates to Quality of Personnel and Customer Rating of Product Technology Benefits, but is more future-oriented. It attempts to measure the customers' perceptions of the ability (not willingness) of the firm to bring value to the customers' future problems. Assuming that a technology is used in a variety of markets, the metric should integrate over the different needs and wants of the customers. A competitor with a strong technology position having broad value will be the most successful at meeting the customer needs in many markets, rather than only a few. A competitor's indifference to a market niche should not be equated with its inability to bring technical strength to bear when desired. Consequently, this

Table 31.1—*Continued*

Metric/Factor	Definition	Comments
		metric probably should not be a weighted average across markets, but an evaluation of the breadth of capability across the entire range of opportunities. A customer survey focused on the competitor's ability to use technology to solve new problems should be used with quantitative rankings.
21. Number and quality of patents/ AVT, PRD		
a. Percent Useful	Percentage of active patents from the company's total patent estate which are incorporated into or used to defend the firm's commercial products or processes.	There is considerable debate about the validity of using the number of patents as an indicator of the strength of the technology or the management of technology for a firm. Research indicates some negative correlation to
b. Value Ratio	Interval rating (1 to 5) for potential strategic value times rating (1 to 5) for strength of protection divided by 25 (maximum attainable value). Yields a number between 0 and 1.	financial performance and patent volume. Also, some firms do not patent extensively as a matter of strategy, and patents are not particularly valuable in other industries. This metric should only be used if patents are considered especially valuable in the industry and competitors follow a vigorous strategy
c. Retention Percent	Percent of granted patents maintained.	of creating a patent "forest." The number of high-quality patents is
d. Cost of Invention	Number of patents from R&D/R&D effort. One can also calculate this just using the number of useful patents from R&D.	probably a more useful metric. Internal assessment of value is one method of determining quality of the patenting activity, which does not require the lag time of citations or sales development, and which should relate to success of the firm's strategy. The metric should be kept for patents in the key technologies of greatest strategic value and in the relevant pacing technologies. All patent metrics should be benchmarked against competitors and compared to exploitation of the current technology position.
22. Sales protected by proprietary position/AVT, VC		
a. % Patent Protected Sales	Percentage of sales protected by patents owned by the company.	Sales of products under patent is another means of assessing the value of the patenting activity. Rigor is required
b. % Proprietary Sales	Percentage of sales protected by patents plus trade secrets and/or other exclusive company know-how or arrangements.	to count only patents that truly protect the sales, and not background patents. This assumes a strategy of first-filing in the U.S. and ignores the potential for protection of sales by longer-lived counterparts in other countries. The assumption is useful for tracking the turnover of the patent portfolio.

Table 31.1—*Continued*

Metric/Factor	Definition	Comments
		The % Proprietary Sales metric allows inclusion of proprietary positions achieved by means other than patents.
23. Peer evaluation/ AVT, PRD		
a. External	Numerical rating (interval rating scale) from 1 to 5 by a panel of external experts on the merits of the firm's technology positioning and technology management practices. Panel selections critical; must be capable and objective. Could include outside directors from technology companies or university science/engineering departments, consultants in technology management, professors, venture capitalists, etc.	Peer evaluation is another method of assessing the strength of the technology of a firm and the management of technology. A variant of this metric (b) is to use an internal panel of experienced scientists, probably technical ladder rather than management people, whose responsibility is to form overall judgments on the technology positioning and management of the firm. In either case, consistency of the panel over time is important in identifying and making improvements.
b. Internal	Same rating scale applied by internal experts; probably staff on company's technical ladder.	
24. Customer satisfaction/ AVT, PRD		
a. External	Average rating by key external customers using a 1 to 5 interval rating scale to evaluate various dimensions regarding product technology or process technology benefits and technical service provided.	Customer ratings of Quality of Personnel, Technical Capability, and Product Technology Benefits speak to external customer satisfaction. Recognizing that the first customer for technology management is internal, customer satisfaction surveys should be used to determine satisfaction from engineering, production and marketing
b. Internal	Same rating approach along dimensions important to key internal customers such as marketing, engineering, manufacturing etc.; dimensions would include, for example, timeliness of developments, competitiveness of solutions, etc.	along axes that are important to them: on-time delivery of technology developments and packages; competitiveness of solutions developed; general satisfaction with the job being done.
25. Development pipeline milestones achieved/PRD		
a. Percent of Project Milestones Achieved	Percent of project milestones achieved within three months of projected achievement date (or within that time appropriate for an industry). Plot as histogram to reveal actual performance, showing percent that beat the target as well as those that miss it.	Reaching milestones in the time predicted is a measure of effective planning and management. Use of milestones correlates with shorter cycle time. A consistent management system is required to track this metric so that variability of the number/quality/ difficulty of the milestones does not cause random fluctuation in the metric.
b. Performance Level at Each Milestone	Percent completion of objectives expected by a milestone date at the milestone date.	

Table 31.1—*Continued*

Metric/Factor	Definition	Comments
26. Customer contact time/PRD	Average hours per researcher spent in direct contact with external (or internal) customers.	One of the best practices identified for technology management is customer contact time by the scientists. This metric provides a quantitative measure; management must also ensure that the time is well spent.
27. Preservation of technical output/PRD	Percent of research project outcomes captured in technical reports.	The value of counting reports and publications is debatable. Positive correlations have been seen with items relating to communications, cross-functional activities, and use of milestones; however, they have shown no relation to financial outputs or to creation of value.
28. Efficiency of internal technical processes/PA, PRD		
a. Project Assessment	The total cost of all commercially successful projects divided by the number of commercially successful projects.	There are many metrics of this type that relate to the efficiency of the daily processes by which research is conducted. Examples are turnaround
b. Portfolio Assessment	The total R&D budget divided by the number of projects with commercial output. Subdivide by projects of similar type (technical service, short-term, long-term) and use in conjunction with project value assessment.	time for customer service requests and dollars spent per customer service request. Internal services such as analytical service, computer services, or library services can be measured in the same way.
29. Employee morale/PRD	Quantitative ratings of key aspects of employee satisfaction and morale as shown by direct employee survey.	Employee morale or job satisfaction is generally thought to be a direct input to motivation and productivity. The level or extent of motivation may be a more useful metric since it is closer to productivity. Motivation can be measured by a direct survey, or more qualitatively through focus group discussions.
30. Goal clarity/PRD	Interval rating scale assessing the extent to which project performance objectives are clearly identified and understood by all participants on the project team.	Internal research in some companies has shown this measurement to be significantly related to successful innovation. Low rankings here would point to a need to improve communication regarding project objectives between R&D and key internal and external customers.
31. Project ownership/ empowerment/ PRD	Interval rating scale assessing the extent to which participants feel they have the support and freedom they need to be successful in the project.	This measurement has also been reported to correlate positively with successful innovation from R&D. It is also likely related to employee morale on a project team.

Table 31.1—*Continued*

Metric/Factor	Definition	Comments
32. Management support/PRD	Interval rating scale assessing the extent to which participants feel they have management's backing and an understanding that failure while learning will not be punished.	This support is also critical to successful innovation. Low ratings in this area point to a breakdown in relations (and perhaps in credibility) between R&D management and the project team and, perhaps, between R&D management and business management.
33. Project championship/ PRD	Percent of projects for which an effective project champion can be identified on the project team.	As with metrics 30–32, the project championship measure has been correlated in some companies with successful innovation. Project champions, whether or not they are identified as project leaders, lend an air of enthusiasm, optimism and urgency to a project that tends to permeate the team.

in the TVP through selection of relevant measures of Portfolio Assessment and Asset Value of the Technology), the need to assess development of the organization (using metrics of the Asset Value of Technology and Practice of R&D Processes To Support Innovation factors) and the need to assess how well the firm's core technical competencies are being developed and protected (Portfolio Assessment, Asset Value of the Technology, and Practice of R&D Processes To Support Innovation factors).

Previous measurement proposals that focused on R&D practices and ignored value creation gave some business managers the sense that R&D was parochial and myopic. The TVP makes it clear that a holistic view is the only comprehensive way to measure and appreciate R&D's contribution to a company. It provides the flexibility for a firm from any industry to assess R&D's true value to that firm.

A powerful feature of the model is its applicability to various organizational formats for managing R&D—centralized, decentralized and hybrid. The centralized format has one laboratory organization, usually a corporate lab supporting all technical activities in the firm, from upstream ap-

plied research and technology development (including basic research) to product and process development and technical service. Decentralized formats have no corporate lab and all technical work is carried out by business unit labs. Hybrids will have a corporate laboratory, usually handling upstream work and new business technology development needs, and perhaps new technical competency development, with business unit labs handling product development, process development and technical service.

By using the TVP model, R&D and business unit leaders can understand the broad responsibilities of R&D and can jointly decide which measures will be appropriate for that business unit. The chief technical officer must work with the CEO and general management to determine which measure will be used to assess broader corporate responsibilities. For corporate labs whose customers are business unit labs, both laboratory managements must agree on the measures they will use to evaluate the productivity and effectiveness of their organization, and this must be understood by both business unit and corporate management.

The TVP model can be used either retrospectively or prospectively, with the time frame that is pertinent to the particular firm. In using the model prospectively, however, one must remember that the projections will only be as good as the assumptions about the probabilities of technical and commercial success that led to them. Thus, prospective measures must always be used cautiously.

Not all measurements are right for all companies, or at all times, for describing or tracking the most important aspects of R&D. However, each company should be able to select a small set of metrics that are appropriate for assessing the value received from R&D and the likelihood of being able to sustain that value. From an R&D perspective, the *critical factors of the moment* are dependent on the situation of the company, the perspective required and the basic dynamics of the model. Let's start from the top.

1. Top Down and Output-Oriented

The TVP model provides a top-down perspective that is output-oriented. Value Creation is the prime driver of overall business returns that are derived from technology-based new products/processes. Metrics that track value are predictors of business growth and (implicitly) a critical input for strategic business reviews. These metrics are used to answer such critical questions as:

- Are we spending the right amount on R&D?
- Are we getting good returns on our R&D?

If the metrics associated with Value Creation are being maintained or going up:

- The corporation has the likely substance to extend a technology-based or innovation-based growth program;
- The investors can expect an extended stream of positive returns from the accumulation of financial payoffs from technology-based innovations; and
- The R&D units enjoy the likelihood of consistent funding to reinvest in technology applications for the near term and base-building for the future.

The key words here are "likely" or "possible." Value Creation is a necessary but not sufficient condition for growth. It is also only a measure of the moment, whether it is looking to the past or to the future. And, any downward movements will predict the difficulties the business will have in achieving solid gains against the competition. These indicators are crucial to assessing the total returns from R&D investments, whether enough is being spent on R&D, and what the likely future value is to the company from a technology perspective.

2. Drivers of "Value Creation"

Value Creation is the result of an accumulation of effort by R&D and the business to produce new ideas and to put the best ones into practice. Due to all of the factors that are involved, momentum is built into these factors over time. They will change, but not rapidly. Change is caused by the drivers of Value Creation. These drivers are strategies that transform the R&D foundations of competencies, know-how, etc. into specific projects and implementation. These are the strategies that are represented and measured by metrics associated with the Portfolio Assessment and the Integration with Business factors. In other words, the portfolio of R&D programs, when reviewed in total, represents the technical strategy of the company, and measures of how well the R&D program is integrated with the business give insight into how relevant the technical strategy is to the business strategy.

When Value Creation is positive, these strategies are most likely working well. When Value Creation is going in the wrong direction, look first to these areas for the cause.

Given this importance to Value Creation, it is no surprise that companies have routinely focused in recent years on methodologies and activities dealing with Portfolio Assessment and Integration with Business issues. These are correctly perceived as the means to the ends that improve the future stream of results from R&D.

Metrics that track Portfolio Assessment describe the state of the various pipelines that run through the R&D enterprise, as well as the targets that are being pursued. They provide a view of how the R&D dollars are being spent in terms of timing, risks and possible returns. Portfolio Assessment is a prime place to look for answers if there are problems with Value Creation, with competitive or market share issues, or with internal satisfaction.

Unlike metrics that track Value Creation, which tend to have modest levels of momentum associated with them and provide significant underpinnings to the business returns, metrics associated with Portfolio Assessment can vary quickly and have little immediate effect on the business. Major effects are cumulative. This dynamic often leads to short-term, risk-averse behaviors that over time undermine Value Creation. Thus, when used properly, measurement of Portfolio Assessment can alert companies to this trend and point to corrective measures before Value Creation turns down. Maintaining an aggressive monitoring of the Portfolio Assessment is extremely important to the long-term support for Value Creation.

Similarly, when problems within the portfolio are corrected, it is necessary to give the solution enough time to work.

If metrics that can be associated with Portfolio Assessment show strategy by *which* categories of R&D and targets are being developed, the metrics associated with integration with Business show a strategy for *how* it is being done, and, consequently, with what level of quality and execution.

Measurements of Integration with Business focus on process, culture, teamwork, and organization. They also touch on many of the aspects of cycle time. The issues addressed by these metrics change slowly and are probably the true pacing items that are applied by the organization to the Portfolio Assessment which, in turn, puts limits on the realizable Value Creation.

When there are difficulties attributable to barriers, for example, that can be removed, then the metrics and the results can be changed relatively rapidly. However, when there are difficulties due to lack of cooperation, lack of contact with the market, lack of good competitive intelligence, or with

a lack of risk taking, then new attitudes and new behaviors are required. This takes time to build into the culture, and the measurements and the results will be slow to change.

3. Business and Technology Leadership

The dynamics of metrics associated with Integration with Business depend on organizational matters, in contrast to those of the Portfolio Assessment, which depend on investment decisions. And, a consideration of organizational *and* investment matters quickly brings the matters of business and technology leadership into the model.

The underlying dynamics do not permit converting management desires for immediate gratification (at low risk) into sustained profitable growth. Presuming a management commitment to doing the right things, there must be allowances made for the application of enough time to link all of the elements on an ongoing basis. Over time, all of the measurements should point to a consistent improvement in the transformation strategies and to the attendant output in Value Creation.

Conversely, monitoring the transformation strategy factors of an otherwise healthy technology-based enterprise will show the early warning signs of any degradation and, to the degree possible, allow timely correction action to be taken before a sound technology foundation begins to crumble.

4. Foundations

The foundations for the strategies represented by the Portfolio Assessment and the Integration with Business factors are built on the Asset Value of Technology and the Practice of R&D Processes To Support Innovation factors. Some of the critical questions regarding these dynamics are:

- Are we becoming more or less productive with our R&D?

- Are we building a strong enough future base of competencies?

- Are we getting an early warning of any declining capability?

How Useful Is the TVP?

The approach advocated in this article was first presented at the Industrial Research Institute Fall Meeting, October 1994 in Williamsburg, Virginia. Following the presentation, workshops explored the acceptance and utility of the Technology Value Pyramid. Two hundred twenty-four attendees from 165 companies completed pre-workshop questionnaires, and 213 attendees from 161 companies filled in post-workshop questionnaires.

Eighty-five percent of the respondents were either R&D directors or chief technical officers. Slightly over half of the respondents were from firms with hybrid R&D organizations, with 32 percent from firms with centralized R&D and 15 percent from firms with decentralized R&D. Except for the engineering/construction industry segment, all segments had at least 10 companies represented, with chemicals leading with 58. The companies were predominantly product oriented (85 percent), with the balance equally split between process and service orientations.

Eighty-five percent of the respondents to the post-workshop questionnaire agreed that the TVP was a useful technique for selecting metrics for managing R&D and communicating with stakeholders.

Respondents considered all managerial fac-tors highly important, with average ratings going from 3.8 to 4.3 (out of 5). No factors of importance were thought to be omitted. There was no significant difference between industries regarding which factors are important, nor which individual metrics are important.

The top 11 metrics from Table 31.1 are listed below in rank order (the rating is not monotonic), and it is interesting to note that all five TVP factors are represented by these top 11 measures.

There was unanimous agreement that it is necessary to consider the interests of different stakeholders in selecting metrics for R&D. While 55 percent of the respondents felt the list of stakeholders was not complete, most suggested modifications were for specific identification of stakeholders that are included in the broad classes already identified. For example, marketing and manufacturing managers can be included in the broad class of "SBU Managers" and, in fact, the menu of metrics does allow one to deal specifically with their interests. Additional stakeholder suggestions worthy of further consideration include both external customers and the public/community. This latter group is of particular concern regarding the environmental performance of the firm and R&D's role in that regard.

Rank*	Metric menu number	Avg. score	Metric
1	1	3.7	Financial Return to the Business.
2	7	3.5	Strategic Alignment with the Business.
3	2	2.6	Projected Value of R&D Pipeline.
4	9	1.9	Sales or Gross Profits from New Products.
5	11	1.7	Accomplishment of Project Milestones.
6	8	1.7	Portfolio Distribution of R&D Projects.
7	24	1.6	Customer Satisfaction Surveys.
8	6	1.6	Market Share.
9	19	1.5	Development Cycle Time.
10	4	1.4	Product Quality & Reliability.
11	5	1.4	Gross Profit Margin.

*Respondents were asked to identify metrics in use, to rank order the top five, and to rank order those thought to be the top five for the firm's CEO.

Foundation factors have the most momentum of any category. They are very slow to change and provide the real rate limitations to growth through technology-based innovation. However, they are also very vulnerable to neglect. They represent significant elements of the culture of the R&D organization and, like most cultural issues, can *deteriorate rapidly* but *improve only slowly*. They need nurturing, leadership and the execution of well-focused technology strategies to become strong elements of a company's growth foundation. And, just as weak transformation strategies degrade Value Creation, they also degrade the foundations.

External Technological Change is the other major dynamic that affects foundations. In these cases, technological breakthroughs undermine traditional technical competencies, bringing in new competitors, toppling current competitors and even redefining the structure of an industry. Although the resulting paradigm shifts take time to develop, R&D and the business are usually both entrenched in the traditional areas and either don't see the changes coming or insist on devaluing the importance until it is too late. The foundations which, if strong, took considerable time to build, must nonetheless be constantly extended and rebuilt to provide options that are the logical growth paths for the business's future. Otherwise, over time, they will erode significantly. Thus, certain of the metrics associated with the Practice of R&D Processes To Support Innovation factor ask for an assessment of either competitors or companies in related markets. Such measures alert R&D leadership to both threats and opportunities within and outside the industry.

Thus, the dynamics of the foundation factors are that they are slow to build and relatively easy to degrade in spite of strong momentum. They are also fundamental to a strong competitive strategy (based on technical core competencies), to R&D creativity, and to productivity. They are a significant contributor to cycle-time reduction, and a rate-limiting source for the options to be taken up by the transformation strategies. The drivers of these factors are the R&D leadership.

GUIDELINES FOR USE

The principal use of the model is for communication and control. This includes communication about, and control of, the overall R&D program, the research program, the business technical programs, and the development of external relationships. The model is not intended to aid project evaluation except as a project would contribute to the improvement of various factors.

Depending on particular needs of stakeholders or of decisions to be taken, different categories and factors should be examined. In addition, time periods should be adjusted for different industries, technologies and types of R&D.

Each of the primary stakeholders will tend to concentrate their attention on different parts of the model. And, while that is logical, it is important to note that all of the factors are connected with time lags by the basic dynamics. It remains for R&D and, in particular, the CTO to make sure that there is awareness of all the factors of the menu, that each stakeholder is reminded of their interconnectedness, and that a consistency is maintained (or that corrective action is taken to achieve consistency) between the collective expectations of the stakeholders and the realities of the model represented by the Technology Value Pyramid.

A HOLISTIC APPROACH

The TVP provides a holistic approach to assessing the value of R&D to the corporation and for guiding improvement efforts when such are deemed necessary. Implementation will require close interaction with the firm's accounting function in order to obtain appropriately detailed sales and cost information. This information is important for developing meaningful calculations of the Value Creation factor. With this information, and with accurate data tracking mechanisms for various R&D activities, the TVP is a powerful tool. By working at the Value Creation level, one can assess value delivered by working retrospectively; prospectively, one can estimate future value creation potential. If the results of these analyses are dissatisfying, one can use the deeper layers of the pyramid to examine potential sources of problems (diagnosis) and to develop appropriate action programs for involvement (prescription). Thus, wise use of the model should allow R&D organizations to improve their innovative output and to develop a means of sustaining innovative output over time.

Use of the model, in and of itself, will not improve an R&D organization; measurements only provide data. We do believe that the set of measurements in the TVP will allow most organizations to select a subset for themselves that can meaningfully report on the effectiveness and productivity of their R&D organization, and the value that it brings to the corporation. Improvement de-

pends on R&D management action consciously taken to improve various measures of performance within the selected set of measurements.

The TVP as currently developed also does not provide a technique for R&D managers to automatically select the measures appropriate for a firm. The ongoing work of the Industrial Research Institute's Subcommittee on Measuring the Effectiveness and Productivity of R&D will provide such a technique. This work is focusing on the development of an expert system-based program that would allow users of the TVP to easily select an appropriate set of measures for their particular circumstances. Future work will also address the utility and validity of the model through use of the TVP concept.

32

Metrics to Evaluate RD&E

JOHN HAUSER AND FLORIAN ZETTELMEYER

OVERVIEW: Metrics affect research decisions, research efforts, and the researchers themselves. From a review of the literature, interviews at ten research-intensive organizations, and formal mathematical analyses, the authors conclude that the best metrics depend upon the goals of the RD&E activity as they vary from applied projects to competency-building programs to basic research explorations. For applied projects, market outcome metrics (sales, customer satisfaction, margins, profit) are relevant if they are adjusted via corporate subsidies to account for short-termism, risk aversion, scope, and options thinking. The magnitude of the subsidy should vary by project according to a well-defined formula.

For RD&E programs that match or create core technological competence, outcome metrics must be moderated with "effort" metrics. Too large a weight on market outcomes leads to false rejection of promising programs. The large weight encourages the selection of lesser-value programs that provide short-term, certain results concentrated in a few business units. This, in turn, leads a firm to use up its "research stock." Instead, to align RD&E with the goals of the firm, the metric system should balance market outcome metrics with metrics that attempt to measure research effort more directly. Such metrics include many traditional indicators.

For long-term research explorations, the right metrics encourage a breadth of ideas. For example, many firms seek to identify their "best people" by rewarding them for successful completion of research explorations. However, metrics implied by this practice lead directly to "not-invented-here" attitudes and result in research empires that are larger than necessary but lead to fewer total ideas. Alternatively, by using metrics that encourage "research tourism," the firm can take advantage of the potential for research spillovers and be more profitable.

Research, Development, and Engineering (RD&E) metrics are important for at least three reasons: First, they document the value of RD&E and are used to justify investments in this fundamental, long-run, and risky venture; second, good metrics enable CEOs and CTOs to evaluate people, objectives, programs, and projects in order to allocate resources effectively; third, metrics affect behavior. When scientists, engineers, managers and other RD&E employees are evaluated on specific metrics, they make decisions, take actions and otherwise alter their behavior in order to improve the metrics. The right metrics align employees' goals with those of the corporation; the wrong metrics are counterproductive and lead to narrow, short-term and risk-avoiding decisions and actions.

The International Center for Research on the Management of Technology (ICRMOT) at MIT's Sloan School of Management is funding an ongoing scientific study of RD&E metrics in order to understand and improve their use in industry. This article is a briefing on that research.

We began the study by interviewing 43 rep-

Reprinted with permission from *Research-Technology Management*, July–August, 1997.

John Hauser is Kirin Professor of Marketing at MIT's Sloan School of Management. **Florian Zettelmeyer** is Professor of Marketing at the Simon Graduate School of Business at University of Rochester.

resentative CEOs, CTOs and researchers at ten research-intensive organizations including Chevron Petroleum Technology, Hoechst Celanese ATG, AT&T Bell Laboratories, Bosch GmbH, Schlumberger Measurement & Systems, Electricite de France, Cable & Wireless plc, Polaroid Corporation, U.S. Army Missile RDEC and Army Research Laboratory, and Varian Vacuum Products [1]. All interviews were conducted in the native languages of the managers. We continued with a comprehensive review of the published literature [2]. We then attempted to abstract and generalize the insights we obtained from these exercises. This resulted in a scientific theory to guide the selection of RD&E metrics [3].

We attempt here to summarize the basic intuition resulting from our research. We hope to highlight the long-term implications of current trends and suggest improvements to current practice. If nothing else, we seek to encourage debate on issues that are fundamental to managing RD&E.

THE TIER METAPHOR

RD&E is a diverse activity. Some applied projects attempt to solve short-term problems faced by a single business unit. There may be little uncertainty as to the outcome of these projects because they are well within the current capabilities of the RD&E organization. At the other extreme, some research explorations seek to build a basic competency in an area of science that is likely to be important to the corporation in the future. These explorations often have a much longer perspective, apply to many business units, and entail considerable risk. Other programs fall somewhere between these extremes. Naturally, the mix of projects, programs and explorations varies by organization. For example, central R&D laboratories often have a greater percentage of basic research programs than RD&E operations within a business unit. Indeed, many individual scientists and engineers have a mix of activities within their own portfolios. However, our observations suggest that every organiza-

tion faces the challenge of integrating these diverse activities.

Best-practice organizations recognize variation and select metrics accordingly. By selecting the right metric for each activity, the firm encourages the right decisions and actions by scientists, engineers and managers. If a firm applies the same metrics throughout the RD&E process, it does not get the most out of its technological efforts. Many of the mistakes that we observed in the field occurred when technology managers attempted to apply the same metrics throughout the process.

To understand better how metrics vary, we introduce a tier metaphor. This metaphor enables us to categorize a diverse continuum of projects, programs and explorations and focus on key characteristics. We define "Tier 1" as basic research that attempts to understand basic science and technology. Tier 1 explorations may have applicability to many business units. Indeed, they may spawn new business units. We define "Tier 2" as those activities that select and develop programs to match or create the core technological competence of the organization. "Tier 3" is defined as specific projects focused on the more immediate needs of the customer, the business unit and/or the corporation. Tier 3 is often accomplished with funding by both business units and the corporation.

For clarity we adopt Lowell Steele's terminology and use the words "objectives" and/or "explorations" for Tier 1 activities, the word "programs" for Tier 2 activities, and the word "projects" for Tier 3 activities [4]. We note, however, that these terms are often used interchangeably in the literature and in practice.

While the word "tier" was used at only a few of the organizations we visited, most firms had a concept that the management of technology varied depending on the stage of the process. (See also [5–8]). For example, the U.S. Army classifies its research as 6.1, 6.2 and 6.3 (and beyond) which corresponds to our metaphor of Tiers 1, 2 and 3. Even the shorthand "RD&E" seems to separate the stages of innovation.

Table 32.1 lists the metrics used by our interviewees, categorized with the tier metaphor. No-

TABLE 32.1
R&D Metrics Reported by Interviewees (from [3])

Category	Qualitative Judgment		Quantitative Measures	
	Metric	Most Relevant	Metric	Most Relevant
Strategic Goals	Match to organization's strategic objectives	Tier 2	Counts of innovations	Tier 2
			Patents	Tier 2
	Scope of the technology	Tier 2	Refereed papers	Tiers 1, 2
	Effectiveness of a new *system*	Tier 2	Competitive response	Tier 3
Quality/Value	Quality of the research	Tiers 1, 2, 3	Gate success of concepts	Tier 3
	Peer review of research	Tiers 2, 3	Percent of goal fulfillment	Tiers 1, 2
	Benchmarking comparable research activities	Tiers 2, 3	Yield = [(quality * opportunity * relevance * leverage)/overhead] *	
	Value of top 5 deliverables	Tier 3	consistency of focus	Tiers 2, 3
People	Quality of the people	Tier 1		
	Managerial involvement	Tiers 2, 3		
Process	Productivity	Tier 3	Internal process measures	Tiers 1, 2
	Timely response	Tier 3	Deliverables delivered	Tier 3
			Fulfillment of technical specifications	Tier 3
			Time for completion	Tier 3
			Speed of getting technology into new products	Tier 3
			Time to market	Tier 3
			Time of response to customer problems	Tier 3
Customer	Relevance	Tier 3	Customer satisfaction	Tier 3
			Service quality (customer measure	Tier 3
			Number of customers who found faults	Tier 3
Revenues/Costs			Revenue of new product in 3 years/R&D cost	Tier 3
			Percent revenues derived from 3- to 5-year-old-products	Tier 3
			Gross margin on new products	Tier 3
			Economic value added	Tier 3
			Break-even after release	Tier 3
			Cost of committing further	Tiers 2, 3
			Overhead cost of research	Tiers 1, 2, 3

tice that some metrics measure incremental profit, some are surrogates for incremental profit (e.g., customer satisfaction and time to market), and some attempt to measure scientific and engineering effort.

We now use the tier metaphor to illustrate how the issues of short-termism, risk aversion, option value thinking, scope, portfolio planning, research spillovers, and research tourism affect RD&E metrics. While these issues are relevant to all RD&E ac-

tivities, the implications and foci of these issues are more intense in some tiers than others. We begin with Tier 3, that is, projects that address shorter-term issues with one or more well-defined customers.

CUSTOMER-DRIVEN RD&E

Many managers, consultants and researchers have argued that in order to succeed, RD&E should be more customer-driven. This viewpoint was reinforced in our interviews. "RD&E has to be developed in the marketplace." "Technical assessment is 'What does it do for the customer?'" In many instances, RD&E managers maintained their budgets by "selling" projects to internal customers, such as the business units.

There is no doubt that the customer is important. In order for the firm to have a good bottom line (profits) it must have a good top line (revenues). Good top-line performance means products and services that fulfill customer needs and satisfy customers [9]. RD&E provides the means with which the firm achieves good top-line performance.

However, our interviewees recognized a downside to a pure customer focus. Instead, many subsidized RD&E projects with central funds. Business units were asked to pay only a fraction of the cost of an R&D project. One CTO stated that the business units were better able to judge an RD&E project if they did not have to pay the full cost. One business unit manager told us about "tin cupping" where she would go around to other business unit managers to ask for contributions to a research project as if she were a beggar with a tin cup. These firms have structured themselves so that the customer is an arbiter, but not the only arbiter, of RD&E funding.

In addition, there is scientific evidence supporting a perspective that all projects should not be entirely customer-driven. Mansfield found that, holding total R&D expenditure constant, an organization's innovative output was driven by the percentage allocated to basic research [10]. Cooper and Kleinschmidt found that adequate resources in RD&E was a key driver separating successful firms from unsuccessful firms [11]. And, based on a survey of 12 chemical firms, Bean suggests that "those chemical firms that exhibited the highest productivity growth in 1987 spent proportionally more for basic research and less for technical service than did firms with lower productivity growth" [12].

In our interviews and analyses we found that both viewpoints have merit. Good project decisions balance customer-driven and research-driven foci. We found that, with the proper central subsidies

and "options thinking," business units could select the RD&E projects that were in the best interests of the firm. However, without well-designed subsidies there was a bias toward short-term, narrow projects with predictable outcomes. We address each of these issues in turn.

SHORT-TERMISM

For an RD&E manager, the costs of RD&E projects occur "today" and are easy to observe, whereas the benefits of RD&E projects are realized many years in the future and may not be attributed to the project. Furthermore, because good scientists, engineers and managers are mobile, they may leave their jobs, or even the firm, before market outcomes such as customer satisfaction, sales or profits can be observed. Even if they stay in the same job, they may not get credit for the benefits. However, scientists, engineers and managers face mortgages, tuition bills and other expenses that occur now and are more predictable than long-term customer-based measures. It is rational for these employees to be more short-term-oriented than the firm.

The impacts of such individual rationality can be dramatic. For example, a ten-year project might be valued by the firm at $100 million given its discount rate, but only $86 million by a business unit manager with a discount rate only 1 percent higher [13]. Thus, the project has a value to the business unit manager that is only 86 percent of the value to the firm. We call this a short-termism ratio of $\gamma = 0.86$.

Furthermore, the short-termism ratio varies by project. The impact of short-termism is more dramatic for projects with long-term payback—the short-termism ratio might be 0.86 for a project with a rapid payback but 0.50 for a project with payback spread over many years. If the firm does not use central subsidies to adjust for this effect, then there will be strong business-unit pressure to fund only the short-term projects. The same phenomenon applies to any rewards, incentives or evaluations of individual scientists, engineers or managers in

RD&E. Because RD&E employees discount the future more than the firm, unadjusted customer-based measures will cause them to favor short-term-oriented projects.

RISK AVERSION

Some projects are more risky than others. A large firm can diversify this risk across many projects, and stockholders can diversify risk across firms. But individual business-unit managers can not diversify risk as easily. If they are risk averse, and most managers are, they will undervalue risky projects. For most situations, we can approximate the effect of risk aversion with a ratio similar to the short-termism ratio. We call this ratio R, for risk aversion [14]. Without central subsidies to counteract this risk factor, business units tend to favor projects with predictable paybacks even though the projects provide less value to the firm.

BENEFITING FROM SCOPE

Often a single business unit funds an applied project, but many business units benefit. For example, Mechlin and Berg illustrate scope by discussing research at Westinghouse on water flows through porous geological formations [15]. This research was done for the uranium mining division, but had substantial additional benefits for heat-flow analyses for high-temperature turbines and below-ground heat pumps, and for the evaluation of environmental impacts (real estate division). Such scope was highlighted many times by our interviewees. The firm, but not the business unit, realizes the benefits of scope.

Thus, we add another ratio—a concentration ratio (α)—to reflect the percent of total benefits that accrue to the business unit for which the RD&E project was completed. (For example, if approximately 40 percent of total benefits accrue to the business unit providing the funding, then $\alpha = 40$ percent.) If the firm does not adjust for such concentration, RD&E customers will favor focused projects over those that have wide applicability.

OPTIONS THINKING

The last concept highlighted by our interviewees was options thinking. The idea is simple: Investing in an R&D project is like buying a financial option to make further investments. If initial investigations justify a further investment, the firm will invest further in the project. (In theory) if further investment is not justified, the firm will abort the project. This means that the value of the project should reflect these investment contingencies—the option value is higher than that which would be calculated if all future investments were locked in. Options thinking implies that an outcome's uncertainty provides an option value. For more discussion, see Mitchell and Hamilton [16] and Faulkner [17].

Options thinking is critical to project evaluation, but it applies equally to the firm and to the business unit. Thus, it does not affect the calculation of central subsidies beyond that already captured by the short-termism, risk aversion and scope indices.

SUBSIDIZING PROJECTS

We found that the magnitude of the subsidy varied by firm and, in some cases, by project. The variation was important because the characteristics of payback-length, risk and concentration varied. Setting the best subsidy for a project was recognized as a critical management challenge. Interestingly, if one interprets "tin-cupping" as an auction in which business units "bid" for RD&E projects, then tin-cupping might be an efficient economic means by which the firm can overcome the short-termism, risk aversion and narrow foci of business unit managers.

We found that firms subsidize RD&E projects in order to align internal customer decisions with those of the firm. When used properly, these subsidies adjust for variations in short-termism, risk aversion and scope. The optimal subsidy (S) is given by the equation: $S = \gamma R \alpha$, where γ is the short-termism ratio, R is the risk aversion index and α is the concentration index.

More importantly, from a research policy perspective, firms should retain central subsidies of research projects. Such subsidies maintain a long-term, wide-scope, balanced mix of RD&E projects. A more subtle message is that S varies by project. Firms that use a single subsidy ratio (and many do) gain in ease of implementation, but they are not achieving the efficiency that is possible with a more flexible process.

METRICS FOR SELECTING TECHNOLOGY

We now consider RD&E activities that attempt to select technology to match or create core technological competence (Tier 2). This is an important function of RD&E. On one hand, these programs are based on the firm's strategic plans and technological capabilities. On the other hand, these programs determine future core technological competence. Our interviewees suggested that it is a critical challenge to decide how heavily customer input should be weighed in the selection of Tier 2 programs. As one of our interviewees said: "The customer knows the direction but lacks the expertise; researchers have the expertise, but lack the direction."

Some firms are attempting to make these decisions customer-driven by using metrics that measure "outcomes"; that is, sales, satisfaction or incremental profit. For example, one popular metric measures RD&E effectiveness by comparing the profit due to new products to the amount spent on RD&E [18]. Many of our interviewees and our analyses suggest that such metrics, when used alone, increase profits in the short-term, but sacrifice the future.

To illustrate this phenomenon, we consider a slightly stylized representation of the process by which RD&E decides to invest in science and technology in order to match or create core technological competence. This stylized representation is based on suggestions by our interviewees that the most critical decision in Tier 2 was to *select* the right programs. Motivating the right amount of scientific, engineering and process effort was impor-

tant, but not as important as program selection. These statements suggest the following process.

Step 1: RD&E chooses one or more programs to develop science and technological capabilities to fulfill (or anticipate) customer needs. Such programs develop resources for competitive advantage.

Step 2: RD&E undertakes initial research to evaluate the program(s). This research determines the potential contribution if the program is successful.

Step 3: RD&E invests scientific, engineering and process *effort* to refine the program into one or more applied projects. This effort matches RD&E's capabilities to the needs of the business units.

The order of the steps is important. Program choice is made before the value of the research is known and before the bulk of the research activity is undertaken. Thus, RD&E faces considerable uncertainty in the outcomes of program choice. If the firm uses metrics that encourage RD&E to choose programs on the basis of anticipated market outcomes, then the impact of risk aversion and short-termism is enormous. The effects are so large that the firm cannot overcome them with program subsidies alone. (This is one way in which programs differ from projects.)

Risk aversion and short-termism lead to false rejection—some programs are rejected that are of value to the firm—and false selection—short-term, certain programs are favored relative to long-term, risky programs that have a larger expected value to the firm. When we mapped the regions for false rejection and false selection, we were surprised at the strength of these effects for program selection. We found that the only way to avoid large regions of false rejection and false selection was to place a low weight on market outcome metrics; that is, low relative to other metrics.

Our interviews and analyses are at odds with recent authors who advocate a simple comparison of market outcomes to research costs (e.g., [19]).

Such metrics could lead to decisions and actions that use up "research stock" by favoring short-term, certain projects. If the firm does not recognize that programs to match or create core technological competence differ from applied projects, then a heavy emphasis on outcome metrics will lead to under-investment in new technologies and science. This, in turn, could lead to long-term ruin.

BALANCING EFFORT AND OUTCOMES

Despite the problem with using market-based outcome metrics for these RD&E activities, we cannot reject outcome metrics entirely. Not only must Step 1 have input from the market, albeit small relative to other metrics, but outcome metrics are critical to Step 3. In Step 3, the firm's goal is to motivate scientists, engineers and managers to allocate the right amount of scientific, engineering and process effort *after* the program has begun. RD&E must incur costs and the people involved must be motivated to work on the projects that are best for the firm. Because these costs are real, individual decisions will only be aligned with the goals of the firm when the rewards are aligned with the rewards to the firm.

To see this another way, consider that costs and individual efforts are easy for scientists, engineers and managers to observe. This means that such costs are given a high implicit weight in any decision. If the firm wants to balance outcomes and costs, it must provide metrics to ensure that RD&E gives the proper weight to outcomes. These arguments imply a larger weight on market-based outcome metrics [18].

We now face a management dilemma. To encourage the right program choice, the firm wants a small weight on market-based outcomes metrics. To motivate the right allocation of effort after program choice, the firm wants a large weight on market-based outcomes metrics.

In theory, a firm can overcome this dilemma if it can find metrics that measure today's RD&E *efforts* without exposing RD&E to the risks inherent in long-term market outcomes. When such metrics are available, the firm can evaluate RD&E on the effort metrics (and thus align potential outcomes with costs). Of course, this means that the effort metrics must correlate with expected, long-term market outcomes. By placing a larger weight on the effort indicators for Tier 2 programs and a smaller weight on market outcomes for Tier 2 programs, the firm attempts to balance the motivations for the right decisions (small weight on market outcomes) and the right effort (larger weight on effort metrics). With the right balance, at least in theory, scientists, engineers and managers will, acting in their own best interests, select the programs that are best for the firm *and* allocate the right amount of effort to completing those programs.

Unfortunately, we found few ideal "effort" metrics. The firms we interviewed attempted to use such metrics as publications, citations, patents, citations to patents, peer review, and other measures as indicators of scientific and engineering effort. However, each metric has potential problems because each metric, taken alone, can be "gamed" by the people involved. No one metric captures all relevant efforts. To overcome these deficiencies, most firms used a combination of metrics.

The recent trend toward a heavy reliance on customer-driven outcome metrics (sales, satisfaction, profit) is counterproductive if it sacrifices long-term benefits in the development of core technological competence. This effect is most pronounced in the choice of science and technology programs. Some "effort" metrics are needed to motivate the right amount of scientific, engineering and process effort. The best metric system uses a combination of outcome and "effort" metrics. The longer term and more risky the research, the lower the weight should be on market-based outcomes. To the extent that they measure effort, we should not reject traditional metrics, such as publications, citations, patents, citations to patents, and peer review. However, each single metric can be "gamed"; hence, these metrics should be used in combination to minimize "gaming" effects.

RESEARCH TOURISM VS. NOT INVENTED HERE

We turn now to the basic research explorations that provide the scientific and technological knowledge upon which Tier 2 programs are based. We call these activities Tier 1 explorations.

Basic research explorations are the most difficult to measure. Not only is the outcome of scientific investigations unknown, but specific business implications are difficult to predict. Furthermore, researchers often have a better idea than management of which explorations will be in the best interests of the firm's core technological strategy.

We found many issues in Tier 1, such as the selection of the best people and the balancing of a high-variance research portfolio, that are covered in the extant literature (see [2] for a review). Here, we focus on how some metric systems encourage "research tourism" and others encourage "not-invented-here" decisions.

By research tourism, our interviewees referred to a common practice among RD&E employees of visiting other laboratories and universities and of entertaining visitors from other laboratories and universities. Attending conferences and reading the literature might also be considered research tourism. The business purpose is to identify and evaluate outside ideas that have the potential to enhance a firm's internal development. The literature calls these outside ideas "research spillovers." If research spillovers are managed correctly, they can be quite profitable. For example, in an econometric study, Jaffe suggests that the indirect effect of research spillovers from competitors is so large that it more than offsets the fact that competitors' RD&E strengthens competitors [20].

However, there is a catch. In order to benefit from research spillovers, a firm must maintain its expertise in the area. The more a firm invests in say, polymers, the more it is able to benefit from outside research activities in that area.

Despite the importance of research spillovers, we found that many firms identify their best people by the internal explorations that these people complete successfully. When researchers are rewarded mostly for internal explorations, they have less incentive to seek outside ideas.

When we analyzed such metric systems we found that the potential for spillovers makes it more profitable for the firm to undertake more *total* (internal and external) basic science explorations than it would in the absence of spillovers. But we found that the most profitable number of *internal* explorations might actually decrease. In other words, a policy of seeking outside ideas could cause internal research empires to shrink. When research tourism is encouraged, and basic science researchers are measured with respect to the outcome of all explorations, whether internal or external, it is possible to align the incentives of the researchers with those of the firm. Researchers will make the decisions that are in the best strategic interests of the firm.

However, we found that if researchers are measured by internal explorations alone, then: (1) They will adopt a "not-invented-here" attitude and spend little or no time on research tourism; (2) they will work on *more* internal explorations than is in the firm's best interests; and (3) the net result will be *fewer* scientific developments. In other words, not-invented-here is the result of the metrics by which research teams are evaluated rather than a generic characteristic of these teams. Poorly designed metrics lead RD&E to spend excessive resources on internal ideas and to devote too few resources to external explorations. Such metrics lead to research empires that are larger than they need be.

Fortunately, many RD&E organizations are recognizing the need to reward explicitly ideas that come from outside the firm. For example, in March 1996, the General Motors Corporation approved a vision statement that included the phrase "Deploy more highly valued innovations, *no matter their source*, than any other enterprise." [Emphasis added.]

RECOMMENDATIONS

Table 32.2 summarizes the recommendations that result from MIT's ongoing research on RD&E met-

TABLE 32.2
Summary of Research Findings

Tier 1
 Basic Research Explorations
  1. Research tourism encourages research spillovers which enhance long-term profitability.
  2. Metrics based on *all* ideas, no matter what their source, match RD&E incentives with those of the firm.
  3. Metrics that reward people only for internal ideas lead to (a) too few ideas, (b) excessive research empires, and (c) "not-invented-here" actions and decisions.

Tier 2
 Programs to Match or Create Core Technological Competence
  1. Metrics must recognize that program decisions differ from decisions on applied projects.
  2. Metrics must recognize that the *choice* of research program is critical and that it is made before most of the scientific, engineering and process effort is undertaken.
  3. Sole emphasis on market-based outcome metrics is counter-productive when *choosing* research programs. Market-based outcome metrics should be used but given a small relative weight.
  4. However, *after* the program is chosen, RD&E must encourage the right amount of scientific, engineering and process effort. This requires *effort* metrics to balance *cost* metrics.
  5. Traditional metrics, such as publications, citations, patents, citations to patents, and peer review, can serve as effort metrics. The best metric systems use a combination of effort metrics and market-outcome metrics to overcome "gaming" issues.

Tier 3
 Applied Projects with or for Business Unit "Customers"
  1. Business units have an important say in the choice of applied projects. However, if they have the only say, then they will choose projects that are shorter-term, less risky, and more focused than is best for the firm.
  2. Subsidies can be used to adjust for short-termism, risk aversion and narrow scope.
  3. To be efficient, the level of subsidy should vary by firm and by project according to the formula $S = \gamma R\,\alpha$.
  4. Options thinking should be used to measure the value of flexibility in decisions to continue projects. (This will lead the firm to accept more uncertainty.)

rics. In this table, we use the tier metaphor to emphasize that a variety of metrics are needed to evaluate and manage RD&E. Metrics that are best for one type of activity may be counter-productive for another type.

REFERENCES AND NOTES

1. Zettelmeyer, Florian and John R. Hauser. "Metrics to Value R&D Groups, Phase I: Qualitative Interviews." Working Paper, International Center for Research on the Management of Technology, MIT Sloan School, Cambridge, MA 02142, March 1995.

2. Hauser, John R. "Metrics to Value R&D: An Annotated Bibliography." Working Paper, International Center for Research on the Management of Technology, MIT Sloan School, Cambridge, MA 02142, March 1996.

3. Hauser, John R. "Research, Development and Engineering Metrics." Working Paper, International Center for Research on the Management of Technology, MIT Sloan School, Cambridge, MA 02142, Jan 1997.

4. Steele, Lowell W. "What We've Learned: Selecting R&D Programs and Objectives." *Research-Technology Management*, March–April 1988, pp. 1–36.

5. Bachman, Paul W. "The Value of R&D in Relation to Company Profits." *Research Management*, May–June 1972, pp. 58–63.

6. Krause, Irv and Liu, John. "Benchmarking R&D Productivity: Research and Development; Case Study." *Planning Review*, January 1993, pp. 16–21.

7. Pappas, Richard A. and Donald S. Remer. "Measuring R&D Productivity." *Research Management*, May–June 1985, pp. 15–22.

8. Tipping, James W., Eugene Zeffren, and Alan R. Fusfeld. "Assessing the Value of Your Technology." *Research-Technology Management*, Sept.–Oct. 1995, pp. 22–39.

9. Hauser, John R., Duncan I. Simester, and Birger Wernerfelt. "Customer Satisfaction Incentives." *Marketing Science*, Fall 1994, pp. 327–350.

10. Mansfield, Edwin. "Basic Research and Productivity Increase in Manufacturing." *American Economic Review*, December 1980.

11. Cooper, Robert G. and Elko J. Kleinschmidt.

"Benchmarking the Firm's Critical Success Factors in New Product Development." *Journal of Product Innovation Management*, Dec. 1995, pp. 374–391.

12. Bean, Alden S. "Why Some R&D Organizations Are More Productive Than Others." *Research-Technology Management*, Jan–Feb 1995, pp. 25–29.

13. For this example, take a time stream of profits (in $millions) consisting of −9, −12, −20, −8, 0, 5, 14, 20, 28, 35, 40, 41, 42, and 43. Use discount rates of 7% for the firm and 8% for the business unit.

14. Technically, R is the ratio of the certainty equivalent of the income stream to the expected value of the income stream. The certainty equivalent is the amount of guaranteed income the business unit would accept in place of the uncertain income stream. For example, if the income stream is relatively uncertain, the business unit manager might accept a guaranteed income stream that is only 90% as large. In this case, $R = 90\%$. (See Ref. 3 for details.)

15. Mechlin, George F. and Daniel Berg. "Evaluating Research—ROI is Not Enough." *Harvard Business Review*, Sept—Oct 1980, pp. 93–99.

16. Mitchell, Graham R. and William F. Hamilton.

"Managing R&D as a Strategic Option." *Research-Technology Management*, May–June 1988, pp. 15–22.

17. Faulkner, Terrence W. "Applying 'Options Thinking' to R&D Valuation." *RTM*, May–June 1996, pp. 50–56.

18. It is possible to derive these implications mathematically with a set of methods known as "agency theory." The basic idea is to set up equations for how scientists, engineers and managers, the "agents," will react to the metric system and then adjust the metrics until the agents, acting in their own best interests, choose those actions and make those decisions that are in the best interests of the firm. For details see (Ref. 3). These equations balance the effects of outcome metrics, cost metrics, risk, and short-termism.

19. McGrath, Michael E. and Michael N. Romeri. "The R&D Effectiveness Index: A Metric for Product Development Performance." *Journal of Product Innovation Management*, 11, 1994, pp. 213–220.

20. Jaffe, Adam B. "Technological Opportunity and Spillovers of R&D: Evidence for Firms Patents, Profits, and Market Value." *American Economic Review*, December 1986, pp. 984–1001.

V

MANAGING INNOVATIVE CLIMATES
IN ORGANIZATIONS

11

Creating Innovative Climates

33

A Skunkworks Tale

THOMAS J. PETERS

Innovation is unpredictable. It thrives in the chaos of "skunkworks," where product champions go scrounging for success.

Yellow Post-it Note Pads have quickly become as commonplace in the American office as paper clips. The product is a $100 million winner for 3M. The idea behind it came from a 3M employee who sang in a choir. The slips of paper he used to mark the hymnals kept falling out, and it dawned on him that adhesive-backed pieces of paper might solve his problem.

The requisite technology existed, and a prototype was soon available. "Great story," you say—but wait, this tale's not quite over yet. Major office-supply distributors thought the idea was silly. Market surveys were negative. But 3M secretaries got hooked on the product once they actually used it. Post-it's breakthrough finally came when 3M mailed samples to the personal secretaries of Fortune 500 CEOs, using the letterhead of the 3M chairman's secretary.

The Post-it story would amount to nothing more than a charming tale were this development

Thomas J. Peters is founder and president of The Tom Peters Group.

process not repeatedly played out at companies across the U.S. The course of innovation (idea generation, prototype development, contact with an initial user, and breakthrough to the final market) is highly uncertain. Moreover, it will always be sloppy, disorganized, and unpredictable, and that is the important point. It's important because we must learn to design organizations that explicitly take into account the unavoidable sloppiness of the process and use it to their advantage rather than fight it.

From America's best-run companies come tales of incredible perseverance, countless experiments, perverse and unusual product-users, five-person "skunkworks" sequestered in dingy warehouses for 60 days, plans gone awry, inventions made in the wrong industry at the wrong time for the wrong reason, and specifications for complex systems scrawled across the backs of envelopes. Innovation just doesn't happen the way it's supposed to.

THE 10 MYTHS OF INNOVATION

The hyperorganized approach leads companies to fall prey to the 10 myths of innovation management, which have already hampered many a firm. These false beliefs must be put to rest, and the sloppy side of innovation must be exploited. The 10 myths are as follows:

1. Specs and a market plan are the first steps to success.
2. Detailed strategic and technological plans greatly increase the odds of a no-surprises outcome.
3. Only a big team can blitz a project, especially if it is a complex one.
4. Contemplation stimulates creativity.
5. Big projects are inherently different from small projects and must be managed differently.
6. An organization must have a rigid hierarchy if would-be innovators are to get a fair hearing.
7. Product compatibility is the key to economic success.

8. Customers will tell you only about yesterday's needs.
9. Technology push is the cornerstone of success.
10. Perfectionism pays off.

Some companies love to make plans more than they love to make profitable new products. In these bureaucratic behemoths, someone's bright idea is turned into a six-month, $2 million study—a paper study. A paper evaluation of the study by various interested parties takes another three months. Some sort of design go-ahead is given, and writing the technical specs, at a cost of $3 million, takes six more months. The specs are evaluated in the four months after that.

During this last stage, a prototype is finally built. It costs $5 million to $10 million and takes four to six months to complete. And guess what? It doesn't work. Throughout the history of successful corporation innovation—from the development of French-fries seasoning at McDonald's to faded jeans at Levi Strauss & Company to the System 360 computer at IBM—neither the first nor the second prototype has *ever* worked. The successful innovators just go back to the drawing board.

But now the people in charge of the project really begin to sweat. By this time, careers are on the line and a lot of time, money, and pride has been invested in the design. So now they enter the "ignore the misfit data and make the damn think work" stage. Meanwhile, the competitors have introduced three or four new products, each with several new features. As time goes by, the plodding planners fall further behind. So they recomplicate the product. "We're going to get it exactly right," they boast. But when they finally get it to the marketplace, it's adorned with so many bells and whistles that it doesn't work well.

This mentality is the antithesis of the Wee Willie Keeler approach. Wee Willie Keeler was a consummate opportunist who played baseball from 1892 to 1910. He once said, "Hit 'em where they ain't," and he proved that strategy's worth by making it into the Baseball Hall of Fame even though he stroked only 34 home runs in 19 seasons. His approach is imitated to a T by firms like Hewlett-

Packard, 3M, McDonald's, Wang Laboratories, PepsiCo, Citicorp, Johnson & Johnson, Digital Equipment, and others like them. This philosophy says, in effect: *Start out by spending $25,000, or even as much as a quarter of a million dollars. Build a prototype, or a big hunk of one, in the first 60 to 90 days. And then poke it to see if it moves.*

Whether projects like these involve aircraft, missiles, or French fries, the results achieved by scores of companies suggest that something can always be built in this length of time. The evaluation of the prototype should take another 60 days. (Even at such an early stage, firms following this approach may decide—explicitly or not—to start up a second team doing roughly the same work as the first, just to get a different look.)

"We're already playing with something tangible," say project leaders at these companies. "Now we take the next little step. We build another new version in 90 days. It's a more developed prototype that will cost a little more, around $100,000 to $200,000. After it's built, we can probably get it, or part of it, into a user's hands—not an average user (that's still years away), but a lead user who's willing to experiment with us. Even an in-house lead user might do the trick." And on goes the process, always involving investments that increase little by little and time-frames that do the same.

At each step the innovators learn a little more, because they set up harsh reality tests with hard products and real users. If something doesn't work, they weed it out quickly before career lock-in and irreversible psychological addiction to hitting home runs take place.

In the aircraft manufacturing industry, one such harsh confrontation with reality is known as the chicken test. Aircraft engines have to be built to withstand possible ingestion of flocks of birds. To determine what would happen in that unlikely event, engineers buy 15 or 20 gross of chickens, stuff some into a cannon with a barrel four feet in diameter, and fire them at engines running full throttle. It's the ultimate pragmatic tests. Rolls-Royce spent several years and several hundred million dollars on a new graphite-material engine. After all that, it failed the chicken test.

What the Wee Willie Keeler, or experimental, approach boils down to is getting your inevitable chicken test out of the way early. Every new product fails a chicken test or two at some point. The burning issue is, when does it fail? At the end of four years, by which time the competitors have a new array of products on the market? Or at the end of 90 days?

BUREAUCRACY UNDER ATTACK

Strategic planning is being attacked on all fronts. Many claim that it is too rigid. Others say it's too bureaucratic. Some believe that corporations should at least decentralize such planning (General Electric and Westinghouse, who were early pioneers in strategic planning, are doing just that). A few even suggest that we get rid of it altogether.

But do we really want to do that? The new "in" terms are *technology* and *production*. "Technology planning" and "manufacturing planning" are the preferred substitutes for strategic planning. Before heading off down a new trail, though, let's look at the record of technology planning. It's hardly spotless.

Think of recent inventions that we're all familiar with. "We do not consider that the aeroplane will be of any use for war purposes," declared the British minister of war in 1910. In the late 1940s, market research predicted that the total sales of mainframe computers would be about a dozen. Even though the robotics industry is crowded with such competitors as United Technologies Corporation, General Electric, Westinghouse Electric, and IBM, the first "intelligent mobile robot" will come from a less-than household name—Denning Systems Inc. of Washington, D.C., a classic three-inventors-in-a-garage operation.

A highly systematic analysis of this phenomenon may be found in the book *The Sources of Invention* by John Jewkes, a professor of economic organization at Oxford University. After studying the development of 58 of this century's major inventions, Jewkes concluded that at least 46 of them occurred in "the wrong place." Note the unusual origins of the following inventions:

- Kodachrome film was invented by a couple of musicians.
- A watchmaker fooling around with brass castings came up with the process involved in the continuous casting of steel.
- The developers of the jet engine were told by reciprocating-aircraft-engine people that it was useless. (They finally peddled their invention not to engine-makers but to airframe-makers.)

According to Jewkes, there is no industry group in which much innovation has taken place as or when it was supposed to. On top of this, "the initial use and vision for a new product is virtually never the one that is ultimately of the greatest importance commercially," reports Jim Utterback, an MIT associate professor of engineering who for more than a decade has studied the development of inventions. He has concluded that users play a special role in this process. To support his point, he recounts the path to success of invention after invention.

His analysis of incandescent lighting is typical. Its first use was on ships, which in retrospect seems natural enough: it's dangerous to keep gas lamps on a seafaring vessel, whose rolling motion can upset them. Thus the incandescent light found its first home in a highly specialized market niche. Then, in a move that *every* market research department could easily have predicted, incandescent lighting spread to—baseball parks! Night games have been with us ever since. From there the invention moved to neighborhoods, where it replaced gas streetlamps, and only 15 years later did incandescent lights begin to make it into homes. As a more recent example of this pattern, transistors were first used for missile guidance systems; their use by home consumers lagged 20 years behind.

The role of corporations in all this is truly frightening. Organizations have an apparently inherent tendency to make exactly the wrong moves in trying to stimulate innovation, according to Utterback, who states, "In 32 of 34 companies, the current product leaders reduced investment in the new technology in order to pour more money into the old."

Not only, then, does the leader *not* embrace the new, he actually reduces his investment in the new to hold on to the old. The problems involved in switching to a new technology are manifold. First, there's scientific hubris (the engineer knows best, he can predict the use of the product most accurately); then comes marketing hubris (how could all those tons of data on the Edsel be wrong?). Jewkes offers three rules of thumb regarding technological planning, all of which are well worth heeding:

- Peering into the future is a popular and agreeable pastime that, if not taken seriously, is also comparatively innocuous.
- There is a great virtue in picking and choosing from a variety of available options.
- The industrial laboratory does not appear to be a particularly favorable environment for the inducement of innovation.

Does this mean that corporations should do away with central planning? Should centralized R&D activity be abolished? The answer is no. First, one does need to make general bets on technological directions: it's important to know the difference between, say, north and northwest. That's fine. What isn't sensible is trying to prespecify the difference between a course of 43 degrees and a course of 46 degrees. As a former managing director at the consulting firm McKinsey & Company liked to argue, "About the best you can hope for is to get the herd heading roughly west." And this is a task that centralized research can do.

"As a regimen or discipline for a group of people, planning is very valuable," notes Fletcher Bryom, the iconoclastic former chairman of Koppers Company. "My position is, go ahead and plan, but once you've done your planning, put it on the shelf. Don't be bound by it. Don't use it as a major influence on the decision-making process. Use it mainly to recognize change as it takes place."

QUICK-AND-DIRTY SOLUTIONS

When the U-2 spy plane emerged as the country's most sophisticated airborne surveillance system 30 years ago, many experts said that it would never fly. It's still doing yeoman service. The developers were a retired aeronautical engineer named Kelly Johnson and a small band of Lockheed Corporation mavericks. They called their off-line group "the Skunk Work"—the original business use of an apt term that (as far as I can determine) may have been coined by Al Capp, who drew the comic strip *Li'l Abner.*

Lockheed is not unique. At GE the same activity is called "bootlegging"; at 3M they label it "scrounging." It would not be difficult to argue that 3M, Hewlett-Packard, Digital Equipment, and Johnson & Johnson are today nothing more than collections of skunkworks.

The finding stands out more and more clearly as the evidence rolls in: whenever a practical innovation has occurred, a skunkwork, usually with a nucleus of six to 25 people, has been at the heart of it. Most skunkworks seem to do things in an incredibly short period of time. While visiting a Westinghouse lab, General Curtis LeMay, then Chief of Staff of the Air Force, found a pencil sketch of what was at the time a beyond-the-state-of-the-art product: a side-mounted radar. He asked if he might have one within 90 days. The next day he sent Westinghouse an airplane to hang it on. He got his device less than 90 days later. In the recent book *The Soul of a New Machine* by Tracy Kidder, Data General's computer-project leader, Tom West, speculates that the company's crucial breakthrough in microcoding may have taken place in less than a week. [*Editor's note:* Microcoding builds into a computer the instructions that make it operate.]

But what happens with a quick-and-dirty skunkwork project? Is the quality as high? Does it ever fit into the rest of the product line? The record shows, delightfully, that the stuff that comes from skunkworks is often of high quality, even though it was invented in a fraction of the so-called normal time.

The creative impetus behind skunkworks boils down to ownership and commitment. In *The Soul of a New Machine,* West describes the phenomenon: "There are 30 guys out there who think they've invented it; I don't want that tampered with." Firms like 3M, Johnson & Johnson, and Hewlett-Packard all agree that in creating the sense of ownership, intense commitment, and unbounded energy that comes from turned-on teams, a surprisingly small group is optimal.

A struggle against others is also important. It, too, engenders feelings of ownership and commitment. Interestingly, its most important form is rivalry with others *inside* the company, not with an outside competitor. Few companies are really familiar with their competitors, but their divisions sure know one another. Constructive internal competition is difficult to manage. There are a great number of subtleties and traps. The net result, however, is almost always positive.

The skunkwork cannot do all things. On the other hand, the empirical indications seem to say, loud and clear, "Ignore this form of organization for innovation at your own peril." The alternative is *de novo* design of the tiniest parts, excessively long product-development cycles, large teams in which ownership and commitment are missing, do-everything-inside attitudes, overcomplexity, and situations in which competing central staffs make the decisions on technical issues or delay them endlessly on the basis of the most tenuous market or financial projections. The show just doesn't get on the road.

HELL-BENT ON SUCCESS

If big, well-orchestrated teams were at the heart of successful innovation, we would expect to find them populated with powerful thinkers who regularly ascended to their mountaintop retreats to look out over the pines. As a result of such reflection, they would accomplish the necessary breakthroughs, presumably on schedule. If, on the other hand, rough-and-tumble skunkworks, hell-bent on outproducing some formal group, were the norm, we would expect to find bleary-eyed folks staring at computer screens or test tubes in dirty, forgotten basement corners.

It does turn out that bleary eyes play quite a large role in innovations. When a year's worth of work is routinely accomplished in five weeks, someone called a "champion" will be found at the heart of the operation. Formal IBM in-house studies of research projects always unearth a champion. National Science Foundation studies suggest that the champion's role in pushing an idea to fruition is crucial. When the brand manager of a consumer-goods company, even in a highly structured system, becomes a determined champion, the odds of success go up tenfold. Looking back over his career in *Adventures of a Bystander,* Peter Drucker, the noted business expert, remarks, "Whenever *anything* is accomplished, it is being done, I have learned, by a monomaniac with a mission."

A crucial corollary is that the corporation that would nourish inventors must also tolerate, even praise, failure. Going through 3M's roster of senior officers with one of the company's executives a couple of years ago, I discovered that virtually every 3M officer had reached the top because he himself had introduced several important new products. Moreover, each story, as it was recounted in conventional form, focused on the rough places in the road: the 10 years of ups and downs when the product was too advanced for the marketplace, when it had to be reformulated, when the manufacturing scale-up didn't work. Setbacks are considered standard operating procedure. Above all else, the winners are those who persist.

SMALL WITHIN BIG

Massive projects like the manned space program or the development of the transistor at AT&T Bell Labs aren't that different from less complex undertakings; they're just bigger. They too can be treated, to a substantial degree, as collections of skunkworks. In an important sense, the principle "small within big" turns out to be essential to the success of big projects. Most of the breakthroughs

in these cases are the results of champions' operating off-line. Charles Brown, chairman of AT&T, said recently, "Today the long-distance network looks like one big, perfectly conceived solution. The reality that we often forget when we think about innovation planning is that the network is a collection of thousands of small breakthroughs that occurred here and there, and certainly not according to schedule or by courtesy of a flawless master plan."

The story of Boeing's recent development of the air-launched cruise missile is even more pertinent. The system is complex. Undoubtedly it should have been developed all at once, with the aid of a 100,000-bubble PERT chart (a "program evaluation and review technique" diagram that indicates the relationships among the phases of a project). The missile-development program was in fact broken down into seven major pieces. Modest-size teams were assembled to deal with the seven projects. Each task was then accomplished in a remarkably short period of time relative to the norm. Each had a champion. Each was in competition with all the others on several vital fronts.

Then what happened? You guessed it. Put the seven pieces together and they don't fit exactly right. So you have to spend some time, as much as a few months, getting the interfaces just right, despite the prior effort that went into interface specifications. (Twice-a-week meetings of a "tie-breaker" group sorted out many of the issues in question.) The final design isn't as technically beautiful as ideals of theoretical perfection suggest is possible. But multiple passes usually take less time and result in the development of simpler, more practical systems than a single everything-at-once pass. (Boeing's cruise missile was delivered more than a year ahead of schedule and well under budget.)

But back to the question of whether big differs from small. There is no question that it does. The Boeing 767 and the French-fries seasoning change at McDonald's are not the same. On the other hand, commitment, championing, small within big, piece-versus-piece competition, the overtight deadline, and the turned-on modest-size group are the keys to breaking down a big, forbidding task into smaller, more manageable ones.

CHARGED-UP TEAMS

The conventional wisdom holds that only a strong functional monolith will keep the engineers' (and innovators') viewpoints to the fore. It's a nice argument on paper, but it doesn't hold much water in practice. What actually happens is that engineers lose out to marketing and finance people in divisional organizations. The divisions are interested only in short-term profit.

By definition, the functional monolith is almost always bureaucratic; it's not oriented to commitment and small-team action. Too many firms force creative people to work on five or six projects that span three or four divisions. But my experience on this one is crystal-clear. No one with one-seventh of the responsibility for anything ever felt committed to it. Peter Drucker's "monomaniacs with missions" were not monomaniacs with *seven different* missions.

Under some forms of management, divisional organizations that grow too big become hopelessly bureaucratic. On the other hand, "the division is the solution" (and the strategy) for Hewlett-Packard, 3M, Johnson & Johnson, Emerson Electric, and the like. Johnson & Johnson constantly creates new divisions. Its corporate watchword is simple: "Growing big by staying small."

These companies carefully monitor the size of their divisions. At HP, divisions are kept to less than a thousand people so that, in president and CEO John Young's words, "the general manager will know all his people by their first names." Bill Gore, chairman of W.L. Gore & Associates, comments, "As the number of people in an organization approaches 200, the group somehow becomes a crowd in which individuals grow increasingly anonymous and significantly less cooperative." The low numbers, whether 200 or 1,000, are all aimed at enhancing ownership and commitment.

Another vital part of the small-team, small-division mentality is the ability to manage, with rel-

atively little muss and fuss, the bureaucratic conflicts that fatally delay much development. As an old hand at skunkworking once said, "Let's be clear about the magnitude of the effects that small teams have. The charged-up team that contains 10 to 50 people isn't in the '10 percent productivity improvement' game. Its results are often 300, 400, even 700 percent beyond those achieved by larger groups."

GET IT OUT THE DOOR

Some firms don't believe in meeting product-release dates. First the date is pushed back three months; then it gets shoved back another 45 days. All the while the bosses are thinking, "We've got to make sure that the software is totally compatible with all the rest of the product family." So the logic goes.

Compatibility is important, particularly in the case of systems-related high-technology products. But sometimes the last 2 percent that's needed for 100 percent compatibility takes 12 months to achieve. Meanwhile, 10 competitors have found a solution to the problem and gotten their products to the marketplace. In such extremely fast-paced markets as data handling, computers, and telecommunications, though, there are literally thousands of entrepreneurs who will fill in the spaces and do the last 2 percent of the work for you.

Digital Equipment's products overlap; users occasionally found that some of its products are incompatible with products that they're supposed to be compatible with. HP's engineers, marketers, and salesmen also lament the incompatibility of some of their products. But companies that wait, trying to achieve the last percentage point of compatibility, may well go belly-up.

The same principle holds true in many other markets, although they show a little less intensity. That's the reason Procter & Gamble, 3M, Mars, and Johnson & Johnson are so insistent about spurring competition among their own divisions and brand managers. Bloomingdale's does the same thing with buying and floor-space assignments in its stores, and Macy's has done extremely well emulating Bloomingdale's. In most markets, new things are happening all the time. The lion's share is often virtually invisible—that is, you frequently don't see it until it's too late. To keep up with the competition, you have to keep getting new items into the market.

Errors of premature release can be (and frequently are) disastrous. Often a product hits the marketplace before the bugs have been worked out. Its technical superiority is blunted by poor reliability or insufficient support. This type of nightmare must be avoided at all costs. But getting that last possible feature, that last degree of complexity (read "overcomplexity"), that last percentage of compatibility, may cost you more of the market than you would have gained by making a perfect product. Unfortunately, the perfectionists tend to get their way because they always use the argument "It'll only take us another 30 days." But we all know that those 30-day projects always seem to take 120 days—if you're lucky.

CUSTOMERS GENERATE IDEAS

The evidence is overwhelming: the great majority of ideas for new products come from the users. Eric von Hippel, a professor at MIT's Sloan School of Management, has studied scientific-instrument equipment manufacturers, and his results are revealing. He reviewed 160 inventions and found that more than 70 percent of the product ideas originated with users. And these weren't just bells-and-whistles ideas, either. Sixty percent of the minor modifications came from users, as did 75 percent of the major modifications. But astonishingly, *100 percent* of the so-called "first of type" ideas for sophisticated devices like the transmission electron microscope were user-generated. According to von Hippel's studies, users that got their ideas across to the producers did a lot more than whisper into their ears. The users came up with the ideas, they prototyped them, they debugged them, and they had them working. Only then did they tap the produc-

ers for their experience in reliable production of multiple copies.

Lead users don't have to be Ph.D.'s or work in germ-free labs. One classic lead user was a housewife whose husband worked at the Corning Glass labs. One day he took home a new glass container that he was going to store acid in. She accidentally used it to heat some food in the oven, and it didn't break. Such is the origin of Pyrex cookware!

Stay in touch with users. It's important in every industry from fast food to computers. Hewlett-Packard has coined the term MBWA—"Management by Wandering Around." Wandering around should mean listening to the user in a direct, not an abstract or shorthand, way. A general manager who designed a major new computer describes a neat trick he pulled off: "I bought my uncle a computer store. I spent nights and weekends working there. My objective was to stay close to the ultimate user, to observe his frustrations and needs firsthand and incognito." What he learned was reflected in the eventual computer design in a thousand little ways and several big ones.

SERVICE AND QUALITY COUNT

"More scientists in bigger labs" seems to be the conventional watchword, along with "Better planning, better tools." The heck with skunkworks. But it's more than skunkworks. It's more than listening to users, too. Service and quality hold as much value as gee-whiz technology—or more.

Recently I talked with the president of a technology company about commodities. He was disturbed by some people's unfortunate tendency to call high-technology products (chips, instruments, personal computers) "commodities." The problem with this is that if you label a specific product a commodity, you'll start to behave as if it is one, neglecting service and quality. For instance, let's take a mundane product: toilet paper. If you go to your local grocery store and purchase a four-roll, 220-square-foot package of one-ply generic-brand toilet paper, the price will be around 79 cents. But if you go to a Seven-Eleven-type grocery, a package of Procter & Gamble's Charmin will cost you $1.99. The difference in distribution channels (Seven-Eleven) and the quality difference (P&G) is obviously enough to add $1.20 to a 79-cent product—or, more accurately, to add $1.20 to a product that cost about a quarter to produce.

Technology push is crucial, but it is not the principal reason that America is undergoing so many industry setbacks. User-unfriendliness, the inability to realize that the customer perceives a product in his own terms, is at least as big a weakness. If you don't believe me, ask 'em in Detroit.

PERFECTIONISTS FINISH LAST

If it weren't for people, 10,000-person research groups would be the most efficient. If it weren't for people, execution via 100,000-bubble PERT charts would be the most efficient. If it weren't for people, huge amounts of money invested in technical forecasting would allow companies to anticipate competition, customer-related problems and technological surprises. If it weren't for users, in-house development of every part that went into every invention would be the best way to assure quality.

Optimization. What's optimal? It's hard to believe, but the "suboptimal" system is often the most truly optimal. Go back to the big-versus-small debate. As a way to do the job, skunkworking is faster, cheaper, and higher-quality than the optimization route. Getting 90 percent compatibility and letting the marketplace do the rest turns out to be optimal, not suboptimal. Getting the last 10 percent may cost you 60 percent of the market.

Tom West of Data General didn't care a whit about building a machine that the "technology bigots" would like. He was interested in people who "wanted to get a machine out the door with their name on it." The stories about the U-2, the missile-development program broken down into seven parts, and the Post-it pads seem to be the same. Committed people, people competing against the market and other corporations and other divisions, those are the people who get the job done. Hail to the skunkworks!

34

The One-Firm Firm
What Makes It Successful

DAVID H. MAISTER

What do investment bankers Goldman Sachs, management consultants McKinsey, accountants Arthur Andersen, compensation and benefits consultants Hewitt Associates, and lawyers Latham & Watkins have in common? Besides being among the most profitable firms (if not *the* most profitable) in their respective professions? Besides being considered by their peers among the best *managed* firms in their respective professions? The answer? They all share, to a greater or lesser extent, a common approach to management that I term the "one-firm firm" system.

In contrast to many of their competitors, one-firm firms have a remarkable degree of institutional loyalty and group effort that is clearly a critical ingredient in their success. The commonality of this organizational orientation and management approach among each of these firms suggests that there is indeed a "model" whose basic elements are transferable to other professions. The purpose of this article is to identify the elements of this model of professional firm success and to explore how these elements interact to form a successful management system.

METHODOLOGY

The information on specific firms contained in this article has been gleaned from a variety of "public domain" sources, as well as selected interviews (on and off the record) at a number of professional service firms, including but not restricted to those named herein. However, none of the information presented here represents "official" statements by the firms involved. As with most professional service organizations, the firms discussed here are private partnerships with no requirement, and with little incentive, to expose their inner workings. Consequently, public information on the management practices (and economic results) of such firms is difficult to obtain.

This situation is regrettable because the professional service firm represents the confluence of two major trends in the U.S. (and worldwide) economy: the growing importance of the service sector and the increasing numbers of "knowledge workers." As a result, any lessons that can be learned about successful management of such enterprises could potentially be of importance not only to the professions but also to other service entities and organizations grappling with the problems of managing large numbers of highly educated employees.

In an attempt to discover the principles of "good management" of professional service firms, I have worked very closely with a broad array of service firms in a variety of capacities. My research has been driven by two propositions: first, that professional service firms are sufficiently different from industrial corporations to warrant special study; and second, that the management issues faced by professional service firms are remarkably similar, regardless of the specific profession under

David H. Maister is President of Maister Associates, a Boston-based consulting firm specializing in the management of professional service organizations.

consideration. I have chosen in this article to concentrate on the second proposition.

WHAT IS MEANT BY "WELL MANAGED?"

The firms chosen for discussion were identified in the following way. In the course of my research and consulting work, I have made it my practice to ask repeatedly the question, "Which do you consider the best managed firm in your profession?" The question is, of course, ambiguous. In any business context, "well managed" can be taken to refer, alternatively, to profitability, member satisfaction, size, growth, innovativeness, quality of products or services, or any of a number of other criteria. The difficulty in identifying "successful" firms is particularly acute in the professions because many of the conventional indicators of business success do not necessarily apply. For example, since there are few economies of scale in the professions,[1] neither size nor rate of growth can be taken as unequivocal measures of success: many firms have chosen to limit both. Even if "per-partner" profit figures were available (which they are not), they would also be unreliable measures, since many professional firms are prepared to sacrifice a degree of profit maximization in the name of other goals such as professional satisfaction and/or quality of worklife. Finally, since "quality" of either service or work product is notoriously difficult to assess in professional work,[2] few reliable indicators of this aspect of success are obtainable.

In spite of these difficulties, it has been remarkable how frequently the same names appear on the list of "well-managed" firms in the professions, as judged by their peers and competitors. The firms discussed here were on virtually everyone's list of admired firms, often together with the comment, "I wish we could do what they do." It should be noted that other firms, not discussed here, were also mentioned frequently. However, as expressed earlier, what makes Arthur Andersen, Goldman Sachs, Hewitt Associates, and Latham & Watkins worthy of some special attention is not only that

they are successful and well respected, but that, in spite of being in different professions, they appear to share a common approach to management (the one-firm firm system) that is readily distinguishable from many of their competitors. This approach is clearly now the only way to run a professional service firm, but it is certainly *a* way that is worthy of special study.

THE "ONE-FIRM FIRM" SYSTEM

Loyalty

The characteristics of the one-firm firm system are institutional loyalty and group effort. In contrast to many of their (often successful) competitors who emphasize individual entrepreneurialism, autonomous profit centers, internal competition and/or highly decentralized, independent activities, one-firm firms place great emphasis on firmwide coordination of decision making, group identity, cooperative teamwork, and institutional commitment.

Hewitt Associates (described along with Goldman Sachs in a recent popular book as one of "The 100 Best Companies in America to Work For")[3] says that, in its recruiting, it looks for "SWANs": people who are *Smart*, *Work* hard, and *Ambitious*, and *Nice*. While emphasis on the first three attributes is common in all professional service firms, it is the emphasis on the last one that differentiates the one-firm firm from all the others. "If an individual has ego needs that are too high," notes Peter Friedes, Hewitt's managing partner, "they can be a very disruptive influence. Our work depends on internal cooperation and teamwork."

The same theme is sounded by Geoffrey Boisi, the partner in charge of mergers and acquisitions at Goldman Sachs: "You learn from day one around here that we gang-tackle problems. If your ego won't permit that, you won't be effective here."[4] By general route, Goldman has achieved its eminance with a minimum of the infighting that afflicts most Wall Street firms. In contrast to many (if not most) of its competitors on the street, Goldman frowns upon anything resembling a star system.

Downplaying Stardom

The same studied avoidance of the star mentality is evidenced at Latham & Watkins. As Clinton Stevenson, the firm's managing partner, points out: "We want to encourage clients to retain the firm of Latham and Watkins, not Clint Stevenson."[5] Partner Jack Walker reinforces this point: "I don't mean to sound sentimental, but there's a bonding here. People care about the work of the firm."[6] The team philosophy at McKinsey, one senior partner explained to me, is illustrated by its approach to project work: "As a young individual consultant, you learn that your job is to hold your own: you can rest assured that the team will win. All you've got to do is do your part."

Above all else, the leaders and, more important, all the other members of these firms view themselves as belonging to an *institution* that has an identity and existence of its own, above and beyond the individuals who happen currently to belong to it. The one-firm firm, relative to its competitors, places great emphasis on its institutional history, broadly held values, and a reputation that all actively work to preserve. Loyalty to, and pride in, the firm and its accomplishments approaches religious fervor at such firms.

Teamwork and Conformity

The emphasis on teamwork and "fitting in" creates an identity not only for the firm but also for the individual members of the firm. This identity, for better or worse, is readily identifiable to the outside world. Reference by others in the profession to members of one-firm firms is not always flattering. Members of other Big-8 firms, particularly those where individualism and individual contributions are highly valued, often make reference to "Arthur Androids." The term "A McKinsey-type" has substantive meaning in the consulting profession— sometimes even down to the style of dress. In the 1950s, I am told by a McKinsey-ite, a set of hats in the closet of a corporation's reception room was an unmistakable sign that the McKinsey consultants were in. The hats have disappeared, but the mentality has not. Goldman Sachs professionals are referred to by other investment bankers as the "IBM clones of Wall Street."

Long Hours and Hard Work

For all the emphasis on teamwork and interpersonal skills, one-firm firm members are no slouches. All of the firms discussed here have reputations for long hours and hard work, even above the norms for the all-absorbing professions in which they compete. Indeed, the way an individual illustrates his or her high involvement and commitment to the firm is through hard work and long hours. Latham & Watkins lawyers are reputed to bill an *average* of 2,200 hours apiece, with some heroic performers reaching the heights of 2,700 hours in some years: this contrasts with a professionwide average of approximately 1,750. At Goldman Sachs, sixteen-hour days are common. It has been said: "If you like the money game, here's [Goldman's] a good team to play on. If you like other games, you may not have time for them."[7] James Scott, a Columbia Business School professor, has commented: "At Goldman, the spirit is pervasive. They all work hard, have the same willingness to work all night to get the job done well, and yet remain in pretty good humor about it."[8] Similarly, McKinsey, Hewitt, and Arthur Andersen are all hard-working environments, above the norms for their respective professions.

Sense of Mission

In large part, the institutional commitment at one-firm firms is generated not only through a loyalty to the firm but also by the development of a sense of "mission," which is most frequently seen as client service. *All* professional service firms list in their mission statement what I call the "3 S's": the goals of (client) Service, (financial) Success, and (professional) Satisfaction. What is recognizable about one-firm firms is that, in their internal communications, there is a clear priority among these.

Within McKinsey, a new consultant learns within a very short period of time that the firm believes that the *client comes first,* the firm second, and the individual last. Goldman Sachs has a reputation for being "ready to sacrifice anything—including its relations with other Wall Street firms—to further the client's interests."[9] At Hewitt Associates, firm ideology is that the 3 S's must be carefully kept in balance at all times; however, client service is clearly number one. None of this is meant to suggest that one-firm firms necessarily render superior service to their clients compared with their competitors, nor that they always resolve inevitable day-to-day conflicts among the 3 S's in the same way. The point is that there *is* a firm ideology which everyone understands and which no one is allowed to take lightly.

Client Service

The emphasis at one-firm firms is clearly one of significant attention to managing client relations. In these firms, client service is defined more broadly than technical excellence: it is taken to mean a more far-ranging attentiveness to client needs and the quality of interaction between the firm and its clients. Goldman Sachs pioneered the concept on Wall Street of forming a marketing and new business development group whose primary responsibility is to manage the interface between the client and the various other parts of the firm that provide the technical and professional services. In most other Wall Street firms, client relations are the responsibility of the individual professionals who do the work, resulting in numerous (and potentially conflicting) contacts between a single client and the various other parts of the firm. Hewitt Associates, alone in its profession, has also pioneered such an "account management" group.[10] At McKinsey, in the words of one partner, "Here everyone realizes that the (client) *relationship* is paramount, not the specific project we happen to be working on at the moment."

The high-commitment, hard-working, mission-oriented, team-intensive characteristics of one-firm firms are reminiscent of another type of organization: the Marine Corps. Indeed, one-firm firms have an elite, Marine Corps attitude about themselves. An atmosphere of a special, private club prevails, where members feel that "we do things differently around here, and most of us couldn't consider working anywhere else." While all professional firms will assert that they have the best *professionals* in town, one-firm firms claim they have the best *firm* in town, a subtle but important difference.[11]

SUSTAINING THE ONE-FIRM FIRM CULTURE

Up to this point, we discussed a type of firm culture, a topic much discussed in recent management literature.[12] Our task now is to try to identify the management practices that have created and sustained this culture. Not surprisingly, since human assets constitute the vast majority of the productive resources of the professional service firm, most of these management practices involve human resource management.

A good overview of the mechanisms by which an "elite group" culture, with emphasis on the *group*, can be created is provided by Dr. Chip Bell,[13] a training consultant, who suggests that the elements of any high performance unit include the following:

- Entrance requirements into the group are extremely difficult.
- Acceptance into the group is followed by *intensive* job-related training, followed by team training.
- Challenging and high-risk team assignments are given early in the individual's career.
- Individuals are constantly tested to ensure that they measure up to the elite standards of the unit.
- Individuals and groups are given the autonomy to take risks normally not permissible at other firms.
- Training is viewed as continuous and related to assignments.
- Individual rewards are tied directly to collective results.

- Managers are seen as experts, pacesetters, and mentors (rather than as administrators).

As we shall see, all of these practices can be seen at work in the one-firm firms.

Recruiting

In contrast to many competing firms, one-firm firms invest a significant amount of *senior* professional time in their recruitment process, and they tend to be much more selective than their competition. At one-firm firms, recruiting is either heavily centralized or well coordinated centrally. At Hewitt Associates, over 1,000 students at sixty-five schools were interviewed in 1980. Of the seventy-two offers that were made, fifty accepted. Each of the 198 invited to the firm's offices spent a half-day with a psychologist (at a cost to the firm of $600 per person) for career counseling to find out if the person was suited for Hewitt's work and would fit within the firm's culture. At Goldman Sachs, 1,000 MBAs are interviewed each year; approximately thirty are chosen. Interviewing likely candidates is a major responsibility of the firm's seventy-three partners (the firm has over 1,600 professionals). Goldman partner James Gorter notes, "Recruiting responsibilities almost come before your business responsibilities."[14] At Latham & Watkins, all candidates get twenty-five to thirty interviews, compared to a norm in the legal profession of approximately five to ten interviews. As a McKinsey partner noted:

> In our business, the game is won or lost at the recruiting stage: we take it very seriously. And it's not a quantity game, it's a quality game. You've got to find the best people you can, and the trick is to understand what *best* means. It's not just brains, not just presentability: you have to try and detect the potentially fully developed professional in the person, and not just look at what they are *now*. Some firms hire in a superficial way, relying on the up-or-out system to screen out the losers. We do have an up-or-out system, but we don't use it as a substitute for good recruiting practices. To us, the costs of recruiting-mistake turnover are too high, in dollars, in morale, and in client service, to ignore.

Training

One-firm firms are notable for their investment in firmwide training, which serves both as a way to add to the substantive skills of juniors and as an important group socialization function.[15] The best examples of this practice are Arthur Andersen and McKinsey. The former is renowned among accounting students for its training center in St. Charles, Illinois (a fully equipped college campus that the firm acquired and converted to its own uses), to which young professionals are sent from around the world. In the words of one Andersen partner: "To this day, I have useful friendships, forged at St. Charles, with people across the firm in different offices and disciplines. If I need to get something done outside my own expertise, I have people I can call on who will do me a favor, even if it comes out of their own hide. They know I'll return it."

Similarly, McKinsey's two-week training program for new professionals is renowned among business school students. The program is run by one or more of the firm's senior professionals, who spend a significant amount of time inculcating the firm's values by telling Marvin Bower stories—Bower, who ran the firm for many years, is largely credited with making McKinsey what it is today. The training program is not always held in the U.S. but rotates between the countries where McKinsey has offices. This not only reinforces the one-firm image (as opposed to a headquarters with branch offices) but also has a dramatic effect on the young professionals' view of the firm. As one of my ex-students told me: "Being sent to Europe for a two-week training program during your first few months with the firm impresses the hell out of you. It makes you think: 'This is a class outfit.' It also both frightens you and gives you confidence. You say, 'Boy, they must think I'm good if they're prepared to spend all this money on me.' But then you worry about whether you can live up to it: it's very motivating." All young professionals are given a copy of Marvin Bower's history of the firm, *Perspectives on McKinsey,* which, unlike many professional firm histories, is as full of philosophy and advice as it is dry on historical facts.

"Growing Their Own" Professionals

Unlike many of their competitors, all of the one-firm firms tend to "grow their own" professionals, rather than to make significant use of lateral hiring of senior professionals. In other words, in the acquisition of human capital, they tend to "make" rather than "buy." This is not to say that no lateral hires are made—just that they are done infrequently, and with extreme caution. "I had to meet with the associates (i.e., not only the partners) before the firm [Latham & Watkins] took me on," Carla Hills [former Secretary of Housing and Urban Development] recalls. "Lateral entry is a big trauma for this place. But that's how it should be."[16]

Avoiding Mergers

A related practice of one-firm firms is the deliberate avoidance of growth by merger. Arthur Andersen, unlike most of the Big-8 firms, did not join in the merger and acquisition boom of the 1950s and early 1960s, in an attempt to become part of a nationwide accounting profession network. Instead, it grew its own regional (and international) offices. Similarly, the decade-long merger mania in investment banking has left Goldman Sachs, which opted out of this trend, as one of the few independent partnerships on the street. In contrast to many other consulting firms, McKinsey's overseas offices were all launched on a grow-your-own basis, initially staffed with U.S. personnel, rather than on an acquisition basis. With one recent exception, all of Latham & Watkins's branch offices were all grown internally.

It is clear that this avoidance of growth through laterals or mergers plays a critical role in both creating and preserving the sense of institutional identity, which is the cornerstone of the one-firm system.

Controlled Growth

As a high proportion of the professional staff shares an extensive, common work history with the one-firm firm, group loyalty is easier to foster. Of course, this staffing strategy has implications for the *rate* of growth pursued by the one-firm firm.

At such firms (in contrast to many competitors), high growth is not a declared goal. Rather, such firms aim for *controlled* growth. The approach is one of, "We'll grow as fast as we can train our people." As Ron Daniel of McKinsey phrases it: "We neither shun growth nor idolize it. We view it as a by-product of achieving our other goals." All of the one-firm firms assert that the major constraint on their growth is not client demand, but the supply of qualified people they can find and train to their way of practicing.

Selective Business Pursuits

Related to this issue is the fact that one-firm firms tend to be more selective than their competitors in the type of business they pursue. It has been reported that an essential element of the Goldman culture is its calculated choosiness about the clients it takes on. The firm has let it be known, both internally and externally, that it "adheres to certain standards—and that it won't compromise them for the sake of a quick buck."[17] At McKinsey, the firm's long-standing strategy is that it will only work for "the top guy" (i.e., the chief executive officer) and, as illustrated internally with countless Marvin Bower stories, will only do those projects where the potential value delivered is demonstrably far in excess of the firm's charges. Junior staff at McKinsey quickly hears stories of projects the firm has turned down because the partner did not believe the firm could add sufficient value to cover its fees. Similarly, while Andersen has been an aggressive marketer (a property common to all the one-firm firms), Andersen appears to have taken a more studied, less "opportunistic" approach to business development than have their competitors.

Consequently, one-firm firms tend to have a less varied practice-mix and a more homogeneous client base than do their more explicitly individualistic competitors. Unlike, say, Booz, Allen, McKinsey's practice is relatively focused on three main areas: organization work, strategy consulting, and operations studies. In the late 1970s, the heyday of "strategy boutiques," many outsiders commented on the firm's reluctance to chase after fast-growing new specialties.[18] But McKinsey, like all of the other one-firm firms, enters new areas "big, or not at all." Andersen's strategy in its consulting work (the fastest growing area for all of the Big-8) has been more clearly focused on computer-based systems design and installation than has the variegated practices of most of its competitors. Goldman has been notably selective in which segments of the investment market it has entered, and has become a dominant player in virtually every sector it has entered.

Outplacement

One of the fortunate consequences of the controlled growth strategy at one-firm firms and the avoidance of laterals and mergers is that these firms, in contrast to many competitors, rarely lose valued people to competitors. At each of the firms named above, I have heard the claim that, "Many of our people have been approached by competitors offering more money to help them launch or bolster a part of the practice. But our people prefer to stay." On Wall Street, raiding of competing firms' top professionals has reached epidemic proportions; yet, this does not include Goldman Sachs. It is said that one of the rarest beasts on Wall Street is an *ex*-Goldman professional: very few leave the firm.

Turnover at one-firm firms is clearly more carefully managed than it is among competitors. Those one-firm firms that do enforce an up-or-out system (McKinsey and Andersen) work actively to place their alumni/ae in good positions preferably with favored clients. McKinsey's regular alumni/ae reunions, a vivid demonstration of its success in breeding loyalty to the firm, are held two or three times a year. In part, due to the "caring" approach taken to junior staff, one-firm firms are able to achieve a very profitable high-leverage strategy (i.e., high ratio of junior to senior staff) without *excessive* pressures for growth to provide promotion opportunities.[19]

Compensation

Internal management procedures at one-firm firms constantly reinforce the team concept. Most important, compensation systems (particularly for partners) are designed to encourage intra-firm coopera-

tion. Whereas many other firms make heavy use of departmental or local-office profitability in setting compensation (i.e., take a *measurement*-oriented, profit-center approach), one-firm firms tend to set compensation (both for partners *and* juniors) through a *judgmental* process, assessing total contribution to the firm. Unique among the Big-8, Andersen has a single worldwide partnership cost-sharing pool (as opposed to separate country profit centers): individual partners share in the joint economics of the whole firm, not just their country (or local office). "The virtue of the 'one-pool' system, as opposed to heavy profit-centering, is that a superior individual in an otherwise poor-profit office can be rewarded appropriately," one Andersen partner pointed out. "Similarly, a weaker individual in a successful office does not get a windfall gain. Further, if you tie individual partner compensation too tightly to departmental or office profitability, it's hard to take into account the particular circumstances of that office. A guy that shows medium profitability in a tough market probably deserves more than one with higher profitability in an easy market where we already have a high market share."

Hewitt Associates sets its partner compensation levels only after all partners have been invited to comment on the contributions (qualitative *and* quantitative) made by other partners on "their" projects and other firmwide affairs. Vigorous efforts are made to assess contributions to the firm that do not show up in the measurable factors. Peter Friedes notes:

> We think that having no profit centers is a great advantage to us. Other organizations don't realize how much time they waste fighting over allocations of overhead, transfer charges, and other mechanisms caused by a profit-center mentality. Whenever there are profit centers, cooperation between groups suffers badly. Of course, we pay a price for not having them: specific accountability is hard to pin down. We often don't know precisely whose time we are writing off, or who precisely brought in that new account. But at least we don't fight over it: we get on with our work. Our people know that, over time, good performance will be recognized and rewarded.

Goldman Sachs also runs a judgment-based (rather than measurement-based) compensation system, including "a month-long evaluation process in which performance is reviewed not only by a person's superiors but by other partners as well, and finally by the management committee. During that review, 'how well you do when other parts of the firm ask for your help on some project' plays a big part."[20] At Latham & Watkins, "15 percent of the firm's income is set aside as a separate fund from which the executive committee, at its sole discretion, awards partners additional compensation based on their general contribution to the firm in terms of such factors as client relations, hours billed, and even the business office's 'scoring' of how promptly the partner has logged his or her own time, sent out and collected bills, and otherwise helped the place run well."[21]

Investments in Research and Development

In most professional service firms, particularly in those with a heavy emphasis on short-term results or year-by-year performance evaluations, any activity that takes an individual away from direct revenue-producing work is considered a detour off the professional success track within the firm and is therefore avoided. This is not the case with one-firm firms.

As the one-firm culture is based on a "team-player" judgment-system approach to evaluations and compensation (at both the partner and junior level), it is *relatively* easier (although it is never easy) for one-firm firms to get their best professionals to engage in nonbillable, stafflike activities such as research and development (R&D), market research, and other investments in the firm's future. For example, McKinsey is noted in the consulting profession for its internally funded R&D projects, of which the most famous example is the work that resulted in the best seller, *In Search of Excellence*. This book, however, was only one of a large number of staff projects continually under way in the firm. An ex-student of mine noted that "at McKinsey, to be selected to do something for the firm is an honor: it's a quick way to get famous in the firm if you succeed. And, of course, you're ex-

pected to succeed. Firm projects are treated as seriously as client work, and your performance is closely examined. However, my friends at other firms tell me that firm projects are a high-risk thing to do: they worry about whether their low chargeable hours will be held against them later on."

Andersen likewise invests heavily in firmwide activities. For instance, it conducts extensive cross-office and cross-functional industry programs, which attempt to coordinate all of the firm's activities with respect to specific industries. In fact, it is rumored, although no one has the statistics, that Andersen invests a higher proportion of its gross revenues in firmwide investment activities than does any other firm.

Goldman's commitment to investing in its own future is illustrated by the firm's policy of forcing partners to keep their capital in the firm rather than to take extraordinarily high incomes. Hewitt's commitment to R&D is built into its organizational structure. Rather than scatter its professional experts throughout its multiple office system (staffed predominantly with account managers), it chose to concentrate its professional groups in three locations in order to promote the rapid cross-fertilization of professional ideas. Significant investments of professional time are made in nonbillable research work under the guidance of professional group managers who establish budgets for such work in negotiation with the managing partner.

Communication

Communication at a one-firm firm is remarkably open and is clearly used as a bonding technique to hold the firm together. All the firms described above make *heavy* use of memorandums to keep everyone informed of what is happening in other parts of the firm, above and beyond the token efforts frequently made at other firms. Frequent firmwide meetings are held, with an emphasis on cross-boundary (i.e., interoffice and interdepartmental) gatherings. Such meetings are valued (and clearly designed) as much as for the social interaction as for whatever the agenda happens to be: people *go* to the meetings. (At numerous other firms I have observed, meetings are seen as distractions from the firm's, or the individual's, business, and people bow out whenever they can.)

At most one-firm firms, open communication extends to financial matters as well. At Hewitt, they believe that "anyone has a right to know anything about the firm except the personal affairs of another individual." At an annual meeting with all junior personnel (including secretaries and other support staff), the managing partner discloses the firm's economic results and answers any and all questions from the audience. At Latham & Watkins, junior associates are significantly involved in all major firm committees, including recruiting, choosing new partners, awarding associate bonuses, and so on. All significant matters about the firm are well known to the associates.

Absence of Status Symbols

Working hard to involve nonpartners in firm affairs and winning their commitment to the firm's success is a hallmark of the one-firm firm and is reinforced by a widely common practice of sharing firm profits more deeply within the organization than is common at other firms. (The ratio between the highest paid and lowest paid partner tends to be markedly less at one-firm firms than it is among their competitors.) There is also a suppression of status differentials between senior and junior members of the firm: an important activity if the firm is attempting to make everyone, junior and senior alike, feel a part of the team. At Hewitt Associates, deemphasizing status extends to the physical surroundings: everyone, from the newest hire to the oldest partner, has the same size office.

The absence of status conflicts in one-firm firms is also noticeable across departments. In today's world of professional megafirms composed of departments specializing in vastly different areas, one of the most significant dangers is that professionals in one area may come to view *their* area as somehow more elite, more exciting, more profitable, or more important to the firm than another area. Their loyalty is to their department, or their local office, and not to the firm. Yet the success of the firm clearly depends upon doing well in all ar-

eas. On Wall Street, different psychological profiles of, and an antipathy between, say, traders and investment bankers is notorious: many attribute the recent turmoil at Lehman Brothers (now Shearson Lehman) to this syndrome. In some law firms, corporate lawyers and litigators are often considered distinct breeds of people who view the world in different ways. In some accounting firms, mutual suspicion among audit, tax, and consulting partners is rampant. In consulting firms, frequently there are status conflicts between the "front-room" client handlers and the "back-room" technical experts.

What strikes any visitor to a one-firm firm is the deeply held mutual respect across departmental, geographic, and functional boundaries. Members of one-firm firms clearly *like* (and respect) their counterparts in other areas, which makes for the successful cross-boundary coordination that is increasingly essential in today's marketplace. Jonathan Cohen of Goldman Sachs notes that out-of-office socialization among Goldman professionals appears to take place more frequently than it does at other Wall Street firms. Retired Marvin Bower of McKinsey asserts that one of the elements in creating the one-firm culture is mutual trust, both horizontally and vertically. This atmosphere is created primarily by the behavior of the firm's leadership, who must set the style for the firm. Unlike many other firms, leaders of one-firm firms work hard not to be identified with or labeled as being closer to one group than another. Cross-boundary respect is also achieved at most one-firm firms by the common practice of rotating senior professionals among the various offices and departments of the firm.

Governance: Consensus-Building Style

How are one-firm firms governed? Are they democracies or autocracies? Without exception, one-firm firms are led (*not* managed) in a consensus-building style.[22] All have (or have had) strong leaders who engage in extensive consultation before major decisions are taken. It is important to note that all of these firms do indeed have leaders: they are not anarchic democracies, nor are they dictatorships. Whether one is reading about Goldman's two Johns (Weinberg and Whitehead), McKinsey's (retired) Marvin Bower and Ron Daniels, Latham & Watkins's Clinton Stevenson, or Hewitt's Peter Friedes, it is clear that one is learning about expert communicators who see their role as preserver of the "true religion." Above all else, they are cheerleaders who suppress their own egos in the name of the institution they head. Such firms also have continuity in leadership: while many of them have electoral systems of governance, leaders tend to stay in place for long periods of time. What is more, the firm's culture outlasts the tenure of any given individual.

Of course, the success of the consensus-building approach to firm governance and the continuity of leadership at one-firm firms is not fortuitous. Since their whole philosophy (and, as I have tried to show, their substantive managerial practices) is built upon cooperative teamwork, consensus is more readily achieved here than it is at other firms. The willingness to allow leaders the freedom to make decisions on behalf of the firm (the absence of which has stymied many other "democratic" firms) was "prewired" into the system long ago, since everyone shares the same values. The one-firm system *is* a system.

CONCLUSION: POTENTIAL WEAKNESSES

Clearly, the one-firm firm system is powerful. What are its weaknesses? The dangers of this approach are reasonably obvious. Above all else, there is the danger of self-congratulatory complacency: a firm that has an integrated system that *works* may, if it is not careful, become insensitive to shifts in its environment that demand changes in the system. The very commitment to "our firm's way of doing things," which is the one-firm firm's strengths, can also be its greatest weakness. This is particularly true because of the chance of "inbreeding" that comes from "growing-your-own" professionals. To deal with this, there is a final ingredient required in the formula: self-criticism. At McKinsey, Andersen, Goldman, and Hewitt, partners have asserted to me that "we have no harsher

critics than ourselves: we're constantly looking for ways to improve what we do." However, it must be acknowledged that, without the diversity common at other professional service firms, one-firm firms with strong cultures run the danger of making even self-criticism a proforma exercise.

Another potential weakness of the one-firm firm culture is that it runs the danger of being insufficiently entrepreneurial, at least in the short run. Other more individualistic firms, which promote and reward opportunistic behavior by individuals and separate profit centers, may be better at reorganizing and capitalizing on emerging trends early in their development. Although contrary examples can be cited, one-firm firms are rarely "pioneers": they try to be (and usually are) good at entering emerging markets as a late second or third. And because of the firmwide concentrated attack they are able to effect, they are frequently successful at this. (The similarity to IBM in this regard, as is much of what has been discussed above, is readily noticeable.)

The one-firm approach is *not* the only way to run a professional service firm. However, it clearly is a very successful way to run a firm. The "team spirit" of the firms described here is broadly admired by their competitors and is not easily copied. As I have attempted to show, the one-firm firm system is *internally* consistent: all of its practices, from recruiting through compensation, performance appraisal, approaches to market, governance, control systems, and above all, culture and human resource strategy, make for a consistent whole.

REFERENCES

1. D. H. Maister, "Profitability: Beating the Downward Trend," *Journal of Management Consulting,* Fall 1984, pp. 39–44.
2. D. H. Maister, "Quality Work Doesn't Mean Quality Service," *American Lawyer,* April 1984.
3. R. Levering, M. Moskowitz, and M. Katz, *The 100 Best Companies to Work for in America* (Reading, MA: Addison-Wesley, 1984).
4. B. McGoldrick, "Inside the Goldman Sachs Culture," *Institutional Investor,* January 1984.
5. S. Brill, "Is Latham & Watkins America's Best Run Firm?" *American Lawyer,* August 1981, pp. 12–14.

6. Ibid.
7. Levering et al. (1984).
8. McGoldrick (January 1984).
9. Ibid.
10. D. H. Maister, *Hewitt Associates* (Boston, MA: Harvard University, Graduate School of Business, HBS Case Services).
11. D. H. Maister, "What Kind of Excellence?" *American Lawyer,* January–February 1985, p. 4–6.
12. See, for example, V. J. Sathe, *Culture and Related Corporate Realities* (Homewood, IL: Richard D. Irwin, Inc., 1985).
13. C. Bell, "How to Create a High Performance Training Unit," *Training*, October 1980, pp. 49–52.
14. McGoldrick (January 1984).
15. D. H. Maister, "How to Build Human Capital," *American Lawyer,* June 1984.
16. Brill (August 1981).
17. McGoldrick (January 1984).
18. See, for example, "The New Shape of Management Consulting," *Business Week*, 21 May 1979.
19. For a discussion of the role of turnover on professional service firm success, see D. H. Maister, "Balancing the Professional Service Firm," *Sloan Management Review,* Fall 1982, pp. 15–29.
20. McGoldrick (January 1984).
21. Brill (August 1981).
22. For a discussion of governance in professional firms, see D. H. Maister, "Partnership Politics," *American Lawyer,* October 1984.

35

3M's Post-it Notes: A Managed or Accidental Innovation?

P. RANGANATH NAYAK AND JOHN KETTERINGHAM

In late 1978, the bleak reports from the four-city market tests came back to the 3M Corporation. The analyses were showing that this "Post-it Note Pads" idea was a real stinker. Such news came as no surprise to a large number of 3M's most astute observers of new product ideas, for this one had smelled funny to them right from the beginning! From its earliest days, Post-it brand adhesive had to be one of the most neglected product notions in 3M history. The company had ignored it before it was a notepad, when the product-to-be was just an adhesive that didn't adhere very well. The first product to reach the marketplace was a sticky bulletin board whose sales were less than exciting to a company like 3M.

But why was this adhesive still around? For five years, beginning before 1970, this odd material kept coming around, always rattling in the pocket of Spencer Silver, the chemist who had mixed it up in the first place. Even after the adhesive had evolved into a stickum-covered bulletin board, and then into notepad glue, there was manufacturing saying that it couldn't mass-produce the pads and marketing claiming that such scratch pads would never sell. So by 1978, when the reports came in from the test markets, it seemed everyone who'd said disparaging things about the Post-it Note Pad was right after all. 3M was finally going to do the merciful thing and bury the remains. At that critical moment, it was only one last try by two highly placed executives, Geoffrey Nicholson and Joseph Ramey, that kept "those little yellow sticky pads" from going the way of the dinosaur.

To understand Silver's persistence with his innovative commercial challenge, it is necessary to go back to his moment of discovery. Silver's role in the development of Post-it Note Pads began in 1964 with a "Polymers for Adhesives' program in 3M's Central Research Laboratories. The company has always had a tradition of periodically reexamining its own products to look for ways to improve them. "Every so many years," said Silver, "3M would put together a bunch of people who looked like they might be productive in developing new types of adhesives." In the course of that "Polymers for Adhesives" research program, which went on for four years, Silver found out about a new family of monomers developed by Archer-Daniels Midland, Inc., which he thought contained potential as ingredients for polymer-based adhesives. He received a number of samples from ADM and began to work with them. This was an open-ended research effort, and Silver's acquisition of the new monomers was the sort of exploration the company encouraged. "As long as you were producing new things, everybody was happy," said Silver. "Of course, they had to be new molecules, patentable molecules. In the course of this exploration, I tried an experiment with one of the monomers to see what would happen if I put a lot of it into the reaction mixture. Before, we had used amounts that would correspond to conventional wisdom." Silver had no expectation whatsoever of what might occur if he did this. He just thought it might be interesting to find out.

In polymerization catalysis, scientists usually

This article is a modified, shortened version of a chapter from P. R. Nayak and J. M. Ketteringham's book *Breakthroughs,* an Arthur D. Little international study of 16 major innovations (Rawson Press, 1986). Published with permission of ADL with additions and modifications made by Professor Ralph Katz, based on his 1996 interviews with Art Fry of 3M.
P. Ranganath Nayak and *John Ketteringham* are Management Consultants at Authur D. Little. *Ralph Katz* is Professor of Management at Northeastern University's College of Business and Research Associate at MIT's Sloan School of Management.

control the amounts of interacting ingredients to very tightly defined proportions, in accordance with prevailing theory and experience. Silver said with a certain measure of glee, "The key to the Post-it adhesive was doing the experiment. If I had sat down and factored it out beforehand, and thought about it, I wouldn't have done the experiment. If I had limited my thinking only to what the literature said, I would have stopped. The literature was full of examples that said you can't do this." Highly regarded publications and experts would have told Silver there was no point in doing what he did. But Silver understood that science is one part meticulous calculation and one part fooling around. "People like myself," said Silver, "get excited about looking for new properties in materials. I find that very satisfying, to perturb the structure slightly and just see what happens. I have a hard time talking people into doing that—people who are more highly trained. It's been my experience that people are reluctant just to try, to experiment—just to see what will happen!"

When Silver went ahead with the "wrong" proportions of the ADM monomers, just to see what would happen, he got a reaction that departed from the predictions of theory. It was what some call an "accident" and what Silver called a "Eureka moment." What Silver experienced was the appearance of what would become the Post-it adhesive polymer. It was the moment for which all scientists become scientists—the emergence of a unique, unexpected, previously unobserved and reliable scientific phenomenon. Each time Silver put those things together, they fell into the same pattern—every time. "It's one of those things you look at and you say, this has got to be useful! You're not forcing materials into a situation to make them work. It wanted to do this. It wanted to make Post-it adhesive," Silver said.

Technically the material was what the research program called for, a new polymer with adhesive properties. But in examining it, Silver noticed among its other curious properties that this material was not "aggressively" adhesive. It would create what scientists call "tack" between two surfaces, but it would not bond tightly to them. Also, and this was a problem not solved for years, this

material was more "cohesive" than it was "adhesive." It clung to its own molecules better than it clung to any other molecules. So if you sprayed it on a surface (it was sprayable, another property that attracted Silver) and then slapped a piece of paper on the sprayed surface, you could remove all or none of the adhesive when you lifted the paper. It might "prefer" one surface to another, but not stick well to either. Someone would have to invent a new coating for paper if 3M were to use this as an adhesive for pieces of paper. But paper? Not very likely, thought Silver, and on this point, at least, everyone agreed with him.

What Silver had done was more than the usual 3M lab synthesis; it was a discovery—the sort of thing a scientist can put his or her name on. When he watched the reaction, Silver was achieving fatherhood, and he was falling in love. He knew he might never again be responsible for so pure and simple a phenomenon. Almost instantly, he personified this viscous goo, calling the stuff "my baby." It may not have been very sticky, but Spencer Silver got very attached to it. As he started to present this discovery to other 3Mers, however, he soon realized that few people shared his views about the beauty of this glue. Interested in practical applications, they had only a passing appreciation for the science embodied in Silver's adhesive. More significantly, they were "trapped by the metaphor" that insists that the ultimate adhesive is one that forms an unbreakable bond! The whole world in which they lived was looking for a better glue, not a worse glue. And like any other sensible adhesives manufacturer, 3M's sights had never wavered from a progressive course of developing stronger and stronger adhesives. Suddenly, here was Spencer Silver, touting the opposite of what was considered normal product virtue.

Although he couldn't say exactly what it was good for, "it had to be good for something," he would tell them. Aren't there times, Silver would ask people, when you want a glue to hold something for a while but not forever? Let's think about those situations. Let's see if we can turn this adhesive into a product that will hold tight as long as people need it to hold but then let go when people want it to let go. From 1968 through 1973, company support systematically slipped away from him. First, the Polymers for Adhesives Program disappeared. 3M had given its researchers a specified time and a limited budget to conduct that program. When the time and money were used up, the researchers were reassigned even though some, like Silver, were just starting to have fun.

"The adhesives program died a natural death," Silver recalled. "The company's business went off, and, in the usual cycle of things, the longer-range research programs were cut. So the emphasis was diminished and we still had invented some interesting materials that we wanted to push." The members of the Polymers for Adhesives group were assigned to new research projects. Left as a team, they might have fought together to keep alive a number of their odd little discoveries. But all those discoveries were shelved, with Silver's one glaring exception, and he got little assistance from his teammates in promoting the survival of his oddball adhesive. So he did what seems to happen frequently at 3M. He shrugged at the organization and he did it himself. He had to wage a battle to get the money just to patent his unique polymer. 3M eventually spent the minimum money possible. Post-it adhesive was patented *only* in the United States. "We really had to fight to get a patent," said Silver, "because there was no commercial product readily apparent. It's kind of a shame. I wish it would change. If 3M commits itself to millions of dollars for research, it ought to allow you to follow up with the money for a patent."

People at 3M, when they fight for something, seem to do it with an understated grace, a politeness that conceals their tenacity. This is true of Silver, who quietly began the arduous struggle to capture the imagination of his colleagues and superiors. Silver's only advantage was that he was, after all, in love. "I was just absolutely convinced that this had potential," Silver said. "There are some things that have a little spark to them—that are worth pursuing. You have to be almost a zealot at times in order to keep interest alive, otherwise it will die off. It seems like the pattern always goes like this: In the fat times, R&D groups appear and we do a lot of interesting research. And then the lean times come just about at the point when you've

developed your first goody, your gizmo. And then you've got to go out and try to sell it. Well, everybody in the divisions is so busy that they don't want to touch it. They don't have time to look at new product ideas with no end product already in mind."

Silver went door-to-door to every division at 3M that might be able to think up an application for his adhesive. The organization never protested his search. When he sought slots of time at in-house technical seminars, he always got a segment to show off his now-it-works, now-it-doesn't adhesive. At every seminar, some people left, some people stayed. Most of them said, "What can you do with a glue that doesn't glue?" But *no one* said to Silver, "Don't try. Stop wasting our time." In fact, it would have violated some very deeply felt principles of the 3M Company to have killed Silver's pet project. Much is made of 3M's "environment for innovation," but 3M's environment is, more accurately, an environment of nonintervention, of expecting people to fulfill their day's responsibilities, every day, without discernible pressure from above. Silver, no matter how much time he spent fooling around with the Post-it adhesive, never failed in his other duties, and so, at 3M, there was no reason whatsoever to overtly discourage his extracurricular activities. The positive side of this corporate ethic is the feeling of independence each worker experiences in doing his job. The disadvantage is that, when you have a good idea that requires more than one person to share the work and get the credit, it can be hard to convince people to postpone their chores and help with yours.

As Silver pursued his lonely quest, his best inspiration for applying his adhesive was a sticky bulletin board, a product that wasn't especially stimulating even to its inventor. He got 3M to manufacture a number of them—through a fairly low-tech and inexpensive process—and they were sent out to the company's distribution and retail network. The outcome was predictable. 3M sold a few, but it was a slow-moving item in a sleepy market niche. Silver knew there had to be a better idea. "At times I was angry because this stuff is so obviously unique," said Silver. "I said to myself, 'Why can't you think of a product? It's your job!' "

Although Silver had overcome the metaphorical trap of always striving for stickier stickum, he, too, became trapped, albeit by a different metaphor. The bulletin board, the only product he could think of, was totally coated with adhesive—it was sticky everywhere. The metaphor said that something is either sticky or not sticky. Something *partly sticky* did not occur to him. More constraining was the fact that, until Silver's adhesive made it possible, there had been no such thing as a self-adhesive piece of note paper. Note paper was cheap and trivial, and the valuable elements used with these bits of paper were their durable fasteners of pins, tacks, tapes, and clips. So silver was immersed in an organization whose lifeblood was tape: Scotch brand tapes like magic tape, cellophane tape, duct tape, masking tape, electrical tape, caulking tape, diaper tape, and surgical tape, to name a few. In this atmosphere, imaging a piece of paper that eliminates the need for tape is almost unthinkable.

In the early 1970s, 3M transferred Silver to its System Research group within the Central Research labs. There he met Oliveira, a biochemist who shared Silver's fascination with things that did stuff you didn't think they could do. Silver and Oliveira kept each other from getting discouraged; they were a duo that eventually presented the adhesive technology to Geoff Nicholson, which in the course of the seemingly accidental nature of Post-it notes, may have been the biggest accident of all. Nicholson was, in 1973, appointed the leader of a new venture team in the Commercial Tape Division laboratory. Now venture teams were open-ended research and development groups formed, when funds are available, to explore new directions in one of 3M's many lines of business and technology. Nicholson had been given a fresh budget and a free hand to develop new products in the company's Commercial Tape Division, whose new product development had grown sluggish. It is a standing policy at 3M that each division must generate 25 percent of its annual revenue from products developed in the last five years, a tall order for any division, especially those in the old, established product lines, and one on which Commercial Tape consistently had been coming up short.

Silver had been to see the people in Com-

mercial Tape at least twice before. Both times they had rejected his adhesive. Two days before Nicholson arrived in Commercial Tape, Silver and Oliveira had been around again, trying to sell the idea to the division's technical director, James Irwin. Irwin sidestepped them by saying there would be a new guy running research projects there in a couple of days. Two days later, Silver and Oliveira were almost the first people in Nicholson's new office. "Here I am, brand-new to the division, and I don't know a lot about adhesives. And here they were talking to me about adhesives," Nicholson recalled. "I'm ripe for something new, different, and exciting. Most anybody who had walked in the door, I would have put my arms around them."

Silver explained his adhesive discovery for the umpteenth time, and Nicholson, who didn't understand half of what he was saying, was intrigued. "It sure sounded different and unique to me," said Nicholson. "I was ripe for the plucking." Finally, Silver's unloved, uncommitted adhesive had a home. Nicholson went about recruiting people for the new venture team; Silver hoped that one of those people would arrive with a *problem* to match his five-year-old *solution*. The one who had the problem was a chemist, a choir director, and an amateur mechanic named Arthur Fry. It was Fry who eventually took the baton from Silver's weary grasp and carried it over a host of discouraging hurdles. Even before joining the new venture team in Commercial Tape, Fry had seen Silver show off his adhesive and had kept the idea turning slowly in the back of his mind. He agreed with Silver that this adhesive was special, although he too wondered what to do with it.

"Then one day in 1974, while I was singing in the choir of the North Presbyterian Church in north St. Paul, I had one of those creative moments," Fry explained. "To make it easier to find the songs we were going to sing at each Sunday's service, I used to mark the places with little slips of paper." Inevitably, when everyone in the church stood up, or when Fry had to communicate through gestures with other members of the choir, he would divert his attention from the placement of his array of bookmarks. One unguarded move, and they either fluttered to the floor or sank into the deep crack of the hymnal's binding. Suddenly, while Fry leafed frantically for his place in the book, he thought "Gee, if I had a little adhesive on these bookmarks, that would be just the ticket." Fry decided to check into that idea the next week at work. What he had in mind, of course, was Silver's adhesive.

What had happened in Fry's ever-searching curiosity was the creative association of two unrelated ideas. When Fry went to work on Monday, he ordered a sample of the adhesive, mixed different concentrations, and invented what he called "the better bookmark." Encouraged by Silver's enthusiasm and Nicholson's push for new products, Fry began to realize the magnitude of his creative activity. "I knew I had made a much bigger discovery," said Fry. "I soon came to realize that the primary application for Silver's adhesive was not to put it on a fixed surface, like the bulletin boards. That was a secondary application. The primary application concerned paper to paper." Fry had also coated only the edge of the paper so that the part protruding from the book wouldn't be sticky. In using these bookmarks for notes back and forth to his boss, Fry had come across the heart of the idea. It wasn't a bookmark at all, it was a note—a system of communication where the means of attachment and removal were built in and did not damage the original surface!

Over the years, Fry has been ordained as the Post-it notes champion, a title which, in ensuing years, has imposed some unusual burdens on him. Today, rather than working side-by-side in a lab with old friends like Silver and Oliveira, Art Fry is ensconced in his own laboratory. To a chemist, this is the equivalent of the corporate corner office— lofty among the echelons of the organization, but such loftiness often makes for a lonely job. On the other hand, Fry is often freed from the splendid isolation of his private lab to speak, as a company spokesman, to large groups of businessmen about the climate for creativity at 3M. He has been interviewed and quoted so often that business writers invariably peg him as the sole Post-it notes product champion. With Fry trapped by this role

and its demands, it's easy to see why Spence Silver seems relieved, perhaps even grateful at the comparatively short shrift given to his role in the Post-it story. Silver is still in 3M's basement, working out of a cramped, windowless office in a large, open, multihood laboratory, a place where experimental ferment still seems to take place. In Silver, the scientific playfulness that gave birth to the Post-it adhesive still seems intact. In fact, without much prompting, he will hold up a glass cylinder of the old Post-it polymer, showing its milky white color in its restful state. He then squeezes the polymer with a plunger and, under pressure, the contents magically become crystal clear. Silver releases the pressure and the adhesive becomes opaque again! Silver doesn't know why it does that. "Isn't that wonderful?" he says. "There must be some way you can *use* that!"

In 1974, after Silver had been making the same exclamation for many years, Fry had provided the first truly affirmative response. But with the "Eureka moment" at the North Presbyterian Church came many other problems. On the bulletin board, Silver's adhesive was attached to a favorable "substrate." It stuck to the bulletin board better than anything else. Move it to paper, however, and it peeled off onto everything it touched. If you couldn't change this property, you still couldn't make a future for Silver's Post-it adhesive. Says Fry, "You had to get the adhesive to stay in place on the note instead of transferring to other surfaces. I think some of the church hymnals have pages that are still sticking together." The two members who invented a paper coating that made the Post-it adhesive work were named Henry Courtney and Roger Merrill. Silver said, "Those guys actually made one of the most important contributions to the whole project, and they haven't received a lot of credit for it. The Post-it adhesive was always interesting to people, but if you put it down on something and pulled it apart, it could stay with either side. It had no memory of where it should be. It was difficult to figure out a way to prime the substrate, to get it to stick to the surface you originally put it on. Roger and Hank invented a way to stick the Post-it adhesive down. And they're the ones who really made the breakthrough discovery, because once you've learned that, you can apply it to all sorts of different structures."

Courtney and Merrill's contribution was the first in a series of actions that definitely were not accidents. Although there was still organizational resistance after Fry's choir book epiphany, every action thereafter, including Courtney and Merrill's research, was directed toward the development, production, and market success of the Post-it note. Fry was a tenacious advocate of the product through all phases from development to production scale-up. While Silver's task had been simply to convince his corporation that his glue was not just a footnote in the obscure history of adhesives, the job Fry assumed was to overcome the natural resistance of people to manufacturing a product differently from their normal experience base. The engineers in 3M's Commercial Tape Division were accustomed to tape, which is sticky all over on one side and then gets packaged into rolls. To apply glue selectively to one side of the paper, and to move the product from rolls to sheets, the engineers would have to invent at least two entirely unique machines. Furthermore, even though 3M is noted for its coating expertise, the company did not have the coating equipment capable of putting the necessary precision on an imprecise surface such as paper. Nor did they have a good way of measuring the coating's weight. Have you ever noticed, for example, that the pads are no thicker at the adhesive layer then at the rest of the pad?

In war and politics, the best strategy is to divide and conquer. In production engineering, the reverse seems to be true. Fry brought together the production people, designers, mechanical engineers, product foremen, and machine operators and let them describe the many reasons why something like that could not be done. He encouraged them to speculate on ways that they might accomplish the impossible. A lifelong gadgeteer, Fry found himself offering his own suggestions. Although the problems bothered the production people, they delighted Fry. "Problems are wonderful things to have if, in overcoming them, you've created a product that is easy for customers to use but difficult for competitors to make."

Inevitably, from these discussions people started thinking of places around 3M where they'd seen machines and parts they could use to piece together the impossible machines they needed to build. And they thought of people who could help. "In a small company, if you had an idea that would incorporate a variety of technologies and you had to go out and buy the equipment to put those together, you probably couldn't afford it, or you'd have to go as inexpensively or as small as possible," said Fry. "At a large company like 3M, we've got so many different types of technology operating and so many experts—guys that really know all about any subject you want—and so much equipment scattered here and there, that we can piece things together when we're starting off. It's the old 80:20 rule; that is, 80 percent of the equipment and materials needed can probably be found within the company and can be scrounged by an 'entrepreneuring' champion."

Then there was Art Fry's basement. He had had arguments with several mechanical engineers about a difficult phase of production, applying adhesive to paper in a continuous roll. He said it could be done; they said it couldn't. Fry assembled a small-scale basic machine in his basement, then adapted it until he'd solved the problem. The machine worked, and it would work even better once the mechanical engineers had a chance to refine it. But the next problem Fry had was worse: the new machine was too big to fit through his basement door. If he couldn't get it out of his cellar, he couldn't show it off to the engineers. Fry accepted the consequences of his genius and did what he had to do. He broke down an external wall in his ground-level basement and delivered his machine by caesarean section!

Within two years, Fry and 3M's mechanical engineers had tinkered their way to a series of machines that, among other things, coated the yellow paper with its "substrate," applied adhesive, and cut the sticky paper into little square and rectangular note pads. All of the machines are unique and proprietary to the company. They are the key to the Post-it Notes' marvelous high-quality consistency and dependability. The immense difficulty of duplicating 3M's machinery is part of the reason few competitors have made it to the market with Post-it note imitations. Fry and the engineers worked on their unique machines and mass-production methods in a pilot plant in the Commercial Tape lab. The project team mapped out every raw material, processing step, test procedure, and intermediate product needed to produce the final output (according to Fry, the quality is so good that there have been fewer than 75 complaints since Post-its were introduced nationwide in 1980). The pilot plant produced more than enough Post-it note prototypes to supply all the company's offices. All the sticky pads went to Nicholson's office. From there his secretary carried out a program of providing every office at 3M with Post-it Notes. Early in the program, secretaries on the fourteenth floor, where the senior managers work, all received Post-it Notes and became hooked. Jack Wilkins, the Commercial Tape Division's marketing director at the time, described the process of discovery that hit people the first time they encountered the Post-it Notes. "Once people started using them it was like handing them marijuana," said Wilkins. "Once you start using it you can't stop."

Strangely enough, the personal enthusiasm of secretaries and marketing people like Wilkins did not impress the people responsible for putting Post-it Notes onto the market. For the division's marketing organization, fear of the unfamiliar repeatedly raised its head and threatened to scuttle the program. The marketing department had got out of the habit of dealing directly with consumers. This is ironic, because that much-heralded 3M hero, William L. McKnight, had established a tradition of direct contact with consumers in 1914. That year, as the company's brand-new national sales manager, the first act performed by the young McKnight was to visit furniture factories in Rockford, Illinois, and find out from workers what was wrong with 3M's mediocre sandpaper, which was then the company's only product. That trip to Rockford was the first instance of an executive from 3M walking in the door, approaching a user, and saying, "Here! Try this! Tell me what you think!"

By 1978 the Commercial Tape Division's marketing department was involved in the introduction of half a dozen new products that met eas-

ily identified needs for clearly defined markets, products like book binding tape for libraries and PMA adhesives for the art market. The Post-it Note was just another new product, and not a high-priority product at that. While the company's marketing people had become mesmerized by Post-it Notes in their own offices, they couldn't imagine that other people would feel the same way. They said you could only sell these things if you gave them away free, because who's going to pay a dollar for scratch paper? Although most of the marketing group had used Post-it Notes, when they created marketing materials to present the new product they included no samples. Instead they wrote brochures describing the note pads, they sent boxes of samples separately—which people would open only if they got excited by the brochures. The 3M marketing group was trapped by its own paradigm. It was their job, as marketing experts, to explain products, not to demonstrate them. And as explainers, they had no words to overcome the "scratch paper" metaphor. If they couldn't explain them, they couldn't sell them.

Nicholson, who had spread Post-it Notes like an infection within 3M, only had limited power to push them outside the company. When the four-city market test failed, he alone might not have had the influence to keep the produce alive. But by this time Nicholson had a heavyweight ally in his own boss Joe Ramey, a Division vice president and General Sales Manager of the Commercial Tape Division. Nicholson and Ramey were curious as to why a product that to them had obvious appeal had bombed so terribly. Had 3M's conventional marketing approach victimized an unconventional product? They were sufficiently curious about the trial to fly to one of the market-test cities—Richmond, Virginia. Ramey had been a marketing troubleshooter and he knew realistically that some market problems are just too far advanced to be saved. Nevertheless, he agreed to go to Richmond because he liked Nicholson, not because he liked Post-it Notes' chances of survival.

If Nicholson and Ramey hadn't gone to Richmond, 3M almost certainly would have ceased pilot production of Post-it Notes, retired the new machinery they'd designed for the job, and let the several hundred thousand note pads dwindle into dusty inventory. 3M had always been a company very skilled at developing new variations from old products and then expanding their range of activities as a result of such developments. But Post-it Note Pad was unique, a product entirely unrelated to anything that had ever been sold by 3M. The reason Nicholson made the extra effort to go to Richmond with Ramey to engineer a market reversal was that they had both used Post-it Notes. They knew how clever and irresistible they were. They also knew that their own marketing people had approached the market tests in the four cities of Tulsa, Denver, Richmond, and Tampa in a traditional style. These were tests that relied heavily on advertising to generate enthusiasm in distributors who did not themselves use Post-it Notes and who saw little sense in exerting sales efforts for a scratch pad that represented both an exorbitant price and a dubious profit margin. Nicholson and Ramey took to Richmond a bit of understanding that had eluded all the marketers and distributors: Post-it Notes were just something you had to *use* to appreciate.

Nicholson and Ramey took the next logical step: they stopped depending on the organization. They went out and did it themselves. To do this, they returned to the two things that had already "sold" Post-it Notes more than once. First, like Spencer Silver shuffling from one 3M division to another with his queer adhesive, Nicholson and Ramey went door-to-door. Second, they gave away the product, which is what they had been doing within 3M for more than a year. Throughout the banks and offices of Richmond's business district, Nicholson and Ramey introduced themselves and handed out little sticky pads of Post-it Notes, saying, "Here, try this." And they watched as all kinds of people, from secretaries to programmers to vice-presidents, did just that. They tested Post-it Notes in the flesh and saw firsthand the excitement and addiction of first-time users. In one day of personal contacts in Richmond, Nicholson and Ramey had obtained vivid assurance, not only that people liked these things, but that they were pleading for 3M to make more and that they were going to tell their friends about them. As was later demonstrated in a massive marketing giveaway program in Idaho, now immortalized in 3M as the "Boise Blitz," people loved the Post-it Notes they got free at first, and if getting more meant they had to pay a dollar a pad, it was well worth the price. Post-it Notes seem to spoil office people forever, for they do something no product ever did before. They convey messages in the exact spot you want with no after marks, dents, or holes. They can be moved from place to place and they come in various sizes (and now in colors) for different kinds of messages. Once you've used them, it's hard to go back to staples and paper clips.

The Boise Blitz was unusual but not unique at 3M. The company had saturated test markets before with products and ads. In addition to spending a small fortune on advertising, promotions, and free Post-it Notes, 3M diverted most of its Office Supply Division sales force and a battalion of temporary employees to the city of Boise in Idaho. The blitz confirmed the appeal of Post-it Notes, revealing that sales inevitably follow the distribution of free samples. Reorders came in at a rate of 90 percent, which is double the rate of any other wildly successful office product. But Boise notwithstanding, the real key to the market breakthrough for Post-it Notes was the first effort in Richmond, when Nicholson and Ramey did what 3M sales representatives had been trained to do since the early days when sandpaper was their only product; they talked directly to the end-user and then they showed distributors and retailers the results.

Recalling the trip to Richmond, Nicholson called it an "accident" and "an act of desperation." Neither he nor Ramey were hopeful that they could rejuvenate a doomed product by an impulsive flight to Richmond to knock on strange doors. "What made me go out into the market was the enthusiasm of Nicholson and Fry," said Ramey. "I just figured for their morale I should get out and find out whether we ought to kill it once and for all. My reaction when I first went out into those markets was that we probably had a dead duck on our hands. Frankly, I thought it was a product that people just wouldn't buy." Nicholson described the Richmond revelation as the last in a series of accidents from the initial invention of the adhesive technology by Silver to the invention of the Post-it Note itself. Fry was around the adhesive and he had a problem that he needed to solve. Had Fry not been in an environment where people were playing around with that adhesive, he never would not have come up with his contribution.

Retrospective writings about Post-it Notes refer effusively to the encouragement provided to creative people by champions and patrons in 3M management. Silver often wonders where all that management encouragement was during the first five years of his struggle to be heard. The 3M organization does not provide interesting soil for new ideas to grow, but until Nicholson listened to a presentation by Silver and his colleague Oliveira, 3M management had given no hint of support for what eventually became the Post-it Notes project. Until then, the flame was borne by researchers from below, acting largely in solitude and occasionally in defiance of the organization's implicit desires. Silver's adhesive (and the sticky bulletin board it spawned) lasted out a half decade of cold shoulders only because 3M has a tradition of "internal selling"; that is, anyone with a product idea can shop it around the company's many divisions for developmental support. This means that inventors never really get stopped at 3M—there isn't any central overseer saying, "Cut that out and get back to work!" Instead, inventors labor in their spare time, experiencing mounting rejections from managers, most of whom do not have the imagination, the patience, or the budget to take a serious look at their ideas. As in other companies, product ideas die at 3M, but their deaths are often more slow and lingering.

Silver and Oliveira were chemists, working at 3M's central R&D lab to develop variations in chemical products. Like other chemists, they worked within specific programs set out by 3M to attain certain results, but they also had encouragement to follow up on interesting, unexpected results—within reason, of course! According to 3M policy, scientists can use up to 15 percent of their time pursuing interests outside their primary assignments. But when asked who keeps track of 3M

researchers' use of the 15 percent rule, and how this is done, the answer is that no one really keeps track. In fact, Fry points out that "No one really has extra time. The 15 percent is time that's put in after 5:00 or in weekends. (The 'bootleg' rule was instituted by McKnight after he had ordered Dick Drew back in 1923 to stop working on what turned out to be masking tape.) It gives us a chance to shape our own careers, for McKnight recognized that people give their best efforts to projects they're most interested in. The reward for the extra effort is that we are soon officially asked to do what we wanted to do all along." Fry goes on to emphasize that the beauty of bootleg projects is that they don't rely on top-down decision making. "If you are going after an established market with existing technology, then top-down decision making is fine, but new-to-the-world things generally require perspectives and information from people scattered within the organization. While innovation starts with the initial idea for a creative product, a lot more creativity and new ideas are needed to build the idea into a business." The creative climate allows one to keep a low profile during the time when the early, tough problems arise that require creative solutions. One of the things that Fry had going for him right away was the support from his immediate lab supervisor, Bob Molenda, to charge expenses to "miscellaneous accounts." This is another of the ways the corporation puts teeth into McKnight's policy of giving freedom to chase new ideas. The company had provided Fry with just enough time and money to get started. "Throwing a lot of money or people at the task not only won't speed it up," says Fry, "it will only cut down on management's ability to afford to be patient. Things can be easily killed before they get a real chance."

Silver also kept the Post-it adhesive alive for a remarkably, and perhaps unreasonably, long time because he also kept busy with other research tasks assigned by the company and didn't devote his entire energy to his funny discovery. He is also a cheerful man with an amazing tolerance for rejection. For more than five years, Silver's adhesive was a really oddball idea that make little sense either technically or commercially. It had no perceptible application; it was a solution looking for a problem. And of all the ways to devise new products, probably the most difficult and inefficient is to invent some substance with novel properties and then search for ways to use it, especially when the goal is to develop a product for which people will pay. Nevertheless, seeing face-to-face the reactions of people in Richmond "playing" with Post-it Notes was so dramatic to Ramey and Nicholson that they finally had all the evidence they need to orchestrate the Boise Blitz.

It's remarkable that Post-it Notes and sandpaper, two of the company's greatest breakthroughs, sixty-six years apart, grew out of a similar style and faith in the wisdom of sitting down with customers and asking questions, without any of the trappings of corporate protocol. It could be just a coincidence, but according to many analysts, Post-it Notes finally succeeded because 3M's corporate culture creates a positive environment for innovation. Although corporate culture is one of those ill-defined and overused business concepts, suffice it to say that there is something in 3M's style that tends to encourage a measure of individual ingenuity among its workers. Fry comments in his talks that "if managers aren't innovative, if they don't provide the climate for creativity, if they can't set aside their carefully laid plans to take advantage of a new opportunity, then intrapreneurs (entrepreneurs within a large established business) have little encouragement."

"3M operates on a simple principle," Forbes magazine once said, "that no market, no end product is so small as to be scorned; that with the proper organization, a myriad of small products can be as profitable, if not more so, than a few big ones." This tolerance of the small-scale certainly helped Spence Silver, and then Art Fry, to keep the company from stomping on the Post-it Notes project before the project had developed a life of its own. But there was also the benefit of bigness. Over the years, 3M has grown into a loosely integrated cluster of divisions, with senior management in the St. Paul corporate headquarters. One of the results of this corporate sprawl is that it permits the clever researcher to hide in the crevices and carry out his

own version of the "15 percent principle." Silver benefited more from this "neglect" than from anyone overtly encouraging him to innovate. Fry also enjoyed this dispensation from scrutiny as he fostered the Post-it project through the touchy and costly labor of product development. Although Fry started out as the team leader, the project's formal coordination passed back and forth between marketing and engineering. "Others were better suited to that function than I," says Fry, "and I needed to be free to focus on technical problems."

A more provocative issue, though, is why people at 3M enjoy such unchecked opportunity to "get away with things." A hasty judgment might be that the company's senior management is consciously fostering and rewarding innovative growth. But there is ample evidence to challenge this assertion. The company tends to recognize its most successful creative people by investing them into the company's Carlton Society or, as in the case of Fry, installing them in private laboratories. After each unexpected invention emerges at 3M, the company tends to follow up by creating new programs for innovation (the latest is called Genesis) and new honors to motivate inventors. 3M also gives "Golden Step Awards" for products that sell $2 million, at a profit, within the first two or three years of national introduction. When Post-it Notes won a Golden Step Award in 1981, 13 other products also won the award. In 1987, 3M had over 50 Golden Step Winners. Yet there seems to have been no desire for trophies, promotions, or rewards in any of the Post-it project principals nor in any of 3M's prior inventors. They were people obsessed with problems, not rewards, and they usually invented their own program in order to get a problem solved.

Extrinsic incentives simply don't explain why 3M gets creativity from its Silvers and Frys. There might be a more credible explanation in the company's origins. Since 1910, 3M has been inextricably linked with the city of St. Paul, and some 80 percent of its employees have historically come from the upper Midwest. One of the striking characteristics of community-linked Midwestern companies like 3M is that company and community have grown up together, and they like to think they know what to expect from each other. This bond among town, corporate management, and workers creates trust, and with trust comes an air of amiability. The ease and unpretentiousness of the highest officials at 3M is different from the formality and status sensitivity of managements in other regions, especially in the East. Nicholson and Ramey, for example, did not need to overcome a lot of deep-seated conditioning in order to go out on the streets and behave like peddlers. Fry himself sold pots and pans and luggage door-to-door while he was in college. At 3M, it is simply not good form for management to watch too closely over the shoulders of its veteran employees. It is equally bad form for employees to violate the trust placed in them by a less than vigilant management. There is an honor system, and it works.

The source of this heartland ethos may lie in the farms that surround St. Paul and the pioneering spirit from which they originated. A midwestern American farm is a place where—for generations—each worker has been expected to complete his daily chores before sitting down to supper. Nobody ever watches him do his chores; if he doesn't do them, the disastrous evidence will become apparent by the next day's dawn. Nobody ever asks him if he did his chores, because he wouldn't be eating if he hadn't. People carry on without permission at 3M because they're trustworthy. And they're trustworthy because trust is a part of the larger culture that has surrounded and affected 3M for eighty-five years. In fact, one thing 3M has shown is that when it gets too structured and self-conscious about managing its innovation, it doesn't innovate any better than any other company. As Nicholson said, Post-it Notes came from accidents, not calculations.

The Post-it note accidents were Spence Silver's polymer discovery, Arthur Fry's bookmark epiphany, and Geoff Nicholson's dragging Joe Ramey off to Richmond. Each accident occurred after one person took an entirely independent course of action from the one assigned by the corporation. Each time, the individual got frustrated by either the indifference or the resistance of the organization. Similar accidents had occurred in the past. In 1956 a researcher spilled a tube full of totally useless fluorocarbon compound on her shoes—and from that accident, chemists Patsy Sherman and Sam Smith created Scotchguard fabric protector. In 1950, after three polite 3M requests to stop wasting money, researcher Alvin W. Boese squeezed synthetic fibers mixed with wood pulp through a makeshift comb and created one of the most successful types of nonwoven decorative ribbon ever devised. Masking tape, cellophane tape, and many other big product successes can trace their origins to a similar sequence of "happy accidents."

These accidents happened because when the organization, or management, discouraged people from doing something, the cancellation order didn't carry much conviction. Ego is not popular at 3M, and it is clear that the people thinking up things often have more room to express their egos than the people who are supposed to be running things. If there is an organizational key to breakthrough at 3M, a significant element of corporate culture, it is the fact that people there don't believe in placing the values of the corporation above the values of the individual. People keep the organization vital by not taking the organization too seriously. As a result, when the creative people, Silver and Fry and Nicholson, inevitably ran into the resistance of the organization, they felt the freedom to say, "Well, okay. Never mind. I'll do it myself." The organization simply did not have an equal measure of persistence in response. 3M gives in to people who are sure of themselves. Just as important, everybody at 3M knows that, if someone's pet project blows up in his face, it isn't the end of the world. If Silver, Fry, or Nicholson had failed, they wouldn't have been dismissed or disgraced. As long as they had their chores done, they always had a place at the table.

36

Making Teflon Stick

ANNE COOPER FUNDERBURG

Discovering it took luck. Developing it into a successful product took all of DuPont's technological resources—and the zeal of one French entrepreneur.

One of the most versatile and familiar products of American chemical engineering, Teflon, was discovered by accident. There are many such tales to be found in the history of industrial chemistry, from vulcanized rubber to saccharin to post-its, all of which were stumbled upon by researchers looking for other things. So common, in fact, are unplanned discoveries of this sort that one might expect would-be inventors to simply mix random chemicals all day long until they come up with something valuable. Yet the circumstances behind the Teflon story show how each step along the way drew on the skills and talents of workers who were trained to nurture such discoveries and take them from the laboratory to the market.

Teflon was developed at DuPont, the source of many twentieth-century chemical innovations. It came about as a byproduct of the firm's involvement with refrigerants. In the early 1930s a pair of General Motors chemists, A. L. Henne and Thomas Midgley, brought samples of two compounds to the Jackson Laboratory at DuPont's Chambers Works in Deepwater, New Jersey. The compounds, called Freon 11 and Freon 12, were chlorofluorocarbons (CFCs)—hydrocarbons in which some or all of the hydrogen was replaced with chlorine or fluorine. GM's research laboratories had developed the family of Freons for its Frigidaire division, which made refrigeration equipment. They were meant to re-

place existing refrigerants such as ammonia, sulfur dioxide, and propane, which were less efficient than Freons and either too poisonous or too explosive for residential use.

Having made the basic discovery, GM teamed up with DuPont to take advantage of the latter's expertise in manufacturing and research and development. The two companies formed a joint venture called Kinetic Chemicals, which by the mid-1930s had isolated and tested a wide range of CFCs and put the most promising ones into mass production. The best seller was refrigerant 114 (later called Freon 114), or tetrafluorodichloroethane (CF_2ClCF_2Cl).

Kinetic had agreed to reserve its entire output of Freon 114 for Frigidaire, so in the late 1930s DuPont was looking for an equally effective refrigerant that it could sell to other manufacturers. One of the chemists assigned to this project was the 27-year-old Roy J. Plunkett, who had been hired in 1936 after completing his doctorate at Ohio State University.

Plunkett was working on a new CFC that he hoped would be a good refrigerant. He synthesized it by reacting tetrafluoroethylene (TFE), a gas at room conditions, with hydrochloric acid. To further this research, Plunkett and his assistant, Jack Rebok, prepared 100 pounds of TFE and stored it in pressure cylinders, to be dispensed as needed. To prevent an explosion or rupture of the cylinder, they kept the canisters in dry ice.

On the morning of April 6, 1938, Rebok connected a canister of TFE to the reaction apparatus he and Plunkett had been using. His standard procedure was to release some TFE into a heated

Reprinted with permission from *American Heritage of Invention & Technology*, vol. 16, no. 1, Summer 2000.
Anne Cooper Funderburg is a freelance writer in Mandeville, LA, and writes about American history.

chamber and then spray in hydrochloric acid, but this time, when he opened the valve on the TFE container, nothing came out. A cursory examination did not reveal anything wrong with the valve. Had the gas somehow leaked out? Rebok and Plunkett weighed the cylinder and discovered that most of the gas was still inside. They fiddled with the valve some more, even using a wire to unclog it, but nothing happened.

A frustrated Plunkett removed the valve completely, turned the canister upside down, and shook it. Some flecks of white powder floated out. Plunkett and Rebok sawed open several of the storage canisters and found that their interior walls were lined with a smooth, waxy white coating. In his lab notebook Plunkett wrote, "A white solid material was obtained, which was supposed to be a polymerized product." This entry shows that he instantly understood what had occurred, even though it was generally believed at the time that a chlorinated or fluorinated ethylene could not be polymerized because previous attempts to do so had failed. Something about the combination of pressure and temperature had forced the TFE molecules to join together in long chains, and the resulting compound turned out to have a most interesting set of properties.

Two days later Plunkett noted some additional characteristics of the intriguing substance: "It is thermoplastic, melts at a temperature approaching red heat, and boils away. It burns without residue; the decompositive products etch glass." He also observed that it was insoluble in cold and hot water, acetone, Freon 113, ether, petroleum ether, alcohol, pyridine, toluene ethyl acetate, concentrated sulfuric acid, glacial acetic acid, nitrobenzene, isoanyl alcohol, ortho dichlorobenzene, sodium hydroxide, and concentrated nitric acid. Further tests showed that the substance did not char or melt when exposed to a soldering iron or an electric arc. Moisture did not cause it to rot or swell, prolonged exposure to sunlight did not degrade it, and it was impervious to mold and fungus.

Plunkett's next step was to duplicate the conditions that had produced the first batch of polymerized tetrafluoroethylene (PTFE). After experi-

mentation he succeeded in re-creating what had occurred by chance inside the canisters. On July 1, 1939, he applied for a patent (which he assigned to Kinetic Chemicals) on tetrafluoroethylene polymers. The patent was granted in 1941.

The patent application ended Plunkett's involvement with his discovery, since at that point the problem shifted from fluorine chemistry, which was his area of expertise, to polymer chemistry and process development. Plunkett was named chemical supervisor of DuPont's tetraethyl lead plant and stayed with DuPont in various positions until his retirement in 1975; he was inducted into the National Inventors Hall of Fame in 1985 and died in 1994.

For about three years DuPont's organic chemicals department experimented with ways to produce TFE, also known as TFE monomer, which was the raw material for PTFE. Plunkett and Rebok had produced small batches for laboratory use, but if PTFE was ever going to find a practical use and be produced commercially, the company would have to find a way to turn out TFE monomer in industrial quantities. When the organic group came up with a promising method, DuPont's central research and development department began looking into possible polymerization processes.

Spontaneous polymerization of TFE can lead to explosive reactions because heat is released in the process, so it had to be carefully controlled. Experiments by the chemist Robert M. Joyce soon led to a feasible but costly procedure. Meanwhile, DuPont's applications group began identifying the properties of PTFE that would be useful in industry, such as its resistance to electric currents and to most chemical reactions. Then came World War II, which gave a large boost to the development of PTFE (and many other technologies).

Scientists working on the Manhattan Project faced the difficult problem of separating the isotope U-235 (which makes up about 0.7 percent of the element uranium in its natural state) from the far more plentiful but inert U-238. The method they settled on was gaseous diffusion, in which a gas is forced through a porous material. Since heavy molecules diffuse more slowly than light ones, multiple repetitions of the diffusion process will yield a gas enriched in the lighter isotopes. Gen. Leslie Groves, director of the Manhattan Project, chose DuPont to design the separation plant. To make it work, the designers needed equipment that would stand up to the highly corrosive starting material, uranium hexafluoride gas, which destroyed conventional gaskets and seals. PTFE was just what they needed, and DuPont agreed to reserve its entire output for government use.

For security reasons PTFE was referred to by a code name, K 416, and the small production unit at Arlington, New Jersey, was heavily guarded. Despite the tight security and DuPont's efforts to control the polymerization process, the Arlington production unit was wrecked by an explosion one night in 1944. The next morning construction workers stood by while Army and FBI investigators looked for evidence of sabotage. Working with DuPont chemists, they found that the explosion had been caused by uncontrolled, spontaneous polymerization that was detonated by the exothermic, or heat-releasing, decomposition of TFE to carbon and tetrafluoromethane. When the investigators left, the construction crews took over, working two 12-hour shifts a day. Within two months the unit had been rebuilt with heavy barricades surrounding it.

The Manhattan Project consumed about two-thirds of Arlington's PTFE output, and the remainder was used for other military applications. It proved to be ideal for the nose cones of proximity bombs because it was both electrically resistive and transparent to radar. It was also used in airplane engines and in explosives manufacturing, where nitric acid would destroy gaskets made of other materials, and as a lining in liquid-fuel tanks, whose cold temperatures could make other linings brittle. When the Army needed tape two-thousandths of an inch thick to wrap copper wires in the radar systems of night bombers, it was painstakingly shaved off a solid block of PTFE at a cost of $100 per pound. The high cost was justified because PTFE did a job nothing else could do.

When peace returned, DuPont decided to go ahead with commercializing PTFE, since its manifold military uses had shown its great industrial potential. With its unmatched knowledge of polymers, the company was in a good position to take advantage of the postwar manufacturing boom. In 1944 the company had registered the trademark Teflon, probably suggested by the abbreviation TFE. The new substance was an ideal fit for DuPont's traditional marketing strategy, which was to shun the manufacture of commodity plastics and specialize in sophisticated materials that could command premium prices. Other materials with some of Teflon's properties were available, but none were as comprehensively resistant to corro-

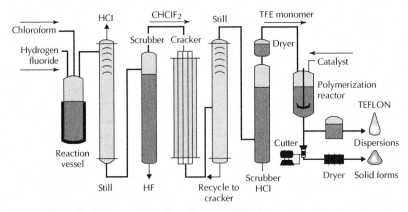

Figure 36.1. Typical process flowsheet for Teflon TFE fluorocarbon resins.

sion, and none of the lubricants or low-friction materials then in use were anywhere near as durable or maintenance-free.

The company faced significant obstacles before it could produce large amounts of Teflon uniformly and economically. Company chemists had developed several ways to polymerize TFE, but the properties of the resulting product varied significantly from batch to batch. And nearly every step of the manufacturing process raised problems that no chemical manufacturer had faced before. Equipment had to withstand temperatures and pressures beyond previous limits. Even a minute quantity of oxygen would react with the gases used as raw materials, fouling the process lines and valves.

After the synthesis was completed, fabricating Teflon into useful articles raised another set of difficulties. Its melting point was so high that it could not be molded or extruded by conventional methods. A further problem was caused by the very properties that had made Teflon so valuable to begin with. Chemistry students like to joke about the inventor who isolates a substance that will dissolve anything, then cannot find a container to hold it. With Teflon, DuPont's chemists faced the opposite problem: How do you make the greatest non-stick substance ever invented bond to another surface?

Research led to the production of Teflon in three basic forms: granules, a fine powder, and an aqueous dispersion. Borrowing the technique of sintering from powder metallurgy, Teflon was compressed and baked into blocks that could be machined into the required shape. In this process the application of heat did not actually melt the Teflon, but it softened the microscopic granules and made them stick together when pressed. Powder could also be blended with hydrocarbons and cold-compressed to coat wires and make tubing. Aqueous dispersions were used to make enamels that could be sprayed or brushed onto a surface and then baked in place.

Another technique involved etching the surface of a piece of Teflon with specially formulated solvents that extracted some of the fluorine atoms. These solvents left behind a thin, carbon-rich surface layer to which conventional adhesives could bond. Yet another solution was to implant fine particles of silica in the Teflon, creating a rough "sandpaper surface" that would also accept adhesives. This method was not as effective as chemical etching, but it was adequate for some purposes. DuPont chemists also developed fluorocarbon resins that would stick to both Teflon and metal surfaces. And of course, sheets of Teflon could be attached to other items with screws, bolts, clamps, and other mechanical fasteners.

Machine parts requiring a uniform coating could be immersed in a "fluidized bed"—a layer of Teflon powder that was agitated with a stream of air until it behaved like a liquid. The item to be coated was first heated to 650 degrees Fahrenheit and then dipped in the fluidized bed for a second or two. After the excess powder was blown off, a film of one to two thousandths of an inch was left behind. As with other methods, repeated applications were often required to get a thick enough film. This method was especially useful with irregularly shaped mechanical components, such as valves and rotors, as well as with small items like ball bearings.

By 1948 DuPont had made enough progress to prepare for full-scale production. Two years later the company's first commercial Teflon plant, designed to produce a million pounds a year, went on line at the Washington Works, on land once owned by George Washington near Parkersburg, West Virginia. DuPont stepped up its efforts to market Teflon for industrial applications, promoting the use of tape and sheets for insulation in many kinds of electrical equipment. Teflon was also used for gaskets, packings, valve components, pump components, bearings, sealer plates, and hoppers. To help users understand the polymer's unusual properties and tricky fabrication requirements, DuPont sent out a team of scientists to advise customers on integrating Teflon into their production processes. Members of the research, manufacturing, and sales staff met regularly to compare notes.

Within a year Teflon was also being used in commercial food processing. DuPont saw the potential for expansion in this field but decided to proceed slowly. In bread manufacturing, rollers were coated with Teflon to keep dough from sticking. Teflon-lined bread pans and muffin tins became standard equipment in many bakeries. Teflon coatings also stopped dough from sticking to cookie sheets and reduced the number of damaged cookies that had to be thrown away. In candy factories Teflon coated conveyor belts, hooks for pulling taffy, and the cutting edges of slicing machines. In all these applications, Teflon proved much more effective than the old method of coating the surface with oil or grease.

A 1953 DuPont television commercial showed a Teflon-coated bread pan and boasted that it had "baked 1,258 loaves of bread and . . . never had a drop of grease in it." The first draft of the script for this ad also predicted that frying pans would be coated with Teflon in the future, but that line was deleted before the commercial was filmed. DuPont was reluctant to market Teflon-coated cookware for home kitchens because of concerns that misuse might lead to injuries and lawsuits. Until the company could be sure that Teflon was absolutely safe in untrained hands, it preferred to stay with industrial users. Nylon, another DuPont product, had become a great success in consumer products, but it was not subjected to the extreme conditions that Teflon cookware would encounter.

DuPont's tests showed that while Teflon could withstand brief exposure to temperatures as high as 1,000 degrees Fahrenheit, it began to soften at 620 degrees Fahrenheit. This was no problem for baking pans, which are rarely subjected to temperatures above 500 degrees, but it could potentially cause problems with pans used on stovetops. Researchers found that at high temperatures, small quantities of gaseous decomposition products were released. Because some of these gases were toxic and might cause temporary flu-like symptoms, adequate ventilation was required. Although the fumes given off by overheated Teflon pans were less toxic than those given off by heated cooking oil or butter, DuPont decided to proceed with caution. Even as late as 1960 the company sold less than 10 million pounds of Teflon per year, with receipts of a piddling (by DuPont standards) $28 mil-

lion. Expanding consumer uses would be the key to boosting sales, but DuPont had to convince itself that Teflon was harmless before selling it to the housewives of America.

While DuPont hesitated, an enterprising French couple took matters into their own hands. Marc Grégoire, an engineer, had heard about Teflon from a colleague who had devised a way to affix a thin layer of it to aluminum for industrial applications. The process involved etching the aluminum with acid to create a microscopically pitted surface, covering the surface with Teflon powder, and heating it to just below its melting point, which caused it to interlock with the aluminum surface.

Grégoire, an avid fisherman, decided to coat his fishing gear with Teflon to prevent tangles. His wife, Colette, had another idea: Why not coat her cooking pans? Grégoire agreed to try it, and he was successful enough to be granted a patent in 1954. The Grégoires were so happy with the results that they set up a business in their home. Starting around 1955, Marc coated pans in their kitchen and Colette peddled them on the street. French cooks, despite their customary reverence for tradition, snapped them up. Encouraged by this reception, the Grégoires formed the Tefal Corporation in May 1956 and opened a factory.

Soon afterward France's Conseil Supérieur de l'Hygiène Publique officially cleared Teflon for use on frying pans. The Laboratoire Municipale de Paris and the École Supérieur de Physique et Chimie also declared that Teflon-coated cookware presented no health hazard. In 1958 the French ministry of agriculture approved the use of Teflon in food processing. That same year the Grégoires sold one million items from their factory. Two years later sales approached the three million mark.

DuPont executives, who were aware of these developments in France, decided to seek the approval of the U.S. Food and Drug Administration (FDA) for wider use of Teflon in cooking and food processing. The company tested frying pans and other cooking surfaces under conditions even more rigorous than those used in France. DuPont's researchers concluded that utensils coated with Teflon were "unquestionably safe" for both domestic and commercial cooking. In January 1960 the company gave the FDA four volumes of data, collected over nine years, on the effects of Teflon resins in food handling. Within a few months the FDA decided that the resins did not "present any problems under the Food Additives Amendment." Despite the favorable FDA decision, DuPont continued to move slowly, since marketing Teflon-coated cookware was not a high priority. Then one man's enthusiasm nudged DuPont into action.

Thomas G. Hardie was an American who admired French culture. After graduating from college, he served in the military, worked for the Marshall Plan in Paris, and became a foreign correspondent for an American newspaper chain. Then he entered his family's business, Nobelt, a Maryland firm that makes textile machinery. During a business trip to France in 1957 or 1958, Hardie met Marc Grégoire at a party on the Left Bank. The Frenchman enthusiastically told Hardie about his business and the factory he was building in a Paris suburb. Hardie was intrigued by Grégoire's tale of the fast-selling cookware.

After Hardie went home to Maryland, he decided that the popular French pans would sell in the United States too. He went back to Paris to meet with Grégoire, who was reluctant to do business with an American because he didn't trust Yankees. But Hardie was very persuasive and eventually won Grégoire's confidence. With visions of quick success, he returned to the States with the rights to manufacture nonstick cookware using Tefal's process.

During the next two years Hardie called on many American cookware manufacturers, trying to persuade them to make Teflon-coated pans. He had no success because the idea of nonstick pans was simply too new. All these rejections turned Hardie's business venture into a personal crusade. Although he had no experience in the import business, he cabled the French factory to ship him 3,000 Tefal pans, which he warehoused in a barn on his sheep farm in Maryland. He sent free sample pans, along with promotional literature, to housewares buyers at 200 department stores. Not one of them placed an order.

Next Hardie met with an executive at DuPont in Wilmington, Delaware. By describing the success of nonstick pans in France, he was able to convince the executive that cookware could be a valuable new market. When the executive objected that the name Tefal was too close to Teflon, Hardie agreed to market his imported French pans under the name T-fal. Later a DuPont salesman was assigned to accompany Hardie on a visit to Macy's in New York City. There, in a tiny basement office, a buyer named George Edelstein placed a small order. Hardie was so excited that he sent a victory cable to the French factory. On December 15, 1960, during a severe snowstorm, the T-fal "Satisfry" skillets went on sale for $6.94 at Macy's Herald Square store. To almost everyone's amazement, the pans quickly sold out.

Shortly afterward Hardie made his second sale when he telephoned Roger Horchow, a buyer for the Dallas department store Neiman Marcus. Horchow agreed to test a sample skillet even though his store didn't have a housewares department. He gave the skillet to Helen Corbitt, a cookbook editor who ran a popular cooking school in Dallas. Corbitt loved it, prompting Neiman Marcus to place a large order and run a half-page newspaper advertisement. The store sold 2,000 skillets in a week. Horchow later recalled, "Skillets were piled up, still in the shipping crates, as in a discount house, with the salesladies handing them out to customers like hotcakes at an Army breakfast." The news spread to other department stores. Buyers jumped on the nonstick bandwagon, and Hardie was swamped with orders.

The inventory in Hardie's barn was quickly exhausted. He phoned France daily to ask for more pans, but the French plant couldn't work fast enough to supply both sides of the Atlantic. Hardie flew to France to press his case with Grégoire. He even lent Tefal $50,000 to expand its facilities, but it still could not meet the American demand. To cope with the avalanche of orders, which reached a million pans per month in mid-1961, Hardie built his own factory in Timonium, Maryland.

Unfortunately for him, around the same time, several major American cookware companies decided that the time was right to start making Teflon pans. Suddenly the market was saturated with nonstick cookware. Because the American companies had no experience with Teflon coatings, much of it was inferior to the French product, and nonstick pans soon acquired a bad name. Just as quickly as the U.S. demand for nonstick pans had soared, it plummeted, and warehouses were filled with unsold stock. Hardie sold his factory and focused on his family's business. (T-fal cookware, the standard of quality in the early 1960s, is still being manufactured and is sold in stores in the United States and abroad.)

Despite the problems with early Teflon cookware, DuPont's managers still believed that it had enormous potential. So the company commissioned some research. Six thousand consumers, along with a sampling of professionals in the cookware business, were asked what was wrong with Teflon products. The respondents overwhelmingly liked the idea of Teflon cookware; the problem lay with faulty production methods that turned out shoddy pans whose coatings scraped off much too easily.

DuPont knew that cookware could be more than just a way to sell lots of Teflon. It could also be an invaluable marketing tool, a vehicle to familiarize vast numbers of consumers with Teflon and its properties. Conversely, low-quality merchandise could only harm the product's reputation. As a result the company established coating standards for manufacturers and initiated a certification program, complete with an official seal of approval for Teflon kitchenware. To verify compliance with its standards, DuPont performed more than 500 tests per month on cookware at its Marshall Laboratories in Philadelphia.

The DuPont certification program was so successful that a marketing survey in the mid-1960s found that 81 percent of homemakers who had purchased nonstick pans were pleased with them. By 1968 DuPont had developed Teflon II, which not only prevented food from sticking to the pans but was also (supposedly) scratch-resistant. Later generations of Teflon cookware, with thicker coatings and improved bonding, would be introduced under the trade names Silverstone in 1976 and Silverstone Supra in 1986.

As Teflon became better known to consumers, rumors began to circulate that it was unsafe. Tales sprang up about how Teflon had caused the mysterious deaths of unidentified workers. In other versions users of nonstick cookware had suffered the flu or seizures after breathing Teflon fumes. Industrial safety bulletins and at least one medical journal warned readers of Teflon's supposed dangers.

Whenever one of these false reports came to DuPont's attention, the company demanded a published retraction. It also published a booklet called *The Anatomy of a Rumor* that summarized the results of research carried out at DuPont and elsewhere. In addition, DuPont tried to set the record straight by acknowledging whatever minor problems could be documented. The company admitted that there had been isolated incidents of "polymer fume fever," which produced symptoms similar to those of influenza for a brief period but had no lasting effects. It also acknowledged at least one case of a worker suffering "the shakes" after smoking cigarettes that might have been contaminated with Teflon dust. In fact, as early as 1954 DuPont had instructed its employees not to smoke or carry cigarettes with them while working with Teflon. However, no serious illnesses or injuries had ever been linked to Teflon.

When Teflon cookware was introduced, many national magazines printed articles about the new products. Most discussed the safety issue, and several mentioned the rumors, but none gave any credence to the gossip. Nevertheless, *Consumer Reports* got so much mail about the rumors after a 1961 article that the editors had to print a second article refuting them again. As late as 1973 *Consumer Reports* was still receiving mail on the "old bugaboo about nonsticks," prompting the editors to publish yet another article emphasizing that they knew of "no consumer illnesses resulting from . . . nonstick cookware in ordinary home use."

As nonstick cookware became accepted, Teflon made the transition from a low-volume specialty material used chiefly in industry to a mass-market consumer item. Today Teflon is used to insulate fabrics in tablecloths and carpets and to coat the surfaces of steam irons. Teflon plumbing pipes and valves can be found in many new homes; Teflon flakes add toughness to nail polish. In fiber form, as part of the fabric known as Gore-Tex, it is beloved by campers and skiers for its ability to insulate while wicking moisture from the skin. It can also be found in pacemakers, dentures, medical sutures, artificial body parts, printed circuits, cables, space suits, and thousands of other manufactured products. The surest sign of the slippery material's success is its adoption as a slang term in political discourse, where *Teflon* is used to describe an officeholder who unaccountably remains popular despite having opinions with which one disagrees.

While the discovery of Teflon was unplanned, the rest of its story is anything but accidental. Plunkett's training in fluorine chemistry allowed him to recognize what he had found and to analyze its properties, a byway he might not have been able to explore in a smaller firm. When the project grew beyond laboratory scale, he knew he could hand it off to other departments with confidence. DuPont had the knowledge base to find ways of producing the monomer cheaply enough, controlling the polymerization, applying the useful but hard-to-handle polymer to industrial use, and making sure that consumer products were durable, safe, and reliable. Large research groups can have their disadvantages, but in the case of Teflon, DuPont's size was a critical ingredient in its success.

12

Maintaining Innovative Climates

37

That's Easy for You to Say

LUCIEN RHODES

An obsession with "corporate culture" can be worse than no culture at all. Just ask the man who wrote the book on the subject.

It all began on Labor Day weekend in 1982. Allan A. Kennedy was sitting in a low beach chair on the shore in front of his cottage on Cape Cod. Next to him was his friend and fellow consultant Tony Merilo. As they relaxed there, watching the sailboats drift across the Cape Cod Bay, drinking beer, and listen-ing to a Red Sox game on the radio, Kennedy turned to Merilo and, with the majestic eloquence suited to great undertakings, said: "Gee, Tony, you know, we ought to start some kind of business together."

This identical thought has, of course, passed between countless friends ever since the discovery of profit margins. Coming from most people, it would have fallen into the general category of loose talk. But Kennedy was not most people. For one thing, he was a 13-year veteran of McKinsey & Co., the management consulting firm, and partner in charge of its Boston office. More to the point, he was the coauthor of a recently published book that offered a startling new perspective on corporate life—one that challenged the whole way people thought about business.

The book was entitled *Corporate Cultures,* a term that was itself new to the language, and it dealt with an aspect of business that, up to then, had been largely ignored. Broadly speaking, that aspect involved the role played by a company's values, symbols, rites, and rituals in determining its overall performance. Citing examples from some of the country's most dynamic companies, Kennedy and co-author Terrence E. Deal showed that these "cultural" factors had a major effect on the attitudes and behavior of a company's employees, and were thus of critical importance to its long-term success.

By any measure, the book was a groundbreaking work, challenging, as it did, the rational, quantitative models of corporate success that were so popular in the 1960s and '70s. But its impact had as much to do with its timing as its content. Published in June 1982, during a period of economic stagnation—with unemployment at 9.5%, the prime over 16%, and trade deficits soaring to record levels—*Corporate Cultures* offered a welcome antidote to the doom and gloom that was abroad in the land. Like *In Search of Excellence,* which appeared a few months later, it suggested that Japan was not the only nation capable of producing strong, highly motivated companies that could compete effectively in the international arena. America could produce—in fact, was already producing—its own.

What the book did not detail, however, was how corporate cultures were actually constructed. The authors could describe a particular culture and demonstrate its effects, but they offered few clues as to how a company might develop a culture in the first place. So the news that Allan Kennedy was going into business was greeted with more than passing interest among the followers of corporate culture. Here was an opportunity to find out how a living, breathing culture could be created, and the creator would be none other than the man who wrote the book.

After an extensive survey of business opportunities, Kennedy and Merilo decided to develop microcomputer software for sales and marketing management. They felt this was their most promising option, given the anticipated growth of the microcomputer market and their own experience as consultants. Acting on that assessment, they resigned from McKinsey and, in February 1983, formally launched Selkirk Associates Inc. with four of their friends.

Kennedy had lofty ambitions for Selkirk. More than a business, he saw it as a kind of laboratory for his theories. He wanted it to function as a society of professional colleagues committed to building a culture and a company that would stress collaboration, openness, decentralization, democratic decisions, respect, and trust. In this society, each individual would be encouraged to devise his or her own entrepreneurial response to the challenges of the business.

For Kennedy, this was not a long-term goal, something that would evolve naturally in the fullness of time. On the contrary, it was a pressing, immediate concern. Accordingly, he focused all his attention on creating such a culture from the start. "I spent lots of time," he says, "trying to think about what kind of values the company ought to stand for and therefore what kind of behavior I expected from people." These thoughts eventually went into a detailed statement of "core beliefs," which he reviewed and amplified with each new employee. In the same vein, Kennedy and his colleagues chose a "guiding principle" namely, a commitment to "making people more productive." They would pursue this ambition, everyone agreed, "through the products and services we offer" and "in the way we conduct our own affairs."

And, in the beginning at least, Selkirk seemed to be everything Kennedy had hoped for. The company set up shop in Boston, in an office that consisted of a large, rectangular room, with three smaller attachments. Each morning, staff members would pile into the main room and sort themselves out by function—programmers and systems engineers by the windows; administrations in the middle, sales and marketing folk at the other end. In keeping with Kennedy's cultural precepts, there were no private offices or, indeed, any physical demarcations between functions.

It was a familial enterprise, informed with the very qualities Kennedy had laid out in his statement of core beliefs. The work was absorbing, the comradeship inspiring. Most mornings, the staff feasted on doughnuts, which they took to calling "corporate carbos," as a wordplay on "corporate cultures." They began a scrapbook as an impromptu cultural archive. Included among the memorabilia was "The Ravin," an Edgar Allan Poe takeoff that commemorated Selkirk's first stirrings in earlier temporary headquarters:

Once upon an April morning,
    disregarding every warning,
In a Back Bay storefront,
    Selkirk software was begun:
True, it was without a toilet,
    but that didn't seem to spoil it.

To strengthen their bonds even further, the staff began to experiment with so-called rites, rituals, and ceremonies—all important elements of a corporate culture, according to Kennedy's book. Selkirk's office manager, Linda Sharkey, recalls a day, for example, when the whole company went out to Kennedy's place on Cape Cod to celebrate their common purpose with barbecues on the beach. "The sun was shining, and we were all there together," she says. "It was a beautiful day. That's the way it was. We didn't use the terms among ourselves that Allan uses in the book. With us, corporate cultures was more by seeing and doing." Sharkey remembers, too, Friday afternoon luncheons of pizza or Chinese food, at which everyone in the company had a chance to talk about his or her accomplishments or problems, or simply hang out.

Kennedy was pleased with all this, as well he might be. "We were," he says, "beginning to develop a real culture."

Then the walls went up.

The problem stemmed from the situation in the big room, where the technical people were la-

boring feverishly to develop Selkirk's first product, while the salespeople were busy preselling it. The former desperately needed peace and quiet to concentrate on their work; the latter were a boisterous lot, fond of crowing whenever a prospect looked encouraging. In fact, the salespeople crowed so often and so loudly that the technicians complained that they were being driven to distraction. Finally, they confronted Kennedy with the problem. Their solution, which Kennedy agreed to, was to erect five-foot-high movable partitions, separating each functional grouping from the others.

In the memory of Selkirk veterans, "the day the walls went up" lives on as a day of infamy. "It was terrible," says Sharkey. "I was embarrassed."

"It was clearly a symbol of divisiveness," says Kennedy.

"I don't know what would have been the right solution," says Reilly Hayes, Selkirk's 23-year-old technical wizard, "but the wall certainly wasn't. It blocked out the windows for the other end of the room. Someone [in marketing] drew a picture of a window and taped it to the wall. The whole thing created a lot of dissension."

Indeed, the erection of the walls touched off a feud between engineering and marketing that eventually grew into "open organizational warfare," according to Kennedy. "I let the wall stand, and a competitive attitude developed where engineering started sniping at marketing. We had two armed camps that didn't trust each other."

As if that weren't bad enough, other problems were beginning to surface. For one thing, the company was obviously overstaffed, having grown from 12 people in June 1983 to 25 in January 1984, without any product—or sales—to show for it. "That was a big mistake," says Kennedy. "We clearly ramped up the organization too fast, particularly given the fact that we were financing ourselves. I mean, for a while, we had a burn rate of around $100,000 per month."

Even more serious, however, was the problem that emerged following the release of the company's initial product, Correspondent, in February 1984. Not that there was anything wrong with the product. It was, in fact, a fine piece of software, and it premiered to glowing reviews. Designed as a selling tool, it combined database management, calendar management, word processing, and mail merge—functions that could help customers organize their accounts, track and schedule sales calls and followups, and generate correspondence. And it did all that splendidly.

The problem had to do with the price tag, a whopping $12,000 per unit. The Selkirk team members had come up with this rarefied figure, not out of greed, but out of a commitment to customer service—a goal to which they had pledged themselves as part of their cultural mission. In order to provide such service, they figured, a Selkirk representative might have to spend two or three weeks with each customer, helping to install and customize the product. Trouble was, customers weren't willing to *pay* for that service, not at $12,000 per unit anyway. After a brief flurry of interest, sales dropped off.

"We just blew it," says Kennedy. "We were arrogant about the market. We were trying to tell the market something it wasn't interested in hearing. We took an arbitrary cultural goal and tried to make it into a strategy, rather than saying we're a market-driven company and we've got to find out what the market wants and supply it." Unfortunately, six months went by before Kennedy and his colleagues figured all this out and began to reduce Correspondent's price accordingly.

By then, however, Selkirk's entire sales effort was in shambles, a victim of its commitment to employee autonomy. Sales targets were seldom realized. Indeed, they were scarcely even set. At weekly meetings, salespeople would do little more than review account activity. "If a salesman said each week for three weeks in a row that he expected to close a certain account, and it never happened," say Merlo, "well, we didn't do anything about it. In any other company, he would probably have been put on probation." As it was, each of the participants entered the results of the meeting in a red-and-black ledger book and struck out once again to wander haphazardly through uncharted territory. "The mistake we made," reflects Merlo, "was using real money in a real company to test hypotheses about what sales goals should be."

Finally, in June 1984, Kennedy took action, laying off 6 people. In July, Correspondent's price was dropped to $4,000 per unit, but sales remained sluggish. In September, Kennedy laid off 5 more people, bringing the size of the staff back to 12.

One of those laid off was the chief engineer, a close friend of Kennedy's, but a man whose departure brought an immediate ceasefire between the warring factions. That night, the remaining staff members took down the walls and stacked them neatly in the kitchenette, where they repose to this day. "We felt," says Sharkey, "like we had our little family back together again."

With morale finally rebounding, Selkirk again cut Correspondent's price in the early fall, to $1,500. This time, sales responded, and, in November, the company enjoyed its first month in the black.

But Selkirk was not yet out of the woods. What remained was for Kennedy to figure out the significance of what had happened, and to draw the appropriate conclusions. Clearly, his experiment had not turned out as he had planned. His insistence on a company without walls had led to organizational warfare. His goal of providing extraordinary service had led to a crucial pricing error. His ideal of employee autonomy had led to confusion in the sales force. In the end, he was forced to fire more than half of his staff, slash prices by 87%, and start over again. What did it all mean?

Merlo had one answer. "We're talking about an experiment in corporate culture failing because the business environment did not support it," he says. "The notion of corporate culture got in the way of tough-minded business decisions." He also faults the emphasis on autonomy. "I don't think we had the right to be organized the way we were. I think we should have had more discipline."

Kennedy himself soon came around to a similar view. "Look in [the statement of core beliefs] and tell me what you find about the importance of performance, about measuring performance or about the idea that people must be held accountable for their performance," he says. "That stuff should have been there. I'm not discounting the importance of corporate culture, but you have to worry about the business at the same time, or you simply won't have one. Then you obviously won't *need* a culture. Where the two come together, I think, is in the cultural norms for performance, what kind of performance is expected of people. And that's a linkage that wasn't explicit in my mind three years ago. But it is now." He adds that, if the manuscript of *Corporate Cultures* were before him today, he would include a section on performance standards, measurement systems, and accountability sanctions.

On that point, he might get an argument from his co-author, Terrence Deal, a professor at Vanderbilt University and a member of Selkirk's board of directors since its inception. Deal does not disagree about the importance of discipline and performance standards, but he questions the wisdom of trying to impose them from above. The most effective performance standards, he notes, are the ones that employees recognize and accept as the product of their own commitment, and these can merge only from the employees' experience. "One of the things that we know pretty handsomely," says Deal, "is that it's the informal performance standards that really drive a company."

In fact, Kennedy may have gotten into trouble not by doing too little, but by doing too much. Rather than letting Selkirk's culture evolve organically, he tried to impose a set of predetermined cultural values on the company, thereby retarding the growth of its own informal value system. He pursued culture as an end in itself, ignoring his own caveat, set down in his book, that "the business environment is the single greatest influence in shaping a corporate culture." Instead, he tried to shape the culture in a vacuum, without synchronizing it with the company's business goals.

In so doing, Kennedy reduced corporate culture to a formula, a collection of genetic "principles." It was a cardinal error, if not an uncommon one. "There are a lot of people," says Deal, "who take our book literally and try to design a culture much as if they're trying to design an organization chart. My experience across the board has been that, as soon as people make it into a formula, they start making mistakes." By following the "for-

mula," Kennedy wound up imposing his own set of rules on Selkirk—although not enough of them, and not the right kind, he now says. The irony is that a real corporate culture allows a company to manage itself *without* formal rules, and to manage itself better than a company that has them.

Deal makes another point. Kennedy, he observes, might be less concerned with performance today if he had not hired so many friends at the beginning. Friends are nice to have around, but it's often hard to discipline them, or subject them to a company's normal sanctions. Over the long run, Deal says, their presence at Selkirk probably undermined the development of informal performance standards.

Kennedy himself may have played a role in that, too. He estimates that, over the past year, he has spent only one day a week at Selkirk. The rest of the time he has been on the road as a consultant, using his fees to help finance the company. In all, he has sunk some $1 million of his own money into Selkirk, without which the company might not have survived. But it has come at a price. "Nobody had to pay attention to things like expenses, because there was a perception of an infinite sink of money," Kennedy says.

The danger of that perception finally came home to him last summer, when three of Selkirk's four salespeople elected to take vacations during the same month. The result was that sales for the month all but vanished. Kennedy had had enough. "I told the people here that either you sustain the company as a self-financing entity, or I will let it go under. I'm unwilling to put more money on the table."

And yet, in the end, it was hard to avoid the conclusion that a large part of Selkirk's continuing problem was Allan Kennedy himself—a thought that did not escape him. "I've got a lot to learn about running a business successfully," he says, "about doing it myself, I mean. I think I know everything about management except how to manage. I can give world-class advice on managing, but—when it comes right down to it—I take too long and fall into all the traps that I see with the managers I advise."

Whatever his shortcomings as a manger, there is one thing Kennedy can't be faulted for, and that is lack of courage. Having drawn the inevitable conclusion, he went out looking for someone who could help him do a better job of managing the company. For several months, he negotiated with the former president of a Boston-based high-tech firm, but the two of them were unable to come to terms. Instead, Kennedy has made changes at Selkirk that he hopes will achieve the same effect. In the new structure, Merilo is taking charge of the microcomputer end of the business, while Betsy Meade—a former West Coast sales representative—has responsibility for a new minicomputer version of Correspondent, to be marketed in conjunction with Prime Computer Corp. As for Kennedy, he will concern himself with external company relations, product-development strategies, and, of course, corporate culture.

Kennedy is full of optimism these days. He points out that, despite its checkered history, Selkirk has emerged with a durable product and an installed base of about 1,000 units. In addition, the company will soon be bolstered with the proceeds from a $250,000 private placement. Meanwhile, he says, some of the company's previous problems have been dealt with, thanks to the introduction of a reliable order-fulfillment process, the decision to put sales reps on a straight commission payment schedule, and the establishment of specific sales targets for at least the next two quarters. "I think we have much more focused responsibility," he says, "and much more tangible measures of success for people in their jobs."

Overall, Kennedy looks on the past three years as a learning experience. "There are times when I think I should charge up most of the zigs and the zags to sheer rank incompetence," he admits. "But then there are other times when I look back and say, 'Nobody's that smart, and you can't do everything right.' In life, you have to be willing to try things. And if something doesn't work, you have to be willing to say, 'Well, that was a dumb idea,' and then try something else." Now, he believes, he has a chance to do just that.

In the meantime, he is in the process of writing another book. He already has a proposal circulating among publishers. In his idle moments, he occasionally amuses himself by inventing titles. One of those titles speaks volumes about where he has been: *Kicking Ass and Taking Names.*

38

Organizational Issues in the Introduction of New Technologies

RALPH KATZ AND THOMAS J. ALLEN

1. INTRODUCTION

More than ever before, organizations competing in today's world of high technology are faced with the challenges of "dualism," that is, functioning efficiently today while planning and innovating effectively for tomorrow. Not only must these organizations be concerned with the success and market penetration of their current product mix, but they must also be concerned with their long-run capability to develop and incorporate in a timely manner the most appropriate technical advancements into future product offerings. Research and development–based corporations, no matter how they are organized, must find ways to internalize both sets of concerns.

Now it would be nice if everyone in an organization agreed on how to carry out this dualism or even agreed on its relative merits. This is rarely the case, however, even though such decisions are critically important to a firm competing in markets strongly affected by changing technology (Allen, 1977; Roberts, 1974). Amidst the pressures of everyday requirements, decision makers representing different parts of the organization usually disagree on the relative wisdom of allocating resources or particular RD&E talents among the span of technical activities that might be of benefit to today's versus tomorrow's organization. Moreover, there are essentially no well-defined principles within management theory on how to structure organizations to accommodate these two sets of conflicting challenges. Classical management theory with its focus on scientific principles deals only with the efficient production and utilization of today's goods and services. The principles of high task specialization, unity of command and direction, high division of labor, and the equality of authority and responsibility all deal with the problems of structuring work and information flows in routine, predictable ways to facilitate production and control through formal lines of authority and job standardization. What is missing is some comparable theory that would also explain how to organize innovative activities within this operating environment such that creative, developmental efforts will not only take place but will also become more accepted and unbiasedly reviewed, especially as these new and different ideas begin to "disrupt" the smooth functioning organization. More specifically, how can one structure an organization to promote the introduction of new technologies and, in general, enhance its longer-term innovation process, yet, at the same time, satisfy the plethora of technical demands and accomplishments needed to support and improve the efficiency and competitiveness of today's producing organization?

Implicit in this discussion, then, is the need for managers to learn how to build parallel structures and activities that would not only permit these two opposing forces to coexist but would also balance them in some integrative, meaningful way. Within the RD&E environment, the operating organization can best be described as an "output-

Reprinted with permission of Plenum Press, "Organizational Issues in the Introduction of New Technologies" in *The Management of Productivity and Technology in Manufacturing*, P. R. Kleindorfer (ed.), © 1985, pp. 275–300.

Ralph Katz is Professor of Management at Northeastern University's College of Business and Research Associate at MIT's Sloan School of Management.   **Thomas J. Allen** is the Howard Johnson Professor of Management at MIT's Sloan School of Management.

oriented" or "downstream" set of forces directed towards the technical support of the organization's current products and towards getting new products out of development and into manufacturing or into the marketplace. Typically, such pressures are controlled through formal structures and through formal job assignments to project managers who are then held accountable for the successful completion of product outputs within established schedules and budget constraints.

At the same time, there must be an "upstream" set of forces that are less concerned with the specific architectures and functionalities of today's products but are more concerned with the various core technologies that might underlie the industry or business environment not only today but also tomorrow. They are, essentially, responsible for the technical health and excellence of the corporation, keeping the company up-to-date and technically competitive in their future business areas.

In every technology-based organization, as discussed by Katz and Allen (1985), the forces that represent this dualism compete with one another for recognition and resources. The conflicts produced by this competition are not necessarily harmful; in fact, they can be very beneficial to the organization in sorting out project priorities and the particular technologies that need to be monitored and pursued, provided there are mechanisms in place to both support and balance these two forces.

If the product-output or downstream set of forces becomes dominant, then there is the likelihood that sacrifices in using the latest technical advancements may be made in order to meet budget, schedule, and immediate market demands. Given these pressures, there are strong tendencies to strip the organization of its research activities and to deemphasize longer-term, forward-looking technological efforts and investigations in order to meet current short-term goals which could, thereby, mortgage future technical capabilities. Under these conditions, requirements for the next generation of new product developments begin to exceed the organization's in-house expertise, and product potentials are then oversold beyond the organization's technical capability.

At the other extreme, if the research or upstream technology component of the organization is allowed to dominate development work within R&D, then the danger is that products may include not only more sophisticated but also perhaps less proven, more risky, or even less marketable technologies. This desire to be technologically aggressive—to develop and use the most attractive, most advanced technology—must be countered by forces that are more sensitive to the operational environments and more concerned with moving research efforts into some final physical reality. Technology is not an autonomous system that determines its own priorities and sets its own standards of performance. To the contrary, market, social, and economic considerations eventually determine priorities as well as the dimensions and levels of performance necessary for successful commercial application (Utterback, 1974).

To balance this dualism—to be able to introduce the new technologies needed for tomorrow's products while functioning efficiently under today's current technological base, is a very difficult task. Generally speaking, the more the organization tries to operate only through formal mechanisms of organizational procedures, structures, and controls, the more the organization will move towards a functioning organization that drives out its ability to experiment and work with new technological concepts and ideas. More informal organizational designs and processes are therefore needed to influence and support true innovative activity, countering the organization's natural movement towards more efficient production and bureaucratic control. These informal mechanisms are also needed to compensate for the many limitations inherent within formal organizational structures and formal task definitions. In the rest of this paper, we will describe three general areas of informal activity that need to take place within an RD&E environment (in parallel with the formal, functioning organization) in order to enhance the innovation process for the more timely introduction of new technologies into the corporation's product portfolio. The general proposition is that these areas of informal activity need to be managed within the

RD&E setting, strengthening and protecting them from the pressures of the "productive" organization in order to increase the organization's willingness and ability to deal with the many advancements that come along, especially with respect to new areas of technology.

2. PROBLEM SOLVING, COMMUNICATIONS, AND THE MOBILITY OF PEOPLE

To keep informed about relevant developments outside the organization as well as new requirements within the organization, R&D professionals must collect and process information from a large variety of outside sources. Project members rarely have all the requisite knowledge and expertise to complete successfully all of the tasks involved in new technical innovations; information and assistance must be drawn from many sources beyond the project both within and outside the organization. Furthermore, if one assumes that the world of technology outside the organization is larger than the world of technology inside the organization, then one should also expect a great deal of emphasis within R&D on keeping in touch with the many advancements in this larger external world. Allen's (1977) 20 years of research work on technical communications and information flows clearly demonstrates just how important this outside contact can be in generating many of the critical ideas and inputs for more successful research and development activity.

At the same time, the research findings of many studies, including Katz and Tushman (1981), Allen (1977), and Pelz and Andrews (1966), have consistently shown that the bulk of these critical outside contracts comes from face-to-face interactions among individuals. Interpersonal communications rather than the formal technical reports, publications, or other written documentation are the primary means by which engineering professionals collect and transfer important new ideas and information into their organizations and project groups. In his study of engineering project teams, for ex-

ample, Allen (1977) carefully demonstrated that only 11% of the sources of new ideas and information could be attributed to written media; the rest occurred through interpersonal communications. Many of these "creative" exchanges, moreover, were of a more spontaneous nature in that they arose not so much out of formal project requirements and interdependencies but out of factors relating to past project experiences and working relationships, the geographical layouts of office locations and laboratory facilities, attendances at special organization events and social functions, chance conversations with external professionals and vendors at conferences and trade shows, and so on. Anything that can be done to stimulate informal contacts among the many parts of the organization and between the organization's R&D professionals and their outside technology and customer environments is likely to be helpful in terms of both technology development and technology transfer.

Since communication processes play such an important role in fostering the creative work activities of R&D members, it would be nice if each individual or project team were naturally willing or always motivated to expose themselves to fresh ideas and new points of view. Unfortunately, this is usually not the case as engineering individuals continue to work in a particular project area or in a given area of technology. In fact, one of the more important assumptions underlying human behavior within organizations is that people are strongly motivated to reduce uncertainty (Katz, 1982). As part of this process, individuals, groups, and even organizations strive to structure their work environments to reduce the amount of stress they must face by directing their activities and interactions toward a more predictable level of certainty and clarity. Over time, then, engineers and scientists are not only functioning to reduce technical uncertainty, they are also functioning to reduce their "personal and situational" uncertainty within the organization (Katz, 1980). In the process of gaining increasing control over their task activities and work demands, three broad areas of biases and behavioral responses begin to emerge. And the more these trends are allowed to take place and become reinforced, the more difficult it will be for the organization to consider seriously the potential, long-term advantages of the many new and different technologies that are slowly being developed and worked on by the larger outside R&D community.

2.1 Problem-Solving Processes

As R&D professionals work together in a given area for a long period of time and become increasingly familiar with their work surroundings, they become less receptive toward any change or innovation that threatens to disrupt significantly their comfortable and predictable work patterns of behavior. In the process of reducing more and more uncertainty, these individuals are likely to develop routine responses for dealing with their frequently encountered tasks in order to ensure predictability, coordination, and economical information processing. As a result, there develops over time increasing rigidity in their problem-solving activities—a kind of functional stability that reduces their capacity for flexibility and openness to change. Behavioral responses and technical decisions are made in fixed, normal patterns; and consequently, new or changing situations that may require technical strategies that do not fit prior problem-solving modes are either ignored or forced into these established molds. R&D professionals interacting over a long period, therefore, develop work patterns that are secure and comfortable, patterns in which routine and precedent play a relatively large part. They come, essentially, to rely more and more on their customary ways of doing things to complete project requirements. In their studies of problem-solving strategies, for example, Allen and Marquis (1963) show that within R&D there can be a very strong bias for choosing those technical strategies and approaches that have worked in the past and with which people have gained common experience, familiarity, and confidence, all of which inhibit the entry of competing tactics involving new technologies, new ideas, or new competencies.

What also seems to be true is that as engineers continue to work in their well-established areas of technology and develop particular problem-solving procedures, they become increasingly committed to these existing methods. Commitment is a function of time, and the longer individuals are asked to work on and extend the capabilities of certain technical approaches, the greater their commitment becomes toward these approaches. Furthermore, in accumulating experience and knowledge in these technical areas, R&D has often had to make clear presentations, showing progress and justifying the allocation of important organizational resources. As part of these review processes, alternative or competing ideas and approaches were probably considered and discarded, and with such public refutation, commitments to the selected courses of action become even stronger. Individuals become known for working and building capability in certain technical areas, both their personal and organizational identities become deeply ensconced in these efforts, and as a result, they may become overly preoccupied with the survival of their particular technical approaches, protecting them against new technical alternatives or negative evaluations. All of the studies that have retrospectively examined the impact of major new technologies on existing organizational decisions and commitments arrive at the same general conclusion: those working on and committed to the old, invaded technology fail to support the radical new technology; instead, they fight back vigorously to defend and improve the old technology (e.g., Cooper and Schendel, 1976; Schon, 1963). And yet, it is often these same experienced technologists who are primarily asked to evaluate the potential effects of these emerging new technologies on the future of the organization's businesses. It is no wonder, therefore, that in the majority of cases studied, the first commercial introduction of a radical new technology has come from outside the industry's traditional competitors.

2.2 Communication and Information Processing

One of the consequences of increased behavioral and technical stability is that R&D groups also become increasingly isolated from outside sources of relevant information and important new ideas. As engineers become more attached to their current work habits and areas of technical expertise, the extent to which they are willing or even feel they need to expose themselves to new ideas, approaches, or technologies becomes progressively less and less. Instead of being vigilant in seeking information from the outside world of technology or from the market place, they become increasingly complacent about external events and new technological developments. After studying the actual communication behaviors of some 350 engineering professionals in a major R&D facility, Katz and Allen (1982) found that as members of project teams worked together, gained experience with one another, and developed more stable role assignments and areas of individual contribution, the groups also communicated less frequently with key sources of outside information. Research groups, for example, failed to pay sufficient attention to events and information in their external R&D community while product development and technical support groups had reduced levels of communication with their internal engineering colleagues and with their downstream client groups from marketing and manufacturing. Such low levels of outside interaction also result in stronger group boundaries, creating tougher barriers to effective communication and more difficult information flows not only among R&D groups but also to other organizational divisions and to other areas outside the organization.

Another set of forces that affects the amount and variety of outside contact that R&D employees may have is the tendency for individuals to want to communicate only with those who are most like themselves, who are most likely to agree with them, or whose ideas and viewpoints are most likely to be in accord with their own interests and established perspectives. Over time, R&D project members learn to interact selectively to avoid messages and information that might be in conflict with their current dispositions toward particular technologies or technical approaches, thereby restricting their overall exposure to outside views and allowing themselves to bias the interpretation of their limited outside data to terms more favorable to their existing attitudes and beliefs. Thus, the organization ends up getting its critical and evaluative information and feedback not from those most likely to challenge or stretch their thinking but from those with whom they have developed comfortable and secure relationships, i.e., friends, peers, long-term suppliers and customers, etc. And it is precisely these latter kinds of relationships that are least likely to provide the inputs and thinking necessary to stimulate the organization's movement into new technical areas.

2.3 Cognitive Processes

One of the dilemmas of building in-house capability in particular areas of technology is that engineers responsible for the success of these technical areas become less willing to accept or seek the advice and ideas of other outside experts. Over time, these engineers may even begin to believe that they possess a monopoly on knowledge in their specialized areas of technology, seriously discounting the possibility that outsiders might be producing important new ideas or advances that might be of use to them. And if this kind of outlook becomes mutually reinforced within a given R&D area or project group, then these individuals often end up relying primarily on their own technical experiences and know-how, and consequently, are more apt to dismiss the critical importance of outside contacts and pay less attention to the many technical advances and achievements in the larger external world. It is precisely this attitude, coupled with the communication and problem-solving trends previously described, that helps explain why many of the most successful firms in a very new area of technology had never participated in the old or substituted area of technology.

This rather myopic outlook within R&D is also encouraged as technologists become increasingly specialized, that is, moving from broadly defined capabilities and solution approaches to more narrowly defined interests and specialties. Pelz and Andrews (1966) argue from their study of scientists and engineers that with increasing group stability, project member preferences for probing deeper and deeper into a particular technological area become greater and greater while their preferences for maintaining technical breadth and flexibility gradually decrease. Without new challenges and opportunities, the diversity of skills and of ideas generated is likely to become progressively more narrow. They are, essentially, learning more and more about less and less. And as engineers welcome information from fewer sources and are exposed to fewer alternative points of view, the more constricted their cognitive abilities become, resulting in a more restricted perspective of their situation and a more limited set of technological responses from which to cope. One of the many signs of obsolescence occurs when engineers retreat to their areas of specialization as they feel insecure addressing technologies and problems outside their direct fields of expertise and experience. They simply feel more comfortable and creative when they can see their organizational contributions in terms of their past performance standards rather than on the basis of future needs and requirements.

Finally, there is not only a strong tendency for technologists to communicate with those who are most like themselves, but it is just as likely that continued interaction among members of an R&D project team will lead to greater homogeneity in knowledge and problem-solving behaviors and perceptions. The well-known proverb "birds of a feather flock together" makes a great deal of sense, but it is just as accurate to say that "the longer birds flock together, the more of a feature they become." One can argue, therefore, that as R&D project members work together over a long period, they will reinforce their common views and commitments to their current technologies and problem-solving approaches. The group not only tries to hire or recruit new members like themselves, thereby exacerbating the trend towards greater homogeneity and consensus and less diversity. Such shared values and perceptions, created through group interactions, act as powerful constraints on individual attitudes and behaviors and provide group members with a strong sense of identity and a great deal of assurance and confidence in their traditional activities. At the same time, however, these shared systems of meaning and beliefs restrict individual creativity into new areas and isolate the group even further from important outside contacts and tech-

nical developments, thereby causing the old technologies to become even more deeply entrenched.

2.4 Mobility of People and the "Not Invented Here" Syndrome

What is implied by all of this discussion is that R&D managers need to learn to observe the strong biases that can naturally develop in the way engineers select and interpret information, in their willingness to innovate or implement radically new technological approaches, or in their cognitive abilities to generate or work with new technical options so that appropriate actions can be undertaken to encourage R&D to become more receptive and responsive to new ideas and emerging technological opportunities. The trends described here are observable; one can determine the extent to which project groups are communicating and interacting effectively with outside information sources, whether project groups are exposing themselves to new ideas and more critical kinds of reviews, or whether a project group is becoming too narrow and homogeneous through its hiring practices.

In the best-selling book, *In Search of Excellence,* organizations are encouraged by Peters and Waterman to practice the Hewlett Packard philosophy of MBWA (Management by Wandering Around). But managers have to know what to look for as they wander around. In particular, technical managers can try to detect the degree to which these different trends are materializing, for the way engineering groups come to view their work environments will be very critical to the organization's ability to introduce and work with new technologies. The more the perceptual outlook of an R&D area can be characterized by the problem-solving, informational, and cognitive trends previously described, the more likely it has internalized what has become known in the R&D community as the "Not Invented Here" (NIH) or the "Nothing New Here" (NNH) syndrome. According to this syndrome, project members are more likely to see only the virtue and superiority of their own ideas and technical activities while dismissing the potential contributions and benefits of new technologies and competitive ideas and accomplishments as inferior and weak.

It is also argued here that the most effective way to prevent R&D groups from developing behaviors and attitudes that coincide with this NIH syndrome is through the judicious movement of engineering personnel among project groups and organizational areas, keeping teams energized and destabilized. Based on the findings of Katz and Allen (1982), Smith (1970), and several other studies, new group members not only have a relative advantage in generating fresh ideas and approaches, but through their active participation, project veterans might consider more carefully ideas and technological alternatives they might otherwise have ignored. In short, project newcomers represent a novelty-enhancing condition, challenging and improving the scope of existing methods and accumulated knowledge.

The mobility of people within the organization is a most fruitful approach for keeping ideas fresh, building insights, and maintaining innovative flexibility. Japanese organizations, for example, assume that the best course of development for capable individuals is lateral rotation across major functional areas of the firm before upward advancement takes place. In a Japanese company, an engineer progressing well may move from R&D into marketing, then into manufacturing, and perhaps back into R&D at a higher level. This is seldom the kind of career track that American firms find appropriate; yet, we all know for sure the kinds of problems one is avoiding as well as the benefits that would accrue over the long run through the greater use of rotation programs even if rotation were limited to between research and development and engineering groups.

In an additional attempt to foster new thinking and to build stronger intraorganizational bridges and communication networks, some companies hold special meetings in which organizational areas report on what they have been doing and on the kind of capability they have. The 3M Corporation, for example, holds a proprietary company fair at which there are presentations of technical papers, exhibits, and demonstrations of projects and prototypes. The fair enables the rest of the people in the company to begin to learn about what is taking place in other divisions or laboratories. The Monsanto Company uses what it calls the Monsanto technical community to bring together technical people, trained in similar disciplines but employed in different divisions of the firm, and it convenes these people in different workshops and groups, encouraging them to exchange ideas and information. Many other companies, such as Procter and Gamble, Corning, and Motorola, have similar kinds of internal forums. These kinds of programs can be very helpful in fostering communication and in stimulating the identification of new technical capabilities as well as the identification of new market and technical needs throughout the firm.

3. ORGANIZATIONAL STRUCTURES

Unlike productivity, which is the efficient application of current solutions, innovation usually connotes the first utilization of a new or improved product, process, or practice. Innovation, as a result, requires both the generation or recognition of a new idea followed by the implementation or exploitation of that idea into a new or better solution. So far, we have discussed organizational processes to the extent that they primarily affect the idea-generation phase of the innovation process. It is just as important, of course, for an organization to plan for the idea-exploitation phase, where exploitation includes the appraisal, focusing, and transferring of research ideas and results for their eventual utilization and application. To say that one is managing or organizing for the introduction of new technologies within the innovation process implies that one is "pushing" the development and movement of new technical ideas and capabilities downstream through the organization from research to development to engineering and even into manufacturing and perhaps some phase of customer distribution.

Innovation, then, is a dynamic process involving the movement and transfer of technologies across internal organizational boundaries. Formal organizational design, on the other hand, is a static concept, describing how to organize collections of activities within well-defined units and reporting relationships, e.g., research, advanced development, product development, engineering, quality assurance, etc. Formal organizational structures tell us what to manage and with whom to interact within certain areas of interdependent activity; they tell us little about how to move information, ideas, and in particular technologies across different organizational areas, divisions, or formal lines of authority. In fact, formal structures tend to separate and differentiate the various organizational groupings, making the movement of ideas and technologies particularly difficult across these groupings, especially if there are no compensating integrating mechanisms in place. And it is in the movement of new technological concepts from research to advanced development to successful product development that we are particularly interested.

The effective organization, therefore, needs to cause the results of R&D to be appropriately transferred. Technically successful R&D, especially if it embraces new radical technologies, is very likely to pose major problems of linkages with the rest of the firm, particularly product development, engineering, manufacturing, marketing, sales, field service, and so on. A company can do a terrific job of R&D and a terrible job of managing the innovating process overall simply because the results of R&D have never been fully exploited and successfully moved downstream. Witness, for example, the problems of Xerox, where the R&D labs have generated and surfaced many major new advances and approaches only to discover that the company has failed to fully exploit and capture benefit from many of them. Other corporations, on the other hand, have benefited extremely well from Xerox's research activities—so many in fact that some have quipped that Xerox's research facilities should be declared a national resource instead of a resource for Xerox (see *Fortune* magazine, September 1983).

Over the past decade or so, Roberts (1979) has been studying the problems of moving R&D results through the organization. From carrying out these studies, he has found that most large organizations have been dissatisfied with the degree of transfer of their own R&D results and feel very uncom-

fortable about how little of their good technical outcomes ever reach the marketplace and generate profitable pay-back for the firm. The R&D labs he studied seemed to have broad enough charters to do almost anything they chose, but ended up being quite narrow as to what they in fact implemented within their own organizations. To enhance the transfer of R&D results across the barriers of organizational structures, Roberts (1979) advocates the building of bridges; and in particular, he recommends three different groups of bridges: procedural, human, and organizational.

The procedural approaches, according to Roberts, try to tie together both the R&D unit and the appropriate receiving units by joint efforts. In the case of new technological concepts, the most immediate receiving unit is typically some advanced development group or some divisional product development organization that receives the output from a centralized research and development lab. The kinds of procedural bridges that have been suggested include joint planning of R&D programs and joint staffing of projects, especially immediately before and after transfer, for those are the most critical phases of the process in which key know-how and information can easily slip through the cracks.

Joint appraisal of results by research, development, and any other appropriate downstream unit or customer is also employed in some labs. From the viewpoint of generating useful information, the best time to carry out joint appraisal of results is when failure has occurred, for there is usually something objective to look at from which one might be able to learn and improve. At the same time, however, this exercise must be done carefully and sensitively to prevent this opportunity from becoming a situation of mutual fingerpointing, showing why the other group is really at fault and how those people caused the failure. In these joint appraisals, the attributions of failure should be centered around substantive issues that can be dealt with behaviorally, structurally, or procedurally; otherwise, intergroup conflicts and differences will be strengthened, which is likely to cause even greater difficulty in future technological handoffs.

Joint appraisal of successes should also not be overlooked, for they can be very helpful in generating the goodwill and trust necessary to strengthen organizational linkages, especially after a history of prior difficulty or failure.

The establishment of human bridges also helps to cope with transfer issues. Interpersonal alliances and informal contacts inevitably turn out to be the basis of integration and intraorganizational cooperation that really matter. The human approaches focus on the relationships that convey information between people, that convey the shift of responsibility from one person to another, and that convey enthusiasm for the project. Roberts argues strongly, in fact, that the building of human bridges is by far the best way to transfer this vital enthusiasm and commitment.

Technology moves through people, and the most effective of these human bridges is the actual movement of people in two directions. Upstream movement of development engineers to join the R&D effort well in advance of the intended transfer is a very important step. This transfers information from the product development areas into the research process, creates an advocate to bring the research results downstream, and builds interpersonal ties for the later assistance that will inevitably be needed as the technology encounters problems. Downstream movement of research individuals will also be helpful in providing the technical expertise necessary for development to build up its own understanding and capability.

In addition to the specific movement of people, human bridges are also built through the interpersonal communication systems that have developed over time through the history of working relationships, rotation programs, task force participation, and other organizational events and activities. Another important device to be considered is the joint problem-solving meeting in which development individuals are asked to sit down with research colleagues to let them explain their difficulties and initial problem-solving thinking. Such meetings are not only helpful in dealing with specific project problems, but will also be useful in building stronger human bridges between the re-

lated R&D areas and may even be helpful in solving additional related problems that were not initially put forth.

The final area for considering the movement of R&D results toward development and eventual commercialization consists of organizational changes and organizational bridges. According to Roberts, these are the toughest kinds to create and implement effectively in an organization. It is far easier to alter procedures or to try to build human bridges across groups than it is to change organizational arrangements and relationships. Nevertheless, several different structural approaches can be effective under different organizational conditions. Some organizations have developed specialized transfer groups, created solely for the purpose of transferring important technical advances or important new processes. Under this approach, the transfer group is like the licensor of a technology who is not just sending equipment and documentation but who is also responsible for training others to work with the technology, for installing the equipment, etc. If used, the specialized transfer group should consist of at least a few of the key technical players. Senior management should not be allowed to argue that they cannot spare the superstars of the research organization to support development or manufacturing engineering.

Another organizational approach is to employ integrators or integrating groups that are given responsibility for straddling the various parts of the RD&E organization. This is a very uncomfortable and a very difficult job to assume because it is extremely difficult to ask someone to take care of an integrating function across two separate suborganizations when he or she does not have responsibility for either the sending or the receiving organization. To perform this function successfully requires someone who can cope with the political sensitivities of multiple groups and who has built substantial informal influence and credibility within the organization.

Finally, a variety of corporate venture strategies can be considered by companies that are concerned with developing new technical approaches, new product lines, or want a stronger emphasis on technical entrepreneurship. Roberts (1980) suggests a large variety of possible venture strategies, ranging from the high corporate involvement of internal venturing to low corporate involvement through venture capital investments in outside firms for the purpose of gaining windows on technology and new market opportunities. Additional venture strategies are also described by Roberts, including the coupling of R&D efforts from both the large corporation and the small independent firm. In general, there is no single best way to organize for the effective introduction of new technologies; but the more informal mechanisms one puts in place to foster both the idea generation and the idea exploitation phases, the more one is likely to be successful at managing the innovation process.

4. ORGANIZATIONAL CONTROLS

All of these organizational attempts at stimulating new technological innovation will fall flat, of course, if organizational controls are not consistent with the innovation process. In looking at many case histories of successful versus unsuccessful innovations based on radical new technologies. Cohen *et al.* (1979) and several other studies have identified a number of factors as being critically important for trying to influence the generation and successful movement of new technologies through the organization.

4.1 Technical Understanding

One of the most important issues in working with new technologies is that the research function must fully understand the main technical issues of the technology before passing it on. Although this point seems obvious, it is often overlooked. The research function must focus not only on the benefit of the new technology in and of itself; it must also deal with the technology's limitations relative to conventional technologies and to other new technological approaches. In the early days of transistors, for example, one large electronics company spent a great deal of money and many years of research effort on understanding the materials and process-

ing problems of germanium for point contacts and junction transistors. Unfortunately, the research organization failed to compare the use of germanium to silicon, whose own development was continuing to make a great deal of progress. Only after many years did the organization finally realize the limitations in the advantages of germanium over silicon and these limitations had less to do with the devices themselves and more to do with device implementation in packaging and circuitry.

It is also important, therefore, to make sure that research understands where the new technology might fit in with respect to the product line or at least what requirements must be met to reach this fit. Research should not waste its time solving problems that do not exist or producing technologies that cannot be sold. Whirlpool, for example, invested substantial research resources in making appliance motors more energy efficient long before the oil crisis, but of course, the marketplace was not yet interested in these kinds of advances. Similarly, GE conducted a great deal of research in environmental concerns in the 1940s but at that time there was very little interest in improving the ecology of our environment. As a last example, DuPont developed Corfam as a synthetic substitute for leather, but unfortunately for DuPont, the public was perfectly satisfied with leather and saw no need for the manmade substitute.

Full understanding also means that research must begin to examine the means of manufacturing, the availability of key materials and technical talents, the ease of use, and so on. Air Products and Chemicals, for example, spent millions of dollars to develop a fluorination process so that textile manufacturers could make fabrics, especially polyesters, more resistant to oil and grease. Unfortunately, textile manufacturers did not want fluorine—a poisonous and corrosive gas—anywhere near their plants and refused to buy the system. Research should also be able to make, at the very least, preliminary cost estimates. One of the most basic elements of a technology is its cost. In fact, a study of technology programs at GE concluded that most of the barriers to the introduction of new technologies (even hardware and software) were cost constraints and not technical feasibility; it was getting the technology to perform capably at a marketable cost.

To help ensure these kinds of requirements, some labs have begun to hire full-time marketing representatives and cost estimators as a regular part of the R&D organization. Previously, corporate R&D organizations were completely dependent on product line divisions for both marketing and sales effort and for business and economic analysis as well. These dependencies, especially the latter, were harmful in getting research projects justified, supported, and accepted by the divisions who were supposed to be the eventual customer of the research results.

4.2 Technical Feasibility

All too often, a technology is transferred before there has been sufficient time within research to demonstrate true feasibility. Such pressures can come from the downstream organization or they can arise from the "unbridled enthusiasm" of the researchers themselves. In either case, it would be more beneficial to discuss what constitutes feasibility and for research to strive to achieve it.

Most new technical concepts do not succeed simply because they must run a gauntlet of barriers as they enter the main part of the functioning organization. In many cases, the new technology is embedded within a system of established technologies. The question then is whether the new technology will offer a sufficient competitive advantage to warrant its incorporation into this interdependent system, perhaps changing drastically the tooling and the overall manufacturing process. Experienced technologists will typically warn you that what you do not yet know about the workings of a new technical advance will probably come back to haunt you. What often appears to be a simple technical issue turns out to be more complicated than we realize. GE discovered a fiber, for example, that looked and behaved more like wool than any synthetic yet known. Unfortunately, the fiber disintegrates in today's cleaning solvents, and the problem has yet to be solved.

4.3 Research and Development Overlap

As previously discussed, it is very helpful to the movement of a new technology if development, or some other appropriate receiving organization, also has a group of technical people who have been getting up to speed on the technology before the actual transfer, e.g., the presence of "ad tech" groups. Such advanced technical activities within development can greatly aid the movement of technology and the smoothing of conflicts.

In a similar fashion, it is also important for research to maintain some activity to support and defend the new technology or to find new ways to extend the technology. Research must not be allowed to feel that it is "finished" at the time of transfer, for if this feeling is present, their willingness and enthusiasm to support the technology will be minimal. Most new technologies are relatively crude at first. Ball-point pens, for example, blotted, skipped, stopped writing all together, and even leaked in consumers' pockets when they first appeared on the market. The first transistors were expensive and had sharply limited frequencies, power capabilities, and temperature tolerances. Such experiences are very typical of new technologies, especially radical new technologies. And the more prepared research is to help "push" the technology, the less likely it will be for the new technology to be dismissed prematurely as a "fad" or as a technology with very limited application.

4.4 Growth Potential

As a related point, all too often a research program sells itself short by being too narrow and not showing a clear path toward technical growth and growth in product applicability. In almost every instance, when the new technology appears on the scene, the old technology is forced to "stretch" itself, often with major advances being achieved in the threatened technology. Under these circumstances, the new technology is in the position of trying to chase or catch a "changing target." Moreover, this new potential in the old technology often holds back the entry of the new technology. Advances in flashbulbs, for example, held off the widespread use of electronic flash for quite some time, while advances in magnetic tape audio and video recording have prevented the emergence of thermoplastic recording. In their well-known study of strategic responses to technological threats, Cooper and Schendel (1976) indicate that in the majority of cases, sales of the old technology did not decline after the introduction of a new technology. To the contrary, sales of the old technology expanded even further. It is for these reasons that the diffusion and substitution of a radical new technology must be viewed as a long-term process and research and development must carefully prepare to argue and demonstrate why the pressured organization should be patient during this time period.

4.5 Organizational Slack and Sponsorship

When an organization pushes too hard for productivity within the RD&E environment, trying to measure and control all aspects of the innovation process, there is little room or slack for experimenting or pursuing novel ideas and concepts. The environment is simply too tightly run and the climate becomes unfavorable for very new or long-term innovation. Engineers and scientists become anxious, restrict the depth of exploration along new paths, and center their attention upon issues closely related to the company's immediate output. Creative innovation, on the other hand, is harder to measure and takes a long period to assess. It requires speculative investments on the part of the firm that wants to nurture the ideas and the experimenting activities that will eventually be worth it.

Given all of the resistance and testing that a new technological idea will eventually encounter from the functioning organization and from operational review committees, strong corporate sponsorship is needed to protect new technological innovations. And the more radical the new technology, the stronger the corporate sponsorship has to be. One of the observations we have made from working and consulting with many technology companies is that most (and in some high technology companies, all) radical new technologies have

had to have well-identified sponsorship at the corporate level in order to succeed.

Another important finding from retrospective studies of radical innovation is that new technologies are not really new! By this, we mean that technological change is a relatively continuous and incremental process which casts shadows far ahead. According to Utterback and Brown (1972), the information incorporated in successful new innovations has been around for roughly 5–30 years prior to its use. They further argue that there are many multiple signals within the external environment that can be used to predict the direction and impact of future technological changes and development. Von Hippel (1983), for example, argues that one can often anticipate future innovations by identifying what he calls "lead users," that is, users whose needs today foreshadow the needs of the general marketplace tomorrow. Nevertheless, even if particular areas of new technology were identified as extremely important, without strong sponsorship it is unlikely that sufficient resources would be diverted to it, that engineers would be isolated from other pressures or tasks to work on it, or that they would be given sufficient uninterrupted time to complete it. One of the reasons why so many new technologies are introduced through the emergence or spin-offs of new firms is that in these situations, the new technology does not encounter resistance from or have to fight against already existing businesses and entrenched technical approaches.

Another benefit of strong sponsorship is that it helps protect the individual risk taker who is willing to take on the entrepreneurial burden of moving the new technology through the organization. No matter how beneficial the new technology appears to be, someone must be willing to sell the effort and make it happen. Schon's (1963) analysis of successful radical innovation is quite clear. At the outset, the new technological concept encounters sharp resistance, which is usually overcome through vigorous promotions by one emerging champion. What is important to recognize here is that these champions are typically self-selected; it is extremely difficult to appoint someone to withstand all of the pressures, hassles, and risks associated with being an idea champion and then to expect him or her to do it excitedly for a long period.

Finally, we also know from research studies that the ultimate use of a new technology is often not known or may change dramatically as the technology becomes further developed. The new technology, moreover, often invades traditional industry by capturing a series of submarkets, many of which are insulated from competition for some extended period. The earliest application of the transistor, for example, was in hearing aids, but its use was not immediately transferred to the organization's defense divisions. Because of these more limited niche markets (and consequently, relatively low sales volume), R&D often concludes that it does not have to work closely with marketing; nor does it want to subject its technological concept to the typical market screens of revenue and volume. Such a conclusion, however, does not help to build the strong harmonious relationship between marketing and R&D that has been shown to be so important for successful commercialization of new innovations (e.g., Souder, 1978). The key to success in these kinds of situations may be to find a pioneering application where the advantages of the new capability are so high that it is worth the risks. This would require the coupling of technical perspective with creative marketing development to identify such pioneering applications. On this basis, early involvement of marketing could be very helpful in providing inputs and market perspective (but not market screens) to the new technological effort.

4.6 Organizational Rewards

Ultimately, we all know that those activities which are measured or get rewarded are those which get done. If the managerial and organizational recommendations and suggestions discussed in this chapter are to be effectively implemented, then the reward systems must be consistent and commensurate with the hoped-for behaviors. One of the most important of these is that research engineers and scientists must come to see that part of their reward system is not just the generation of publi-

cations of new technological concepts and advances, but that part of their responsibilities is also the successful transfer of their work. A few high-technology companies we know have been making such reward systems explicit within their corporate labs, and although it has taken some time to take hold, it has been quite effective in moving technology through the development cycle. It has also resulted in research seeking more joint sponsorship of its activities, especially with the development divisions—all of which has helped to strengthen the communications and bridging mechanisms within the corporation.

Finally, in most areas of day-to-day functioning, productivity rather than creativity is and should be the principal objective. Even where innovation and creativity are truly desired and encouraged, activities that are potentially more creative may be subordinated to those activities of higher organizational priority or more closely tied to identified organizational needs. Nevertheless, organizations exhibit simultaneous demands for routinization and for innovation. And it is in the balance of these countervailing pressures that one determines the organization's true climate for managing and encouraging the introduction of new technological opportunities.

REFERENCES

Allen, T. J. (1977). *Managing the Flow of Technology,* MIT Press, Cambridge, Massachusetts.

Allen, T. J., and Marquis, D. G. (1983). Positive and negative biasing sets. The effect of prior experience on research performance, *IEEE Trans Eng. Manag.* 11, 158–162.

Cohen, H., Keller, S., and Streeter, D. (1979). The transfer of technology from research to development, *Res. Manag.* May, 11–17.

Cooper, A. C., and Schendel, D. (1976). Strategic responses to technological threats, *Bus. Hor.* February, 61–69.

Katz, R. (1980). Time and work: Toward an integrative perspective, *Res. in Organ. Behav.* 2, 81–127.

Katz, R. (1982). The effects of group longevity on project communication and performance, *Admin. Sci. Q.* 27, 81–104.

Katz, R., and Allen, T. J. (1982). Investigating the not

invented here (NIH) syndrome, *Res. Dev. Manag.* 12, 7–19.

Katz, R., and Allen, T. J. (1985). Project performance and the locus of influence in the R&D matrix, *Acad. Manag. J.* 26, 67–87.

Katz, R., and Tushman, M. (1981). An investigation into the managerial role and career paths of gatekeepers and project supervisors in a major R&D facility, *Res. Dev. Manage.* 11, 103–110.

Pelz, D. C. and Andrews, F. M. (1966). *Scientists in Organizations,* Wiley, New York.

Roberts, E. B. (1974). A simple model of R&D project dynamics, *Rev. Dev. Manage.* 5, 1–15.

Roberts, E. B. (1979). Stimulating technological innovations: Organizational approaches, *Res. Manage.* 22, 26–30.

Roberts, E. B. (1980). New ventures for corporate growth, *Harv. Bus. Rev.* July–August, 58, 134–142.

Schon, D. D. (1963). Champions for radical new inventions, *Harv. Bus. Rev.* 41, 76–84.

Souder, W. E. (1976). Effectiveness of product development methods, *Ind. Market. Manage.* 7, 299–307.

Utterback, J. M. (1974). Innovation in industry and the diffusion of technology, *Science* 183, 620–626.

Utterback, J. M., and Brown, J. W. (1972). Monitoring for technological opportunities, *Bus. Hor.* 15, 5–15.

von Hippel, E. (1983). *Novel Product Concept from Lead Users: Segmenting Users by Experience,* Massachusetts Institute of Technology Working Paper No. 1476–83.

39

Implementing Radical Innovation in Mature Firms

The Role of Hubs

RICHARD LEIFER, GINA COLARELLI O'CONNOR, AND MARK RICE

EXECUTIVE OVERVIEW: There is increasing evidence of the importance of radical or breakthrough innovation to long-term firm success in the competitive marketplace today. Although this recognition has permeated many established companies, there is uncertainty about how to accomplish such innovation. This article is based on a six-year longitudinal study of 12 radical-innovation projects in 10 large, mature companies. The life cycle of radical-innovation projects is unlike those of incremental projects, because of an abundance of uncertainties and discontinuities. These characteristics require that radical-innovation projects be managed quite differently from incremental ones. Seven key strategic imperatives are offered for successfully implementing radical innovation.

WHY RADICAL INNOVATION IS IMPORTANT

The contemporary competitive landscape has been and continues to be driven by technological revolution, globalization, hypercompetition, and extreme emphasis on price, quality, and customer satisfaction, requiring an increased recognition and focus on innovation as a strategic competence. While there has been an emphasis on incremental innovation in the past decade, there has been less emphasis on radical or breakthrough innovation. Consequently, a great deal is known about implementing incremental innovation, but implementing radical innovation is poorly understood.[1] This is true even though the importance of radical or breakthrough innovation has been underscored by a number of consultants and business scholars.[2]

Radical or breakthrough innovations transform the relationship between customers and suppliers, restructure marketplace economics, displace current products, and create entirely new product categories. They provide the engine for long-term growth that corporate leaders seek. Unfortunately, recognizing the importance of radical innovations and developing and commercializing them are two different things.

Companies that have succeeded over the long haul punctuate ongoing incremental innovation with radical innovations that create new markets and business opportunities.[3] While it is clear that radical innovation is important to firms concerned with long-run growth and renewal, it is also clear that large, established firms have difficulty managing the radical-innovation process. Large, established firms have grown excellent at managing operational efficiencies, and at introducing next-generation products. However, the chaos and uncertainty that come with commercializing new technologies for markets that may not yet exist require vastly different competencies.

In 1995, we embarked on a study to learn how radical-innovation projects are managed in large,

Reprinted with permission of *Academy of Management Executive*, "Implementing Radical Innovation in Mature Firms: The Role of Hubs," Richard Leifer, Gina Colarelli O'Connor, and Mark Rice, vol. 15, no. 3, 2001.

Richard Leifer and ***Gina Colarelli O'Connor*** are Professors of Management at the Lally School of Management & Technology at Rensselaer Polytechnic Institute. ***Mark Rice*** is a Professor of Entrepreneurial Studies at Babson College.

established U.S.-based firms. (See Appendix.) Our expectation was that, by first observing and describing the nature of radical innovation, we could draw some insights into how management of radical innovation might be improved. Our observations led us to suggest a set of seven strategic initiatives for developing and sustaining an organizational radical-innovation competency.

THE NATURE OF RADICAL INNOVATION

A radical innovation is a product, process, or service with either unprecedented performance features or familiar features that offer significant improvements in performance or cost that transform existing markets or create new ones. Examples include computerized tomography (CT) and magnetic resonance imaging (MRI), personal computers, pagers, and cellular telephones.

We came to a pragmatic definition of radical innovation after understanding the theoretical work of others and engaging representatives from the Industrial Research Institute (IRI), a professional association of the technology leaders of *Fortune* 1000

companies located in Washington, D.C.[4] We agreed that we would consider only formally established projects with explicit budgets and organizational identities. The definition of radical innovation that emerged from our study included one or more of these criteria: an entirely new set of performance features, at least a five-fold improvement in known performance features, and a significant (30-percent or greater) reduction in cost. A listing of the companies and a brief description of their projects can be found in Table 39.1.

UNCERTAINTIES IN THE RADICAL-INNOVATION LIFE CYCLE

Radical innovation has traditionally been defined as that arena where technical and market uncertainties are high.[5] Technical uncertainties refer to questions about the validity of the underlying scientific knowledge, whether the technology will work, technical specifications of the product, and ramping-up issues. Market uncertainties include issues related to customer needs and wants—either existing or latent forms of interaction between the

TABLE 39.1
Companies and Project Descriptions

Company	Projects
Air Products Corporation	• An ionic transport membrane (ITM) for separating oxygen from air.
Analog Devices, Inc.	• A micro-electromechanical system (MEMS) accelerometer, a small microchip capable of detecting changes in speed, initially targeted for automobile airbag acutators.
DuPont	• A new material that emitted light that made it attractive in electronic-display applications.
	• An environmentally friendly polyester film that could be recycled or decomposed.
General Electric	• A digital x-ray imaging system that would replace existing film-based x-ray systems.
General Motors	• A hybrid vehicle capable of drawing power from both electrical and conventional engines.
IBM	• A new generation of communication chips using silicon germanium (SiGe). This innovation aimed to increase switching speeds and greatly reduced power requirements.
	• Development and integration of a high-density display, and memory and battery technologies to help create an electronic book.
Nortel Networks (and its spinoff, NetActive)	• A technology allowing digital content to be rented over an Internet link between the consumer and a NetActive server.
Polaroid	• The creation of low-cost, high-capacity computer memory-storage devices.
Texas Instruments	• A Digital Micromirror Device capable of creating a screen image by bouncing light off 1.3 million microscopic bidirectional mirrors squeezed onto a one-square-inch chip for business conference projection systems and large-screen movie theaters.
Otis Elevator Division of United Technologies	• An elevator that could move vertically and horizontally for solving the problem of moving people within extremely tall buildings.

customer and proposed products, and methods of sales and distribution. However, we found two other sources of uncertainty critical for radical-innovation project success. Organizational and resource uncertainties stemming from the conflict between mainstream organizations and radical-innovation teams more often caused projects to stall. Among the organizational uncertainties were questions about the capabilities of the project team; recruiting the right people; managing relationships with the rest of the organization; dealing with variability in management support; overcoming the short-term, results-oriented orientation of operating units, and their resistance to products that might jeopardize existing product lines; and counteracting vested interests in the current business model.

Resource uncertainties also claimed an unexpectedly large share of the teams' attention. Teams needed to find out what funding and competencies were required to complete the project, whether there were sources other than those allocated through the normal corporate budgeting process, who the right partners were, and how to manage their partnerships most effectively. Coping with these uncertainties is essential to managing radical-innovation projects and underlies the dynamics of the radical-innovation life cycle.

DYNAMICS OF THE RADICAL-INNOVATION LIFE CYCLE

We constructed timelines for the 12 projects that show that the radical-innovation life cycle is long term (often a decade or longer), unpredictable, sporadic (with stops and starts, deaths and revivals), non-linear, and stochastic (with unpredictable exogenous events). (See Figure 39.1 for an example.) Radical-innovation projects are also context dependent in that corporate culture and informal relationships accelerate or retard progress. These characteristics contrast with the course of incremental innovation, which follows a more linear, orderly process with far fewer organizational and resource uncertainties.[6] As a result, managers of the two processes must take strikingly different paths.

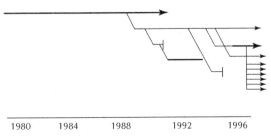

1980   1984   1988   1992   1996

Figure 39.1. DuPont's biodegradable polymer project. Note: Arrows refer to applications pursued, thickness of arrows to resource commitment. A vertical line indicates the end of an application development.

The life cycle of DuPont's biodegradable polymer project, pictured as Figure 39.1, illustrates the impact of these uncertainties on a radical innovation's time horizon. Each solid horizontal line represents a market application that was pursued, the thickness of those lines indicates the level of resource commitment, and the short vertical lines mark discontinuities that were unsuccessful market applications, any of which could have killed the project.

The four types of uncertainty discussed above were found in this project. Technical uncertainties emerged from different potential applications. For example, how can the degradability, biodegradability, and manufacturability questions be answered for each potential application explored, since each application required a different polymer characteristic?

Market uncertainties abounded and accounted for most of the project discontinuities. The original application, disposable diapers, disappeared when the OEM that requested the application development withdrew its interest in the project. The project went into hibernation until another potential application was identified. Unfortunately, this application also proved a dead end. The project was also beset with other market uncertainties—dead-end applications, environmental regulations that were rumored but never enacted, and improper management of beta trials at potential customer sites. This set of application investigations and subsequent dead ends characterizes the major form of the project life-cycle diagram. Organizational uncertainties included three different project sites and a changing cast of scientists, technicians, and project champions. Resource uncertainties included fluctuating financial support, which at one point supported only two part-time people.

The greatest opportunity for enhancing the possibility of radical-innovation success, we believe, is to expend energy on managing resource and organizational uncertainties. These factors are, in fact, under managerial control. If firms learn to reduce these uncertainties in a systematic way— through leadership and organizational and managerial approaches—then radical-innovation project teams would be better able to address the less controllable and more chaotic market and technical uncertainties.

Based on patterns observed in all 12 projects and on feedback from workshops, seminars, and discussions with almost 40 other companies, we isolated seven key strategic imperatives for developing and driving radical-innovation projects to success. While none of the participating companies demonstrated a competence in all of these strategic imperatives, the full range of implementing these imperatives can result in greater quantity, shorter project life cycles, and increased project success of radical-innovation projects.

IMPERATIVE NO. 1: BUILD A RADICAL-INNOVATION HUB

A radical-innovation hub can oversee and help nurture projects by reducing uncertainty without increasing bureaucracy. A radical-innovation hub can serve as a repository for cumulative learning about managing radical innovation, and is a natural home base for those who play pivotal roles in making radical innovation happen: the idea hunters and gatherers, internal venture capitalists, members of evaluation and oversight boards, and corporate entrepreneurs experienced in the realm of high uncertainty.

At Air Products, for example, a business development manager in corporate R&D helped formulate cost estimates and worked on developing full-system concepts with project teams. At DuPont, a director of new-business development connected technological discoveries with market opportunities and built project teams to help explore the market potential for new technologies. At Polaroid, a new-business division was set up to handle innovative efforts that could not be nurtured in one of the company's two major business units. Finally, at Nortel Networks, a Business Ventures Group was structured to receive, evaluate, and help develop novel ideas that could not gain the attention of business-unit management. They also helped build the project teams and advisory boards for each project.

Effective radical-innovation hubs perform several functions for reducing organizational and resource uncertainties. They capture radical-innovation ideas by recruiting and training idea hunters and gatherers and establishing skilled early-evaluation boards. They build and train radical-innovation project teams and serve as mentors during the incubation period, advising project teams about resource acquisition, market-learning methods, project-evaluation criteria, and management of interfaces with existing business units and senior management. They organize and recruit project advisory boards, decide when a project should transition to a receiving unit (a currently existing operating unit, a newly formed one, or a spinout venture), and organize a transition team to facilitate this transition. Hubs also recruit and develop those who thrive in the radical-innovation environment of risk, uncertainty, and potentially high payout. Radical-innovation hubs thus provide oversight and management from a project's inception to its commercialization and build and accumulate expertise in overseeing a portfolio of radical-innovation projects for the firm. Finally, they create performance benchmarks for senior management. They assemble and update a knowledge-management system that shows how long it takes for radical innovations and markets to develop, and how much money is required. An example of how this might play out is illustrated with the hub at Nortel Networks.

The hub at Nortel Networks was called the

Business Ventures Group. In addition to issuing a request for proposals, the hub organization developed a Web site that helped employees get started on articulating ideas. The initial contact person in the hub engaged the idea generator before the idea was fully formulated to help develop the concept. Hub personnel judged the potential attractiveness of the opportunity. Finally, the hub decided whether or not to fund the concept development, leading to a business case for evaluation by senior management. If the business-case evaluation resulted in a sanctioned project by senior management, the hub was able to provide project-incubation services: finding the right people, putting the team together, and providing business-development skills, marketing, and financial-management assistance to the team.

IMPERATIVE NO. 2: DEPLOY HUNTERS AND GATHERERS

Good ideas can come from everywhere—business units, R&D, senior managers, bench scientists, even people outside the organization. Idea generators need a place to take their ideas for a quick back-of-the-envelope assessment, for help with extending or redirecting their thinking, or with articulating the idea's potential. To move radical ideas forward, firms need to first recognize opportunities. The technical and market uncertainties associated with radical innovations, however, often make this difficult. In 10 of our 12 cases, the individuals who generated the ideas did not recognize the opportunities. In incremental innovation, opportunities are more easily recognized. They are typically derived from analytically based market research and are built on known technologies. But when markets do not yet exist, and there are many alternative directions that the technology-development path can take, opportunity recognition is much more challenging. The boundaries of the firm's current markets and organizational structure further complicate recognition; it is sometimes difficult to recognize and pursue an opportunity that one knows will seriously challenge parts of the es-

tablished organization. Those who do recognize potential breakthrough opportunities have the market knowledge and organizational perspective to connect ideas with applications. They can think broadly about potential scientific connections, social trends, markets, and customers.

Opportunity recognition depends more on individual initiative than routine practices and can be either reactive or proactive. Gatherers are alert and ready to react to promising radical ideas, while hunters take responsibility for actively seeking out ideas with business potential.[7]

Opportunity gatherers are receivers of ideas. They have the experience, skills, judgment, and motivation to be alert to activities that are going on in R&D or that appear from other sources. They have the technical and market sophistication to assess what they encounter. In most of our cases, the first-line or midlevel research managers and senior scientists played the gatherer role. They viewed their responsibility as helping the idea generator acquire the resources needed to develop the idea. The gatherer at DuPont played a critical role in the discovery of a new kind of fiber.

A researcher at DuPont was working on characterizing materials, and noticed unexpected properties in the fiber he was working with. Under certain conditions, the fiber emitted light at an unusually high speed. When his project was coming up for a technical review, his boss said: "This might be of interest to the new-business development guys." He went to the director of corporate business development (the gatherer) and said: "You should come to this review." In our interviews, the director commented: "I get invitations all the time, but on this one, they grabbed my attention and would not let me go." Once he saw the presentation, he was impressed. He called his buddy in the electronic materials division, who indicated he would inherit the project if and when it was ready for commercialization.

Opportunity hunters take a more active approach. They go out into the organization, asking questions to uncover latent ideas, and making connections. One respondent told us that, after sniffing new ideas out of the lab, he would tell scientists: "I know you just invented this yesterday, but, wow! Can I see a market for this!" Like gatherers, hunters have technical training, but are also experienced in marketing or business development. Perhaps as important, a successful hunter knows how to articulate the opportunity in compelling terms that gain the attention of higher management, a skill few bench scientists have.

IBM hired an individual to play the role of hunter. His job was to troll for ideas and evaluate them at a very early stage. He described his job this way: "I started looking through our research organization to uncover intellectual property that I could leverage into the marketplace. I was actively scanning and knew [that one scientist] had been running around evangelizing [the technology] for two or three years. He hadn't been able to build a case that got it recognized and funded, which is what I did." The hunter worked with this scientist to develop a business plan that articulated the opportunity in a way that senior management could understand. Together, they were successful in gaining high-level attention and support for the development of the silicon-germanium chip technology that became one of IBM's promising new ventures.

Hubs Can Help in Generating Ideas

Unlike these ad hoc, nonsystematic approaches, hubs can take responsibility for fuzzy front-end functions, imposing continuity and system on a catch-as-catch-can set of activities. Hubs create a network of idea generators, hunters, and gatherers, and actively develop their skills. Figure 39.2 is a diagram of a hub structure, portraying hunters and gatherers as interfacing with various organizational units, as well as working in those units, but reporting to the radical-innovation hub that coordinates them.

The venture-development organization at Nortel Networks was a radical-innovation hub. It used a request for proposals to stimulate idea generation and a Web site to help collect them. The initial hub contact person would help the idea generator articulate and develop ideas into business concepts, and would help sort out the known from the uncertain. Once the idea was submitted, the hub

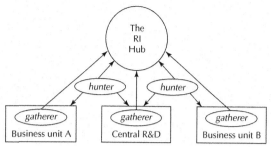

Figure 39.2. A radical-innovation hub with hunters and gatherers.

team performed a preliminary screening. If there was a compelling opportunity, a business-development specialist worked with the idea generator to further develop it. Eventually, a three-person team judged the attractiveness of the opportunity and decided whether to fund it. Funding at that point did not sanction a formal project, but rather a commitment to develop a business proposal for senior management. That commitment ranged from several weeks to several months. According to the hub director: "The overall competency that we brought as a team was the ability to take technology and translate it into compelling business propositions."

IMPERATIVE NO. 3: MONITOR AND REDIRECT PROJECTS

The inevitable uncertainties of radical-innovation projects contribute to ad hoc, crisis-oriented management practices. What is needed is an uncertainty-reduction mindset quite different from that held by incremental project managers. Radical-innovation managers must relinquish the traditional control-to-task mentality in favor of a monitor-and-redirect mindset.

The four kinds of uncertainties in radical-innovation environments make control-to-task impractical. After the head of GE Medical Systems (GEMS) decided that the investment in digital x-ray project technology was a distraction from the short-term budget, the project manager was forced to seek funding from outside the corporation. At the same time, his manufacturing partner failed to meet expectations, forcing him to bring manufacturing into R&D. These resource-uncertainty problems were compounded with organizational ones. Since the head of GEMS was more concerned with cost cutting than innovation, he was not ready to market the project. The project languished in R&D until he was replaced with a more innovation-minded president who brought the project to commercialization within six months. Technical-uncertainty issues continued to beset the project as difficulties were experienced going from a small demonstration unit in a laboratory to a commercially viable product. Market uncertainties were also present. Although customers were identified—GEMS knew all users of x-ray machines worldwide—the substantial cost increase of the new machine required educating a price-sensitive market on how this innovation would still be cost effective—by eliminating film and film storage, for example.

Of course, not all uncertainties can be confronted simultaneously. By cataloguing uncertainties, the team can focus on some while deferring or even outsourcing others. The natural tendency is to confront the uncertainties with which the team is most comfortable and put the others on the shelf. This is a particular problem for teams dominated by scientists or engineers who may prefer to focus on technical challenges.

Progress can be monitored by checking off assumptions as they are tested. The learning that results can then be documented, along with the decisions made as a consequence of that learning. According to one respondent, radical-innovation evaluation was based on the amount that was learned for the amount of money invested in the project, rather than tracking task completion against budget and schedule.

Besides managing internal progress, the project manager must manage interfaces with the mainstream organization. This requires gaining legitimacy for the project, preparing the organization to assimilate it as a mainstream activity, and securing resources to continue incubating the project. Our

research suggests that when radical innovation is incubated in the mainstream organization, both the project and the organization benefit from the mutual learning. They can achieve greater success than do skunk-works projects, which develop in isolation from the rest of the organization without the wealth of resources that the mainstream organization has to offer. For example, the IBM project manager was able to borrow a chip fabrication facility during a slow period to test his ideas. The IBM book project worked closely with IBM Solutions to identify a beta test partner. And at DuPont, connections to the internal network caused a senior technical researcher in a division other than where the project was working to identify a potential market opportunity.

Successful managers used several practices to integrate radical-innovation projects into the mainstream. They maintained regular and frequent communication with the mainstream organization. They capitalized on the desirability of preempting other innovators, and introduced testimony from potential customers as ways to get an operating unit to recognize the importance of the innovation. And they assembled influential advisors. A board of brand-name people provided internal legitimacy as it contributed the insights and support of its members.

Hubs Can Reduce Uncertainty

A hub can proactively engage radical-innovation project teams and provide experts and mentors who help projects understand the nature of the radical-innovation life cycle and how to reduce uncertainties. A hub establishes radical-innovation project-management systems, refines them through cumulative experience, and then helps teams implement them. It teaches teams new market-learning approaches and implements resource-acquisition strategies. A hub can be most effective in helping the project leader manage the interface between the project and the rest of the organization; using its own informal network to complement that of the project leader, it acts as a conduit for money, human resources, advice, facilities, and legitimacy.

IMPERATIVE NO. 4: DEVELOP A RESOURCE-ACQUISITION SKILL SET

Radical-innovation projects typically outstrip available research resources. Getting money, facilities, and people is universally difficult for radical innovators; they must spend an inordinate percentage of time and energy chasing resources.

When a radical-innovation project is established, it is generally given a small formal budget. The time required for a technology or market to mature may take many years. Investments in radical-innovation projects take a long time to demonstrate any tangible returns. Furthermore, radical-innovation projects based on new technologies often require development funds far beyond normal budget limits. Because a radical-innovation life cycle may span many years, a project can expect to see its supporters and funding sources change two or three times during its lifetime. Consequently, project leaders must approach a variety of potential funding sources and even reorient their projects to suit whoever holds the purse strings. Acquiring resources is a dynamic process. In all 12 projects in our study, the persistence of project champions in acquiring resources was critical.[8] The price paid for this persistence was substantial, since time spent chasing money was time not spent on project development.

After three years of effort, part of the hybrid-vehicle development effort at General Motors was in danger of losing its funding. In an effort to save it, the project team staged a technology demonstration for GM's corporate executives. Though the demonstration was impressive, it failed to produce continued funding. "You put on a good show," the director of GM's development group informed the project team, "but we've received orders to close you down. Our budgets can't handle you." A month later, a GM manager found himself sitting on an airplane next to a U.S. Department of Energy (DOE) official, whom he had known for some time. The GM manager described the hybrid-vehicle project to the DOE official, who liked what he heard, and agreed to recommend the project for federal funding. Approximately one year later, that funding was authorized. GM participated in the Partnership for a New Generation of Vehicles (PNGV), a consortium of the Big Three U.S. automakers, each of which was working on a hybrid vehicle. The GM project was significantly increased in scope by the funding, and effort within and by the original development group was consolidated under one project heading. In the view of the project team members and manager, the project would surely have folded without this funding.

Hubs Can Facilitate Resource Acquisition

The hub has an important role in helping teams develop a resource-acquisition capability. First, it can assign someone to the team who has a track record of acquiring resources for radical-innovation projects. Alternatively, it can train the project manager or a team member in that skill set, or assign a resource-acquisition specialist, who can identify internal and external financing and develop funding proposals on an as-needed basis.

At Nortel Networks, the project manager was assigned the role of obtaining resources. As one member of the project told us: "I think [he] has made a very, very good decision in terms of insulating the organization from the business of going out and getting money. And the reason I say that is because I worked at an Internet startup for 13 months before I joined [this project], and what happened is they did IPOs and money getting, and all that sort of thing, but they got everybody really excited and it was the sexy thing to do in the company. So what happened was that they took their entire general management staff, the CFO, the CEO, the COO, and, for six months, they went away (to get money), so the company basically went to hell in a handbasket."

The hub can also explore alternative venture-capital models. During the six-year period of our study, we witnessed firms' experimenting with several approaches to venture-capital funding. These initiatives were directed at internal ventures, potential spinoffs, and external startup ventures of strategic interest to the firm. In some cases, the firm opted to develop an internal venture-capital investment capacity. In others, the firm chose to partner with external venture-capital firms. Firms that make a strategic decision to pursue radical innovations can learn from the experience of others and choose an approach to funding that is consonant with their objectives.

In addition, the hub can assemble an appropriate decision-making board for radical-innovation investments. The higher level decision-making process required for ramp-up and transition-funding decisions should involve people with sufficient experience, capacity, and independence from the operating units that they can make objective and effective decisions. There is no more certain way of making a venture-funding board irrelevant than staffing it with managers who are driven by the short-term interests of their own business units, or who lack the skill and judgment to make appropriate decisions. Senior management must be involved in making decisions on radical-innovation projects because of strategic considerations, but the decision-making team can and should include people experienced with radical innovations and the technologies in question, and with venture-capital background.

Some decision makers may come from outside the organization. This was the case at Lucent, which hired a venture capitalist to run its new division, whose purpose was to incubate internal technologies and opportunities. The venture board was composed of the president of the venture-capital division, three vice presidents, and industry experts. Funding came from the corporate budget for investment purposes and additional funding was sought from outside investors.

IMPERATIVE NO. 5: ACCELERATE PROJECT TRANSITION

Projects cannot stay in R&D forever. At some point they must be transitioned to a receiving unit in operations for ramp-up and market introduction. The transition from project to operating business presents its own set of hurdles. After conquering technical,

market, organizational, and resource uncertainties, this last hurdle turns out to be quite difficult.

During transition, market and technical issues continue to beset the project. Early adopters are often willing to accept a prototype and work with the innovating firm to define the form and function of the new product, but customers buying a commercial product expect development to be fully completed. The new radical-innovation product is sufficiently different from current products that potential customers and the sales force need to be educated. Technical specifications that were adequate for the prototype stage require substantial revisions, since the new product is customized for specific applications. A project team and an operating unit often have quite different perceptions of a project's readiness for transition.

When the project team transitioned its product at DuPont, the operating-unit manager exclaimed: "I can't believe they sent it to me this early." In his mind, the technical and market uncertainties had not been sufficiently resolved. Because additional applications-development work was required before significant production could be undertaken, the project team remained involved even after the transfer. A project manager in the business unit was assigned the task of completing the technical and market development. Unfortunately, after he had only begun the effort he was promoted. The project was at a standstill for almost a year until a new product manager was assigned and brought up to speed. The second project manager forced his team to assess more than 30 leads and to focus on four. The search for commercial applications continued until, in frustration, the manager sent the project back to R&D. If the project had, in fact, been ready for transition, with a clear market opportunity identified, the confusion in the operating unit could have been avoided, time to market reduced, and revenues realized. Since it was transitioned too early, none of this occurred.

In most cases, neither the receiving operating unit nor the project team should be expected to possess the competencies needed to accelerate the project through this transition. A transition team—formed and supported by the radical-innovation hub—can be a more effective organizational approach. Although this requires two handoffs rather than one—from project to transition team and from transition team to operating unit—it is easier to bridge two small gaps than one substantial gap.

Hubs Can Smooth the Transition Process

As we saw at DuPont and GE, transitioning is where several projects really stalled. A series of actions will make the transition process work more effectively. First, transition readiness needs to be assessed from both the project and operating-unit perspectives. A key output of the transition-readiness assessment is the transition plan. This also facilitates buy-in by key leaders from R&D, the operating unit, and corporate levels, creating a strategic push to consummate the transition. Second, a transition team should be created, composed of personnel from the project team and the receiving operating unit, transition-management experts, market-development specialists, and a special oversight board. Funding should come from a corporate fund, because neither R&D nor the operating unit feels an ownership of the project at that point in its development. Hub staff, through repeated assessment exercises, eventually develops a competency in expediting this process and ensuring the quality and usability of the outcome.

The transition team develops a detailed transition plan that defines tasks, timetable, roles, and responsibilities. A senior-level-management transition champion gives the transition process the high priority it needs to be successful. The transition team oversight board is a useful organizational mechanism for concentrating the power of senior-management supporters. It also provides a natural mechanism for reviewing the progress of the transition team and ensuring cooperation among the various stakeholders.

The ultimate goal of any radical-innovation project is a successful business. From a market-development perspective, that goal can be reached via several paths. It is difficult, but critically important, to set realistic expectations in the transition plan about the likely evolution of the market. At IBM, the program manager in the microelectronics unit,

which received the silicon-germanium project, told us that the project almost did not make his unit's list of priority projects because projections of market size were too long-term to be credible. The projections were predicated on the telecommunications market, but the immediate market application, satellites, was much smaller and required too much work to develop for enough short-term revenues.

IMPERATIVE NO. 6: FIND PEOPLE WHO DRIVE RADICAL INNOVATION

Radical innovation will not happen without the right people. People with risk-taking propensity, drive, and out-of-the-box thinking were involved in every project we followed. Nevertheless, we saw few deliberate attempts to recruit, develop, and retain such people; they either emerged or had the maverick personality that is attracted to radical innovation. In some cases, people volunteered onto project teams when they heard about them through the grapevine. Large companies will have a difficult time retaining these people, as they are often entrepreneurial and ambitious, and often at odds with the organizational framework within which they work. Developing a reward system is therefore a critical organizational problem. A few organizations, including Procter & Gamble, 3M, and Lucent, experimented with implementing appropriate reward systems for radical innovators, but even these firms were dissatisfied with their approaches.

Cross-functional teams formed for incremental innovation typically include a technical guru, an engineer, a designer, a manufacturing expert, a marketing specialist, and even a financial person. Radical innovation requires a core group of multifunctional individuals, particularly in the early stages. They may be technical people first and foremost, but their value to the team is enhanced if they understand marketing enough to think broadly about application possibilities, are interested in the financial impacts of alternative courses of development, appreciate the consequences of development choices on manufacturing, and build connections to internal and external partners.[9]

None of the companies we followed had developed human-resources strategies for coping systematically with the personnel dimensions of radical innovation. Virtually all the managers we interviewed understood that people who drive radical innovation have different characteristics from those in more traditional roles, but had not translated those realizations into organizational policies. Recruiting, developing, and providing career opportunities for people who can drive radical innovation appears to be a major gap in the radical-innovation competencies of firms.

Hubs Can Search for Talent

Firms need to learn how to attract, develop, reward, and retain people who carry out radical innovation. The first step is to discover the radical-innovation types already within the firm. The hub serves as a magnet, drawing them out of the corporate woodwork, and can also actively search the organization for radical-innovation talent. Going beyond the hub, dynamic leadership, an effective innovation-support infrastructure, and deep reservoirs of technology, competencies, knowledge, and talent all create a corporate culture that appeals to the innovators and entrepreneurs who are most likely to promote radical innovation.

IMPERATIVE NO. 7: MOBILIZE THE MULTIPLE ROLES OF LEADERS

Since senior management has a great impact on the capacity of an organization to succeed or fail at radical innovation, it follows that senior-management turnover produces disruption, both positive and negative, in the radical-innovation culture of organizations. At one company in our sample, the retirement of a senior officer resulted in the dismantling of its radical-innovation hub. At Texas Instruments, the sudden death of the CEO who instigated and supported the Digital Light Processor threw the young project into turmoil. Fortunately, the vice chairman and most of the senior staff had voiced support for the project, which had achieved formal status in the organization, with a budget and personnel. But the project was not out of the woods until the new president also demonstrated support. We identified three ways that senior managers can champion radical innovation.

Executives as Patrons

Just as the powerful and rich have supported and protected artists throughout history, a system of patronage works in corporate innovation. In all firms in our study, one or more senior executives played the role of patron, variously providing organizational protection, resources, and encouragement to maverick innovators. We found across most projects that the patron had faith in a project champion because of the champion's personal characteristics, a lengthy relationship between the two, or the champion's track record in bringing other important projects to fruition.

To be an effective patron, the senior executive must be accessible, especially to the middle managers who generate many radical innovation opportunities. The senior executive must also have a passion or personal liking for the project, and must give sustained support, or pass that support on to another executive. In half the companies studied, the executive who followed a project's departed patron either slowed or killed the project.

Executives as Provocateurs

In 90 percent of our sample companies, executives played an active role in driving radical innovation by issuing a call to arms that stimulated innovation. The CEO of Air Products voiced his concern that the company had missed an earlier game-changing innovation, and declared: "By God, we're not going to miss the next one." At one of his regularly scheduled monthly meetings with his top managers, Otis Elevator's CEO challenged his team to produce ideas for an elevator in an imaginary mile-high building. An elevator car that used electric motors rather than conventional cables was the result. Such executives stimulated both the level of activity and its direction, impelling their organizations to launch major new efforts.

Executives as Shapers of Culture

An executive's greatest contribution to radical innovation is to shape the organizational culture in ways that make it natural, accepted, and valued. Ray Stata put his own stamp on the culture of the company he founded, Analog Devices. Early in the history of Analog Devices, Stata wanted to acquire a company that offered the opportunity to expand into integrated circuits, a technology he believed would become important over time. Although his board of directors disagreed, Stata used his founder's stock to acquire the company, which has since become a cornerstone of Analog's success.

This story is frequently cited as the basis of an entrepreneurial culture that justifies breaking the rules to pursue an attractive business opportunity. Stata reinforced this culture when he came back from retirement to help support and protect the solid-state accelerometer project. The project did not have a home at the time, and was passed from operating unit to operating unit. Largely because of Stata's support, a new division was established to develop, manufacture, and sell the innovation. One of its applications allowed the replacement of three separate airbag initiators with a single solid-state accelerometer. Subsequent applications emerged in areas as diverse as medical devices and video games.

DEVELOPING A MATURE RADICAL-INNOVATION CAPABILITY

Radical-innovation maturity comes when an organization has systematically implemented processes for initiating, supporting, and rewarding radical-innovation activities. Depending on an organization's level of radical-innovation competency, there are different approaches to solving the major challenges identified in this article. Moving from lower to higher maturity is not easy. However, failure means that firms must rely on a combination of luck and extraordinary individual effort. In firms with a mature radical-innovation capability, radical-innovation hubs can play a supporting role in some tasks and take the lead in others. An integrated diagram of the hub and its relationship to the project team and the larger organization is contained in Figure 39.3. The hub sits at the interface between the radical-innovation project team and key internal and external stakeholders. Internal stakeholders include various operating units, R&D, and senior management. External stakeholders include early-adopter partners, manufacturing partners, technology-development partners, and funding partners. The hub acts as a source of radical-innovation expertise and facilitates constructive relationships between the radical-innovation project and its stakeholders.

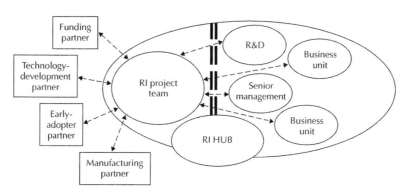

Figure 39.3. The radical-innovation hub's role in the organization.

As a firm builds a more mature radical-innovation capability, its leadership sets expectations, creates an innovative, supportive culture, establishes a radical-innovation hub, and develops appropriate goals and reward systems surrounding the hub's activities. Since radical innovation is critical to long-term organizational success, such organizational arrangements are absolutely critical. Without them, radical innovation will remain a haphazard and ad hoc activity.

APPENDIX

A multiple case-study design was used, since there is little research concerning new-product development for radical innovation. Case-study research is especially appropriate for this type of exploratory research, with a focus on documenting a phenomenon within its organizational context, exploring the boundaries of a phenomenon, and integrating information from multiple sources.

This research was conducted in cooperation with the Industrial Research Institute (IRI), with the financial support of the Sloan Foundation. All the participating companies were members of the IRI, and volunteered projects on the basis of the definition of radical innovation provided in this article. We lost no companies over the course of the research. The findings represent the results of a longitudinal (since 1995), multidisciplinary study of the management of radical innovation. Six researchers represented the management disciplines of entrepreneurship, marketing, operations, product design, organizational behavior, and technology management. All six participated as interviewers, and met regularly to review and interpret the data. We collected interview data, as well as company records, in real time rather than retrospectively to control for the history effects that weaken case research. Data were gathered at multiple times, at least once per year, as each project moved forward. To learn about each project, we interviewed senior management (including directors and vice presidents of R&D and corporate development), project managers, and individual team members. Using multiple interviewees reduces the risk of undue influence that an individual interview may have on the case study and develops a richer portrait of each case.

ENDNOTES

1. See, in particular, Foster, R. 1986. *Innovation: The attacker's advantage*, New York: Summit Books; Utterback, J. M. 1994. *Mastering the dynamics of innovation*. Boston: Harvard Business School Press; Christensen, C. 1997. *The innovator's dilemma*. Boston: Harvard Business School Press; and Hamel, G. 2000. *Leading the revolution*. Boston: Harvard Business School Press. This article builds on research reported in our book, Leifer, R., McDermott, C. M., O'Connor, G. C., Peters, L., Rice, M., & Veryzer, R. W. 2000. *Radical innovation: How mature companies can outsmart upstarts*. Boston: Harvard Business School Press. However, our conclusions and recommendations reflect new thinking beyond those presented in the book.

2. Numerous other writers have recognized the difficulty of managing radical innovation in large, established firms. Christensen, ibid., indicates that it is highly uncommon for firms that manage established lines of business well to anticipate and respond effectively to a disruptive technology coming from an external agent, much less to commercialize one themselves. Also see Leonard-Barton, D. 1995. *Well-springs of knowledge: Building and sustaining the sources of innovation.* Boston: Harvard Business School Press; Katz, R. & Allen, T. 1985. Organizational issues in the introduction of new technologies. In P. R. Kleindorfer, (Ed.), *The management of productivity and technology in manufacturing*: 275–300. New York: Plenum Press; Kanter, R. M. 1989. Swimming in newstreams: Mastering innovation dilemmas. *California Management Review*, Summer: 45–69: Tushman, M. & O'Reilly, C. 1997. *Winning through innovation: A practical guide to leading organizational change and renewal.* Boston: Harvard Business School Press; Dougherty, D. 1992. Interpretive barriers to successful product innovation in large firms. *Organization Science*, 3(2): 179–202; Block, Z. & MacMillan, I. 1993. *Corporate venturing: Creating new businesses within the firm.* Boston: Harvard Business School Press, provide actionable prescriptions to enhance an organization's readiness to commercialize radical innovation.

3. See Morone, J. 1993. *Winning in high tech markets*. Boston: Harvard Business School Press, and Tushman & O'Reilly, ibid.

4. Abdul, A. 1994. Pioneering versus incremental innovation: Review and research propositions. *Journal of Product Innovation Management.* 11: 56–61; Lee, M. & Na, D. 1994. Determinants of technical success in

product development when innovation radicalness is considered. *Journal of Product Innovation Management*, 11: 62–68; March, J. G. 1991. Exploration and exploitation in organizational learning. *Organization Science*, 2(1): 71–87.

5. See Ansoff, H. 1957. Strategies for diversification. *Harvard Business Review*, September–October, 35: 113–124; and Booz-Allen & Hamilton, Inc. 1982. *New product management for the 1980s*. New York: Booz-Allen & Hamilton.

6. For an excellent discussion of stage-gates, see Cooper, R. G. 1990. Stage-gate systems: A new tool for managing new products, *Business Horizons*, May–June: 44–54. See also Cooper, R. G. 1993. *Winning at new products: Accelerating the process from idea to launch*, 2nd ed. Reading, MA: Addison-Wesley.

7. Rosabeth Kanter's concepts are most similar to ours. She refers to "scouts" and "coaches." A scout is the receiver of new ideas and has corporate money to allo-

cate to them, and the coach is the more active role, helping refine the idea and create the sales pitch to help senior management understand its importance and eventually find a funding sponsor. See Kanter, R. M. 1989. *When giants learn to dance*. New York: Simon & Schuster.

8. See also Dougherty, D. & Hardy, C. 1996. Sustained product innovation in large, mature organizations: Overcoming innovation-to-organization problems. *Academy of Management Journal*, 39(5): 1120–1153; and Angle, H. L. & Van de Ven, A. H. 1989. Suggestions for managing the innovation journey. In A. H. Van de Ven, H. L. Angle, & M. S. Poole, (Eds.), *Research on the management of innovation: The Minnesota studies*. New York: Harper & Row.

9. Leavitt, H. & Lipman-Blumen, J. 1995. Hot groups. *Harvard Business Review*, 73: 109–116; Wenger, E. C. & Snyder, W. M. 2000. Communities of practice: The organizataional frontier. *Harvard Business Review*, January–February: 139–145.

40

Dreams to Market

Crafting a Culture of Innovation

KAREN ANNE ZIEN AND SHELDON A. BUCKLER

In 1993, while at Polaroid, we were troubled by this critical question: *How can a mature company keep its innovative spirit alive—or, if necessary, rekindle and reinvigorate it?*

To find the answer, we organized the Polaroid Invention and Innovation Research Project. Our cross-functional team videotaped more than 140 in-depth interviews with technical, marketing, and management personnel from 12 leading U.S., European, and Japanese companies—companies like Sony, Hewlett-Packard, Toshiba, Club Med, and Polaroid that have introduced a stream of financially successful new products founded on new business concepts and new inventions. The complete list of study companies is shown in Exhibit 1.

What we found was a strikingly consistent model of how companies craft and sustain a culture in which innovation is not only tolerated but nurtured, rewarded, and even demanded. We learned about mavericks and champions, corporate storytelling, the Fuzzy Front End—and most critical of all, the role of Top Management.

In an earlier article, we emphasized one aspect of how these companies invigorate and sustain an innovative culture: leaders at all levels of these highly successful, mature enterprises tell and retell compelling stories of innovative experiences and exploits [3]. These stories, myths, and legends are the organization's teaching parables and are ubiquitous throughout highly successful enterprises.

One story we retold in our earlier article was Marv Patterson's description of a moment in the development of Hewlett-Packard. In this story a leader's spontaneous human reaction (Bill Hewlett's direct response to an unwieldy pen plotter prototype) spurred a salient invention (Larry La Barre's elegant microsprocket drive design), which helped to change the nature of Hewlett-Packard's business (to a major provider of diverse drafting plotters for a variety of key markets), and set the stage for the innovative printers and computers of the '80s, now HP's biggest business:

Exhibit 1: Invention and Innovation Research Project—Study Companies

United States	3M
	Apple Computer
	Hewlett-Packard Company
	Polaroid Corporation
Japan	Sony Corporation
	Toshiba
	Two other leading consumer electronics companies
Europe	Club Méditeranée
	Kenwood Electronics Italia
	Océ
	SmithKline Beecham

Reprinted from the *Journal of Product Innovation Management*, July 1997, vol. 14, no. 4, pp. 274–287. Reprinted with permission.

Karen Anne Zien is a senior innovation systems anthropologist and principal at Apogee, an international consulting firm, and is co-founder of the Creativity and Innovation Lab in Cambridge, MA. **Sheldon A. Buckler** is Chairman of Commonwealth Energy Systems and is a well-known technologist and business leader who worked closely with Edwin Land at Polaroid to create breakthrough film technology.

... The bottom line was that we made that decision back in 1978, the plotter was introduced in January of '81, and by the end of 1982, we owned 60% of the marketplace that had grown 60% because of this product introduction. We literally revolutionized the drafting plotter marketplace at the time. In the next 4 or 5 years we dominated that marketplace. Absolutely dominated it. And by 1985, any plotter that was competitive used grit wheels. So to me that's a story of innovation.[1]

In this article we will describe how these companies strive to keep their innovative spirit vibrant and robust by employing *principles* of innovation. In keeping with our cultural anthropological approach, we will rely once again on our interviewees' narrative descriptions and on the patterns and themes that occurred across all the interviews. Exhibit 2 describes our method.

SEVEN PRINCIPLES

Innovative companies, regardless of differences in industry and geographic culture, share a set of characteristics, qualities, and behaviors that differentiates them from other less innovative companies. In the course of our research, there emerged a remarkably consistent pattern, a well-crafted fabric, that has kept the innovative force strong and functional in the companies we studied, despite changes in leadership and direction, industry structure, the marketplace, and the passage of time. We, and our study companies, were completely surprised by this finding, since we had expected to see vastly different patterns in the form of regional models for Japan, Europe, and the U.S.[2]

We identified seven key principles at work in innovative companies across all these geographic

---

Exhibit 2: Interview and Analysis Methods Borrowed from Cultural Anthropology

Interviews

We asked each company to identify 2 or 3 very innovative products and some participants from marketing, technical/invention, and mid-level management who were active in the *earliest* stages of the product's conception. Experience taught us to request interviews with corporate "storytellers." In addition, we met with heads of companies, divisions, research labs, and subsidiaries. We visited both headquarters and divisions of these global companies. We also interviewed Polaroid employees worldwide.

Our visiting interview teams at each company consisted of 3 people, also representing the same cross functional mix of marketing, technical/invention, and mid-level business management points of view. Our methods for interviewing, observing, and analyzing came from cultural anthropology [3: "Exhibit 2 Cultural Anthropology" and "Table 1 Comparison of Two Methodological Paradigms"].

Therefore, our guidelines for each visit were: to suspend our earlier judgments and hypotheses; pay attention to the larger context; look at the situation from many angles (a 360 or systemic perspective): notice the process we experienced entering each company; and learn from our own intuition (the tool for "discovery") and perceptions during the visits. We probed and listened to interviewees, following their path, rather than directing the interviews. When we encountered problems (for example, Océ set our interviews up jointly with marketing and technical partnerships, though our request was to speak with these representatives separately) or opportunities (such as the chance to follow our intuition and interview storytellers), we tried to understand if such problems or opportunities pointed to something important about the company (Océ was remarkable for the quality of marketing and technical partnerships) or the topic of invention and innovation. We discussed such problems or opportunities with our interviewees in order to jointly learn more with them during the interview or visit. During each company visit we were seeking qualitative information, diversity of insight, intensity of contact, and personal experiences more than quantitative data. We had thoroughly researched each company and the literature on invention and innovation before scheduling our visits.

We also became cultural anthropologists in the way we observed experiences within our own company. The two inventors on our project team were astonished to detect that people at Polaroid reacted to the idea and activities of our research project with the same patterns of behavior that our corporate culture exhibited when relating to their new technological ideas at that time. We were a microcosm of the culture of innovation at Polaroid; we even had to go underground for a while.

Analysis

Cultural anthropology was also the source of our methods of analysis. We further developed anthropological methods to fit our corporate circumstances. The visiting interview teams of 3 selected the most representative segments of video from the company interviews. The full team could then fruitfully engage in interpreting the data and experiences gathered from each company, almost as if we had all been on the visit. As we went along, we took time to invest the full team and our key sponsors with the learnings from each subsequent company visit.

At the end of all the visits to study companies, one of our sponsors and the team reviewed all the videotapes and interview transcripts. Then, small cross-functional partnerships of 3 team members from diverse backgrounds identified the most striking themes as they were expressed in each transcript. Multiple diverse views of the same

data is a method verified by experiences in market research as well as cultural anthropology [6].

We then worked as a full group to identify the patterns and themes that were present in all the companies within each of the three regions. Only at this point could we document that all the companies, regardless of geography or nation of headquarters, shared one set of patterns, themes, or principles.

We did not "report" our team's findings to our sponsors or senior management, in the strict sense of the word. Rather we invited them to attend "Creative Engagements" or symposia. We all viewed the representative videotapes from each company and engaged in further discussion and interpretation of the meaning of these for our company. At each symposium we asked participants: (1) What did you discover about invention and innovation from the videotaped interviews? What stands out? What surprises you? (2) Why are these key learnings and surprises important to you? What does this mean to you personally and your company?

Our extensive database of videotaped narrative interviews, qualitative analysis, and further interpretations by leading company practitioners, continues to grow as we conduct similar symposia in our study companies and work with other companies. Some of our study companies have responded to our findings with yet another round of innovative insights and activities. Observing this we learned again that innovative companies craft and *continually* take new actions to sustain their innovative cultures!

---

regions. These are listed in Exhibit 3. The principles are a set. We found them *all present throughout and woven together* in these highly innovative companies. As far as we could determine, these companies had not learned them from each other or a common source. Rather, the principles arise spontaneously as a pattern in companies that continually hone a culture of innovation.

While the principles are universal, we found that each company's implementation "formula" is particular and specific to that company. Each company customizes the principles for their own cor-

porate culture by *systematically* and *systemically* implementing a set of practices or approaches throughout the whole enterprise. The innovative companies in our study take a few provocative approaches to these principles, hone them, work them fully into their enterprise, and repeatedly refresh and renew them by eagerly adding new approaches to sustain their innovative character.

"Innovation" is the whole spectrum of activities, from dreams to market introduction and maintenance, necessary to provide new value to customers and a satisfactory financial return to the

---

Exhibit 3: Seven Key Principles

Principle #1: Sustain faith and treasure identity as an innovative company.

Principle #2: Be truly experimental in all functions, especially in the front end.

Principle #3: Structure "really real" relationships between marketing and technical people.

Principle #4: Generate customer intimacy.

Principle #5: Engage the whole organization.

Principle #6: Never forget the individual.

Principle #7: Tell and embody powerful and purposeful stories.

---

company. Figure 40.1 is the conceptual model we developed as we interviewed.

Everyone in these companies is engaged in the dynamic creation and recreation of a culture of innovation. Innovative companies carefully craft and continually take new actions to sustain their innovative cultures. We found that innovation occurs throughout the whole system we call the enterprise, and there are three micro climates or micro cultures—the "Fuzzy Front End," the Product Development Process, and Business Operations—as represented in our conceptual model. All three micro cultures are essential to the creation of new value. The apparent dilemma is that each of these micro cultures has characteristics and requirements that seemingly are incompatible with the others, as we have indicated in Table 40.1. In the innovative companies we studied, senior leadership develops

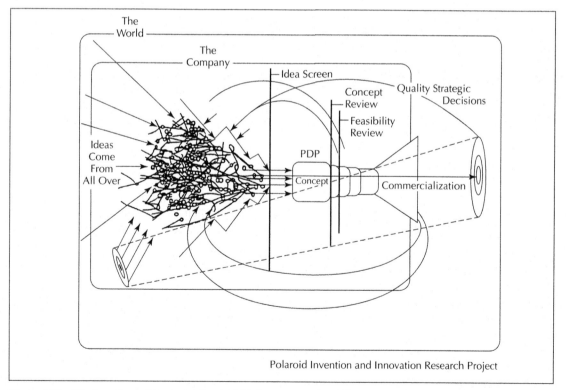

Figure 40.1. The Invention and Innovation Research Project conceptual model.

TABLE 40.1
Three Micro Cultures of Innovation

Fuzzy Front End	Product Development Process	Market Operations
Experimental	Disciplined	Seeks predictability and order
High tolerance for ambiguity and uncertainty	Focus on numerous quantitative goals and measurements	Strong financial orientation
Chaotic	Commitment to the goal	Commitment to established values and businesses
"Unreasonable"	Schedule oriented	Oriented to rules and routine
Enjoyment of the Quest itself	Urgency	Slow to change
		Highly organized
Unpredictable	Trained—generally not receptive to new ideas	Does not welcome revolutionary ideas
Much individual activity	Teamwork of paramount importance	Large in size compared to FFE or PDP

and sustains a context that simultaneously fortifies all three micro cultures and includes strong, vital feedback loops and connections among them.

Managing the paradox represented by these three micro cultures is what we mean when we say innovative and transformational leaders create the context and craft a culture for continual innovation. For these leaders innovation is the core of *all* value creation in business organizations. The inception of innovation is the "Fuzzy Front End," as we have described it. In most mature business organizations the fuzzy front end culture becomes intolerable because it is so "unbusinesslike." But these leaders of continually innovative enterprises have learned to love the fuzzy front end and to keep it central to the enterprise.

THE SEVEN PRINCIPLES AT WORK IN HIGHLY INNOVATIVE COMPANIES

Principle #1: Sustain Faith and Treasure Identity as an Innovative Company

Leaders of highly innovative companies demonstrate in every decision, action, and communication that innovation propels profitability. The emphasis is on developing whole new business concepts, product platforms, mapping generations, and systematically "destroying one's own." Continual in-

novation is their soul business. All other business concerns flow from this single overriding purpose.

Often the leader's role in innovation is to set challenging targets. At Sony the leadership set ambitious schedules for both the CD and the 3.5 inch floppy disk projects:

> . . . not only clear, but also challenging targets— that is the very key for innovation. Yes, a sort of 'voice from above' that does not question the possibilities and absolutely believes it is possible. The discussion of 'if there is possibility' will go into a negative spiral. . . . So, I conducted an orchestra of friends and colleagues I had known for 25 years in order to make the schedule.[3]

The most successful companies are those that create inventive and innovative work "at the margin," continually bring such new ventures into the core business, and drop others out, often by cannibalizing and obsoleting their own. As another key player in Sony's CD project told us:

> . . . There were an awful lot of stakeholders in black vinyl, but my nose told me, 'This [CD] technology will build up very steadily and constantly and will invite something new for the record industry and the hardware industry.' I just caught some scent of the flavor of it by myself. Of course, I asked a lot of questions of Mr. Tsurushima [the

CD Project Leader], and Dr. Nakajima [the Technical Leader and inventor]. Then, gradually, I generated my own confidence for myself.[4]

3M not only treasures their identity as an innovative company, they emphatically advertise it. All the innovation "self-talk" at 3M is truly reinforcing to the efforts of their people. Slogans and mottos are not at all trivial to the task of crafting their culture of innovation, but provide an ongoing sense of identity and self reference. We found that 3Mers talked about measurements the most, though all companies had worked to find the most effective metrics to motivate innovative activities.

HP moved away from "break even time" to a more immediate feedback metric, a "slip rate" measurement that is married to a running analysis and interpretation of the causes of "slips" in a new product development project schedule. They tally these on a *pareto* chart that also documents the causes of slippage in each phase of new product development, from the initial idea investigation phase on into the market introduction phase. Then, improvement programs focus first on the most frequent causes of slippage.

In all the companies we studied, such corporate-wide metrics and goals are baked into business goals and performance measures for groups at all levels of the enterprise and are central to the personal goals and performance reviews for individuals. Such measures also contribute to an innovative identity and the ability to maintain self reference as great changes occur.

Our study companies measure the internal "success rate" of research projects and programs, including starts and failures or ends of programs in development. This is a reminder to senior leadership and is a measure of how many good ideas need to be worked on in order to have even one market success. One hundred investigations for every market success was a common rule of thumb.

HP teams know the amount of revenues to be earned or lost for each day a program may enter early or late into the market. Using this dollar amount as a decision criterion for development costs, a team chooses resources to shorten time lines, stick to the critical path, or unjam a bottleneck. If the team members need to ask the Program Manager, s/he has one day to respond, then the decision is bumped to the next higher level each day. In 4 days the CEO is deciding. HP applied this decision-making approach first to an integrated set of software development projects and then broadcast the results and approach company-wide by a series of HP educational videos.

Eventually we came to understand that such metrics, and the ensuing visibility of improvement programs, are more than just good business process and discipline. They do much to sustain employees' "faith" in the innovation process, keep motivation high, and, as a result, spur even more initiatives. A robust and productive product development process similarly sustains faith and makes innovation worth it to thousands and thousands of creative people. Without business process disciplines, measurements, and continual improvements to all phases of the invention and product delivery process, good ideas would remain just that—ideas. Creative and inventive people fervently work to see that their ideas come to fruition! The crafting of an innovative culture requires creating an environment of faith and trust that good ideas have a likely chance to become great products.

Principle #2: Be Truly Experimental in All Functions, Especially in the Front End

Successful product and service innovation organizations are truly experimental in the front end. For example: 3M now requires 30% (up from 25%) of sales to come from *new* products introduced in the last 4 years (down from 5 years). And they carefully define what they mean by "new." They designate 7 classes of central and divisional technical work. "New" means "Class 5" technical work at 3M. Occasionally "Class 4" is considered "new," but only if "Class 4" results in taking a product, related to existing business, into a totally new market.

> . . . We're putting together a portfolio of new programs in my division that we want to sponsor and support. We call these "Class 5" programs. Class

5 are "Unrelated Business programs." There's a multiple class system. Class 1 is what's called Sales Support/Tech Service. Class 2 is Production Support. Class 3 is Support for Current Business and Process Development. Class 4 is Related Business Product Development. Class 5 is Unrelated Business Product Development. Class 6 is Research. Class 7 is Basic Research. Those are the 7 classes of how 3M Corporation divides up technical work.

Each laboratory has their portfolio. I know how many dollars go into and how much of our time is spent in Class 1, 2, 3, 4, 5. And I can give you all the statistics for my lab. Class 3 is Current Business and Process Development. That's the core of my business. Class 1 is Sales Support. That's Tech Service. Every time a salesperson calls, a customer's got a complaint or a problem in the field, I've got to have Class 1 resources. And so without losing my handle on what I need to do to support current business, I dabble in Class 4 and Class 5.

And we in 3M have a norm now of every technical director sponsoring at least one Class 5 program, because Class 5s are the future growth opportunities. They are the future potential for multi-hundred million dollar businesses, because Class 5 is unrelated business.[5]

All the companies we studied create "safe havens" for small amounts of highly experimental work by all functions. 3M's 15% rule allows people 15% time to work on self-defined innovations. The 15% rule was at first only applied to technical people. This is still the more usual, but we found references throughout the culture and recommend it for all functions. Fifteen percent means 15% of a person's time, 15% of a colleague's time added to your time, 15% of machine run time for experiments, 15% of a budget in key functions or in a business, depending on how an enterprise is organized.

Based on one person's initiative, 3M has recently committed a full-time resource to spend a year with Eric von Hippel at MIT to develop a practitioner's version of his "lead user method" to identify new markets. Such fellowships have been frequently developed between 3M and MIT, but more usually in technical areas. In this case preliminary research by the "visiting fellow," the year's on site experience in von Hippel's lab, and the integration of the 3Mer's learning with skillful training resources have created an accessible and serviceable curriculum resource for marketing and new product teams at 3M.[6]

Sony has Tokyo laboratories for "next generation" exploration that are located *in*conveniently away from ongoing business activities, in locations where "a bowl of soup will get cold if carried from headquarters to the lab."[7] Dr. Nakajima, winner of an Imperial Award for Technology, attributes the success of his work on digital sound, at a time when the industry and Sony were committed to analog, to the location of his lab and the protection of his supervisor, who contained his excitement until Nakajima's ideas were developed enough to take forward for presentation.

For similar reasons HP has a concept of "G-jobs" ("government jobs") that are projects undertaken by individuals, using company equipment and resources for unauthorized, even personal work. They have tracked the relationship of this personal work, such as building a pump for a small pond, to such company breakthroughs as a team's work to design a gas spectrometer.

Sony and Toshiba both have small organizations, with projects budgeted for 6 months at a time, to take new products from the Concept Phase to Sales Development in small quantities. This approach to produce planning is called "Darwinian selection" by Dorothy Leonard-Barton [10]. The point is to keep small scale and run things all the way into the market before investing a lot in order to: help define the market, create a technological standard, and identify an initial group of early adopters. This requires experimental and early participation from a variety of functional groups, not just R&D.

> . . . Not only in our company, but also in other companies we are facing some problems, and we're thinking about the efficiency of R&D, specifically how to introduce the products of R&D into business areas and create new markets, especially in the case of totally nosy concepts, where there is no business group yet formed to support

the project. So the development laboratory takes a project from research to the market.

In the current system, with an existing business, after research and development has a good result, then we move the project to the business group efficiently. So we directly support the current state of the business.

The question is if we have a new idea that doesn't match those existing product lines, what happens to the new idea? And in this case, this development laboratory is able to take a brand new idea to the market. If we try to get the consensus with the existing business groups, that project would be crushed. Business R&D groups are funded by each business group and corporate laboratories are funded by corporate. The potentially new product that does not fit the existing businesses would be separately funded in the development lab.[8]

Another approach to business ideas that fall outside the current structural definitions is 3M's. 3M awards $30K "genesis" grants to fund the early stages of innovation programs that don't fit the current business structure. These funds are allocated by a panel of colleagues, fellow scientists.

All our study companies have people dedicated to keeping (computerized or other) active databases to collect ideas and create access to them. If a project does not go ahead at first, they assure and secure documentation of all ideas and investigations for better timed opportunities later. Océ, a maker of computer peripherals and specialized CAD applications and products, calls this their "refrigerator" for ideas.[9]

Xerox Palo Alto Research Center (Xerox PARC), where we conducted a preliminary visit and interviews, developed a product concept while working to support collaboration in new product idea generation and development activities among their own staff. They have a prototype of a computerized community "whiteboard," to promote the collaboration most innovations require [15].

Their concept was "WYSIWIS" (pronounced "whizzy whiz") or "What You See Is What I See." One person can manipulate the words and symbols of another, leading to true "co-creation" of ideas and concepts. Everyone can see ideas as they are generated and this often spurs their own thinking. The software makes it easier to participate and contribute: people from very different backgrounds and locales can comment on and share in constructing an idea. Ideas can be expressed, captured, built upon, sorted, and categorized without interrupting the verbal flow of conversation or necessitating colocation [14].[10]

Being truly experimental in the front end means understanding that some new ideas are significant enough to redirect the strategy of the enterprise. More than 25 years ago at Polaroid the SX70 camera, the integral film and camera system that emerges from the camera, self-times and self-develops, without anything to peel apart, was begun as "Special Experiment #70" (SX70), but became the basis for all the new instant photographic product systems since. The more recent beginning of digital photography at various photographic companies was similarly designated as "special experiments," worked "off the beaten track" of new product development projects.

Principle #3: Structure "Really Real" Relationships Between Marketing and Technical People

High performance product and service innovation organizations structure strong, *direct* interaction between visionary business/marketing people and technical inventors. These *really real* relationships, not just structured work assignments, operate both formally and informally. People know where to go with radically new concepts to get nurturing for their ideas and help from people in other functions.

In addition, these relationships, when brought to bear early in the idea and concept generation phase, serve to resolve the traditional tensions between cost control and experimentation. Now the emphasis can be on strategic fit and building alliances for extending or redirecting strategy. "Otherwise, the technology people always try to take control and the marketing people are too conservative."[11]

3M people speak in apparent code, about a "3-legged stool." As we listened to story after story of invention and innovation at 3M, each one told of early seemingly "automatic" connecting between marketing, technical, and manufacturing people. This occurred informally at first, with people from the other two elements of this "3-legged stool" joining the first to share an idea. The connections, it was explained, are useful just to bounce off ideas, and add informal resources to early investigations.

Sony makes especially effective use of the Japanese practice of rotating managers for 2–3 year assignments in marketing, product planning & development, manufacturing, and finance. Many people refer to Sony marketing and product planning people as "technically astute and [their] technical people have marketing savvy." The experiences gained and relationships developed help keep the views of the Chairman [Akio Morita] active in practice.

> . . . Our Chairman says, "Science alone is not technology and technology alone is not innovation." Innovation needs creativity in technologies, product planning and marketing, not only in technologies. For example, our Walkman did not need any innovative technologies, just the existing technologies. But innovation of the concept itself, use of the existing technologies, product planning, and marketing were required. That's also innovation in products.[12]

Océ distinguishes itself as a "market driven company," not a "Marketing driven company." Océ assigns all new product work to small partnerships or groups of co-located, cross-functional people. Other work is also assigned to multifunctional partnerships or teams, so relationships are deepened.

Our study companies look for and implement a multitude of ways to create "connective tissue" and "a campus vs. commuting college" atmosphere among their people. They invest in and sustain all of these, rather than look for only one method to create relationships across functions [7].

Principle #4: Generate Customer Intimacy

Generating customer intimacy is very closely related to being truly experimental in the front end of all functions. "Consumers are clever. They need

something totally different."[13] Innovative companies develop many ways to interact with customers and users. There is no single "best way" to structure these innovative interactions with customers. All the innovative companies we studied have "sensor" people and a great variety of "sensing" activities that maintain their antennae in the outside scientific/technical and customer worlds.

Marketing and technical people together engage in a host of formal and informal relationships with customers and end users at the front end. They are especially active *before* having a particular product concept to explore. Connecting-with-the-customer-activities occur at all levels in the enterprise and continue through all stages of the product creation process. What is remarkable in these enterprises is their *up front* involvement with customers.

An example of such involvement is HP's "scanning teams." To identify new opportunities and new ideas, HP has scanning teams—partnerships of two or so people, equally balanced with technical and marketing expertise—who go out and about to scan the *environments of use* of their current, former, and potential customers; to gather "premonitions" of future scientific and industry trends; and to acquaint themselves fully with competitor products and direction. Supporting the scanning teams' efforts is systematic background research from other HP groups. From this research and the scanning team's intelligence gathering the teams routinely make "maps" of innovation targets that match customer and market opportunity areas with emerging internal technical capabilities.

People on the scanning team are able to follow a new concept by becoming part of the investigation and new product development team. Once a project begins, HP monitors each effort against 10 factors the company has found crucial to product success. These include understanding user needs; alignment with HP and divisional strategy; competitive analysis and product positioning; technical risk assessment; priority criteria as to cost, schedule, and features; regulatory compliance; distribution channel considerations; continuing commitment to the market; endorsement by upper management; and total organizational support.

An internal company study led by Edith Wilson found that successful projects performed well on *all* 10 factors while unsuccessful projects had been inadequate on one or more factors [17]. Of 20 projects studied, no project succeeded if it ignored the single most important factor, the necessity to understand customer needs.[14]

HP has been actively moving from being a technology driven company to becoming a customer driven company, since 1991. The Corporate Market Research group, the Medical Group, and others have looked into various new approaches to generating more connection with customer environments and interviewing customers, such as "voice of the customer," "concept engineering," and our cultural anthropological approach.[15]

The general manager of the Medical Group makes a point of visiting customers (frequently those who are *not* currently HP customers) on a worldwide basis at least once a month. Senior leaders are specially trained so they can better "probe and listen," an anthropological approach, rather than "direct" the interview or attempt to "make a deal" with current and potential key customers. In addition, every new engineer in the Medical Group attends a course in physiology and a 2-week internship at Boston University Hospital, where s/he follows residents, nurses, and doctors on their rounds at University Hospital.[16]

HP's senior level customer interviews are part of gathering customer requirements for what HP calls "Phase Zero," before any product concepts are considered. Once a project is in Phase Zero, partnerships of a marketing and a technical representative go together to customer sites, similar to the scanning teams described above, completing some 25–50 interviews worldwide. It is important that the interviews be conducted in person in the customer or potential customer's environment. This allows the team to observe and collect contextual, implicit and qualitative data that would be overlooked in other methods of data gathering.

Concept engineering is a quite structured customer centered process of data collection, reflection, and clarification. It was jointly developed from 1990 to 1992 by professors at MIT and several founding members of the Center for Quality Management (Bolt Beraneck and Newman, GenRad, Analog Devices, Bose, and Polaroid) [5].[17]

In addition to using an anthropological approach for the interviews, concept engineering uses a special "KJ" methodology (based on the "affinity diagram" developed by Kawakito Jiro, a Japanese anthropologist) to distill images and requirements of the customer environment. Later analytical steps use more quantitative techniques to measure the intensity and frequency of perceived needs.

An extension of this kind of approach is "empathic design" named by Dorothy Leonard-Barton, from her observations of highly innovative designers. It can be used for adapting known technologies to users' needs, such as the functioning of doors on automobiles, or for radically new technologies and products, such as Personal Digital Assistants (PDA's).

In addition to standard market research and traditional interviews providing all the information they can, technologists accompanied by someone knowledgeable in the marketplace observe and videotape various users' practices in detail, such as the opening and closing of automobile doors in the streets of New York or people's actual behaviors as they utilize a wide spectrum of electronic devices that may all be integrated into a PDA. Having sophisticated technologists in the potential customer's environment of use builds empathy and reveals opportunities for product uses that the potential customer would not conceive and therefore would not identify in a more traditional market research data collection process. In this way the designers are able to identify elements of design that will make the product more "empathic" for customers [10,11].

Using this approach, HP Medical Products observers found and solved logistical problems with complex video directed surgery equipment. The doctors and nurses neither articulated nor perceived these problems. They were identified by having HP expert trainers present in operating rooms while doctors and technicians were completing their final qualifications to become certified in using specific HP equipment.[18]

Apple and HP both have developed creative and useful ways to package and communicate the rich qualitative data that can be gathered using a variety of these customer focused approaches. They each produced internal videos, "The Office of the Future" and the "Hospital of the Year 2000," respectively. The HP video is very realistically but futuristically staged around specific clinical cases. It shows these being managed by technologies that leading edge hospitals anticipate 5–10 years from now. The totality of the hospital scenes depicted comprise a holistic vision of HP technologies integrated and linked together in innovative ways to support the customer and the patient.

Through such processes of "being in the customer's shoes" or "having the customer on the team," leaders of successful, continually innovative companies focus on a common purpose and create a heightened sense of opportunity, urgency, and strategic targeting to stimulate their individual innovators, diverse specialists, and collaborating organizations [2].

Principle #5: Engage the Whole Organization

Leaders of successful, continually innovative companies create a sense of community across the whole organization. In these companies everyone connects with a common inspiring purpose, knows why they are working together, and participates in innovation as the fundamental way the company creates and brings basic new value to customers. Each person experiences his/her role as key to the company's innovative performance, and has found a way, with the organization's encouragement, to align the purpose of their life-work with their work-work.

The organization *as a whole* is called to high performance in product and service innovation. Invention and innovation activities are not delegated to one or only a few people, groups, or functional areas. This includes a highly visible encouraging role for top management—it is not necessary to be the technical inventor or innovator, but to be the innovation leaders calling for, recognizing, and acknowledging innovative results in others. For example at Club Med:

> . . . We spend more time going there, talking to the people, gathering the people together, than writing papers, sending memos, sending videos, taping things. This is against our culture. Our president is going to retire this year, he is 73 years old. Every single time he has a decision to make, he'd rather go there by plane, physically be there with his team, and share his idea or decision. And that's something that we all do in management. Jose spends all summer going to the resorts, to see how each and every team is matching with the need of the customers. I spend more time going and doing conferences about how the company is doing, our innovations, our new products, what we're going to do tomorrow, than just signing the catalogue with a nice little note saying this is what it's all about. So we believe very strongly in passing on those messages person to person.[19]

3M's abundance of rewards and recognition (big, little, daily, weekly, annually), such as "The Circle of Technical Excellence," reinforce that people are the most important resource for innovation. At the time we were visiting 3M they were discussing how to encourage "champions" of innovative people and their ideas. As with other desired behavior sets, they decided to establish a new award, publish and broadly communicate the criteria (the desired behaviors) so that peers could review the criteria and nominate awardees. Their published criteria will also motivate people to exhibit these behaviors and others to recognize and emulate them. In this way, 3M's expansive reward and recognition system develops people in new roles as well as rewards successful results.

HP, Apple, and Club Med are all explicit that good ideas come from sources internal and external to the enterprise. One of our Japanese study companies is adamant that the best ideas come from the middle of the organization, where people are closely linked with customers. "You need passion. Any top-driven product with too many people [politically] involved will not succeed."

Océ, 3M, Sony, and Apple all have formal, regular "meetings" (some electronic) to compile and interpret what is learned from customers and to make sure everyone in the organization knows what "we" know. They work at engaging the whole system. HP, Sony, and Apple also make video tapes for internal information, process, and idea sharing.

Principle #6: Never Forget the Individual

All of the interviewees in our study reflected a sense of adventure and wonder. These individuals provide curiosity, commitment, and courage. Carol Steiner reminds us of Heidegger's philosophy of human nature and its applicability to innovative enterprises: It is human nature to be *practically involved in a complex world* rather than rationally involved with a conceptually simplified world; it is human nature to be authentic, i.e., *unconventional*, at least some of the time; and it is human nature to be *collaborative* [8,16].

People spoke of feeling energized, saying, "Success breeds success." At Sony, inventive and innovative people were brought back from retirement and assignments in other (related) companies to interview with us. Sony people remarked that their partner company in some of the CD development work had no idea where to locate the highly inventive people who had been on that team in the 1970s and early '80s. The Japanese companies where we interviewed reward inventive and innovative people at two points in time. The first is when an idea is evaluated as worthy to try to implement. The second is when some commercial success based on that idea is realized.

Many interviewees described their experience as "fun." We were asked by the senior editor of a very distinguished business magazine, "Aren't managers concerned about 'burn out' for their employees if everyone is involved in such self-motivated innovative activities?" We replied that we observed this high energy and productivity for self and the company to be much preferable to the "snuff out" of people in organizations that do not encourage and support the intrinsic motivation of individuals to be creative and innovative. While it is never easy to bring a new idea to market, working for a company where new ideas are encouraged gives our interviewees a sense of direction and pos-

sibility, of having the freedom to think for themselves and for the future of their company.

Actively helping individuals create a linkage between their "work life" and longer term "life work" is a crucial step in generating an environment where innovation and high productivity flourish together. Management cannot control this process, or direct it in detail. But it can actively energize and support many linkages between individual purpose (intrinsic motivation) and organizational purpose (extrinsic motivation) [1].

At ICI Polyester (Imperial Chemicals Industries), despite the company's long history of past innovations, Global Melinex Chief Executive Jim Alles and a leadership group became convinced that its people's innovativeness was in danger of becoming limited. He assembled 75 people from top marketing, technical and leadership positions worldwide to develop an initiative to improve innovativeness. The core of this initiative was to begin discussing the goals that were meaningful to these individuals in their own lives and to take a fresh look at their organizations' work in terms of those goals. Alles' directive for the project became to "create an environment where people will work at what they are best at doing and what they like doing best."

During the meeting, as the participants—including executives such as Alles—began to discuss how their own life purposes might connect more fully to the innovation needs of ICI in its business climate, the participants soon noted a prevailing disconnection and use of "we-they" terms in describing critical organizational relationships. This discussion led to three important steps: (1) active management of these disconnecting attitudes during the meeting itself, (2) the establishment of a long-term process to open discussions to better connect life-work and work-work at all levels, including qualitative ways to establish progress, and (3) a commitment by all the executives to fully participate themselves in each step they asked employees to undertake.[20]

Principle #7: Tell and Embody Powerful and Purposeful Stories

The stories told in highly innovative organization support and reinforce the principles and practices of innovation. Innovative organizations *treasure* their identity and support their faith with an abundance of stories, teaching parables, myths, and legends that foster and align the myriad innovative activities in highly successful innovation organizations.

In our work with study companies and others since conducting our worldwide study, we have demonstrated that collecting and understanding the organization's *founding* stories is important as a first element of intervention and rejuvenation [3]. Often the founding stories of these companies include novel business concepts, not just technical inventions or service breakthroughs. Perhaps the story of the founding of Club Med is new to you, as it was to us:

> The man who created Club Med years ago was Gerard Blitz from Belgium. He was a swimming champion and a very good water polo player with the Belgian team at the Olympics. He had been contacted to help with the reintegration of former prisoners of war [after World War II]. People knew that they couldn't just take the prisoners of war back to their families right away. They had to help them start step-by-step, to give them back first good health, and some rest, and some food, and to have some medical assistance, and everything. They created this kind of place in Switzerland, in the mountains. And he was in charge of this entire development.
>
> After the war, and after this entire project, lasted a year or two, I think, he said, "This is a great system: to have the people all together and to have their families visiting them at the same time. It would be great to do this as a kind of vacation. To have a place where everybody was working and playing together: cleaning their rooms, and cooking, and playing volleyball, and relaxing, walking, having a little hike in the mountains. It would be nice if we could have a team taking care of all those things, and having people come and vacation with us." He wanted to go with his friends and live in a place where they could have fun.
>
> So they decided to create the first Club Med village in the Balearic Islands. It was very focused on the ocean, on spear fishing, they all were good swimmers. So they went there to have fun, do some sports, play volleyball, and play some music at night, and dance—like you organize a party in your country [the USA]. So this is what they decided to do. They created this first Club Med village, and this is how it all came about.[21]
>
> . . . And the timing was also perfect. In 1950— I don't know about in America, but in Europe very few people had been to the ocean. Few people had actually swum in the water. Vacations were for wealthy people. You had to have money to vacation at the ocean. I mean, if you had to have a house, or money go to the hotel and to pay for the hotel room, that was an expensive type of vacation. So most people either didn't take vacations, or when they did they would go to their families', which were most of the time in the countryside. They went camping a lot.
>
> In between luxury and nothing, there was a big vacation gap. This gap was filled by guys like Gerard Blitz, who also said, "Because we're going to share the cost of sports equipment, sport gear and so on, it's going to be less expensive for everybody. We're going to buy a water-ski boat, we're going to buy some spear-fishing equipment, some scuba tanks, and it's not going to cost too much because everybody's going to use it." And at the same time, the idea of this club was 'Let's share together!'[22]

At HP one member of the PR department views her job as collecting and shepherding the publication of HP stories in external media so that HP people will hear/read them, get ideas, and put them into practice. Being mentioned in a story of innovation, published or not, is a significant non-monetary reward in all these companies.

RICH AND PROFUSE NETWORKS OF RELATIONSHIPS

Two other conditions found in highly innovative organizations are: (1) total *interconnectivity* among remotely located individuals and (2) *permeable boundaries* across all systems and work groups, even those external to the enterprise itself. Knowledge, not necessarily people, is what needs to cross organizational boundaries. Hence, creating productive knowledge interactions among individuals is where the emphasis is needed [12].

The innovative productivity of a collaboration comes from the *differentness* of the individuals in a group, not their sameness. "Connectedness" requires wide ranging interests and a profuse network of interactions with others. This takes time, conscious reaching out to different people and skills, and a culture that rewards lateral participation.

ALIGNED PURPOSE

How do leaders within an enterprise deal with the highly diverse backgrounds, disaggregated organizations, and multiple locations generally found in our global innovative activities? Innovative collaboration demands strong alignment at three levels: (1) to the individual's own goals in life (for creativity and motivation), (2) to others in the enterprise (for collaboration), and (3) between individuals in the enterprise and the larger society (for creating value).

The process starts by focusing equally on the overarching organizational purpose that bonds members of the enterprise *and* on talented people's intrinsic desire to collaborate with others in order to create something truly new and of value in the world. Successful leaders tap into these intrinsic motivations and align extrinsic opportunities and incentives to encourage members of the enterprise. Both *clarity* of purpose and *alignment* of organizational and personal purpose are vital to sustaining the passion and commitment of a culture of innovation. Active alignment of purposes can lead to much more creative outcomes and a self renewing process both within the enterprise and between the enterprise and its external partners and stakeholders [9,13].

Such highly innovative systems cannot be planned and directed in detail, they cannot be "rolled out" but must be "co-created" and nurtured by all the members in the whole organization. The specific practices of other companies give us some ideas, but we must create *our* environment of in-

novation so that practices, that are true to the whole set of innovation principles, are also consistent with *our* corporate culture. Crafting a culture of innovation is a "story of connections" between one person and other employees; between employees and external partners; between employees and the organization's purpose.

Other members of the Polaroid research team included: Carole Uhrich (Co-Sponsor), Ed Chan, Patrick Flaherty, David Hinds, Suzanne Merritt, Phil Norris, Leigh Shanny, and Michael Zuraw. Dr. Barbara Perry, a cultural anthropologist who works with organizations, assisted the team. She provided a framework for our qualitative research and analysis. Carolyn J. Sullivan has been a skillful research associate throughout the project and the writing of our articles. We are grateful to the more than 140 interviewees who told us their personal stories of invention and innovation experiences.

REFERENCES

1. Amabile, Teresa. Motivational Synergy: Toward New Conceptualizations of Intrinsic and Extrinsic Motivation in the Workplace. *Human Resource Management Review* 3:3 (1993).
2. Bailetti, Antonio J. and Guild, Paul D. Designers' Impressions of Direct Contact Between Product Designers and Champions of Innovation. *Journal of Product Innovation Management* 8(2):91–103 (June 1991).
3. Buckler, Sheldon A. and Zien, Karen Anne. From Experience: The Spirituality of Innovation: Learning from Stories. *Journal of Product Innovation Management* I 3(5):391–405 (September 1996).
4. Collins, James and Porras, Jerry I. *Built to Last: Successful Habits of Visionary Companies*. New York: HarperCollins, 1994.
5. *Concept Engineering*. Document 71, Center for Quality Management. Boston, MA. September 1992.
6. Falk, Dennis Robert. *The Effects of Perspective Taking on Heterogeneous and Homogeneous Problem Solving Groups*. Ph.D. dissertation, University of Minnesota, December 1974.
7. Harryson, Sigvald J. From Experience: How Canon and Sony Drive Product Innovation Through Networking and Application-focused R&D. *Journal of Product Innovation Management* 14(4): (July 1997).
8. Heidegger, Martin. The Question Concerning Tech-

nology. In: *The Question Concerning Technology and Other Essays*, trans. William Lovitt. New York: Harper and Row, 1977. 3–35.

9. Hershock, Robert J., Cowman, Charles O. and Peters, Douglas. From Experience: Action Teams That Work. *Journal of Product Innovation Management* 11(2):95–104 (March 1994).

10. Leonard-Barton, Dorothy. Empathic Design and Experimental Modeling: Explorations into Really New Products. In: Report No. 94-124. Marketing Science Institute. Cambridge, MA, 1994, pp. 19–21.

11. Leonard-Barton, Dorothy. *Wellsprings of Knowledge*. Boston: Harvard Business School Press, 1995.

12. Nonaka, Ikujiro. The Knowledge-Creating Company. *Harvard Business Review* (November-December 1991). pp. 96–104.

13. Nonaka, Ikujiro and Yamanouchi, Teruo, Managing Innovation as a Self-Renewing Process. *Journal of Business Venturing* 4(5):299–315 (September 1989).

14. Quinn, James Brian, Baruch, Jordan and Zien, Karen Anne. Software-Based Innovation. *Sloan Management Review*. 37(4):11–24 (Summer 1996).

15. Schrage, Michael. *No More Teams! Mastering the Dynamics of Creative Collaboration*. New York: Currency Doubleday, 1995. Originally published as *Shared Minds*. Random House, Inc., 1990.

16. Steiner, Carol J. A Philosophy for Innovation: The Role of Unconventional Individuals in Innovation Success. *Journal of Product Innovation Management* 12(5):431–440 (November 1995).

17. Wilson, Edith. Product Definition: Assorted Techniques and Their Marketplace Impact. Institute of Electrical and Electronics Engineers: Engineering Management Conference. 1990, pp. 64–69.

ENDNOTES

1. Patterson, Marv. Formerly, Director, Research and Development Operations, Hewlett-Packard Company. Now, President, Innovation Resultants International. San Diego, CA. Interview at Hewlett-Packard, Palo Alto, CA. April 21, 1993.

2. Yamada, Toshiyuki. President of Sony's Research Center, was the first representative from our study companies to call after receiving our findings. Such an omnipresent pattern, he said, was very important to Sony at the juncture of designating their new leadership, the first from outside Sony.

3. Tanaka, Yoshinori, General Manager, R&D Coordination Department, corporate Technology Group. Interview at Sony Corporation, Tokyo, Japan. June 17, 1993.

4. Suzuki, Akira. Director, Tohokushinsha Film Corporation. Interview at Sony Corporation. Tokyo, Japan. June 17, 1993. Mr. Tsurushima was the CD Project Leader and Dr. Nakajima, now retired, was given an Imperial Award for his invention and technical leadership in the development of the CD.

5. Narayan, Sankar, Technical Director. Interview at 3M. Minneapolis, MN, May 11, 1993.

6. Author's consultations with Mary Sonnack, Division Quality Associate. Brookline, MA. August 1995 and 3M presentation: Denais, Claude, "Risky Decision Making: Project Prioritization at 3M." Conference: "Defining New Products and Services." Product Development Roundtable/Management Roundtable, Orlando, FL. April 2, 1996.

7. Nakajima Heitaro, Executive Technology Advisor. Interview at Sony Corporation. Tokyo, Japan, June 17, 1993.

8. Tanaka, Yoshinori, General Manager. R&D Coordination Department. Corporate Technology Group. Interview at Sony Corporation, Tokyo, Japan. June 17, 1993.

9. Dupont, Jean Pierre. Vice President, Marketing and Strategic Planning, Engineering System Division, Océ. Interview in Créteil Cedex, France. September 2, 1993.

10. Interviews at Xerox Pare. Palo Alto, CA. February 1994. Descriptions of electronic tools for collaborative innovation from Xerox Parc and others are also described in Schrage, Michael. *No More Teams! Mastering the Dynamics of Creative Collaboration*. Currency Doubleday. New York, N.Y. 1995.

11. Yajima Toshio, Senior Manager, Advanced Information Group. Interview at Toshiba Corporation. Tokyo, Japan. June 15, 1993.

12. Tanaka Yoshinori. General Manager, R&D Coordination Department, Corporate Technology Group. Interview at Sony Corporation, Tokyo, Japan. June 17, 1993.

13. Yajima Toshio, Senior Manager, Advanced Information Group. Interview at Toshiba Corporation. Tokyo, Japan. June 15, 1993.

14. Author's conversation with Edith Wilson. October 21, 1992.

15. HP was the second recipient of The Product Development Management Association's "Outstanding Corporate Innovator Award," presented in 1989. Much of the work in the Medical Group at that time was important to HP's application.

16. Halloran, Mark. Group R&D Manager. Interview at Hewlett-Packard Medical Products Group, Andover, MA. April 1991.

17. Professor Shoji Shiba and Ph.D. student Gary Burchill of MIT. Founding companies participated in a variety of ways. For example, Polaroid contributed a

New Product Delivery Support staff member for almost 2 years with additional research and administrative support to assist in the generation of the method and to develop training materials.

18. Halloran. Interview. April 1991.

19. Pello, Jean-Louis. Responsable de la Communication Interne, Club Méditerranée. Interview in Paris, France. September 3, 1993.

20. Alles, Jim. CEO, Global Melinex, Imperial Chemicals Industries.

21. Aliel, Jose. Directeur Animation, Club Méditerranée. Interview in Paris, France. September 3, 1993.

22. Pello, Jean-Louis. Responsable de la Communication Interne, Club Méditerranée. Interview in Paris, France. September 3, 1993.

VI

THE MANAGEMENT OF ORGANIZATIONAL PROCESSES FOR INNOVATION

13

Decision-Making Processes

41

Managing Organizational Deliberations in Nonroutine Work

WILLIAM A. PASMORE

For a few organizations that exist in stable environments, inflexible thinking isn't a problem; but for most, the ability to change and adapt is critical to success.[1] Flexible thinking is especially important in R&D, marketing, and management. In fact, flexible thinking is a good thing in any part of an organization in which creating or applying knowledge is an important element in task completion. In these parts of the organization, work is nonroutine in nature. That is, tasks are frequently changing and rarely performed in exactly the same way

twice. Routine tasks are repetitive, requiring little learning or creativity once they have been mastered. Nonroutine tasks demand some degree of both creativity and expertise; people may be professionally prepared to undertake nonroutine tasks, but still find their completion challenging or impossible. Medical researchers may be trained to look for a cure for the AIDS virus, for example, but still experience difficulty in finding one.

The fact that nonroutine work involves human thinking rather than machine performance makes it

Reprinted with permission from the author.

William A. Pasmore is Professor of Organizational Development at Case Western Reserve University's Weatherhead School of Management.

fascinatingly unpredictable and subject to all sorts of human foibles. In the wonderful and wacky world of nonroutine work, whether you like the person you are working with can make the difference between succeeding at a task or failing; doubling the number of people working on a project may just slow things down; and relying on what you think you know may only get in the way. It's as if nonroutine work takes place in some science fiction-inspired other dimension in which the laws of normal work don't hold. "Do it right the first time" and "zero defects" make no sense at all; goals can change in midstream; productivity is virtually immeasurable until the task is finished, and even then people argue about what has been accomplished. The history of product development is replete with examples of failures turned into successes and accidental inventions that appear brilliant only in retrospect. More than a few ulcers have resulted from people betting their careers on the outcomes of work that at its very core is inherently unpredictable and nonroutine.

R&D ORGANIZATIONS AS SOCIOTECHNICAL SYSTEMS

In traditional sociotechnical systems theory, it is held that organizations should be designed around their core production process; in R&D organizations, the core process is clearly that of producing ideas about products or processes. This work is knowledge-based; the availability and utilization of knowledge determine to a large extent the quality and viability of the ideas that are eventually conceived. Thus to improve R&D organizations we must understand the ways in which social and technical systems influence the development and utilization of knowledge. This emphasis on knowledge in R&D systems is made clearer when one contrasts the nature of work in routine (manufacturing) versus nonroutine (R&D) units. Table 41.1 shows some of the major differences between routine and nonroutine work.

Nonroutine work is, by definition, partially undefined at its commencement. How the work will

be performed and even by whom the work will be carried out is determined along the way as new information influences thinking about the task. Given the lack of prior experience regarding the current task, many decisions are made intuitively rather than on the basis of complete data and logic. Moreover, many methods can be employed to work toward objectives, which themselves are often in conflict: Consumers desire products that are relatively inexpensive, but they demand high quality as well; trade-offs are made between such things as product durability and weight as different materials are considered; working toward a major breakthrough is played against utilizing familiar development techniques. Finding the correct solutions to such trade-offs is complicated by the fact that information regarding consumer tastes, competitors' plans, and many other crucial inputs may be lacking and difficult to obtain. Even if such information is available, forecasting how long it will take to complete the development task or how much the product will ultimately cost to manufacture is often more like reading entrails than running a railroad. Thus, in nonroutine environments like R&D, the inputs that would allow logic, data, and science to guide plans for the use of resources in task completion are often missing. In the absence of this information, choosing the right path to the desired outcome is necessarily an intuitive and political process.

Partly because of this uncertain information and partly because of the nature of R&D itself, success is defined differently in R&D than in traditional manufacturing operations. In manufacturing, repetitive performance of the same tasks allows micromeasurement of routinely observable manufacturing parameters. Often these measures are referred to as indicators of manufacturing "efficiencies." The emphasis in manufacturing is on doing the same thing better and faster with less waste and higher quality—efficiency. In R&D, managers are more concerned with results than in process measures. Achieving the goal is of paramount importance, not how efficiently material or labor is utilized. Productivity is difficult to measure in the R&D environment. How does one calculate the

TABLE 41.1
Differences Between Routine and Nonroutine Systems

	Routine	Nonroutine
Nature of work	• Defined • Repetitive • One right way • Clear, shared goals • Information readily available • Forecasting helpful	• Undefined • Nonrepetitive • Many right ways • Multiple, competing goals • Information hard to obtain • Forecasting difficult
Nature of success	• Efficiency • Technical perfection • Productivity measurable • Physical technology • Standard information	• Effectiveness • Human perfection • Productivity unmeasurable • Knowledge technology • Nonstandard information
Nature of decision making	• Rules applicable • Experience counts • Authority-based • Complete operational specs • Authority by position	• Rules inhibiting • Experience may be irrelevant • Consensus-based • Incomplete operational specs • Authority by virtue of expertise
Nature of context	• Short time horizon • Stable environment • Predefined outcomes	• Long time horizon • Unstable environment • Emergent outcomes
Nature of variances	• Obvious	• Hidden

productivity of an engineer who makes only one major breakthrough in his career, but with it ensures the company's future?

Because of these difficulties in measuring R&D performance, surrogate measures are often applied. Project reports are written and submitted regularly as an indication that "we're working on it"; budgets are measured against what other companies are spending or what we spent last year; headcount is changed in the vain hope that the number of people in the lab somehow relates to the number of new ideas produced there. More appropriately, efforts are made to recruit and develop high-caliber people who have the right education or a good track record. In manufacturing, the emphasis is on finding the best machine for the job; in R&D, the focus is on getting the best people. It

follows that decisions about personnel are a key to the evolution of knowledge in R&D systems.

The way decisions are made in R&D organizations also differs from the approach in manufacturing organizations. In manufacturing, technical understanding of the process is nearly complete, leading to tight specifications that dictate how the system is to be operated. Past experience is directly applicable to future operations, so that those with experience are accorded greater authority in decision making. As these people rise in the system, the combination of their expertise and the tight operating specifications for the system lends strength to the hierarchy. Many decisions are made in a top-down fashion; little input is required from those below in order to reach decisions of acceptable quality. In R&D organizations, expertise is widely distributed throughout the organization. Past experience may not be applicable to the development of new products and in fact may inhibit creative thinking. There are few rules to govern the creative process, yet the pieces of complex products must somehow fit together before they are handed to manufacturing. This leads to consensus-based decision making in R&D, as specialists who often understand more about their work than their supervisors meet together over coffee and doughnuts to work out the bugs in promising designs.

Time is measured differently, as well. In manufacturing, timeclocks are punched and minutes of downtime are recorded. In R&D, quarters or years are the appropriate unit of time for most discussions. R&D specialists can go for months without receiving feedback on their performance and even longer before they know whether the product they have designed is a success.

By the time products reach manufacturing and facilities have been constructed for their fabrication, someone has already determined that there is at least a semistable consumer demand for the product. This means that the environment of the manufacturing organization is relatively stable compared to that of the R&D Unit. In R&D, projects are started (and stopped) on the basis of incomplete information about consumer demand for the product, projections of manufacturing costs, evidence of unintended environmental impacts, the departure of key scientists, or the whims of the R&D manager. Hence R&D units tend to be more subject to their environments than are routine operations. They are also more sensitive to the loss of knowledge that accompanies the departure of even a few individuals, if those figures have played key roles in the development of important products.

Finally, there is an important difference between manufacturing and R&D operations regarding the variances or problems that affect the quality or quantity of outputs from the system. In manufacturing, variances tend to be visible and repetitive. With some detective work, they can be tracked down and identified. This in turn makes them relatively easy to control. In R&D, variances are often hidden and sometimes go undetected un-

til the product is in manufacturing or even later. The variances are hidden because they tend to occur in people's heads when they are thinking about how to design the product. Incorrect assumptions, mistakes, guesses, misinformation, misunderstandings, and trade-offs are an integral part of virtually every complex product development process. Often it is difficult to trace these variances to their source or to catalogue them for future reference. It is even difficult to recognize variances after they have occurred, for they gradually become an accepted part of organizational life. Since delays, misunderstandings, and miscommunications are taken to be par for the course, they are consequently ignored when people discuss what can be done to improve the system. Understanding what variances look like in nonroutine environments and how they can be controlled requires new sociotechnical systems thinking.

TYPICAL VARIANCES IN R&D OPERATIONS

Since the core product of R&D organizations is knowledge, it follows that variances in R&D systems will be knowledge-related. Here are some quotes from a first-rate Research and Development organization in a very respected company that characterize the kinds of variances often encountered in R&D:

> Product development teams don't like to report bad news upward, sometimes they will even hide unfavorable data. When you raise issues, you get an arrow through the heart. Upper management doesn't want to hear the truth.

Does that affect the speed of product development when it happens? You bet!

> Managers are aligned with their *functional* accountabilities. They can go away from a meeting and have very different perspectives on what was agreed upon.

Does that affect cooperation among their people on projects later on? You bet!

We need to find ways to pass on decisions and earlier work. Usually, when a new person representing a new function comes onto the team, they want to challenge the work done previously.

Does it slow things down when you hand off work between departments? You bet!

People we hire think that working in industry is going to be just like graduate school. It isn't. It takes us years to teach them to work in teams instead of as individuals and to learn to ask the right questions instead of just answering a question raised by the professor.

Does socializing and integrating professionals into an organization's technical culture take time? You bet!

There's a great deal of uncertainty associated with early test data. Market research often tries to make it look like science—almost to fool people.

Do departmental boundaries create mistrust? You bet!

Turning over our ideas to marketing is like giving the keys to a race car to a three-year-old.

Do departmental boundaries get in the way of communication and teamwork? You bet!

A problem in the organization is that everybody can say "no" and nobody can say "yes."

Does the hierarchy sometimes slow things down? You bet!

ORGANIZING FOR KNOWLEDGE WORK

It used to be that the way to manage knowledge work was a combination of brute force and wishful thinking. The best people were hired or recruited to a large team. They were given space, equipment and resources; told to go to work; and then flogged regularly for not producing results fast enough. The nature of knowledge work, being inherently nonroutine, made for lots of flogging. As Albert Einstein said, "Science is a wonderful thing as long as one doesn't have to make a living at it."

But what other choice did the manager of knowledge workers have? He or she could get involved directly, but since the subordinates know more than the manager, the result of direct involvement was a lot of time spent in educating the boss just enough to screw up the decision-making on the project. Leaving people alone didn't feel very comfortable either, particularly with someone higher up the chain of command breathing down the manager's neck. Something must be done, action must be taken, people need leadership! They need management! So lead! The manager would lead, if he or she could just figure out how, when neither he nor others know exactly what is supposed to be done or how to do it.

In knowledge work organizations especially, the brightest people often do get promoted. Even so, being prompted doesn't make them the brightest person in everything, in every specialty, in every situation. But the pressure is on to demonstrate leadership and that means making decisions. Or, more precisely, making certain the right decisions are made. Bradford and Cohen state that, in order to feel competent, managers often feel compelled to live up to the following myths:[2]

- A good manager knows at all times what is going on in the department.

- A good manager should have more technical expertise than any subordinate.

- A good manager should be able to solve any problem that comes up.

- A good manager should be the primary person responsible for how the department is working.

The answer, it turns out, is neither to jump in nor to stand back. It's to make certain that people and teams are adequately prepared for the tasks they have been given, that the problem has been framed properly, and to help people organize themselves to answer the critical questions they have identified. What does organizing for knowledge work entail?

It's here that Calvin Pava's breakthrough in perceiving knowledge work as a series of deliberations, rather than discrete decisions, provides important clues.[3] Pava's argument, essentially, is that by the time people are ready to decide something, the knowledge work is over. Therefore, all of the attention that has been placed on organizational decision-making is in fact *mis*placed. The real knowledge work goes on long before the meeting at which the decision is made; and it tends to be a very messy, disorganized process, open to the full negative force of all the human foibles and social dynamics described earlier. By the time the decision is framed, the battle is lost; it's classic garbage in–garbage out.

Instead, Pava suggests that we trace the process of developing knowledge concerning key decisions over time, in all the various forums in which learning and discussion take place. Some of this work is individual in nature, some of it is collective. Some work is open for all to see, other work goes on in the minds of one or two people or behind closed doors in private sessions. To understand how knowledge is invented, processed, and eventually made available for decisions, we need to trace the evolution of knowledge about a topic through what Pava labeled *deliberations* to distinguish them from what we typically think of as meetings or decisions. Deliberations are times when people think about a topic, when they learn what they think they know, when they share their learnings with others, and when, ultimately, sense is made of that knowledge and concretized into a decision. If the quality of decisions can't be affected much once the knowledge work that goes into the decision is over, it follows that the point of optimal intervention in knowledge work is *while the knowledge is still being developed.* Organizing for knowledge work and *managing knowledge workers* means making the process by which people in the organization learn and influence each other much more explicit, and then working to improve that process over time, as its mechanisms are better understood and experiments with new ways of approaching knowledge tasks are conducted.

Typically, deliberations are spontaneous, un-planned events that we regard as practically inconsequential, yet they are the essence of knowledge work. Deliberately planning and managing them allows us to enhance the quality of knowledge work; it's almost like putting a power tool in the hands of someone who has been doing a task manually. Since knowledge work is made up of a series of deliberations, if we improve individual deliberations, we eventually improve the overall knowledge output and application. The secret is in recognizing what knowledge work is, and how to improve it at the microlevel. What have we learned from past examples of effective and ineffective deliberations? Well, some characteristics of effective deliberations are listed in the box entitled Characteristics of Effective Deliberations.

The first characteristic, knowledge highly developed and available, is determined primarily by effective preparation. The next two, knowledge utilized fully and without bias and apolitical discussion, have more to do with the dynamics that occur during the deliberation process. Having the right people present is a matter of deliberation planning. Often, we forget to include people with knowledge relevant to a problem or, worse, actively exclude them because they are not from our department, not at the proper "level" to be part of the discussion, or have opinions contrary to our own. Disruptive people are those who use their position power to influence deliberation outcomes even when they know nothing about the content being discussed. Planning and holding discussions at key choice points eliminate some of the problems associated with people who want to work alone and commit others to their decisions without the others' input. The next four items have to do with preparation and opportunity framing: goals clear and shared, challenging but realistic time frames, decision-making procedures clear, and appropriate attention to the external environment. The last item is a reminder that the purpose of planning deliberations is not to add bureaucracy or to slow things down, but rather to try to make certain that knowledge work is performed as effectively as it possibly can be.

Ineffective deliberations, by contrast, proceed

Characteristics of Effective Deliberations

- Knowledge highly developed and available
- Knowledge utilized fully and without bias
- Apolitical discussion of facts and alternatives
- People with most knowledge present
- Disruptive/inappropriate people absent
- Discussion held at key choice points
- Goals clear and shared
- Challenging but realistic time frames
- Decision making procedures clear
- Appropriate attention to external environment
- Minimum bureaucracy

without highly developed knowledge, resulting in the wrong decisions being made or decisions constantly being postponed. Knowledge that is available may be set aside, as in the *Challenger* incident, for a variety of social or political reasons. Parties to ineffective deliberations may enter dialogue with one another from an adversarial or defensive posture, leading to incomplete or misleading information in key decisions. Disruptive parties can undo months of careful work as powerful individuals inject opinions or push personal agendas. The avoidance of deliberations at key decision points makes it impossible for people with relevant knowledge to influence the course of knowledge work in a timely fashion. Goals not clearly set forth cause knowledge workers to diverge their thinking and then to become set in the paths they have chosen. Unrealistic time frames cause people to forego adequate preparation, rush through deliberations, exclude important external parties, and settle for lower quality than desired. Unclear procedures for making decisions leave the door open for one-on-one lobbying, subgroup influence, majority voting, and a host of other threats to quality decision-making. Inadequate attention to the external environment, as in the case of Xerox, creates self-

sealing truths, and an insistence on internal decisions being the only logical way to look at the world. Too much structure constrains flexible thinking and causes people to avoid the deliberation process altogether. Based on my consulting and research experiences, here's a more detailed list of the kinds of ineffective deliberations that I have been able to observe:

Lack of Knowledge

Perhaps the most obvious variance consists of a lack of knowledge needed to complete a task appropriately. Typically the lack of relevant knowledge is demonstrated in wrong decisions or in decisions being delayed or avoided altogether. When R&D professionals lack relevant knowledge, they are usually quite aware of it—making this one of the simpler variances to detect in nonroutine systems. Controlling this variance, however, may be difficult or even impossible. Sometimes the knowledge is too costly to develop or simply beyond our capacity to create; under these circumstances, risks are assessed, guesses are made, and consequences are measured somewhere down the road. But often this variance can be corrected by exposing the area of relevant uncertainty and involving people with appropriate expertise to answer the questions being posed.

Failure to Use Knowledge

More difficult to detect is the failure to utilize knowledge that already exists within the system. In common parlance, we generally refer to the results of this variance as "mistakes." In contrast to the preceding variance—lack of knowledge, which is viewed as beyond a person's control—the failure to use existing knowledge to make a proper decision is clearly attributable to human error. People sometimes "forget" to check with the appropriate authority on an issue or forget what they themselves know. Alternatively, they may fail to communicate what they know to others who need it because they assume that the others either would not understand it or "wouldn't value their input." Detecting this variance usually requires an analysis of past project performance, including interviews to discover why existing knowledge was not utilized when it should have been.

Lack of Cooperation

When parties who possess knowledge relevant to the tasks of others deliberately withhold it, incorrect decisions often result. People may withhold knowledge if they view their relationships with others as fundamentally competitive rather than cooperative, if they feel wronged by the other party in the past, if they are working toward opposing objectives, if they stand to lose political power by making the other look good, or if they feel that their help is not desired. Lack of cooperation is also apparent when one group shows reluctance to adopt or evaluate fairly the ideas generated by another group. People tend to work on their own ideas and to resist input from others. This tendency is so common in R&D that it has become known as the "Not Invented Here Syndrome." Its occurrence frequently results in delays or even barriers to innovation.

Many organizations are designed in ways that heighten the chances that competition and politics will overshadow cooperation and mutual support. Raises and promotions are limited in number; tall pyramidal hierarchies and clear functional boundaries interfere with natural tendencies by those at lower levels to help others in need; those who play political games are often the most highly rewarded; cooperation is given little recognition. All too often, those who cooperate feel as if they do so at their own peril and almost in violation of the wishes of their superiors.

Missing Parties in Key Discussions

One of the beliefs underlying hierarchical forms of organization is that the most qualified decision makers rise to the top. When this belief is not challenged during key discussions, those in positions of authority may trust themselves to make the choices without soliciting the opinion of others. A notorious example of this type of variance occurred at Morton Thiokol, where lower-level engineers tried to stop the launch of the *Challenger* but were excluded from key discussions by their superiors. This variance also occurs when colleagues in other functions are excluded from discussions in which their ideas would be crucial. The old "toss it over the transom" relationship between R&D and production units is a classic expression of this variance. Production people are explicitly excluded from R&D discussions that will affect production because R&D people tend to view production people as inflexible, anti-innovation, and too narrowly focused on efficiencies rather than the "pull" of the product in the marketplace.

Wrong Parties in Key Discussions

This variance is the obverse of the previous one. Just as excluding certain people can bias decisions, so can including others. Some people are invited to important meetings because they always have been invited, not because they possess technical information or experience relevant to the choices at hand. Nevertheless, because of their status or verbal dexterity, they are allowed to influence decisions in disastrous ways. Many organizations will never know the missed potential of ideas shot down by those who did not know what they were talking about. Instead, we have only reminders of a persistent few who risked their careers to prove that what others thought was impossible was indeed feasible.

No Key Discussions at All

These days, everyone hates meetings. Scientists and engineers may hate them more than most people. Given the choice of attending a meeting to have her ideas ripped apart by others or trusting her own professional judgment, a scientist may not hesitate to press on alone. The same is true of managers; if there is a chance that an idea may be shot down, many would prefer to make the decision on their own and live with the consequences than to take the time to have it reviewed. There may at times be powerful reasons for avoiding discussions of one's work with others. But when discussions are not held, knowledge cannot be transferred or developed among decision makers.

Lack of Goal Clarity

Even in single projects, efforts often proceed in multiple directions because goals are unclear.

Should the product be elegant or cheap? Should it be developed quickly or incorporate all the latest discoveries? Should it be considered high priority or low priority? In many projects, the answers to such questions are unclear. Even when goals are clearly stated at the beginning of a project, they may change as the project proceeds. Add to this the fact that most R&D organizations need to juggle multiple projects simultaneously, and the goals of the individual engineer or scientist become even more confused. In the face of this confusion, knowledge crucial to the project's success may not be developed at all, while knowledge that is in reality much less important is developed fully.

Another variance in this genre is goal displacement. In the search for certainty in an uncertain world, some managers of R&D organizations place more emphasis on paperwork being completed properly than on actual project results. When secondary tasks occupy the time needed to work on knowledge development, knowledge output suffers.

Time Frame Too Short or Too Long

The most common form of this variance occurs when insufficient time is allotted to develop critical knowledge on a project. Under the pressure of arbitrary deadlines, knowledge development is sacrificed for the sake of expediency, thereby precluding the expected level of success. At times, however, time frames can be too long. When project deadlines are set too far into the future, they may be regarded as nonexistent. When those trying to develop knowledge for the project approach others for assistance, they may find that attention is riveted on projects with tighter deadlines. In a world where there is never enough time to do everything one would like, short deadlines are used to focus attention on certain projects and away from others.

Procedures Unclear or Nonexistent

When procedures are not clearly stated for such things as project review sessions, the allocation of resources, or project selection, the informal system is allowed to drive decision making. Often this

works out well because the informal system may utilize knowledge more effectively than the formal system does. But the informal system also has a way of making decisions that optimize local benefits and short-term gains. Procedures help to specify who should be involved in making key decisions and what knowledge they should use. In the fight against all things bureaucratic, some R&D organizations neglect opportunities to ensure that sound decisions are made more consistently.

Inadequate Attention to External Environment

By now, the joke about what the customer requested versus what the engineer designed has become trite. Nevertheless, it remains true far too often. Contact with the customer, or with other relevant segments of the external environment, is frequently less than it should be. As a result, critical information never enters the design process or is overlooked once the process has begun.

Too Much Bureaucratic Structure

Earlier we noted that a lack of procedures may prevent the utilization of knowledge in allocating resources. At the same time, we recognize that adding too much structure to an R&D organization can kill it. Numerous levels of hierarchy, an overabundance of rules and regulations, a flood of trivial paperwork—such things sap energy that would otherwise be available for the creation and application of knowledge. Tom Peters has claimed repeatedly that "skunkworks" outperform major research labs, ostensibly because the research and organizational bureaucracies get in the way of people doing what they would naturally do.[4] But if one reads accounts of how highly successful skunk work groups actually function (such as the Data General Eagle computer group described by Kidder, 1981),[5] it becomes clear that a very definite structure is in place. The difference is that the structure is largely self-generated and appropriate to the challenge at hand. Structure is a liability when it interferes with knowledge generation and utilization. It adds little or no value to the outcome and serves mainly to calm the nerves of frightened administrators.

IMPROVING DELIBERATIONS

There are many ways to improve the quality of deliberations. Knowledge workers can form *learning action teams* which are the parallel of quality action teams in routine work settings. Learning action teams conduct regular brainstorming, analysis, and action planning about how to improve knowledge availability, knowledge utilization, and the flexibility of organizational thinking processes. Learning action teams can be formed within a single function, such as marketing, or across functions, such as marketing and R&D. Like quality action teams, learning action teams are authorized to try experiments to improve performance within limbs imposed by a steering committee. Proposals from learning action teams can range from simple changes in data gathering procedures to changes in organizational design.

Deliberation quality also can be improved by studying past deliberations and performing a critical review of what went well and what went poorly. Particularly if outsiders are invited to join in the discussion and ask innocent questions regarding the culture of deliberations, the learnings can be eye opening.

Alternately, deliberations can be improved by planning them in advance, making certain that the right people are involved, that they come prepared with the right knowledge, and that rules are agreed upon before interaction begins. Planning important deliberations in advance makes sense, particularly in these days of impossibly busy calendars and multiple responsibilities.

Ultimately, however, the most important intervention to improve deliberation quality is to redesign the organization so that effective deliberations take place naturally, rather than fighting against improper structural influences. The organization design for effective deliberating takes into account the need to constantly realign knowledge with authority, yet integrate the outcomes of separate deliberations. What kind of organization design allows both flexibility and integration? Clearly, the design can't be based upon hierarchy. Hierarchy is one means of achieving integration,

but it is not the only means and hierarchy destroys flexibility and almost guarantees that knowledge will not be aligned with authority. So the first principle is that the *design must be nonhierarchical.* The examples of ineffective deliberations illustrate a fundamental problem associated with knowledge work in hierarchical organizations: *The people with the most knowledge about something are often the least empowered to make decisions.* Even when they are invited to be part of the discussion, as in the case of Roger Boisjoly and the *Challenger* tragedy, their testimony is often discounted or their opinion isn't even asked. Why do we manage knowledge workers this way?

When I ask managers this question, their most frequent answer is, "There's too much at stake." But what's the logic here? That the more important the decision is, the less important it is to have knowledgeable people making it? When managers say, "There's too much at stake," what are they really saying? There's too much at stake for the company? Or there's too much at stake in their own careers? Or is it that they just don't trust others to make the same decisions they would make themselves? Don't knowledge workers also have a lot at stake? Don't their continued employment and well-being depend on the organization's finding the right answers to the tough questions?

The second principle is that people with knowledge must be capable of working with others who need that knowledge in order to complete their thinking about an interdependent task. Maximum freedom of movement is achieved by reducing role and boundary restrictions. So, the second principle is that *the organization should maximize freedom of movement.*

Third, because it is extremely important for everyone to know what knowledge exists in the system and how to access it, the organization must be holographic; that is, everyone needs to know a little about what is going on in the rest of the organization. So, the third principle is that *knowledge must be widely shared and easily accessible.*

Fourth, in order to provide integration and direction, there must be ways for people to agree on goals and strategies. Based on these goals and strategies, further choices regarding projects and teams can be made. Since it is important for people with knowledge to be involved in setting direction, the fourth principle is that *the organization must involve people with knowledge in goal setting and integration activities.*

Finally, because the development and utilization of knowledge require the development of people who are knowledgeable and highly committed, *the organization must be designed to encourage, support, and reward learning.* Together, these five principles point toward a form of organization which is quite different from the norm in use today.

REFERENCES

1. Pasmore, W. *Creating Strategic Change: Designing the Flexible, High Performing Organization.* New York: Wiley & Sons, 1994.
2. Bradford, D. and Cohen, A. *Managing for Excellence.* New York: Wiley & Sons, 1984.
3. Pava, C. *Managing New Office Technology: An Organizational Strategy.* New York: The Free Press, 1983.
4. Peters, T. *Liberation Management: Necessary Disorganization for the Nanosecond Nineties.* New York: Alfred Knopf, 1992.
5. Kidder, T. *Soul of a New Machine.* Boston: Little Brown, 1981.

42

Speed and Strategic Choice

How Managers Accelerate Decision Making

KATHLEEN M. EISENHARDT

Strategy making has changed. The carefully conducted industry analysis or the broad-ranging strategic plan is no longer a guarantee of success. The premium now is on moving fast and keeping pace. More than ever before, the best strategies are irrelevant if they take too long to formulate. Rather, especially where technical and competitive change are rapid, fast strategic decision making is essential.

But how do people make fast choices? Conventional wisdom suggests several strategies. One strategy is to skimp on analysis. That is, managers could look at limited information, consider only one or two alternatives, or gather data from only a few sources. Yes, this is fast. But the obvious problem is that such skimping seriously compromises the quality of the choice. A more subtle concern is whether decision makers will actually have enough confidence to make major choices with so little information and analysis to bolster their decisions.

Another strategy suggested by conventional wisdom is to limit conflict. Conflict drags out decision making, and the more powerful the combatants, the longer this conflict is likely to persist. So, minimizing conflict seems likely to accelerate choice. But, how can managers actually go about repressing real conflict among key executives? And if it can be suppressed, will managers support decisions if their opinions have been ignored? Most importantly, is it possible to make high-quality decisions without conflict? A wide spectrum of research indicates that high conflict yields more innovative, thorough decision making.

Conventional wisdom also suggests a third strategy to accelerate choices. Be an autocrat—make bold and rapid unilateral moves. While such a leader can move quickly, the era of swashbucklers is over. Such leaders often become isolated. This means poor information for making important choices, lack of support once those choices are made, and disabling anxiety which plagues people attempting to make major decisions alone.

Thus, at first glance, conventional wisdom offers strategies which appear to accelerate choices. More often, they are ineffective because they fail to deal with important realities. How can decision makers formulate high-quality choices when information and analysis are limited? How can they maintain a committed group if conflict and debate are suppressed? How can they avoid the natural tendency to procrastinate, especially when information is poor and stakes are high? The purpose of this article is to explore how managers actually do make fast, yet high-quality, strategic decisions.

RESEARCH BASE

The ideas described here partially rest on data which I collected with a colleague, Jay Bourgeois. Our motivation was to study how executives coped with strategic decision making in fast-moving,

Kathleen M. Eisenhardt is professor of Industrial Engineering and Engineering Management at Stanford University.

high-technology environments. Past research on choice processes had neglected such environments in favor of studying large bureaucracies in stable settings. We tracked decision-making processes in 12 microcomputer firms. We relied on extensive interviews with each member of the top management team of every firm, plus questionnaires, observations of group meetings, and various secondary data. I then followed up the microcomputer study with contacts with numerous Silicon Valley firms and their key executives.

These field data suggest striking differences in the pace of strategic decision making across firms. Some decision makers are fast. They make decisions on critical issues such as product innovations, strategic alliances, and strategic redirection within several months. Others are slow. They spend 6 months, more often 12 to 18 months, on decisions that the fast decision makers can execute in 2 to 4 months.

The ideas in this article are bolstered by recent psychological research. Writings on artificial intelligence and problem solving under time pressure are useful for understanding how people accelerate cognitive processing by more efficient use of information.[1] Work on the effects of emotion on decision making is also germane. Particularly relevant are findings about how individuals cope with anxiety and stress when dealing with high uncertainty. Finally, the psychological literature provides insight into how groups build cohesive interactions and ensure perceptions of equity when resolving conflict situations.[3] Thus, the combination of field study plus related psychological research led to the portrait of fast, yet high-quality, decision making that follows.

Overall, fast decision makers use simple, yet powerful tactics to accelerate choices (see Table 42.1). They maintain constant watch over real time operating information and rely on fast, comparative analysis of multiple alternatives to speed cognitive processing. They favor approaches to conflict resolution which are quick and yet maintain a cohesive group process. Lastly, their use of advice and integration of decisions and tactics creates the self-confidence needed to make a fast

choice, even when information is limited and stakes are high.

At the other end of the spectrum, slow decision makers become bogged down by the fruitless search for information, excessive development of alternatives, and paralysis in the face of conflict and uncertainty.

TRACKING REAL TIME INFORMATION

One of the myths of fast strategic decision making is that limiting information saves time. That is, slashing the amount of information, the number of information sources, and the depth of analysis accelerates choice. But, is this what fast decision makers actually do? The answer is "no." They do just the opposite. They use as much, and sometimes more, information than do their slower counterparts.

However, there is a crucial difference in the kind of information. Slow decision makers rely on planning and futuristic information. They spend time tracking the likely path of technologies, markets, or competitor actions, and then develop plans. In contrast, the fast decision makers look to real time information—that is, information about current operations and current environment which is reported with little or no time lag.

Fast decision makers gather real time information in several ways. One critical source is operational measures of internal performance. Fast decision makers typically examine a wide variety of operating measures on a monthly, weekly, and even daily basis. They prefer indicators such as bookings, backlog, margins, engineering milestones, cash, scrap, and work-in-process to more refined, account-based indicators such as profitability. The key finance manager often has a critical role in the fast decision-making organization. This executive typically is charged with providing this "constant pulse" of what is happening. In comparison with the classic big-company view, fast decision makers keep the key financial manager close to operations, and not in a watch-dog, staff role.

For less quantitative data, fast decision mak-

TABLE 42.1
Fast versus Slow Strategic Decision Making

Fast	Implications
• Track real time information on firm operations and the competitive environment	• Acts as a warning system to spot problems and opportunities early on • Builds a deep, intuitive grasp of the business
• Build multiple, simultaneous alternatives	• Permits quick, comparative analysis • Bolsters confidence that the best alternatives have been considered • Adds a fallback position
• Seek the advice of experienced counselors	• Emphasizes advice from the most useful managers • Provides a safe forum to experiment with ideas and options • Boosts confidence in the choice
• Use "consensus with qualification" to resolve conflicts	• Offers proactive conflict resolution which recognizes its inevitability in many situations • Is a popular approach which balances managers' desires to be heard with the need to make a choice
• Integrate the decision with other decisions and tactics	• "Actively" copes with the stress of choice when information is poor and stakes are high • Signals possible mismatches with other decisions and tactics in the future

Slow	Implications
• Focus on planning and futuristic information, keeping a loose grip on current operations and environment	• Can be time-consuming to develop • Quickly obsolete in fast changing situations
• Develop a single alternative, while moving to a second only if the first fails	• Obscures real preferences • Limits confidence that the best alternatives have been considered • Eliminates a fall back position
• Solicit advice hapharzardly or from less experienced counselors	• Fails to take best advantage of the experienced executives
• Use of consensus or deadlines to resolve conflicts	• Consensus is often wishful thinking in complex business decisions • Deadlines may not exist and so decisions can be postponed indefinitely
• Consider the decision as a single choice in isolation from other choices	• Increases stress by keeping the decision in the abstract • Risks the chance that the decision will conflict with other choices

ers emphasize frequent operational meetings—2 or 3 such meetings per week are not unusual. And, the intensity of such meetings is high, with each being a "must" on all calendars. Typically, these meet-ings cover "what's happening" with sales, engi-neering schedules, releases, or whatever comprises the critical operating information of the organiza-tion. But, these meetings are not limited to internal information. Fast decision makers also relay to each other external real time information such as new product introductions by competitors, competition at key accounts, and technical developments within the industry.

A good example is Zap Computers (a fictitious name for an actual firm). Zap's top management team is known for rapid decisions. Typically, they execute, in 2 or 3 months, decisions which elsewhere often drag on for a year or more. How do they do it? The popular press highlights their "laid back" and "fun-loving" California culture. A closer inspection reveals slavish dedication to real time information.

Zap executives claim to "over-MBA it," to "measure everything." They come close. Zap executives review bookings, scrap, inventory, cash flow, and engineering milestones on a weekly and sometimes daily basis. The monthly review is more comprehensive, emphasizing ratios such as revenue per employee and margins. Firm executives maintain fixed targets for margins and key expense categories. These targets themselves are not so unusual, but what is striking is the number of people who can recite them. Zap executives also attend three regularly scheduled operations meetings each week. One is a staff meeting for general topics while another is for products and the third is a review of engineering schedules. The tone is emotional, intense, and vocal.

The Zap top management team plays an important role in gathering real time data. The VP Finance, described as "having a good understanding of business" and years of experience, oversees more than just the usual treasury and accounting functions. He is responsible for the financial model of the firm which is run at least weekly. The model itself is simple, but it allows Zap executives to translate possible decisions into their impact on basic operating results. His group also provides updated operational data, usually available on a daily basis.

Other executives are also essential to the real time information network at Zap. For example, the VP Marketing is charged with tracking the moves of competitors as they occur. This means constant phone calls and frequent travel. The VP R&D also

works the phone to maintain a complex web of university and business contacts which keep him cognizant of the latest technical developments.

Zap executives also favor electronic mail or face-to-face meetings. As they described it, "We e-mail constantly." They are also frequently in and out of each other's offices. On the other hand, Zap executives avoid time delayed media such as memos. They are seen as too slow and too dated. Overall, dedication to real time information gives Zap executives an extraordinary grasp of the details of their business.

In contrast, slow decision makers have a much looser grip on current operations and the competitive environment. For example, decision makers in the slower firms track few operational measures and they review them less frequently than do their faster counterparts. Their emphasis is on future, not current, information.

The lack of real time information is also evident in the use of group meetings. In the microcomputer study, slow decision makers had few, if any, weekly operations meetings. Several firms did not have a VP Finance or else relied on a less experienced executive. For example, one used an ex-engineer, claimed by all to be "weak" in financial matters. At another firm, the VP Finance had left the firm and there were no current plans to replace him.

Instead of real time data, these executives prefer planning information. For example, executives at one corporation spent close to a year doing a technology study of various operating systems for microprocessors as a prelude to a new product decision. At another firm, executives responded to a performance decline by spending 6 months developing a technology forecast for the industry. The elapsed decision-making time in both firms was over a year.

Why does real time information speed decision making? An obvious reason is that the continual tracking allows managers to spot opportunities and problems sooner. Real time information acts as an early warning system so that managers can respond before situations become too problematic. When crises do arise, such managers can go right to the problem, rather than groping about for relevant information.

However, a more subtle explanation comes from artificial intelligence. Research on the development of intuition suggests that the basis of intuition is experience.[4] For example, chess players develop their inuition by playing chess over and over again. This repeated practice allows the chess player to play the game using what lay people term "intuition." In fact, intuitive chess players have actually learned to process information in patterns or blocks. Because they recognize and manipulate information in blocks, they can process information much faster than others who think only in single items of information.

Consistent with this view, managers who track real time information are actually developing their intuition. Aided by intuition, they can then react quickly and accurately to changing events. Indeed, in the microcomputer study, the executives who were most attuned to real time data were also those most described as intuitive. For example, one executive was described as a "numbers" person and claimed to "over-MBA it." Yet, he was also described by his colleagues as "intuitive," "a lateral thinker," and as having "the best sense of everything in the business." Another also claimed to be a "numbers guy." His VP Finance praised "the quality of his understanding" and frequent use of operating data. This executive was also described as having "an immense instinctive feel" and a superior "grasp of the business."

In contrast, slow strategic decision makers emphasize planning and forecasting information. They look to the future and attempt to predict it. Or, they hope that, by waiting, the future will become clear. Yet, their faster counterparts maintain that this is foolish. They claim that extensive planning wastes time. Why? It's difficult to predict what will happen and impossible to predict who will do it and when. As one fast-moving executive claimed, "No company can know how things will evolve. You can only monitor the outside world and direct the evolving strategy at what you see." Overall, it appears that real time information—which gives executives an intimate knowledge of their business—speeds choice, but planning information—which attempts to predict the future—does not.

BUILDING MULTIPLE, SIMULTANEOUS ALTERNATIVES

A second myth is that fast decision makers save time by focusing on only one or two alternatives. The underlying logic is that fewer alternatives are faster to analyze than more. But, in fact, fast decision makers do the reverse. They explicitly search for and debate multiple alternatives, often working several options at once. At the extreme, some fast decision makers will support alternatives that they oppose if doing so furthers debate, and they will even introduce alternatives which they do not actually support.

A good example of multiple alternatives occurred at one of the microcomputer firms. Here, a new CEO was faced with improving the lacklustre performance of the firm. New products were slow in coming out and there was pressure from investors to do something. This new CEO launched a fact-gathering exercise. Almost simultaneously firm executives began to develop a rough set of alternatives. As the fact-gathering continued, so did the shaping of alternatives. In less than 2 months, the executives had developed 4 options. They considered selling some of their technology—there were willing buyers, especially from overseas. They also considered a major strategic redirection of the firm which would involve using the base technology to enter a new market. A third option involved various tactical changes in the form of redeployment of some engineering resources and adjustments to the marketing approach. The final option was extreme—liquidation of the firm.

Executives at the firm admitted that this decision strategy was ambiguous and complicated. Why did they do it? As one executive claimed, they and the CEO, in particular, liked to have multiple options, "a larger set of options than most people do." There was a preference for "working a multiple array of possibilities instead of just a couple."

At another firm, the problem was cash flow. The business was prospering, but the cash flow was not keeping pace. These executives also developed multiple alternatives. One executive negotiated with banks to extend credit lines. Others developed several strategic alliance alternatives with both U.S. and foreign firms. A third set of executives planned a major equity financing. With rough details of each option in hand, firm executives chose the strategic alliance for flexibility and marketing reasons. When the first choice alliance partner backed out, firm executives quickly cut a deal with the second. The credit and equity plans were waiting on the shelf, if the alliance option had failed.

In contrast, surprisingly, slow decision makers work with fewer, not more, alternatives. They typically develop and analyze a single alternative, and only seriously consider other alternatives if the first becomes infeasible. Thus, slow decision makers favor a highly sequential approach to alternatives—and, one which emphasizes depth of analysis over breadth of options.

A new product decision at one of the microcomputer firms illustrates this process. These executives wanted to develop a new product which made greater use of VLSI technology. The rationale was that increased integration would lower product costs. Firm executives explored the possibility of in-house development for several months. When they concluded that the firm lacked sufficient expertise, they then migrated to a second alternative, a strategic alliance with a major U.S. firm. They spent about six months getting to know the personnel of their could-be partner and negotiating the terms of the deal. However, the deal fell through when the two parties could not reach a final agreement. At this point, firm executives then had to search several months for another partner. After further delay, they eventually closed the deal with this second firm.

This same pattern of single, sequential alternatives is characteristic of many slow decisions. At another firm, executives were also interested in developing a new product. They too explored the in-house route to a new product for several months. When they determined that in-house development would be too slow, they belatedly went outside for a product source. It took several more months to locate a suitable partner. At another firm, executives noted increasing competition in their marketplace. They spent almost a year deciding whether they needed a new strategy. Only after they decided that the old strategy was no longer workable did they then seriously consider what that new strategy should be.

Why are multiple, simultaneous alternatives fast? One reason is that multiple alternatives (at least, in the range of 3 to 5) are faster to analyze than 1 or 2. Why? The reason is comparison. Comparative analysis sharpens preferences. For example, car buyers often find it difficult to understand their preferences in the abstract. Rather, actually driving cars and then comparing across cars helps prospective buyers to decide whether they prefer a leather or plush interior, a standard or rally suspension package, one make over another, and so forth. Comparative analysis is also fast because it allows decision makers to use rankings to assess alternatives. The superiority of an alternative is often apparent in comparison, even if its superiority cannot be readily quantified.

Multiple alternatives also speed decision making because they are confidence building. With multiple alternatives, decision makers are more likely to feel that they have not missed a superior alternative. Again, to use the car buying example, it is difficult to buy a car without seeing others because buyers often cannot overcome the feeling that they may be missing something better elsewhere.

Finally, multiple alternatives are fast because they provide a fall back position. When one option falls through, decision makers can quickly move on to a second. Although a first choice option sometimes prevails, situations can change rapidly and dramatically. So, the odds are good that adjustments will be needed. In the illustrations above, the firms which pursued multiple alternatives simultaneously had fall back positions when one or more options proved infeasible. In contrast, the firms pursuing sequential alternatives lost time because they waited until an option failed before looking for a new one.

Overall, there is a fundamental difference in how fast and slow decision makers treat alternatives. Fast decision makers develop multiple alternatives, but analyze them rapidly. They rely primarily on quick, comparative analysis, which reveals relative rankings and sharpens preferences. Theirs is a "breadth-not-depth" strategy. Recent laboratory research indicates that this is the efficient cognitive processing approach when the decision maker is under time pressure.[5]

In contrast, slow decision makers emphasize depth of analysis. They analyze few alternatives, but do so in greater depth. And so, they often conduct a similar amount of analysis, but without gaining the confidence in their choice that multiple alternatives bring and without gaining the advantage of fall back positions.

RELYING ON THE ADVICE OF COUNSELORS

A third critical aspect of fast strategic decision making is the judicious use of advice. Most fast decision makers rely on a two-tier advice process in which all executives offer some advice, but the key decision maker focuses on the advice provided by one or two of the most experienced executives in the group, who are termed "counselors." By contrast, slow decision makers typically have no one in this counselor role.

What do these counselors do? Typically, they work in the background advising the key decision maker about a wide range of issues. They also serve as an early and confidential sounding board for ideas. For example, at one firm, the counselor played an important role in a new product decision. The situation was triggered by an unexpected new product introduction by an important competitor. In private, the counselor alerted the CEO to the imminent introduction. The two then conferred, with the counselor helping the CEO to shape and test alternatives. Even after the CEO brought the issue to the attention of the entire top management team, the counselor continued to work behind the scenes with the CEO. As the CEO described, "Our interaction is more general than just sales . . . When I talk with Joe it's often about company issues."

A striking feature of the counselor role is the consistent demographic profile of these individuals. The counselor is typically an older and more experienced person, who is recognized as "savvy" or "street smart" by colleagues. For example, the counselors at one firm were 10 to 20 years older than the rest of the top management team. One of these executives had been a senior manager at a major, international firm. He was described as "the best manager" on the top management team. The other had been a senior executive at two important firms in the industry. He was credited with being the "most knowledgeable about the outside world." Counselors are also frequently on a career plateau, with their aspirations no longer centering on the fast-track to the top. Rather, they relish the personal challenge. As one counselor claimed, "It's fun to build an organization again."

What do fast decision makers do when they have no colleague who fits the counselor profile? In one such case, a consultant was hired to play the role. This consultant had extensive industry contacts, had been a senior executive at two other firms, and had known the CEO for many years. This executive was credited with an important advisory role to the CEO as well as to several other executives.

In contrast, slow strategic decision makers typically have no executive who acts as a counselor. These decision makers usually do not develop any kind of a close, advisory relationship with another colleague. Or, if they do, that colleague often is a poor choice. For example, the counselor to a CEO in a slow decision making firm was considered to be "bright," but "young." He was in his early 30s and had only functional staff experience. Given his modest background, his ability to be an effective counselor was limited.

Why do experienced counselors accelerate choices? Clearly, one reason is that these individuals can provide high-quality advice to decision makers more readily than less experienced colleagues. They have simply seen more and done more. Not surprisingly, they can usually assess situations more rapidly and offer better advice than less experienced people.

Second, they are excellent sounding boards. They combine strong experience with a trustworthiness that comes with limited personal ambition and often long acquaintance with the key decision maker. These are people who understand discretion and the subtle exercise of power.

Perhaps most importantly, counselors can boost the confidence of decision makers to decide. One of the highest barriers to fast decision making is anxiety. Big stakes decision making with high uncertainty is stressful, and so it is extraordinarily tempting simply to procrastinate. However, conversations with an experienced confidante can counteract this tendency to delay by bolstering decision makers' confidence to make difficult choices.

RESOLVING CONFLICT

Another myth of fast strategic decision making is that conflict slows down the pace of choice. Obviously, conflict can have this effect. But, fast decision makers know how to gain the advantages of conflict without extensive delays in their decision process. The key is conflict resolution.

Fast decision makers typically use a two step process, termed "consensus with qualification" by one executive, to resolve deadlocks among individuals. This process works as follows. First, executives talk over an issue and attempt to gain consensus. If consensus occurs, the choice is made. However, if consensus is not forthcoming, the key manager and most relevant functional head make the choice, guided by the input from the rest of the group. As one executive told us, "Most of the time we reach consensus, but if not Randy makes the choice."

A description of decision making at Forefront (a fictitious name for an actual firm) serves an an illustration. Forefront was faced with a major challenge in its principal market from an important competitor. This firm had unexpectedly announced a new machine which appeared to challenge Forefront's leadership in its primary area of business. Forefront executives confronted the problem, and substantial disagreement was apparent. Several executives wanted to shift R&D resources to counter this competitive move. The price was diverting sig-

nificant engineering talent from a more innovative product currently in design. Others argued that a simple extension of an existing product was appropriate. Under this plan, Forefront would simply repackage an existing product with a few new features from its stable of modest technical improvements. A third set of executives perceived that the threat was not all that important and that Forefront should continue with current plans, making no response.

The team held a series of meetings over several weeks. Consensus was not in the cards. Given the stalemate, the CEO and his marketing VP simply made the choice. Not all agreed with their selection, but everyone had a voice in the process. As the CEO claimed, "The functional heads do the talking . . . I pull the trigger."

The approach to conflict used by fast decision makers, such as the executives at Forefront, constrasts markedly with that used by the slow decision makers. Sometimes slow decision makers wait for consensus. They forage for an option which satisfies everyone. However, since conflict is common in decision making, the search for consensus often drags on for months. For example, the decision makers at one firm debated the specifications of a new product for about a year. Finally, consensus came—after several executives who opposed one of the options left the firm.

Sometimes slow decision makers wait for deadlines, which then energize them to make a choice. For example, the annual meeting triggered a decision at one firm. The CEO had worked for almost a year on a proposal to develop a new market. Others in the group felt that such a project would stretch sales and engineering resources too much. The CEO was unwilling to do nothing and yet also unwilling to decide. So, he continually refined his proposal in the hopes of gaining others' agreement. What was the result? Each refinement improved the proposal, but also stiffened the opposition. This pattern might have dragged on indefinitely except for the annual meeting. Frustrated by repeated rejections and facing the impending deadline, the CEO came up with a new alternative and as he claimed, "shoved it down their throats."

Why is consensus with qualification rapid? One reason is that it takes a realistic view of conflict. Conflict is seen as natural, valuable, and almost always inevitable. Therefore, fast decision makers recognize that choices must be made even if there is disagreement. The other reason that consensus with qualification is rapid is its popularity. Managers like it. Most people want a voice in the decision-making process, but are willing to accept that their opinions may not prevail. Consensus with qualification gives people this voice, and goes one better by giving them added influence when the choice particularly affects their part of the organization.

In constrast, slow decision makers are stymied by conflict. They delay in the hopes that uncertainty will magically become certain. Or, they look for consensus. But unfortunately, consensus is often wishful thinking in most complex business situations. People are likely to have differing opinions, expecially regarding big and important choices. Although consensus sometimes emerges, often it does not. Rather, as one executive described, "We found that operating by consensus essentially gave everyone veto power. There was no structure. Nothing was accomplished." Overall, many managers dislike a strictly consensual approach to choice and prefer simply to "get on with it."

INTEGRATING DECISIONS AND TACTICS

The final key to fast decision making is the integration of the focal decision with other key choices and tactical plans. In effect, fast strategic decision makers fit any single decision into a web of interlocking choices. This decision integration does not imply any sort of elaborate planning. In fact, frequently there is no written plan. Rather, fast decision makers maintain a cognitive map which they can readily describe or sketch on a piece of paper. At most, fast decision makers stitch together a 5 to 10 page document describing the relationship among choices and tactics.

A good example of decision integration occurred at Triumph (a fictitious name for an actual firm). The decision began with the arrival of a new CEO. The firm was struggling in the wake of the highly mercurial, former CEO. The new CEO spent several weeks learning about people and products. He and other executives also began developing options for how to energize the firm. In the process of defining and refining these options, the executive group also decided on the specifications for a new product, scheduled three new product releases, reprioritized engineering assignments, and rebudgeted the firm for the year. All of this occurred in about 2 months.

These decisions contrast with a similar one executed by a slower team. These executives also faced deteriorating financial performance. However, their response was a technology forecasting project. This project was completed several months later. Firm executives then spent the next several months debating whether to change the firm strategy or to execute the existing approach more effectively. Key firm executives were seriously split on the issue. Finally, after several opponents left the firm and the financial situation had deteriorated to the point that the existing strategy could no longer be salvaged, the CEO chose to alter the strategy. Only then did the executives think about what the new strategy would be. Five more months passed before the new strategic direction was set. And, there were still tactical plans such as engineering assignments to be made.

The contrast with the Triumph case is striking. Triumph executives made a similar decision on strategic redirection—plus they chose a new product, scheduled 3 new product releases, reassigned engineering priorities and rebudgeted the firm—in less than 2 months, compared to the 18-month period of the second company.

Why does decision integration accelerate decision making? On the surface, it appears time-consuming to link together decisions and tactics. However, this surface view neglects the value of decision integration for building the confidence of decision makers. Anxiety is a major impediment to fast choice. Making choices, when information is poor and stakes are high, is paralyzing. The psychological literature indicates that a key to efficacy in such stressful situations is proactive and structuring behavior—that is, formulation of concrete action steps to structure one's unstable world.[6] Such "active coping" enhances feelings of competence and control which, in turn, boost the confidence to decide. Consistent with this view, managers who integrate decisions and tactics are actually engaging in active coping. Aided by enhanced feelings of competence and control, they can make choices more quickly and confidently in high stress and information poor situations.

Secondly, decision integration does more than simply give a psychological illusion of control. The process also provides better understanding of alternatives and potential conflicts with other decisions. By linking together decisions and tactics at the outset, managers can avoid many of the delays that occur when executing one action has unanticipated consequences for other actions.

In contrast, slow decision makers treat each decision as a separate event, detached from other major choices and from tactics of implementation. In effect, they employ a linear view of decision making. Unfortunately, such an approach does nothing to diminish anxiety. Decisions remain in the abstract, unattached to other activities within the organization. Evidence from several firms confirms that anxiety looms large for slow decision makers. For example, one slow decision maker worried that "we don't know if we have the confidence to do it." A second executive lamented, "Maybe we saw too much mystery. Maybe we needed more gut." His conclusion was simple: "You don't know any more even though you wait." Overall, slow decision makers see decisions as very large, discrete, and anxiety-provoking events whereas fast decision makers see individual decisions as a smaller part of an overarching pattern of choices.

MANAGERIAL IMPLICATIONS

Strategy making is changing. Strategies that may have been viable in the past are no longer feasible if they take too long to formulate. The field data echo this point. For example, the microcomputer study revealed that fast strategic decision makers led either high performing organizations or organizations that achieved performance turnarounds.[7] These fast decision makers also explicitly linked the speed of their strategic decision making to success. They claimed: "you have to keep up with the train," "you've got to catch the big opportunities," "simply do *something*," and so on.

In contrast, the slow strategic decision makers managed mediocre organizations, some of which have since failed. These decision makers usually recognized that speed was important, but they did not understand how to be fast. As a result, they missed opportunities and lost the learning that comes with making frequent choices. As one described, "The company wound up doing a random walk. Our products were too late and they were too expensive."

How do managers actually make fast, yet high-quality, strategic decisions? This article has identified five key tactics for accelerating decision making. They involve simple behaviors which decision makers in a variety of settings can use. To summarize:

- Before decisions arise, track real time information to develop a deep and intuitive grasp of the business. Focus on both operating parameters and critical environmental variables to hone your intuition.

- During the decision process, immediately begin to build multiple alternatives using your intuitive grasp of the business. Be certain to analyze the alternatives quickly and in comparison with one another. Possibly even begin execution of several before settling on a final choice.

- Ask everyone for advice, but depend on one or two counselors. Be selective in your choice of counselors. Look for savvy, trustworthy, and discreet colleagues.

- When it's time to decide, involve everyone. Try for consensus. But, if it doesn't emerge, don't delay. Make the choice yourself or better yet, with the others most affected by the decision. Delaying won't make you popular and won't make you fast.

- Ensure that you have integrated your choice with other decisions and tactical moves. You'll feel more confident and you will have avoided many of the headaches of mismatched decisions down the road.

CONCLUSION

Previous scholarly research on decision making has ignored speed in favor of topics such as the breakdown of rationality and the difficulty of identifying goals. It has also emphasized the study of large bureaucracies in stable settings, rather than the high velocity environments which many decision makers actually face.

However, most managers have recognized that speed matters. A slow strategy is as ineffective as the wrong strategy. So, fast strategic decision making has emerged as a crucial competitive weapon. But, knowing how to be fast is difficult. The process involves accelerating information processing, building up the confidence to decide, and yet maintaining the cohesiveness of the decision-making group. Should managers learn how to be fast? One executive summarized the prevailing reality in many industries: "No advantage is long-term because our industry isn't static. The only competitve advantage is in moving quickly."

REFERENCES

1. J. Hayes, *The Complete Problem Solvers* (Philadelphia, PA: Franklin Press, 1981); H. Simon, *"Making Management Decisions,"* Academy of Management Executive (1987); J. Payne, J. Bettman, and E. Johnson, "Adaptive Strategy Selection in Decision Making," *Journal of Experimental Psychology* (1988).
2. R. Gal and R. Lazarus, "The Role of Activity in Anticipating Stressful Situations," *Journal of Human Stress* (1975); E. Langer, "Illusion of Control," *Journal of Personality and Social Psychology* (1975).
3. P. C. Earley and E. A. Lind, "Procedural Justice and Participation in Task Selection: The Role of Control in Mediating Justice Judgments," *Journal of Personality and Social Psychology* (1987).
4. Simon, op. cit.
5. Payne et al., op. cit.
6. Gal and Lazarus, op. cit.
7. K. Eisenhardt, "Making Fast Strategic Decisions in High Velocity Environments," *Academy of Management Journal,* 28/3 (1989).

43

Vasa Syndrome
Insights from a Seventeenth-Century New Product Disaster

ERIC H. KESSLER, PAUL E. BIERLY III,
AND SHANTHI GOPALAKRISHNAN

The Swedish ship *Vasa* was one of the most spectacular warships ever built. On its maiden voyage in August of 1628, after going less than one mile, the vessel keeled over and sank 110 feet to the bottom of the Stockholm harbor. Fifty crewmembers went down with the ship. It was truly a disaster—and an excellent example of a failure in the new-product development process. In this article, we show how insights gleaned from the *Vasa* incident are relevant to contemporary organizations. Seven potential problems in new-product development are examined. Together, these problems comprise the *Vasa* syndrome—a complex set of challenges that can ultimately overwhelm an organization's capabilities. Each problem provides an opportunity to develop managerial competencies in understanding these problem areas, linking these problems to failures described in the *Vasa* case and contemporary organizations, and determining how to avoid or minimize these problems in the new-product development process. The *Vasa* case and examples from contemporary organizations demonstrate how history continues to repeat itself in the process of new-product development, and we provide guidelines on how to avoid falling prey to the *Vasa* Syndrome.

On August 10, 1628, the Swedish Navy launched one of the most spectacular warships ever built: the *Vasa*. This magnificent ship, which cost over 5 percent of Sweden's GNP, was built to symbolize Swedish strength and beauty and to put fear into the hearts of Sweden's enemies. Lavishly equipped with 700 sculptures and ornaments, including carvings of lion heads, angels, mermaids, and grinning devils by some of the best woodcarvers in the world, the *Vasa* was considered to be the mightiest vessel of its time, armed with 64 heavy guns and 300 soldiers. After sailing less than a mile, the *Vasa* keeled over and sank 110 feet to the bottom of the Stockholm harbor. Fifty crewmen went down with the ship.

Because the failures of the past are often ignored, researchers have found it valuable to draw from historical events such as Admiral Nelson's strategy at Trafalgar and Napoleon's retreats from Russia to illustrate critical lessons for today's managers.[1] In this article, we use the story of the *Vasa* to vividly and dramatically illustrate common failures that continue to haunt new-product development. We discuss seven key problems whose root cause is an over-ambitious approach to developing new products that exceeds an organization's resources and capabilities. We call this condition the *Vasa* Syndrome, and provide numerous examples from modern-day organizations to illustrate that,

Reprinted with permission of *Academy of Management Executive*, "*Vasa* Syndrome: Insights from a 17th Century New Product Disaster," Eric Kessler, Paul Bierly III, and Shanthi Gopalakrishnan, vol. 15, no. 3, 2001.

Eric H. Kessler is a Professor of Management at Pace University's School of Business. *Paul E. Bierly III* is Professor of Management at James Madison University's College of Business. *Shanthi Gopalakrishnan* is a Professor of Management at New Jersey Institute of Technology.

even though more than 300 years have passed since the *Vasa* sank, the underlying problems are still being repeated.

THE *VASA* STORY IN BRIEF[2]

The *Vasa* disaster occurred at a tumultuous time when Sweden was engaged in naval battles with Denmark, Russia, and Poland. In 1625, 10 Swedish navy ships ran aground while on patrol in the Bay of Riga. Because of this loss, development of the *Vasa* was accelerated. Discovering that Denmark planned to build a larger and better-armed ship than the *Vasa*, King Gustavus Adolphus ordered specifications altered to add a second gundeck and more cannons than originally planned. These design changes caused the *Vasa* to far exceed the size that could be supported by its ballast. Additionally, the death of master shipbuilder Henrik Hybertson left the ship's construction under the supervision of a weaker manager.

In the summer of 1628, a stability test was conducted by Admiral Klas Fleming and Captain Sofring Hansson. Thirty men ran from one side of the ship to the other. After the third crossing, the ship was heeling so violently that the test was halted. Even so, Fleming decided not to postpone the commissioning of the ship, commenting that "the shipbuilder has built ships before and he should not be worried." Less than a month later, the *Vasa* put to sea. Hannson sailed the ship with open gunports, an unusual practice, to show off its powerful armaments. Despite calm conditions, the ship keeled over. Inquiries were unable to identify the root causes of the failure and no one was formally blamed. In reality, many can be blamed: the King because of his unrealistic demands, Hybertson because of the *Vasa*'s poor design, Fleming because of his inaction following the stability test, and Hannson because of his poor seamanship. More fundamentally, an overly ambitious product-development process appears to have sunk the *Vasa*.[3]

LESSONS LEARNED

The *Vasa*'s story highlights the need to maintain strategic realism and managerial realism in the development of new products. We contend that an organization's goals must be appropriately matched to its capabilities, or at a minimum to its accessible skills and knowledge sets. For this to occur, strategic decision makers must have access to accurate and unbiased information, and organizations must develop processes that facilitate the transfer of tacit and explicit knowledge throughout the organization.

We propose seven basic dimensions of the *Vasa* Syndrome and suggest lessons for product development. (See Table 43.1.) These elements are organized by problem area and include conceptual foundation, examples from the *Vasa* case, and insight for managers, with current examples from business organizations. Although the problem areas are conceptually distinct, they are not mutually

TABLE 43.1
Seven Problems and Insights for New-Product Development

Problem Area	Key Insights
1. Lack of external learning capability	Don't imitate ideas unless you fully understand them.
2. Goal confusion	Clarify and focus the project goal on essential user needs.
3. Obsession with speed	Use fast-paced development as a means, not an end, without being hasty.
4. Feedback system failure	Keep an open mind and foster flexible problem solving.
5. Communication barriers	Facilitate vertical- and horizontal-information flows. Share and integrate knowledge.
6. Poor organizational memory	Document and catalog critical areas of expertise. Create a knowledge bank.
7. Top-management meddling	Don't micromanage projects. Set objectives at the top levels and give teams the resources and autonomy to achieve them.

exclusive. A failure in one area may be interrelated with other areas. Accordingly, the seven problems need to be viewed as interrelated components of a complex process. Consistent with a systems view, the focus on only one or two problem areas could still have tragic consequences.

Problem 1: Lack of External Learning Capability

An important strategic choice for firms is to determine the balance of internal and external sources of technology that best meets their needs and fits their resource base.[4] Internal sourcing enables firms to develop core competencies and better understand the tacit knowledge associated with a new product. External sourcing enables a firm to develop a broader knowledge base and keep abreast of cutting-edge technologies. Yet successful external sourcing can occur only if the firm has already internally developed learning capability so that it can effectively understand, interpret, and know how to apply the external knowledge.[5] This is particularly important during dominant design shifts, because leaders in the old technology often fail to become leaders in the new technology.[6]

Leaders in the old technology have tangible and intangible resources invested to the extent that it is difficult for them to abandon historical areas of strength. Often they are not knowledgeable about newer technologies and lack the capability to learn how to use them.

Learning Failures on the **Vasa.** The builders tried to imitate the design of the Danish *Sancta Sophia* with two gundecks and more cannons, but lacked general technical knowledge and procedures for developing a ship with these specifications. They had no theories or mathematical methods for calculating stability and other critical measures, and did not use any standard methods for limiting the overall weight. The second gundeck, placed high on the ship, without proper ballast in the lower part of the ship, created instability and was one of the primary reasons the *Vasa* capsized.

Core Insight for Managers. Don't try to imitate ideas unless you fully understand them. Man-

agers should not imitate ideas from other firms unless they know and understand the basic principles underlying the design decisions, how and why the other product was changed, and how similar changes will affect their product. Managers need to assess their operational readiness by analyzing the needs and capabilities of their firm. Generally speaking, if the product and underlying technology are outside organizational capabilities and this gap cannot be reasonably bridged, then the firm should not blindly pursue such a project. However, if this new technological area is seen as strategically important and a basis of future competition, then the firm should never allow itself to be in such a vulnerable position. It must develop and maintain a competence in the given area, even if it means relying heavily on external partners.

A current example of poor external learning capacity is the attempt of Greyhound Lines, Inc. to design and implement a new computerized reservation system, called Trips.[7] The initial plan was to mimic an airline computer reservation system, such as American Airlines' SABRE system. But the bus reservation system was much more complicated, managing thousands of buses and 10 times the number of vehicle stops per day.

Contributing to Greyhound's failure was the rigid management style of its top managers, who came to Greyhound from unrelated industries and lacked specific knowledge about the bus industry. Nevertheless, they conceived and announced their plan for the computerized reservation system only two months after taking on their new positions. They replaced most of the current regional executives, further reducing the company's core of management knowledge. Several managers, including the project director, challenged the overall plan, but top management refused to budge and demanded that the system be implemented on schedule. It was an unqualified failure. Since the system data bank did not include many minor routes, clerks had to frequently revert to a manual process. The system repeatedly crashed, and, even when operational, it took clerks twice as long on average to issue tickets. Eventually, after huge losses, the project was scrapped and the top managers were forced to leave the company. Greyhound's ambition clearly outpaced its resources and abilities. As with *Vasa*'s builders imitating the Danes, Greyhound managers copied a product they didn't understand and that was quite different from what they needed. The results were also a disaster.

Problem 2: Goal Confusion

Uncertainty about organizational objectives is common when the environment changes dramatically. This change can be a major technological breakthrough, a dramatic change in customer needs, or a change in the competitive rules of the game. Each of these environmental jolts may require a new set of attributes and a realignment of the company's objectives. This internal transition is rarely smooth and almost always incurs resistance to change from employees who have been comfortable with and committed to the old way of doing things. As a result, the direction of an organization's new projects and development activities may be placed in doubt.

Poor coordination can also make goals confusing. This affects new-product development at the strategic level and the operational level.[8] At the strategic level, a common problem is creeping elegance or features creep, which is a failure to freeze product-performance specifications. Instead, some firms change specifications to incorporate new technological advancements as they become available, often resulting in feature changes not being completely or adequately communicated to the technical personnel.[9] At the operational level, the goals of one department may even work against the goals of other departments and, hence, cause uncertainty. This is especially damaging in new-product development, which requires much cross-functional and interdepartmental coordination.[10]

Goal Failures at **Vasa.**   There was an overemphasis on the ship's elegance and firepower and reduced importance on its seaworthiness and stability, which are the more critical issues. There was goal confusion regarding the many changes in the ship's features and components without a clear understanding of how each affected the functioning of the whole. A ship originally designed for 36 guns

was put out to sea with 64 bronze cannons. The *Vasa* also featured many sculptures and carved ornaments, causing some to observe that it was more a work of art than a serious sailing vessel. Armaments and artwork were very heavy and added to the instability of the ship. Resources and management attention focused on the artwork (form) probably came at the expense of the basics of shipbuilding (functionality).

There was also goal confusion concerning how the ship should be designed for combat. The *Vasa* was built when the rules of engagement for naval warfare were dramatically changing. Ships had been designed for close-in combat; typically, only a few shots would be fired before the warships would approach each other and soldiers would board the other ship. The new strategy for naval warfare was to keep a distance from enemy ships and use powerful artillery to sink them. The *Vasa* was built with all possible features. For close combat, the ship was designed to carry 300 soldiers and had a high stem platform to fire down at the enemy. The ship also had tremendous firepower to sink other ships at a distance. This lack of focus and ambiguous performance goals required a more complex ship design, which contributed to the difficulty in maintaining ship stability.

Core Insight for Managers.   Clarify and focus the project goal. More specifically, adopt a clear perspective by focusing on essential customer and end-user needs. Do not tinker with product design in an attempt to please all internal groups. Firms should not get caught up in continually changing details. They should beware that what seems logical for the specific part of the product does not undermine the overall design integrity and functionality of the product. Though competitors may force improvements by releasing a superior product, managers must anticipate the cumulative effects of these changes to the larger system. A holistic perspective should be applied to projects, such as through the use of teams. Teams can represent the broader interests of the organization, which improves each member's understanding of the entire project and helps achieve a customer- and value-based delivery focus, instead of a compilation of functional agendas. This is especially true when cross-functional representation enables teams to become holographic (i.e., the parts mirror the whole) so that the many different facets of the development process are included in the smaller team unit, helping reconcile trade-offs towards shared overall objectives.[11] For example, manufacturing could identify design flaws before marketers commit to it, or marketing can warn against poor demand for a feature before engineers dedicate too many resources to it.

More recent examples of goal confusion are General Motors' unsuccessful electric car development project and Motorola's Iridium project. GM invested $1.5 billion to develop the EV1, the first mass-produced electric vehicle in the U.S.[12] Unfortunately, GM was more intent on meeting government clean-air mandates than consumer demands.[13] GM focused primarily on having a zero-emission car, rather than making the car more practical by increasing its range per electric charge. Meanwhile, Toyota and Honda are the leaders in the development of hybrid electric and gasoline cars, which are more consumer-oriented. In addition, GM was not clear on which new technology to pursue, and discontinued production three years after its launch.

Iridium, the now defunct $5 billion global-satellite phone system designed and built by Motorola and 18 other electronics and telecommunications companies, was a colossal failure because goals that addressed customer needs were neglected. The handsets were large and heavy, did not work well in cars or buildings, and cost too much. Few customers valued world-wide calling capability more than the convenience and low cost of the competing cellular phones. Motorola also had different goals from its partners, namely to develop technological competence in satellite-communication technology. Motorola pushed this new, risky technology to market before many major technical problems were resolved.

The Vasa, GM, and Iridium examples reinforce the point that firms should control their temptation to tinker and modify designs only when it adds to the functionality of the product. A useful managerial tool for regulating this activity is the house of quality, originated, ironically, at Mitsubishi's Kobe shipyards, which helps project teams make important design decisions by getting them to think together about what users want and how to get it to them most effectively.[14] First, one identifies product requirements, and weights their relative importance from the users' perspective. Figure 43.1 shows that, in the case of the Vasa, staying afloat was by far the most important attribute of the ship. Second, one identifies the engineering attributes in such measurable terms as number of gundecks and proportion of ballast. Third, one describes the direction and strength of the relationships between the user requirements and engineering attributes (body of the house) as well the relationships among the attributes themselves (roof of the house). For the Vasa, the body of the house shows that floatability is influenced positively by proportion of ballast and negatively by number of gundecks and elegance of ornamentation. This is because ballast is extremely critical in stabilizing the ship whereas top-heavy payload can destabilize it, not to mention divert time and energy in the manufacture of guns and artwork. The roof also shows that more gundecks would require a bigger ship, which would reduce the proportion of ballast below needed levels. Notwithstanding, ballast was traded for power and aesthetics in favor of less critical attributes such as looks impressive.

Problem 3: Obsession with Speed

There tends to be a positive bias for speed in product development.[15] The aura of first-mover advantages and heroic pioneer surrounds the efforts to accelerate innovation processes and produces what some might characterize as an obsession with speed to market. However, there is a growing counterbalance in the literature to the assumption that fast product development is universally desirable. Some researchers have pointed out the disadvantages of pioneering new technologies and concluded that it is sometimes better to go slower and be a follower instead of a pioneer.[16] Others discuss some of the hidden costs of speed, such as more

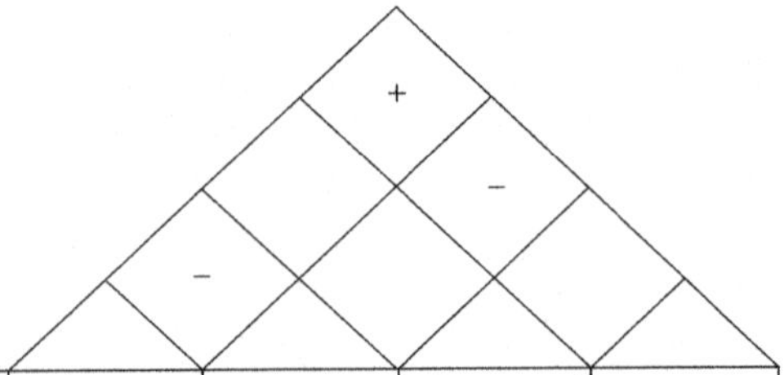

Engineering Attributes	Importance (maximum = 10)	Number of gun decks	Proportion of ballast	Elegance of orientation	Size of ship
Customer Requirements Firepower for close and distant combat	6	++			+
Looks impressive	2	+		++	+
Stays afloat	10	–	++	–	
Completed quickly	5	–		–	–

++ = Strong positive relationship   – = Negative relationship
+ = Positive relationship   (Blank) = Weak relationship

Figure 43.1.   House of quality adapted for the *Vasa*.

mistakes, heavy use of resources, and disruptions in workflow.[17] It is also important to consider that development speed may be purposefully slower under certain conditions, such as when products affect the health and safety of their users (e.g., pharmaceuticals, transportation),[18] and faster in others, such as when a firm has monopoly-like control in an industry (e.g., Microsoft) and can afford to release an imperfect product with the promise to fix it later.

Rapid development may be less advantageous under conditions of high uncertainty.[19] This is due partly to increased forecasting error (less information is available) and increased noise (shifting rules of the game). Additionally, speed is less functional for radical innovations because these types of products are very risky and need more time for testing and control. Rushing big projects is especially risky because they are often based on new, less familiar areas of technical expertise.[20] It is often less advantageous to go fast if you can't see clearly and could be going in the wrong direction. Managers should pursue speed most vigorously when the technology is known and markets can be forecast.[21]

Speed Failures on the **Vasa.**   There was a definite overemphasis on the ship's speed of development to the detriment of its quality. The *Vasa* was a radically new design and embodied new technologies, which contributed to high levels of uncertainty. Because of the many design changes and other activities in the shipyard, the work on the *Vasa* was forced into an overly compressed time schedule. Steps were skipped and there was a general lack of checks and balances and quality control. Military activities with Poland, naval losses in Riga, and Denmark's activity increased the time pressure and accelerated the schedule.

Core Insight for Managers.   Use fast-paced development as a means, not as an end. Do not compromise the integrity of the product in the interest of blindly speeding it to market. For example, in its desire to gain quick market share, Black Rock Corp. rushed its long-handled Killer Bee golf clubs to market. The firm failed to get the sport's governing authority, the USGA, to confirm that the design conformed to standards. Although initial sales were dramatic, and the cofounders took the company public, profits were wiped out when the USGA ruled that the clubs' markings violated its rules. The company was forced to refund money to almost 10,000 customers, for a total loss of $500,000. Although the USGA later reversed its decision, the company missed out on the key selling months and lost credibility with retailers and golfers. This resulted in a $2.7 million loss and the company was forced to file for bankruptcy.

Sometimes even the most conservative organizations fall into the trap of focusing too much on innovation speed, often because of an internal pressure to meet an arbitrary deadline. For example, in October 2000 the Federal Aviation Administration (FAA) attempted to upgrade the computer software of its air-traffic control radar at its Los Angeles Center in the Mojave Desert. However, the new system was unable to read flight data from the less advanced Mexican air traffic system. The FAA's new system failed and shut down air traffic across the southwestern United States, grounding or delaying about 2,200 flights. The entire U.S. airline system quickly became gridlocked. Critics argued that the FAA hastily implemented this upgrade without adequately testing it or properly training air traffic controllers in its use.[23] Luckily, this failure did not result in a catastrophic accident, only massive chaos and inconvenience.

However, there are instances where fast market introduction is appropriate, even if the product is not perfect. For example, Netscape's strategy to launch a new product every six months (versus traditional industry cycles of 12 months) forced it to forego a standard practice of using major beta test sites. Netscape simply released prototypes onto the Internet and waited for users to give them feedback. This provided Netscape with an army of debuggers who quickly refined its prototype into a finished product.[24] Put simply, there is a difference between speed and haste. Speed is gained from fundamentally improving the strategic and operational nature of the new-product development process.[25] Haste infects development processes when steps are haphazardly skipped (e.g., prototyping, testing, approval from appropriate authorities) and traditional processes are pushed past their functional limita-

tions.[26] Therefore, organizations are usually best served when they use fast-paced development practices as a means to better focus teams, reduce delays, control costs, and meet demand, but not as ends in themselves.

Problem 4: Feedback System Failure

Throughout the new-product development process, there should be continuous testing to ensure the product meets customer needs. Feedback must be accurate and passed on to all key people involved. In many cases, numerous iterations of a design-build-test cycle are used to maximize learning and develop the best final product.[27] During the design stage, clear objectives for the project are established and key problems are framed. During the build stage, working models are built to ensure that the design is feasible. These models may be physical prototypes or computer simulations developed in a computer-aided design (CAD) workstation.[28] During the test phase, the working models are tested to ensure the product meets the goals outlined in the design stage. To get good information out of the testing phase, it is imperative that the tests are carefully designed and performed to ensure accuracy and reduce bias. It is also important that all affected participants analyze the results. The overall process should be an integrated problem-solving exercise. For example, someone is frequently designated to fulfill the role of critic in the new-product development process, to openly question the established assumptions and ensure that valuing team harmony over good decision making does not push a faulty product forward.[29]

Feedback Failures on the **Vasa***.* There were numerous incomplete product tests, managers ignored test results, and there was an overly optimistic approach despite warnings to the contrary. Even though this type of large project is not conducive to multiple design-build-test cycles, an initial high quality small-scale prototype would have identified many of the problems with the *Vasa* early in the process. Such a prototype would also have communicated the overall vision to everyone involved. There is no indication that a prototype was built. As mentioned earlier, a stability test was conducted and, after the ship failed, the admiral chose to ignore the results. The key persons involved were willing to be silent as the price for career advancement and job security.[30] The admiral also did not tell the test results to others, such as the shipbuilder, who were in a better position to interpret the results. In one observer's words, the admiral was a coward because it takes courage to admit mistakes to your boss.[31] This is particularly important in engineering and product development, where faulty specifications can have disastrous results. Faulty reasoning may have also played a part of the admiral's poor decision, as reflected in his comments that he was not worried because the shipbuilder had built ships before. Given the difficulty of seeing with the naked eye how top-heavy the *Vasa* was, an unbiased read of the test results was very important. The issues of honesty to oneself (scientific objectivity) and honesty to other persons (ethical practices) are intertwined in the handling of the test results.

Core Insight for Managers. Keep an open mind and foster flexible problem solving throughout the development process. Don't rationalize away bad news in the process of finding solutions to problems. Unfortunately, the *Vasa* team fell victim to the confirmation trap, seeking information consistent with what they think or want to be true and deemphasizing critical but contradictory information.[32] There are several ways for managers to overcome this bias, such as dedicating themselves to such overarching principles as safety, reliability, and quality assurance. Additionally, managers need to take negative feedback seriously and act on it, especially in cases such as the *Vasa* where human lives are at risk.

A modern example of this type of failure is Lever Brothers' European launch of Persil Power. Lever Brothers' Persil had a large market share in the detergent industry in the early 1990s and was one of the strongest brands in the European grocery sector.[33] After spending 10 years and $300 million on research and advertising, a new product called Persil Power was launched, containing manganese,

which was believed to have superior cleaning properties. However, Procter & Gamble, Lever Brothers' primary competitor, aggressively initiated a negative advertising campaign, claiming Persil Power damaged clothes and fabrics. Independent researchers confirmed P&G's claims and Persil's market share dropped dramatically. Within a year, the product was withdrawn from the market. Although scientists had known for years about the problems of using manganese in detergents, managers and marketers at Lever Brothers overruled the R&D people and disregarded their warnings,[34] in part, because they were overly concerned with quickly countering P&G's new product, Ariel Future.

Problem 5: Communication Barriers

An effective communication system includes open and efficient channels across vertical levels as well as across divisions and job descriptions. Such a system ties the employees to the major ongoing activities, decreases uncertainty by clarifying issues and providing a forum where misunderstandings can be aired and information clarified, and creates coordination, cooperation, and a culture that is open and conducive to change.[35] Some barriers to effective communication are selective perception, where individuals only see what they want to see, and projection, where one person or group's motives are mirrored onto another.[36]

Communication Failures on **Vasa.** There were three sets of official specifications for the *Vasa*, drawn up by the king, Hybertson, and another shipbuilder. These parties did not communicate well and never clearly resolved critical differences. The king and Hybertson disagreed about keel length and bottom width, causing both to operate under different assumptions about the amount of weight the *Vasa* could accommodate. Hybertson's long illness harmed communication between himself and his assistants. The shipbuilders were not at the stability test, causing them to receive only filtered, biased feedback from the admiral. The king gave many orders, but did not get proper advice from experts. Thus communication failed within and across the project, as well as up and down the hierarchy. When communication is forced to pass through multiple, insulated layers, critical shortcomings will almost certainly be known only in retrospect.[37] The king, in providing design additions, and Admiral Klas Fleming, in interpreting the results, seemed to employ selective perception.

Core Insight for Managers. Organizations must promote the sharing and integration of knowledge within and across project teams, and develop a system that creates a culture with a positive attitude towards reconciling divergent viewpoints.[38] Open communication enables boundaryless coordination across vertical, horizontal, and external dimensions. A supportive communicator does not engender defensiveness, because an overly defensive person no longer listens and contributes, but is more concerned with self preservation.[39]

Disney is an example of a company that has removed internal communication barriers to facilitate new-product development.[40] Historically, different divisions were mostly independent, trying to maximize their own creativity. CEO Michael Eisner saw that all divisions had to be closely coordinated, enabling creative content to be exploited in numerous, mutually reinforcing ways. He created a Synergy Group, headed by a vice president in charge of synergy, with all other divisions reporting to it. A remarkable success of Eisner's vision was the *Lion King*, with world box office sales of $766 million, and retail merchandise sales of $1.5 billion from over 1,000 *Lion King* products—books, toys, games, clothes, lunchboxes, sheets, and toothbrushes. Toys "R" Us carried 200 *Lion King* products the week before the movie opening and the Disney theme park started planning related attractions a year before the movie was finished. The *Lion King* soundtrack became the best-selling album in the U.S. and its video became the best-selling video ever, making a profit of $300 million. A successful Broadway play followed the movie. Clearly, Disney was able to profit in all of these *Lion King* ventures because it was highly coordinated, which enabled cross-promotion and the development of a very strong brand.[41]

3M Corporation is another example of a company that has successfully removed external communication barriers to facilitate boundaryless new-product development. Its lead user program, initially developed by Eric von Hippel of MIT, is a systematic process for identifying key individuals, called lead users, who experience needs far ahead of the other average users in the market segment.[42] When 3M's Medical-Surgical Markets Division wanted breakthrough ideas in surgical drapes, it formed a lead user group consisting of a surgeon, an antimicrobial pharmacologist, a disease control expert, a chemist, a biologist, a Hollywood make-up artist, and a veterinarian surgeon.[43] Each of these experts possessed specialized knowledge critical to the development of new products. The team provided three-concrete new-product recommendations and also suggested a new business strategy.[44]

Problem 6: Poor Organizational Memory

An organization's memory is the stored information from its history that can be brought to bear on present decisions.[45] As individuals share and integrate knowledge, the knowledge is stored in the organizational structure, systems, routines, and procedures.[46] When individuals leave an organization, some of their knowledge also leaves, but, depending on the effectiveness of the organizational memory, some also remains within existing institutionalized practices. This knowledge can be tacit or explicit. Explicit knowledge can be stored in files and formal procedures, and dissemination can be improved by the use of computer systems and organizational intranets. Many large consulting organizations, such as McKinsey, PricewaterhouseCoopers, and Arthur Andersen (now known as Accenture) have made significant strides in collecting, codifying, disseminating, and reusing explicit knowledge through databanks that can be readily accessed by employees anywhere in the world.

Tacit knowledge is difficult to codify and must be shared, stored, and retrieved by direct, personal communications and sharing of experiences. Tacit knowledge is hard to articulate, and is dependent on more subtle transfer techniques, such as common organizational stories and rituals.[47] Maintaining interaction across different organizational levels and groups, and sharing experiences through apprenticeship programs and job rotations, also enables effective storage of tacit knowledge. During the new-product development process, tacit knowledge stored in the organization's memory must be converted into explicit knowledge that can be understood by individuals lacking experience in a specific area. This is difficult when there is high personnel turnover and the organization fails to adequately store information.[48]

***Memory Failures on* Vasa.** The master shipbuilder, Henrik Hybertson, became ill and died a full year before completion of the ship. His individual knowledge was not documented or transferred to others to become organizational knowledge. His assistant, Hein Jacobsson, had little management experience, and much less knowledge about the project. Jacobsson was not fully knowledgeable about important attributes of the *Vasa*, such as its armament and overall weight.

Core Insight for Managers.   A culture that values learning better prepares managers to learn from past experiences and adapt to future contingencies. In its handling of the Hubble telescope, NASA repeated many of the mistakes in product development and project management that caused the Challenger disaster. Established procedures and behavioral patterns interfered with their ability to learn from past mistakes.

The Hubble and Challenger incidents both involved failures in complex systems that relied on chains of events that were expected to trigger one another deterministically.[49] The knowledge gained from the Challenger was not stored in the organization's memory and could not be retrieved. Instead, NASA's complex structure and culture institutionalized faulty project paradigms and failed to create adequate corrective mechanisms.[50]

In contrast, Cisco Systems built up detailed organizational memory on managing acquisitions and the integrating of acquired companies. Cisco derives almost 40 percent of its revenue from these acquisitions. Once an acquisition is consummated, Cisco uses a documented and repeatable process for integration.[51] The R&D and product divisions of each acquired company are grafted onto the product side of the organization, and manufacturing, sales, and distribution are integrated into Cisco's functional organization. All computer equipment is immediately made compatible with Cisco's information system. The process is so well institutionalized that most acquired products can be fully integrated within 60 to 100 days.

Advances in computer and telecommunication technology afford organizations powerful tools for storing and accessing lessons from past projects. Some firms have developed expert lists that enable any project member to access the individuals in their organization who have knowledge in specific problem areas. Middle managers play a critical role in most organizations in transferring and storing knowledge throughout the organization.[52] When companies downsize and eliminate many of the middle-manager positions, they risk destroying knowledge flow by eliminating critical communication nodes and storage banks.

Problem 7: Top-Management Meddling

Delegating decision making can improve new-product development because it diffuses the power necessary to go against the status quo, increases workers' involvement in and awareness about a project, and subsequently strengthens workers' commitment to it.[53] Decision autonomy also provides a buffer against excess outside interference, reduces frequent, mandated changes in the product, and limits the number of bureaucratic snags and necessary approvals.[54] A lack of autonomy may also hurt product development over the long run, because people who are not allowed to make decisions cannot learn from these experiences.[55] This relates to empowerment, which is delegating decision making to the lowest level where a competent choice can be made. Empowerment necessitates sharing information and knowledge in order to ground people's decisions, authorizing them to make these decisions, and designing performance-based compensation systems to reward their decisions.[56] When these principles are embraced, organizations can reap the advantages of a more committed, informed, proactive, and flexible workforce. However, many organizations have yet to realize the promise of empowerment and remain vulnerable to information overload, detached decision making, and poorly leveraged worker expertise.[57]

Top managers need to be careful not to become too over-bearing and interfere with the new-product development process. Their value-added is greatest early in the process by providing vision and leadership during the concept formulation and basic design phase. Later, it is usually better for them to step back and let others handle the more specific issues. However, some top managers do the opposite: they do not get involved in the early stages, tend to become very reactive, and pay most attention at the end of the development process when the product is close to being commercialized.[58] Unfortunately, at this point, they usually have little ability to improve the process or influence the outcome. An exception is when top managers have strong engineering skills and can legitimately execute the chief engineering role (e.g., Bill Gates of Microsoft and Andy Grove of Intel). Of course, few executives, including King Gustavus Adolphus, can boast such credentials.

Meddling Failures on Vasa.

King Gustavus Adolphus was a one-person top-management team, and he tended to get overly involved in areas far beyond his expertise. The changes in the size and armament of the ship were made under intense pressure from the king. He had suffered military losses and was pushing the development team to deliver the Vasa so that it could be put into action immediately. He approved the ship's dimensions but continuously requested more design changes. The master shipbuilder and his team had little influence over the amount of ornamentation or firepower. On several occasions the master shipbuilder cautiously tried to dissuade the king, but the king would not listen. Some speculate the extreme vanity of the king was to blame for his overly grandiose goals and over-involvement.

Core Insight for Managers.

Top management should resist the temptation to micromanage projects from an unreasonable distance and instead focus its energy on setting broad-based objectives, while giving teams the resources and autonomy to achieve them. Organizations need to staff projects with good people and let them do their jobs. This is consistent with the notion that top managers should set overall directions for development, then release it to the project team and capture it only after the creative and technical work is complete.[59] Top-management influence has its place in upstream goal setting and downstream implementation, but, as in the Vasa example, bureaucratic interference can harm organic project learning and execution. A more recent example of top-management meddling comes from the example of Greyhound Lines, where domineering leaders meddled in the project and did not fully listen to critics.

Two examples of appropriate top-management involvement are Toyota and Cisco. In 1983, when Toyota sold mostly inexpensive cars, Chairman Eiji Toyoda decided the firm should enter the luxury market with the Lexus. He challenged his engineers to "develop the best car in the world," provided general specifications, and then released the project to his team.[60] These specifications were so challenging that they required the development team to make breakthroughs in aerodynamics, noise dampening, suspension, and engine design.[61] Toyoda provided the engineers resources and patients; he invested over $500 million during the six-year development period. He was actively involved in providing the project's vision, but was careful to let the engineers figure out how to make it happen.

John Chambers of Cisco Systems created an overarching system for integrating acquisitions, while providing a lot of freedom to the management and personnel of the acquired companies. When Cisco acquired Cerent in 1999, Chambers assured its CEO that personnel decisions about Cerent's workers would be made jointly by the two executives. This attitude of collaboration and give and take enabled Cisco to leverage Cerent's cutting-edge technology, which is critical to linking the Internet and telephone system industries.[62]

PREVENTING THE *VASA* SYNDROME

The *Vasa* Syndrome can severely and negatively affect the success of new-product development, but it is certainly not inevitable. Organizations can prevent the *Vasa* Syndrome by realistically deciding what to pursue and how to pursue it in the development of new products and technologies. Although speed is very important for competition, managers should not be obsessed with speed to the extent that other functions like product safety, quality, and reliability are neglected. In general, managers must take care to appropriately match current goals to current capabilities, or, at a minimum, to the skill and knowledge sets to which they have access. Executive decision makers must generate, circulate, and collect accurate and unbiased information. Other implications of the *Vasa* tragedy suggest the need to develop and institutionalize processes that facilitate the transfer of explicit and tacit knowledge, incorporate it in organizational memory, and foster organization-wide awareness of such capabilities.

If we listen, the *Vasa* tragedy speaks to us across the centuries and offers specific insights for today's managers. Put a priority on clear, customer-focused project goals and ensure that your organization has the capacity to understand and apply critical knowledge. Value speed but do not be hasty. Work to integrate all phases of the development process and facilitate the sharing of project-relevant information across functional disciplines and managerial levels. Institutionalize learning and leverage experiences through a knowledge bank. Finally, after setting the strategic direction for development, resist the temptation to micromanage it from the executive level. The *Vasa* case illustrates a violation of all these maxims, but this is not to say that guarding against one or two of them would have saved the ship. Often these problems occur simultaneously and are mutually reinforcing. The Greyhound example shows how multiple shortcomings can systematically affect an organization, whereas other examples illustrate the consequences of a single problem.

The *Vasa* was not the first new product to sink, and it surely won't be the last. But heeding its lessons can reduce vulnerability to the *Vasa* syndrome and subsequently lessen the probability that it will occur.

ENDNOTES

1. Pringle, C. D., & Kroll, M. J. 1997. Why Trafalgar was won before it was fought: Lessons from resource-based theory. *The Academy of Management Executive*, 11(4): 73–89; Kroll, M. J., Tooms, L. A., & Wright, P. 2000. Napoleon's tragic march home from Moscow: Lessons in hubris. *The Academy of Management Executive*, 14(1): 117–128.

2. Details from the *Vasa* story are gathered from a collection of sources, including Borgenstam, C., & Sandstrom, A. 1995. *Why* Vasa *capsized*. Stockholm: AB Grafisk Press; Ives, B. 1994. *The* Vasa *capsizes*. *http://virtualschool.edu/*; Manning, W. H. 1998. *Test, technology, and the tender ship*. Santa Monica, CA: Graduate Management Admissions Council; Mayol, D. E. 1996. *The Swedish ship* Vasa's *revival*. Miami, FL: University of Miami; Squires, A. 1986. *The tender ship*. Boston: Birkhauser; Vasa Museum homepage: *http://www.vasamuseet.se*.

3. Nearly half a century later, in 1961, the *Vasa* was salvaged. Subsequent analysis of the remains has provided clues to explain why the *Vasa* capsized. The *Vasa* has been restored and is displayed in the Vasa Museum in Stockholm.

4. Bierly, P., & Chakrabarti, A. 1996. Generic knowledge strategies in the U.S. pharmaceutical industry. *Strategic Management Journal*, 17 (winter special issue): 123–135.

5. Cohen, W. M., & Levinthal, D. A. 1990. Absorptive capacity: A new perspective on learning and innovation. *Administrative Science Quarterly*, 35: 128–152.

6. Bower, J. L., & Christensen, C. M. 1995. Disruptive technologies: Catching the wave. *Harvard Business Review*, January-February: 43–53.

7. Picken, J. C., & Dess, G. G. 1998. Right strategy—wrong problem. *Organizational Dynamics*, 27(1): 35–49.

8. Kessler, E. H., & Chakrabarti, A. K. 1996. Innovation speed: A conceptual model of context, antecedents and outcomes. *Academy of Management Review*, 21(4): 1143–1191.

9. Gupta, A. K., & Wilemon, D. L. 1990. Accelerating the development of technology-based new products. *California Management Review*, 32(2): 24–44; Stalk, G. & Hout, T. M. 1990. *Competing against time: How time-based competition is reshaping global markets*. New York: The Free Press.

10. Daft, R. L. 1998. *Organization theory and design*. Cincinnati, OH: South-Western College Publishing.

11. Brown, S. L., & Eisenhardt, K. M. 1995. Product development: Past research, present findings, and future directions. *Academy of Management Review*, 20: 343–378; Van de Ven, A. H. 1986. Central problems in the management of innovation. *Management Science*, 32: 590–607.

12. Electric car drives factory innovations. *The Wall Street Journal*, 27 February 1997, B-1.

13. Electric cars get jolt of marketing. *Marketing News*, 18 August 1997, 1.

14. Hauser, J. R., & Clausing, D. 1988. The house of quality. *Harvard Business Review*, May-June: 63–73.

15. Crawford, C. M. 1992. The hidden costs of accelerated product development. *Journal of Product Innovation Management*, 9: 188–199; Kessler & Chakrabarti, op. cit.

16. Lounamaa, P. H., & March, J. G. 1987. Adaptive coordination of a learning team. *Management Science*, 33: 107–123; Von Braun, C. F. 1990. The acceleration trap. *Sloan Management Review*, 32(1): 49–58; Golder, P. N., & Tellis, G. J. 1993. Pioneering advantage: Marketing logic or marketing legend. *Journal of Marketing Research*, 30: 158–170; Lieberman, M. B. & Montgomery, D. B. 1988. First-mover advantages. *Strategic Management Journal*, 9: 41–58.

17. Crawford, op. cit.; Von Braun, op. cit.

18. Kessler & Chakrabarti, op. cit.

19. Meyer, M. H., & Utterback, J. M. 1995. Product development cycle time and commercial success. *IEEE Transactions on Engineering Management*, 42(4): 297–304; Kessler, E. H., & Bierly, P. E. Is faster really better? An empirical test of the implications of innovation speed. *IEEE Transactions on Engineering Management*, in press.

20. Tushman, M., & Anderson, P. 1986. Technological discontinuities and organizational environments. *Administrative Science Quarterly*, 31: 439–465.

21. Kessler & Bierly, op. cit.

22. Macht, J. Shortcut derails maker of long golf clubs. *Inc.*, August 1999.

23. Shiver, J. FAA software flaw spotlights malady of digital age aviation: The glitch in Palmdale that delayed air traffic is blamed on coding and on insufficient testing and controller training. *Los Angeles Times*, 27 October 2000, C-1.

24. Eisenhardt, K., & Brown, S. L. 1998. Time pacing: Competing in markets that won't stand still. *Harvard Business Review*, March-April.

25. Kessler & Chakrabarti, op. cit.

26. Wheelwright, S. C., & Clark, K. B. 1992. *Revolutionizing product development*. New York: The Free Press.

27. Ibid.

28. Iansiti, M. 1995. Shooting the rapids: Managing product development in turbulent environments. *California Management Review*, Fall.

29. Bazerman, M. H. 1990. *Judgment in managerial decision making*. New York: Wiley.

30. Manning, op. cit.

31. Ives, op. cit.

32. Bazerman, op. cit.

33. 26 million pounds sterling power behind Persil. *Super Marketing*, 18 March 1994, 8.

34. Persil's power failure. *Marketing*, 19 January 1995, 10.

35. Daft, op. cit.

36. Dearborn, D. C., & Simon, H. A. 1958. Selective perception. *Sociometry*, 21: 140–143.

37. Manning, op. cit.

38. Pearson, C. M., & Mitroff, I. I. 1993. From crisis prone to crisis prepared: A framework for crisis management. *The Academy of Management Executive*, 7(1): 48–59.

39. Whetton, D. A., & Cameron, K. S. 1995. *Developing management skills*. New York: Harper Collins.

40. Rayport, J. F., Knoop, C., & Reavis, C. 1999. *Disney's The Lion King: The Synergy Group (A & B)*, HBS Cases 9-899-041 and 9-899-042. Boston: Harvard Business School Publishing.

41. Ibid.

42. For a detailed description of the lead-user method, see: von Hippel, E., Thomke, S., & Sonnack, M. 1999. Creating breakthroughs at 3M. *Harvard Business Review*, September-October: 47–57.

43. Thomke, S., & Nimgade, A. 1999. *Innovation at 3M (A)*, HBS Case 9-699-012. Boston: Harvard Business School Publishing.

44. Ibid.

45. Walsh, J. P., & Ungson, G. R. 1991. Organizational memory. *Academy of Management Review*, 16: 57–91.

46. Spender, J.-C. 1996. Making knowledge the basis of a dynamic theory of the firm. *Strategic Management Journal*, 17 (winter special issue): 45–62.

47. Nonaka, I. 1994. A dynamic theory of organizational knowledge creation. *Organization Science*, 5: 14–37.

48. Huber, G. P. 1991. Organizational learning: the contributing processes and the literatures. *Organization Science*, 2: 88–115.

49. Stein, B. S., & Kanter, R. M. 1993. Why good people do bad things: A retrospective on the Hubble. *The Academy of Management Executive*, 7: 58–62.

50. Ibid.

51. Nolan, R. L., & Porter, K. A. 1998. *Cisco Systems, Inc.* Harvard Business School Case #9-238-127. Boston: Harvard Business School Publishing.

52. Nonaka, op. cit.; Floyd, S. W., & Wooldridge,

B. 1994. Dinosaurs or dynamos? Recognizing middle management's strategic role. *The Academy of Management Executive*, 8(4): 47–57.

53. Damanpour, F. 1991. Organizational innovation: A metaanalysis of effects of determinants and moderators. *Academy of Management Journal*, 34: 555–590.

54. Ancona, D. G., & Caldwell, D. 1990. Improving the performance of new product teams. *Research Technology Management*, 33:25–29; Deschamps, J. P., and Nayak, P. R. 1992. Competing through products: Lessons from the winners. *Columbia Journal of World Business*, 27(2): 38–54; Stalk & Hout, op. cit.

55. Eisenhardt, K. M. 1989. Making fast strategic decisions in high velocity environments. *Academy of Management Journal*, 32: 543–576.

56. Ford, R. C., & Fottler, M. D. 1995. Empowerment: A matter of degree. *The Academy of Management Executive*, 9(3): 21–29.

57. Forrester, R. 2000. Empowerment: Rejuvenating a potent idea. *The Academy of Management Executive*, 14(3): 67–90.

58. Hayes, R. H., Wheelwright, S. C., & Clark, K. B. 1988. *Dynamic manufacturing*. New York: The Free Press.

59. Spender, J.-C., & Kessler, E. H. 1995. Managing the uncertainty of innovation: Extending Thompson (1967). *Human Relations*, 48: 35–56.

60. Taylor, A. Here come Japan's new luxury cars. *Fortune*, 14 August 1989, 62–66.

61. Ibid.

62. Meet Mr. Internet. *BusinessWeek*, 13 September 1999, 128–140.

14

Organizational Practices, Policies, and Rewards

44

Winning the Talent War for Women

DOUGLAS M. MCCRACKEN

Nine years ago, the professional services firm of Deloitte & Touche realized too many of its talented women were walking out the door. Stopping them was urgent—but it took a deeper change in the organization than anyone expected.

Nine years ago, we came to grips with the fact that women at Deloitte were on the march—out the

door. In 1991, only four of our 50 candidates for partner were women, even though Deloitte & Touche—America's third largest accounting, tax, and consulting firm at the time—has been heavily recruiting women from colleges and business schools since 1980. Not only that. We also found that women were leaving the firm at a significantly greater rate than men.

To be frank, many of the firm's senior partners, including myself, didn't actually see the exodus of women as a problem, or at least, it wasn't *our* problem. We assumed that women were leaving to have children and stay home. If there was a problem at all, it was society's or the women's, not Deloitte's. In fact, most senior partners firmly believed we were doing everything possible to retain women. We prided ourselves on our open, collegial, performance-based work environment.

How wrong we were, and how far we've come.

Over the next few years, we analyzed why women were leaving and worked to stop the outflow. At first, the program was largely our CEO's idea; unlike many of us, he saw women's leaving as a serious business matter that the firm could and should fix.

These days, you'd be hard-pressed to find partners within the firm who disagree. It took a cultural revolution, but Deloitte now has a radically different approach to retaining talented women. Based on six principles, it is an approach that other companies might well consider, for its results speak for themselves.

Today 14% of our partners and directors are women. While we aren't yet where we want to be, this percentage is up from 5% in 1991 and the highest in the Big Five. The number of women managing partners has increased dramatically, and we've eliminated the gender gap in our turnover: women now stay on at about the same rate as men each year. The firm's annual turnover rate as a whole fell from around 25% in the early 1990s to 18% in 1999, despite an intensifying war for talent. Besides saving us $250 million in hiring and training costs, lower turnover has enabled Deloitte to grow faster than any other large professional services firm in the past several years.

A TWO-STAGE PROCESS

Deloitte's Initiative for the Retention and Advancement of Women grew out of a 1992 task force chaired by Mike Cook, then CEO of Deloitte & Touche. A number of women partners initially wanted nothing to do with the effort because it implied affirmative action. But Cook, along with a handful of partners—women and men—insisted that high turnover for women was a problem of the utmost urgency. In professional services firms, they argued, the "product" is talent, billed to the client by the hour; and so much of our firm's product was leaving at an alarming rate. Cook made sure that both women and men were part of the task force and that it represented a broad range of views, including outright skepticism.

Once in place, the task force didn't immediately launch a slew of new organizational policies aimed at outlawing bad behavior. Instead, it approached the problem methodically, just as we would approach a consulting assignment. Thus, it first investigated the problem and gathered the data necessary to make a business case—not a moral or emotional one—for change. Then it prepared the groundwork for change by holding a series of intensive, two-day workshops for all of our management professionals. These sessions were designed to bring to the surface the gender-based assumptions about careers and aspirations that had discouraged high-performing women from staying.

Only then did the firm announce a series of policies aimed at keeping women. A major component of these policies was to first get all the firm's offices to monitor the progress of their women professionals. The head of every office received the message that the CEO and other managing partners were watching, and in turn, women started getting their share of premier client assignments and informal mentoring. Other policies, designed to promote more balance between work and life for women and men, also helped. These efforts have opened up our work environment and our culture in ways we never expected.

PREPARING THE WAY FOR CHANGE

Along the way, we've learned a series of lessons. Other companies, with different traditions and operating environments, may well follow other paths to achieve equitable treatment of men and women. But we think our lessons will apply to a great many organizations.

Make Sure Senior Management Is Front and Center

Despite its name, the Women's Initiative was always driven by the managing partners—it never became an "HR thing" foisted on the firm. Like other organizations, we were used to having new personnel programs every so often, just one more thing added to an already full plate. I'm sure most of our partners felt initially that the focus on women was the latest "program of the year"; we would try our best and then move on to something else. But from the start, senior management signaled that the initiative would be led by the partners. Cook named Ellen Gabriel, a star partner, as the first leader of the initiative.

Cook's own leadership involved no small investment and risk. In a firm like ours, where the partners are also owners, leadership is not top-down. He took charge of the effort personally and visibly, and with every step, we all got the sense that change was a high priority for him. In Cook's case, a reputation for toughness helped to give this initiative credibility.

Make an Airtight Business Case for Cultural Change

The task force prepared the firm for change by laying a foundation of data, including personal stories. Deloitte was doing a great job of hiring high-performing women; in fact, women often earned higher performance ratings than men in their first years with the firm. Yet the percentage of women decreased with each step up the career ladder, in all practices and regions, and many women left the firm just when they were expected to receive promotions. Interviews with current and former women professionals explained why. Most weren't leaving to raise families; they had weighed their options in Deloitte's male-dominated culture and found them wanting. Many of them, dissatisfied with a culture they perceived as endemic to professional services firms, switched professions. And all of them together represented a major lost opportunity for the firm.

These facts made for a sobering report to the senior partners on the firm's management committee in 1993. As Cook summarized, "Half of our hires are now women, and almost all of them have left before becoming partner candidates. We know that in order to get enough partners to grow the business, we're going to have to go deeper and deeper into the pool of new hires. Are you willing to have more and more of your partners taken from lower and lower in the talent pool? *And* let the high-performing women go elsewhere in the marketplace?

Let the World Watch You

With the endorsement of the management committee, the firm moved forward. It held a press conference to launch the Women's Initiative, but it also went further and named an external advisory council. Chaired by Lynn Martin, former U.S. secretary of labor, the council comprised business leaders with expertise in the area of women in the workplace. Besides reviewing the initiative's progress, the council brought visibility to the effort. As the task force realized, going public would put healthy pressure on the partners to commit to change and deliver results. And that's what happened, particularly with slow-moving offices in the organization. Local managers received prodding comments from their associates like, "I read in the *Wall Street Journal* that we're doing this major initiative, but I don't see big change in our office."

The council has held the firm's feet to the fire in a variety of ways: an annual report on the initiative; periodic voicemail updates from Lynn Martin to the entire firm; and full-day meetings of the council with the firm's senior executives. The council defines the challenges we still face, and it lets senior management know they're not off the hook.

Along with helping the task force think about gender, the council has opened the firm's eyes to broader issues. In 1994, the council was meeting with a group of eight professionals—four men and four women—identified by their managers as rising stars at Deloitte. At the end of the meeting, one member of the council asked, almost as an afterthought, "How many of you want to be partners next time we see you?" Only one of the eight said yes. Stunned, the council asked for an explanation.

They were surprised to find that young men

in the firm didn't want what older men wanted; they weren't trying to buy good enough lifestyles so that their wives didn't have to work. At the time, the average partner at Deloitte was making $350,000 and working 80 hours a week, but these young people—men and women both—would've been happy working 60 hours a week for $250,000. They believed they were good enough, and they weren't willing to give up their families and outside lives for another $100,000. One council member recalls, "When we asked if they wanted to be partners, we thought they were going to salute and thank us and hope we put nice letters in their files. Instead they looked at us and said, 'Perhaps.'"

Begin with Dialogue as the Platform for Change

The task force had found that women at Deloitte perceived they had fewer career opportunities than men, but no one could point to any specific policies as the culprits. We had to tackle our underlying culture to fix the problem. Accordingly, the firm held special two-day workshops designed to explore issues of gender in the workplace. We needed to begin a dialogue: in our view, the key to creating cultural change in the firm was to turn taboo subjects at work into acceptable topics of discussion.

During 1992 and 1993, nearly every management professional at Deloitte & Touche—5,000 people, including the board of directors, the management committee, and the managing partners of all of our U.S. offices—attended the workshop in groups of 24. Cook personally monitored attendance; as one partner puts it, "Resistance was futile." Many harbored doubts. I myself saw it as just one more thing to do, and I had always been skeptical of HR-type programs. I'm sure I wasn't the only partner calculating in my head the lost revenue represented by two days' worth of billable hours, multiplied by 5,000—not to mention the $8 million cost of the workshops themselves.

I was dead wrong. The workshops were a turning point, a pivotal event in the life of the firm. Through discussions, videos, and case studies, we began to take a hard look at how gender attitudes

affected the environment at Deloitte. It wasn't enough to hear the problems in the abstract; we had to see them face to face. Sitting across a table from a respected colleague and hearing her say, "Why did you make that assumption about women? It's just not true," I, like many others, began to change.

The lightbulbs went on for different partners at different times. Many of us had little exposure to dual-career families but did have highly educated daughters entering the workforce. A woman partner would say to a male counterpart, "Sarah's graduating from college. Would you want her to work for a company that has lower expectations for women?" Suddenly he'd get it.

Case studies were useful for bringing out and examining subtle differences in expectations. Drawing on scripts provided by outside facilitators, people in the workshops would break into groups, discuss cases, and share solutions with the full group. A typical scenario would have partners evaluating two promising young professionals, a woman and a man with identical skills. Of the woman, a partner would say, "She's really good, she gives 100%. But I just don't see her interacting with a CFO. She's not as polished as some. Her presentation skills could be stronger." The conversation about the man would vary slightly, but significantly: "He's good. He and I are going to take a CFO golfing next week. I know he can grow into it; he has tremendous potential." Beginning with these subtle variations in language, careers could go in very different directions. A woman was found a bit wanting, and we (male partners) couldn't see how she would get to the next level. As one woman summed up, "Women get evaluated on their performance; men get evaluated on their potential."

Another scenario had two members of a team arriving late for an early-morning meeting. Both were single parents, one a father and one a mother. The team joked about and then forgot the man's tardiness but assumed the woman was having childcare problems. After the meeting, the team leader, a woman, suggested that she think seriously about her priorities.

Scenarios like these lent realism to the workshop discussions, and hard-hitting dialogue often ensued. One partner was jolted into thinking about an outing he was going to attend, an annual "guys' weekend" with partners from the Atlanta office and many of their clients. It was very popular, and there were never any women. It hadn't occurred to him to ask why. He figured "no woman would want to go to a golf outing where you smoke cigars and drink beer and tell lies." But the women in the session were quick to say that by not being there, they were frozen out of informal networks where important information was shared and a sense of belonging built. Today women are routinely included in such outings.

Work assignments got a lot of attention in the workshops. Everyone knew that high-profile, high-revenue assignments were the key to advancement in the firm. Careers were made on big clients; you grew up on the Microsoft engagement, the Chrysler engagement. But the process of assigning these plum accounts was largely unexamined. Too often, women were passed over for certain assignments because male partners made assumptions about what they wanted: "I wouldn't put her on that kind of company because it's a tough manufacturing environment," or "That client is difficult to deal with." Even more common, "Travel puts too much pressure on women," or "Her husband won't go along with relocating." Usually we weren't even conscious of making such assumptions, but the workshops brought them front and center.

The workshops also highlighted one of the worse aspects of these hidden assumptions: they were self-fulfilling. Say a partner gets a big new client and asks the assignment director to put together a team, adding, "Continuity is very important on this engagement." The assignment director knows that women turn over more rapidly than men and has the numbers to prove it. So the thinking goes, "If I put a woman on this account, the partner will be all over me—and that's who evaluates me." In the end, John gets to work on the big account and Jane works "somewhere else." After a while, Jane says, "I'm not going anywhere here. I'm never going to get the big opportunities," so she leaves. And the assignment director says, "I knew it."

The task force realized the workshops were risky; the firm was opening a can of worms and couldn't control the results. Indeed, a few of the workshops flopped, disintegrating into a painful mixture of bitterness and skepticism. Some people dismissed the experience as a waste of time. But ultimately the workshops converted a critical mass of Deloitte's leaders. The message was out: don't make assumptions about what women do or don't want. Ask them.

PUTTING THE NEW ATTITUDES TO WORK

The workshops generated momentum, but the dialogue had to be followed with concrete operational steps if we were going to bring about real change. The task force had clear expectations: more of our qualified women should be promoted, and the turnover rate for women should fall. But the firm had to be careful not to set quotas or seem to give women all the plum assignments. The key was to send a clear, powerful message for change while still giving heads of local offices some discretion.

Use a Flexible System of Accountability

Since the fastest way to change behaviors is to measure them, the task force started by simply asking for numbers. Beginning in 1993, in the midst of the workshops, local offices were asked to conduct annual reviews to determine if the top-rated women were receiving their proportionate share of the best assignments. Some offices resisted, questioning the usefulness of this time-consuming exercise or fearing that the initiative would lead to quotas. However, a few pointed phone calls from the CEO prodded the laggards. The reviews confirmed our suspicions: women tended to be assigned to projects in nonprofit, health care, and retail—segments that generally lacked large global accounts—while men received most of the assignments in manufacturing, financial services, and highly visible areas like mergers and acquisitions.

The reviews had their intended effect. Like many other managing partners, I began routinely discussing assignment decisions with the partners in charge of project staffing to make sure women had opportunities for key engagements. Most offices began tracking the activities of their high-performing women on a quarterly basis. To complement the connections that men naturally made with one another, we began hosting regular networking events for women—for example, panel discussions where women partners discussed their careers and leadership roles, followed by networking receptions. We also started formal career planning for women partners and senior managers. This planning proved so helpful that women suggested men also be included, thus giving rise to Deloitte Consulting's current Partner Development Program.

Only after the operational changes had percolated through the organization did the task force introduce clear accountability for the changes that were being made. It offered offices a menu of goals derived from the Women's Initiative—such as a recruiting hit rate or a reduction in the gender gap in turnover—yet left it up to the offices to pick the goals best suited for their particular situations. Office heads started including their choices among the objectives that drove their year-end evaluations and compensation. And the firm made sure that results on turnover, promotion, and other key numbers for each office were circulated widely among man-

agement, feeding a healthy internal competitiveness. Low-performing offices got calls or visits from task force members to push for better progress. Today partners know that they will not become leaders of this organization if they have not demonstrated their commitment to the Women's Initiative.

IT'S NOT JUST ABOUT WOMEN

Moving toward equality in career development was fundamental. But as people began to discuss gender issues in workshops, meetings, and hallways, what started out as a program for women soon began to affect our overall corporate culture.

Promote Work-Life Balance for Men and Women

We discovered that work-life balance was important to everyone. On paper, we had always allowed temporary, flexible work arrangements, but people believed (rightly, at the time) that working fewer hours could doom an otherwise promising career. In 1993, only a few hundred people were taking advantage of the policy. So now we said that opting for flexible work wouldn't hinder advancement

---

Lessons from Deloitte's Women's Initiative

1. Make sure senior management is front and center. To overcome the resistance of partners, the CEO actively led the Women's Initiative. He put his own reputation on the line.

2. Make an airtight business case for cultural changes. Emotional appeals weren't going to be enough. We had to document the business imperative for change before we could justify the investment and effort that the initiative would require.

3. Let the world watch you. We appointed an external advisory council and told the press about our plans. They wouldn't let the initiative be another "program of the year" that led nowhere.

4. Begin with dialogue as the platform for change. We required everyone to attend intensive workshops to reveal and examine gender-based assumptions in mentoring and client assignments.

5. Use a flexible system of accountability. We first required local offices to measure their efforts with women professionals. Next, we worked with the office heads to select their focus areas for change under the initiative.

6. Promote work-life balance for men and women. Policies for flexible work arrangements and lighter travel schedules not only eased the strain on busy professionals but also helped open our corporate culture.

in the firm, though it might stretch out the time required for promotion. Use of these arrangements became one more benchmark of an office's progress with the initiative. And when a woman was admitted to the partnership in 1995 while on a flexible work arrangement, people really began to get the message. By 1999 more than 30 people on flexible work arrangements had made partner, and in that year, the total number of people on flexible schedules had doubled to 800.

We also reexamined the schedule that all of us work, especially within the consulting practice. A grinding travel schedule had long been an accepted part of the macho consultants' culture. Typically, a consultant was away from home five days a week, for up to 18 months at a time. In 1996, we started a new schedule, dubbed the 3-4-5 program. Consultants working on out-of-town projects were to be away from home three nights a week, at the client site four days a week, and in their local Deloitte offices on the fifth day.

The 3-4-5 schedule hasn't been feasible on all projects—for example, those with tight deadlines like Y2K-driven system implementations. In fact, many of us were concerned initially that the program would compromise client service. But most clients embraced our new program. It turned out that employees from the client's regional offices were exhausted, too, by traveling to meet Deloitte's team at their home offices all week long. One day each week without the Deloitte consultants at their sites was a relief, not an inconvenience! By breaking the collective silence about the personal price everyone was paying, we made everyone happier. We now expect the vast majority of all projects to conform to 3-4-5.

As a result of these and other changes, we've transformed our culture into one in which people are comfortable talking about aspects of their personal lives, going well beyond client assignments and career development. Teams are getting requests like "I want to talk to my kids every night at 7:00 for half an hour," or "I'd really like to go to the gym in the morning, so can we start our meetings at 8:30 instead of 7:30?" This more open environment not only helps us keep our rising stars but also makes us more creative in a variety of areas.

A NEW OUTLOOK

The changes at Deloitte are by no means complete. For many years, women have made up one-third to one-half of Deloitte's recruits, so we need to make sure the percentage of women partners and directors rises well above 14%. And we face new challenges. Now that more women are becoming partners, how can we make sure they continue to develop and advance into positions of leadership? In an increasingly global firm, how can we extend the values of the initiative while respecting local cultural differences?

Still, we have transformed our work environment, even in the smallest details. When a visiting speaker—even a client—cracks a joke at women's expense, none of us laughs, not even politely. One partner turned down an invitation to join a premier lunch club in Manhattan when he learned it excluded women. And we've opened our eyes to differences in style that go beyond gender to include culture. For example, on a recent client engagement, the project manager described an Asian consultant on his team as "shy" and therefore not ready to take on more responsibility. But another partner pushed the project manager for details and suggested that consultants could still be successful even if they didn't "command a room" or raise their voices when speaking in meetings.

We've not only narrowed the gender gap; we've narrowed the gap between who we think we are and who we truly are. Now when I say ours is a meritocracy, I'm speaking about men and women. It's not easy to manage a diverse group of people; we have to be creative and flexible in developing coaching and mentoring capabilities. Although the Women's Initiative has made managing more complicated, the benefits are substantial: greater creativity, faster growth, and far greater performance for our clients.

45

Danger: Toxic Company

ALAN M. WEBBER

The problem isn't that loyalty is dead or that careers are history. The real problem, argues Stanford's Jeffrey Pfeffer, is that so many companies are toxic—and that they get exactly what they deserve.

According to Jeffrey Pfeffer, when it comes to the link between people and profits, companies get exactly what they deserve. Companies that treat their people right get enormous dividends: high rates of productivity, low rates of turnover. Companies that treat their people poorly experience the opposite—and end up complaining about the death of loyalty and the death of talent. These are "toxic workplaces," according to Pfeffer, 52, the Thomas D. Dee Professor of Organizational Behavior at the Stanford Graduate School of Business and the author of *The Human Equation: Building Profits by Putting People First* (Harvard Business School Press, 1998).

Pfeffer disputes much of the conventional wisdom in the current conversation about work and business. Loyalty isn't dead, he insists—but toxic companies are driving people away. There isn't a scarcity of talent—but there is a growing unwillingness to work for toxic organizations. Pfeffer also disputes the idea of the end of the career. "I don't believe that people are looking to go flitting from one job to the next," he says. "People are looking for the opportunity to have variety in their work and to tackle challenging assignments. The best companies are figuring out how their employees can have both opportunities—without leaving." When *Fast Company* interviewed the plain-talking, provocative Pfeffer in his Palo Alto office, he offered the following observations about the primacy of people in the new economy and about how you can detoxify your workplace.

THE ONE GUARANTEED WAY TO GET A 30% TO 40% PRODUCTIVITY GAIN

It mystifies me that so many companies think they can get a cheap competitive advantage by purchasing something on the open market! Anything that you can purchase on the open market is also available to your competitors. So the question is, How can you distinguish yourself in a world in which your competitors can copy everything you do?

The answer is, all that separates you from your competitors are the skills, knowledge, commitment, and abilities of the people who work for you. There is a very compelling business case for this idea: Companies that manage people right will outperform companies that don't by 30% to 40%. This principle even applies to the current IPO market: IPO firms that value their people have a much higher five-year survival rate than those that don't. Similar studies of the steel industry, the oil-refining industry, the apparel industry, and the semiconductor industry all demonstrate the enormous productivity benefits that come with implementing high-performance, high-involvement management practices.

Most people immediately understand this point. It's not as though I've discovered some mysterious black magic. There is conclusive evidence that holds for all industries, regardless of their type,

Reprinted from *Fast Company Magazine*, November, 1998, pp. 152–156. Used with permission.
Alan M. Webber is the editor of *Fast Company Magazine*.

size, or age. The results are the same. If you don't believe me, look at the numbers.

"WELCOME TO THE TOXIC WORKPLACE! WE FIRE AT WILL!"

There is a lot of turnover in Silicon Valley, because there are so many toxic workplaces in Silicon Valley. These are companies that create the conditions that they deplore. Companies say to me, "Nobody who comes to work for us stays for any length of time. Loyalty is dead." Let's accept that premise for a moment—even though it's wrong. But if we do accept that premise, the question becomes, If loyalty is dead, who killed it?

Companies killed loyalty—by becoming toxic places to work! Start with the interviewing and recruiting process. What happens on a new employee's first day? The company asks the employee to sign an at-will employment contract that gives the company the right to fire the person at any time and for any reason. The document was prepared by a lawyer; the company tells the employee to have it reviewed by a lawyer before signing it.

Think about it: It's your first day on the job, and I've already told you that you don't have a permanent employment relationship with me, that your job is based on a contractual relationship—and then I wonder why, on your second day, you're not approaching me with a long-term perspective and a feeling of trust!

A PLACE WHERE PEOPLE COME TO WORK TO GET RICH ENOUGH TO QUIT

Here's another example of a practice that creates the conditions that companies deplore: stock options. David Russo, the head of human resources for SAS Institute, gave a talk to my class. My students were dumbfounded to learn that this successful software-development company doesn't offer its people stock options. David said, "You must know lots of people who have gotten stock options. Why do they want them? Explain the logic."

Finally one of my students raised his hand and said, "I can tell you the logic. For most people, stock options are like the lottery. People are hoping to strike it rich and then quit." David smiled and said, "What an interesting thing! We've built an organization in which your motivation for coming to work is to make a lot of money—so that you can get the hell out of the organization."

To me, that's an operational definition of a toxic workplace: It's a place where people come to work so they can make enough money so they can leave. Dennis Bakke, of Applied Energy Services Corp. (AES), likes to point to a photo of the top 20 people at his company. The photo was taken more than a decade ago, and today 17 of those 20 people are still there. They're all plenty wealthy—because AES has grown tremendously. They all could have quit. The fact that they didn't says a lot about that company.

TOXIC FLEXTIME: "WORK ANY 18 HOURS YOU WANT"

Another sign that a company is toxic: It requires people to choose between having a life and having a career. A toxic company says to people, "We want to own you." There's an old joke that they used to tell about working at Microsoft: "We offer flexible time—you can work any 18 hours you want."

A toxic company says, "We're going to put you in a situation where you have to work in a style and on a pace that is not sustainable. We want you to come in here and burn yourself out—and then you can leave." That's one thing that SAS manages brilliantly: When you take a job there, you don't have to ask yourself, "Am I going to be a successful and effective SAS employee, or am I going to know the names of my children?"

WHAT'S THE DIFFERENCE BETWEEN A FACTOR OF PRODUCTION AND A HUMAN BEING?

Another sign of a toxic workplace is that the company treats its people as if they were a factor of production. At a toxic workplace, the managers can reel off all of the various economic factors: "We've got capital that we invest, we've got raw material that we use, we've got the waste from the manufacturing process that we recycle—and, in the same category, we've got our people." It's a workplace that doesn't see people as people, but rather sees them as factors of production. And that's ironic, because what we celebrate as a competitive, capitalistic practice actually reflects a Marxist orientation: People are seen as a factor of production, from which a company has to extract an economic "surplus."

There is a huge difference between that perspective and the way AES, for example, looks at its people. Dennis Bakke even objects to the term "human resources." Dennis says that fuel is a resource—but that people aren't. Underlying this difference in language is a difference in philosophy that guides much of what a company does. If a company looks at you merely as a factor of production, then every day it must calculate whether your marginal revenue exceeds your marginal cost. That's how it decides whether or not to keep you.

IF YOUR COMPANY IS SO GREAT, WHY DOESN'T ANYONE WANT TO WORK THERE?

You hear a lot about the shortage of talent. The thing to remember is that, for great workplaces, there is no shortage of talent. Companies that are short on talent probably deserve to be! Anyone who is smart enough to work in a high-tech company is too smart to work in a toxic workplace. And if they do work in one, as soon as they have a choice, they choose to leave.

For example, according to David Russo, SAS Institute had a 3% voluntary-turnover rate in 1997. SAS almost never loses one of its people to a competitor, he says. When it does lose people, it's usually because of a lifestyle change or because someone at the company has to move to a place where there is no SAS facility.

That kind of thing is happening across the economy. Hewlett-Packard has lower turnover than many of its competitors. The Men's Wearhouse has lower turnover than many other companies in the retail industry. Starbucks has comparatively lower turnover than other companies in the fast-food business. Of course, none of these companies is perfect. But a company that says, "We want to create a place that attracts people, that makes them want to stay," will have lower turnover than places that say, "We don't care about our people's well-being or about whether they stay." And then, when these toxic companies conduct themselves in this way, they wonder why people leave.

WHICH IS BETTER BUSINESS—PAYING SIGNING BONUSES OR TREATING PEOPLE RIGHT?

High turnover costs big money. First of all, it costs money to go out and replace all of the people you've lost. If the companies in Silicon Valley that are losing people would stop paying $50,000 signing bonuses, and instead do what's necessary to keep the people they've got, they would be much better off economically. Along with incurring replacement costs, when you lose people, you lose knowledge, you lose experience, and you lose customer relationships. Every time a customer interacts with your company, he or she sees a different person. I like to go to my branch bank because I know that I'll always make a new friend there: The turnover is so high, I'm always meeting new people!

There is nothing soft and sentimental about this part of the argument. This is simple economics. David Russo did a calculation in my class one day: A student asked him why SAS does so much family-friendly stuff. He said, "We have something like 5,000 employees. Our turnover rate last year was 3%. What's the industry average?" Somebody said 20%. Russo replied, "Actually, 20% is low, but I don't care. We'll use 20%. The difference between 20% and 3% is 17%. Multiply 17% by 5,000 people, and that's 850 people. What does turnover cost per person? Calculate it in terms of salary." The students estimated that the cost is one year's salary and that the average salary is $60,000. Russo said, "Both of those figures are low, but that doesn't matter. I'll use them. Multiply $60,000 by 850 people, and that's more than $50 million in savings."

That's how Russo pays for the SAS gymnasium, for on-site medical care, for all of the company's other family-friendly items. "Plus," he said, "I've got tons of money left over." If you can save $50 million a year in reduced turnover, you're talking about real financial savings. This is not tree-huggery. This is money in the bank.

WHEN YOU LOOK AT YOUR PEOPLE, WHAT DO YOU SEE—EXPENSES OR ASSETS?

You've got to ask a question that gets back to an old cliche: Do you walk the talk? It's easy for a company to say, "We invest in people. We believe in training. We believe in mutual commitments between the managers and the workforce. We believe in sharing information widely with our people."

Many organizations say those things—but in their heart of hearts, they don't believe them. Most managers, if they're being honest with themselves, will admit it: When they look at their people, they see costs, they see salaries, they see benefits, they see overhead. Very few companies look at their people and see assets.

In part, it's because of the financial-reporting systems that we've got. The fact is, your salary is an expense. If I buy a computer to replace you, I can capitalize the computer and then depreciate its useful life over many years. If I hire you, I take on an expense.

But there are other things that companies can measure. Whole Foods Market Inc. and AES, for instance, not only do employee surveys; they also take them seriously. I know of managers at Hewlett-Packard who were fired because they received such poor reviews from their employees.

WHY NOTHING CHANGES #1: WISHING DOESN'T MAKE IT SO

Everybody knows what to do, but nobody does it. For example, a lot of companies confuse talk with action. They believe that, because they've said it, it's actually happened. One of my students did a

research project on the internship program at a large Wall Street securities company. Under the program, the firm hired interns right out of college; then, after a few years, the interns would go back to business school. But the program was catastrophic. The firm treated the interns like dog doo, and that had two bad consequences. First, the interns didn't go back to work for the firm after business school, so the two or three years that the firm had invested in them were wasted. Second, and even worse, when the interns arrived at Harvard, Stanford, Chicago, or Northwestern, they would tell all of the other business-school students that the firm was horse manure—which made its recruiting difficult, to say the least. At some level, people at the firm understood these problems. So the senior leadership asked my student, who had interned there, to help the firm fix its program.

After she had done a bunch of interviews, she told them, "You have a model that says you're going to treat people with respect and dignity. Let me tell you the 30 things—and that's a low number—that you do to your interns that violate your model."

When the top people at the firm heard her report, they said, "This can't be. The core values of this firm are respect for the individual, treating the individual with dignity, and teamwork. We believe in these values." In effect, they were saying, "Because we believe it, it must be happening." These were not bad people. They just thought that their wishes had become reality.

WHY NOTHING CHANGES #2: MEMORY IS NO SUBSTITUTE FOR THINKING

There's another reason why companies don't do what they know they should: They fall prey to the power of precedents. They do something once, and then they get trapped by their own history: This is the way we do it because this is the way we've always done it. They substitute memory ("We did it this way before") for thinking ("Is this a sensible way to do it?").

Not long ago, I went to a large, fancy San Francisco law firm—where they treat their associates like dog doo and where the turnover is very high. I asked the managing partner about the turnover rate. He said, "A few years ago, it was 25%, and we're now up to 30%." I asked him how the firm had responded to that trend. He said, "We increased our recruiting." So I asked him, "What kind of doctor would you be if your patient was bleeding faster and faster, and your only response was to increase the speed of the transfusion?"

But what this law firm knew how to do was recruit. Over the preceding five years, it had created 162 partners—and it had lost 163 people in the same period. It preferred undertaking lavish recruitment efforts to dealing with the root causes of the turnover. People do what they know how to do. This law firm knows how to recruit—so it steps up recruiting. But what it hasn't thought about for five seconds is how to solve the underlying problem.

HOW TO MAKE SOMETHING CHANGE: START WITH YOU

Where do you start? You start with a philosophy, and the rest follows from that. If you believe in training and developing people, you don't necessarily need a huge training budget. You begin by imparting knowledge in various ways—by holding meetings, by talking to people, by coaching them, by mentoring them. If you believe in reciprocal commitments, you start by building those commitments with the people you work with. If you believe in information sharing, you share information with the people you have the most contact with. In other words, you begin in your immediate sphere of influence. You start with your own behavior.

46

Managing Dual Ladder Systems in RD&E Settings

RALPH KATZ AND THOMAS J. ALLEN

Organizations employing professional specialists, engineers, and scientists in particular face the dilemma of establishing reward systems that are both stimulating to the professional and productive for the organization. This problem stems in part from the notion that specialist groups bring to the organization a set of attitudes and career aspirations that are in conflict with the organization's work requirements and established career paths. It is often argued that many Research, Development, and Engineering (i.e., RD&E) professionals are socialized into their technical occupations with values and definitions of success that differ significantly from those prevailing in the traditional managerial setting [1]. In the typical organization, for example, management expects authority to be aligned with the hierarchical principle, discharged through a progression of well-ordered job positions. Technical professionals, on the other hand, value the freedom to pursue their own technical interests, the responsibility for making judgments in their areas of expertise, and the exercise of organizational control through knowledge, logical arguments, and collegiality.

For many years now, much has also been written about how professional needs clash with those organizational incentives normally available to managers [16, 22]. In theory, RD&E professionals are supposedly motivated by a desire to contribute to their disciplines and to establish a credible or distinguished reputation among their technical colleagues. In a sense, they are strongly oriented toward work in their professions, developing strong commitments to their specialized skills and outside professional reference groups [8]. And it is often this more "cosmopolitan" orientation that causes one to be seen as being less organization-loyal. For if publications, the production of knowledge for its own sake, or independence are the primary goals of scientists and engineers, then they will feel thwarted in achieving these goals when confronted by the economic realities and practical needs of the company. Managers, in contrast, desire upward mobility in the organizational hierarchy. In a sense, they are more committed to developing their own "local" organizational careers. They do this by focusing more on the achievement of company objectives and the acquisition of organizational approval and promotion. As one R&D professional recently framed it for us: "To have my ability recognized rather than my authority is far more rewarding." To the true professional, then, upward mobility in the managerial hierarchy is of little importance compared to autonomy in the practice of one's technical specialty. Success is, therefore, defined independent from managerial advancement. In short, the argument is that professionals acquire status and define success from the perspectives of their technical colleagues while managers build these same attributes from the perspectives of their organizational superiors.

These conflicts between professional and business-related goals can be very problematic for an organization as scientists and engineers try to uphold professional standards in the face of strong pressures for commercializable results [20]. Not all

Reprinted with the permission of the authors. Earlier versions of this research were published in T. J. Allen and R. Katz, "The dual ladder: motivational solution or managerial delusion?" *R&D Management,* vol. 16, no. 2, pp. 185–197, 1986; and T. J. Allen and R. Katz "Age, education, and the technical ladder," *IEEE Transactions on Engineering Management,* vol. 39, pp. 237–245, 1992.

Ralph Katz is Professor of Management at Northeastern University's College of Business and Research Associate at MIT's Sloan School of Management. *Thomas J. Allen* is the Howard Johnson Professor of Management at MIT's Sloan School of Management.

scientists and engineers are alike in their orientations toward career success, however. Whether engineers and scientists are more interested in peer recognition than they are in organizational advancement has been the subject of much debate. While some studies have suggested that engineers are very different in their professional and organizational orientations than their more scientific counterparts [1, 17], other studies question whether one can truly generalize within any professional occupation [6, 13]. Generally speaking, research studies have substantiated an important distinction between professionals who are either more "cosmopolitan" or more "local" in their career orientation [8, 7]. In contrast to the previously described cosmopolitan stereotype, many technical professionals are local in their overall perspectives and are, in fact, interested in working on the application of technology that achieves the business aims of the company. Local professionals, as a result, are more involved than their cosmopolitan counterparts in establishing organizational identities and careers through the successful commercialization of technical accomplishments. In a sense, the loyalties and work focus of locals are more strongly aligned with their organizations than with their professional peers. In any organization, there is generally some proportion of professionals who prefer technical problem solving and for whom management has very little attraction and vice versa.

THE DUAL LADDER REWARD SYSTEM

Despite these purported "professional/managerial" or "cosmopolitan/local" differences, the highest rewards in most business organizations are conferred on those who assume additional managerial responsibility. Advancement up the managerial ladder secures increases in status, recognition, salary, influence, and power. For many professionally oriented technologists, movement into management becomes the most viable career strategy simply because their opportunities to achieve success without undertaking such managerial responsibilities are very limited. As a result, many productive engi-

neers and scientists feel frustrated as they are "pressured" to take on managerial and administrative roles they really do not want in order to attain higher salary and more prestige.

The "dual ladder system of career advancement" is an organizational arrangement that was developed to solve these individual and organizational problems by formalizing promotions along two parallel hierarchies: one provides managerial progression, while the other provides opportunity for professional advancement [3]. The dual ladder system promises equal status and rewards to equivalent levels in the two hierarchies. In providing the more professionally oriented specialists with opportunities and incentives to remain active in their fields, without having to shift to management, the dual ladder aims to secure for the firm a highly motivated and creative pool of technical talent. It tries to establish a viable career track that maintains the productivity of highly innovative individuals who either see themselves or who are viewed by others as less interested in or less capable of carrying out managerial responsibilities by rewarding them with increased levels of prestige, freedom, and appropriate job perquisites as they are promoted up the technical ladder.

Although dual ladders have now been in use for some time, their success has been the focus of much debate [12]. Three generic kinds of problems seem to underly the concept. First, most cultures automatically associate prestige with managerial advancement. Titles of Department Head and Vice President convey images of success, while titles of Senior Researcher and Lead Engineer are considerably more ambiguous and therefore more subject to skepticism. Many organizations exacerbate these differences by not living up to their promised commitments of creating equal status, perquisites, resources, and other financial and symbolic rewards to those of equivalent levels in the managerial and professional hierarchies. Frequently too, management does a poor job of publicizing the technical ladder and little observable change takes place either in work activities or responsibilities after technical promotion. An additional problem arises when technical promotions are debated through justifications of past contributions while managerial promotions are more positively discussed in terms of future promise and potential.

A second set of problems concerns the nature of incentives associated with each ladder. Movement up the managerial ladder usually leads to positions of increased influence and power within the organization. The number of employees under a manager typically increases with promotion and such resources can be mobilized more easily to carry out the manager's needs and demands. In sharp contrast, advancement up the technical ladder usually leads to increased autonomy in the pursuit of one's technical interests but often at the expense of organizational influence and power. Neither the number of subordinates nor any visible means of power increase, fostering perceptions that the technical ladder might really be less important. The issue of relevance becomes even more difficult as the organization grants professionals enough freedom to select their work with little linkage between their activities and company objectives, returns, or paybacks. Such conflicts are aggravated even more as the organization chooses to either eliminate or deemphasize certain areas of interest. As a result, supervision of individual contributors becomes more difficult and feelings of isolation from the organization become more pronounced. The risk is that the technical side becomes a "parking lot" for bright technologists whose abilities to generate ideas easily outstrips the capability of the organization for dealing with them. The rewards of freedom and independence can also bring with them feelings of rejection and disconnection.

Finally, there is the inevitable tendency to "pollute" the technical side of the dual ladder. In addition to rewarding outstanding technical performers who choose to remain in the organization as individual contributors, the technical ladder becomes a repository for less successful, unnecessary, and even incompetent managers. Over time, the criteria for technical promotion are gradually corrupted to encompass not only technical contributions but also organizational loyalty, rewarding those individuals who have been "passed over" for managerial positions. Another common practice is to use the professional ladder primarily for pacifying in-

dividuals who are technically competent and who deserve to be rewarded, but who lack diplomatic skills or management ability. When any of this is done, it can make the technical ladder into a consolation prize, demotivating individuals who interpret technical promotions not as a reward but simply as a signal that they are "not good enough to be a manager." Certainly, such misuses undermine the integrity of the dual ladder system.

Much has been done over the past few years to improve the formal structures of dual ladder systems to alleviate these problems. Using internal and external peer reviews, organizations have begun "policing" their technical ladders to protect their purity and prevent the "dumping ground" abuses. They have tried to strengthen their commitment to the technical side through increased publicity, recognition, career counseling, and information dissemination; through making the ladders more comparable in numbers of people, and perquisites at equivalent hierarchical levels; through clearer job descriptions, qualifications, responsibilities, performance standards, and reporting relationships; and though greater involvement in organizational decision-making and in influencing technical strategy.

Despite these changes, there is still very little empirical research regarding the reactions of professionals for whom the technical ladder was originally designed. The present research, as a result, revolves around two key questions. First, what proportion of a laboratory's technical staff will find the technical ladder career an attractive one and to what extent do their perceptions change over time? This temporal investigation is critical since most technical professionals do not graduate with concrete notions of career success; instead they develop and change their orientations and perspectives over time as they encounter different work experiences [7, 19].

The second research question focuses on the characteristics of those scientists and engineers who indicate a preference or predisposition for the technical ladder. Were they to show a consistent set of characteristics, then knowledge of that fact could provide management with important guidance regarding the appropriate use of the dual ladder system for those types of people or situations. It is our initial contention that career orientations of technical professionals are strongly influenced by the nature of the work in which they are engaged and by their level of education.

One of the most important factors affecting one's work experiences lies in the nature of one's task assignments. Within an RD&E setting, projects can be categorized along a continuum ranging from research to development to technical service [21]. More importantly, empirical studies have revealed substantial problem-solving and information-processing differences among projects engaged in these kinds of activities [10]. Technical professionals with a more cosmopolitan orientation are more likely to be engaged in research projects since their problems and solution approaches are more universally defined and less constrained by organizational circumstances and boundaries [2]. Research projects are also more likely to involve the advancement of knowledge and are therefore less pressured by the business goals of the company. Development and technical service projects, however, require a more local outlook since the development or design of technology is without meaning if the technology is not successfully commercialized or otherwise put to use. As emphasized by Ritti [17], development or design projects are ultimately tested by competition in the marketplace and not by exposure to or review by one's technical peers and colleagues. We would expect to find, therefore, that technical professionals working in development and technical service projects are less cosmopolitan in their career and work orientations than their research colleagues. Based on this logic, the following hypothesis is suggested:

H1: The degree of preference for a technical ladder career will vary systematically with the nature of the work performed by an individual scientist or engineer, viz., those performing basic research will exhibit a stronger preference than those engaged in applied research, development, and technical service. Those doing applied research will show a stronger preference than those engaged in development or technical service, and so on.

Additional research also indicates that educational background strongly affects the values and career interests of scientists and engineers [17, 13, 6]. According to these studies, technical professionals with Ph.D.'s are not only more likely to prefer research work but they are also more likely to be interested in scientific and technical accomplishment rather than promotion within their organization. This is in sharp contrast to non-Ph.D.'s who have been shown to place greater emphasis on salaries and organizational advancement. Scientists and engineers with a Ph.D. degree have been shown to stress the importance of technical communication with their outside professional community while also striving to achieve greater autonomy within their organization [17, 1]. One explanation for these effects is that university education at the Ph.D. level is substantially more specialized and emphasizes more explicitly the values of fundamental knowledge and discovery. As a result of this exposure, Ph.D.'s often seek and expect upon graduation to be in positions that will allow them to continue to contribute to their technical disciplines. Our supposition is that the university experience of Ph.D.'s provides them with expectations and values that fit more directly with the goals of cosmopolitans than with local types of professionals. It is likely, therefore, that Ph.D.'s will develop a stronger preference for the technical side of the dual ladder than their non-Ph.D. colleagues. These educational differences lend themselves to the following hypothesis:

H2: The degree of preference for a technical ladder career will vary systematically with the level of education of the individual engineer or scientist, viz., those with a Ph.D. degree will exhibit a stronger preference than those with a Master of Science degree who in turn will show a stronger preference than those with a Bachelor of Science.

RESEARCH METHOD

The data were collected in a study of about 2,500 scientists, engineers and managers in nine U.S. and two European organizations. The selection of participating organizations could not be randomized, but they were chosen to represent several distinct sectors and industries. Two of the organizations are government laboratories, and three are not-for-profit firms doing most of their business with government agencies. The six remaining organizations are in private industry: two in aerospace, one in electronics, two in the manufacture of industrial equipment, and one in the food industry.

In each organization, short meetings were scheduled with the respondents to solicit their voluntary cooperation and to explain the purposes of the study. Each scientist or engineer received an individually addressed questionnaire at this time. The questionnaire included the usual demographic questions plus several questions about the ways in which the respondent viewed his or her future career and the ways in which the organization structured its reward system around career factors. There were also several questions addressing the way in which engineers viewed their jobs and the importance that they attached to various features in their jobs. The present paper is developed around the central questions shown in Table 46.1. These ques-

TABLE 46.1
Format of the Principal Questions

To what extent would you like
your career to be:

		Not at all		Somewhat			To a great extent	
a)	a progression up the technical professional ladder to a higher-level position?	1	2	3	4	5	6	7
b)	a progression up the managerial ladder to a higher-level positon?	1	2	3	4	5	6	7
c)	the opportunity to engage in those challenging and exciting research activities and projects with which you are most interested, irrespective of promotion	1	2	3	4	5	6	7

tions ask engineers the degree to which they would prefer each of three alternative careers. They were asked to choose between progression on either the managerial or technical ladders or in lieu of these, the opportunity to engage in challenging and exciting projects irrespective of promotion. Using questions and definitions developed by the National Science Foundation and Pelz and Andrews [15], individuals also indicated how well the categories of research, development, and technical service represented the activities in which they were engaged.

Individuals were asked to complete their questionnaires as soon as possible. They were provided with stamped return envelopes so they could mail completed forms to the investigators directly. These procedures not only enhance data quality since respondents must commit their own time and effort but they also increase the response rate. The response rate across organizations was extremely high, ranging from 82 percent to 96 percent. A total of 2,199 usable questionnaires were returned.

RESULTS

Respondents (including staff technologists, managers, and professionals already promoted to technical ladder positions) ranged in age from 21 to 65, with a mean of 43 years. Individuals within this overall respondent sample were initially classified as being oriented toward a technical, managerial, or project-centered career if their response on one of the three scales exceeded the response on the other two by at least one scale point. Those who reported equally favoring any two of the three options were left out of the analysis. A total of 1,495 respondents indicated a preference for one of the three options. Of these, 488 (32.6 percent) preferred the managerial ladder over the two alternative career paths, 323 (21.6 percent) preferred the technical ladder and a surprising 684 (45.8 percent) reported a preference for having the 'opportunity to engage in those challenging and exciting research activities and projects with which (they) are most interested, irrespective of promotion.'

Such a large proportion of respondents preferring a somewhat nontraditional form of reward arouses suspicions that the questionnaire's wording may have made the alternative more attractive than was intended. It would seem reasonable that, were this the case, the induced preference would not be as strongly felt as preferences based on a more substantial conviction. Increasing the margin of preference required in defining orientation did not, however, decrease the proportion of those preferring interesting projects. In fact, the proportions of respondents indicating a stronger preference for this third option over the managerial and technical ladder promotional options went from 45.8 percent to 48.4 percent to 51.4 percent as the margin of preference in scale points went from 1 to 2 to 3 (although the absolute numbers of individuals with this preference did of course decrease from 642 to 393 to 213 as the margin of scale preference was increased from 1 to 2 to 3).

Orientation as a Function of Age

Career preferences, as one might expect, are significantly related to age (F = 18.25; df = 2, 1399;

$p < 0.001$). The proportion of engineers citing a preference for interesting projects increases almost monotonically with age (Figure 46.1). This may be due, partially, to a realization that advancement opportunities along the two traditional ladders are diminishing with age. This can be only partially true, since such a high proportion of those in their twenties indicate this preference. In fact, it is the most preferred alternative for all engineers, save those from 25 to 30.

The technical ladder career attracts the smallest proportion of engineers in all ages. The proportion indicating this preference hovers around 20 percent showing only a mild peak among those in their thirties. The proportion preferring a managerial career peaks in the late twenties and declines steadily thereafter.

Career Preference as a Function of Position

As one might expect, managers report a marked preference for a managerial career. There is some diminution with age (Figure 46.2) with a con-

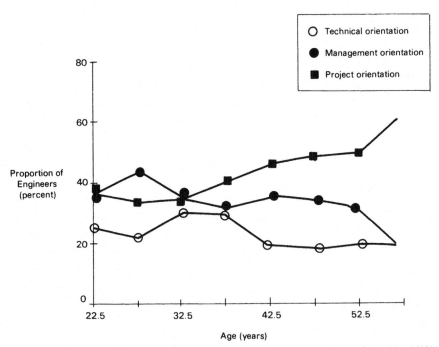

Figure 46.1. Career preferences of engineers in nine organizations as a function of age (N = 1402).

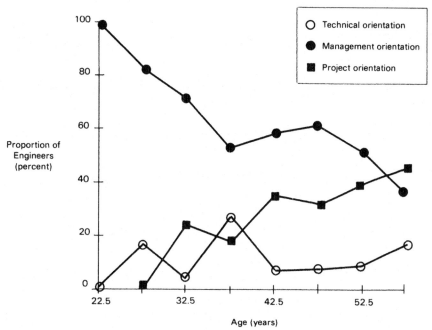

Figure 46.2. Career preferences of managers as a function of age (N = 374).

comitant increase in preference for interesting projects. Only for a brief period in their late thirties do managers show any interest in the technical ladder.

From the data displayed in Figure 46.3, it appears that most of the engineers who are on the technical ladder prefer one of the other two alternatives. The younger ones tend to show a slight preference for management over their own technical ladder. But the older technical ladder engineers indicate a strong preference for interesting work.

This increasing concern for interesting work among the older engineers in all of the figures is a very important and often misunderstood issue. Work assignments for older engineers are often made with the implicit assumption of the inevitability of technical obsolescence. That inevitability has been seriously questioned in recent years. Furthermore, such an assumption leads to work assignments that are inherently less challenging and thereby creates a self-fulfilling prophecy, guaranteeing obsolescence. Recent research [4] shows that instead of age being the cause of obsolescence, the failure of management to

provide challenging work and to emphasize the need for technical currency are the more likely causes. If older engineers seek more challenging work but seldom find it, can there be any wonder that they often allow themselves to sink into obsolescence? Our findings reinforce the importance of career growth for older engineers to prevent their movement into a career stage of stabilization [9]. Older engineers can be challenged by modifying job assignments and thereby forcing the acquisition of new knowledge. That they are interested and concerned about maintaining this type of challenge is quite evident in the data.

TECHNICAL LADDER ORIENTATION

Although career orientations vary with age, our data also show that a substantial proportion of the technical professionals have a preference for technical advancement within their organizations. The next question in our research, therefore, is to identify the characteristics of the respondents in this popula-

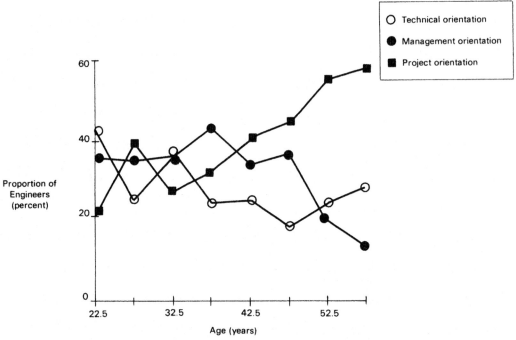

Figure 46.3. Career preferences of technical ladder engineers as a function of age (N = 351).

tion. Toward that end, a three-way analysis of variance was performed on the data, with nature of the work (basic research; applied research; development; technical service), education level (B.S., M.S., Ph.D.) and position (manager or not)[1] as independent variables and the degree of technical ladder preference as the dependent variable. Since age was a significant factor influencing career orientation, it is used as a covariate in the analysis. Surprisingly, among this set of respondents, type of work does not affect the degree of technical ladder preference (F = 0.78; N.S.). The first hypothesis is not supported by the data (Figure 46.4).[2] Education level, on the other hand, has a significant effect (F = 8.06; p < 0.001), as does managerial position. As hypothesized, Ph.D.'s have a much stronger preference for the technical ladder than engineers without a doctorate. Managerial position, as one might expect, has an effect on attitudes toward the technical ladder (managers lose interest in technical ladder careers) so this variable is included as a control in the analysis. It is interesting that even among

the managers, education has an effect. While the average for all managers was below the population mean in desire for a technical ladder career, those managers with a Ph.D. degree scored much higher than those without (−0.02 vs. −0.29) and were very close to the overall mean.[3] There is no indi-

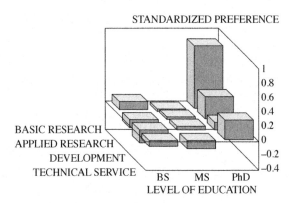

Figure 46.4. Degree of technical ladder preference as a function of education and the nature of the technical work performed.

cation of significant interaction among any of three categorical variables. The effect of the covariate (Age) was not significant ($F = 3.76$; $p > 0.05$).

The analysis shows that the more education an individual has, the more likely that person is to choose or prefer a technical ladder career. This is particularly true of those with a Ph.D. degree. This information should be very helpful to organizations, since it indicates what circumstances would justify the cost of a dual ladder system and which people would feel more rewarded by promotion onto the technical ladder.

Since it is primarily those with a doctoral degree who fall into this class, it will be interesting to see in which other ways their motivations differ from those of their colleagues. If this were known, it could provide a deeper understanding of the reasons for choosing a technical ladder career and provide organizations with better guidance for its appropriate use.

One way of assessing an individual's work goals is to ask that individual for indicators of successful work outcomes. Fortunately, our questionnaire included a series of questions[4] that asked people to rate several possible outcomes on the degree to which, in their work, they would consider them measures of success. To simplify the analysis and reduce the number of variables, a factor analysis is performed on the responses to the questions, reducing them to two factors, one of which describes what one might consider academic/scientific measures of success, the other describing product-related commercial success (Table 46.2).[5]

An ANOVA was then performed using educational attainment (i.e., Ph.D. or non-Ph.D.) as the independent variable and each of the two factor scores as dependent variables (Figure 46.5). Age was again controlled as a covariate. It is no surprise that those with a Ph.D. degree attach importance to academic success criteria and are significantly less interested in commercial success. Adherence to success criteria, such as these, represent to some degree the way in which these individuals expect and want to be evaluated. Just who the anticipated evaluators are is not completely clear, nor does it matter. It is the internal self-

TABLE 46.2
Factor Analysis of Questions Concerning Perceived Measures of Success

Question	Loading on factor 1	Loading on factor 2
Publishing a paper which adds significantly to the technical literature.	0.79	
Developing new theoretical insights or solutions.	0.84	
Developing concrete answers to important technical problems.	0.55	
Contributing to a product of high commercial success.		0.79
Contributing to a product of distinctly superior technical quality.		0.80
Coming up with a highly innovative idea or solution.	0.60	

evaluation that is important for our purposes. That these individuals evaluate themselves against particular external success criteria should provide insight into the nature of their career orientation and underlying value system. Those who are more inclined toward an academic career will measure their success according to appropriate criteria such as publication or theorizing (factor 1). Those inclined toward an industrial career will also choose appropriate criteria, in that case participating in the development of a successful product (factor 2).

It is clear from Figure 46.5 that educational

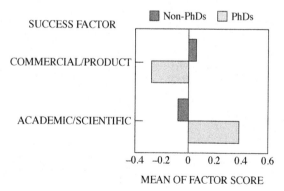

Figure 46.5. Factor scores relating to success criteria as a function of education.

level strongly influences the choice of success criterion. Those individuals with a Ph.D. degree are much more inclined toward the academic/scientific criteria ($F = 52.36$; $p < 0.001$) and less toward the commercial/product-oriented criterion ($F = 23.38$; $p < 0.001$) than are their colleagues who do not hold a doctoral degree. This is certainly understandable, although perhaps not desirable.

The long time which those with a Ph.D. degree spend in graduate school allows a degree of socialization into academic values that apparently persists even after these individuals have worked in industry for quite some period. In a sense one might argue that occupational socialization for these people is much stronger than their organizational socialization.

SOCIALIZATION AND RESOCIALIZATION

It should be interesting to see how long the effect of academic socialization persists. This can be examined by plotting the degree to which respondents cite, as a function of their age, the two types of success criteria. These plots are shown for those without a Ph.D. degree in Figure 46.6 and for those with the Ph.D. in Figure 46.7. It is startlingly clear from these figures that the effect of academic socialization is very persistent. It occurs for engineers and scientists, regardless of education level. They

all enter industry with a much stronger orientation toward academic/scientific goals. For those without a doctorate, however, the commercial/product goals gradually increase in importance becoming more dominant at about age 30. This is a surprisingly long period of accommodation and would appear even more unusual, were it not for the situation among the Ph.D.'s, who appear never to reach a reasonable accommodation with industrial goals. Although the orientation toward commercial/product goals also increases in importance for those with a Ph.D., the magnitude of this success factor always falls below that for academic/scientific success.

To be certain, once again, that it is educational level and its concomitant socialization process that causes the effects shown in Figures 46.6 and 46.7, the same data were replotted after separating people who were working on basic and applied research activities from those working on development and technical service projects. Divided in this manner, the comparative plots for each of the two subgroupings were not very different in their overall patterns from those seen in Figures 46.6 and 46.7. Those without a Ph.D. degree begin their industrial careers with a stronger academic than commercial orientation. After a few years of experience, they shift and become more commercially oriented. The Ph.D.'s begin similarly but never lose their academic orientation. Deep into their industrial careers,

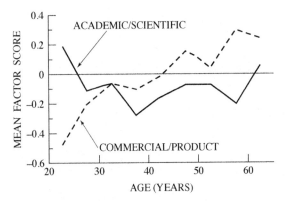

Figure 46.6. Success criteria as a function of age for non-Ph.D. engineers and scientists.

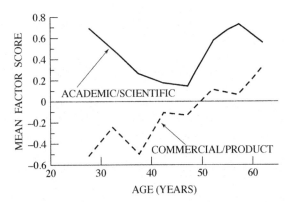

Figure 46.7. Success criteria as a function of age for Ph.D. engineers and scientists.

they are still concerned about academic success measures. This phenomenon appears for those with a Ph.D. degree whether they are working in research or in development or technical service.

RELATIONSHIP TO TECHNICAL LADDER PREFERENCE

To investigate further the interrelationships between educational attainment and perceptions of success, a two-way ANOVA was performed on preference for a technical ladder career, with education and success factors as independent variables. Although both effects are significant, that of education is greater[6] (Figure 46.8). What is more important, the standardized means clearly indicate that it is the combination of a Ph.D. degree and an academic/scientific orientation toward success that produces the strongest preference for the technical ladder. Non-Ph.D.'s with a commercial/product orientation have the lowest performance. It is also important to note that Ph.D.'s with a commercial/product orientation and non-Ph.D.'s with an academic/scientific orientation respond in a similar fashion concerning their preference for the technical ladder. Both are very close to the overall mean of zero.

To investigate the degree to which academic or commercial success criteria are associated with education and managerial position, a logistic regression and categorical data analysis were performed on the data. This analysis shows Ph.D.'s, who are not managers, to be most likely to be academically oriented. Managers, who do not have a Ph.D. degree, are the most likely to have a commercial orientation toward success. Finally, those with neither a Ph.D. degree nor a managerial position are 1.32 times more likely to have a commercial orientation than are Ph.D. managers. Education, in other words, has a greater effect than organizational position on the chances of adopting a particular orientation.

If, in fact, the population groupings differ sharply in how they view the technical ladder, then it is important to know if they differ greatly in other work-related ways. Toward this end, a series of questions, previously used and described by Pelz and Andrews [15] were included in the study. These questions measure respondents' perceptions of important work-related opportunities and problem-solving approaches. The standardized mean responses to these items for the two extreme[7] subgroups are shown in Table 46.3. Not only do these subgroups differ sharply in their preference for the technical ladder, but they also differ significantly in the way in which they relate to their work environments.[8] Those who most prefer the technical ladder also prefer to work more conceptually and in greater depth on problems that are more important to their professional disciplines. In contrast, those who least prefer the technical ladder want to work on more immediate solutions to problems that are more relevant to the organization. Similarly, those preferring the technical ladder also value freedom and independence and prefer to work less collaboratively.[9] The groups did not differ on their need to work on challenging tasks or with competent colleagues.

DISCUSSION

Companies recruit engineers and scientists with a Ph.D. degree for their level of education and for their demonstrated intelligence and perseverance, having survived a long and sometimes arduous ed-

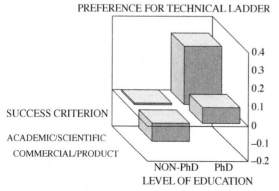

PREFERENCE FOR TECHNICAL LADDER

SUCCESS CRITERION

ACADEMIC/SCIENTIFIC
COMMERCIAL/PRODUCT

NON-PhD   PhD
LEVEL OF EDUCATION

0.4
0.3
0.2
0.1
0
−0.1
−0.2

Figure 46.8. Preference for technical ladder as a function of education and criteria for success.

TABLE 46.3
Contrasting Perceptions of Work-Related Issues

	Means of standardized scores for:		
Work opportunities and problem solving approaches	Ph.D.'s with academic/scientific orientation	Non-Ph.D.'s with commercial/ product orientation	Difference
Preference for working toward immediate concrete solutions	−0.41	0.20	0.61*
Importance of working on organizationally significant tasks	−0.18	0.21	0.39*
Preference for working in collaboration with others	−0.21	0.14	0.35*
Importance of working on difficult and challenging assignments	0.18	−0.01	0.19
Importance of working with technically competent colleagues	0.14	0.02	−0.12
Importance of pursuing one's own ideas	0.16	−0.05	−0.21
Importance of having freedom to be creative	0.23	−0.16	−0.39*
Importance of working on professionally significant tasks	0.28	−0.13	−0.41*
Preference for deep probing of narrow areas	0.32	−0.19	−0.51*
Preference for working with general established principles	0.38	−0.14	−0.52*

*$p < 0.001$.

their demonstrated intelligence and perseverance, having survived a long and sometimes arduous educational career. They are also frequently recruited for their independence of thought. To the degree that this latter goal exists, industry may be getting more than they bargained for. Few firms can truly afford to support employees whose principal goals are publishing and theory development. Even in industry these may be fine as secondary goals. But the primary goal must be developing products that will allow the company to remain in business. Those without a doctoral degree are much quicker to see this. They begin their careers with a similar academic orientation, but after a few years they reorient themselves appropriately and become more commercially oriented. A real danger exists in that the Ph.D.'s are much more likely to be promoted onto the technical ladder. The risk is that they may have their academic values therefore reinforced and

never become adequately socialized into the goals necessary to keep the firm in business. The Ph.D.'s in our study maintained a very strong academic and scientific orientation throughout their careers.

Education

If there were ever any doubt that the technical ladder reward system is better received by those educated to the level of the Ph.D., the present research should certainly remove that doubt. Over the several years required to achieve a doctoral degree, students are very strongly socialized into an academic system which has its own values and rewards. These are distinctly different from those of industry. The technical ladder reward system was originally designed to be more aligned with that view of the world. It attempts to emulate the academic reward system, based on a belief in the importance of peer recognition for technical professionals.

without the Ph.D. degree, in contrast, are not as thoroughly socialized into the academic system. The technical ladder consequently does not have the same appeal or value in their eyes. The failure to recognize this fact has led to many problems in implementing the dual ladder system. It has also led to frustration on the part of personnel officers who cannot understand why so many engineers fail to see the wonderful benefits of their technical ladder.

About 80 percent of the 278 Ph.D.'s in our sample had a stronger academic/scientific focus than a commercial/product one. Based on the results of our study, it is these individuals who are especially likely to opt for technical ladder careers. It is important to recognize that this group of individuals differs from their organizational counterparts in many other important ways particularly with respect to work-related goals and problem-solving approaches. The establishment of two formal parallel career ladders creates added differentiation within the organization [14]. If the two ladders are also staffed with individuals who are significantly different from each other not only in terms of educational background but also in terms of values, attitudes, and work-related preferences, then the organization runs the risk of differentiating itself even more.

One of the more important forces affecting interaction patterns within organizations is the tendency for individuals to communicate more frequently with those who are most like themselves and whose ideas and viewpoints are most likely to agree with their own interest and perspectives [10]. Based on this notion of selective exposure and the strong differences emerging from our dual ladder study, it is not surprising that communication between technical and managerial groups may be severely strained. Promotional dynamics may even be exacerbating this problem. Prior research by Katz and Tushman [11] showed that individuals promoted on the technical ladder communicate less often and are significantly more isolated from organizational peers than those promoted to management. The results of our study indicate that organizations may be compounding this problem by promoting to the technical side individuals who not only have weaker communication ties to begin with but who also

claim "they cannot do their best work in collaboration with others." They prefer, instead, the freedom to work independently and pursue their own ideas. Clearly, organizations need to build forces for integration that compensate for the structural and staffing differentiation that accompanies the dual ladder. Without establishing strong bridging mechanisms [18] to overcome the problems of coordination and communication, those on the technical ladder are likely to become decoupled from the rest of the organization. And as described by Allen [2] and others, such segmentation can be a strong barrier to effective technology transfer and innovation.

CONCLUSION

Although the Ph.D.'s, through their longer and more intense exposure, are more thoroughly indoctrinated in the academic values, all young people coming through the university system are to some degree affected. The views expressed by the younger people in the present sample show this. They feel that publication, theory building, and specialization are important. This culminates, after a few years on the job, in an attraction toward a technical ladder career. Shortly after that, however, reality begins to set in. They begin to understand that industry needs management as much as technology—that theories and publications don't put bread on the table and that commercially important projects are not necessarily those of the greatest scientific interest. This awakening occurs in the early to mid-thirties and results in a pronounced shift away from the technical ladder and increased interest in management. The initial state is found both among the research Ph.D.'s who are most interested in the technical ladder and among the other engineers and scientists not so predisposed. The degree to which the two groups adapt as time goes on differs considerably, however. Over time, the strength of the commercial focus among the non-Ph.D.'s greatly exceeds the strength of their academic interests. Although the commercial focus of Ph.D.'s also increases over time, it is always exceeded by their concern for academic and scientific success.

While many alternative personality and situational explanations could account for this strong difference, it is our contention that the separate organizational experiences of the two groups provide a stronger explanation of the observations than do differences in personality. Perhaps through the nature of their work assignments, their reporting relationships, or even the location of their offices, non-Ph.D.'s become more socialized into the value system of the organization and its management. The organizational socialization encounters and interactions of Ph.D.'s, on the other hand, may be very different. Perhaps they are given more independent activities, or research tasks that require little interaction, or they are co-located with each other, or they are assigned only to supervisors with similar academic values. Whatever the reasons, it may be that the organizational socialization experiences of these individuals are very different from their less highly educated colleagues.

Since whatever happens during organizational socialization dramatically affects one's performance, career, communication networks, and overall perspective, future research is clearly needed to understand and compare the organizational socialization process for engineers and scientists from differing backgrounds and educational environments. If the dual ladder is to work effectively in organizations, we must learn how to better organize and structure the early experiences of engineers and scientists to create better working relationships between those promoted technically and managerially, rather than estranging them from each other.

NOTES

1. Managerial position was included as an independent variable to control for its effect, since it would be expected to influence attitudes toward the technical ladder.

2. Although the means of the standardized preference scores of engineers in different types of technical work differ in the predicted direction, the differences do not reach statistical significance.

3. The overall mean is zero, since the data have been standardized.

4. Developed by Pelz and Andrews [15].

5. In line with our previous discussion, these two factors closely parallel the differences described by the local-cosmopolitan distinction.

6. For Education, $F = 17.72$, $p < 0.001$; for Success Orientation, $F = 5.80$, $p = 0.02$.

7. Ph.D.'s with an academic/scientific vision of success and non-Ph.D.'s with a commercial/product vision of success.

8. The items are listed in Table 46.3 by the magnitude of the disparities between the two groupings, not in the order in which they appeared in the questionnaire.

9. Interestingly enough, for each of the items, the mean standardized responses for the non-congruent (i.e., commercial/product Ph.D.'s and academic/scientific non-Ph.D.'s) groupings fell within the ranges reported in Table 46.3 and were not significantly different from the overall population means.

REFERENCES

1. T. J. Allen, "Distinguishing engineers from scientists," in *Managing Professionals in Innovative Organizations: A Collection of Readings*, R. Katz, Ed. New York: Harper Business, 1988, pp. 3–19.

2. T. J. Allen, *Managing the Flow of Technology*. Cambridge, MA: MIT Press, 1984.

3. T. J. Allen and R. Katz, "The dual ladder: motivational solution or managerial delusion?" *R&D Management*, vol. 16, no. 2, pp. 185–197, 1986.

4. T. J. Allen and R. Katz, "The treble ladder revisited: Why do engineers lose interest in the dual ladder as they grow older?" *International Journal of Vehicle Design*, vol. 12, nos. 5/6, 1991.

5. T. J. Allen and R. Katz "Age, education, and the technical ladder," *IEEE Transactions on Engineering Management*, vol. 39, pp. 237–245, 1992.

6. L. Bailyn, *Living with Technology: Issues in Mid-Career*. Cambridge, MA: MIT Press, 1980.

7. G. W. Dalton and P. Thompson, *Novations: Strategies for Career Management*. Glenview, IL: Scott Foresman, 1985.

8. A. W. Gouldner, "Cosmopolitans and locals—towards an analysis of latent social roles," *Administrative Science Quarterly*, vol. 2, pp. 281–306, 1957.

9. R. Katz "Managing creative performance in R&D

teams, in *The Human Side of Managing Technological Innovation*, R. Katz, Ed. New York: Oxford University Press, 1996.

10. R. Katz and T. J. Allen, Investigating the not invented here (NIH) syndrome," *R&D Management*, vol. 12, pp. 7–19, 1982.

11. R. Katz and M. L. Tushman, "A longitudinal study of the effects of boundary spanning supervision on turnover and promotion in R&D," *Academy of Management Journal*, vol. 26, pp. 437–456, 1983.

12. R. Katz, M. L. Tushman, and T. J. Allen, *Managing the dual-ladder: A longitudinal study*. Greenwich, CT: JAI Press, 1991.

13. S. Kerr and M. A. Von Glinow, "Issues in the study of 'professionals' in organizations: The case of scientists and engineers," *Organizational Behavior and Human Performance*, vol. 18, pp. 329–345, 1977.

14. P. R. Lawrence and J. M. Lorsch, *Organization and Environment*. Boston: Harvard Business School Press, 1967.

15. D. C. Pelz and F. M. Andrews, *Scientists in Organizations*. New York: Wiley, 1976.

16. J. Raelin, *Clash of Cultures*. Boston, MA: Harvard Business School Press, 1985.

17. R. R. Ritti, *The Engineer in the Industrial Corporation*. New York: Columbia University Press, 1971.

18. E. B. Roberts, "Stimulating Technological Innovations: Organizational approaches," *Research Management*, vol. 22, pp. 26–30, 1979.

19. E. H. Schein, "How 'Career Anchors' hold executives to their career paths," in *Managing Professionals in Innovative Organization: A Collection of Readings*, R. Katz, Ed. New York: Harper Business, 1988, pp. 487–497.

20. H. A. Shepard, "The dual hierarchy in research," *Managing Professionals in Innovative Organization: A Collection of Readings*, R. Katz, Ed. New York: Harper Business, 1988, pp. 177–187.

21. M. L. Tushman, "Technical communication in R&D laboratories: The impact of project work characteristics," *Academy of Management Journal*, vol. 20, pp. 624–645, 1977.

22. M. A. Von Glinow, *The New Professionals: Managing Today's High-Tech Employees*. Cambridge, MA: Ballinger Press, 1988.

15

Managing Across Functions for Rapid Product Development

47

A Six-Step Framework for Becoming a Fast-Cycle-Time Competitor

CHRISTOPHER MEYER

The ongoing ability to deliver a quality product or service quicker than the competition yields a sustainable competitive advantage. At the expense of the U.S. Post Office, Federal Express created an entire business based on this principle. Citibank became the leader in home mortgage originations in part because it offered loan commitments within 15 minutes! Compaq computer established itself as more than a quality IBM clone manufacturer when it brought its 386 and 486 machines to market *be-* *fore* IBM. The bottom line is that when customers decide they want to buy, the first supplier who can fill that need with a quality product or service will flourish. These companies operate in Fast Cycle Time (FCT).

The benefits for FCT competitors are substantial. The first entrant into a market typically dominates that market in both share and profit margins. Pricing pressure does not exist when there is no competition. FCT leaders reinforce their posi-

Reprinted with permission from Christopher Meyer.

Christopher Meyer is Chairman of the Strategic Alignment Group, a management consulting firm in Menlo Park, CA.

tion because they set the standards which others must follow. They secure the prime distribution channels which create additional entry barriers for the competition.

Becoming a FCT competitor is not easy. *FCT requires a systemic integration of new values, structures, and rewards into the core work process.* One cannot simply accelerate the work pace without negative impact. First, people will make the same mistakes they always have, only quicker. Second, management will rapidly burn out the organization's most important resource: people. An image employees often have when they first hear about reduced cycle time is a cardiac stress test. They equate reducing cycle time to speeding up the organizational treadmill. Regrettably, they are often correct.

FCT is the ongoing ability to identify, satisfy and be paid for meeting customer needs faster than anyone else. There are several key words in this definition. The first one is *ongoing*. Although useful, single shot cycle time reductions do not provide a sustainable competitive advantage. In a competitive environment, the race is never over. Competitors who improve continuously will pass those who pause to relax. The next key word is *identify*. FCT is the responsibility of all organization functions from the start of the business cycle through the end. Some incorrectly consider cycle time an exclusively manufacturing or engineering issue. The firm that identifies the customer's need first has a head start in filling that need. *Satisfy* means that one cannot sacrifice quality for time. The old rule was that if you required a product or service quickly, it would cost more and the quality couldn't be guaranteed. That thinking is dead. World class competitors such as Toyota have clearly demonstrated that speed does not have to sacrifice quality or cost. *Paid* refers to the attention FCT companies place on completing the business cycle. For example, while Toyota was able to reduce its manufacturing cycle time to 2 days, it still took 17 days to sell and deliver its cars. FCT companies view their organization as a value delivery system. As a system, the slowest sub-cycle limits the overall system's total cycle time. *Meet-*

ing customer needs declares that products or services which do not meet customer needs are not acceptable. And last, *faster than anyone else* reflects the reality of increasing competition. If there is a foreign or domestic competitor who is faster, it is only a matter of time before they will dominate that market. Detroit and the semiconductor industry have both learned the hard lesson of ignoring international competition.

Reducing product development cycle time requires a systematic strategy. This article defines the six key steps to becoming a FCT competitor and outlines how to implement them.

Step 1: Understand what your end customer regards as added value, and reflect that in every job and level within the organization.

An FCT strategy focuses the entire organization on work that adds value to the end customer while concurrently trying to eliminate anything that does not. Thus, product development managers in FCT companies structure and manage their organizations as value delivery systems focused on adding value for their customers. In order to do this, all employees must know who the end customers are and what is added value to them. Only people who *pay* for the product are end customers.

There may be more than one end customer. For example, consumer products have several end customers along the distribution chain, starting with the distributor and ending at the customer. To distributors, packaging may be a value added component of the product, whereas for the consumer, packaging adds little value. The main lesson here is to understand what is value added for each customer you serve.

This end customer view contrasts with the notion of internal customers popularized by quality programs. The internal customer concept suggests that each job has an upstream supplier and a downstream customer. For example, manufacturing is the customer of engineering's designs. While this approach improves understanding of mutual dependencies, calling internal groups customers can cause people to incorrectly equate internal definitions of value added with those of the end customer.

The difference is critical: End customers generate revenue while internal customers generate cost. For example, internal customers create 99 percent of the paperwork in organizations. Paperwork rarely adds value to end customers. Using the end customer's definition of value added exposes non-value added time and activities. Motorola, for example, no longer encourages the internal customer perspective.

After defining the end customer, one has to define what is value added in their eyes. A rule of thumb for determining value added is whether or not the end customer is willing to pay for the product, service or feature. If they are not willing to pay for it, then it is probably not value added. The information to make this determination comes from one place: the customer.

Traditionally, we have relied on sales and marketing to channel the customer's definition of value added into the organization. While efficient in the use of people, this approach limits the direct contact other functions have with the customer. It is increasingly evident that expanding the breadth of organizational contact with the end customer sharpens all employees' understanding of what is value added, as well as their motivation to deliver it.

For example, a leading manufacturer of electronic test equipment conducted a focus group in which customers compared their equipment to a competitor's. Invited to the focus group were several young engineers from the development team. Standing behind a one-way mirror, the engineers saw that most of the customers were attracted to their company's product before it was turned on. But after it was turned on, the customers drifted en masse to the competitor's product. Why? Simply because the display on the competitor's product was easier to read. Because display readability was not an issue for the engineers' young eyes, they had dismissed customer complaints. Seeing their competitor's "inferior" product surrounded by customers quickly changed their minds.

In another example, it was a Du Pont development technicians' visit to Reebok that generated a competitive response to Nike's "air cushion" heel. The technicians were there for another purpose; yet when they heard of this problem, they devised a solution involving implanted rubber tubes—made, of course, by Du Pont.

These examples illustrate that making customer needs visible to those who have traditionally been isolated or removed from the end customer can quickly reorient functional activities toward outcomes that are truly value added.

While end customers are the ultimate source of defining what is or is not value added, top management is responsible for defining the organization's value added focus and allocating resources accordingly. It does this by defining a *value proposition* for the organization. The value proposition is unique to each organization and defines that organization's value adding strategy.

A well-defined value proposition that is aligned with customer needs is critical for ensuring competitive advantage. For example, prior to 1989 Quantum Corporation, a computer disk manufacturer, had a value proposition that stressed quality and performance. Quantum's position in the marketplace grew accordingly. In contrast, a new competitor, Conner Peripherals, had an explicit value proposition to be first to market with their products. Conner's philosophy of "sell, design, build" focused on being first to market with products designed specifically for their customers' needs. In contrast, Quantum had been designing disk drives to meet market standards rather than a specific customer's needs. By carefully picking leading computer manufacturers as its customers, Conner was able to leverage the initial sales and development effort into a broader market opportunity. This was most graphically seen in Conner's relationship with Compaq Computer. Compaq was one of the initial investors in Conner as well as being the initial customer. Compaq's success in the PC market established a presence for Conner's drives.

As Conner's success grew, Quantum reexamined its value proposition of quality and performance relative to Conner's focus on custom designs done fast. Quantum determined that product performance was important, but availability of new products was even more so. In short, quality and performance value were insufficient without time to market. Even though many of Conner's customers complained about the quality of early models, one should not forget that these complainers were purchasing their drives from Conner, not Quantum. This is not to suggest that quality should be ignored in favor of developing products faster, but, rather, that a fanatical devotion to technical elegance can often undermine a company's ability to compete against those who can deliver their products faster. In sum, the value proposition focuses behavior and resources.

A good value proposition clarifies what is value added from what is not. As simple as this may appear, many organizations have value propositions that are not clearly stated. Management often assumes the value proposition is obvious and understood by all. Therefore, it concentrates on managing operations. Each element of the organization may have its own operating definition of the value proposition and act accordingly. The net result is that value adding efforts do not build on, but actually subvert, one another. Ultimately, the customer becomes confused.

In the service sector, SAS Airlines provides another good example of how a clear value proposition guides behavior. Jan Carlzon, President of SAS, led that company's revival by defining SAS' value proposition as being the airline of the business traveler. SAS tuned its schedule, route structure and service to the needs of the business traveler. Since there are few business travelers on the weekends, SAS limits flights on Saturday and Sunday. During the initial implementation of this value proposition, employees frequently suggested ways to increase weekend aircraft utilization. Since aircraft utilization is a traditional measure of performance, this behavior was easily understandable. Under the new value proposition, however, weekend flights did not add value to the business traveler. Carlzon rejected these ideas and used weekends for maintenance and training to improve service during the week.

An organization must be aligned around its value proposition. When it is not, people may be working hard but the net force of their collective efforts will be significantly less. Alignment is more than a function of common understanding. Structural elements such as reward systems, policies, cultural norms, and organization design must be in alignment as well. To foster alignment, management should continuously communicate and test whether the value proposition is incorporated into each job. Like management, each employee has his or her own mental model of how they should add value for the customer. Without ongoing education efforts, employees may use outdated or conflicting models. Educational efforts minimally include exposing all employees to the corporate mission and value proposition on a regular basis. Additional methods may include:

- One-way briefings such as corporate video newscasts, "all hands" meetings, annual reports, and internal newsletters.
- Staff meetings where senior managers explicitly employ the value proposition as a criterion for decisions.
- Special meetings to describe *and* test the value proposition with key constituencies, including project reviews, new employee orientation, buzz groups, brown bag lunches, etc.
- Company symbols and giveaways such as desk accessories, T-shirts and coffee cups.

Step 2: Focus the entire organization on work that adds value to the end customer.

As we have stated, there are two types of work: value added and non-value added, the former being work for which the customer is willing to pay. For example, painting a car a specific color is value added work, whereas testing the paint for durability is not something a customer is willing to pay for. Surprised? Many might argue that untested paint could fade quickly, thus upsetting the customer. No question about that, but consider your reaction to a car's price sticker that showed below the $699 for optional leather seats a $45 charge for paint testing!

Non-value added practices such as testing are required because we do not fully trust the process being tested. This may be because the process is not well understood or developed or it may be due to poor operating practices. In either case, testing is a stop gap measure until the process is made stable. In the language of quality experts, testing quality into the product is inferior to designing it in. While one may argue as to when a process is stable enough to eliminate testing, the goal of doing so should always be present.

The only way to identify which work adds value is to study the organization's value delivery system. Constructing a high-level map can provide a macroview of the entire organization's value delivery system. Figure 47.1 is a generic example of such a map.

By limiting the amount of detail, it is easier to identify which steps are the most important value adding processes. For some, it may be steps in the process engineering or development process, while for others it could be the testing or prototype stages. Once identified, critical processes can be exploded in greater detail by flow charting and tracing the multi-functional processes and interdependencies,

as described in Meyer (1), chapter 7. At the least, this process map should identify the critical players, key tasks and the time required to complete them. In addition, it is also useful to specify the inputs and outputs of critical steps.

The map should accurately reflect how the process works today. During the mapping exercise, there is a strong urge to incorporate how the process ought to look or be changed. It is important to resist this urge and to defer these discussions until the map is completed. Until there is agreement on how the work flows are currently conducted, such discussions can become irrelevant. However, reaching agreement is not necessarily easy. Everyone has their own model of how the value delivery system works. Defined by personal experience, these models differ for each individual.

Similarly, senior management's models are frequently built on past experience that may no longer be valid. For example, a high technology company wanted to reduce new product development time and therefore mapped the entire development process. Management was shocked to dis-

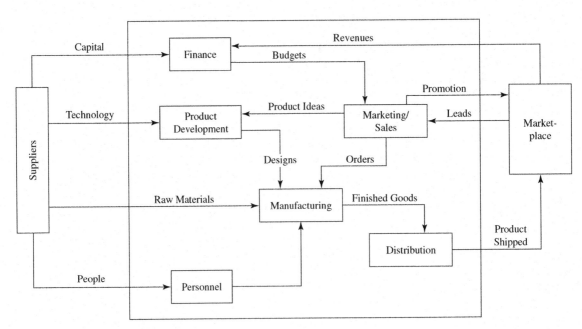

Figure 47.1. A value delivery process map.

cover that the product definition phase often took as long as 12 months. Its mental model was based on experience gained when the company was much smaller and products were simpler. Management had lost contact with the impact of many small process changes made over time.

The mapping process does not require everyone to agree on each and every step. Rather, strive for general agreement that the map accurately reflects the most critical process steps. Once this is achieved, review the map and identify value added and non-value added steps. Test each non-value added task to eliminate or compress it. First efforts typically result in a "shrinkage" of the current process. A shrinkage has the same basic structure as the current process with many non-value activities removed. Shrinkages will often cut cycle time by 50 percent or more. But one should not stop here. Shrinkages are the results of picking the low-hanging fruit. These are efficiency improvements which do not fundamentally restructure the value delivery process or yield a *substantial* competitive advantage. Major breakthroughs come from insights that fundamentally restructure the core elements of the value delivery process. Clearly understanding the current delivery process, combined with seriously questioning and entertaining creative alternatives, generates such insights.

To become an FCT competitor, one must understand the value delivery process sufficiently to define and focus everyone's attention on value added work. It is curious that American business leaders love to invoke sports metaphors as competitive models yet rarely include the attention sports professionals pay toward process analysis and improvement. Specifically, world champions such as the San Francisco 49ers spend 100 hours off the field for every hour played. They use non-game time to analyze films about their value delivery process or for training. Yet in an informal survey that we conducted during a series of California Institute of Technology FCT seminars, we never found more than two executives per class who spend more than one day a month examining their organizations' work processes. One cannot expect dramatic breakthroughs without putting in more time.

Step 3: Redesign your organization so that it is flat and based on multi-functional teams, with blurred boundaries—inside and out.

By definition, large hierarchical organizations can never be quick. Every time an approval is required, there is a delay as the request is communicated, considered and responded to. The further the approval level is from the point of origination, the greater the time delay. When functionally organized, companies divide customer problems into pieces and channel each piece to the appropriate functional group. Inevitably, some needs fall in the cracks between functional specialties while others are miscast. In any event, internal functions never experience the complete problem or its ramifications as the customer does. Instead, all they see and attempt to remedy is a sub-set of the problem.

A further difficulty with this process is how solutions are defined. For the most part, solutions are *internally* defined within each function. To wit, engineering has its own design standards just as manufacturing has its quality standards. These standards are isolated from each other and most importantly, the customer. The customer is concerned about the total product or service while the functions are concerned with their piece of the action. The functional paradigm practically ignores the inherent interdependence between functions. The responsibility for cross-functional integration is concentrated in a single individual—the boss, who, in simpler times, might have been able to make all the integration choices wisely. Today's complex technologies and business conditions contain too many unknowns to expect any one person to know all the answers. The functional organization creates expectations that the boss should provide the answer when, in fact, he or she may be the least qualified to do so.

The multifunctional, team-based organization is an alternative. Composed of people from different functional disciplines, the teams have responsibility for the core value-added work processes of the business. Rather than operating as a coordinating body on top of the existing functional organization, the team focuses on delivering the organization's value proposition, and is allocated resources accordingly. Teams are typically organized around major product or market groupings.

Business teams are different from a classic matrix organization primarily in the breadth and scope of their responsibility. Theoretically, a matrix divides the power equally between project and functional bosses. In practice, the functional bosses wield more than 50 percent of the power. Because the core value adding process is within the team, the team-based organization keeps at least 51 percent of the power in the team.

Overly simplified, the teams become the line organization and the functions become staff. In contrast to the team itself, the team-based organization resists easy depiction. The supporting cast and connections to other teams are in constant flux. Teams are nodes in an organic network that continually adjusts itself in response to new customer demands. The role of the functions is to support the teams in the value-adding processes as defined by the teams.

Placing the core value delivery process into teams requires a fundamental shift in the organization's power structure. If management allows functional dominance to continue, employees will accurately perceive the teams as nice but not essential. The success of a multi-functional, team-based organization depends on creating a new organization *architecture*. The worst thing one could do would be to throw a group of people from engineering, operations and marketing together and call them a multi-functional team. They would find themselves struggling to define their role as they simultaneously tried to shift the organization's functional orientation. Well-designed team-based structures define the roles and responsibilities for those off the team as well as those on the team. Issues to be considered in the architecture include:

• Initial definition of team goals.
• Team charter.
• Team member responsibilities.
• Boundary conditions/limits of the teams.

• Linkages to other teams, functions and management.
• Team and personal rewards.
• Team support requirements from functions.

To work effectively, senior management must make certain that clear goals and chartered responsibilities are part of the team's design. This is management's control mechanism, and implementing the newly defined architecture is senior management's responsibility. They must demonstrate through their behaviors (and measures, as discussed later) the importance of teams. During implementation, the reality of the power shift becomes palatable as teams struggle to take control while some functional elements inevitably resist. Old habits will continue to dominate if senior management fails to publicly support the teams during the transition. Within well-defined parameters, teams need the power to take whatever actions are necessary to serve customers. And if teams contain the functional expertise required, there should be few cracks for issues to fall into.

In the same way that multi-functional teams blur boundaries within the organization, FCT competitors attempt to do the same thing between suppliers and customers. FCT companies consider suppliers and customers as partners in the value delivery effort. The fewer boundaries that exist in space or time, the more effective the value delivery process can be.

For example, Quantum Corporation designs application-specific integrated circuits (ASIC) for chip vendor manufacturers. Quantum now includes FCT in its vendor selection criteria. The net result is that a recent supplier placed two engineers plus their own engineering workstations at Quantum to speed the design of an ASIC. The cost to Quantum: nothing.

At the other end of the spectrum, customers are being included in the design phase of the value delivery process. For example, Conner Peripherals begins a product development effort only after they have a customer commitment for that product. Their senior technologist spends much of his time on the road meeting *and selling* customers.

The message is clear: to become fast, redesign your organization for speed. Just as one should not try to turn an oil tanker into a ski boat, one should not try to speed up a large, hierarchical organization. Overcoming the inherent structural limitations of centralized control and specialization is not possible. Flat, multi-functional, team-based organizations provide the architecture that enables locally informed and quick decision making.

Unfortunately, companies are often far more successful at launching these teams than they are at getting the teams *and* the functional areas to work together effectively. This occurs because the design of any performance-measurement system should reflect the basic operating assumptions of the organization it supports. But if the organization changes and the measurement system doesn't, then the latter is likely to be less than effective—perhaps even counterproductive. At many companies that have moved from control-oriented, functional hierarchies to a faster and flatter team-based approach, traditional performance-measurement systems not only fail to support the new teams; they often *undermine* them.

Ideally, a team-based management system should help get members from different functions on a team to focus their efforts and speak a common language. In many measurement systems today, however, each relatively independent function has its own set of measures, whose chief purpose is to keep top managers informed about its own performance. Marketing tracks market share, operations monitors inventory, finance watches costs, and so on. Such outputs or *"results measures"* tell an organization where it stands in achieving its goals but not how it got there or what it should do differently. Most results measures track what goes on within a function, not what happens across them, and most cross-functional results measures (such as gross margins or revenues) are limited to being financial in nature. What are needed, therefore, are *"process measures"* to track the various tasks and activities throughout the organization that are a responsible for producing a given result.

Teams must create the measures that support their mission, or they will not fully exploit their ability to perform the process faster and in a way that is more responsive to customer demands. A process measure that a product-development team might use, for example, is one that tracks staffing levels to make sure that the necessary people are on a given team at the right time. Another might be the number or percentage of new or unique product parts. One must also be careful not to burden and constrain teams with excess measures. All too often, I have seen organizations react to intensifying competition by piling more and more measures on their operations in the hope of encouraging them to work harder. The end result, however, is that team members often spend too much time collecting data, monitoring activities, and discussing minutia, rather than managing their projects and concentrating on what they really need to do together. Trying to run a team without a good, simple guidance system is like trying to drive a car without a dashboard. You might do it in a pinch but not as a matter of practice (see Meyer [2] for further discussion of "dashboards" and process measurement systems).

Although useful for keeping score on the performance of a business, results measures like market share, profits, and cost do not really help a multifunctional team monitor the activities or capabilities that enable it to perform a given process. Nor do these measures tell members what they must do to improve their team performance. Tracking staffing levels or years of experience in certain job categories during the course of a project, for example, might help explain why a program is 3 months late or over budget. A 10 percent rise in service costs combined with an 8 percent drop in quarterly profits does not tell a team what its service technicians should do differently on their next call. But knowing that the average time spent per service call rose 15 percent last month and that, as a result, the number of late or missed calls rose 10 percent might help explain why costs had risen and profits and customer satisfaction had gone down.

Finally, a team's reliance on traditional measures often causes its members to forget the team's goal and revert to their old functional way of working. Consider the case of a well-known automobile company during the development of its 1991 luxury model. The project was one of its first attempts to use multifunctional teams for faster product development. Unfortunately, the team's measurement system was still a collection of the individual measures that each functional area had used for years.

Shortly before they were to freeze the clay styling model and begin engineering the new body, a controversy erupted over a new door handle design. Purchasing and finance argued that it was too expensive, while the design and body engineering members responded that the new shape was integral to their styling theme and was affordable. Each came to the next meeting armed with data to support their respective functional perspectives. Design and body engineering presented their styling analysis and engineering model along with a vendor bid that showed the new handle would cost no more to manufacture than the old one. Purchasing and finance presented their figures that showed the new handle's warranty costs had not been taken into account. The current door handle was highly reliable and its warranty cost was extremely low. They asked body engineering to prove that the new handle could match that warranty cost. Body engineering responded that the part was not inherently any more complex than existing handles, but this didn't satisfy finance. Purchasing added that, while the bid came from an approved vendor, other approved vendors had historically achieved better warranty costs and presented their ranking list to prove it. After a short shouting match, design and body engineering gave up. The redesign minimally cost a week because going back to the old handle required changing its placement, which in turn required significant changes to the clay model.

While the above story is familiar to anyone who's worked within a traditional, functional organization, isn't this exactly what the multifunctional team was designed to address and if so, what happened? By using only their individual functional measures within the team, both parties quickly typecast each other in familiar functional roles and reverted to old, familiar behavior. Finance and purchasing was viewed as concerned only about cost and styling, and engineering was seen to care only about performance. Regardless of who was right or wrong, because functional measures were the only ones used, they quickly channeled the discussion into a polarized, functional debate, rather than a team decision. *During the discussion, neither side asked the critical question: Would the new handle increase the car's ability to compete in the marketplace?* When a team hasn't created an integrated measurement system that tracks its overall goal, it's highly likely that during conflict, people will revert to their familiar roles. When cross-functional teams are established, organizations have to institute not only a measurement system that protects its functional excellence but also one that supports the company's basic strategy, prevents unpleasant and unexpected surprises, and truly empowers the teams to make appropriate decisions and trade-offs.

Step 4: Pursue process development as avidly as product or service development.

Process improvements can provide tremendous leverage. A single improvement can ripple across the product delivery system. For example, one of the major dilemmas in developing and manufacturing technology-intensive products is the ability to test them for functionality, reliability and quality. Frequently in their zeal to develop the product, development engineers do not focus on test development until late in the development process. In order to improve and innovate testing, the product development process must be changed. In short, test development must come earlier.

At Quantum, for example, managers realized that the bottlenecks that occurred downstream during testing were a significant impediment to rapid introduction of new products. Typically, drives were tested at the back end of the process using specially designed test equipment. The test hardware and software was developed after the product was well along in development. Early prototypes were typically tested by engineering and did not utilize any production test equipment because it was not readily available. In addition, there was always the issue of keeping the test equipment that was used up to date. Frequently, new drives would fail tests not because the drive was bad but because the test software was often a revision or two older than the drive it was testing.

Another factor that contributed to delays was the cost and number of testers. Testers represent one of the largest capital outlays required in drive development and manufacturing. For example, the more volume produced, the more testers that are required. Thus, testers came to be viewed as a significant bottleneck. And, when orders dropped, testers represented a very expensive and underutilized investment.

Quantum's approach to this problem was to incorporate "self-test" procedures within the drive itself. This required development engineers to design the test procedure into the product. It was an elegant approach. A more conventional approach would have been to persuade development engineers to pay more attention to test development earlier in the product development process. However, this approach has no real hook that grabs the development engineer. In contrast, development engineers had no choice but to focus their creative energies on process development when they were faced with the challenge that the product needed to be delivered without having to go through testing in downstream stages.

Why don't more organizations pay attention to process development? Three reasons stand out:

1. Because significant process improvements take time to design and implement, instant results are rare. In addition, it may take time for employees to learn and familiarize themselves with new work processes.

2. The second reason is embodied in the old development axiom that what you measure is what people pay attention to. The vast majority of organization performance measures are end results. While these measures tell us what we have accomplished, they provide little insight into how we did it. Because we don't measure the process itself, people don't pay much attention to improving it.

3. The third reason is closely related: We focus whatever measures we *do* use on those items that are easiest to measure. The dilemma created is

similar to the drunk who looks for his keys only under the lamp post because that is where the light shines. Most organizations limit their process measures to the tangible, linear work processes such as those found in manufacturing. Those processes are much easier to measure precisely than the nonlinear work process found in design engineering, R&D, marketing, or sales. Non-linear work process measures are not as precise, nor do they need to be. What is most important is to develop measures that aid cycle time improvement efforts. Relative measures that compare current cycle time to past are sufficient.

Process improvement begins with allocating time to it in conjunction with the establishment of performance measures. Simple as this sounds, focusing on the "what" is so familiar that it takes effort to shift attention to the "how." Making process improvement routine requires improvement goals, dedicated time, appropriate rewards, forums for process improvement work, and skills to do it. This starts with top management since people take their cues from them.

An excellent starting point for the top team is to develop the macro map of the value delivery system that we described earlier. For each critical process, select a process champion from the top team. The champion is responsible for ensuring the continuous improvement of that process. This includes developing improvement goals and performance measures. Incorporating these goals into each executive's personal objectives and performance review makes process improvement very real.

Process champions do not work alone. They must involve those intimately familiar with the process in the diagnostic and improvement effort. A favorite tactic uses this enlarged group to create two additional maps. The first is the detailed map of the targeted process and the second proposes a new process architecture. The new architecture map becomes the foundation of the process improvement plan.

The mapping process creates a forum where people can engage in a dialogue around process improvements. Some firms find it useful to continue to segregate process improvement work from the mainstream in order to gain momentum. Others devote time to process improvement within their existing meetings. The point is that one must create, communicate and use clear forums for such process improvement dialogues.

Step 5: Set "stretch" cycle time goals and measure progress publicly.

Initial FCT goals should seek a minimum 50 percent improvement in cycle time. Setting goals lower than this does not achieve the cycle time reduction available or required to compete. Merely working harder within the existing process can reduce cycle time by 20 to 30 percent. When management sets a more aggressive goal, it signals the organization that everyone must consider new ways of working. This challenge in turn stimulates learning.

In the past, cycle time reductions were confined to manufacturing and were viewed in terms of incremental cost or efficiency improvements rather than as a competitive strategy. Knowledge work has received little attention, yet this is where the greatest leverage is (2). For example, the average design cycle of U.S. auto makers has been six years while production cycle time is less than a day. Until international competition in the auto industry made it clear that U.S. development cycles were double the Japanese, managers didn't realize that breakthroughs in excess of 50–200 percent were achievable. The reality is such improvements are now the minimum required merely to catch up!

When communicating stretch goals, expect people to respond with skepticism. Their response is entirely rational since they are being asked to make dramatic improvements without knowing exactly how they will do it. For this reason, it is essential that management have a strong FCT improvement vision that they agree to. Hewlett-Packard's CEO John Young asked all employees to cut in half the time it takes to break even on a new product. This vision provides a picture of the future to sustain people along the hard journey of getting there. Since the vision is only as effective as management's belief and enthusiasm in it, people will test it for holes as soon as it is communicated.

Defining the beginning and end of the cycle and sub-cycles is the first step in developing measures. This can be more difficult than it appears, since defining when a customer's need first exists is not easy or precise. Recognize that beginning and end points are inherently arbitrary choices. One can always mount a rational argument for other points. Select points that make sense for your industry and organization. If there are any sub-cycles that you will be focusing on, such as new product development, define that cycle as well. Once completed, establish the organization's initial FCT goal by benchmarking total current cycle time against your best competitor worldwide. Recognize that their current capability is the *minimum* target you can set since they won't be standing still. Only if the gap between you and them is enormous should an interim goal be selected.

Hewlett-Packard starts the new product development cycle when applicable technology exists within H-P Labs. The end of the cycle is when the profits from the product equal the total investment in the product. They call this break-even time (BET). The primary reason for using break-even time is to discourage quick introduction of products that fail to meet customer's needs. Quantum begins its new product development cycle with the first approved specification and ends at the passage of a process maturity test. Pick what works for your business.

Defining the cycle and sub-cycles permits internal comparison to past efforts and external comparison to competitors. For example, pick the last three products in a given area and see how long they took to develop. In doing so, it is quite easy to examine what contributed to FCT and what impeded it. This can also occur with key components required to build those products. Quantum has analyzed the time it has taken to develop firmware on every disk drive the company has made. The beauty of this approach is that it quickly moves the FCT conversation to the operating level and stimulates quality improvement discussions.

It is highly advantageous to distribute and display cycle time measures. A lack of feedback regarding results limits the possibility for initiating improvements. Research demonstrates that major breakthroughs frequently come from people who are less familiar with the process. As functions and teams develop their own FCT goals, display them. Graphics are far superior to tables and words.

One cannot understate the importance of aggressive FCT goals and publicly displayed measurements. Perhaps a colleague of mine describes this best using what he calls Management's Apparent Interest Index. Employees throughout an organization take action based on what they believe interests management. If management frequently discusses FCT, sets clear goals and consistently updates public measures, employees will recognize and act on management's attention. Without these factors, management's apparent interest will not be visible.

Step 6: Create an environment that stimulates and rewards continuous learning and action.

Increasing the rate of organizational learning is the heart of an FCT competitive strategy. As a relatively new area, organizational learning deserves special attention. Learning in organizations is a reiterative process by which individual insights and discoveries are converted into organizational knowledge (3). Thus, organizational learning involves the building of a knowledge base that can be made accessible to others, shared, evaluated, and codified—rendering the knowledge independent of the inventor or person who originally generated the insight. Moreover, access to a shared knowledge base in organizations is critical; a lack of knowledge can result in major product errors and delays (4).

By reducing the time it takes to learn, fast-cycle organizations internally benefit from a parallel reduction in the time it takes them to detect and correct errors in the product development process. Many development efforts have been delayed, or have failed, because errors in the early stages of development were either ignored or went undetected.

Failure to detect errors in product development is costly, since problems are usually not seen until prototypes have been scaled up or advanced into manufacturing—usually with disastrous results (5). Similarly, product development efforts are thwarted when problems take too long to fix. Knowing what caused a problem (error detection) is useful only when someone acts to prevent its recurrence (error correction). Hence, the ability to correct errors in a timely manner once they have been detected is also critical to organizational effectiveness. Fast cycle companies are intent on increasing the rate of organizational learning.

Organizational learning is an organizational *process,* as opposed to merely the collection or summation of individual insights. Further, there are two critical differences between organizational and individual learning. First, organizations have a collective purpose, and to be useful, learning in an organization must be in line with that purpose. Second, organizations are social systems. While individual learning can be entirely personal, organizational learning requires that employees at large have access to the learning process and understand the new knowledge created.

Learning in public presents major dilemmas, however. Consider the following: Organizations reward technical competence, which drives people to speak out only when they know the right answer. According to this definition, learning is equated with competence. Thus, if the behavioral norms of the organization reward people who demonstrate competence, it would be wise to be quiet if one didn't know something. *Yet, if people cannot be public about what they don't know, how can we expect them to learn?* The logic of this model is as frightening as it is clear. FCT organizations and leaders begin by rewarding the right questions as much as they do the right answers.

A further dilemma is that the knowledge we have gained to date blocks the path to organizational learning. These established mental models are deeply entrenched and act as walls: they support what we are doing today *and* they act as barriers to thinking and doing differently (6,7). The older our organization is, the more we must "unlearn" in order to learn. The process of learning requires letting go of existing beliefs about current organizational practices, technologies and existing power relationships. To become a FCT competitor, it is essential that senior management embrace organizational learning as a strategic objective. The executive's job is to define a *strategy and system architecture* that increases the speed of learning throughout the firm. Because organizational learning is a social activity, it requires an architecture that creates forums where ideas and experience can be exchanged. The multi-functional team creates such a forum for task-related interactions between people of diverse background and experience. When combined with a clear team goal and rewards, individuals enthusiastically engage in both dialogue and discussions of differences that functional organizations usually evade or ignore. Insights that could not have been attained by individuals acting alone inevitably result from these interactions. In sum, team members gain invaluable experience by being actively engaged in the process of organizational learning.

While organizational learning is a new area of strategic focus for corporations, there are no hard and fast templates to follow. Leading FCT competitors are writing the rules right now by defining both a strategy and an architecture that supports learning in their organizations. The shift to a process-sensitive, learning organization requires time and effort. New values and structures take hold only as old ones are retired. FCT yields a sustainable competitive advantage because it is woven into the cultural fabric of the entire organization's value delivery process. World-class quality has become the "ante" required to be a global competitor, but it does not ensure leadership. Any organization that couples world-class quality with an FCT capability will have a competitive edge.

REFERENCES

1. Meyer, Christopher. *Fast Cycle Time: How to Align Purpose, Strategy, and Structure for Speed.* New York: The Free Press, 1993.
2. Meyer, Christopher. "How the right measures help teams excel." *Harvard Business Review* 72, No. 3 (1994): 95–103.
3. Nonaka, Ikujiro. "The Knowledge-Creating Company." *Harvard Business Review,* November–December (1992): 96–104.
4. Shrivastava, Paul. "A Typology of Organizational Learning Systems." *Journal of Management Studies* 20, No. 1 (1983): 9–25.
5. Purser, Ronald, Pasmore, William, and Tenkasi, Ramkrishnan. "The influence of deliberations on learning in new product development teams." *Journal of Engineering and Technology Management* 9 (1992): 1–28.
6. Purser, Ronald. "Redesigning the Knowledge-Based Product Development Organization: A Case Study in Sociotechnical Systems Change." *Technovation* 11 No. 7 (1991): 403–416.
7. Senge, Peter. *The Fifth Discipline: The Art and Practice of Organizational Learning.* New York: Doubleday, 1990.

48

Shortening the Product Development Cycle

PRESTON G. SMITH AND DONALD G. REINERTSEN

OVERVIEW: As techniques like cross-functional development teams and concurrent engineering become widespread, these approaches to shortening development cycles lose their competitive edge. Decisive advantage is likely to come from the techniques competitors are *not* using. The authors explain that there are other untapped sources of cycle time reduction for R&D managers to exploit. These include opportunities to accelerate the "fuzzy front end," in which half of a typical development cycle vanishes before the team even starts work. The authors caution against structuring a development process around a company's largest projects because this can excessively delay smaller ones. They also question the use of phased development systems, which often cause delays in their attempts to standardize control of projects.

The demands on product developers have never been greater. Product life cycles have shortened and continue to shrink. Product technologies, particularly in the electronics and materials areas, are changing faster than ever. As a consequence, the pressure is on to shorten product development cycles.

Frequently, the R&D manager becomes the focal point of efforts to cut development time. To deal effectively with this challenge, the technical manager must look beyond popular but limited solutions to see the breadth of the problem. Such techniques and tools as quality function deployment

(QFD), simultaneous engineering, and computer-aided design (CAD) still leave many time-saving opportunities untouched (1).

Great reductions in cycle time are obtainable by applying various techniques blended to suit a particular company's needs. Fortunately, the R&D manager is in a strong position to initiate and foster many time-saving approaches. This article covers 10 such approaches in which such managers play a particularly important role.

1. BE FLEXIBLE ABOUT PROCESS

There are many sound approaches to managing R&D projects. Each method has its advantages and shortcomings. The correct approach can only be selected when one has a clear vision of which advantages are critical in a particular situation. For example, consider the tradeoff between managing development time and technical risk. Most product development systems, such as phased systems (see Authors' Addendum at end of article), are designed to monitor and control technical risk. Such systems are effective and appropriate where reducing technical risk is the paramount concern. Yet managing technical risk is not always the prime objective. Speed can be more important when trying to head off an emerging competitor, and cost can dominate our concerns in a mature market.

The most effective organizations have different development systems available and tuned to suit these distinct objectives. Without alternative processes, all projects tend to get sent through the same process, a common denominator that suits no

Reprinted with permission from *Research-Technology Management,* May–June, 1992, pp. 44–49.
Preston G. Smith is a management consultant in West Hartford, CT. ***Donald G. Reinertsen*** is a management consultant in Redondo Beach, CA.

objective well. In practice it usually errs on the side of minimizing technical risk at the expense of speed. This one-size-fits-all mentality usually creates a system tailored to the largest, most complex projects to the detriment of simpler ones. If the R&D manager is successful in minimizing product complexity, simpler systems can be used.

2. LET ECONOMICS BE YOUR GUIDE

You need a yardstick to decide which development goal to stress for a particular project, and in business the time-tested yardstick is marked in dollars. It is both simple and valuable to develop financial yardsticks for the development process. Such yardsticks tell you the relative financial impact of project delay versus a product cost overrun. They guide you toward choosing the most productive development goal and applying a development process that facilitates this goal.

In addition to these strategic decisions, there are countless daily tactical decisions where yardsticks help to make accurate, fast, low-level decisions on tradeoff issues. For example, is it worth spending $100 on air freight to get a sample into a customer's hands for evaluation two days sooner? How about buying an extra microscope for $2000 if it will cut a week off of the schedule? Or $50,000 for temporary tooling that will allow you to start production two months early while permanent tooling is being made? Without sound decision rules these decisions are likely to be made both incorrectly and slowly.

The financial model that provides the yardsticks is not hard to build (2), but it requires cross-functional effort. The finance department may have the greatest expertise in financial model building, and marketing has important data. But R&D should probably initiate this activity because it will obtain the greatest benefit from the model. For instance, it is not unusual to discover that the product's development expense has far less impact on its life cycle profitability than development delay. When this is the case, R&D managers who spend much of their time massaging the budget are concentrating on a low-leverage area. (See Editor's Addendum for additional discussion.)

3. WATCH OUT FOR COMPLEXITY

The degree of complexity in a project determines the effort needed and thus the length of the development cycle. Although this is not surprising, product development people are frequently startled to discover how quickly complexity mounts.

For example, consider the experience of a company that makes industrial process controls. They made ambitious plans to use a microprocessor in their product for the first time, figuring that this is hardly a new technology anymore. To make full use of the new capability, they tapped a new market they had been unable to serve previously, and they added new product features that had been unavailable to them before. In retrospect, they realized that complexity had multiplied on them. Microprocessors may have been familiar components to others, but this firm had to learn how to procure them, design for them, test and assemble them—everything. In addition, they had to master the new features and establish the new market. This project was slow and fraught with other difficulties.

Complexity is insidious because it multiplies quickly and its effects are indirect and often not apparent. It increases the risk—both technical and market (discussed later). It increases workloads because many more interactions among elements must be considered. It tends to draw in more people, often specialists, which complicates communication and decision making. All of this necessitates a more complicated—thus slower—management process.

The way to get new products out quickly is to minimize complexity by moving in short, simple steps, sampling customer response along the way by selling intermediate models (2). This incremental innovation strategy involves two roles for the R&D manager. First, as manager of technical people who generally enjoy experimenting with new technologies, the R&D manager needs to temper others' desires to put the latest technologies into new products by applying his or her accumulated wisdom. In particular, stress the use of carryover parts and standard components, such as fasteners and connectors. Emphasize inelegant but clean architectures (2). Rechannel the technical brainpower toward solving those problems that will provide more substantial benefit for the customer.

These days the Japanese are often held up as innovation leaders, particularly companies like Honda. In fact, these companies control their pace of innovation carefully. Even their "all new" products are often far from all new. For instance, the initial Acura automobile models which made their debut in 1986 were advertised as "new automobiles . . . designed and engineered from the driver out." However, inspection of the Acura Integra revealed that the skin was indeed new, but many functional components—the most highly engineered ones like the engine, brake, door latches, and panel instruments—were carryovers from Honda Civics and Accords.

4. MANAGE THE INVENTION PIPELINE

Complexity is minimized by moving into new areas in a planned and evolutionary way, as just covered. This does not mean that newness is avoided. Quite the contrary, newness and invention must be embraced and managed.

Invention presents a dilemma to rapid product development. On the one hand, invention is essential to innovation: continual repackaging is a dead-end strategy. On the other hand, invention is a notoriously unpredictable activity. It cannot be scheduled into a normal project, must less an accelerated one. Any attempt to schedule this wild card into a project just adds uncertainty to the schedule, and in some markets schedule uncertainty is more detrimental than a longer but certain schedule.

Resolving this dilemma falls to the R&D manager. The solution is to invent off-line in a separately scheduled program that is tightly integrated with your market and product plans. Companies like Canon, Honda and Sony are innovation leaders because they devote considerable resources to maintaining a storehouse of developing technologies basic to their businesses. Both the invention track and the product development track are market driven, and both are given resources adequate to keep projects moving swiftly. The difference is that the former is loosely scheduled while the latter is tightly scheduled. When a technology reaches the point that much of its schedule uncertainty is eliminated, it switches tracks.

Consider the two types of failures that occur when this type of system is short-circuited. Many companies try to avoid the invention track by integrating it with the product development track. Then, schedule uncertainty is high in development projects, which ultimately causes both employees and customers to have little regard for development schedules. Every project proceeds at its own pace, unable to be accelerated.

The other failure is even worse. In this case a company simply does not invent. Its development schedules are predictable, but so is its demise.

5. AVOID THE "THINKING STAGE" TRAP

Other departments are often quick to blame R&D for slow product development, but the fact is that half of the typical product development cycle has vanished before development is even authorized (2). What we call the fuzzy front end is frequently one of the largest and cheapest opportunities to shorten the development cycle.

The front end of a development project starts when the need for a new product is first apparent, whether the company acts on it or not. Product need could be mandated by the enactment of a new government regulation, the emergence of a new technology, or certainly, the appearance of a competing product. The front end terminates when the firm commits significant human resources to development of the product.

We are not suggesting that the front end is unimportant; it is more like the heavens—mostly empty. Some crucial decisions are made during this period regarding the size of the market opportunity, the target customer, alignment with corporate strategy, and availability of key technologies and resources. In fact, research on the market success of new products suggests that products fail because companies don't do enough of this "homework" (3). Nonetheless, front-end time is still mostly a vacuum, largely because managers who haven't calculated the dollar value of development delay believe that time is free until people are assigned to the project.

As an R&D manager, your role in this phase is to be hard-nosed about using your people on product concepts. Resist the attempts of marketing or general management to have one of your people "look at" an idea in their "spare time." Remind them that delay erodes product profitability, and offer to assign one of your people immediately at full- or half-time to reach a certain decision by a definite date. If you aren't this serious about using your resources, then the company isn't serious about the product concept.

6. STAFF TEAMS ADEQUATELY

Our experience, and that of many others, suggests that product development proceeds most quickly and effectively with a team of six to ten full-time members. Although some products, like automobiles, computers or aircraft, require more effort than this, far more common is the development project that seems too small to justify this level of commitment. It receives perhaps a full-time person, a couple of part-timers, and a flock of bit-part participants. Given the heavy load of projects underway at a typical company, this appears to be the best that can be done. No single project has enough importance to command adequate resources.

The solution to this situation is simple in principle. If each project requires a certain number of person-years of effort, consider doubling the staff on half of the projects and complete them in half the time. Then do the other half of the projects similarly. Fewer projects will be underway at any point in time, but the same number will be completed each year.

Although the annual output is the same, the shorter, more intense project option has several benefits. The projects started first get to market sooner, giving them a competitive advantage and a longer sales life. The ones started later are completed no later than before, but they enjoy the advantages of a late start, such as better market information and more recent technologies. Both the early starts and the late starts reap the advantages of a short cycle: fewer opportunities for the market or the project objectives to shift, which means less redesign.

The shorter, more intense option is a valid model if project pacing is primarily dependent on labor availability, that is, project tasks typically sit waiting for people to work on them. In our experience, this is a frequent occurrence. Occasionally, project pacing depends primarily on outside events, like tooling lead time or prototype testing time, in which case it may not be possible to save appreciable time through heavier staffing.

A common objection here is that a more intense effort suggests large teams and thus a greater communication burden, which negates part of the anticipated benefit. However, we can obtain the extra effort without extra people by staffing teams with full-timers rather than part-timers.

Although it is possible to overstaff a team, our experience suggests that American development teams suffer much more from fragmented understaffing than from overstaffing.

7. STAFF WITH GENERALISTS

Often teams are fragmented, having many part-time members, because people are viewed too narrowly, and they in turn often mold themselves as narrow specialists. Some people indeed must be highly trained in a specific technical area in order to advance the state of the art, but the need for such skills is limited in most product development, which instead stresses application and integration. Having a narrow person on a development team causes the R&D manager three problems:

• First, it is difficult to keep such people fully occupied on the project. They tend to split their time commitment among one or more other projects. They drift in and out of a project as their particular skills are needed or as they have the time. They may shift to another project when the going gets tough on a particular project. Consequently, the team leader is consumed by simply keeping the team together and communicating with the part-timers, not on the primary task of developing the product.

• The second difficulty with specialization on the team is that good products require balance to provide value to the customer. This balance is achieved most quickly and with the least communication burden if everyone involved has a solid appreciation of the customer, the economics, the various technologies involved, and the manufacturing methods.

• Third, a high degree of specialization inhibits the manager's ability to redeploy people within a development team to match the workload, which leads to queues and delay.

There are a couple of implications for the R&D manager. First, staff teams with generalists or those willing to become generalists. This will give the team a more comprehensive view, which will allow them to move quickly and precisely. It will also strengthen the team if team members can shift to various secondary team jobs rather than dropping off of the team for a while.

Second, encourage and develop generalists. We recently saw a manufacturing engineer take on the company's cost accounting system when the manufacturing costs for his product weren't coming out to his liking. He didn't overhaul the corporate system, but he did negotiate a more equitable way of costing his product, a more creative one than a cost accountant was likely to have proposed. He learned a lot about cost accounting in the process, and he is now more valuable to his organization.

Deliberately expose people to new areas, either by transferring them to new departments or through outside training. Send your engineer to an accounting course or your draftsperson to a production machine programming course; encourage a marketing person to enroll in a microwave fundamentals course. Just by staying on a development team from start to finish, people will broaden, but this process can be accelerated through deliberate training.

8. LET THE TEAM MANAGE THE TEAM

Product development is just a succession of problems to be solved, so development speed depends on the speed of the problem-solving process, which in turn depends on how tightly problem-solving loops are connected. Every time the team has to go outside of itself to obtain a decision, additional delay is incurred. The farther it has to go, geographically or organizationally, the greater the delay is likely to be.

A Boston-area computer peripherals firm attacked the problem-solving-loop issue directly. The vice president observed that their development team was wasting time because designers weren't getting good enough guidance at their weekly meetings. They would design what they thought was desired, only to find out a week later that they were off track and most of the week had been wasted. So the vice president organized short, daily team meetings at the ten o'clock coffee break. Not only was there less waste of design talent, but everybody moved faster and more surefootedly because progress was now measured on a daily basis.

Just as important as the fivefold shortening of the loop is the vice president's role in this process. He recognized the problem, got the group to meet daily, and even attended many of the meetings. But he didn't run the meetings or participate in their content. He only made sure the group got together daily and left each meeting with a clear idea of what they would be doing next. The team ran the team, and if the company had had a team leader, the vice president's involvement would have been unnecessary.

Getting the team to run itself involves a couple of difficult organizational challenges. In typical organizations, managers of R&D and other functions typically control pieces of development projects, which are completed as these managers coordinate their efforts—often slowly. These roles must shift as the team assumes control of the project. The functional managers now act as advisors and coaches, assisting their colleagues on the team but not using the colleagues as conduits to carry information back to the functional manager for a decision. If this occurs, the organization has just installed a group of puppets as an additional level in the decision-making process. The R&D manager will still have plenty to manage, but his or her role changes with respect to a fast development team.

Frequently the team is also uncomfortable with its new role of making final decisions. If it tries to toss the decision back to management, management must simply toss it back to the team. Before long the team will rely on management as a source of decision-making information, not as a source of decisions.

9. MANAGE BOTH TECHNICAL AND MARKET RISK

Risk management has always been a large part of the technical manager's job. As cycle length shrinks, this job becomes even more essential. Faster projects employ a greater amount of task overlapping, which creates loose ends and, in turn, more opportunities for key steps to be omitted accidentally (2). If the team is managing itself, as just suggested, there is less formal opportunity for the management hierarchy to apply its considerable experience to averting past mistakes. Finally, there is less time available in a compressed schedule to recover from problems.

Fortunately, there is a great deal the R&D manager can do to manage risk in an accelerated project. Sensitivity to risk comes in part from years of experience, which managers are more likely to possess than are the members of the team. Through frequent informal interactions with the team, management can see potential pitfalls and inject insight to cope with them, all without infringing on the team's charter to make its own decisions.

One area where the manager's experience is most valuable is in balancing testing and analysis. Many technical people are prone to analyze an issue profusely before building something and testing it. Just making a model seems like an unprofessional expedient, but expedients are often just what we are looking for as we try to shrink tasks. Others, who may lack the analytical skills or discipline, do the opposite. They build and test repeatedly before thinking much about what the underlying issues may be, so they waste time in resolving risk, too.

The trick is in knowing when to test and when analysis would get the answer faster, or better yet, how test and analysis can be blended to get the best of both. The R&D manager's accumulated wisdom can be invaluable in raising and helping to resolve these issues. The manager also must make sure that analytical and testing resources, such as an open lab, are easily available to the team for this hands-on work.

Risk is of two types: technical risk, which is the inability to satisfy the product specification, and market risk, which is the inability to sell the product assuming it meets specifications. We tend to concentrate on technical risk, ignoring market risk, because we have better techniques for resolving technical risk, it is easier to identify and measure, and its symptoms usually appear sooner.

The R&D manager's job here is to teach the rest of the organization that market risk is just as real as technical risk and that the same general risk management techniques apply to it, although the two should be managed independently (2).

10. DEVELOP A RESERVE

We have saved the toughest topic until last. As suggested in 6, above, development projects are slow largely because they spend most of their lives waiting to be worked on. Projects are abundant but resources are tight. One reason for this predicament is that we use the popular development funnel concept where it doesn't apply.

For some products, often chemical products, the concept of a development funnel does make sense. The failure rate in the initial feasibility stages of a project is high, and the cost of these stages is low. So we start lots of projects at the top of the funnel, and a few winners flow from the bottom through a natural selection process.

Ironically, the development funnel doesn't work well for many products because the failure rate isn't high enough. Such projects are more likely to succumb to market causes either before or after development than to fail on technical grounds during development. Nevertheless, companies load the funnel with plenty of new-product ideas, and marketing is in fact encouraged to overstock the funnel (2). Because few projects actually fail, projects languish in the funnel awaiting resources. R&D managers must discourage application of the development funnel mentality where it does not apply. Applying it under low failure-rate circumstances generates a glut and demoralizes technical people whose perfectly acceptable projects get shelved in midstream for lack of resources.

However, eliminating just the glut is not going far enough. There actually has to be some slack because unplanned new product ideas will arise unexpectedly. The time-competitive firm needs some reserve development capacity to respond to these customer needs quickly, just as they retain reserve manufacturing capacity to fill unanticipated production orders responsively.

This presents a difficult challenge for the R&D manager. At planning time, don't accept a full load and then a bit more to cover fallout. Instead, leave some unused capacity for the really new projects.

This completes our tour of 10 areas where the R&D manager can shorten development time dramatically. You obtain the greatest benefit by making all of these improvements, because they all reinforce one another. But this is a long-term goal because no company we know of does all of these things well yet. So get started with some of them, perhaps by using a pilot rapid development project to initiate several of the changes immediately (2). Finally, get the non-R&D parts of the company in-

volved compressing development time too. Even those apparently removed from the process, like corporate planners, have essential parts to play (4).

REFERENCES

1. Reinersten, Donald (1991); "Outrunning the Pack in Faster Product Development," *Electronic Design*, 39, 1, pp. 111–118.
2. Smith, Preston, G., and Donald G. Reinersten (1991); *Developing Products in Half the Time* (New York: Van Nostrand Reinhold).
3. Cooper, Robert G. (1988); *Winning at New Projects* (Reading, Mass: Addison-Wesley).
4. Reinertsen, Donald G., and Preston G. Smith (1991); "The Strategist's Role in Shortening Product Development Cycles," *The Journal of Business Strategy*, July/Aug. pp. 18–22.
5. Rosenau, Jr., Milton D. (1990); *Faster New Product Development* (New York: AMACOM).
6. Reinertsen, Donald G. (1983); "Whodunit? The Search for the New-Product Killers," *Electronic Business*, July, pp. 62–66.
7. Reinertsen, Donald G. (1992); "The Mythology of Speed," *Machine Design*, March, pp. 95–98.

AUTHORS' ADDENDUM

Beware of Phased Development

The concept of dividing a project into phases and funding each phase only if it satisfies certain prerequisites would appear to be a good management tool. Yet, as consultants, we see an oil-and-water relationship between phased systems and accelerated development.

Developed by NASA as the PPP (phased project planning) process, phased development systems are designed to control technical risk. But when speed is important, market risk becomes more critical: even if the product is designed according to spec, there is a significant possibility for market failure if it is introduced late. When technical risk must be balanced with market risk, a monolithic PPP-type system is no longer the clear choice. Adaptation and balance are needed in the project management system, and the balance shifts toward empowering the people and away from depending on formal control systems.

The question is really where the balance should lie. We advise some companies, usually fast-growing small ones, to formalize their review systems because with the product line and the staff growing rapidly, more formal management checks are needed to avoid technical failures. Yet in the majority of cases we see phased control systems that are overly cumbersome for a firm's needs. Sometimes a company will just adopt another's phase process, as we once observed when a 100-person instrument manufacturer adopted Hewlett-Packard's phase process, figuring that HP was also an instrument producer—a very good one indeed. But HP's process, fine-tuned for a large company with dozens of divisions, was excess baggage for this small firm.

Fundamental Limitations

Phased systems have a number of fundamental limitations that restrict the ability to shorten development cycles. They often preclude employing one of the most fundamental time-shrinking tools: overlapping activities. Often a particular stream of activities could be overlapped to advantage, but a phase review breaks the chain by requiring that all activities be finished up for review before the next phase can start.

Moreover, the very act of discovering overlapping opportunities requires a new attitude for an organization long indoctrinated in a sequential phased review process. It is that much more difficult to get people thinking creatively about overlapping dissimilar activities when their mindset is built around established phase gates.

Overlapping is enabled by employing partial, fragmentary information that evolves in a stream (2). Phased systems fight against partial information, providing credit and passage to the next phase only when the information package is complete. Here again, the time-saving opportunities must be discovered in a particular circumstance by thinking creatively about inching forward with the information at hand. A tidy phase framework discourages these discoveries.

Exploiting system architecture opportunities is another means of compressing development cy-

cles (2). By dividing a product into subsystem modules with relatively clean interfaces and ample performance margins, these modules can be developed concurrently by different teams. For example, a transceiver might be divided into a power supply, a transmitter, a receiver, and an audio amplifier. There is no reason to believe that these four modules will have the same timing. One module might require more conceptual design or technology exploration, while another might need a great deal of prototype testing. Putting all modules in a lockstep phase review process stretches the overall cycle.

Moreover, a phased system encourages queues. Queue reduction is a huge and inexpensive opportunity to shorten cycles simply because most development projects spend the majority of their time sitting in queue somewhere. With phases, queues build up in preparation for a review. (Remember that only a complete package is acceptable for a review, so some items wait while the package is completed.) Then, when the review is complete, a flood is released into the next stage of the system, swamping it.

For example, upon final approval of the design, purchasing may be faced with simultaneously ordering a million dollars of capital equipment from a dozen suppliers. Or in one case we observed, the chief engineer signed and released into manufacturing nearly 500 drawings in a single day when his project passed a milestone. You could actually observe the glut passing through the manufacturing transition process.

A phase process also causes problems when, as is often the case in practice, "the product" is really a line of products in different sizes, materials or colors. Then, forcing all variants into synchronization for convenience in review creates both pre-review delays and post-review gluts unnecessarily.

Responsibility Belongs to People

In short, phased approaches are attempts to build judgment into the process rather than into the people. It is reactive, it is slow, and it removes the responsibility from the people doing the work, where the responsibility belongs.

Shifting away from a phased process is difficult. There is a great deal of management comfort involved in taking a thorough, formal look at a project periodically and making an explicit decision whether or not to proceed. Unfortunately, the cost can be high when time is at stake. A balance must be struck between comfort and speed, and all too often, comfort wins out even when speed is the key competitive factor.

Some rapid product development specialists suggest that the phased approach should remain the foundation but that the phases should be compressed and "dead time" between phases should be eliminated (5). Our observations of how development projects actually proceed in industry suggests that the greatest opportunities for improvement lie in eliminating the delays associated with synchronization and queueing. This requires a fundamental departure from the phased approach, not fine-tuning it.—D.R. and P.S.

EDITOR'S ADDENDUM

In Reinertsen's (6) original publication of the trade-off between speed and costs, he emphasized that the key to winning lies in a company's flexibility in adapting a product's development to its specific market conditions and strategy. Based on a relatively simple economic model of new product development and commercialization, Reinertsen compared the effects of a product delay, a product-development expense overrun, and a product cost increase on total life cycle profits in two different kinds of market situations: (1) a high-growth, price-eroding market and (2) a low-growth, non-price-eroding market. After examining the comparative results of his model, Reinertsen drew the following three conclusions:

1. Development expense overruns have little effect on total life cycle profits. In both high- and low-growth market conditions, overrunning development cost budgets by 50 percent had less than a 4 percent impact on total life cycle profits before tax.

2. Speed is not always the most critical factor. In a high-growth market with short product-life cycles, shipping a product six months late can decrease its cumulative profits by 33 percent. On the other hand, in a slow-growth market with long product-life cycles, a six month delay creates only a 7 percent decline in overall profits.

3. In many market situations, the single most important factor may be product cost. A product cost overrun of only 9 percent reduced total life cycle profits by 22 percent in a fast-growth market and resulted in a devastating 45 percent decrease in profits in a slow-growth market.

In general, Reinertsen stresses that the art of managing product development depends on making well-informed trade-offs between the four possible objectives of (1) development speed, (2) product cost, (3) product performance, and (4) development expense. He urges managers to build some simple economic models and tools to show the profit impact of the trade-offs they are making, often even unconsciously, between these objectives in their specific competitive situation. Based on his experience working with companies to improve and speed up their development processes, Reinertsen (7) offers a number of additional suggestions as follows:

1. Don't wait until all plans and schedules are in place. Design can always begin before planning is complete. Do just enough planning to begin design and then complete it early enough to support key design decisions.

2. Be careful in drawing up elaborate schedules with very long lists of supposedly predictable activities that then force developers to waste effort explaining why they're off schedule. Engineering is stochastic and is, therefore, extremely hard to schedule. Microscheduling can be a huge drain on design resources. Focus instead on a handful of critical high-level milestones. Find ways to set and communicate priorities so that top management does not have to intervene to accelerate one program at the expense of others.

3. Don't get bogged down waiting until all the

specifications are complete, accurate, and frozen. Keep the specs adaptable throughout the design process. Progressively finalize them after compromises have been made on features. Concentrate on making product specifications fit customer requirements to control risk.

4. Don't believe that the key to rapid product development is designing things right the first time, and that too much time is spent correcting mistakes. Early testing is critical for rapid development. Instead, quickly implement a partial solution and then keep working to test and improve it. Pay more attention to how fast you can arrive at an answer, not just the number of mistakes made along the way.

5. Don't implement control and measurement systems that end up disempowering people. Rapid development relies on motivating teams rather than imposing top-down plans.

6. Too much time is often spent planning and rehearsing for formal review meetings. Instead, improve informal monitoring of development programs and strengthen informal channels of communication. Manage by "wandering around" for faster and more timely information.

7. Don't concentrate on defining the perfect development process that avoids all possible problems. It encourages the compliance of people rather than their initiative to find new ways of solving old problems. Well-trained people armed with information and the means to choose and implement appropriate approaches can outperform most systems in a fast-changing environment.

8. Be careful of the "do-it-try-it-fix-it" mentality. While managers get frustrated with too many analyses, make sure that analysis plays a strong role, as it is the only practical means for uncovering problems in complex systems.

9. Establish autonomous teams carefully to avoid the disadvantages of not using the organization's existing designs, solutions, and technologies.

10. The slowest portion of most organizational processes is deciding to get started. Running fast is good, but starting early is better!

49

First to Market, First to Fail?

Real Causes of Enduring Market Leadership

GERARD J. TELLIS AND PETER N. GOLDER

Managers and entrepreneurs frequently adhere to the motto of being first to market. But the authors have discovered that many pioneers fail, while most current leaders are not pioneers. Using a historical method, the authors try to determine why pioneers fail and early leaders succeed. They have found that market leaders embody five factors critical to success: vision, persistence, commitment, innovation, and asset leverage.

Be first to market. This principle is one of the most enduring in business theory and practice. Entrepreneurs and established giants are always in a race to be first. Research from the 1980s that shows that market pioneers have enduring advantages in distribution, product-line breadth, product quality, and, especially, market share underscores this principle. For example, several studies of the PIMS (profit impact of market strategies) database show that mean market shares over a large cross section of businesses are around 30 percent for market pioneers, 19 percent for early followers, and 13 percent for late entrants.[1] Similar estimates emerge from a study of the completely different ASSESSOR data.[2] That two independent databases collected by different methods and researchers should yield similar results is impressive. In addition, PIMS data show that more than 70 percent of current market leaders are market pioneers, while Urban et al. are unaware of any pioneers in their sample that failed.[3] Further evidence in support of a pioneer advantage comes from an *Advertising Age* study that shows that, of twenty-five market leaders in 1923, nineteen were still market leaders in 1983, and all were still in the top five.[4] The belief in enduring pioneer advantage grew so strong that some authors even suggested that firms preannounce a product's introduction to claim the advantages that accrue to the pioneer.

But as most people came to believe in the strong advantages of market pioneering, some researchers warned of potential problems with the studies on which this belief was based.[5] Kerin, Varadarajan, and Peterson state, "The belief that entry order automatically endows first movers with immutable competitive advantages and later entrants with overwhelming disadvantages is naive."[6] In particular, the data that support the advantages of pioneers suffer from three key limitations.

First, PIMS and ASSESSOR data are collected by surveying surviving firms. Thus they include only survivors and not failures. Failures may hold important lessons, and their inclusion may change the statistics. Second, these two databases determine the market pioneer by surveying a current employee in a responding firm. Such surveys may be biased because current market leaders may see or promote themselves as pioneers, especially if the market is old, the managers are new, and the firm has been successful. Third, while all studies define the concept of a pioneer as the first to enter a market, the operational definition has been im-

Reprinted from *MIT Sloan Management Review*, vol. 37, no. 1, Winter, 1996, pp. 65–75. Reprinted with permission.

Gerard J. Tellis is Neely Chaired Professor of American Enterprise at the University of Southern California's Marshal School of Business. ***Peter N. Golder*** is Marketing Professor at New York University's Stern School of Business.

precise. PIMS, in particular, uses the definition, "one of the pioneers in first developing such products or services." This promotes ambiguity in data and analysis.

We reexamined the rewards of pioneering using an entirely different method, historical analysis, that avoids the limitations we described.[7] Our main conclusions, based on fifty consumer product categories, are:

- The failure rate of market pioneers is 47 percent. This failure rate is higher for durables than for nondurables, but does not vary by categories starting before and after World War II.

- The mean market share of pioneers is 10 percent.

- Market pioneers are current leaders in only 11 percent of the categories. The rate is even lower for categories starting after World War II.

- The results on market pioneers are found despite sampling categories that are more favorable to pioneers. If all fifty categories had been selected randomly, the results may have been more unfavorable to pioneers.

- Another class of firms labeled "early leaders" has a minimal failure rate, an average market share almost three times that of market pioneers, and a high rate of market leadership. Early leaders are firms that enter after pioneers but assume market leadership during the early growth phase of the product life cycle.

Is the difference between pioneers and early leaders merely semantic? We emphasize that it is not but, rather, of strategic significance.[8] While firms often rush products to market to get a lead time of months or weeks, early leaders enter an average of thirteen years after pioneers, yet are much more successful. Therefore, the questions we address in this paper are:

1. Why are early leaders so successful?

2. Why do pioneers fail so often?

3. What are the real causes of enduring market leadership, if not order of entry?

We try to answer these questions using the historical method. First we clarify definitions and explain our method.

DEFINING PIONEERS AND CATEGORIES

Three pitfalls are typical in defining a pioneer: too loose a definition, too narrow a definition of the product category, and use of hindsight. First, we define a pioneer as "the first to sell in a new product category." While more narrow than the PIMS definition, this clearly differentiates the order of market entry to determine if being first by itself has any enduring advantages. It distinguishes the class of firms that might enter after the pioneer and learn from its errors, yet still enter early enough to shape and dominate the product category.

Second, we define a product category as "a set of competing brands that consumers perceive as close substitutes." Examples include microwave ovens, copiers, and disposable diapers. Our definition is consistent with research that considers customers in determining product categories.[9] Several authors argue that a broad category definition is preferable to a narrow one because of the competitive advantage that accrues from thinking broadly.[10] At the other extreme, too narrow a definition allows any firm with a loyal following to be designated a pioneer of the market segment it currently dominates. Narrow definitions of categories lead firms to be smugly satisfied in their current success and bear no insights into past success or future options. While we avoid extremely narrow category definitions, we do allow for category extensions. A category extension is "a subset of a category consisting of competing brands that a target segment perceives as close substitutes for each other, but not for brands in the parent category." Examples are light beer, liquid detergent, and color TVs. Category extensions differ from the parent category in at least one key attribute and develop their own character, distinct from product, competitive, and consumer characteristics.

Third, we avoid hindsight when identifying pioneers and defining categories. Every successful, dominant firm has reached that position by doing something right. In a loose sense, that firm has "pioneered" a new concept or a new market segment. However, such a loose definition of a pioneer based on results is circular: if successful firms are labeled pioneers, then pioneers must be successful. The question is, how did the firm appear at the time it entered the market? When a firm enters a market, the only certainty is its order of entry, whether first, second, etc. In this strict definition of pioneer as first to market, pioneering rarely leads to long-term leadership.[11] Therefore, the question becomes, what are the real causes of enduring market leadership?

LEARNING FROM HISTORY

In the study of market pioneering, no sophistication in survey method or econometric modeling can substitute for the insights of historical analysis. The historical method involves analyzing reports written when the market evolved—in other words, it uses accounts of market definitions and entry contemporaneous with their occurrence. Data collection consists of screening for (1) willing, (2) able, and (3) reliable reporters who provide (4) corroborating evidence. Following our earlier approach, we used these four criteria to analyze the same fifty product categories from our 1993 study.[12] We substantially broadened the data collection and analysis to address the questions of this study by examining about 1,500 articles in 25 different periodicals. We also analyzed information from approximately 275 books that document individual product categories and brands.

The historical method relies on induction to determine causality. For the purpose of this study, the method identifies patterns in the failures and successes of the many entrants in each market. While the approach is labor intensive, it promises fresh insights because it recaptures forgotten details in market evolution. Cross-sectional databases cannot provide the same perspective, because a matrix of numbers loses the richness of history while introducing potentially serious biases. The historical method clearly lacks the apparent objectivity

and sophistication of previous authors' econometric models. However, the objectivity of significance tests and sophistication of complex models may lull us into a false sense of security in precise estimates of causes that are really spurious. In contrast, research time invested in piecing together history, from our vantage point, can reveal patterns and insights that actors at the time could not see and current analysts have long forgotten.

Many researchers have called for a longitudinal analysis similar to ours, but few have chosen this method, probably because it is difficult, time consuming, and not clearly charted.[13] To appreciate the insight from the historical method, consider these quotes from *Financial World* about a company in the restaurant business:

- "World's biggest chain of highway restaurants."
- "Pioneer in restaurant franchising."
- "Most strongly entrenched factor and highest quality investment."
- "Most fabulous success story in restaurant chains."[14]

While these quotes may bring McDonald's to mind, they refer to Howard Johnson's restaurants. Reports from the 1960s give a forgotten view of how reporters regarded Howard Johnson's at the time.

The disposable diaper market provides another example of how the business world easily relabels successful firms as pioneers. In 1991, Procter & Gamble (P&G) celebrated the thirtieth anniversary of its entry in the disposable diaper market. For the millions of parents who benefited from the disposable diaper, it was indeed an occasion for celebration. P&G claimed it "literally created the disposable diaper business in the U.S."[15] The truth is that disposable diapers were available in the United States as early as 1935 (e.g., the Chux brand).[16] While some may argue that Chux was a small obscure brand, in 1961, *Consumer Reports* evaluated "nationally available" disposable diapers brands and ranked Chux as clearly the best, ahead of Sears and Montgomery Ward; it didn't even mention Pampers.[17] Moreover, Chicopee Mills, a unit of the very capable Johnson & Johnson, owned

Chux. A few years later, Pampers and Chux were both ranked as best buys, indicating the arrival of Pampers and its perception as Chux's competitor. Thus the passage of time, Pamper's success and P&G's promotion of its achievements have led to a reinterpretation of the history of this market.

The personal computer market is an example of how quickly the business world forgets failures and showers undeserved praise on successes. The first personal computers were sold in 1975.[18] By the early 1980s, the business press referred to Apple Computer as the pioneer of personal computers.[19] However, several studies, including one supported by the National Science Foundation, found that MITS (Micro Instrumentation and Telemetry Systems) was the actual market pioneer.[20] In fact, MITS was so dominant that, in 1976, *Business Week* referred to it as the "IBM of home computers."[21] Further, it reported that MITS's "early lead has made its design a de facto standard for the industry." MITS's initial success certainly did not lead to long-term market leadership.

Because our method does not yield a model, statistics, or effects sizes for the causes of enduring leadership, it runs counter to the dominant approach in marketing. It does, however, yield new case histories gleaned from the corroborated evidence of willing, able, and reliable reporters. The reporters' accounts are publicly available and verifiable. The synthesis and interpretation are our own. We hope that readers will consider our explanation of enduring leadership as an alternative to econometric models of survey data.[22]

THE LESSONS OF EARLY LEADERS

Our study of the fifty product categories from their inception suggests five factors that drive the superior performance of early leaders: a vision of the mass market, managerial persistence, financial commitment, relentless innovation, and asset leverage. Similarly, pioneers' inability or neglect to implement these factors often leads to their failure. The first four factors are related components of a mind-set that seems to fire the success of early lead-

ers, especially in newly emerging categories. The last factor comes into play especially in category extensions. These factors are probably the real causes of enduring market leadership.

Envisioning the Mass Market

The mass market has a bad image in marketing, especially for mature products. Theorists stress the importance of segmentation and differentiation to better serve market niches and increase prices and profits. The mass market is synonymous with cutthroat competition, low prices, and low profits. However, the opposite is true in new markets. Typically, when a new product is first commercialized, product quality is low, prices are high, and applications are few. The product does not seem like an attractive buy, and sales are limited. At this time, it takes a visionary to see the mass market and find the way to open that market. Tapping the mass market provides economies of scale and experience that can overcome low quality, high prices, and limited features. The mass market is a means to exploit the full potential of the new product.

For example, Ampex pioneered the video recorder market in 1956 and was the leading supplier for several years. At $50,000 each, initial recorder sales were limited. RCA and Toshiba, the only competitors, were way behind, so Ampex had almost a monopoly in sales and R&D. However, Ampex managers did little to improve quality or lower costs; ironically, they sought to reduce Ampex's dependence on video recorder sales, and pursued audio products, computer peripherals, and other diversifications.

In contrast, three other companies, Sony, JVC, and Matsushita, inspired by their consumer orientation, saw the mass-market potential and made a concerted research effort to bring the video recorder to that market. In the mid-1950s, Masaru Ibuka, Sony's cofounder, set a target price for its video recorder of $5,000, which he later reduced to $500, 1 percent of Ampex's initial price. At JVC, Yuma Shiraishi, manager of video recorder development, provided just a few guidelines to his engineers: develop a machine that could sell for $500, while using little tape and retaining high picture quality. Such stringent directives were difficult to meet. Each of these companies researched for twenty years to realize its goal. But the directives were clear and firmly focused on the mass market. When their efforts were successful in the mid-1970s, these companies catapulted to dominant positions in the electronics industry. In the fifteen years following 1970, video sales went from $2 million to almost $2 billion at JVC, from $6 million to $3 billion at Matsushita, and from $17 million to almost $2 billion at Sony, but total sales increased from $296 million to only $480 million at Ampex. Ampex started anew in the early 1970s to cater to the mass market for video recorders, but this effort quickly faded in the face of technological hurdles, and Ampex ceded the market it had pioneered.[23]

The disposable diaper market provides another example. Although Chux was probably the best product in the early 1960s, it was relatively expensive, so sales were limited to wealthy households or for use while traveling. Indeed, *Consumer Reports* recommended that Chux be used when traveling.[24] However, P&G's experience in grocery marketing and its early research with Pampers prompted it to pursue the far greater potential of the mass market. Its major problem was to develop a product that was soft yet strong and absorbent, with a dry inner surface, all at an affordable price. P&G launched a concerted research drive to achieve this. In the first five years, its lowest price was 10 cents per diaper, when diaper services cost 3.5 cents and home washing about 1.5 cents. P&G engineers persisted and finally came up with a sophisticated machine that produced 400 diapers a minute at 5.5 cents each. At that price, Pampers' national rollout in 1966 was a huge success, expanding the market from $10 million to $370 million in seven years. P&G had unlocked the mass market for disposable diapers.[25]

Another example is the photographic film market that developed over many years. During most of the 1800s, photography was restricted to professionals and serious amateurs because there was no network of professional film developers. George Eastman had the vision to see the mass market that would develop if film processing were made easy. He designed a system whereby consumers took pictures with a returnable camera, mailed in the camera with the exposed film for developing, and received the developed pictures and a reloaded camera. The company's slogan, "You press the button, we do the rest," convinced consumers that photography was finally available to amateurs. This innovation initiated Kodak's long quest to expand the mass market for cameras and photographic film. It continued this effort by producing low-priced cameras, such as the Brownie line, which made picture taking affordable for everyone.[26]

Many other markets came to be dominated by nonpioneers that entered early or even late but transformed the industry by targeting the mass market. Examples include Timex in watches, Gillette in safety razors, Ford in automobiles, and L'eggs in women's hosiery.

Managerial Persistence

New products are often described as "breakthrough" inventions by tinkering scientists, but that it not the way that market leaders achieved their dominance. Actually, successful products are the fruit of small, incremental innovations in design, manufacturing, and marketing over many years. Management must maintain a commitment to the brand over a long period of slow progress. For example, P&G persisted with research for ten years to make Pampers affordable for the mass market. "P&G spent more time and money on developing and testmarketing Pampers than Henry Ford plowed into his first automobile or Edison spent on inventing the first incandescent bulb," according to *Forbes*.[27]

Sony, JVC, and Matsushita spent more than twice as much time and encountered far greater difficulties than P&G, but their commitment was unwavering. For example, after Ampex commercialized the first video recorder in 1956, Sony, then a small company, took almost seven years to introduce the PV-100, a two-headed, helical scanner model. While a technological wonder, the product was still too expensive and lacked the features and color to appeal to the mass market. After another thirteen years, Sony introduced the Betamax, which finally appealed to consumers. During those years, Sony engineers had worked to include color, reduce the weight by 77 percent, increase recording density eleven times, and, at the same time, reduce the price by 88 percent.[28] The Betamax was not a sudden breakthrough, but the evolutionary result of persistent research over two decades to solve diverse design problems.

Similar efforts to develop a video recorder were so slow at JVC that director Shiraishi chose to keep the project secret from his superiors for fear of losing support. Shiraishi's steadfast direction finally paid off, and JVC introduced the VHS in fall 1976, twenty-one years after starting research, twenty years after Ampex pioneered the market, and at least one year after Sony's Betamax. The VHS actually succeeded better than the Betamax primarily because of its longer tape limit (two hours to one for the Betamax) and Matsushita's early strategy to be an original equipment manufacturer for RCA.[29] During the twenty years after 1956, Ampex made some sporadic attempts to develop a product for the mass market, but it had neither a plan nor the persistence to succeed.[30] Thus, in the video recorder market, order or timing of market entry was irrelevant. A vision of the consumer market and persistence to design a product to meet consumer needs was critical.

RCA pioneered color television in 1954, yet sales did not come easily since the vast majority of programs were broadcast in black and white. In fact, black-and-white TVs received broadcasts better than color TV sets did, so consumers had little incentive to purchase color sets.[31] Even in the early 1960s, *Consumer Reports* recommended that people not purchase color sets.[32]

RCA needed to make a strong long-term commitment to ensure that this product gained widespread acceptance. It accomplished this goal in two ways: First, it started a program of technical research, which led to the needed improvements in quality, protected with patents that RCA licensed profitably for years.[33] Second, it committed to broadcasting color television programs through its subsidiary, NBC, at a time when the vast majority of consumers still owned black-and-white sets. During the 1960 to 1961 season, NBC broadcast fifteen to twenty times as many color programs as CBS did, while ABC did not broadcast in color at all.[34] Better quality products and more color broadcasts finally convinced consumers that color TV had come of age, and the market took off in the mid-1960s. RCA's persistence over ten years was rewarded with long-term market leadership of color TVs and is still symbolized by its famous trademark, the NBC peacock.

Many other markets came to be dominated by firms that persisted over many decades to finally establish the mass markets for their products. Kellogg, Hershey, Crisco, and Wrigley are household names, though their efforts in marketing and R&D are little known. In all these cases, persistence rather than order of entry or sudden breakthroughs was a key to success.

Financial Commitment

Because market dominance of a category requires vision and persistence over many years against great research and marketing odds, firms need to commit finances to last through this struggle, especially when revenues do not cover costs. Specifically, they need to address two aspects: access to financial resources and willingness to use those resources. Failure in either area can lead to a loss of promising markets. Enduring market leaders are firms that commit resources, either their own or those entrusted to them.

For example, most people consider Miller Lite to be the pioneer of light beer because it has dominated the market since its entry in the mid-1970s. However, several brands of light beer were introduced without much success, starting in the early 1960s. Rheingold Brewery of New York made the largest effort. During 1967, it spent $5.5 million to introduce Gablinger's light beer in the eastern United States.[35] Unfortunately, in the same year, it also achieved lower sales of its regular beer. These two factors produced a loss for Rheingold that led the company's directors to fire top management in favor of more profit-conscious managers. The new management dropped support for Gablinger's to make a quick return to profitability.[36] The company's directors would not maintain the financial commitment to Gablinger's, even though it had already achieved distribution in most eastern states.[37] Had immediate profits not been of so much concern, Gablinger's might well have established and dominated the light beer category.

Rheingold's behavior contrasts markedly with the financial resources Philip Morris used to introduce Miller Lite. In 1975, *Business Week* reported that Lite's advertising expenditures averaged $6.50 per barrel, while the industry average was only $1.[38] Further, Philip Morris was willing to forgo any profits from all Miller brands for five years in order to build market share and establish the light beer category.[39] Its effort was rewarded by long-term leadership in light beer.

Some have suggested that the Lite brand name was an important factor in Miller's success. However, the Lite trademark was originally used by Meister Brau, another early failure in light beer. Miller acquired and used the Lite trademark successfully, largely because of Philip Morris's financial commitment.

MITS pioneered personal computers in 1975, followed by many startups during the next few years. However, only one emerged as a dominant player—Apple Computer. While contemporary reports attribute Apple's success to the role of Steven Jobs and Stephen Wozniak, its success may be due as much to a largely unrecognized third partner, Mike Markkula. He was a former executive of Fairchild and Intel who became wealthy from stock options earned at those companies. When Jobs and Wozniak each invested $6,000 in their new venture, Markkula contributed $91,000 for a one-third interest. Markkula helped Apple receive a bank line of credit and venture capital from various sources, including a fund financed by the Rockefeller family.[40] The strong financial backing enabled the firm to continue developing new products and quickly expand its market nationally. Contemporary reports do not mention Markkula's crucial role in bringing early financial strength to the startup.

Two other companies shared Apple's early dominance of the personal computer market: Tandy and Commodore. Each did so because it had financial strength and was willing to commit resources to compete in this rapidly expanding national market. Other examples in which financial commitment was very important in achieving long-term leadership are Seagram and Bartles & Jaymes in wine coolers, Matsushita and Sony in video recorders, RCA in color television, and Crest in toothpaste.

Relentless Innovation

Market pioneering generally follows an invention. However, changes in consumer tastes, technology, and competition require firms to keep improving their products. Long-term leadership requires continuous innovation. Three factors hinder companies from investing and following through with innovations. First, they fear cannibalizing established products. For example, IBM stymied its development of minicomputers and workstations to protect mainframe sales, even though competitors kept making inroads into the mainframe market. Second, they are satisfied with their progress. Ampex's failure to bring video recorders to the home market was caused partly by management's satisfaction with sales to the professional market. Third, large bureaucracies either discourage innovations or are slow in bringing them to market. Despite their technological strength and financial resources, GM and IBM both were slow to bring out new products because of their bureaucratic approval process.

Gillette's history shows how companies must overcome these three hindrances. Prior to 1903, men used straight razors or expensive safety razors. In 1903, with an eye on the mass market, King Gillette introduced a safety razor with low-priced disposable blades.[41] The Gillette company popularized its razors and blades by successfully targeting the U.S. government to buy them for troops during the two world wars. Because of patents on its initial products, persistent pursuit of the mass market, some new innovations, and few competitive threats, Gillette dominated the market for a half-century, with a peak share of 72 percent in 1962. Success may have led to complacency.

In 1962, Wilkinson Sword of Britain introduced a stainless steel blade that lasted three times longer than Gillette's carbon steel blade. With imitators of Wilkinson entering the market, Gillette's share fell to 50 percent in little over a year. Gillette was shaken. The irony was that Gillette had been aware of the new stainless steel technology all along. Indeed, later on, Wilkinson had to license some of this technology from Gillette, which held the patents. But Gillette's introduction of stainless steel would have rendered obsolete much of its manufacturing capacity for carbon steel blades. Wilkinson Sword did not destroy Gillette because it lacked the financial resources to fully exploit this innovation.[42]

The Wilkinson experience galvanized Gillette to innovate even at the cost of cannibalizing its own established products. Gillette introduced the Trac II twin-head razors in 1972 and saw its older brands slowly yield to the twin-blade technology. While Trac II was still in its prime, Gillette introduced Atra pivoting-head razors in 1977, knowing this product would cannibalize Trac II sales. In 1976, Gillette was threatened by the imminent entry of Bic disposable razors. Gillette continued to innovate by introducing the Good News twin-blade disposable razor, even though the cheaper disposable cut into short-term profits. Each of these moves were expensive but well rewarded, as Gillette retained dominance of the razor and blade market and even expanded its share. In 1989, Gillette introduced the next big innovation, Sensor, a razor with twin blades that move independently. This innovation gave such a good shave that it actually began to reverse the loss in share from less profitable disposables.[43]

The atmosphere at Gillette reveals how it succeeds at innovation.[44] There is a passion for the product, while innovation is almost an obsession. At any time, Gillette has at least twenty shaving products planned. Each day, about 200 employees personally test new shaving technology. Technicians test the edge, the guard, the angle of the blade, various aspects of the razor, and even the cut

whiskers. They study facial hair, skin chemistry, and skin follicles. This emphasis on research probably promotes Gillette's dominance of the market, which is evident in its profit figures. It has diversified into numerous other markets, so that razors and blades now account only for one-third of sales but almost two-thirds of profits.[45]

Other examples of relentless innovation include Polaroid's efforts in the instant camera market, Kodak's dominance of the photographic film market, and Johnson & Johnson's steady introduction of new drugs.

Asset Leverage

Late entrants are often able to become leaders in some categories if they hold dominant positions in a related category. This ability depends on shared economies across categories, typically due to brand-name recognition, but also due to strength in distribution, production, or managerial expertise. For example, IBM's huge success in mainframe computers provided it with instant brand recognition and a strong distribution advantage to enter the PC market. Other examples where the ability to leverage assets has led to long-term success are found in diet colas, liquid laundry detergents, and wine coolers.

During the 1950s, diet cola was sold to people with special dietary needs (e.g., diabetics). In 1961, Royal Crown (RC) achieved great success when it redirected diet cola toward the mass market. However, it was virtually powerless to prevent Diet Pepsi and Coca-Cola's Tab from gaining leadership in this market, even though RC spent millions to support its brand.[46] The ability to leverage assets was further demonstrated when Diet Coke captured market leadership within one year of its entry in 1982.

As for liquid laundry detergent, Wisk enjoyed a commanding position from the time it pioneered the category in 1956 through the mid-1980s. Even P&G was unable to dislodge Wisk with its new brand, Era. However, P&G was able to capture market leadership by leveraging its dominance in regular detergent with the introduction of Liquid Tide.

In the wine cooler market, the market pioneer,

California Cooler, enjoyed great success until the dominant firms in the wine and distilled spirits industries decided to enter this category. A few years after entry, Gallo's Bartles & Jaymes brand and Seagram established leadership in wine coolers. Their established distribution networks and managerial expertise contributed to their success.

Other examples of successful asset leverage include Matsushita in camcorders and Coca-Cola's Sprite in lemon-lime soft drinks. The proliferation of brand extensions into related categories is evidence of marketers' attempts to leverage this asset. However, two factors are necessary for successful asset leverage. First, the firm should dominate the original category in at least one dimension, such as distribution, R&D, production, or especially brand recognition. This dominance provides the asset to be leveraged. Second, the new category should be closely related to the original category so that the asset is relevant and can be easily transferred. These examples show how the emergence of a category extension provides an ideal opportunity for a dominant firm to enter late yet still assume leadership by leveraging assets.

Interpretation of Study Results

Our study has raised several questions about our method and findings. Some of these are limitations that qualify our findings, others are opportunities for future research, while still others are misinterpretations we would like to clarify.

• Does the initial leadership of pioneers provide adequate rewards to pioneering? On average, early leaders to not enter markets until about thirteen years after pioneers.[47] During that time, pioneers may well enjoy rewards from market pioneering. However, the high cost and risk of pioneering new markets usually requires long-term rewards in order to be successful. Therefore, pioneers do not seek rewards that are limited to the early years of a category for the following reasons: First, sales increase exponentially following the tapping of the mass market (e.g., recall the sales growth of Sony, JVC, and Pampers). Pioneers that fail in the early stage forgo potentially huge rewards (e.g., Ampex and Chux). Second, pioneers fail at this stage despite their efforts to succeed, because they lack vision, persistence, commitment, or innovation. So their failure is the result of poor strategy rather than a planned withdrawal. Third, some pioneers fail so quickly that they do not realize any significant rewards (e.g., MITS in personal computers and Trommer's Red Letter in light beer). Fourth, even though early leaders enter long after pioneers, other firms compete against the pioneer in the interim. This competition prevents the pioneer from collecting the rewards of a monopolist. Finally, it is important to remember that our analysis considers only successful product categories. The rewards of pioneering are much less when failed product categories are included.

• Do the conclusions change if we focus on profits rather than on market leadership? Our study does not explicitly address profits because data are difficult to obtain. Profit considerations may change the conclusions a little, but not dramatically. The main reason is that the increase in sales following the opening of the mass market is so large and rapid that firms that make this breakthrough enjoy huge profits, even when discounted over the relevant time frame. Pioneers suffer a huge opportunity loss by failing at or before this stage. Moreover, financiers, venture capitalists, and shareholders who invest in pioneers do so with the hope that their targets will be hugely successful rather than merely meet some limited short-term profit goals. The emphasis on short-term profits itself may be inappropriate. The marketing concept suggests that enduring leadership may be a more appropriate goal for a firm than short-term profits.[48] Indeed, a harvesting strategy that focuses on short-term profits has been suggested as a hedge against declining markets, rather than an end in itself, and certainly not for growth markets.[49]

• Do moderators influence the role of the five factors of enduring leadership? Our research indicates that the first four factors (vision, persistence, commitment, and innovation) are more important for entirely new categories, while asset leverage is more important for category extensions. Also, the importance of innovation probably increases as products become more technically complex. The importance of assets, such as brand names, increases as product attributes are more ambiguous and consumers rely on reputation. Attribute ambiguity itself may be negatively related to technical complexity. Thus a new category, an ambiguous attribute, and technical complexity appear to be moderators of the factors of enduring leadership.

• Are the five factors of enduring leadership related? The five factors may well be structurally related in a causal chain. Thus the vision of the mass market may inspire persistence and the willingness to commit huge finances; finances may provide the resources to innovate, while innovation may provide the solutions to achieve and maintain leadership. These factors may emanate from different individuals or departments in a firm, or they may be embodied in a single leader, which is more likely for new categories. As such, the relationship among the factors could occur more because they are part of an individual's drive for market leadership. George Eastman, Henry Ford, Yuma Shiraishi, Edwin Land, and Bill Gates are some examples of leaders who set the tone that drove their organizations.

• Can the relative importance of the five factors be quantified? Survey researchers could develop a specific scale for each cause, measure current organizations with these scales, and test a formal model of our theory. However, we feel that the strength of our thesis lies in the details of the cases and the longitudinal nature of the analysis. If we apply scales and models, we may lose much of the richness while yielding few new insights. With the limited knowledge we had at the beginning of the study and the data we collected, analyzing cases seems to be more fruitful than testing models.

STRATEGIC IMPLEMENTATION

Managers should not misinterpret our findings by concluding that following is better than pioneering. Rather, managers should conclude that the five factors of enduring leadership that we have identified

are much more important than pioneering in determining long-term leadership. Being first does not automatically endow an advantage; it only provides an opportunity.

One of the reasons for short-lived leadership in new markets is that the magnitude of any leadership advantage is proportional only to the size of the current market. Since initial sales are often quite low, any advantage for leading these markets is also low. Therefore, a critical variable for managers to assess is how "new" the new product really is. Building significant consumer demand in most new markets often takes many years. Producer-based advantages (e.g., experience effects or economies of scale) or consumer-based advantages (e.g., brand equity, switching costs, network effects, or industry standards) provide meaningful barriers to entry only after markets grow to a reasonable size. The five factors of enduring leadership are critical for firms to build primary demand for new products. Therefore, the issue for managers is how to promote the five factors in their own organization.

First, a product champion who passionately believes in the potential of a new market and has the freedom, resources, and responsibility to realize that potential is important. Indeed, the histories of many of the successful firms in our sample show that a single product champion shepherded the product from introduction to market leadership: George Eastman for Kodak cameras, William Kellogg for Kellogg's cereals, William Wrigley for Wrigley's chewing gum, Yuma Shiraishi for JVC's video recorder, and Bill Gates for Microsoft software. In contrast, the absence of vision, persistence, and commitment for Ampex's video recorder project may be directly attributed to the inconsistent leadership for the project.

Second is the establishment of an organizational structure, which we call enhanced autonomy. This structure provides the independence to pursue new ideas with vision, persistence, and relentless innovation while also supporting them with financial resources and assets to leverage. This arrangement is often achieved through decentralization. When large corporations are too centralized, senior leadership may impose an obsolete formula for authorizing and developing innovations. Such an atmosphere may stifle the vision of new markets and the pursuit of relentless innovation that are critical to success. Additionally, senior management's involvement in the market decisions of individual business units may slow new product introduction and growth. This situation is especially likely if the innovation threatens established products in the same or related markets. For example, IBM's reliance on mainframe computers may have diminished its efforts in workstations and minicomputers. Autonomous business units may be the solution to this problem. Opportunities to develop an innovation should be entrusted to new business units with independent responsibility for the innovation. The leader of that unit could then serve as its champion, while the unit itself may work as a small, nimble, entrepreneurial company that can draw on the resources of a large corporation.

CONCLUSION

Market pioneering is neither necessary nor sufficient for long-term success and leadership. Instead, enduring market leaders embody five principles more critical to success than pioneering. A strategy for market entry is much like a battle plan. A first strike may be desirable, but careful preparation for attack, counterattack, penetration, and consolidation are critical for success. Such preparation requires answers to five questions:

1. How will the entrant exploit the potential market? Many an innovation initially appears too crude, costly, and unappealing. Market leaders are firms that can *envision* the mass market for these primitive innovations. Firms that can define that vision can assemble resources and inspire people for the task ahead.

2. Can the entrant stick it out? The road to success is rarely easy, otherwise many a competitor would have taken it earlier. Technological blocks, legal constraints, consumer misperceptions, and competitive threats are some of the obstacles that new entrants face. Market leaders *persist* through these challenges in quest of the vision.

3. Does the entrant have the resources to put to the task? The mass market cannot be tapped cheaply. Many innovations hinge on new product technologies, expensive process technologies and large-scale productions, or massive promotion. Market leaders are firms that can *commit* the resources to the vision of the mass market, when sales are a trickle but high costs loom.

4. Can the entrant change, even at the cost of current position? Markets, consumers, competitors, and technology change constantly. Stagnation in this environment leads to erosion of share or quick failure. Market leadership belongs to firms that *innovate* relentlessly even at the risk of cannibalizing or rendering obsolete their own products.

5. Can the entrant transfer its strengths to the new market? Leaders in a mature category often dominate with a well-known brand, extensive distribution, or unique expertise. These strengths constitute a relatively easy means to enter and dominate a category extension. Market leaders are firms that can nurture these strengths while *leveraging* them to dominate new markets.

Should you be first to enter new markets? Without these five factors, a first entrant is merely an alarm for competitors; by embodying these factors, a late entrant can outpace a lethargic pioneer. An earlier entry along with the factors is surely an advantage. But being first to market by itself is neither necessary nor sufficient for enduring market leadership.

REFERENCES

The authors thank Robert Fisher, Dennis Rook, Richard Staelin, and participants at the Fall 1993 ORSA/TIMS conference for their comments. This study was partly supported by a grant from the Marketing Science Institute.

1. R.D. Buzzell and B.T. Gale, *The PIMS Principles: Linking Strategy to Performance* (New York: Free Press, 1987); M. Lambkin, "Order of Entry and Performance in New Markets," *Strategic Management Journal*

9 (1988): 127–40; W.T. Robinson, "Sources of Market Pioneer Advantages: The Case of Industrial Goods Industries," *Journal of Marketing Research* 25 (1988): 87–94; and W.T. Robinson and C. Fornell, "Sources of Market Pioneer Advantage in Consumer Goods Industries," *Journal of Marketing Research* 22 (1985): 305–317.

2. G.L. Urban, T. Carter, S. Gaskin, and Z. Mucha, "Market Share Rewards to Pioneering Brands: An Empirical Analysis and Strategic Implications," *Management Science* 32 (1986): 645–659.

3. Ibid.

4. "Study: Majority of 25 Leaders in 1923 Still on Top," *Advertising Age*, 19 September 1983, p. 32.

5. D.A. Aaker and G.S. Day, "The Perils of High-Growth Markets," *Strategic Management Journal* 7 (1986): 409–21; A.D. Chandler, "The Enduring Logic of Industrial Success," *Harvard Business Review*, March-April 1990, pp. 131–139; A. Glazer, "The Advantages of Being First," *American Economic Review* 75 (1985): 473–480; M.B. Lieberman and D.B. Montgomery, "First-Mover Advantages," *Strategic Management Journal* 9 (1988): 41–58; and M.J. Moore, W. Boulding, and R.C. Goodstein, "Pioneering and Market Share: Is Entry Time Endogenous and Does It Matter?," *Journal of Marketing Research* 28 (1991): 97–104.

6. R.A. Kerin, P.R. Varadarajan, and R.A. Peterson, "First-Mover Advantage: A Synthesis, Conceptual Framework, and Research Propositions," *Journal of Marketing* 56 (1992): 48.

7. P.N. Golder and G.J. Tellis, "Pioneer Advantage: Marketing Logic or Marketing Legend?," *Journal of Marketing Research* 30 (1993): 158–170.

8. Ibid.

9. G.S. Day, A.D. Shocker, and R.K. Srivastava, "Customer-Oriented Approaches to Identifying Product-Markets," *Journal of Marketing* 43 (1979) 8–19; and S. Ratneshwar and A.D. Shocker, "Substitution in Use and the Role of Usage Context in Product Category Structures," *Journal of Marketing Research* 28 (1991): 281–295.

10. For example, see T. Levitt, "Marketing Myopia," *Harvard Business Review*, July-August 1960, pp. 45–56.

11. Golder and Tellis (1993).

12. Ibid.

13. Aaker and Day (1986); G.S. Day and R. Wensley, "Marketing Theory with a Strategic Orientation," *Journal of Marketing* 47 (1983): 79–89; Lieberman and Montgomery (1988); Kerin, Varadarajan, and Peterson (1992); and Urban et al. (1986).

14. *Financial World*, 20 May 1964, p. 5; 5 April 1967, p. 6; 5 April 1967, p. 6; 8 September 1965, p. 5.

15. Procter & Gamble, Annual Report, 1977.

16. "For Babies Only—Chux Throw-away Diapers," *Delineator*, 127 (1935): 6–7.

17. "Disposable Diapers," *Consumer Reports*, March 1961, pp. 151–152.

18. P. Freiberger and M. Swaine, *Five in the Valley: The Making of the Personal Computer* (Berkeley, California: Osborne/McGraw-Hill, 1984); and A. Gupta and H.D. Toong, "The First Decade of Personal Computers," in *Insights into Personal Computers*, ed. A Gupta and H.D. Toong (New York: IEEE Press, 1985), pp. 17–36.

19. For example, see "Personal Computers: And the Winner is IBM," *Business Week*, 3 October 1983, pp. 76–95.

20. Freiberger and Swaine (1984); Gupta and Toong (1985); and G.E. Nelson and W.R. Hewlett, "The Design and Development of a Family of Personal Computers for Engineers and Scientists," in Gupta and Toong (1985), pp. 37–54.

21. "Microcomputers Catch on Fast," *Business Week*, 12 July 1976, p. 50.

22. See, for example: Robinson and Fornell (1985); and Urban et al. (1986).

23. Ampex, Annual Reports, 1969–1976, 1980; S. Ghoshal and C.A. Bartlett, "Matsushita Electric Industrial (MEI) in 1987" (Boston: Harvard Business School, Case 9-388-144, 1988); R.Y. Lurie, "The World VCR Industry" (Boston: Harvard Business School, Case 9-387-098, 1987); Matsushita, Annual Reports, 1971–1973, 1975, 1977, 1979, 1981, 1985; R.S. Rosenbloom and M.A. Cusumano, "Technological Pioneering and Competitive Advantage: The Birth of the VCR Industry," *California Management Review* 29 (1987): 51–76; R.S. Rosenbloom and K. Freeze, "Ampex Corporation and Video Innovation," in *Research on Technological Innovation, Management, and Policy*, ed. R.S. Rosenbloom (Greenwich, Connecticut: JAI Press, 1985); and Sony, Annual Reports, 1973–1974, 1976–1982, 1984.

24. *Consumer Reports* (1961).

25. "The Great Diaper Rash," *Forbes*, 15 December 1970, p. 24; M.E. Porter, "The Disposable Diaper Industry in 1974" (Boston: Harvard Business School, Case 380-175, 1980); H. Tecklenburg, "A Dogged Dedication to Learning," *Research Technology Management* 33 (1990): 12–15; and "The Great Diaper Battle," *Time*, 24 January 1969, pp. 69–70.

26. B. Coe, *Cameras: From Daguerreotypes to Instant Pictures* (New York: Crown Publishers, 1978); E.S. Lothrop, Jr., *A Century of Cameras* (Dobbs Ferry, New York: Morgan & Morgan, 1973); and B. Newhall, *The History of Photography* (Boston: Little, Brown, 1982).

27. *Forbes* (1970).

28. Rosenbloom and Cusumano (1987).

29. "A Flickering Picture for Video Recorders," *Business Week*, 21 August 1978, p. 28.

30. Ampex, Annual Reports, 1972 and 1974; and Rosenbloom and Cusumano (1987).

31. "Personal Business," *Business Week*, 4 June 1960, p. 129.

32. "Color TV," *Consumer Reports*, November 1961, pp. 612–613.

33. D.J. Collis, "General Electric—Consumer Electronics Group" (Boston: Harvard Business School, Case 9-389-048, 1988).

34. *Business Week* (1960).

35. "IRS Approval of Low-Calorie Claim Clears Way for Gablinger's Expansion," *Advertising Age*, 17 February 1969, p. 37.

36. "Gablinger's Goes to Gumbinner; Grey Gets Knickerbocker as 2 Shops Resign," *Advertising Age*, 27 November 1967, p. 2; and "Intro of Gablinger's Leads to Top-Level Shift at Rheingold," *Advertising Age*, 1 April 1968, p. 6.

37. "Meister Brau Wins Amylase Patent Suit," *Advertising Age*, 30 March 1970, p. 62.

38. "How Miller Won a Market Slot for Lite Beer," *Business Week*, 13 October 1975, p. 116.

39. "Turmoil among the Brewers: Miller's Fast Growth," *Business Week*, 8 November 1976, pp. 58–67.

40. Freiberger and Swaine (1984); and M. Lynch, "Investors Buying Nautilus for Shares in Apple Are Told to Watch for a Worm," *Wall Street Journal*, 8 October 1980, p. 10.

41. R.B. Adams, Jr., *King C. Gillette: The Man and His Wonderful Shaving Device* (Boston: Little, Brown, 1978).

42. L. Ingrassia, "Gillette Holds Its Edge by Endlessly Searching for a Better Shave," *Wall Street Journal*, 10 December 1992, p. 1.

43. S.N. Chakravarty, "We Had to Change the Playing Field," *Forbes*, 4 February 1991, pp. 82–86; and Ingrassia (1992).

44. Ingrassia (1992).

45. Chakravarty (1991).

46. "Sales Bubble for Diet Drinks," *Business Week*, 27 June 1964, pp. 88–92; and F. Sinclair, "Diet-Rite Budget Boosted: It's 'Not Fad, but Forever,' " *Advertising Age*, 28 October 1963, pp. 1–2.

47. Golder and Tellis (1993).

48. P.F. Anderson, "Marketing, Strategic Planning and the Theory of the Firm," *Journal of Marketing* 46 (1982): 15–26.

49. D.A. Aaker, *Developing Business Strategies* (New York: John Wiley, 1988); and D.F. Abell and J.S. Hammond, *Strategic Market Planning: Problems and Analytical Approaches* (Englewood Cliffs, New Jersey: Prentice-Hall, 1979).

VII

MANAGING THE INNOVATION PROCESS IN ORGANIZATIONS

16

Managing Cross-Functional Relationships to Enhance New Product Development

50

Managing Relations Between R&D and Marketing in New Product Development Projects

WILLIAM E. SOUDER

INTRODUCTION

Research and development (R&D) and marketing personnel depend on each other for the creation of new product innovations. Yet R&D and marketing departments have frequent misunderstandings and conflicts.

Many managers have first-hand experience with R&D/marketing interface problems and behaviors between R&D and marketing groups have been carefully studied [3,5,7–11]. However, much more information is needed about this complex and important topic. This paper examines the R&D/marketing interface conditions found at 289 new product development innovation projects. Based on these findings, strategies and guidelines are presented for improving the relationships between R&D and marketing groups.

Reprinted by permission of the publisher from "Managing Relations Between R&D and Marketing in New Product Development Projects," by William E. Souder, *Journal of Product Innovation Management*, 1988, Vol. 5, pp. 6–19. Copyright 1988 by Elsevier Science Co., Inc.

William E. Souder is Professor of Management at the University of Alabama Business School.

About the Data Base

Data: Life cycle data were collected on 289 new product development innovation projects at 53 consumer and industrial product firms [7–10]. The data collection focused on project events, with detailed attention given to organization structures, environments, climates, behavioral processes and project success/failure factors. The ultimate objective was to understand development processes for new product innovations.

Sample of Firms: Using published statistics, an industry by industry compilation was made of firms with significant new product activities in either consumer or industrial goods. Target firms were then randomly selected from this list, based on a compromise design that carefully considered the cost of traveling to distant sites and the need to maintain representivity on several important dimensions [7–10]. Approximately five firms were selected from each of the following ten industries: metals, glass, transportation (includes automotive and mass transit), plastics, machinery, electronics (includes computers and instruments), chemicals, food, aerospace and pharmaceuticals.

Sample of Projects: Using carefully specified definitions, the population of new product innovation projects initiated during the preceding five years was assembled at each firm. A random sampling of equal numbers of success and failure outcome projects was taken from these populations at each firm, while maintaining a range of types of technologies, types of innovations, degrees of difficulty of projects, central vs. divisional R&D efforts, and several other important dimensions [7–10]. Several ongoing projects whose success or failure outcomes were unknown at that time were intentionally included in this sampling. Following these procedures, approximately 10% of each firm's portfolios were selected into the 289 project sample studied here.

Data Collection: A total of 27 instruments, numerous telephone interviews and 584 in-depth face-to-face interviews were carried out on each project to record the life cycle histories and extract the relevant data on each of the 289 projects [7–10]. A cascading interview procedure was used to cross-validate information collected from each marketing, R&D and other subject on each project [7–10].

Methodology of This Study

This study was carried out on a comprehensive data base of life cycle information on 289 new product development innovation projects. The data were collected through ten years of intensive field research at 56 consumer and industrial products firms [7–10]. The box "About the Data Base" and Table 50.1 present the methodology and the project outcome measurement scales used in collecting that data base.

The 289-project data base contains numerous detailed descriptions and ratings of key events, activities, attitudes and behaviors of the R&D and marketing personnel who worked on each project. As part of the content analyses, statistical reduction and factor analyses of this large data base [1,2,4], these items were reduced to 42 attitudinal and behavioral descriptors of the R&D/marketing interface. Some examples of these descriptors are "Frequency of Joint Meetings," "Frequency of Joint Customer Visits," "Degree of Perceived Need to Interact," and "Degree of Regard for the Other Party's Competency."

Each of the 289 projects was rated on each of these descriptors. Some of these ratings came directly from the instruments, while others were developed through content analyses of the interviews [1,2,7–10]. Redundancies were built in at several points. For example, the "Frequency of Joint Meetings" was primarily measured by questionnaire items that asked for the number of times per year that joint meetings were held. Details about these joint meetings were solicited during the interviews. Differences greater than 10% in the questionnaire responses of the marketing, R&D and other personnel on the same project were reconciled during the interviews. As another example, the "Degree of Perceived Need to Interact" was primarily measured by asking pointed questions of the respondents during the interviews. The information from the interviews was then checked against the Likert-type scale ratings supplied on the questionnaires. Apparent disparities were resolved by returning to the subjects to clarify their responses and ratings. This type of multi-method, multitrait measurement approach is commonly used to maximize the validity of social science measurements [4].

A profile of ratings was thus developed for each of the 289 projects. Some of these profiles appeared to be very similar; others appeared to be very dissimilar. Statistical cluster analyses techniques were then applied to the profiles in order to exhaustively cluster the projects by the various types of profiles that were found [2,4].

SEVEN R&D/MARKETING INTERFACE STATES

Using a 95% statistical significance level, seven different clusters were found from the cluster analysis. Each cluster was then labeled according to its observed items. For example, a review of the items in one cluster showed that it was characterized by a low frequency of meetings between the R&D and marketing personnel, highly specialized and organizationally separated R&D and marketing functions, and a low degree of perceived need to interact. Therefore, the label "Lack of Interaction" was coined to describe this R&D/marketing interface state of affairs. Twenty-two of the 289 projects, or 7.6% of the sample, exhibited this state. Similarly, the other states and percentages shown in Table 50.2 were found and accordingly labeled.

TABLE 50.1
Project Outcome Measurement Scale

| | Success outcomes | |
| | Descriptors | |
Degrees of success	Technical outcomes	Commercial outcomes
High	Breakthrough	Blockbuster
Medium	Enhancement	Above expectations
Low	Met the specs	Met expectations

| | Failure outcomes | |
| | Descriptors | |
Degrees of failure	Technical outcomes	Commercial outcomes
Low	Learned a lot	Below expectations
Medium	Gained some technology	Protected our position but lost money
High	Complete dud	Took a bath we won't forget

Other outcome
SE = Stopped the effort early due to poor progress

TABLE 50.2
Incidence of Harmony and Disharmony States

States	Percentage of projects experiencing each state
Mild Disharmony	
Lack of interaction	7.6%
Lack of communication	6.6
Too-good friends	6.3
Subtotal	20.5
Severe Disharmony	
Lack of appreciation	26.9
Distrust	11.8
Subtotal	38.7
Disharmony total	59.2
Harmony	
Equal partner	11.7
Dominant partner	29.1
Harmony total	40.8
Overall total	100%

Several firms that experienced the Lack of Interaction, Lack of Communication and Too-Good Friends problems on the projects studied here avoided these states on some subsequent projects. Follow-up studies with these firms showed that they overcame these states through modest efforts. These efforts included more frequent joint meetings, joint involvements in planning proposed projects and increased sharing of information. Moreover, though these problems often lowered the organization's new product development effectiveness, they were not totally disruptive and they seldom led to major project failures. Therefore, as shown in Table 50.2, these problems were labeled 'mild.' By contrast, the Lack of Appreciation and Distrust problem states were labeled 'severe.' Follow-up studies showed that these types of problems were not easily overcome, they usually caused operating disruptions, consumed many hours of managerial talent in moderating disputes, delayed key actions and important decisions and led to project failures.

Many other projects were found that did not exhibit either mild or severe disharmonies. As shown in Table 50.2, these projects were considered to be in a 'harmony' state.

CHARACTERISTICS OF THE MILD DISHARMONY STATES

Lack of Interaction

In this state of affairs, there were very few formal and informal meetings between the R&D and marketing personnel. Both parties were deeply concerned with their own narrow specialties and neither saw any reason to learn more about the other's work. Neither party saw the need for close interaction. R&D expected marketing to use whatever they gave them, and marketing expected R&D to create useful products.

This state resulted more from simple neglect than from any strong animosities between the parties. For example, one subject noted: "You get busy and you don't stop to think about whether or not they should know about this or that. . . . when you have to get your part of the job done." Another subject said: "If you don't get used to seeing each other you don't miss each other, and if you don't think about each other you don't make any effort to get together. And you always have to make an effort." It may be noted that several projects experiencing the Lack of Interaction state were in older, commodity product firms that were attempting to develop new product lines. Most of these firms had no histories of close R&D/marketing interactions.

Lack of Communication

In this state, the two parties purposely maintained verbal, attitudinal, and physical distances from each other. R&D purposely did not inform marketing about their new technologies until very late in the development cycle. Marketing purposely did not keep R&D informed about market needs. This occurred because neither party felt the other had much information of significant value. And neither felt it was important to inform the other of the details of their own work. This state was aptly summed up in the comments of one respondent. "If we told them all this, they wouldn't know what to do with it. . . . We know more about it than they do. Our best source of information comes from right here, from ourselves." Note how this state of affairs is different from the above Lack of Interaction syndrome, where the perceived urgency of pursuing their own activities caused the parties to neglect each other. Here, both parties harbored negative feelings about the worth of the other that stood in the way of interaction.

Though various causes of the Lack of Communication state were observed within the data base, two experiences repeatedly lead directly to this problem. One was the perceived theft of credit. When either party took what the other thought was undue credit for meritorious project achievements, this inevitably led to a Lack of Communication problem. The impression that the other party had taken unfair advantage was long remembered. Another experience that frequently led to the Lack of Communication state was top management's uneven use of accolades. If top management praised one party and did not praise the other, rivalry invariably developed that shut off some future communication. As one subject noted: "If we don't tell them anything, they can't go to management and take credit for it."

Too-Good Friends

In this state of affairs, the R&D and marketing personnel were too friendly and maintained too high a regard for each other. They enjoyed each other's company so much that they frequently met socially, outside the work environment. These social affairs often included the individual's families, e.g., family picnics and Sunday afternoon socials were common. In most of the Too-Good Friends cases, work and social aspects were commingled, e.g., joint visits to customer facilities might also involve a round of golf and the Sunday afternoon socials always included some informal discussion of business. Each party felt that the other had their own area of exclusive expertise, and that the other was beyond reproach. This inhibited each party from challenging the other's assumptions and judgments. Consequently, important information and subtle observations were overlooked that were significant for the project.

What factors led to this type of problem? Surprisingly, past successes sometimes led the team members to become too-good friends. Teams of R&D and marketing personnel who had worked together successfully for long periods of time often became complacent. Their potency appeared to decline once they had achieved complete harmony. Apparently, they needed some conflicts or the challenge of building harmonious relationships to maintain their alertness. A related factor was a kind of blind faith in the correctness of the counterpart person. As one respondent observed: "You are always sort of reluctant to challenge and question what your colleague tells you. He's the expert in that area. And you don't expect that he'll play politics with you, so there's no reason to question his integrity. And you figure he's the best man you've got, so he probably won't steer you wrong."

A detailed examination of the other clusters of projects showed that past successes and great faith in each other also characterized effective R&D/marketing interfaces, i.e., the Equal Partner Harmony state in Table 50.2. What were the distinguishing factors? The answer appears to be a matter of interpersonal dynamics. The parties to an effective interface always challenged and penetratingly questioned each other. They appeared to enjoy and thrive on this aspect, sometimes with impish good humor. When one partner found a gap in the other's logic, both partners were suddenly energized to close that gap. Such experiences further strengthened their relationship. The partner who committed the logic gap never seemed to suffer any loss of prestige in the other's eyes. Rather, the ambience was described by one partner as "a climate where we look for flaws, and it's not important who committed the flaw. We just want to find it and work together to fix it." This is clearly a different climate from the above Too-Good Friends state.

It should be noted that such professional disagreements and challenging behaviors, that often characterize effective R&D/marketing interfaces, may give the outside observer the mistaken impression of disharmony and strife. Professional disagreement appears to be a very healthy and enlightening climate for its members. At times, such disagreements may seem to become very heated and destructive. Yet if these discussions are confined to the issues and do not become personally threatening to the participants, they can actually strengthen the R&D/marketing interface. Thus, it is the lack of professional disagreement (Too-Good Friends) that constitutes disharmony, and not its presence.

CHARACTERISTICS OF THE SEVERE DISHARMONY STATES

Lack of Appreciation

This state was characterized by strong feelings that the other party was relatively useless. Marketing felt that R&D was too sophisticated, while R&D felt that marketing was too simplistic. Marketing felt that R&D should be prohibited from visiting customers because they would talk over their heads. R&D felt that marketing did not have a good grasp of the market needs. In this state, the marketing groups often purchased their R&D work outside the firm rather than use the in-house R&D group. R&D often independently moved ahead with its own ideas, by-passing marketing and attempting to launch their own new products. These efforts seldom succeeded, and the failures were usually rationalized by the R&D personnel as marketing's fault for failing to assist them!

What caused the Lack of Appreciation? No single cause was identified. Some cases had long remembered histories of ineffectiveness by one party, e.g., R&D failed to develop the promised product or marketing failed to correctly identify the market. Sometimes, the organizational climates fostered a lack of appreciation. For example, several respondents indicated that they "never see any signals from management that collaboration is desired." Other respondents noted that "management has not indicated that we are expected to cooperate with them." It is interesting that management must make a special effort to encourage cooperation: it does not seem to be automatic.

The organization of R&D and marketing into separate departments with separate budgets and op-

erations often fostered a lack of appreciation. As evidence, consider the following sampling of statements from personnel at five firms in the Lack of Appreciation state. "We don't have any inputs into their plans and budgets." "They have their own operations and so do we." "We get our rewards from doing our things and they get theirs from something else." "No one is responsible for how it all comes together." "We just go our separate ways."

Distrust

Distrust is the extreme case of deep-seated jealousies, negative attitudes, fears and hostile behaviors. In this state of affairs, marketing felt that R&D could not be trusted to follow instructions. R&D felt they were blamed for failures, but marketing was credited for successes. Several R&D groups in this state feared that marketing wanted to liquidate them. R&D lamented that marketing often attempted to dictate exactly what, where, when and how to do the project, allowing no room for rebuttal and no tolerance for their suggestions. Marketing lamented that when R&D got involved the project disappeared and they never saw it before it was completed, at which point it was seldom what they wanted. Several cases were found where R&D initiated many projects and kept them secret from marketing "so marketing wouldn't kill them before they gained enough strength to move along on their own." Cases were found where marketing brought R&D into the picture only after the product specifications had been finalized "in order to avoid any arguments from R&D about how to do it."

What caused the Distrust state? Though no single cause was found, several important contributing factors were isolated. All the Distrust cases began as either a Lack of Appreciation or a Lack of Communication problem that evolved into Distrust. Many of the Distrust cases were characterized by personality conflicts that top management had allowed to exist for a long time. In some cases, these conflicts had become so institutionalized that even personnel who had not been involved harbored feelings of Distrust. As an example, note the following quote from one respondent, referring to his counterpart in another department. "He once did some things to us. I'm not sure what they were. It all happened before I came into this group. So, you see, you really have to watch out for him." This type of institutionalized Distrust was found surprisingly often.

CHARACTERISTICS OF THE HARMONY STATES

Equal Partner Harmony

In this state, each party appeared to share equally in the work loads, activities and rewards. Each party felt free to call joint meetings on almost any issue. These meetings were characterized by an open given and take of facts, opinions and feelings. No issues were left unresolved and consensus was sought by everyone. Study committees and task forces with joint memberships were common, with the task force chairmanships rotated between the R&D and marketing personnel. Moreover, it was part of the Equal Partner culture to involve R&D and marketing personnel jointly in all customer visits, customer follow-ups, customer service, new product planning and forecasting, project selection and product strategy formulation activities.

Three features were common to all the Equal Partner cases. One, the marketing personnel were technically trained. They all had undergraduate degrees in science or engineering. Two, the marketing personnel had prior careers in R&D. Thus, personnel were often successfully exchanged or rotated between the R&D and marketing functions. Three, the R&D and marketing personnel had a strong sense of joint partnership. As evidence, note the following sampling of quotes collected from R&D and marketing personnel in Equal Partner states. "We couldn't get along without them." "We're on the phone with each other constantly." "I feel like I've known them a long time." "We've been through 'thick and thin' together."

Dominant Partner Harmony

In this state, one of the parties was content to let the other lead. Both R&D-dominant and marketing-dominant cases were found. For example, one R&D

subject in a marketing-dominant case noted: "We have no idea at all what the market needs are. But if they'll tell us what they want and supply the specs we can sure make it for them." A marketing respondent in an R&D-dominant case said: "We can usually sell what R&D gives us. We don't really know what they are able to come up with. They know what it takes to make a good performing product better than we do."

It may be noted that the dominant partner cases seldom involved complex technologies, exacting customer needs or large R&D efforts. Most of these cases involved developmental efforts as opposed to research efforts. This reinforces the notion that problems at the R&D/marketing interface escalates as the technology or the user's environment become more complex.

INCIDENCE, SEVERITY, AND CONSEQUENCES OF DISHARMONY

As the percentages in Table 50.2 show, a surprisingly high incidence of R&D/marketing disharmony was found. Nearly two-thirds (59.2%) of the projects studied here experienced some type of R&D/marketing interface disharmony. Moreover, it is especially disconcerting that over one-third (38.7%) of the projects studied here experienced severe disharmonies. These results are statistically significant at the 99.9% level of confidence (using the binomial statistical test [6]). That is, a statistically significant number of projects were found to be experiencing disharmonies. And a statistically significant number of these projects had severe disharmonies.

But is disharmony disruptive to project success? Table 50.3 responds to this question. Most of the Harmony projects succeeded. Partial success characterized the Mild Disharmony projects. And most of the Severe Disharmony projects failed. As noted in Table 50.3, these results evidence a statistically significant relationship between the degree of harmony/disharmony and the degree of project success/failure. This relationship is significant at greater than the 99.9% confidence level. Thus,

TABLE 50.3
Distribution of Project Outcomes by Harmony/Disharmony States

States	Percentage of projects in each state exhibiting each outcome[a]		
	Success	**Partial success**	**Failure**
Harmony	52%	35%	13%
Mild disharmony	32	45	23
Severe disharmony	11	21	68

χ^2 statistic = 88.84, significant at <.001

[a]The following definitons are used, based on Table 46.1:

Success =	High plus Medium Degrees of Commercial Success (Blockbuster plus above Expectations)
Partial Success =	Low Degree of Commercial Success plus low Degree of Commercial Failure (Met Expectations plus Below Expectations)
Failure =	Medium plus High Degrees of Commercial Failure (Protected Our Position But Lost Money, plus Took a Bath We Won't Forget)

these results demonstrate that the quality of the R&D/marketing interface affects the degree of success of new product development efforts.

A case-by-case examination of the data base revealed many informative details underlying the results in Table 50.3. In many of the projects experiencing the Too-Good Friends problem, important information was overlooked that severely diminished the effectiveness of the end products. In many of the projects experiencing the Lack of Communications problem, the new products either did not match the market needs or failed to meet some important customer specification. In about half of the projects with Lack of Interaction problems, the end products either did not perform as originally planned or arrived too late to capture a rapidly changing market. Thus, Mild Disharmonies generally depreciated the degree of success of the end products. But they seldom resulted in dismal product failures.

By contrast, in a majority of the projects experiencing Lack of Appreciation problems, the end products either failed to perform or they were not cost-effective. In many of the projects where Dis-

trust occurred, the products did not perform at all. Thus, Severe Disharmonies resulted in a high frequency of rather dramatic failures. Moreover, it should be noted that Severe Disharmonies were very difficult to overcome. Attempts by management to ameliorate them through negotiation, reorganization, bargaining or personnel transfers often left deep scars and sowed the seeds for a renewed outbreak of similar problems elsewhere. Thus, the prognosis for firms experiencing Severe Disharmonies is unusually pessimistic. Once they appear, their persistence can doom the firm's new product success rate for a long time.

Thus, these results show that the incidence and seriousness of R&D/marketing interface problems are distressingly high. Moreover, many of these problems are chronic, persistent, difficult to correct and seriously detrimental to new product success. These results are both surprising and disappointing. In spite of previous awareness and study of these problems [3,5,7–11], they still persist.

The reader is cautioned to use some care in interpreting these results. As mentioned above in connection with the discussion of the characteristics of the various R&D/marketing interface states, disharmony is a complex facet of human behavior. Professional disagreements, that may appear disharmonious to a casual observer, are often a sign of a very healthy and harmonious interface. The strong statistical relationships found here between disharmony and success do not mean that every disagreement and all apparent disharmonies are bad. One must be very careful in defining what constitutes real disharmony. In fact, the results show that a lack of professional disagreement (Too-Good Friends) may indicate disharmony. Thus, the reader is cautioned to use these results in the context of the definitions of the R&D/marketing states set forth here.

EIGHT GUIDELINES FOR OVERCOMING DISHARMONY

An analysis of the projects in the data base revealed eight practices that alleviated R&D/marketing interface problems. These practices are summarized in the Box of guidelines.

Guidelines for Improving Relations Between R&D and Marketing

1. *Break Large Projects into Smaller Ones.* Three-fourths of the projects with nine or more persons assigned to them experienced interface problems. By contrast, projects with five or fewer persons assigned to them seldom experienced problems. The smaller number of individuals and organizational layers on the small projects permitted increased face-to-face contacts, increased empathies and easier coordination.

2. *Take a Proactive Stance toward Interface Problems.* In those cases where potential interface problems were avoided and actual problems were overcome, the parties maintained a posture of aggressively seeking out and facing such problems head-on. They openly criticized and examined their behaviors. As one individual noted: "We don't treat it like a social disease and sweep it under the rug. If we got it, we want to know about it so we can get rid of it."

3. *Eliminate Mild Problems before They Grow into Severe Problems.* All the cases of severe (Lack of Appreciation and Distrust) problems studied here began as mild problems at some earlier points in time. As noted elsewhere in this paper, severe disharmonies were extremely difficult to eliminate. Mild disharmonies were much easier to overcome. Thus, it is wise to eliminate mild problems while they are still mild.

4. *Involve Both Parties Early in the Life of the Project.* Much has been said and written about the benefits from participation and early involvement of the R&D and marketing parties in decision processes [3,5,10,11]. The results here reinforce the conclusion that when R&D and marketing are joint participants to all the decisions, from the start of the project to its completion, Lack of Appreciation and Distrust are lessened.

5. *Promote and Maintain Dyadic Relationships.* A dyad is a very powerful symbiotic, interpersonal alliance between two individuals who become intensely committed to each other and to the joint pursuit of a new product idea [10,11]. Dyads are fostered any time persons with complementary skills and personalities are assigned to work together and given significant autonomy. Dyads are worth promoting not only because they encourage innovation in particular cases, but because they can become the kernel of a much wider circle of interrelationships between R&D and marketing. A successful dyad composed of an R&D person and a marketing person will draw other R&D and marketing personnel onto their bandwagon.

6. *Make Open Communication an Explicit Responsibility of Everyone.* This was dramatically illustrated by the Open Door policy at one of the firms in the data base that had a history of poor R&D/marketing interfaces. This policy consisted of quarterly information meetings between R&D and marketing, day-long and week-long exchanges of personnel, periodic gripe sessions, and the constant encouragement of personnel to visit their counterparts. Every employee was formally charged with the responsibility of playing a role in this Open Door policy. Moreover, each employee's success in meeting this responsibility was formally evaluated at the end of each quarter. The open-door policy survived the initial skepticism that surrounded it, and the examples set by a few diligent individuals eventually spread.

7. *Use Interlocking Task Forces.* A vivid illustration of the use of interlocking task forces was provided by one firm in the data base. The top-level task force or steering committee consisted of the company president, the vice presidents of R&D, marketing, and finance, the project coordinator, the R&D task force leader and the marketing task force leader. The marketing and R&D task force memberships changed as the project metamorphosed over its life cycle. In the early stages of the project, phenomenological research work was carried out by Ph.D. scientists. Application oriented scientists gradually replaced them as the project aged. Finally, engineering personnel replaced them. This interlocking task force structure was repeatedly successfully used by this firm to foster R&D/marketing harmony and new product development success.

8. *Clarify the Decision Authorities.* The decision authority is a kind of charter between R&D and marketing. It governs and guides the R&D/marketing venture by detailing who has the right to make what decisions, under which circumstances. For example, at one firm the policy specified that marketing had the sole authority and responsibility for defining the user's needs.

R&D had the ultimate authority and responsibility for selecting the technical means to meet these needs. R&D and marketing were given the joint responsibility for deciding when an adequate product had been defined. Complaints and appeals to top management could not be made unilaterally by either party. Top management only entertained an audience composed of both parties. A decision authority policy, as well as the group process of developing such a policy, can contribute enormously to clarifying the roles between R&D and marketing. Well-developed decision authority policies were observed at several firms. They fostered a sound foundation for the avoidance of many time-consuming conflicts.

Each of these eight practices reflects an actual experience of one or more firms in the data base. The users contended that the practice significantly increased the harmony of their R&D/marketing interface. In every case, these contentions were borne out by the data. The firm's interface became more harmonious after the practice was implemented.

Each guideline in the Box is effective for managing innovations because it pushes the R&D and marketing parties into a more collaborative, partnership role. The guidelines create conditions in which disharmonies are discouraged and harmonious behaviors are encouraged. Unfortunately, the guidelines do not provide much information about where and when they should be used. For example, when is it best to use guideline #8 (decision authority clarification)? Should guideline #8 always be used, on every project? In order for managers to intelligently apply the guidelines, a framework is needed for analyzing the role needs of the situation and for selecting the guidelines that best meet that need. An attempt to present such a Customer-Developer-Conditions (CDC) framework can be found in Souder (10) and in the original source publication of this article.

SUMMARY AND CONCLUSIONS

Nearly two-thirds of the 289 projects in the data base examined here experienced one of five types of R&D/marketing disharmony. The severity of disharmony was found to be statistically significantly related to the degree of success of innovation projects. Since the data base showed that severe disharmonies were extremely difficult to overcome, it is essential to prohibit their formation.

The results of this research indicate that R&D and marketing managers should jointly work together to help avoid disharmonies in seven ways. First, they should make all their personnel aware that R&D/marketing interface problems naturally occur. Second, they should encourage their personnel to be sensitive to the emergence of R&D/marketing interface problems by watching for the appearance of any characteristics of five types of disharmonies, as discussed above. Early detection is the key to their elimination. Third, managers should be especially careful to give equal credit and public praise to their R&D and marketing personnel in order to eliminate jealousies that might form a basis for severe disharmony. Fourth, R&D and marketing managers must make special efforts to reinforce in words and deeds their desire that the R&D and marketing parties collaborate. They must constantly send signals to their personnel that cooperation is essential. Fifth, managers should use teams of R&D and marketing employees at every opportunity. This will help avoid the natural impression that R&D and marketing are two separate organizational entities and cultures. Sixth, managers must not let personality clashes and other problems remain for so long that they become institutionalized into extremes of distrust. Finally, managers must also be aware that there is such a thing as too much harmony: R&D and marketing personnel can become too complacent with each other.

Outdated role concepts appear to be a major obstacle to achieving R&D/marketing harmony. This study encountered a surprising number of organization structures, organization behaviors, organizational reward systems, product strategies and new product development processes that emphasized a clear separation of roles and specialization of functions between R&D and marketing. This separation was only effective for handling simple technologies, simple markets and well defined customer needs, i.e., the Dominant Partner Harmony case. To successfully develop many types of new product innovations, R&D and marketing must work closely together. In some cases, they must work jointly with the customer in a trial and error fashion, trying various prototypes as a means to discovering the customer's real needs and the appropriate product. In other types of innovations, a true creative process is required in which new information and concepts are generated on the basis of the information shared between members of the R&D/marketing team. In still other cases, it is essential that the parties feel a strong sense of joint responsibility for setting new product goals and priorities, generating and selecting new product ideas, researching and analyzing customer wants, setting product performance requirements, and defining the new product's performance and cost trade-offs.

It appears that the institutionalized roles between R&D and marketing must be radically changed before new product development success rates can significantly increase. The only effective means to permanently avoid disharmonies is for the R&D and marketing parties to fully understand and appreciate their reciprocal roles, and to play out these roles in a true team setting. Moreover, it is essential that the R&D and marketing parties establish a team relationship that permits them to flexibly swap roles in response to evolving technologies, markets and customer needs. Unfortunately, there is as yet no recipe for such role swapping. Each R&D/marketing team must discover what works best for them. The point is: this discovery process can only unfold when the R&D and marketing parties act like a true team.

These conclusions and recommendations are all too familiar. Many firms are not implementing the team approaches and organizational techniques that this research has once again shown to be effective. Disharmonies between R&D and marketing continue to be surprisingly prevalent, chronic and disruptive to successful new product development. These findings are discouraging, in view of the obvious importance of the topic and an emerging awareness of it.

As noted above, the lack of detailed experimental knowledge of R&D/marketing interface problems remains a barrier to their prevention. Far too little is known about what constitutes real disharmony, the distinctions between professional disagreement and disharmony, how to alter the institutionalized roles between R&D and marketing and how to implement new team approaches between R&D and marketing personnel. It is hoped that this broadbased, ex post exploratory field study may provide a convincing basis for more advanced experimental research. Perhaps these results can serve as a basis for deriving empirically based propositions and operational hypotheses, that can then be tested through interventions and administrative experiments in real organizations. The results from these experiments should eliminate the last barrier to informed actions for reducing R&D/marketing interface problems.

REFERENCES

1. Berelson, B. *Content Analysis in Communication Research.* New York: Free Press, 1952.
2. Crollier, D. J. *Pattern Recognition Methods for the Social Sciences and Economics.* Cambridge Press: Cambridge, 1986.
3. Gupta, A. K., Raj, S. P. and Wilemon, D. The R&D-marketing interface in high-technology firms. *Journal of Product Innovation Management.* 2: 12–24, March 1985.
4. Kerlinger, F. N. *Foundations of Behavioral Research.* New York: Holt, Rinehart and Winston, 1973, pp. 514–535, 659–692.
5. Shanklin, W. L. and Ryans, J. K. *Marketing High Technology.* Lexington Books: Lexington, MA, 1984.
6. Siegel, Sidney. *Nonparametric Statistics for the Be-*

havioral Sciences. McGraw-Hill: New York, 1956, pp. 36–42, 175–179, 196–202.

7. Souder, Wm. E., et al., *An Exploratory Study of the Coordinating Mechanisms Between R&D and Marketing as an Influence on the Innovation Process*. Final Report, National Science Foundation Grant 75-17195 to the Technology Management Studies Institute, August 26, 1977.

8. Souder, Wm. E., et al., *A Comparative Analysis of Phase Transfer Methods for Managing New Product Developments*, Final Report, National Science Foundation Grant 79-12927 to the Technol-ogy Management Studies Institute, August 15, 1983.

9. Souder, Wm. E. *Technology Management Studies Institute Field Instruments Package*. Technology Management Studies Institute, University of Pittsburgh, Pittsburgh, PA 15261, 1987 edition.

10. Souder, Wm. E. *Managing New Product Innovations*. Lexington Books: Lexington, MA, 1987.

11. Young, H. C. *Product Development Setting, Information Exchange, and Marketing-R&D Coupling*. Unpublished Ph.D. dissertation. Northwestern University, Evanston, IL 1973.

51

Examining Some Myths About New Product "Winners"

ROBERT G. COOPER

OVERVIEW: Too many myths prevail about how to manage new product projects. This article probes what truly separates winners from losers in the new product game. It reports the results of a study of 103 new product cases from 21 major firms and divisions in four countries. Key success factors are identified—factors that distinguish the successful projects from the commercial duds.

New products continue to fail at an alarming rate. Although numerous studies have probed the reasons for failure, or what distinguishes winners from losers, many pundits appear to have ignored their conclusions and prescriptions, clinging to old beliefs, even myths, about how new products ought to be managed. Could some of these traditional beliefs underlie what is wrong with new product management? Consider the following 8 myths:

1. First into the market wins!

2. Analysis means paralysis . . . let's just get out there and do it: "Ready, fire, aim!"

3. Company reputation, a strong brand name, and a good selling effort will make almost any new product a success.

4. Having a low price is critical to winning.

5. We just can't afford the time to do market studies, customer tests, and a trial sell. Speed is of the essence!

6. If you're large, powerful, and strong enough, you don't need synergy to win. Diversify any-

where . . . any product, market, or technology arena is "fair game."

7. Don't try to pin down the definition of the product before development begins. This thwarts the creativity of scientists.

8. The competitive situation makes all the difference between winning and losing. If you don't succeed, blame it on a highly competitive market.

You may not believe all of these, but if one looks at the way many new product initiatives are managed, it is clear that these myths have many adherents. Our research team set out to gather data that would either prove or disprove these and other popular tenets of new product management (1). To do so, we selected one industry—the chemical industry—with a long tradition of science-based product development, and one in which there are many recognized leaders in innovation. Twenty-one major firms and divisions in four countries, including companies such as Du Pont, Dow, Shell-UK, Exxon Chemicals, ICI and Rohm & Haas, provided 103 case histories of significant and recent new products for in-depth study. New products covered a wide range of specialty products, including polymers (e.g., a modified polyethylene, a clarity film resin, an extrusion coating resin and a silicone rubber acoustic), as well as chemicals (e.g., a non-ionic surfactant, a pigment dispersant, an industrial heat transfer fluid, and an inorganic ultraviolet absorber).

The results reported here are from the New-Prod studies, an ongoing research investigation into

For an earlier version of this paper and for more discussion of the issues covered, see Cooper, R. G. "Debunking the myths of new product development." *Research-Technology Management*, July–August, 1994, pp. 40–50; and Cooper, R. G. *Winning at New Products*, 2d ed. Reading, MA: Addison-Wesley, 1993, respectively.

Robert G. Cooper is Professor of Industrial Marketing and Technology Management at McMaster University's Business School.

factors that separate new product winners from losers (2–6). A conceptual framework or model of new product outcomes formed the basis for the study (7,8). Thirteen blocks of variables or *dimensions*—such as Synergy, Market Attractiveness, and Product Advantage—were identified from this model, along with 95 variables that comprised these dimensions. Of the 103 new product cases, 68 were commercial successes and 35 were failures. All were fairly major products of their type; all had been launched and had been on the market for three years. Data on each product were collected via detailed questionnaires administered to members of the project teams. For each project, the 95 descriptor variables were measured on 10-point anchored scales along with a number of key performance measures: profitability, market share, and others.

By observing what separated the winners from the losers in this sample, we were able to shed light on what makes a winner, and which myths really are founded on fact. Although undertaken in only one industry, the results appear to have face validity generally; hence, they are likely to apply to a broader range of moderate-to-high technology industries.

SUCCESS VS. FAILURE

New product success was measured in a number of ways, including profitability, impact on the firm, current sales and market share, and timeliness. Some performance measures were in percentages, others in dollars and many gauged on zero-to-ten scales. Timeliness measures included time efficiency (whether the project was done in a time-efficient manner) and adherence to the time schedule (both were scaled, perception measures).

Two-thirds of the projects were rated as successes, based on profitability. Indeed, this two-thirds did much better on every performance dimension we measured save one. Winners did not differ from losers in terms of costs. The mean development cost for both successes and failures was about $2.25 million.

What then were the main drivers of new product success? To answer the question, we developed *major dimensions* or *indexes*, each based on a number of different measures. New product projects were then split into thirds—the top, middle and bottom third—on each dimension or index; we then looked more closely at the performance of each third. Here are the keys to new product success:

1. PRODUCT SUPERIORITY IS NUMBER ONE

Having a high-quality, superior product that delivered real value to the customer made all the difference between winning and losing. The most superior products—the top third on this dimension—achieved a success rate of 90.6 percent, 61 percentage points higher than the 29.6 percent success rate of products in the bottom third, most of which were "me-too" types of products. These superior products scored high on seven key items:

- Excellent *relative product quality*—relative to competitors' products, and in terms of how the customer measures quality.
- Good *value for money* for the customer.
- Superior *price/performance* characteristics for the customer relative to competitors' products.
- Superior to competing products in terms of *meeting customer needs*.
- Product *benefits* or attributes easily perceived as being *useful by the customer*.
- *Unique attributes* and characteristics for the customer—not available from competitive products.
- Highly *visible benefits*—very obvious to the customer.

There are two messages for managers:

1. Use these seven key elements of product advantage as screening criteria. If your new product projects don't score high on these items, then maybe you should be spending your money elsewhere!

2. These seven items become project objectives. No effort should be spared building these key elements into your next new product. The first step is to listen to the voice of your customer; only then can you fashion a product that truly does deliver these seven winning elements. Key customer questions include:

- What is "quality" in the customer's eyes?

- What is "value" to him?

- What are desired price/performance characteristics?

- What are her needs, wants and preferences?

- What are useful and unique product benefits, attributes and characteristics?

- What benefits will be highly visible—will really jump out at the customer?

Answers to these questions are central to developing that superior product that yields a 91 percent success rate.

That Product Superiority is the number one success factor should come as no surprise. Apparently it does to some people, though, including many project teams in these large, well-run chemical companies.

About one-third of these project teams developed and took *very mediocre new products* to market—products that scored poorly on this Product Superiority index. The results were predictable: This one-third had a 70 percent failure rate!

One element noticeable for its absence is having a low price. Low price was measured, but it was not related to either new product performance, or to the items that comprise Product Superiority. A low-price strategy is not key to winning at new products in the chemical industry.

2. QUALITY OF MARKETING ACTIONS IS CRITICAL

How well the marketing activities were executed from idea through to launch was the second key success factor. Note that "marketing activities" include a lot more than merely the launch. In fact, five marketing actions had almost equal impact on success. In rank order of impact, these activities were:

1. Undertaking *customer tests* or field trials of the product proficiently: the right number and location of test sites, good controls, appropriate metrics, etc.

2. Building in a *trial sell* or test market phase, and executing it well—where the product was sold to a limited number of customers in order to test the product, production and marketing, and to confirm market acceptance.

3. Executing the *launch* well: a solid launch plan, properly resourced, and executed proficiently.

4. Undertaking a *detailed market study* before Development begins: face-to-face interviews with potential customers/users to determine needs, wants, preferences, likes and dislikes, competitive weaknesses, and purchase intent.

5. Carrying out a *preliminary market assessment*—a quick scoping of the market in the *earliest* stage of the project.

Those new chemical products, for which these five marketing activities were executed—in a quality way—were far more successful. They had a success rate of 88.6 percent (versus only 37.5 percent for projects where these actions were poorly handled or not taken at all); and they also scored significantly higher on our other measures of performance (see Cooper [10] for actual data tables).

Sadly, many projects were found lacking in these five pivotal marketing actions. For example, there were many doubtful omissions: 57 percent of projects featured no detailed market study; 46 percent omitted the trial sell; and 28 percent did not even have a *formal* launch. Further, one-third of projects had quality-of-execution ratings for marketing actions below 5.0 out of 10—a dismal score. Not surprisingly, these same projects suffered a failure rate of 63 percent!

No doubt there are good and valid reasons why certain commonly recommended actions may be omitted. Not every project needs a market test or trial sell, for example. But the frequency of omission of too many activities was substantial, certainly more than one would have expected from the occasional skipping over a step in the development and commercialization process. Moreover, most of the excuses for omission were fairly lame. For example, we heard from marketing that "they had a limited window of opportunity, so they had to move fast and that meant cutting out a few of the normal steps." From R&D we heard, "We didn't do a user study because we didn't have the budget—and besides, there was nobody to do it—the marketing folks were too busy doing other things." Project members would also claim that, "We don't usually do a detailed financial analysis prior to development—the numbers really aren't too reliable." Or, "We didn't do 'beta tests' or field trials because we didn't want competitors to find out about the new product." While confidentiality is important, it is also imperative to make sure that the product really does perform under live field conditions in a manner acceptable to the customer.

It is also important to point out the impact of these marketing actions on cycle time. Projects that featured well-executed marketing activities fared much better on both measures of timeliness—time efficiency and adherence to the timeline. Contrary to myth, taking a little extra time to execute the customer test, trial sell, product launch, and detailed market study in a quality fashion *does not add extra time*; rather, it pays off, not only with higher success rates but in terms of staying on schedule and achieving better time efficiency.

The message is clear: Marketing actions are critical to both new product success and cycle time reduction. If these actions are not done, or are carried out in a sloppy fashion, then watch out: Expect a drop in success rates, profitability, market share, and company impact—and watch time efficiency and time-lines suffer. Marketing actions, executed in a quality fashion, must be an integral facet of your firm's product development game plan. Unfortunately, they often are not!

3. DON'T SKIP UP-FRONT HOMEWORK

Homework undertaken before the project proceeds into Development is critical. Projects that boasted superb up-front homework achieved a 43 percent higher success rate and were rated significantly more profitable; they were more likely to be successful technically, and they had a greater impact on the company. Most important, better homework reduced cycle time; such projects were undertaken in a more time-efficient manner, and stayed on-schedule.

Five critical activities comprise the homework phase of the project. In rank order (some of which overlap the marketing actions above) that precede the Development phase are:

1. Initial screening: the first decision to get into the project (the idea screen).
2. The detailed market study or marketing research (described above).
3. The business and financial analysis held just before the decision to "Go to Development." (Some people call this "building the business case.")
4. Preliminary market assessment—the first and quick market study.
5. Preliminary technical assessment—the first and quick technical appraisal of the project.

While the wisdom of doing these tasks may be apparent, good homework certainly was missing in too many new product projects. Indeed, one-third of the projects had a homework quality-of-execution rating of less than 4.5 on the ten-point scale—an indictment of the quality of homework here.

Once again the message is obvious: Don't skimp on the homework. If you find yourself making the case that, "We don't have time for the homework," you are heading for trouble on two counts: First, cutting out the homework drives your success rate way down; second, cutting out homework to save time today will cost you in wasted time tomorrow. It's a "penny-wise, pound-poor" solution to saving time. Make it a rule: No significant project should move into the Development phase without the five actions described above completed, and in a quality way.

4. PICK ATTRACTIVE MARKETS

This sounds a bit like saying, "Buy low and sell high"; except that with new products, the choice is much more apparent—there are *evident market characteristics to look for*, characteristics that most often result in success.

Here are the ingredients of this winning market situation:

- The product type (category) represented an essential one for the customer;
- The market was growing quickly;
- There was a positive economic climate for the new product;
- The market demand for this type of product was stable over time (as opposed to cyclical and unstable);
- Potential customers were innovative adopters, amenable to trying new products;
- Potential customers were relatively price-insensitive;
- The market was a large one;
- Potential customers themselves were very profitable.

Note that few of these characteristics on their own were predictive of success, but when taken together as an "index of market attractiveness," they were strongly linked to performance. For example: Those projects in the top third on this index achieved an enviable success rate of 87.5 percent; they were rated significantly more profitable; they had a higher market share (by 20 share points); and they were executed in a more time-efficient manner.

The implications for management are evident: Create an "index of market attractiveness," perhaps using the eight characteristics cited above, and use this index in scoring or rating projects when determining your project priorities.

5. GET PRODUCT DEFINITION RIGHT—FIRST

Defining the product sharply before proceeding to the Development phase closely parallels the need for homework. Sharply defined projects were decidedly more successful: 85 percent successful versus only 47 percent for those without good, early definition. Sharply defined projects also had significantly higher profitability ratings.

Good definition leads to success in other ways too: technical success, impact on the company, and market share (20 share points higher). Staying on schedule was also tied to good definition. Projects that lacked sharp, early definition had dismal adherence-to-schedule ratings.

Management must make certain that significant projects are clearly defined when they are released for development. Here are the ingredients of this definition (again, in rank order of impact):

1. The product benefits to the customer are clearly defined.
2. The target market is precisely spelled out.
3. The product's requirements, features and specs are clearly defined.
4. The product concept—what the product will be and do—is conceptually laid out.
5. The positioning strategy—how the product will be positioned in the minds of users and vis-a-vis competitors' products—is mapped out.

Unless these five items are clearly defined, written down and agreed to by all parties prior to entering the Development phase, then your project will face tough times downstream: Your odds of failure have just skyrocketed by a factor of three!

6. PLAN AND RESOURCE THE LAUNCH PROPERLY

Whoever said, "Build a better mousetrap and the world will beat a path to your door" was a poet, not a businessman. Not only must the product be a superior one, but it must also be launched, marketed and supported in a strong and proficient manner. This was not always the case for the sample of chemical products studied, and most often the result was failure.

The "goodness of launch" index consisted of eight elements. In rank order of impact, they were:

1. **Service quality:** the quality of the service and technical support aimed at the customer (e.g., the right people, qualified, responsive, etc.).

2. **Reliability of product delivery**—on-time shipments.

3. **Product availability**—the product supply was adequate.

4. **Sales force quality:** quality of the selling effort (the right people, properly trained, etc.).

5. **Promotional quality:** quality of the promotional effort (trade shows, events, etc.).

6. **Promotion magnitude**—enough promotional effort.

7. **Service magnitude**—enough support resources.

8. **Sales force magnitude**—enough sales people and effort.

Although advertising, both quality and magnitude, was measured, it did not impact strongly on success; consequently, it was dropped from this index of "goodness of launch." Apparently advertising is not a critical component of the launch of products (at least in this sample) in the chemical industry; it may, however, be more critical in other industries such as consumer goods.

The launch clearly had an impact on performance. New products scoring in the top third in terms of "goodness of launch" achieved an admirable performance: 78 percent success rate, versus only 41 percent for the poorly launched products. They also had significantly higher ratings on profitability, technical success, and impact on the company; and closer adherence to the time schedule (although one is not sure what is the cause and what is the result here; perhaps an "on-schedule" project resulted in a better launch).

The message is this: Don't assume good products sell themselves, and don't treat the launch as an afterthought. Never underestimate the importance of this final step in the process. Plan for the launch early (some chemical firms' divisions require a *preliminary launch* plan to be delivered as part of the "business case" before the Development phase even begins), and make sure that sufficient resources are allocated to this launch.

7. SYNERGISTIC PRODUCTS DO BETTER

The old adage, "Attack from a position of strength" certainly applies to these new chemical products. Where synergy with the base business was lacking, new products fared poorly on average. Synergistic products, by contrast, achieved an 81 percent success rate (versus only 49 percent for the non-synergistic one-third of products); they had significantly higher profitability ratings and impacts on the company; and they achieved a higher market share—about 19 percentage points higher than for products without synergy.

In the context of the chemical industry, here are some of the more important ingredients of a "synergistic" new product. There was a strong fit between the needs of the new product project and the resources, skills and experience of the firm/division in terms of:

• Management capabilities;

• Technical support and customer service skills/resources;

• Market research and market intelligence skills/resources;

• Selling (sales force) skills/resources;

• R&D (product development) skills/resources (for example, the new product could leverage internal, existing technical skills);

• Manufacturing skills/experience;

• Distribution skills/resources.

These seven synergy ingredients become obvious checklist items in a scoring or rating model to help prioritize new product projects. If your synergy score is low, then there must be other compelling reasons to proceed with the project.

8. NATURE OF PURCHASE HAS STRONG IMPACT

The *nature of the purchase* is often overlooked; yet it constitutes an important body of literature in the field of buyer behavior, namely literature on the *adoption-of-innovation process.*

Our study concludes that many of the characteristics that capture the nature of the purchase—specifically, the level of risk, ease of adoption, and purchase importance from the customer's perspective—are indeed important considerations in the success equation. These adoption characteristics, in rank order of importance, are:

1. Projects where customer tests or trials could be used as valid predictors of ultimate product performance (low risk to the customer);
2. Products that represent important purchases to the customer, with a significant impact on her operation;
3. Products whose adoption holds little risk for the customer;
4. Products for which the customer is certain about the outcome of the purchase;
5. New products that the customer can test or try out easily and inexpensively before adoption;
6. New products that require little change to the customer's own product or process.

Taken together, these characteristics indicate the likelihood of product adoption, and hence have a profound impact on new product performance. For example, new products in the top third on this "adoption likelihood index" achieved an admirable success rate (79 percent); they had a higher profitability rating; they were more likely to be technical successes; and they had a greater sales and profit impact on the firm. Although these purchase characteristics are unfamiliar to many product developers, they deserve to be built into your checklist of items to consider when evaluating the odds of winning.

9. RIGHT ORGANIZATIONAL DESIGN IS AT HEART OF SUCCESSFUL PROJECTS

"Rip apart a badly developed project and you will unfailingly find 75 percent of slippage attributable to (1) 'siloing' or sending memos up and down vertical organizational 'silos' or 'stovepipes' for decisions, and (2) sequential problem solving," according to Peters (9). Our study concurs: Good organizational design was strongly linked to success. Projects that lacked good organizational design—the bottom third on this dimension—fared poorly, with much lower success, profitability, technical success, and timeliness ratings. Good organizational design in this study meant (in rank order of impact) projects:

1. Organized as a cross-functional team (as opposed to each function doing its own part independently);
2. Where the team was dedicated and focused (i.e., devoted a large percentage of their time to this project, as opposed to being spread over many projects);
3. Where the team was accountable for the entire project from beginning to end (as opposed to being accountable for only one phase);
4. Led by a strong champion;
5. With top management committed to, and strongly supporting, the project.

While the ingredients of "good organizational design"—"good O.D."—should be familiar, it is surprising that many projects lacked good O.D. For example, one-third of the projects scored below 6.5 on this ten-point metric or dimension. This is not a particularly admirable score, given that a dedicated, accountable cross-functional team approach is such a well-known success ingredient!

Interestingly, performance for the top third and the middle third of projects in terms of O.D.

was essentially the same; it appears that there are diminishing returns to O.D., and that only when O.D. is poor (bottom third) are the effects felt.

The implication is that careful thought must be given to how the project team is structured and led. Strive toward a cross-functional, dedicated and accountable team, led by a champion and supported by top management. Although some of these concepts are not new, it is reassuring to find concrete evidence that this team approach *really does deliver better results*; it is equally provocative to find that despite the pleas to move to a team approach, many firms/divisions have yet to get the message.

10. STEP-OUT PROJECTS TEND TO FAIL

Some projects took the firm into unfamiliar territory: a new product category, new customers, unfamiliar customer needs served, new competitors for the firm, unfamiliar technology, new sales force, channels and servicing requirements, and an unfamiliar manufacturing process. These step-out projects into unfamiliar territory had a lower success rate (by 26 percentage points); achieved a lower market share (by 8 percentage points); and were rated lower in terms of both profitability and impact on the company.

The encouraging news is that the negative impact here was not as strong as for most factors. New and unfamiliar territory certainly results in lower success rates and profitability, on average; but the rates were not dramatically lower. The message from our study is that sometimes it is necessary to venture into new and unfamiliar markets, technologies or manufacturing processes. Do so with caution, and be aware that success rates will suffer; but note that the odds of disaster are not so high as to justify not making the move altogether.

11. QUALITY OF EXECUTION PAYS OFF

The great majority of new product resources—people and money—goes to technological activi-

ties in these new chemical products. Not surprisingly, how well these activities were executed also impacted on success. As for other activities, we developed a quality-of-execution index for technical actions: preliminary technical assessment; product development; in-house or lab testing; pilot or trial production; and commercial production start-up.

Projects where these five technical activities were well executed out-performed the rest, and by a considerable margin on some performance dimensions. The top third of projects, in terms of quality-of-execution of technical actions:

- Had a higher success rate (73 percent versus only 49 percent for the bottom third);
- Had a higher technical success rating (but not a significantly higher profitability rating);
- Did significantly better on the two time metrics: time efficiency and adherence to the time schedule.

While quality-of-execution of these technical activities did not have the dramatic impact on new product performance that either marketing or homework activities did, the good news is that proficiency was much better here: The top third of projects' technical actions scored over 8.0 on this ten-point quality-of-execution index, considerably higher than for other activities; even the middle third scored reasonably well—a score of 6.8 or better.

12. NON-PRODUCT ADVANTAGES HAVE LESS IMPACT

Some firms seek competitive advantage via elements other than product advantage; for example, through superior customer service, a strong company reputation or a better sales force. Does such a non-product strategy work in the new product arena? Yes, but not as well as gaining advantage via the product itself.

Non-product advantage was gained from six key elements (in rank order of impact):

1. Superior customer service and technical support for the new product;

2. High level of technical competence (as perceived by the customer) for this type of product;

3. Superior sales force (e.g., larger, better qualified);

4. Positive company image or reputation;

5. Faster or more reliable product delivery;

6. A well-known brand name.

One other possible element of non-product advantage included advertising and promotion, but it failed to impact on success, and hence was dropped from this list.

Non-product advantage, as gauged by these six elements, certainly did influence new product outcomes, but not nearly as dramatically as for the other success factors described so far. For example, the top third of products on this dimension had a success rate of 80 percent (versus 57 percent for the bottom third); and they also had significantly higher profitability and impact-on-the-company ratings.

The message is evident: By all means, strive for advantage via non-product elements—superior service and technical support, a reputation for technical competence, a quality sales force, product availability, and a positive company image and brand name. Every advantage helps. But don't pin your hopes on these elements alone; whenever you hear yourself saying "Our company's reputation, brand name or sales force will make this product a winner," be on guard. If these are your only elements of advantage, you may be overestimating your chances of winning!

The competitive situation has surprisingly little impact on new product outcomes: 64 percent of new products were successful in the highly competitive markets while 74 percent were successful in the less competitive markets. Other performance measures showed similar tendencies. The one exception was timeliness: Highly competitive markets meant closer adherence to the time schedule.

One message is that the markets for these chemical products were *all quite competitive*, and there was not much dispersion along this dimension. In short, there is no such thing as a "non-competitive, comfortable market"—they are all tough! Hence, a difficult competitive situation is most often a given. Second, look to other factors as the key to success: New products succeed *not so much because of their external environment, but because of what project teams and leaders do*—because they conceive and develop superior products, execute pivotal activities in a superb fashion, do solid up-front homework, have a strong customer focus, get sharp and early product definition, and plan and execute a good launch.

13. NATURE OF INNOVATION HAS SURPRISING IMPACT

Innovativeness can be measured in one of two ways: whether or not the product is truly new to the marketplace, and how new the product is to the developing firm. The new products studied fit into one of five categories of product innovativeness (percent breakdowns are given):

1. Innovation: a totally new product to the world (18.8 percent of cases);

2. New to market, new to company—a totally new product to the company that also offered new features to an existing market (33.7 percent);

3. New line: a totally new product or line to company, but an existing market and similar products in that market (16.8 percent);

4. A new item in an existing product line for the company (11.9 percent);

5. A modification of an existing company product (18.8 percent).

Surprisingly, success and failure were not strongly connected to the nature of the innovation. For most measures of performance, there was no significant connection, but there were trends and some significant impacts, including some surprising U-shaped impacts (see Table 51.1):

TABLE 51.1
Impact of the Nature of the Innovation

Nature of innovation	True innovation	New to company and to market	New line for company	New item in existing line	Modification
Category:	1	2	3	4	5
Percent breakdown	18.8%	33.7%	16.8%	11.9%	18.8%
Percent successful	63.2%	68.6%	47.1%	83.3%	70.0%
Tech success rating (0–10)	8.32	7.69	6.39	7.75	7.56
Impact on company (0–10)	6.11	5.46	5.72	7.08	6.37

- New items in an existing company line (category 4 above) had the highest success rate (83 percent), while new lines for the firm (category 3) had the highest failure rate (53 percent). Note that highly innovative products had a respectable 63 percent success rate.

- Technical success ratings were greatest for true innovations (category 1), and least for new lines.

- Closer-to-home new products and highly innovative products had the greatest sales and profit impact on the company: new items in an existing line (category 4) followed by true innovations (category 1). Products *new to the company* had the least impact.

The message is that highly innovative products (category 1), in spite of all their perceived risks and pitfalls, achieved an admirable track record in this study. Perhaps there is less to be feared from being bold and innovative than we had imagined. Note that *innovativeness* is not the same dimension as developing step-out products (item 10 above). The latter is measured *relative to the firm*; products that represented new territory for the company did poorly. Product innovativeness, by contrast, is measured *relative to the marketplace and competition*; new products high on this dimension did very well.

Near the other end of the innovativeness spectrum, much less innovative products—new items in an existing company line, category 4—also achieved a high performance, indeed the best *overall*: the highest success rate, highest technical success rat-

ing, and greatest sales and profit impact on the firm. By contrast, the "middle of the road" or "fairly safe" strategy of launching a new product line, where there is already a market with very similar products in it (category 3), may not be so safe after all. Such products yielded the lowest success rates (47 percent) and the lowest impacts on the firm.

ORDER OF ENTRY MAKES LITTLE DIFFERENCE

One of the drives of today's quest for cycle time reduction is the belief that "being first into the market spells success." The evidence from our study does not support this contention, however. True, products that were first into the market were more successful than those that followed (71 percent versus 60 percent), but this difference was not significant; nor were the profitability rating differences significant. Being "first-in" should not be one's ultimate goal. Many of our new product failures were first into their markets—"fast failures"; an ill-conceived product, developed in haste, often leads to disaster. The lesson is that it may be better to be a fast second with a superior product that has real customer benefits than simply being first.

SO MUCH FOR MYTHS!

None of the 13 success factors discussed in this article come as a total surprise. Indeed, many have been reported or hinted at in previous studies or articles. The success factors we identified are totally consistent with our previous NewProd conclusions (6). Therefore, these results and the ensuing message for managers can likely be applied to a wider range of industries. The disconcerting evidence is that many managers and project teams apparently have failed to get the message.

I began the article with eight well-known myths. Here's how they stack up against the evidence:

1. Being first into the market is only marginally more successful. Being "best in" is far more profitable.

2. Up-front homework really does pay off, not only in terms of higher profits and success rates, but it saves time as well.

3. While a company reputation, a strong brand name, and a good selling effort do help, they are not nearly as decisive as gaining advantage via the product itself.

4. Low price is not the key to winning; rather, good value for money and superior price/performance characteristics are.

5. You cannot afford not to do market studies, customer tests, and a trial sell. High-quality marketing actions yield a double payoff: they drive up profitability and drive down the cycle time.

6. Even for large and powerful firms, synergy is important. The ability to build on in-house resources and capabilities is central to success.

7. Pin down the definition of the product before development begins; it may limit the scope of the developer or designer somewhat, but it certainly drives profits up and time-to-market down in a major way.

8. New products succeed in spite of the market's competitiveness; the competitive situation per se is not that strongly linked to winning and losing. You cannot blame competitiveness for your ills.

So much for myths!

REFERENCES

1. The research team consisted of the author, Professor Robert G. Cooper, and his colleague, Professor Elko Kleinschmidt, together with graduate student research assistants. Professor Kleinschmidt has co-authored a number of research reports on the New-Prod studies with Cooper.

2. Cooper, R. G. "Why new industrial products fail." *Industrial Marketing Management*, 4, 1975, pp. 315–26.

3. Cooper, R. G. "The dimensions of industrial new product success and failure." *Journal of Marketing*, 43, Summer 1979, pp. 93–103.

4. Cooper, R. G. and E. J. Kleinschmidt. "An investigation into the new product process: steps, deficiencies and impact." *Journal of Product Innovation Management*, 3, 2, 1986, pp. 71–85.

5. Cooper, R. G. and E. J. Kleinschmidt. "New products: what separates winners from losers." *Journal of Product Innovation Management*, 4, 3, 1987, pp. 169–184.

6. Cooper, R. G. and E. J. Kleinschmidt. *New Products: The Key Factors in Success*. Chicago: American Marketing Assoc., 1990, monograph. See also: Cooper, R. G. "New products: what distinguishes the winners." *Research-Technology Management*, Nov.–Dec. 1990, pp. 27–31.

7. Cooper, R. G. and E. J. Kleinschmidt. "Major new products: What distinguishes the winners in the chemical industry." *Journal of Product Innovation Management*, vol. 2, no. 10, March 1993, pp. 90–111.

8. Cooper, R. G. and E. J. Kleinschmidt. "New product success in the chemical industry." *Industrial Marketing Management*, vol. 22, no. 2, 1993, pp. 85–99.

9. Peters, Tom. *Thriving on Chaos*. New York: Harper & Row, 1988.

10. Cooper, R. G. "Debunking the myths of new product development." *Research-Technology Management*, July–August, 1994, pp. 40–50.

11. Cooper, R. G. *Winning at New Products*, 2d ed. Reading, MA.: Addison-Wesley, 1993.

52

The Rules of Innovation

CLAYTON CHRISTENSEN

Innovation is widely considered a black art—but is it? A leading business thinker lays out four essential rules designed to maximize the chances disruptive technologies will succeed.

Two decades ago, when I was just out of graduate school and working in the automotive industry, I got my first introduction to the statistical process-control chart. We used this laborious technique to make sure the machines employed in our manufacturing process did not drift out of control. Composed of three parallel horizontal lines, the "SPC" chart has long been an important tool in quality management. The center line represents the targeted value for the critical performance parameter of a product being manufactured. The lines above and below it represent the acceptable upper and lower control limits. If the product were, say, an axle, workers would plot the thickness of each piece they made on the chart. When I asked why there was typically a scatter of points around the target, my managers cited the randomness inherent in all processes.

The "Quality Movement" of the 1980s and '90s subsequently taught us that there *isn't* randomness in processes. Every deviation of the actual value from the target has a cause. It appears to be random when we don't know the cause. The Quality Movement developed methods for identifying those additional factors—and we discovered that if we could control or account for all of them, the result would be perfectly predictable, and there would be no need to inspect products as they emerged from manufacturing.

The management of innovation today is where the Quality Movement was 20 years ago, in that many believe the outcomes of innovation efforts are unpredictable. The raison d'être of the venture capital industry is belief in the unpredictability of new businesses. A few ventures will succeed; most won't, the VCs say. They therefore place a portfolio of bets, extracting premium prices for their capital in order to earn the high return required to compensate for the risk that unpredictability imposes. I believe, however, that innovation *isn't* random. Every undesired outcome has a cause. Those outcomes *appear* to be random when we don't understand all the factors that affect successful innovation. If we could understand and manage these variables, innovation wouldn't be nearly as risky as it appears.

The good news is that recent years have seen considerable progress in identifying important variables that affect the probability of success in innovation. I've classified these variables into four sets: (1) taking root in disruption, (2) the necessary scope to succeed, (3) leveraging the right capabilities and (4) disrupting competitors, not customers.

Of course, building successful businesses is such a complicated process, involving subtle interdependencies among so many variables in dynamic systems, that we're unlikely ever to make it perfectly predictable. But the more we can master these variables, the more we will be able to create new companies, products, processes and services that achieve what we hope to achieve.

TAKE ROOT IN DISRUPTION

The startling conclusion suggested by the research that led to my writing *The Innovator's Dilemma*

From Christensen, Clayton, "The Rules of Innovation," *Technology Review*, June 2002, pp. 33–38. Reprinted with permission.
Clayton Christensen is a Professor at the Harvard Business School and a former Rhodes scholar.

was that many successful companies stumble from prominence not because they're badly managed but precisely because they are *well managed*. They listen to and satisfy the needs of their best customers, and they focus investments at the largest and most profitable tiers of their markets. Mastering these paradigms of good management gives established companies, as a group, an extraordinary track record in producing *sustaining* innovations that bring better products to established markets. It matters little whether the innovation is incrementally simple or radically difficult, as long as it enables good companies to make better products that they can sell for higher margins to their best customers in attractively sized markets. The companies that had led their industries in prior technologies led their industries in adopting new sustaining technologies in literally *100 percent* of the cases we studied.

In contrast, the leading companies almost always were toppled when *disruptive* technologies emerged—products or services that weren't as good as those already used in established markets. Disruptive innovations don't initially perform well enough to be sold or used successfully in mainstream markets. But they have other attributes— most often simplicity, convenience and low cost— that appeal to a new, small and initially unattractive (to established firms) set of customers, who use them in new or low-end applications.

The chances a new company could become successful if its entry path was a *sustaining* strategy—trying to make a better product than the incumbents and selling it to the same customers— were about six percent in our study. The chances of success for firms that entered with a disruptive strategy were 33 percent. The disparity stems from the motivation and position of the leading firms. They have far more resources to throw at opportunities than entrants do. When newcomers attack customers and markets attractive to the leaders, the leaders overwhelm them.

All companies are burdened with "asymmetric" motivations in that they must move toward markets that promise higher profit margins and the most substantial and immediate growth and cannot move down market toward smaller opportunities

and profit margins. When new entrants take root with customers in markets that are unattractive to the leaders, they are safer—and it has nothing to do with how much cash or proprietary technology they have. They are safe because the incumbents are motivated to ignore or even exit the very markets that the entrants are motivated to enter. Taking root in disruption, therefore, is the first condition that innovators need to meet to improve the probability of successfully creating a new growth business. If they cannot or do not do this, their odds of success are much smaller.

There are two tests to assess whether a market can be disrupted. At least one of these criteria must be met in order for an upstart to be disruptively successful. If a new growth business can meet both, the odds are even better.

1. Does the innovation enable less-skilled or less-wealthy customers to do for themselves things that only the wealthy or skilled intermediaries could previously do?

When an innovation fulfills this condition, even if it can't do all the things existing offerings can, potential customers excluded from the market tend to be delighted. For example, many people loved the first personal computers, no matter how clunky the booting process and limited the software the machines could run, because the alternative to which they compared the PC wasn't the minicomputer—it was no computer at all. Filling such a void reduces the capital commitments and technological achievements required for an innovation to become viable and creates new growth markets. I call the process of finding and nurturing these opportunities *creative creation*. After a technology takes root in new markets, and after new growth is created, disruption can invade the established market and destroy its leading firms.

Even if innovators succeed in cramming disruptive technology into an existing market application, the incumbents typically win. Digital photography, online consumer banking and hybrid-electric vehicles are examples of potentially disruptive technologies that were deployed in such a sustaining fashion. Billions were spent on these innovations to beat out already acceptable and habitual technology; little net growth resulted, as sales of the new products cannibalized sales of the old; and the industry leaders maintained their rule.

2. Does the innovation target customers at the low end of a market who don't need all the functionality of current products? And does the business model enable the disruptive innovator to earn attractive returns at discount prices unattractive to the incumbents?

Wal-Mart, Dell Computer and Nucor are examples of disruptive companies that attacked the low ends of their markets with business models that allowed them to make money at discount prices. Wal-Mart started by selling brand-name products at prices 20 percent below department store prices and still earned attractive returns because it turned inventory over much more frequently. Such a disruptive strategy can create new growth businesses but does not create new markets or classes of consumers. It has a high probability of success because the reported profit margins of established companies typically improve if they get out of low-end, low-margin products and add in their stead high-margin products positioned in more-demanding market segments. By assaulting the low end of the market and then moving up, a new company attacks, tier by tier, the markets from which established competitors are motivated to exit.

PICK THE SCOPE NEEDED TO SUCCEED

The second set of variables that affects the probability that a new business venture will succeed relates to its degree of "integration." Highly integrated companies make and sell their own proprietary components and products across a wide range of product lines or businesses. Nonintegrated companies outsource as much as possible to suppliers and partners and use modular, open systems and components. Which style is likely to be successful is determined by the conditions under which companies must compete as disruption occurs.

In markets where product functionality is not yet good enough, companies must compete by making better products. This typically means making products whose architecture is interdependent and proprietary, because competitive pressure compels engineers to fit the pieces of their systems together in ever more efficient ways in order to wring the best performance possible out of the available technology. Standardization of interfaces (meaning fewer degrees of design freedom) forces them to back away from the frontier of what is technologically possible—which spells competitive trouble when functionality is inadequate. This helps explain why IBM, General Motors, Apple Computer, RCA, Xerox and AT&T, as the most integrated firms during the not-good-enough era of their industries' histories, became dominant competitors. Intel and Microsoft (raps about the latter's supposed lack of innovation aside) have also dominated their pieces of the computer industry—compared to less integrated companies such as WordPerfect (now owned by Corel)—because their products have employed the sorts of proprietary, interdependent architectures that are necessary when pushing the frontier of what is possible. This also helps us understand why NTT DoCoMo, with its integrated strategy, has been so much more successful in providing mobile access to the Internet than nonintegrated American and European competitors who have sought to interface with each other through negotiated standards.

When the functionality of products has overshot what mainstream customers can use, however, companies must compete through improvements in speed to market, simplicity and convenience, and the ability to customize products to the needs of customers in ever smaller market niches. Here, competitive forces drive the design of *modular* products, in which the interfaces among components and subsystems are clearly specified. Ultimately, these coalesce as industry standards. Modular architectures help companies respond to individual customer needs and introduce new products faster by upgrading individual subsystems without having to redesign everything. Under these conditions (and only under these conditions), outsourcing titans like Dell and Cisco Systems can prosper—because modular architectures help them be fast, flexible and responsive.

LEVERAGE THE RIGHT CAPABILITIES

Innovations fail when managers attempt to implement them within organizations that are incapable of succeeding. Managers can determine the innovation limits of their organizations quite precisely by asking three questions: (1) *Do I have the resources to succeed?* (2) *Will my organization's processes facilitate success in this new effort?* (3) *Will my organization's values allow employees to prioritize this innovation, given their other responsibilities?*

Beyond technology, the resources that drive innovative success are managers and money. Corporate executives often tap managers who have strong records of success in the mainstream to manage the creation of new growth businesses. Such choices can be the kiss of death, however, because the challenges confronting managers in a disruptive enterprise—and the skills required to overcome them—are different from those that prevail in the core business. Many innovations fail because managers do not know what they do not know as they make and implement their plans. That is, they assume that the same strategies and customer needs that apply in mature, stable markets will apply in disruptive ventures. But this is not the case, and by making such assumptions, managers close themselves off from opportunities to discover what customers really find useful in new, disruptive products.

Innovators must avoid two common misconceptions in managing the other key resource, money. The first is that deep corporate pockets are an advantage when growing new businesses. They are not. Too much cash allows those running a new venture to follow a flawed strategy for too long. Having barely enough money forces the venture's managers to adapt to the desires of actual customers, rather than those of the corporate treasury, when looking for ways to get money—

and forces them to uncover a viable strategy more quickly.

The second misconception is that patience is a virtue—that innovation entails large losses for sustained periods prior to reaping the huge upside that comes from disruptive technologies. Innovators should *be patient about the new venture's size but impatient for profits.* The mandate to be profitable forces the venture to zero in on a valid strategy. But when new ventures are forced to get big fast, they end up placing huge bets at a time when the right strategy simply cannot be known. In particular, they tend to target large, obvious, existing markets—and this condemns them to failure. Most of today's envisioned business opportunities for wireless Internet access, for example, involve big applications such as stock-trading and multiplayer gaming that have already found homes on wired, desktop computers. Billions are being sunk into new wireless ventures committed to taking over these markets before innovators have a chance to learn what applications wireless is really best at delivering.

Resources such as technology, cash and technical talent tend to be flexible, in that they can be used for a wide array of purposes. Processes, however—the central element in our second question—are typically inflexible. Their purpose is not to adapt quickly but to get the same job done reliably, again and again. The fact that a process facilitates certain tasks means that it will not work well for very different tasks. Failure is frequently rooted in the forced use of habitual but inappropriate processes for doing market research, strategic planning and budgeting.

Sony, for example, was history's most successful disruptor. Between 1950 and 1980 it introduced 12 bona fide disruptive technologies that created exciting new markets and ultimately dethroned industry leaders—everything from radios and televisions to VCRs and the Walkman. Between 1980 and 1997, however, the company did not introduce a single disruptive innovation. Sony continued to produce sustaining innovations in its product businesses, of course. But even the new businesses that it created with its PlayStation and Vaio notebook computer were great but late entries into already established markets.

What drove Sony's shift from a disruptive to a sustaining innovation strategy? Prior to 1980, all new product launch decisions were made by co-founder Akio Morita and a trusted team of associates. They never did market research, believing that if markets did not exist they could not be analyzed. Their process for assessing new opportunities relied on personal intuition. In the 1980s Morita withdrew from active management in order to be more involved in Japanese politics. The company consequently began hiring marketing and product-planning professionals who brought with them data-intensive, analytical processes of doing market research. Those processes were very good at uncovering unmet customer needs in existing product markets. But making the intuitive bets required to launch disruptive businesses became impossible.

A company's values—the focus of question three—determine the necessity of spinning out separate organizations for new ventures. Values are even less flexible than resources. *Everyone* in an organization—executives to sales force—must put a premium on the type of business that helps the company make money given its existing cost structure. If a new venture doesn't target order sizes, price points and margins that are more attractive than other opportunities on the organization's plate, it won't get priority resources; it will languish and ultimately fail.

Nor is it just the values of the innovating company that matter, because suppliers and distributors have values too, and *they* must put the highest priorities on opportunities that help *them* make money. This is why, with almost no exceptions, disruptive innovations take root in free-standing value networks—with new sales forces, distributors and retailing channels.

DISRUPT COMPETITORS, NOT CUSTOMERS

The fourth factor in successful innovation is minimizing the need for customers to reorder their lives. If an innovation helps customers do things they are already trying to do more simply and conveniently, it has a higher probability of success. If it makes it easier for customers to do something they weren't trying to do anyway, it will fail. Put differently, innovators should try to disrupt their competitors, never their customers.

The best way to understand what customers are actually trying to do, as opposed to what they say they want to do, is to *watch* them. For example, when interviewed by the college textbook industry, students say they would welcome the ability to probe more deeply into topics of interest that textbooks just touch on. In response, publishers have invested substantial sums to make richer information available on CDs and Web sites. But few students actually use these innovations, and little growth has resulted. Why? Because what most students *really* are trying to do is avoid reading textbooks at all. They say they would like to delve more deeply into their subjects. But what they really *do* is put off reading until the last possible minute—and then cram.

To make it simpler and more convenient for students to do what they already are trying to do, a publisher could create an online facility called Cramming.com. Like all disruptive technologies, it would take root in a low-end market: the least conscientious students. Semester after semester, Cramming.com would then improve as a new "cramming-aid" growth business, without affecting textbook sales. Conscientious students would continue to purchase textbooks. At some point, however, learning the material online would be so much easier and less expensive that, tier by tier, students would stop buying texts. This path of innovation has a much higher chance of success than a direct assault that pits digital texts against conventional textbooks.

The observed probabilities of success in innovation are low. But these statistics stem from the sum of sustaining and disruptive strategies, many of which are attempted in organizations whose resources, processes and values render them incapable of succeeding. Many innovators draw lessons from observing other successful companies in very different circumstances and attempt to succeed with just one or a few links in a chain of interdependent values. And many fail after assuming that what customers *say* they want to do is what they actually *would* do.

Hence, the observed probabilities of success don't necessarily reflect what the true likelihood of success can be, if the critical variables in the complex and dynamic process of innovation are understood and managed effectively. Indeed, success may not be as difficult to achieve as it has seemed.

17

Managing User Innovation for New Product Development

53

Product Concept Development Through the Lead-User Method

ERIC VON HIPPEL AND RALPH KATZ

The concepts and examples reported in this selection address an important problem facing all innovative organizations, which is, how can one effectively determine user needs for developing new products and services in markets that are strongly affected by rapid changes in technology? This selection begins by exploring some of the difficulties faced by traditional methods of market research. It then discusses the "lead user" methodology that von Hippel and his associates have developed and tested as a useful managerial solution for dealing with this problem.

Reprinted with permission of the authors.

Eric von Hippel is Professor of Management at MIT's Sloan School of Management. ***Ralph Katz*** is Professor of Management at Northeastern University's College of Business and Research Associate at MIT's Sloan School of Management.

ROOT OF THE PROBLEM: MARKETING RESEARCH CONSTRAINED BY USER EXPERIENCE

One important function of marketing research is to accurately understand user needs for potential new products. Such understanding is clearly an essential input to the success of the new product development process. Nevertheless, users selected to provide input data to consumer and industrial market analysis have an important limitation: their insights into new product and service needs and potential solutions are constrained by their real-world experiences. Users steeped in the present are, therefore, unlikely to generate novel product concepts that conflict with the familiar.

The notion that familiarity with existing product attributes and uses interferes with an individual's ability to conceive of novel attributes and uses is strongly supported by research into problem solving. Extant studies have shown, for example, that when experimental subjects are familiarized with a complicated problem-solving strategy, they are unlikely to devise a simpler one even when this is appropriate. Moreover, subjects who use an object or see it used in a very normal and familiar way are strongly blocked from using that object in a new or novel manner. In fact, the more recently these objects or problem-solving strategies were used in a familiar way, the more difficult it was for the subject to employ them in a more innovative way. In an R&D setting, Allen and Marquis showed that the success of a research group in solving a new problem was strongly dependent on whether the solutions and experiences it had used in the past fit the demands of the new problem.[1] All of these research studies suggest that typical users of existing products—the type usually chosen in market research—are poorly situated with regard to the difficult problem-solving tasks associated with assessing unfamiliar product needs.

The constraint of users to the familiar pertains even in the instance of sophisticated marketing research techniques such as multi-attribute mapping of product perceptions and preferences. Such methods typically frame user information and responses in terms of known attributes; they do not offer a reliable or valid means of going beyond the experiences of those queried or interviewed. First, most users are not well positioned to accurately evaluate novel product concepts or accurately quantify unfamiliar product attributes. Secondly, there is no mechanism in traditional market research to induce users to identify all product attributes potentially relevant to a product category, especially attributes that lie outside the range of their real world experiences. As a result, product development groups in many companies aren't able to deliver the stream of breakthrough products that allow their companies to achieve the rather ambitious growth goals announced by business leaders. There is simply no effective system in place to guide and support new product developers in these kinds of breakthrough efforts.

LEAD USERS AS A SOLUTION

In many product categories, the constraint of users to the familiar does not lessen the ability of marketing research to evaluate needs for new products by analyzing typical users. In the relatively slow-moving world of steels and autos, for example, new models often do not differ radically from their immediate predecessors. Therefore, even the "new" is reasonably familiar and the typical user can thus play a valuable role in the development of new products.

Contrastingly, in high technology industries, the world moves so rapidly that the related real-world experience of ordinary users is often rendered obsolete by the time a product is developed or during the time of its projected commercial lifetime. For such industries, we propose that lead users, who *do* have real-life experience with novel product or process needs, are essential to accurate marketing research. Although the insights of lead users are as constrained to the familiar as those of other users, lead users are more familiar with conditions that lie in the future and, so, are in a position to provide accurate data on needs related to such prospective conditions.

Lead users of a novel or enhanced product, process, or service are defined as those who display two characteristics with respect to it:

1. Lead users face needs that will be general in a marketplace, but they face them months or years before the bulk of that marketplace encounters them, *and*

2. Lead users are in a position to benefit significantly by obtaining a solution to those needs.

Each of the two lead user characteristics provides an independent contribution to the type of new product need and solution data that such lead users are hypothesized to possess. The first specifies that a lead user will possess the particular real-world experience that the manufacturers must analyze if they are to accurately understand the needs that the bulk of the market will have "tomorrow." Users "at the front of the trend" exist simply because important new technologies, products, tastes, and other factors related to new product opportunities typically diffuse through a society over many years rather than impact all members simultaneously.[2]

The second lead user characteristic is a direct application of the hypothesis that the greater the benefit a given user expects to obtain from a needed novel product or process, the greater his investment will be in obtaining a solution. Users who expect high returns from a solution to a need they are experiencing should have been driven by these expectations to attempt to solve their need. This work in turn will have produced insight into the need and perhaps useful solutions that will be of value to inquiring market researchers.

In sum, then, lead users are users whose present strong needs will become general in a marketplace months or years in the future. Since lead users are familiar with conditions that lie in the future for most others, it is hypothesized that they can serve as a need-forecasting laboratory for marketing research. Moreover, since lead users often attempt to fill the need they experience, it is also hypothesized that they can provide valuable new product concept and design data to inquiring organizations in addition to need data. As a result, lead users may have a great deal more to contribute than data regarding their unfilled needs; often, they may

contribute insights regarding solutions as well. Such "solution" data can range from rich insights to actual working and tested prototypes of the desired novel product, process, or service.

This lead user method was developed by von Hippel, based on his twelve-year study of the innovation process.[3] Von Hippel's research traced the role of users in product innovation. The most striking finding was that *users* were often the actual developers of original prototype solutions to what eventually became successful commercial products. In some industries, users had been responsible for most of the important product or process innovations. Users were found, for example, to be the actual developers of 82% of all commercialized scientific instruments studied and 63% of all semiconductor and electron subassembly manufacturing equipment innovations studied. These findings go against conventional wisdom, which holds that manufacturers are typically the developers of new products. It is also a major challenge to the common belief that users can provide market researchers only with data on market need. What von Hippel's research evidence demonstrates convincingly is that often innovative users also have valuable new product information to offer design engineers and product developers.

Consider how an automobile manufacturer would apply the lead user process. If the company wanted to design an innovative braking system, it might start by trying to find out if any innovations had been developed by drivers with a strong need for better brakes, such as auto racers. It wouldn't stop there, however. Next, it would look to a related but technologically advanced field where people had an even higher need to stop quickly, such as aerospace. And, in fact, aerospace is where innovations such as ABS braking were first developed: military and commercial aircraft pilots have a very high incentive to stop their vehicles before running out of runway.

TESTING THE METHOD

To test the usefulness of the lead user concept, a prototype lead user market research study was undertaken in the rapidly changing field of computer-aided-design (CAD) products.[4] (Over forty firms compete in the $1 billion market for CAD hardware and software. This market grew at over 35% per year over the period 1982 to 1986 and the forecast is for continued growth at this rate for the next several years.) Within the CAD field, von Hippel and Urban decided to focus specifically on CAD systems used to design the printed circuit (PC) boards used in electronic products, PC-CAD.

The method used to identify lead users and test the value of the data they possess in the PC-CAD field involved four major steps: (1) identify an important market or technical trend, (2) identify lead users with respect to that trend, (3) analyze lead user data, and (4) test lead user data on ordinary users.

Identifying an Important Trend

Lead users are defined as being in advance of the market with respect to a given important dimension that is changing over time. Therefore, before one can identify lead users in a given product category of interest, one must specify the underlying trend on which these users have a leading position.

To identify an "important" trend in PC-CAD, we sought out a number of expert users. We identified these by telephoning managers of the PC-CAD groups of a number of firms in the Boston area and asking each: "Whom do you regard as the engineer most expert in PC-CAD in your firm?" "Whom in your company do group members turn to when they face difficult PC-CAD problems?"[5] After discussions with expert users, it was qualitatively clear that an increase in the density with which chips and circuits are placed on a board was, and would continue to be, a very important trend in the PC-CAD field. Historical data showed that board density had in fact been steadily increasing over a number of years. And the value of continuing increases in density was clear. An increase in density means that it is possible to mount more electronic components on a given size printed circuit board. This in turn translates directly into an ability to lower costs (less material is used), to decreased product size, and to increased speed of circuit operation (signals between components travel shorter distances when board density is higher).

Very possibly, other equally important trends exist in the field that would reward analysis, but it was decided to focus on this single trend in the study.

Identifying Lead Users

To identify lead users of PC-CAD systems capable of designing high-density printed circuit boards, von Hippel and Urban identified that subset of users: (1) who were designing very high-density boards now and (2) who were positioned to gain especially high benefit from increases in board density. They decided to use a formal telephone-screening questionnaire to accomplish this task and strove to design one that contained objective indicators of these two hypothesized lead user characteristics.

Printed circuit board density can be increased in a number of ways and each offers an objective means of determining a respondent's position on the trend toward higher density. First, the number of layers of printed wiring in a printed circuit board can be increased. (Early boards contained only one or two layers but now some manufacturers are designing boards with twenty or more layers.) Second, the size of electronic components can be decreased. (A recent important technique for achieving this is surface-mounted devices that are soldered directly to the surface of a printed circuit board.) Finally, the printed wires, vias, that interconnect the electronic components on a board that can be made narrower and packed more closely. Questions regarding each of these density-related attributes were included in our questionnaire.

Next, von Hippel and Urban assessed the level of benefit a respondent might expect to gain by improvements in PC-CAD by means of several questions. First, they asked about users' level of satisfaction with existing PC-CAD equipment, assuming that high dissatisfaction would indicate expected high benefit from improvements. Second, they asked whether respondents had developed and built their own PC-CAD systems rather than buy the commercially available systems such as those offered by IBM or Computervision. (They assumed, as noted previously, that users who make such innovation investments do so because they expect high benefit from resulting PC-CAD system improvement.) Finally, they asked respondents whether they thought their firms were innovators in the field of PC-CAD.

The PC-CAD users interviewed were restricted to U.S. firms and selected from two sources: A list of members of the relevant professional engineering association (IPCA) and a list of current and potential customers provided by a cooperating supplier. Interviewees were selected from both lists at random. Von Hippel and Urban contacted approximately 178 qualified respondents and had them answer the questions on the phone or by mail if they preferred. The cooperation rate was good: 136 screening questionnaires were completed. One third of these were completed by engineers or designers, one third by CAD or printed circuit board managers, 26% by general engineering managers, and 8% by corporate officers.

Simple inspection of the screening questionnaire responses showed that fully 23% of all responding user firms had developed their own in-house PC-CAD hardware and software systems. Also, this high proportion of user-innovators that we found in our sample is probably characteristic of the general population of PC-CAD users. The sample was well dispersed across the self-stated

scale with respect to innovativeness: 24% indicated they were on the leading edge of technology, 38% up-to-date, 25% in the mainstream, and 13% adopting only after the technology is clearly established. This self-perception is supported by objective behavior with respect to the alacrity with which our respondents adopted PC-CAD.

The researchers next conducted a cluster analysis of screening questionnaire data relating to the hypothesized lead user characteristics in an attempt to identify a lead user group. The two cluster solution is shown in Table 53.1.

Note that this analysis does, indeed, clearly indicate a group of respondents who combine the two hypothesized attributes of lead users and that, effectively, all of the PC-CAD product innovation is reported by the lead user group.

In the two-cluster solution, what is termed the lead users cluster is ahead of nonlead users in the trend toward higher density. That is, lead users report more use of surface-mounted components, use of narrower lines, and use of more layers than do members of the nonlead cluster. Second, lead users appear to expect higher benefit from PC-CAD innovations that would allow them even further progress. That is, they report less satisfaction with their existing PC-CAD systems (4.1 vs. 5.3, with

TABLE 53.1
Cluster Analyses Revealing Lead and Nonlead User Groups

	Two-cluster solution	
	Lead Users	Nonlead Users
Indicators of user position on PC-CAD density trend		
Use surface mount?	87%	56%
Average line width (mils)	11	15
Average layers (number)	7.1	4.0
Indicators of user-expected benefit from PC-CAD improvement		
Satisfaction[a]	4.1	5.3
Indicators of related user innovation		
Build own PC-CAD?	85%	1%
Innovativeness[b]	3.3	2.4
First use of CAD (year)	1973	1980
Number in cluster	38	98

[a]7-point scale—high value more satisfied.
[b]4-point scale—high value more innovative.

higher values indicating satisfaction). Strikingly, 87% of respondents in the lead user group report building their own PC-CAD system (vs. only 1% of nonlead users) in order to obtain improved PC-CAD system performance.[6] Lead users also judged themselves to be more innovative (3.3 vs. 2.4 on the four-statement scale with higher values more innovative), and they were in fact earlier adopters of PC-CAD than were nonlead users. Note that 28% of our respondents are classified in this lead user cluster (which is a far higher percentage than has been found in other lead user studies). A discriminant analysis also indicated that building one's own system was the most important indicator of membership in the lead user cluster.

Analyzing Lead User Insights

The next step in the analysis was to select a small sample of the lead users identified in the cluster analysis to participate in a group discussion to develop one or more concepts for improved PC-CAD systems. Experts from five lead user firms that had facilities located near MIT were recruited for this group. The firms represented were Raytheon, DEC, Bell Laboratories, Honeywell, and Teradyne. Four of these five firms had built their own PC-CAD systems. All were working in high-density (many layers and narrow lines) applications and had adopted the CAD technology early.

The task set for this group was to specify the best PC-CAD system for laying out high-density digital boards that could be built with current technology. (To guard against the inclusion of "dream" features impossible to implement, the study conservatively allowed the concept the group developed to include only features that one or more of them had already implemented in their own organizations. No one firm had implemented all aspects of the concept, however.)

The PC-CAD system concept developed by our lead user creative group integrated the output of PC-CAD with numerically controlled printed circuit board manufacturing machines; had easy input interfaces (e.g., block diagrams, interactive graphics, icon menus); and stored data centrally with access by all systems. It also provided full functional and environmental simulation (e.g., electrical, mechanical, and thermal) of the board being designed and could design boards of up to 20 layers, route thin lines, and properly located surface-mounted devices on the board.

Testing Product Concept Perceptions and Preferences

From the point of view of marketing research, new product need data and new product solutions from lead users are only interesting if they are preferred by the general marketplace.

To test this matter, we decided to determine PC-CAD user preferences for four system concepts: the system concept developed by the lead user group, each user's own in-house PC-CAD system, the best commercial PC-CAD system available at the time of the study (as determined by a PC-CAD system manufacturer's competitive analysis), and a system for laying out curved printed circuit boards. (This last was a description of a special-purpose system that one lead user had designed in-house to lay out boards curved into three-dimensional shapes. This is a useful attribute if one is trying to fit boards into the oddly shaped spaces inside some very compact products, but most users would have no practical use for it. In our analysis of preference, we think user response to this concept can serve to flag any respondent tendency to prefer systems based on system exotica rather than practical value in use.)

To obtain user preference data regarding the four PC-CAD system concepts, the study designed a new questionnaire that contained measures of both perception and preference. First, respondents were asked to rate their current PC-CAD system on seventeen attribute scales. (These were generated by a separate sample of users through triad comparisons of alternate systems, open-ended interviews, and technical analysis.) Each scale was presented to respondents in the form of five-point agree-disagree judgment based on a statement such as "my system is easy to customize."[7] Next, each respondent was invited to read a one-page description of each of the three concepts that had been generated (labeled simply, J, K, and L) and rate them on the same scales.

All concepts were described as having an identical price of $150,000 for a complete hardware and software workstation system able to support four users. Next, rank-order preference and constant-sum paired comparison judgments were requested for the three concepts and the existing system. Finally, probability-of-purchase measures on an 11-point Juster scale were collected for each concept at the base price of $150,000, with alternate prices of $100,000 and $200,000.

A second questionnaire was sent to 173 users (the 178 respondents who qualified in the screening survey less the 5 user firms in the creative group). Respondents were called by phone to inform them that a questionnaire had been sent. After telephone follow-up and a second mailing of the questionnaire, seventy-one complete or near-complete responses were obtained (41%) and the following analyses are based on these.

CONCEPT PREFERENCES

The analysis of the concept questionnaire showed that respondents strongly preferred the lead user group PC-CAD system concept over the three others presented to them: 78.6% of the sample selected the lead user's creative group concept as their first choice. It was strongly preferred by users over their existing systems. Respondents maintained their preferences for the lead user concept even when it was priced higher than competing concepts. The effects of price were investigated through the probability of purchase measures collected at three different prices for each concept. Even when the lead user concept was priced twice as high as that of competing concepts, the lead user concept was still strongly preferred and more likely to be purchased.

The needs of today's lead users are typically not precisely the same as the needs of the users who will make up a major share of tomorrow's predicted market. Indeed, the literature on diffusion suggests that in general the early adopters of a novel product or practice differ in significant ways from the bulk of the users who follow them. However, in this instance, the product concept preferences and

the probability of purchase measures of lead users and nonlead users were very similar. Furthermore, a comparison of the way in which lead and nonlead users evaluated PC-CAD system attributes showed that this similarity was deep-seated.

Even though this field experiment supports the usefulness of the lead user method, there are certain problematic issues that must be further explored. One problem is accurate trend identification. Currently we rely on a skillful analyst to select an important trend on the basis of judgment (much as product attributes for use in multi-attribute analysis are selected by market research analysts on the basis of judgement and qualitative data). Clearly, it would be useful to improve this method. A second problem is the tacit assumption that the product perceptions and preferences of lead users are or will be similar to non-lead users as a market develops. When this is true, evaluation of the eventual appeal of a lead user product or product concept is straightforward. But what if lead users like the product and non-lead users do not? In this case there are two possibilities: (1) The concept is too novel to be appreciated by non-lead users—but it will later be preferred by them when their needs evolve to resemble those of today's lead users; (2) the concept appeals only to lead users and will never be appreciated by non-lead users even after they "evolve."

And finally, this study focused on the identification and study of naturally occurring lead users. Perhaps lead users can also be created? It is possible for manufacturers to stimulate user innovation by acting to increase user innovation-related benefit. If they can also place users in environments which they judge to foreshadow future general market conditions, they may be able to create lead users. Market research studies which allow users to experience prototypes of proposed new products and then test their reactions are a possible step in this direction.

IMPLEMENTING THE LEAD USER CONCEPT

Over the years, von Hippel and his colleagues have conducted successful lead user studies in many different industries both in the U.S. and in Europe. For example, at Hilti, a leading European manufacturer of components, equipment, and materials used in the construction industry, a lead user study was undertaken to develop a concept for a novel "pipe-hanger" system.[8] This is a type of fastening system used to hang pipes on walls and ceilings in commercial and industrial buildings. By working with lead users, Hilti personnel developed a new pipe-hanger system that was commercially very successful and that also won them an industry achievement award for outstanding product development work. Listed below are the steps they went through in their study.

The Hilti research team first identified a few important need-related trends by conducting telephone interviews with experts in the field of study. Based on their trend analysis they came up with these three market needs:

1. Pipe hangers that are very easy to assemble (the reason being that education levels among installers are going down in many countries);

2. A more secure system of connecting hanger elements together and attaching them to walls and ceilings (the reason—more stringent safety requirements in the industry);

3. Pipe-hangers made from lighter, noncorrodible materials (e.g., plastics rather than commonly used steel elements).

Next they found twenty-two expert users by surveying cooperating firms throughout Europe. The users were all tradesmen who had actually designed, built and then installed hangers that were not commercially available. The list of twenty-two was paired down to twelve lead users who had the richest information to offer. They participated with three Hilti engineers and the marketing manager in a three-day creative problem-solving workshop to develop a system for a novel pipe-hanging system that had the characteristics identified in the trend analysis. The final step in their study was to ask a small sample of "routine" users to evaluate the concept that came out of the workshop. The majority of those surveyed preferred the new concept and

indicated they would be willing to pay a 20% higher price for it, relative to existing systems.

As another example in a much different industry, Lee Meadows and von Hippel undertook a lead user study for a major manufacturer of food products. The company was seeking a new kind of snack. In this study, lead users, nutrition experts, and internal scientists developed a concept for a performance-enhancing cookie designed to appeal to the amateur athlete market. Prior to the study, the client company's market research groups had identified several trends that suggested opportunities for new snack foods. One, for example, was a trend towards more interest in ethnic foods. Another was a growing public interest in healthy foods. A third trend was an increasing interest in workout activities and sports by "weekend athletes." Based on the interests expressed in discussions with Management, it was decided to focus on the trends toward an increasing interest in nutrition and weekend sport activities.

At the start of the study, it was known that nutrition was obviously connected in some ways with athletic performance, but they did not know if nutrition in the form of "snacking" could actually help performance in any way. Thus, they began their work by reading a range of sport magazines (e.g. those aimed at serious amateur runners and weight lifters) and articles by "sport nutrition" experts to determine if there was evidence of a link between certain forms of snacking and improved performance. In their reading they found there was a real basis for the performance-enhancing value of eating some kinds of snacks before, during and/or after athletic activities (e.g. eating certain nutrients after athletic performances could speed recovery of muscles).

Telephone interviews were then conducted with a number of elite athletes, prominent coaches and nutrition scientists to identify an appropriate group to participate with internal scientists in a concept development workshop. Some of those interviewed were olympic athletes, coaches, and scientists associated with their training. The workshop group they assembled included a nutrition scientist who was involved in work with elite Navy fighters, a competitive bike racer, and a winner of national events in weight lifting. Through the interviews, it was found that knowledge about performance enhancing snack foods was segmented between the nutrition scientists and athletes. The scientists knew what ingredients the snacks should contain and something about dose timing. The athletes knew how the cookie should be formulated for easy consumption. To be able to clearly focus on first one and then the other type of information, they decided to run two concept development workshops. One was composed mainly of nutrition scientists and the other was made up primarily of elite athletes.

Workshop participants succeeded in developing a concept for an "olympic cookie" which specified what it should contain and how it should be formulated and packaged. Of course, lead users and the nutrition scientists could only comment on what they cared about and knew. They did not care much how the cookie tasted so the company experts added the taste dimension based on previous studies of more routine snack consumers. Management of the company was very pleased with the concept that came out of the study and planned to introduced it in a line of "healthy snacks."

Typically, lead user studies are carried out by a team composed of both technical and marketing people. It is done before the start of formal product development work. The research process includes four stages, with each stage defined by activities that move the research team forward to a more refined understanding of market needs *and* possible solutions to these needs. The essence of the four stages is described below:

Stage 1: Planning the Lead User Project—Management selects the focus of the innovation effort and determines the resources needed to carry out the lead user study.

Stage 2: Identifying an Important Trend—The core research team identifies a promising need-related trend (by scanning trade literature and interviewing experts in the field of study).

Stage 3: Screening for Lead Users—The team locates a small group of sophisticated users (e.g. 4–6) who "lead" the trend.

Stage 4: Learning from Lead Users—The lead user group and in-house experts co-develop a concept for a new product in a one- or two-day workshop and the appeal of the concept is tested in the general market.

It is our strong belief that there are certain advantages of the lead user method over traditional research approaches. By "traditional" approaches, we mean methods designed to collect data on the *needs* of typical users prior to product development. First, we see the lead user method as enabling research teams to collect more accurate and richer data on future market needs than is possible to get with traditional methods. Recall that by definition, lead users have product needs that are ahead of the general market. Thus, by focusing on this type of user, teams are likely to uncover information on *emerging* product needs—ones that are latent for the majority of the market. The limitation of traditional methods stems from the fact that the "average" user—the type usually selected for analyses—does not have the real work experience required to know or even imagine what products others will want in the future. Lead users are also limited in a similar way. However, because they lead the market in terms of their experiences with existing products, they are in a better position than most users to have insights into future market needs.[9]

Secondly, lead user research typically enables manufacturers to speed up the new product development process. One reason for this is that the output of a lead user study is an actual new product concept, thus cutting down the work required by design engineers. In addition, the whole innovation process is more integrated because technical and marketing people are working collaboratively throughout the lead user study and they maintain their interaction with the external environment as well. As a result, there is much less chance of having product development slowed up later in the process because of misjudging or disagreeing about what the customers want. This heavy involvement with external people is very different from what commonly goes on during front-end market research. Other than one or two focus groups, there is usually no *systematic* interaction with users or experts. Moreover, R&D personnel are typically not directly involved in gathering needs data from users. Even after product development is under way, user contact is likely to be hit and miss, at best. The net effect of the holistic approach employed in a lead user study is to greatly reduce the total expense of new product development. For example, management from Hilti estimated that they were able to cut in *half* both the expense and time of new concept development with the lead user method, versus when they used their traditional methods.

Despite our confidence in the lead user method, we recognize that, for many managers and research teams, it is a very new and therefore, suspect way to go about market research. The "newness" isn't so much in the specific techniques. We find that often teams are familiar with the procedures involved in doing a lead user study. The main difficulty many marketing and R&D personnel have is one of *mindset* about how market research "should" be done. One key aspect of the lead user method that is unfamiliar to many personnel is the idea that sophisticated users can be a source of design data and product ideas, as well as needs and information. Doing a lead user study also requires some managers of the innovation process to make a shift in resource allocation for pre-development market research. Often, managers are skeptical of the value of in-depth market research at the front end of an innovation project, the proverbial *fuzzy* front end, and therefore, do not want to invest much time and money in it. A typical market research reaction is "We can't afford to tie up our best people for this." Unfortunately, many managers still see the "real and important" work as starting with formal product development. They fail to realize that in-depth exploratory market research done early in an innovation project is one of the key ingredients of new product winners.

A RECENT COMPANY EXAMPLE AND ASSESSMENT

The lack of a system to guide product developers who are seeking to create new breakthrough-type products is a problem even for a company like 3M,

long known for its success with innovation. Traditionally, 3M has fostered innovation through its cultural encouragement of exploratory efforts with aphorisms such as: "*Make a little, sell a little*" or "*It's better to seek forgiveness than to ask for permission.*" This cultural reliance led to the creation of a long historical list of profitable breakthrough products, including sandpaper, Scotch tape, Post-it Notes and Thinsulate. But more recently, the company's top managers became concerned that too much of the company's growth was coming from modifications and improvements within existing product lines. Breakthroughs were becoming more rare as safe, predictable benefits from incremental product changes were confining the company too much to its current product mix. To help counter this trend, management set a bold objective—30% of sales should come from products that had not existed four years earlier.

To meet this challenge, the R&D function within many 3M businesses knew it would have to change how it did its work. With the help Mary Sonnack, a 3M Divisional Scientist who had previously spent a year at M.I.T. studying and working with Professor Eric von Hippel, several product development teams in 3M's Medical-Surgical Markets Division became the company's first groups to try to implement the lead user process. The teams were charged with creating breakthrough products. After one year of working with von Hippel's lead user approach, the teams in this business had generated proposals for two new line extensions, one breakthrough product, and a whole new strategic approach for treating infection. The teams were also showing senior management that the lead user process could possibly systematize the company's development of breakthrough-type products.

Learning from Lead Users

Processes designed to come up with new product ideas generally begin with information collected from users. As previously discussed, what differentiates the lead user method is the kind of information collected and from whom they collect it. Product development teams usually work with information from users at the center of their target market, trying to learn what '*typical*' customers might need. The underlying assumption is that users provide information about their needs and R&D's job is to use this data to help brainstorm new product ideas. The lead user process is fundamentally different in that it tries to collect information about both needs and solutions from the *leading edges* of the target market and from markets that face similar problems in more extreme situations. The underlying assumption here is that lead users outside the company have already come up with innovations and that R&D should take advantage of this by tracking down promising lead users and see if they can adapt these ideas and innovations to their own markets' needs.

To find these lead users quickly, the surgical-medical project teams used telephone interviews to *network* their way to experts on the leading edge of their target markets. Team members began by briefly explaining their problem to individuals whom they believed had relevant expertise—for example, authors who had written about the topic. They also asked for referrals to others who might have even more relevant knowledge. Networking is effective because people with a serious interest in any topic tend to know of others who know even more than they do. Based on this iterative approach, it is usually not long before a team reaches lead users at the front edge of what might become the target market. The next step, however, was to continue networking until lead users were found in markets and fields that faced *similar* problems but in *different* circumstances—often industries or conditions in which lead users are able or forced to experiment. Such individuals can then help product development teams uncover truly novel solutions to needs and problems in the target market.

To illustrate, the members of a medical imaging lead user project knew that a major trend in their business was the ability to detect increasingly small features, including very early-stage tumors. The team networked to the leading edge by identifying several radiologists working on some of the most challenging medical-imaging problems. The team discovered that some of these researchers had developed imaging innovations that were way ahead of commercially available products. The team then asked these radiologists for the names of people in *any* field who might be even further ahead in some important aspect of the imaging problems on which they were working. The radiologists identified many such individuals, including pattern recognition specialists and engineers working on fine detail images in semiconductor chips. Having identified lead users from both the target market and from industries and disciplines with analogous problems, the team was able to collect a great deal of new information from which to create more breakthrough-type products. Much of this information was gathered through telephone interviews and on-site visits but more useful information was derived when the team hosted a two-day workshop that included a range of lead user experts as well as key people inside 3M—product developers, marketing specialists, manufacturing engineers, etc. Although the new product concept is based on insights gained from the lead users, it had to be adapted and modified to the business's particular market, manufacturing, regulatory, and financial conditions.

Another interesting and potentially more lucrative example occurred in the surgical drape business, an area in which traditional market research had provided abundant data but had not pointed developers toward a breakthrough. As Rita Shor, a senior 3M product specialist, explained to Mary Sonnack: "Our business unit has been going nowhere. We may be number one in the surgical drapes market, but we're stagnating. We need to identify new customer needs. If we don't bring in radically new ways of looking for products, management may have little choice but to sell off the business." Surgical drapes were an extension of 3M's innovations with masking tape and they work by isolating the area being operated on from other sources of potential infection, including the patient's own body, the operating table, members of the surgical team, etc. But the growing range and resistances of microbes as well as the development of new medical procedures were providing new challenges—for example, surgical drapes don't easily cover catheters or tubes inserted into the patient.

Although 3M's surgical drape products were bringing in more than $100 million in annual sales, the business had not had a breakthrough product for almost a decade. Technological prowess was not the major issue, for in the early 1990s the division had developed technologically advanced disposable gowns for surgeons that would keep them more comfortable while also safeguarding them and their patients by allowing water vapor but not viruses to pass through microscopic pinholes in the gowns' fabric. Unfortunately, this technological feat was developed just as managed care was becoming more pervasive in U.S. hospitals. Surgeons may have loved the fabric, but insurers wouldn't pay for it and so sales were disappointing. In short, the division saw little room for growth in existing markets, margins were declining on current products, and there was little opportunity to expand into less developed countries given the drapes' high costs. Under these circumstances, a cross-disciplinary surgical drape lead user team was assembled to "Find a better type of disposable surgical draping."

The team spent the first few months learning more about the cause and prevention of infections by researching the literature and by interviewing experts in the field. They then held a workshop with management to discuss what they had learned and to set parameters for acceptable types of breakthrough products, including market targets and the kinds of innovations desired by key stakeholders. During the next six or so weeks, the team tried to get a better understanding of important trends in infection control, for one cannot specify what the leading edge of a target market might be without first understanding the major trends in the heart of that market. Much of the team's initial research focused on understanding what doctors in more developed countries needed. As the group asked more questions and talked to more experts, the team realized it didn't know enough about the needs of surgeons and hospitals in countries in which infectious diseases are still major killers. The team broke up into subgroups and traveled to hospitals in Malaysia, Indonesia, Korea, and India to learn how people in less than ideal environments attempted to keep infections from spreading in the

operating room. They were shocked when they saw how many of these surgeons combated infection by using cheap antibiotics as a substitute for disposable drapes and other, more expensive asepsis measures.

The team quickly realized that a potential crisis was germinating in the surgical wards of developing countries. These doctors' reliance on cheap antibiotics to prevent the spread of infection in the short run would eventually result in more drug resistant infections. Furthermore, even if 3M could radically cut the cost of surgical drapes to help overcome this use of drugs, most hospitals in developing countries would simply not be able to afford them. These insights led the team to redefine its goal: find a much cheaper, much more effective way to prevent infections from starting or spreading that does not depend on antibiotics *and might even not depend on surgical drapes.*

After returning from their field visits, the team networked their way with innovators at the leading edge of the trend toward cheaper, more effective infection control, many of whom turned up in surprising places. The team, for example, discovered that specialists in some leading veterinary hospitals were keeping infection rates very low despite facing difficult conditions and cost constraints. As one leading edge veterinarian surgeon explained it, "Our patients are covered with hair, they don't bathe, and they don't have medical insurance, so the infection controls that we use can't cost much." Another surprising leading edge source was Hollywood, where some make-up artists had become experts in applying materials to skin that are non-irritating and easy to remove when no longer needed, attributes that are very important to the design of infection control materials that are also applied to skin.

As a final but important step, the team invited several lead users to a workshop to see how one *"could find a revolutionary, low cost approach to infection control."* Although the group struggled and floundered for a while, the participants pulled together some very creative ideas and solution alternatives toward the end of the session. In fact, the workshop generated concepts for six new product lines, as well as a radical new approach to infection control. The lead user team chose to present to senior management the three product line concepts they felt were the strongest. The first involved a more economical line of surgical drapes that could be made with existing 3M technology, and although this would not be a breakthrough-type product, it would probably be very welcomed in the cost-conscious developing world. The second product concept was for a "skin doctor" line of hand-held devices that would layer antimicrobial substances onto the skin during an operation and vacuum up blood and other liquids during surgery. This product would provide a significant advance in user functionality and could be developed based on existing 3M know-how. The third new product proposal was for an "armor" line that would coat catheters and tubes with antimicrobial protection. This, too, could be created with existing 3M technologies and promises to open up major new market opportunities for the business. Previously, 3M had focused only on products designed to prevent surface infections; this armor line would allow it to enter the $2 billion market aimed at controlling blood-borne, urinary tract, and respiratory infections.

When lead user teams do substantial probing, they often gather information and ideas that point toward the need for a change in strategy. This did in fact take place in this surgical drape team example. In addition to the above novel product introductions, the team uncovered a revolutionary approach to infection control—an approach that would require a change in the division's overall strategy. Previously, the division had always focused on products that were essentially the same for everyone—in a sense, one size fits all. All patients, regardless of the circumstances that brought them there, would get the same degree of infection prevention from the same basic drapes. In the course of their research, the team learned that some people enter a hospital with a significantly greater risk of contracting infection, perhaps because of malnutrition or because they were diabetic. Doctors wanted a way of treating individual patients according to their individual needs through "upstream" containment of infections. They wanted ways to treat people *before* surgery in order to reduce the likelihood of their contracting a disease during an operation. Although the exact nature of this new approach involves proprietary information and therefore cannot be discussed in detail here, the team has recommended this strategic shift to 3M's senior management in this business, who have bought into this approach; it is currently being developed for the marketplace.

A Comprehensive Assessment

Finally, in a recent, somewhat independent attempt to assess the performance of the lead user method for generating new product ideas, a natural experiment was conducted within the 3M Company (see Lilien et al., 2002, for details).[10] Within several participating business divisions, the lead user method was implemented to develop major new product lines, which it succeeded in doing not only in the surgical-medical business as previously described, but also in other business areas, including: (1) electronic test and communications equipment that for the first time would enable isolated field service and equipment repair workers to carry out their problem-solving work as a team; (2) a new approach, implemented though novel equipment, that cuts the application time of commercial graphics films on advertising vehicles such as buses from what used to be 48 hours to less than 1 hour; and (3) a new approach to packaging fragile items in shipping cartons to replace current foamed materials. The new packaging product lines are more environmentally friendly and faster and they are more convenient for both shippers and package recipients than present products and methods.

The study's published findings show that product ideas generated by the lead user process have sales forecasts over the next five years that are eight times greater than the sales projected from a carefully selected sample of comparative funded 3M projects ($146 million annual sales on average versus $18 million). The lead user products and services also have significantly higher forecasted market shares by year five (68% on average versus 33%). And the research also reports that projects funded from the lead user approach were judged to be more novel, more original, and more "new to the world" than ideas from more conventional methods.

Clearly, the lead user methodology is not meant to replace all other means and ways of developing new products through R&D and market research efforts. However, even though it can never guarantee success, the lead user method does open up new avenues for dealing with the difficult problem of developing successful new products and services. It tries to take product development teams and their companies in directions that they might not have considered or imagined. It tries to provide a process that is more systematic to the generation of breakthrough concepts—a process that has been absent in most companies. Perhaps Bill Coyne, 3M's head of R&D, said it best when he reported, "We have now tested the lead user method in eight of our fifty-five divisions. Corporate management is very enthusiastic about the process, and the line of 3M people interested in learning the method from Mary Sonnack's group extends out her office door and around the block."

NOTES

1. T. J. Allen and D. G. Marquis. "Positive and Negative Biasing Sets: The Effects of Prior Experience on Research Performance." *IEEE Transactions on Engineering Management* EM-11, no. 4 (December 1964): 558–614.

2. For example, when Edwin Mansfield (*The Economics of Technological Change* [New York: Norton, 1968], 134–35) explored the rate of diffusion of twelve very important industrial goods innovations into major firms in the bituminous coal, iron and teel, brewing, and railroad industries, he found that in 75% of the cases it took over twenty years for complete diffusion of these innovations to major firms. Accordingly, some users of these innovations could be found far in advance of the general market.

3. Much of this research is reported in E. von Hippel, *The Sources of Innovation*, New York: Oxford University Press, 1988. As recently demonstrated by the open source software movement, it is possible for innovative product users to far outnumber a company's developers. Thousands of users, for example, have participated in the development of very successful products such as the Apache Webserver software, the Perl language, and the Linux operating system. This approach—to mobilize and use the power of thousands of innova-

tive users—has now spread to the development of games for Sony's Playstation and even to the pharmaceutical industry, where Lilly is trying to involve thousands of independent researchers in the synthesis of promising new drugs.

4. See G. Urban and E. von Hippel, "Lead User Analyses for the Development of New Industrial Products," *Management Science*, 1988.

5. PC-CAD system purchase decisions are made primarily by the final users in the engineering department responsible for CAD design of boards. In this study we interviewed only these dominant influencers to find concepts and test them. If the purchase decision process had been more diffuse, it would have been appropriate to include other important decision participants in our data collection.

6. The innovating users reported that their goal was to achieve better performance than commercially available products could provide in several areas: high routing density, faster turnaround time to meet market demands, better compatibility to manufacturing, interfaces to other graphics and mechanical CAD systems, and improved ease of use for less experienced users.

7. The seventeen attributes were: ease of customization, integration with other CAD systems, completeness of features, integration with manufacturing, maintenance, upgrading, learning, ease of use, power, design time, enough layers, high-density boards, manufacturable designs, reliability, placing and routing capabilities, high value, and updating capability.

8. C. Herstatt and E. von Hippel, "From Experience: Developing New Product Concepts via the Lead User Method: A Case Study in a 'Low-Tech' Field," *Journal of Product Innovation Management*, Vol. 8, 1992, pp. 213–221.

9. Most lead users are quite willing to give detailed information to manufacturers simply because they would much rather have someone else take the steps of producing and marketing their innovations on a broad scale and they do not see themselves suffering financial loss by revealing what they know. Most had created an innovative solution for their own use because an acceptable solution to their problem was not available. Obviously, when lead users generate an innovation that gives them a competitive advantage, they will be less willing to share for free what they know. In interviewing lead users or in conducting lead user workshops, product development teams should be up front that their company has commercial interests in the ideas under discussion.

10. G. Lilien, P. Morrison, K. Searls, M. Sonnack, and E. von Hippel, "Performance Assessment of the Lead User Idea-Generation Process for New Product Development," *Management Science*, 2002, Vol. 48, pp. 1042–1059.

54

Shifting Innovation to Users Through Toolkits

ERIC VON HIPPEL AND RALPH KATZ

INTRODUCTION

Research has consistently shown that new products and services must accurately respond to user needs if they are to succeed in the marketplace. However, it is often a *very* costly matter for firms to understand users' needs deeply and well. Need information is very complex, and conventional market research techniques only skim the surface. Techniques that probe more deeply, such as ethnographic studies, are both difficult and time-consuming. Further, the task of understanding user needs is growing ever more difficult as firms increasingly strive to learn about and serve the unique needs of "markets of one" and as the pace of change in markets and user needs grows ever faster. Indeed, firms at the leading edge of these trends are finding that conventional solutions are completely breaking down and that a whole new approach is needed if they are to be able to continue to produce products and services that accurately respond to their users' needs.

Fortunately, an entirely new approach to this problem is being developed in a few high-tech fields. In this emerging new approach, manufacturers actually *abandon* their increasingly frustrating efforts to understand users' needs accurately and in detail. Instead, they outsource key *need-related* innovation tasks to the users themselves after equipping them with appropriate "toolkits for user innovation."

Toolkits for user innovation are coordinated sets of "user-friendly" design tools that enable users to develop new product innovations for themselves. The toolkits are not general purpose. Rather, they are specific to the design challenges of a specific field or subfield, such as integrated circuit design or software product design. Within their fields of use, they give users real freedom to innovate, allowing them to develop producible custom products via iterative trial and error. That is, users can create a preliminary design, simulate or prototype it, evaluate its functioning in their own use environment, and then iteratively improve it until satisfied.

Toolkits for user innovation first emerged in a primitive form in the 1980s in the high-tech field of custom integrated circuit design and manufacturing. In this field, as IC products grew increasingly large and complex, the costs of not understanding user needs precisely and completely at the start of a product design work had grown to punishingly high levels. Many errors due to incomplete or inaccurate specification of user needs were occurring, and the cost of correcting even a single error found late in the design process or during user testing could involve literally months of delay and hundreds of thousands of dollars of extra engineering charges. The introduction of the toolkits approach to the custom semiconductor field has reduced development time by two-thirds or more for products of equivalent complexity (von Hippel 1998). Semiconductor manufacturers' sales of user-designed chips were 15 billion dollars in 2000 (Thomke and von Hippel 2002).

Although toolkits for user innovation are now only applied to the development of a few types of custom industrial products and services, we pro-

A version of this paper was published by *Management Science* in July 2002, pp. 821–833.
Reprinted with permission of the authors.

Eric von Hippel is Professor of Management at MIT's Sloan School of Management. **Ralph Katz** is Professor of Management at Northeastern University's College of Business and Research Associate at MIT's Sloan School of Management.

pose that they will eventually be a valuable product development method for all product types characterized by heterogeneous user demand. As we will see in this selection, the economics of sticky information make toolkits desirable under many conditions, while technical advances in computerization are making them increasingly practical in many fields.

In this selection we begin by explaining the benefits of shifting need-related design activities to users. We then explore how this can be achieved via "toolkits for user innovation" and detail the elements such a toolkit should contain. Finally, we discuss the relationship of toolkits for user innovation to other development methods, and where they can be most effectively applied.

TOOLKITS AND STICKY INFORMATION

The toolkits approach to product and service development involves transferring *need-related* product development tasks from manufacturers to users and equipping the users with tools to carry out those tasks. To understand the utility of such a transfer consider that, to solve a problem, needed information and problem-solving capabilities (also a form of information) must be brought together at a single locus. The requirement to transfer information from its point of origin to a specified problem-solving site will not affect the locus of problem-solving activity when that information can be shifted at no or little cost. However, when it is costly to transfer from one site to another in useable form—is, in our terms sticky—the distribution of problem-solving activities can be significantly affected.

The stickiness of a given unit of information in a given instance is defined as the incremental expenditure required to transfer it to a specified locus in a form useable by a given information seeker. When this cost is low, information stickiness is low; when it is high, stickiness is high (von Hippel 1994). A number of researchers have both argued and shown that information required by technical problem-solvers is indeed often

costly to transfer for a range of reasons. Information stickiness can be due to attributes of the information itself such as the way it is encoded (Nelson 1982, 1990; Pavitt 1987; Rosenberg 1982), and/or it can be due to attributes of information seekers or providers. For example, a particular information seeker may be less able in acquiring information because of a lack of certain tools or complementary information—a lack of "absorptive capacity" in the terminology of Cohen and Levinthal (1990). Also, specialized personnel such as "technological gatekeepers" (Allen 1984; Tushman and Katz 1980; Katz 1997) and specialized organizational structures such as information transfer groups (Katz and Allen 1988) can significantly affect information transfer costs between and within organizations.

In the case of product development, sticky information needed by developers is generated at both product manufacturer and product user sites. Generally, a manufacturer has information regarding solution possibilities and its production process, while users have information about needs and the setting of use. The toolkits approach to product and service development reduces sticky information transfer costs by repartitioning the overall product development task into subtasks, each primarily requiring information from either the user or the manufacturer site. Then, it assigns each subtask to user-based or manufacturer-based problem solvers as appropriate.

Repartitioning of innovation process tasks for this purpose can involve fundamental changes to the underlying architecture of a product or service. Consider, for example, how semiconductor manufacturers shifted to the new toolkits paradigm for custom chip development. Traditionally, manufacturers of custom semiconductors had carried out all chip design tasks themselves, guided only by need specifications from users. And, since manufacturer development engineers were carrying out all design tasks, those engineers had typically incorporated need-related information into the design of both the fundamental elements of a circuit, such as transistors, and the electrical "wiring" that interconnected those elements into a functioning circuit.

The brilliant insight that allowed custom integrated circuit design to be partitioned into solution-related and need-related subtasks was that the design of the chip's fundamental elements, such as its transistors, could be made standard for all custom digital circuit designs. This subtask required rich access to the manufacturer's sticky solution information regarding how semiconductors are fabricated, but did not require detailed information on specific user needs. It could therefore be assigned to manufacturer-based chip design and fabrication engineers. It was also observed that the subtask of interconnecting standard digital circuit elements into a functioning integrated circuit required only sticky, need-related information about chip function—for example, whether it was to function as a microprocessor for a calculator or the voice chip for a robotic dog. This subtask of "wiring" the circuit was therefore assigned to users—the parties already in possession of the relevant need-related information. In other words, this new type of chip, called a "gate array," had a novel architecture created specifically to separate problem-solving tasks requiring access to a manufacturer's sticky solution information from those requiring access to users' sticky need information. Tasks involving sticky solution information were then assigned to chip manufacturers, while those involving sticky need information were assigned to users.

The same basic principle can be illustrated in a less technical context—food design. In this field, manufacturer-based designers have traditionally undertaken the entire job of developing a novel food, and so they have freely blended need-specific design into any or all of the recipe-design elements wherever convenient. For example, manufacturer-based developers might find it convenient to create a novel cake by both designing a novel flavor and texture for the cake body and designing a complementary novel flavor and texture into the frosting. However, it is possible to repartition these same tasks so that only a few draw upon need-related information, and these can then be more easily transferred to users.

The architecture of the humble pizza illustrates how this can be done. In the case of the pizza,
many aspects of the design, such as the design of the dough and the sauce, have been made standard, and user choice has been restricted to a single task only—design of toppings. In other words, all need-related information that is unique to a given user has been linked to the toppings-design task only. Transfer of this single design task to users can still potentially offer creative individuals a very large design space to play in (although pizza shops typically restrict it sharply). Any edible ingredients one can think of—from eye of newt to edible flowers—are potential topping components. But the fact that need-related information has been concentrated within only a single product design task makes it much easier to transfer design freedom to the user.

Once problem-solving and the sticky information needed to perform it have been co-located, the development of new products and services can proceed much more rapidly and effectively. To understand why this is so, consider that problem-solving in general, and development of a new product or service in particular, proceeds via an iterative process of trial and error (Baron 1988; von Hippel and Tyre 1995). User or manufacturer-based designers begin by designing what they think they want. Then, they test the initial solution, find drawbacks, and try again. This iterative process is sometimes called "learning by doing" (Arrow 1962; Rosenberg 1982). When tasks have been subdivided so that the sticky information required to solve them and the problem-solvers are co-located, the need to shift problem-solving back and forth *between* user and manufacturer during the trial-and-error cycles involved in learning by doing is eliminated. Iterative learning by doing is still carried out, but the trial-and-error cycles for each subtask are carried out entirely *within* a user or manufacturer firm.

To appreciate the major advantage in problem-solving speed and efficiency that concentrating problem-solving within a single locus can create, consider a familiar, everyday example: the contrast between conducting financial strategy development with and without "user-operated" financial spreadsheet software.

- Prior to the development of easy-to-use financial spreadsheet programs such as Lotus 1-2-3 and Microsoft's Excel, a CFO might have carried out a financial strategy development exercise as follows. First, the CFO would have asked his or her assistant to develop an analysis incorporating a list of assumptions. A few hours or days might elapse before the result was delivered. Then the CFO would use her rich understanding of the firm and its goals to study the analysis. She would typically almost immediately spot some implications of the patterns developed, and would then ask for additional analyses to explore these implications. The assistant would take the new instructions and go back to work while the CFO switched to another task. When the assistant returned, the cycle would repeat until a satisfactory outcome was found.

- After the development of financial spreadsheet programs, a CFO might begin an analysis by asking an assistant to load up a spreadsheet with corporate data. The CFO would then "play with" the data, trying out various ideas and possibilities and "what-if" scenarios. The cycle time between trials would be reduced from days or hours to minutes. The CFO's full, rich information would be applied immediately to the effects of each trial. Unexpected patterns—suggestive to the CFO but often meaningless to a less knowledgeable assistant—would be immediately identified and followed up, and so forth.

It is generally acknowledged that spreadsheet software that enables expert users to "do it themselves" has led to better outcomes that are achieved faster (Levy 1984; Schrage 2000). The advantages are similar in the case of product and service development. Thus, when custom integrated circuit design is carried out by entirely by manufacturers, users cannot engage in learning by doing with respect to their need and their use environment until a chip has been completely designed by the manufacturer and sample chips have been made available. At that late stage, as was noted earlier, it can cost months and hundreds of thousands of dollars for a manufacturer to incorporate modifications requested by users based upon learning by doing. In contrast, users can learn to identify and correct need-related design errors early, rapidly and at a very low cost if they are equipped with an appropriate toolkit for user innovation. Learning by doing via trial and error still occurs, of course, but the cycle time is much faster because the complete cycle of need-related learning is carried out at a single user site earlier in the development process.

TOOLKITS—A WAY TO TRANSFER DESIGN CAPABILITY TO USERS

In principle, then, when "need-related" design tasks are assigned to users and "solution-related" tasks are assigned to manufacturers, times and costs are compressed, and learning by doing is more effectively integrated into the design process. But users are not design specialists in the manufacturer's product or service field. So, how can one expect them to create sophisticated, producible custom designs efficiently and effectively? Manufacturers who have pioneered in this field have solved the problem by providing users with kits of design tools that can help them to carry out the design tasks assigned to them (von Hippel 1998).

Toolkit development involves "unsticking" manufacturer solution and production information relevant to the development work of user-innovators and incorporating it into a toolkit. This can be done because the stickiness of a given unit of information is not immutable. Rather, it can be reduced by investments made to that end. For example, firms may reduce the stickiness of a critical form of technical expertise by investing in converting some of that expertise from tacit knowledge to the more explicit and easily transferable form of a software "expert system" (Davis 1986), and/or they may invest in reducing the stickiness of information of interest to a particular group of users by encoding it in the form of a remotely accessible computer data base. This is what the travel industry did, for example, when it invested substantial sums to put its various data bases for airline schedules, hotel reservations, and car rentals "on-line" in a user-accessible form.

The incentive to invest in reducing the stickiness of a given unit of information will vary *according to the number of times that one expects to transfer it*. As an illustration, suppose that to solve a particular problem, two units of equally sticky local information are required, one from a user and one from a manufacturer. In that case, there will be an equal incentive operating to unstick either of these units of information in order to reduce the cost of transfer, other things (such as the cost of unsticking) being equal. But now suppose that there is reason to expect that one of the units of information, say the manufacturer's, will be a candidate for transfer n times in the future, while the user's unit of information will be of interest to problem solvers only once. For example, suppose that a manufacturer expects to have the same technical information called on repeatedly to solve n user product application problems, and that each such problem involves unique user information. In that case, the total incentive to unstick the manufacturer's information across the entire series of user problems is n times higher than the incentive for an individual user to unstick its problem-related information.

In the case of the problem-solving work of product and service development, the situation just described is the one encountered when user needs for a given product type are heterogeneous. Under these conditions, manufacturers specializing in a given product type attempt to adapt the same basic approach to the diverse application problems of many users. For example, manufacturers of adhesives will attempt to solve diverse user fastening-related problems with specialized adhesives, while manufacturers of mechanical fasteners will attempt to solve such problems with specialized screws and bolts. The commonality in solution approach means that the sticky information required from a manufacturer to solve each novel application problem tends to be the same, involving such things as the properties and limitations of the solution type. In contrast, the diversity in applications means that sticky information required from users tends to be novel or have novel components. Thus, the higher

the heterogeneity of user needs faced by a manufacturer, the higher its incentive to invest in unsticking problem-related information relevant to user problem-solvers and transfer that information to users in the form of a toolkit for user innovation (von Hippel 1998).

ELEMENTS OF A TOOLKIT

Toolkits for innovation are not new as a general concept—every manufacturer equips its engineers with a set of tools suitable for designing the type of products or services it wishes to produce. Toolkits for users also are not new—many users have personal toolsets that they have assembled to help them create new items or modify standard ones. For example, some users have woodworking tools ranging from saws to glue which can be used to create or repair furniture. Others may have software tools to write or modify software. What is new, however, are integrated toolsets to enable users to create and test designs for custom products or services that can then be produced "as is" by manufacturers.

We propose that effective toolkits for user innovation will enable five important objectives. First, they will enable users to carry out complete cycles of trial-and-error learning. Second, they will offer users a "solution space" that encompasses the designs they want to create. Third, users will be able to operate them with their customary design language and skills—in other words, well-designed toolkits are "user friendly" in the sense that users do not need to engage in much additional training to use them competently. Fourth, they will contain libraries of commonly used modules that the user can incorporate into his or her custom design—thus allowing the user to focus his or her design efforts on the truly unique elements of that design. Fifth and finally, properly designed toolkits will ensure that custom products and services designed by users will be producible on manufacturer production equipment *without* requiring revisions by manufacturer-based engineers.

Learning by Doing via Trial and Error

As was mentioned earlier, it is important that toolkits for user innovation enable users to go through complete trial-and-error cycles as they create their designs. Such cycles begin with the design of a possible solution. The solution is then built (or simulated on a computer), tested and evaluated. If evaluation shows that improvements are needed, the cycle is repeated. For example, suppose that a user is designing a new custom telephone answering system for her firm, using a software-based CTI design toolkit provided by a vendor. Suppose also that the user decides to include a new rule to "route all calls of X nature to Joe" in her design. A properly designed toolkit would allow her to temporarily place the new rule into the telephone system software, so that she could actually try it out (via a real test or a simulation) and see what happened. She might discover that the solution worked perfectly. Or, she might find that the new rule caused some unexpected form of trouble—for example, Joe might be flooded with too many calls—in which case it would be "back to the drawing board" for another design and another trial.

In the same way, toolkits for user innovation in the semiconductor design field allow the users to design a circuit that they think will meet their needs and then test the design by "running" it in the form of a computer simulation. This quickly reveals errors that the user can then quickly and cheaply fix using toolkit-supplied diagnostic and design tools (Thomke 1998). For example, a user might discover by testing a simulated circuit design that he or she had forgotten about a switch to adjust the circuit—and make that discovery simply by trying to make a needed adjustment. The user could then quickly and cheaply design in the needed switch without major cost or delay.

One can appreciate the importance of giving the user the capability for trial-and-error learning by doing in a toolkit by thinking about the consequences of not having it. When users are not supplied with toolkits that enable them to draw on their local, sticky information and engage in trial-and-error learning, they must actually order a product and have it built to learn about design errors—typically a very costly and unsatisfactory way to proceed. For example, custom furniture makers allow customers to select from a range of options for their furniture—but they do not offer the customer a way to learn during the design process and before buying. The cost to the customer is unexpected learning that comes too late: "That style of couch and swatch of fabric did look great in the showroom. But now that the couch has been delivered, I discover that it makes the room feel crowded, and that the color of the fabric clashes with the wallpaper!"

An Appropriate "Solution Space"

Economical production of custom products and services is only achievable when a custom design falls within the pre-existing capability and degrees of freedom built into a given manufacturer's production system. We may term this the "solution space" offered by that system. A solution space may vary from very large to small, and if the output of a toolkit is tied to a particular production system, the design freedom that a toolkit can offer a user will be accordingly large or small. For example, the solution space offered by the production process of a custom integrated circuit manufacturer offers a huge solution space to users—it will produce any combination of logic elements interconnected in any way that a user-designer might desire, with the result that the user can invent anything from a novel type of computer processor to a novel "silicon organism" within that space. However, note that the semiconductor production process also has stringent limits. It will only implement product designs expressed in terms of semiconductor logic—it will not implement designs for bicycles or houses. Also, even within the arena of semiconductors, it will only be able to produce semiconductors that fit within a certain range with respect to size and other properties. Another example of a production system offering a very large solution space to designers—and, potentially to user-designers via toolkits—is the automated machining center. Such a device can basically fashion any shape out of any machinable material that can be created by any combination of basic machining operations such as drilling and milling. As a consequence, toolkits for user innovation intended to create designs producible on automated machining centers can offer users access to that very large solution space.[1]

Large solution spaces can typically be made available to user-designers when production systems and associated toolkits allow users to manipulate and combine relatively basic and general-purpose building blocks and operations, as in the examples above. In contrast, small solution spaces typically result when users are only allowed to combine a relatively few special-purpose "options." Thus, users who want to design their own custom automobile are restricted to a relatively small solution space: They can only make choices from lists of options regarding such things as engines, transmissions and paint colors. Similarly, purchasers of eyeglasses produced by "mass-customization"[2] production methods are restricted to combining "any frame from this list" of pre-designed frames with "any hinge from that list" of pre-designed hinges, and so on.

The reason producers of custom products or services enforce constraints on the solution space that user-designers may use is that custom products can only be produced at reasonable prices when custom user designs can be implemented by simply making low-cost adjustments to the production process. This condition is met within the solution space on offer. However, responding to requests that fall outside of that space will require small or large additional investments by the manufacturer. For example, an integrated circuit producer may have to invest many millions of dollars and rework an entire production process in order to respond to a customer request for a larger chip that falls outside of the solution space associated with its present production equipment.

"User-Friendly" Toolkits

Toolkits for user innovation are most effective and successful when they are made "user friendly" by enabling users to use the skills they already have and work in their own customary and well-practiced design language. This means that users don't have to learn the—typically different—design skills and language customarily used by manufacturer-based designers, and so will require much less training to use the toolkit effectively.

For example, in the case of custom integrated circuit design, toolkit users are typically electrical engineers who are designing electronic systems that will incorporate custom ICs. The digital IC design language normally used by electrical engineers is Boolean algebra. Therefore, user-friendly toolkits for custom IC design are provided that allow toolkit users to design in this language. That is, users can create a design, test how it works and make improvements all within their own, customary language. At the conclusion of the design process, the toolkit then translates the user's logical design into a different form, the design inputs required by the IC manufacturer's semiconductor production system.

A design toolkit based on a language and skills and tools familiar to the user is only possible, of course, to the extent that the user *has* familiarity with some appropriate and reasonably complete language and set of skills and tools. Interestingly, this is the case more frequently than one might initially suppose, at least in terms of the *function* that a user wants a product or service to perform—because functionality is a face that the product or service presents to the user. (Indeed, an expert user of a product or service may be much more familiar with that functional "face" than manufacturer-based experts.)

Thus, the user of a custom semiconductor is the expert in what he or she wants that custom chip to *do* and is skilled at making complex trade-offs among familiar functional elements to achieve a desired end. Thus: "If I increase chip clock speed, I can reduce the size of my cache memory and . . ." As a less technical example, consider the matter of designing a custom hairstyle. In this field there is certainly a great deal of information known to hairstylists that even an expert user may not know such as how to achieve a given look via "layer cutting" or how to achieve a given streaked color pattern by selectively dying some strands of hair. However, an expert user is often very well practiced at the skill of examining the shape of his or her face and hairstyle as reflected in a mirror and visualizing specific improvements that might be desirable in matters such as curls or shape or color. In addition, the user will be very familiar with the nature and functioning of everyday tools used to shape hair such as scissors and combs.

A "user-friendly" toolkit for hairstyling innovation can be built upon on these familiar skills and tools. For example, a user can be invited to sit in front of a computer monitor and study an image of his or her face and hairstyle as captured by a video camera. Then, she can select from a palette of colors and color patterns offered on the screen, can superimpose the effect on her existing hairstyle, can examine it, and repeatedly modify it in a process of trial-and-error learning. Similarly, the user can select and manipulate images of familiar tools such as combs and scissors to alter the image of the length and shape of her own hairstyle as projected on the computer screen, can study and further modify the result achieved, and so forth. Note that the user's new design can be as radically new as desired, because the toolkit gives the user access to the most basic hairstyling variables and tools such as color and scissors. When the user is satisfied, the completed design can be translated into technical hairstyling instructions in the language of a hairstyling specialist—the intended "production system" in this instance.

Module Libraries

Custom designs are seldom novel in all their parts. Therefore, libraries of standard modules that will frequently be useful elements in custom designs are a valuable part of a toolkit for user innovation. Provision of such standard modules enables users to focus their creative work on those aspects of their design that are truly novel. Thus, a team of architects who are designing a custom office building will find it very useful to have access to a library of standard components, such as a range of standard structural support columns with pre-analyzed structural characteristics, that they can incorporate into their novel building designs. Similarly, designers of custom integrated circuits find it very useful to incorporate pre-designed elements in their custom designs ranging from simple operational amplifiers to complete microprocessors—examples of "cells" and "macrocells," respectively—that they draw from a library in their design toolkit. And again similarly, even users who want to design quite unusual hairstyles will often find it helpful to begin by selecting a hairstyle from a toolkit library. The goal is to select a style that has some elements of the desired look. Users can then proceed to develop their own desired style by adding to and subtracting from that starting point.

Translating User Designs for Production

Finally, the "language" of a toolkit for user innovation must be convertible without error into the "language" of the intended production system at the conclusion of the user design work. If this is not so, then the entire purpose of the toolkit is lost—because a manufacturer receiving a user design essentially has to "do the design over again." Error-free translation need not emerge as a major problem—for example, it was never a major problem during the development of toolkits for integrated circuit design, because both chip designers and integrated circuit component producers already used a language based on digital logic. On the other hand, in some fields, translating from the design language preferred by users to the language required by intended production systems can be *the* problem in toolkit design. To illustrate, consider the case of a recent Nestle USA's FoodServices Division toolkit test project developed for use in custom food design by the Director of Food Product Development, Ernie Gum.

One major business of Nestle FoodServices is production of custom food products, such as custom Mexican sauces, for major restaurant and take-out food chains. Custom foods of this type have been traditionally developed by or modified by chain executive chefs, using what are in effect design and production toolkits taught by culinary schools: restaurant-style recipe development based on food ingredients available to individuals and restaurants, processed on restaurant-style equipment. After using their traditional toolkits to develop or modify a recipe for a new menu item, executive chefs call in Nestle FoodServices or other custom food producers and ask them to manufacture the product they have designed—and this is where the language translation problem rears its head.

There is no error-free way to "translate" a recipe expressed in the "language" of a traditional restaurant-style culinary toolkit into the "language" required by a food manufacturing facility. Food factories can only use ingredients that are obtainable in quantity at a consistent quality. These are not the same as and may not taste quite the same as ingredients used by the executive chef during recipe development. Also, food factories use volume production equipment, such as huge, steam-heated retorts. Such equipment is very different from restaurant-style stoves and pots and pans, and it often cannot reproduce the cooking conditions created by the executive chef on his stovetop—for example, very rapid heating. Therefore food production factories cannot simply produce a recipe developed by or modified by an executive chef "as is" under factory conditions—it will not taste the same.

As a consequence, even though an executive chef creates a prototype product using a traditional chef's toolkit, food manufacturers find most of that information—the information about ingredients and processing conditions—useless because it cannot be straightforwardly translated into factory-relevant terms. The only information that can be salvaged is the information about taste and texture contained in the prototype. And so, production chefs carefully examine and taste the customer's custom food prototype, and then try to make something that "tastes the same" using factory ingredients and methods. But executive chef taste buds are not necessarily the same as production chef taste buds, and so the initial factory version—and the second and the third—is typically not what the customer wants. So the producer must create variation after variation until the customer is finally satisfied. In the case of Nestle, this painstaking "translation" effort means that it often takes 26 weeks to bring a new custom food product from chef's prototype to first factory production.

To solve the translation problem, Gum created a novel toolkit of food "precomponent" ingredients to be used by executive chefs during food development. Each ingredient in the toolkit is the Nestle *factory* version of an ingredient traditionally used by chefs during recipe development: That is, it is an ingredient commercially available to Nestle that had been processed as an independent ingredient on Nestle factory equipment. For example, a toolkit designed for Mexican chefs (the first one designed by Nestle) contains a chili puree ingredient processed on industrial equipment identical to that used to produce food in commercial-sized lots. (Each precomponent also contains traces of materials that will interact during production— for example, traces of a tomato "carrier" are included in the chili puree—so that the taste effects of such interactions are also included in the precomponent.)

Chefs using the toolkit of Nestle precomponents to develop new product prototypes do find that each component differs slightly from the fresh components he or she is used to. But these differences are discovered immediately via "learning by doing," and the chef then immediately adapts and moves to the desired final taste and texture by making trial-and-error adjustments in the ingredients and proportions in the recipe being developed. When a recipe based on precomponents is finished, it can be immediately and precisely reproduced by Nestle factories—because now the user-developer is using the same language as the factory for his or her design work. In the Nestle case, testing shows that adding the "error-free translation" feature to toolkit-based design by users can potentially shorten the time of custom food development from 26 weeks to 3 weeks by eliminating repeated redesign and refinement interactions between Nestle and its custom food customers.

DISCUSSION

To this point we have explored why toolkits for user innovation can be valuable, and have developed the contents of a toolkit. We now conclude by discussing the relationship of toolkits to other product development methods, where toolkits will offer the most value, how toolkits can be developed, and the competitive value of toolkits for manufacturers.

Relationship to Other Product Development Methods

Toolkits for user innovation improve the ability of users to innovate for themselves. Users with sufficient incentive to do so can apply toolkits to design products and services that fit their own needs precisely at a lower cost than would otherwise be the case. "Product configurators" used by producers of mass-customized products are similar in intent but less capable than toolkits. They invite product purchasers to configure their own unique product by selecting from lists of options that have been predesigned by the mass customizer. For example, Dell Computer invites visitors to its website to "design your own computer" by making choices among lists of computer components on offer such as monitors and disk drives.

Market research techniques conventionally used for product design such as multiattribute techniques and conjoint analysis have a very different basic purpose. These are used to collect and analyze need and preference information from many individual users. The information is used by manufacturer-based product developers to design standard products that will bring the greatest satisfaction to the greatest number of customers. Products are not designed by users themselves.

Lead user idea generation methods are similar to conventional market research methods in purpose, but allocate idea generation to lead users rather than to in-house developers. Thus, lead user studies begin with market and trend analyses to determine the nature and direction of migration of user preference. Then, potential design solutions are sought from lead users located at the leading edge of important market trends identified. The goal is to incorporate one or a few of these solutions into standard products that will address the preferences of as many as users as possible to the

greatest extent possible (see Selection 53 and/or von Hippel, Thomke, and Sonnack 1999).

The toolkit for innovation approach is complementary to the lead user approach in an interesting way. Some of the users choosing to employ a toolkit to design a product precisely right for their own needs will be "lead users," whose present strong need foreshadows a general need in the marketplace. Manufacturers can find it valuable to identify and acquire the generally useful improvements made by these lead users, and then supply them to the general market. The business model of Stata Corporation illustrates this pattern. Stata Corporation sells a software package for performing complex statistical analyses. The package offers the functions of a toolkit for user innovation, and Stata encourages its customers to create and share new software code for executing novel statistical techniques. The company then selects user developments of interest to many users and adapts and incorporates these into its next product release.

Where Toolkits Offer the Most Value

Toolkits for user innovation are applicable to essentially all types of products and services where heterogeneity of user demand makes custom, "precisely right" solutions valuable to buyers. As market researchers have long known, many markets have high heterogeneity of demand (Franke and Reisinger 2002). The toolkits for the user innovation approach are becoming more attractive in such fields as advances in both computerized design and computerized production technologies progressively reduce the fixed costs associated with the design and production of novel products.

The fixed costs of design are being steadily reduced by the refinement and increased application of computer-aided design tools (CAD). These design tools have sharply reduced the costs of designing a unique product for product producers. When they are simplified and transferred to users in the form of "user-friendly" toolkits described in this selection, they do the same for users. The fixed costs of tooling have been sharply reduced by the introduction of "mass-customized" production methods. These methods involve various combinations of computerized production machines that can be adjusted to produce different outputs near-instantly and at low cost, modular product design and flexible assembly techniques. Manufacturers using mass-customized production can often make even single-unit quantities of custom products at a cost that is reasonably competitive with the costs of manufacturing similar items by traditional mass-production methods (Pine 1993).

We should note that toolkits are not the appropriate solution for all product needs, even in highly heterogeneous markets. They do allow greater scope for users to apply their understanding of a need more directly and thus will generally result in products that "fit the need" better. On the other hand, toolkits will not be the preferred approach when the highest achievable performance on other dimensions is required, because they incorporate automated design rules that cannot, at least at present, translate designs into product or software with the same skill as can a human designer. For example, a design for a gate array generated via toolkit will typically take up more physical space on a silicon chip than would a full-custom design of similar complexity. Even when toolkits are on offer, therefore, manufacturers may continue to design certain products (those with difficult technical demands) while customers take over the design of others (those involving complex or rapidly evolving user needs).

We should also note that the design freedom provided by toolkits for user innovation may not be of interest to all or even to most users in a market characterized by heterogeneous needs. A user must have a high enough need for something different to offset the costs of putting a toolkit to use for that approach to be of interest. Toolkits may therefore be offered only to the subset of users who have a need for them. Or, in the case of software, toolkits may be provided to all users along with a standard "default" version of the product or service, because the cost of delivering the extra software is essentially zero. In such a case the toolkit capability will simply lie unused in the background unless and until a user has sufficient incentive to evoke and employ it.

Development of Toolkits

We have said that manufacturers that offer toolkits for user innovation to their customers are freed from having to know the details of their customers' needs for new products and services. On the other hand, the manufacturer does still have to know the solution space his customers need to be able to design the novel products or services they want. For example, Nestle has to know which 30 ingredients to put into its Mexican sauce design toolkit, even if it does *not* have to know anything about a specific customer's need or anything about the attributes of the sauce that the customer hopes to make.

Fortunately, determining solution dimensions a toolkit must offer does not take superhuman insight on the part of manufacturer experts. Manufacturer-based developers can create a first-generation toolkit by analyzing existing customer products and determining the dimensions that were required to design those. Alternatively, manufacturers can simply modify existing in-house design toolsets to make them more user-friendly, and distribute these as a first-generation toolkit for user innovation. All that is required for initial success is that a first-generation toolkit offer enough functionality to make it valuable to interested users relative to other existing options. As users begin to apply the toolkit to their projects, the more advanced among them will "bump up against the edges" of the solution space on offer and then request the additional capabilities they need to implement their novel designs. Manufacturers can then improve their toolkits by responding to these explicit requests for improvement, and/or they can wait until impatient lead users actually create and test and use the toolkit improvements they need for themselves. Toolkit improvements that prove to be of general value can then be incorporated into the standard toolkit and distributed to the general toolkit-using community just as product improvements developed by lead users can be distributed to the general community of users.

Competitive Value of Toolkits for Manufacturers

Toolkits can create competitive advantages for manufacturers first to offer them. Being first into a marketplace with a toolkit may yield first-mover advantages with respect to setting a standard for a user design language that has a good chance of being generally adopted by the user community in that marketplace. Also, manufacturers tailor the toolkits they offer to allow easy, error-free translations of designs made by users into their own production capabilities. This gives originators a competitive edge even if the toolkit language itself becomes an open standard. For example, in the field of custom food production, customers often try to get a better price by asking a number of firms to quote on producing the prototype product they have designed. If a design has been created on a toolkit based on a Nestle-developed language of precomponents that can be produced efficiently on Nestle factory equipment by methods known best to that firm, Nestle will obviously enter the contest with a competitive edge.

Toolkits can impact existing business models in a field in ways that may or may not be to manufacturers' competitive advantage in the longer run. For example, consider that many manufacturers of products and services appropriately benefit from both their design capabilities and their production capabilities. A switch to user-based customization via toolkits can affect their ability to do this over the long term. Thus, a manufacturer that is early in introducing a toolkit approach to custom product or service design may initially gain an advantage by tying that toolkit to his particular production facility. However, when toolsets are made available to customer designers, this tie often weakens over time. Customers and independent tool developers can eventually learn to design toolkits applicable to the processes of several manufacturers. (Indeed, this is precisely what has happened in the custom integrated circuit industry. The initial toolsets revealed to users by producers of custom integrated circuits were producer-specific. Over time, however, specialist tool design firms such as Cadence developed toolkits that enabled users to make designs producible by a number of vendors.) The end result is that manufacturers that previously benefited from selling their product design skills and production skills can be eventually forced by the shifting of design tasks to customers via toolkits to a position of benefiting from production skills only.

However, manufacturers who project long-term disadvantages that may accrue from a switch to a toolkit-based innovation process will not necessarily have the luxury of declining to introduce one. If any manufacturer introduces the toolkits approach into a field favoring its use, customers will tend to migrate to it, forcing competitors to follow. Therefore, a firm's only real choice in a field where conditions are favorable to the introduction of toolkits is the choice of leading or following.

We conclude by proposing, as we did at the start of this article, that toolkits for user innovation will eventually be adopted by many manufacturers facing heterogeneous customer demand. As toolkits are more generally adopted, the organization of innovation-related tasks seen today especially in the field of custom integrated circuit production will spread, and users will increasingly be able to get *exactly* the products and services they want— by designing them for themselves.

NOTES

1. Note, however, that current computer-aided design and manufacturing software (CAD-CAM) is not equivalent to a toolkit for user innovation. It does not, for example, offer users the ability to conduct trial-and-error tests of the functional suitability of the designs they are constructing.

2. "Mass-customized" production systems are systems of computerized process equipment that can be adjusted instantly and at low cost. Such equipment can produce small volumes of a product or even one-of-a-kind products at near mass-production costs (Pine 1993).

REFERENCES

Allen, T. (1984). *Managing the Flow of Technology.* Cambridge, MA: MIT Press.

Arrow, Kenneth J. (1962). "Economic Welfare and the Allocation of Resource of Invention," in Richard R. Nelson (Ed.), *The Rate and Direction of Inventive Activity: Economic and Social Factors,* A Report of the National Bureau of Economic Research. Princeton University Press, Princeton, NJ, pp. 609–625.

Baron, Jonathan. (1988). *Thinking and Deciding.* Cambridge University Press, New York.

Cohen, Wesley M., and Daniel A. Levinthal. (1990). "Absorptive Capacity: A New Perspective on Learning and Innovation," *Administrative Science Quarterly,* 35, 1, March, pp. 128–152.

Davis, Randall. (1986). "Knowledge-Based Systems," *Science,* 231, 4741, February 28, pp. 957–963.

Franke, Nik, and H. Reisinger. (2002). "Remaining Within-Cluster Variance in Cluster Analyses: A Meta-Analysis." Working Paper, University of Vienna and Vienna University of Economics and Business Administration.

Katz, Ralph. (1997). *The Human Side of Managing Technological Innovation.* Oxford University Press, New York, Chapters 18 and 35.

Katz, Ralph, and Thomas J. Allen. (1988). "Organizational Issues in the Introduction of New Technologies," in Ralph Katz (Ed.), *Managing Professionals in Innovative Organizations.* Harper Business, Cambridge, MA, pp. 442–456.

Levy, Steven. (1984). "A Spreadsheet Way of Knowledge," *Harpers,* 269, 1614, November, pp. 58–64.

Nelson, Richard R. (1982). "The Role of Knowledge in R&D Efficiency," *Quarterly Journal of Economics,* 97, 3, August, pp. 453–470.

Nelson, Richard R. (1990). "What Is Public and What Is Private About Technology?" Consortium on Competitiveness and Cooperation, Working Paper No. 90-9, Center for Research in Management, University of California at Berkeley, Berkeley, CA, April.

Ogawa, Susumu. (1998). "Does Sticky Information Affect the Locus of Innovation? Evidence from the Japanese Convenience-Store Industry," *Research Policy,* 26, pp. 777–790.

Pavitt, Keith. (1987)."The Objectives of Technology Policy," *Science and Public Policy,* 14, 4, August, pp. 182–188.

Pine, Joseph B. II. (1993). *Mass Customization: The New Frontier in Business Competition.* Harvard Business School Press, Cambridge, MA.

Rosenberg, Nathan.(1982). *Inside the Black Box: Technology and Economics.* Cambridge University Press, New York, p. 131.

Schrage, Michael. (2000). *Serious Play: How the World's Best Companies Simulate to Innovate.* Harvard Business School Press, Cambridge, MA, p. 38.

Thomke, Stefan. (1998). "Managing Experimentation in the Design of New Products," *Management Science,* 44, 6, June, pp. 743–762.

Thomke, Stefan. and Eric von Hippel. (2002). Customers as Innovators: A New Way to Create Value," *Harvard Business Review,* April, pp. 74–81.

Tushman, Michael, and Ralph Katz. (1980). "External Communication and Project Performance : An Investigation into the Role of Gatekeepers," *Management Science,* 26, 11, November, pp. 1071–1085.

von Hippel, Eric. (1994). "Sticky Information and the Locus of Problem Solving: Implications for Innovation," *Management Science* 40, 4, April, pp. 429–439

von Hippel, Eric. (1998). "Economics of Product Development by Users: The Impact of 'Sticky' Local Information," *Management Science,* 44, 5, May, pp. 629–644

von Hippel, Eric, and Marcie, Tyre. (1995). "How 'Learning by Doing' Is Done: Problem Identification in Novel Process Equipment," *Research Policy,* January, pp. 1–12.

von Hippel, Eric, Stephan Thomke, and Mary Sonnack. (1999). "Creating Breakthroughs at 3M," *Harvard Business Review,* September–October, pp. 47–57.

18

Organizational Frameworks for Innovation and New Product Development

55

Modular Platforms and Innovation Strategy

MARC H. MEYER AND ALVIN P. LEHNERD

INTRODUCTION

This selection focuses on the relationships among product architecture, product strategy, and the organization of product development. You should be convinced by the end of this selection that these three elements are powerfully related and have at your disposal a management framework to define that integration for your own company.

In the first section, we will look at the common subsystems that may be shared between products and across product lines, referred to by some as *modularization* and by others as *platforms. Modular, shared subsystems are product platforms*. We describe how modular platforms apply to physical, assembled and nonassembled products. We will also show how the concept applies to nonphysical products such as software and services.

Reprinted with permission of the authors.

Marc H. Meyer is the Sarmanian Professor of Entrepreneurial Studies at Northeastern University's College of Business Administration. *Alvin P. Lehnerd* is an executive in residence at Northeastern University after a successful career at Black & Decker and Steelcase. Together, Meyer and Lehnerd have applied these methods to many corporations across a wide range of industries.

Flowing from this discussion then comes the strategic implications of modular platforms. Many management teams use acquisitions to achieve growth, especially when they fear they can't meet publicly announced revenue or growth targets. However, if a management team believes that the same end can be better achieved through organic development, e.g., from within, then it can be much faster and more cost effective to leverage existing subsystem technology into new products and services, as well as new market applications for those technologies. In the second part of this selection, we will therefore examine strategy implications of modular platforms. We will consider new products and services. We will also consider new markets and, further, new business models. Modular platforms, if carefully managed, allow the firm to be innovative in many ways.

We will then focus, in the third part, on organizing the firm for new product development. Here, we propose an approach that blends teams focused on creating modular platforms with other teams that integrate these platforms and add value to them to create specific products and services. Unfortunately, companies are far too often organized *not to share* anything between product development teams, or between those teams and other functions such as marketing. We would further caution that mass customization can be profitless unless built upon a robust foundation of modular platforms in engineering and manufacturing.

And in the final portion of the paper, we will discuss how to begin to apply this thinking to your own company through a series of questions that a management might ask itself in the areas of modular architectures, new product strategy, and organization design. Also, if you should have questions or wish to learn more, please visit Meyer's Web site (www.marchmeyer.com) for links to more examples, planning methods, and additional relevant articles.

DEFINING MODULAR PLATFORMS

The first step is to define product line architecture in a way that is useful for considering the design and evolution of technology intensive products, systems, and services:

> A product line architecture is that combination of subsystems, the interfaces between these subsystems, and the interfaces to external systems that collectively serve as the foundation for a stream of specific products.

Product lines may be thought of as having distinct generations of architecture, where subsystems and their interfaces are either added or removed to improve overall functionality, and where all subsystems and interfaces are refreshed with newer technology created by the firm, its partners, or formulated as standards within the industry.

Product line architecture is most often expressed in the form of a higher level "block diagram" that shows the major subsystems and subsystem interfaces with the current generation product line. Not all architectural diagrams are alike in quality. Good ones show robustness, meaning a combination of subsystems that can serve as the basis for a host of derivative products targeting specific market applications. Good architectures also show those subsystems where the company seeks proprietary competitive advantage, e.g., specific subsystems and interfaces where unique and potentially patented solutions prevail. It also shows how the products and systems of other companies can connect to the architecture or how other companies can build value-added products and services. Good product line architectures such as these are merely the reflection of sound and strong minds, the crown jewels of the technology-intensive corporation.

With this definition of product line architecture, we now consider the subsystems within product line architectures:

> A product platform is a subsystem or interface that is physical or in some other tangible way expressly used in more than one product. The most effective platforms are those modules that are common to and leveraged across several or more distinct product line architectures.[1]

The conceptual distinction between a platform-focused firm and one that is not may be best visualized by contrasting Figure 55.1 and Figure 55.2. Figure 55.1 characterizes a company with multiple product lines. Each product line (i.e., A, B, C & N) has its own distinct product line architecture (those subsystems and interfaces common to all the products within the product line). Further, as we look across most of these product lines, most share some logical subsystems or modules within their respective product line architectures. Yet, as an inventory of the actual implementation of these subsystems is performed, it becomes clear that each product line has its own implementations of what should be common modules.

Figure 55.1 reflects the reality of most large corporations. Research has shown that successful companies tend to diversify into closely related product and service categories, as opposed to unrelated holding company type diversification.[2] Executives also want to build powerful channels of distribution through which to sell the corporation's expanding mix of products and services. And yet, when it comes to the engineering of the products, systems, and services with that mix, each product line is typically treated as a separate business. Teams are typically housed in different locations, and rarely are provided opportunities to interact in substantive ways. Their inter-product line engineering meetings focus on issues other than the internals of their own respective product line architectures, for to do that might sacrifice individual control over design decisions, and therefore, potentially delay product development cycles. Discussions of longer-term R&D requirements and engineering management processes are far safer. The result is that a substantial amount of what might be common between product lines *is not!*

Figure 55.2, on the other hand, shows a firm that leverages common subsystems not only within product lines, but also across them. Each product line has its own distinct architecture, defined as the

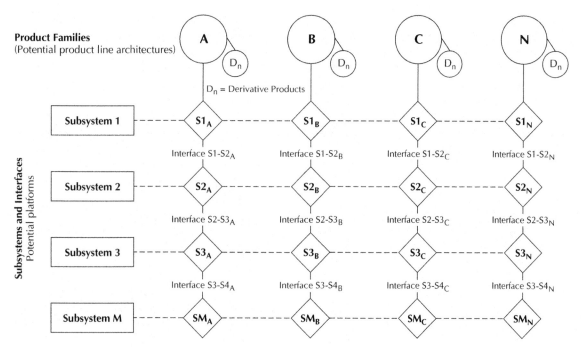

Figure 55.1. Modularization: commonality of subsystems and interfaces.

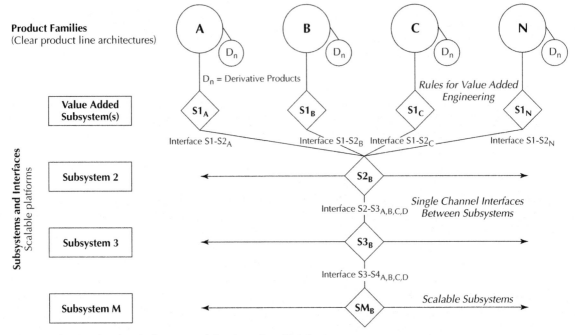

Figure 55.2. Creating shared subsystems and focusing value added developments.

major subsystems and interfaces that are the foundations of all the specific products within that product line of business. Over time, that architecture evolves to meet new user requirements and embrace new core technology. Each version of that architecture may be referred to as a generation of the product line. In well-managed firms, that architecture changes with a frequency that is at least abreast of the product line generational changes of competitors. In poorly managed firms, product architectures are not changed regularly and become dated, incapable of accommodating advances in either subsystem or interface technology. Also, in well-managed firms, the value added engineering performed for individual products is a portion of the total engineering effort, because the majority of the technology for single products comes from common existing platforms.

Within Figure 55.2, the power of the platform discipline occurs when a substantial percentage of all the modules within and between product line architectures are common. When this occurs, the sub-systems themselves become *the platforms*. An interface between subsystems can also become a platform if that interface is used widely across multiple products as an inherent part of the product line architecture. This can be a very powerful technology strategy if the interface is sufficiently robust to allow engineers to migrate to increasingly powerful generations of subsystem technology without violating the overall product line architecture. Think of successful software product companies. The interfaces, either to the user, to other systems, or internally between subsystems, define the strength and durability of that company's products over time. Microsoft is the most visible example. It has standardized on graphical user interfaces and programming interfaces across its Office applications, operating systems, and programming/database tools. This has allowed the company to migrate its various offerings from desktop, to Internet, to repository-centric modalities without significant loss of market share.

We have seen instances where the degree of

commonality is 100%, and variation for different groups of customers is achieved through *services* provided with the product. The more typical scenario, however, is that 70% to 90% of all the engineering required for a new product development within an established product line is actually the integration of pre-existing subsystems based on pre-existing interfaces. Over time, the functionality, performance, and cost of these common subsystems are themselves improved by other parts of the corporation and, frequently, by the firm's key suppliers. The remaining 10% to 30% of a specific new product development effort is the creation of new value-added technology. The more complex the system, the more advantageous the sharing of modular platforms can be, simply because complex systems tend to involve more subsystems and subsystem interfaces than simpler systems. Increasing reuse through modular platforms reduces cost more dramatically. And, as our opening case suggests, modular platforms provide a focused avenue for development teams to increase performance by using the latest available technologies in the design of their product platforms.

The italicized words in Figure 55.2 represent three driving concepts behind this way of thinking. The first of these labels is *scalable designs* for the shared subsystems. By scalability, we mean the range of performance and functionality that the subsystem is designed to provide. That range will generally improve as the subsystem evolves over time. The design targets for that range are not merely grabbed from thin air. Rather, the engineering team carefully defines these requirements by looking at the range of market applications for which the shared subsystem is intended for integration with end-use products.

A very simple example would be cooling subsystems for a line of computing equipment. In a non-platform focused company, the workstation, mid-range server, and high-end server cooling system designers all design their own unique cooling subsystems. In contrast, a platform-focused company has one team designing a cooling subsystem that can serve all three market applications. Its storage team works in the same way, designing scal-

able disk array devices. Electronics are designed as a package to support all three product lines. And, the company's operating system team develops or adapts an operating system that functions and integrates across all three types of computers. One will find that all leading computer manufacturers have transitioned to the shared subsystem approach for new product development. For each, however, maintaining that discipline in the face of new, emerging market requirements remains a constant battle.

Similar issues are in the second set of italicized labels in Figure 55.2: *single channel interfaces to and from subsystems*. By this we mean that there is one clear method or process for connecting to that subsystem for any other subsystem that is clearly documented and strongly enforced. Unfortunately, as engineers add new functionality to other subsystems, they tend to build new, custom interfaces between these subsystems. Over time, this multiplicity of interfaces becomes a tremendous impediment to both new subsystem innovation and value added product development. For example, let us imagine a storage systems vendor that needs to add a new networking adapter to handle Internet Protocol enabled SCSI (iSCI). Rather than change just one subsystem (and introduce the new code on one printed circuit board), numerous subsystems, their microcode, and multiple printed circuit boards need to be changed. Development time cycles and costs skyrocket because even the simplest product enhancements become hard to do. To prevent this undesirable state of affairs, a clear interface discipline needs to be developed and enforced across all teams.

The third set of italicized labels, *set value added engineering constraints*, refers to the management of resource allocation for new product development. Do rules exist that enforce a discipline of using existing platforms? Or, are engineering teams allowed a free hand to create new products "fresh," from top to bottom, without any constraints on using the corporation's established product platforms? If constraints do exist, then there is a process by which product teams can demand periodic enhancements from the platform teams so that appropriate subsystems can have common components across all product lines? Basic ground rules are essential. We have seen some executives set the "75%" rule, e.g., 75% of all components and content in a division's products will be pooled from the repository of common subsystems and interfaces. This ground rule will vary industry by industry. Using common platforms, product engineering teams then create specific products and services by adding value in the form of additional functionality, service, financial packaging, or by forming relationships with external firms for ancillary "plug-in" technology or services.

Together, these three approaches to managing architecture—scalable subsystems, single subsystem interfaces, and rules for enforcing commonality—will drive new product development teams toward the modularization of their product lines.

MODULAR PLATFORMS IN ACTION

We have created a *design matrix* that provides a conceptual framework for integrating specific features within product platforms to target market applications. This is shown in Figure 55.3. Again, it is important to remember that we view the subsystems as the product platforms—elements to be shared across different products—and that we refer to the overall assembly of these pieces of subsystem technologies as the product line architecture.

At the top part of Figure 55.3 are the firm's target market applications. On the left axis of the market segmentation grid are levels of good, better, best, or, more specifically, different types of buying behaviors and preferences. The *value buyer* places cost above all else; the *premium buyer* places performance and functionality ahead of price; and the *standard buyer* is one who looks for the blissful combination of good performance and reasonable price. Across the other axis of the market segmentation grid are the potential groups of users within different vertical markets and/or different geographies.

When defining product strategy, it is particularly important to have an expansive view on this dimension of the segmentation grid. Experience has shown that enterprises achieve sustained growth and profitability by consistently *leveraging their skills and technologies into new market applications*. Witness Honda's application of small engine technology to motorcycles, automobile, lawn mowers, generators, and marine applications, among others. Or, in systems, witness Hewlett Packard's historical march from industrial process control equipment, to medical equipment, to computers and servers, to computer peripherals, to computer services. In services, corporations also sustain growth by finding new types of users for core services domestically and abroad. State Street, for example, first managed the back-office bookkeeping for security purchases by pension funds in the United States. It then applied these processes to mutual funds and, next, to both pension and mutual fund managers around the world.

In the bottom part of Figure 55.3 is a multidimensional grid where each axis stands for one of the major subsystem platforms within the product line architecture. Each subsystem typically has different levels of performance or functionality. In systems, for example, we think of this as the scalability of a subsystem or as the additional functionality created through "plug-ins" or "add-ons." Some subsystems may have only one or several levels of functionality. Other subsystems, by the very flexibility of hardware and software, may have a dozen distinct levels of functionality. Today, most engineering managers in most technology-focused industries have structured subsystems to feature a gradation of functions and performance. The framework described here is a natural extension of that discipline.

The method of applying modular platform to quickly exploit new opportunities is to integrate these two parts of Figure 55.3. In the specifics of the figure, we show four hypothetical point integrations. Point A is of a price-driven buying group in one particular vertical market. The ideal solution for that type of customer is to combine entry-level versions of all four subsystems within the bottom part of the figure. We therefore take Level 1 for Subsystem 1, Level 1 for Subsystem 2, Level 1 for

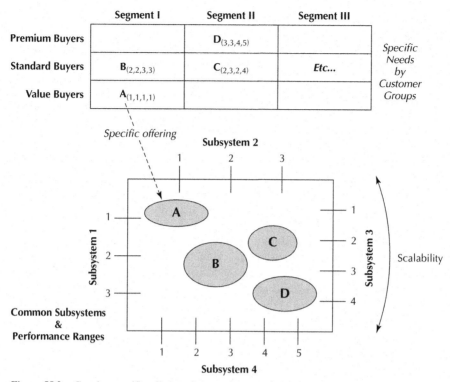

Figure 55.3. Creating specific offerings from a subsystem design matrix.

Subsystem 3, and Level 1 for Subsystem 4 from the design matrix and apply these to the target market application. The second target application, a "standard buyer" expressed as Point B, requires higher levels of functionality across the different subsystems (Level 2 for Subsystem 1, Level 2 for Subsystem 2, Level 3 for Subsystem 3, and Level 3 for Subsystem 4). We include Point C here to illustrate that even users with similar buying behaviors may vary in specific feature preferences across target vertical markets. In other words, the market applications of Point B and Point C are similar but different. Last, a premium buyer group is expressed in Point D. The high levels of functionality (and usually price) for all four subsystems are factored into the composite design of that market application. All target applications are then conveniently displayed on the market segmentation grid with their respective design matrix values in parenthe-

ses. Various individuals involved in product strategy and product management can use this framework to see the "whole picture" for their respective product lines.

There are, of course, a number of strategic and market related research efforts required to uncover and prioritize target market applications. Market growth rates with each segment, competitive dynamics, and available channels for distributing new products are the major types of research needed to stage its attack plan across the market segmentation grid. Other processes are required to elicit user needs correctly and validate these across the broader segment. The design matrix comes into play as both a feasibility assessment technique (if we were to build it, do we have the technologies in place?) and also as an execution tool for managing the design of specific products once strategic direction is established.

Several wonderful examples of corporations applying modular platforms exist in the current literature. One of my favorites is that of Scania trucks, provided by Johnson and Broms.[3] Over the years, engineers at Scania have created a wide variety of trucks based on four basic subsystems: the engines, the transmission, the truck cab, and the chassis. Engineers designed these subsystems initially with performance levels matching target market applications. For example, the truck engine had to meet the needs of hauling requirements (payload) and driving conditions (grades and distances) of the European truck fleets. Over the course of time, Scania enhanced its engine subsystem to meet new needs for power, fuel economy, and emissions and to achieve these new features with the minimum number of engine configurations. This discipline was also applied to transmissions, cabs, and chassis. The results were just 3 basic versions of the engine subsystem, 4 for the transmissions, 4 for the cabs, and 12 chassis. These were the *scalable shared* subsystems that we identified in earlier in Figure 55.2.

Standardized interfaces—the *single channel interfaces* to and from subsystems described in Figure 55.2—were also essential for Scania. Once again, the subsystem engineering teams designed single design interfaces that were nonetheless flexible. The cab would have its designed interfaces. If the engine team designed new shapes or castings for a new generation of its engine, that engine still had to conform to the cab team's mounting interfaces. In this way, each subsystem team could advance its own subsystem to meet new customer needs without violating the overall architecture of the entire truck system. With these underlying platforms, Scania was able to build an incredible variety of trucks with far fewer components than its competitors, which by one benchmarking study was shown to have 50% fewer components. Buying fewer components, but in far greater volume, obviously impacts of cost of goods. At the same time, Scania's engineers could focus all their energies on making the shared subsystem platforms the best building blocks for truck development in the world.

Setting value added engineering constraints is also an important part of this discipline. Scania identifies where a customer's ideal truck design fits into its current design matrix. Because of the quality and scalability of its subsystems, most customers' requests are served with current technology. Value added comes in the form of various accessories, comfort features, and tracking systems that might be added to the cab subsystem. If a customer has a request that exceeds the performance of an existing subsystem, such as the engine, Scania adds that functionality to the "next generation" of that subsystem and makes it available to all customers. A company such as Scania can also say "no" to a particular request, deeming it too unique, the size of the order too small, and hence, a distraction from profitable growth. Or, if the company is in the systems or software business, it can handle the unique request through its professional services group, pricing on a completely business model.

The larger purpose of the framework shown in Figure 55.3 is to *create and sustain value-cost leadership*. Using "off the shelf" scalable subsystems allows the company to serve its customers and also minimize cost of goods. Modular platforms allow a company to provide distinctive value at competitive, distinctive cost. That cost might not be the lowest cost for the particular product category; for example, in the case of Scania, another manufacturer might produce less expensive trucks. Arguably, however, there is no company in the world that can make trucks as powerful and as capable as Scania and offer them at the price points provided by Scania.

Modularization also has downstream benefits over the life cycle of a product line. At the simplest level, when the product or system fails, the user and engineer can work together more effectively to determine the source the problem and fix it. At a more complex level is the cost of maintaining the product and keeping it useful over extended periods of time. This life cycle cost advantage can be seen in the case of leading medical equipment manufacturers, who must upgrade very expensive medical imaging devices without replacing the entire system (no one wants to discard a million dollar piece of hardware into the garbage bin!). The way these device manufacturers achieve upgrades is through modularization of subsystems. In ultrasound devices, for example, manufacturers can replace image acquisition elements (transducers) with newer and better ones through simple plug-in interfaces. Software upgrades to take advantage of these probes are provided on disks or as microcode on printed circuit boards that can be inserted into computer processing subsystem of the device. To the user, the upgrade is incremental, even though the functionality may be an order of magnitude better.

There is even greater reward by taking modular platform design beyond single product lines. *Modular subsystems that are then leveraged across products within a product line, and then across product lines themselves, buy you a lot.* These advantages include lowering cost of goods for manufactured products, significant reductions in product development cycle times, and greater product variety from common assets.

Another rich example lies in how Black and Decker revolutionized the manufacturing of power tools by treating motors, armatures, power cords, and switches as its platforms.[4] A single "universal" motor, for example, was designed to serve the needs of all consumer power tools up to 650 watts in power. It also had a plug-in interface to the rest

of the motor assembly, and the entire assembly itself was designed for automated balancing. Prior to these innovations, the firm had over 120 different motors for its various consumer power tools, and these were attached manually to the motor assembly and also balanced through operator assisted machinery. Figure 55.4 shows the before and after snapshots of the motor design, together with the scalability of the new design of changing length while maintaining width. This subsystem allowed the manufacture of a wide range of motors in a single production process, using the same materials and the same quality control process. That motor is the classic product platform, combining a scalable engineering design with automated manufacturing processes to drive down cost of goods and improve quality. This approach was applied to all the other major subsystems of the respective product architectures for drills, sanders, circular saws, and jigsaws. The effect of the platform approach was to dramatically lower Black and Decker's cost of goods. Management used this advantage to bring a tremendous price discontinuity to the market. The vast majority of the company's competitors were driven off the shelves. Black and Decker continues this discipline today, sharing many subsystems between its consumer power tool line and its higher powered DeWalt line.

During the 1990s, IBMs revitalized its hardware business by rigorous implementation of the

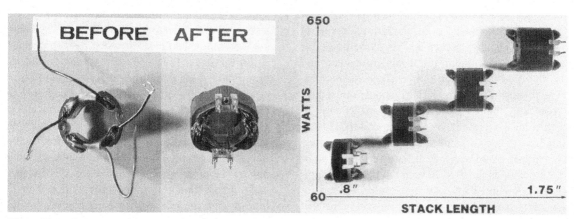

Figure 55.4. The power of scalable subsystem design.

platform discipline that we have described above.[5] IBM was in total financial crisis at the beginning of that decade, reporting a loss of approximately $8 billion in 1993. Senior management challenged the different hardware "server" groups to reduce operating costs through a program of standardization, complexity reduction, and commonality in parts. Reuse of common parts hardly existed at the start of this initiative. For example, a study of 1500 server products revealed 12,000 features, 103,000 bills of materials, and 540,000 active part numbers! The systems running the business were largely older, legacy systems that could not share data readily. There were 59 different floor management systems in the factories, 25 different material logistics systems, and 12 different purchasing systems.

IBM went about the process of platform renewal. Each major server group still developed its own product line architecture for its product families (mainframes, workstations, PC servers, and central storage systems). IBM refers to these common architectures as "reference architectures." However, strict processes and rules were imposed on teams within each product family to use common electrical, semiconductor, and other computer-related components within different reference architectures. Over time, this reuse had a telling effect. IBM reported a 3% reduction in base manufacturing cost for its different computer lines and enjoyed a 15% reduction in component procurement costs as a result of this initiative, savings that continue to provide billions of dollars in increased earnings. It was also spending 42% less on new product development even though its revenues were considerably higher and its cycle times for developing new mainframes, minicomputers, and workstations had dropped from years to months. The application of the platform concept to reduce costs and improve cycle times was central to achieving the tremendous turnaround at IBM! Having always been organized in the past as a separate functional group serving all hardware lines, marketing was also moved in with the hardware engineering groups. This was an essential change for enabling the company to build new systems that would accommodate the rapidly evolving E-business needs of its core customers groups in financial services, manufacturing, and wholesale/retail. Today, IBM is working toward a single, scalable product line architecture as the foundation for all of its server product lines. There is perhaps no more dramatic example of shared subsystems and interfaces within industry today.

THE ROLE OF MODULAR PLATFORMS IN FORMULATING A GROWTH STRATEGY

The integration of methods for segmenting markets with methods for the design and development of modular, reusable modules can be central to a company's formation of a competitive strategy. In this section, we will discuss this connection between platforms and competitive strategy by presenting examples drawn from traditional products, computer systems, software, and services.

The larger issue here is the constant challenge of redefining the firm's *strategic focus* to embrace growth opportunities. We define strategic focus in very pragmatic terms: the types of customers that it targets and the product and service offerings provided to them. Strategic focus must also be seen as dynamic, but at the same time, not chaotic. To grow, firms must consider new types of customers and new types of product and service solutions, balancing the cost of developing and marketing these solutions with the upside of potential new revenues. At the same time, constantly changing from one focus to another serves only to dilute resources and confuse staff and customers. Growth means leveraging technologies, products, and services to new, related market applications in an orderly, step-by-step manner.

Firm maturity is also an important consideration. For new firms, defining that focus in terms of customer groups served and the technologies used to serve them can be particularly problematic. Technological entrepreneurs have a tendency to take any customer that "fogs up the mirror," without regard for the dilution that this lack of focus has on the development of branding, distribution, and support capabilities.[6] In other words, the ambitious entrepreneur wishes to become a multi-market, multi-product company all at once, and perhaps raised his or her venture capital with a plan based on such a promise. Strategic focus requires a discipline in planning and control that many new firms only acquire years after startup.

Older firms, on the other hand, often have the opposite problem. Having focused for decades on specific types of customers, with specific types of solutions, management is reticent about venturing into new, related market applications. The organization bureaucracy, efficient at creating incremental variations on existing product architectures and manufacturing processes, is equally reticent about retooling its products and processes to create modular platforms that are needed for timely and cost-effective development of new market applications. This lack of flexibility in both thinking and in execution leads to several undesirable outcomes that stand in the way of enterprise growth. The first is that traditional market leaders find themselves suffering the classic competitive "squeeze"—attacked "from above" by new entrants employing better product technologies and redefined premium product expectations in the market and attacked "from"

below" by existing competitors who have forcefully driven down cost of goods with scalable product and process platforms that have redefined the low-cost "value" segment in the market. Most often, the traditional market leader sees its market slowly erode, not enough year by year to sound the bells of alarm, but over the course of a decade the cumulative effects are potentially disastrous.

The thinking process for considering the role of modular platforms in the context of growth strategy starts with the simple market segmentation grid, an example of which is shown in Figure 55.5. Such a grid should show not only those customer groups that the firm presently targets, but also groups that it could target. The new technology firm uses the segmentation grid to decide *who not to sell to*, whereas the established corporation uses the grid to consider *where to grow*. The example shown in Figure 55.5 is one developed by a company participating in the construction products industry. If you examine the figure closely, you can see that the major market segments identified by

	New Home Building			Remodeling & Repair			Commercial			
	Custom Builder	Volume Builder	Small Volume Builder	Full Service Remodeler	Self Repair	Specialty Repairs	Retail	Schools	Multi-Family	Industrial
Units (in millions)	0.6	0.2	0.2	1.1	0	0.2				
Growth Rate	3%	3%	4%	3%	1%	2%				
Market Leaders	Fragmented	Fragmented	Fragmented	Firm C	Fragmented	Fragmented	2.1 / 1% / Firm K, Firm L	3.6 / 3% / Firm K, Firm L	1.4 / 5% / Firm K	0.7 / 3% / Firm L
Units (in millions)	1.5	0.5	0.4	3.5	1.8	0.8				
Growth Rate	−4%	Flat	Flat	Flat	Flat	Flat				
Market Leaders	Firm A	Firm A	Firm A	Firm A	Firm A	Firm G				
Units (in millions)	0.6	0.7	0.3	2	1.3	1.1				
Growth Rate	4%	1%	Flat	3%	Flat	1%				
Market Leaders	Firm B	Firm A, Firm B	Firm A, Firm B	Firm B	Firm B	Firm B	5.8 / 1% / Firm K, Firm L	6.0 / 3% / Firm K, Firm L	2.5 / 5% / Firm G	1.4 / 3% / Firm K, Firm L
Units (in millions)	1.1	1.2	0.5	4	2.6	2.1				
Growth Rate	2%	8%	4%	Flat	−1%	4%				
Market Leaders	Firm D	Firm E, Firm D	Firm D	Firm F	Firm F	Firm F				
Units (in millions)	1.9	5.8	1.9	8.6	5.5	6.2	6.5	2.1	1.6	2
Growth Rate	−1%	−3%	−1%	2%	1%	2%	1%	3%	5%	3%
Market Leaders	Firm H	Firm H	Firm H	Firm H, Firm J	Firm H, Firm J	Firm H, Firm I	Fragmented	Fragmented	Fragmented	Fragmented

Figure 55.5. Market segmentation and new product strategy.

the company were home construction by professional builders, home construction and repair (including by "do-it-yourselfers"), and light industrial construction. The vertical axis is arrayed in bands of price-performance. This firm, noted as Firm A, was a traditional market leader in the high end of the market place. Over time, however, its high position was eroded by new entrants who made more featured filled products. Also, its overall market share position had eroded due to low cost suppliers working in the bottom three bands of the price-performance. Perhaps most striking is the fact that the niches focused on by Firm A were in slow or negative growth. Its substantial spending on R&D, if not rechanneled to growing segments, will probably not add to the corporation's revenues!

Armed with this comprehensive strategic picture of the market in its entirety, management decided that "business as usual" was no longer acceptable. Executives took decisive steps to broaden the company's focus toward the home construction repair market. At the same time, a study of its product designs and production processes showed incredible complexity and redundancy. Senior management launched a modular platform design initiative to simplify that complexity and create common parts and pieces across all its products. It also created a new window architecture, the first iteration of which could be implemented in the faster growing, lower price segment of new home construction. The product launch exceeded expectations and market share improved. The company then began to leverage the new, simpler window architecture into the home repair market. The advanced technologies group also began prototyping how to leverage the new window architecture into light industrial construction applications. In Figure 55.5, you should be able to follow these strategies by the shading of the cells. While change takes time in any mature business, this company is proactively providing its competitors with product innovation.

There are many ways to segment groups of customers. For the purposes of using market segmentation to drive product and service strategy, my preference is to array distinct customer groups on the horizontal axis, e.g., as market segments. Then,

within each group, the firm can identify different value drivers on the part of users and buyers. In many cases, there is a clear continuum of performance and price, e.g., the good, better, and best. In other cases, larger phenomena drive buyers' decisions.

Derick Abel's *Defining the Business* remains a wonderful resource for thinking about how to segment markets.[7] Rather than simple tiers of price-performance, Abel's presentation developed matrices of customer functions and customer groups, and then tracked successive generations of technology passing through these dimensions. For example, he considered IBM's market segmentation as evolving along three clear axes: vertical markets on one axis, customer functions on another axis (storage, to general purpose computer hardware, to applications software and tools, to networking, to computer-related services), and types of core technology on a third axis (IBM proprietary, Wintel, and open systems). This three-dimensional array allows managers to clearly articulate the essential "who, what, and how" for a new product strategy.

We can think about Microsoft's new product strategy in a similar manner: vertical industrial and various consumer markets on one axis, product uses on another (operating system, database, office applications, etc.), and types of core technology on a third (desktop, Internet/distributed, and repository [.NET]). And, as both Microsoft and IBM show, each dimension is dynamic. To grow, a firm must change its new product strategy in step with significant changes to any of the segmentation dimensions, whether it be a new emerging customers groups, a new product use, or a new disruptive technology.

A company does not have to undergo crisis in the form of lost market share to come to the conclusion that it is desirable to leverage its core technology into new market applications. Enlightened management teams can choose to build and improve modular platforms from the start, and systematically find new market applications and *technology and market partners* for these platforms.

A classic example is Hewlett Packard's (HP's) extension of its ultrasound imaging systems into new clinical applications during the 1990s. (HP's medical product division became part of Agilent in the late 1990s, and was then purchased by Philips for its medical imaging business.) Traditionally used for highly complex imaging of blood flow volume and valve performance within the heart, this phased-array ultrasound machine was extended into new, related market applications. These new uses included blood vessel imaging and soft tissue radiology applications. The Hewlett Packard team made this happen by building a plug-in hardware interface between the central processing unit and the data acquisition device, typically highly complex transducers. The image analysis software was then modified to show the different body parts and the flow of fluids within them. This growth into new market segments is shown in Figure 55.6, comprising three distinct generations of the product line architecture throughout the 1980s and 1990s. The figure shows the evolution of the company's product strategy in this market space. It drove the cost of its devices downward to build market share with new generations of technology and lowered component procurement costs by virtue of expanding into new market applications.[8]

The ultrasound example shows the strategy of *extending product lines through the work of others.* Some of the transducers used with the ultrasound machines were made by companies other than HP. Without these firms, HP would not have entered several niche applications. These firms could build new data acquisition devices that worked with ultrasound machine because HP de-

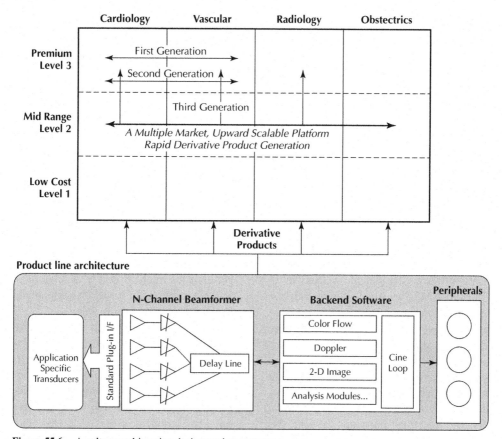

Figure 55.6.   An ultrasound imaging device product strategy.

veloped and marketed a general purpose hardware interface into its machines and created a library of software routines and algorithms that allowed other programmers to make the necessary connections. Market leaders in technology-intensive industries are famous for providing the hardware and software interface architecture and development tools needed by other companies to build complementary products and services. This can explode the breadth and depth of a firm's total solution set. It is a competitive strategy that challenges the "not invented here" syndrome of traditional R&D bureaucracies.

This strategy drives the software industry. For example, Quark has been a market leader in high-end graphics desktop publishing software for several decades. It sells sophisticated layout and design software for publishers. Like other desktop software companies, Quark started off at the desktop, but then expanded to enterprise versions where users can create repositories for their work to be shared with other users. As powerful as Quark's software has proven to be, it assumes even greater functionality because the user can also purchase specialized software modules created by independent software companies around the world. These modules allow users, for example, to more rapidly create different types of industrial-grade labeling used in new packaging designs. There are hundreds of these modules, covering a great range of specific publishing needs. To build this following, Quark created and marketed a software development toolkit and has helped them reach the market through regional distributors around the world. Quark has also developed new interfaces for the Web, such a software filter that allows users to import and export XML formatted files to and from its own products.

The company MathWorks is another strong example of this strategy put to work. This company develops technical computing software for technical professionals, and hundreds of thousands of students and professionals have used the firm's languages for modeling and simulating the function of mechanical and electrical parts and devices. The modeling of controls and digital signal processing

systems are two key market applications. The MathWorks family of products consists of two core products, MATLAB and Simulink. Each of these programs serves as a platform for product development, and together they support more than 50 additional MathWorks products. These languages, MATLAB and Simulink, are useful products by themselves—used by students and engineers to build new models and simulations. However, these languages also serve as the company's common product platforms. In software, languages and the development libraries associated with those languages are often the key enabling platforms for applications development and business growth.

Like other successful software companies, MathWorks is always trying to improve its platforms and add new subsystems to them. One example is the development of a software module that allows models developed in MATLAB to gather and analyze data in real time from sensors and input sources. The 50 products that are available with the MathWorks two languages are toolkits that help engineers apply these languages to a host of specific applications. This might be the design of an automotive subassembly, an aircraft wing, or a process control device used for chemical manufacturing. All of these toolkits are developed either with MATLAB or Simulink, the two common platforms. Taking the strategy one step further, a number of these toolkits have been developed by third parties and licensed back to MathWorks. This flexibility has allowed to the company's solution set to be domain-specific in areas in which it has limited in-house capability. As the company targets new market segments, it works with "lead users" or "lighthouse" customers in target domains to either jointly develop or participate in the distribution of newly created toolkits [see Selections 53 and 54 in this volume].

The challenges faced by established, traditional corporations are an interesting contrast to these software companies. The traditional corporation wishes to become more entrepreneurial and innovative, too, but is bound by organizational bureaucracies and decision-making processes nurtured in incrementalism. As previously discussed, many traditional market leaders find themselves attacked from above by new entrants using new technologies and from below by competitors who are copying and discounting established product lines. In the example of the window manufacturer described earlier, companies may seek to create new architectures that simplify complexity and remove cost. The difficulties of changing production and other related infrastructures may be great, but are nonetheless necessary if the firm wishes to stop the gradual erosion of its market share. This addresses attacks "from below."

An effective response to new premium product competitors can be equally challenging. Here, the traditional leader must take nothing for granted in terms of its understanding of user needs. It must re-engage in talking to existing and prospective customers, seeking to connect its own products to how these products are used to produce end-solutions or results. The company must also try to find needs and problems that cannot be addressed by existing competitive products, including its own. Such latent needs, based in the current frustrations of end users and buyers, are the secret sauce of designing the next generation of a product line, where the goal is to recapture the "high end" of an established market. Again, these are difficult principles to put into practice, for they imply looking at users with fresh eyes, finding the unarticulated latent needs, and building those needs into new products.

While many traditional strategists believe that a firm must choose either a low-cost or a differentiated strategy, we believe that the great power of product line architecture and modular platforms allows a firm to consider both of these strategies in parallel. Its basic, low-cost offerings have the interfaces or connections through which additional levels of functionality can be easily added, either during manufacturing or later in the field. Since many products today are increasingly systems, where software provides the brains behind the brawn, premium solutions will be created from entry-level systems by dynamically downloading specialized software modules that adapt the systems to specific usage requirements. This is but one way that technology challenges the conventional wisdom of business strategy development.

APPLICATIONS TO TECHNOLOGY-INTENSIVE SERVICES

We find that most of these concepts apply equally as well to services businesses. To illustrate this, we will describe the service innovations undertaken by the reinsurance division of Lincoln National Corporation (which is now a part of SwissRe).[9] At the turn of the millenium, Lincoln National's reinsurance division was the largest life reinsurer in the United States, with approximately $6 billion in annual revenues and more than $130 billion in assets under management. A reinsurer provides insurance to insurance companies. Typically, the insurance company sets a retention level on individual policies. The dollar amount of any policy that exceeds that level is ceded to a reinsurer, who then bears that additional risk a cost per thousand dollars of the amount being reinsured. Lincoln National had developed distinctive competence in assessing the risk of any new life insurance policy application. It would share that understanding with its insurers, e.g., its customers, not only providing them with reinsurance, but also telling them which borderline applications not to take or how much more should be charged to cover the additional embodied risk.

Through many decades, Lincoln National's reinsurance group prospered with this strategy. However, during the 1990s, many large financial corporations desired to enter the life reinsurance market. They began to eat away not only at Lincoln's market share, but also at its profitability by offering substantial discounts to Lincoln's current customers. In fact, many insurance companies would maintain relationships with several reinsurers, comparing costs and coverages on a case-by-case basis. Senior management at Lincoln National's reinsurance division might have chosen to fight back by lowering its prices, too. However, that approach would only reinforce the competitors' advantage by destroying Lincoln National's traditional brand of high-risk management quality. Instead, senior management brought together a multifunctional team of highly experienced professionals and asked that team to create a new reinsurance offering that would change the terms of competition in the industry.

The team took several bold initiatives. First, it created a new market segmentation framework for the company. Visualize a vertical axis that has simply large firms and small firms as the categories. On the horizontal axis, also visualize two categories: traditional insurers and banks (or other financial services firms). Then, within each of the quadrants, the team differentiated between knowledge/service-driven customers and price-driven customers. The initial target for the new service design became large, knowledge-driven insurers and financial services firms.

Then, the team examined the core subsystems of its traditional reinsurance service. These subsystems included "deal negotiation" (since each reinsurance treaty must be individually negotiated with an insurance company to take into account the demographics of its own target market), underwriting assistance, and administration of business on the books. Since each of the core processes was applied to individual insurers and modified to meet particular needs, they were truly "platforms" just as in any physical product. Lincoln's team also added an entirely new subsystem that met a latent, unfilled need. A number of insurers were selling insurance products on which they were not making money due to the terms, conditions, and pricing of those products. So, Lincoln developed an expert system that would integrate the insurer's desired product type (term insurance versus whole life, for example) together with characteristics of an insurer's target population (such as age and income) to produce a profitable product. In short, Lincoln got into the business of designing their customers' own unique insurance products.

Management then decided to deploy this new reinsurance offering through multifunctional teams, with a sales person, an actuary, and an underwriter all working together to solve the customer's problems. Prior to this restructuring, salespersons would work through functionally organized bureaucracies.

To complete the renewal process, management then reconsidered the value of this new reinsurance offering, one that would help customers design a profitable insurance product as well as reinsure it. Could Lincoln change it pricing, its business model, and escape the price pressures of low-cost reinsurers? Rather than charge on a price per thousand over a certain predetermined retention level, Lincoln National asked for—and received—a rather significant percentage of the insurer's premium revenues starting on the very first dollar of those revenues for individual life policies. This premium-sharing model also placed Lincoln directly into partnership with its customers. These decisions rather quickly produced new sales and higher profits. Five years later, SwissRe, one of the largest low-cost life reinsurers, found the recipe too hard to implement on its own and made an offer to acquire Lincoln's reinsurance business.[10]

THE ORGANIZATIONAL IMPLICATIONS OF MODULAR PLATFORMS

Bold strategies for product and service innovation often fail due to the lack of complementary innovation in the development and marketing organizations of the firm. The architectural and strategic approaches that we have described also have fundamental implications for the organization of new product development within the enterprise. As executives in the medical products group at HP have remarked, *"Products have an uncanny way of looking like the organizations that make them."* If development groups operate in geographically and organizational distinct entities—the classic stovepipe scenario—then it is unlikely that the products or systems made by these groups will share a significant percentage of subsystems and components within those subsystems. Similarly, if new product development processes are focused largely on stage gate checkpoints for individual products, with little in the way of process to integrate architecture and engineering between those individual products, then it is also unlikely that common architecture and shared subsystem platforms will be realized.

Many companies that develop systems with specific market applications organize themselves with almost complete decentralization of R&D to specific product-market business units. This decentralization of R&D came to pass with the frustration by many of the slowness and lack of market focus of centralized R&D that was dominant during the 1960s and 1970s. Also, many corporations have adopted highly complex stage gate management systems where executive teams review and pass judgment on single products, product by product, within the proposal, development, manufacturing ramp-up cycle, and commercialization cycle. Such processes were generated on a foundation of management research whose unit of analysis was *single* product successes and failures and then identified success factors on such an idiosyncratic basis. The generation of shared platforms and the continuous renewal of architecture underpinning a stream of products over time was not part of that thinking. Yet, as we have argued in the pages above, renewal of architecture and the creation of

shared subsystems are essential for sustained success in technology-intensive industries.

Figure 55.7 provides an approach to "platforming" the development organization that we have seen used by a number of technology-intensive manufacturers within the telecommunications, computing, and consumer electronics. The figure shows three basic organization entities: market teams (at the top of the figure), product engineering teams (in the middle), and platform teams (at the bottom).

The platform teams are research and development staff who have been assigned responsibility for each of the major subsystems that comprise the shared platform. These platform teams develop the core technologies and processes within their particular subsystem focus. The responsibility of these platform teams is two-fold: first, to maintain, test,

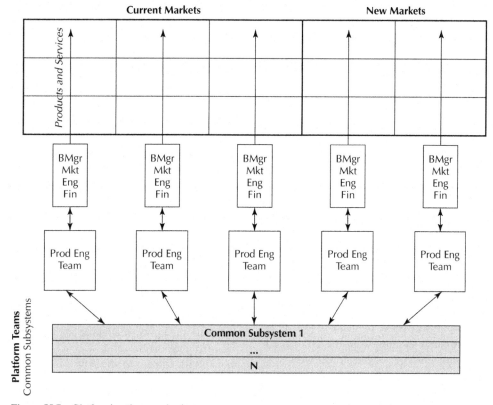

Figure 55.7. Platforming the organization.

and incrementally improve *current* generation processes and technologies, and second, to obsolete these with *next* generation processes and technologies, either through breakthroughs developed internally or by other companies, large or small. To as great an extent as possible, these persons should be co-located to facilitate effective communication.

The market teams are shown at the top part of Figure 55.7. These teams can be treated as full profit and loss centers, developing the product and families, marketing them, and held responsible for growth in their target segments. Or, management can think of these teams as "product management groups," defining new product and service engineering requirements, channels and promotion, and product launch strategies.

The product engineering teams are responsible for developing and integrating the specific product, systems, and services required for the solutions defined by the market teams. The focus of the product engineering teams must be to use as much common subsystem or platform technology as is possible and available within the firm or its partners. By focusing its value added engineering work in this way, the product engineering team can work faster and provide a more advantageous cost of goods position.

We have found that a systematic, calendar-based process must be implemented to channel structured communications between market teams, specific product engineering teams, and common platform teams. Whether a P&L center or not, each market team is responsible for understanding the specific needs of target customers and stating these in a product requirements roadmap which is then effectively passed to the product engineering groups. In turn, these product engineering groups must pass their subsystem requirements to the common platform teams who must then assess what they can do with the resources at hand, request further resources if necessary, or make tradeoffs on the various requests coming from the product engineering teams. Last, these teams must work with their business unit managers to determine what may, and may not, get done during a given cycle of development. Senior management must arbitrate emerging disputes—while strongly nurturing the discipline of using modular platforms. As noted earlier, we believe that this is best done by imposing limits on the amount of value added market teams can engineer into their respective product lines.

For large corporations, where business units house multiple product lines, the organization may be extended to include business teams which oversee both product line teams and the platform teams. IBM is a strong example of this. First, it has *Portfolio Management Teams* with one Portfolio Management Team per division. These teams are comprised by the functional heads in each of the product divisions, including the vice presidents of marketing, engineering, operations, and distribution along with the CIO and CFO, i.e., the top decision-makers needed to reallocate resources to new market opportunities. The division General Manager chairs the Portfolio Management Team meetings. Reporting to the Portfolio Management Team is a specialized team called the *Product Line Planning Team*. Product line planners include the division's best systems architects and business development personnel. Collectively, they plan and map the product platforms and the product line roadmaps that will be used across the division's product lines. These teams are also responsible for performing market research in target market requirements and competitive offerings. Also reporting to the Portfolio Management Team are *Product Development Teams* who are responsible for seeing specific products developed and introduced successfully into manufacturing.[11]

GET STARTED BY ASKING A FEW QUESTIONS

These ideas lead directly to questions that senior management must ask itself in areas of developing growth strategies, managing architecture, and organizing the firm.

With respect to strategy:

- Are low-cost competitors taking share—either by copying the company's designs and discounting prices or with new product line architectures that are simply more cost effective? Does the corporation's response to these competitors threaten its own profitability?

- Alternatively, are new entrants taking leadership in defining the high end of the "good, better, best" spectrum in the company's product categories? Are these new entrants commanding premium prices for their endeavors, and what is that doing to the corporation's ability to charge premium prices and to its profitability?

- Is the overall market growing but the corporation's own sales largely flat because it is not effectively addressing new faster growing segments within the market? Is the failure to respond effectively a product development problem, a distribution problem, or both?

Similarly, with respect to architecture:

- Does the firm lack a clear definition of its product platforms or a strategy for leveraging these across product lines?

- If the firm produces systems of one form or another, is it missing a well-developed API (Application Programming Interface) that allows not only its own engineers but customers and third parties to build and integrate value added modules for those systems?

- When the firm attacks new emerging market applications, does it create entirely new product architectures and entirely new subsystems within those architectures—as opposed to leveraging architecture and modular platforms for those new applications?

With respect to the organization of development, if the answer to any of the questions above is "Yes," the very way that development teams are organized and the processes by which they manage work and communicate with different parts of the company may be impediments to effective innovation.

- Are there specific groups that are responsible for the development of core, shared subsystems that

are to be shared between products and product lines?

- Is there a tendency for development teams to follow their own understanding of user requirements? Or, do market and sales drive new product functionality?

These questions, if answered candidly, reflect the challenges facing technology-intensive companies working in dynamic markets. Not all management teams succeed at forming solutions to the issues raised by these questions. Many executives still perceive platforms as only technical phenomena, and therefore, platform development is left only to the engineers to decide and execute. However, developing new product line architectures based on modular platforms first requires *market thinking*, if for no other reason than the corporation must first identify the current and near-future market application targets for common technology. That market thinking will also include performing competitive benchmarking, examining channels of distribution, and thoughtfully considering the impact of brand on customer decision-making for new products and services.

Similarly, the company's senior executives may still be focused on sequential product development and have no experience in building common, modular platforms and leveraging them into products and services for new markets. Further, the "new" product development system that they have just installed may be a single-product stage-gate system that reinforces sequentially for individual product developments. Beyond this, perhaps most management teams are still largely consumed with completing the very next product. When the product launches, everyone breathes a collective sigh of relief before racing off onto the next product. The fire drill of getting the next new product "out the door" overrides the desire to build sustainable platforms and create a plan for a stream of derivative products.

Calmer heads must prevail. Growth relies on a stream of value-rich products and services based on common product platforms and production processes, and the knowledge that goes into the creation of these assets. We believe that the greatest challenge, however, is to get the people who understand this logic working together on empowered teams. For only then will the firm create the dynamic products and services it needs for substantial growth.

NOTES

1. Meyer, Marc H., and Lehnerd, Alvin. 1997. *The Power of Product Platforms.* New York: The Free Press.

2. Rumelt, R., 1974, *Strategy Structure and Economic Performance.* Boston, MA: Harvard Business School; Meyer, M.H.; and Roberts, E.B. 1988. "Focusing New Product Strategy for Corporate Growth," *Sloan Management Review,* 29:4, pp. 7–16.

3. Johnson, H. Thomas, and Broms, Anders. 2001. *Profit Beyond Measure.* New York: The Free Press.

4. Meyer, Marc H., and Lehnerd, Alvin. 1997. *The Power of Product Platforms.* New York: The Free Press.

5. Meyer, Marc H., and Mugge, Paul. 2001. "Make Platform Innovation Drive Business Growth," *Research Technology Management,* January-February 2001, pp. 25–39.

6. Meyer, M.H., and Roberts, Edward, 1988. "Focusing New Product Strategy for Corporate Growth," *Sloan Management Review,* 29:4, pp. 7–16.

7. Abell, Derick, 1980. *Defining the Business.* Englewood Cliffs, NJ: Prenctice-Hall.

8. Meyer, Marc H., Tertzakian, Peter, and Utterback, James. 1997. "Metrics for Managing Product Development within a Product Family Context," *Management Science,* 43:1, January 1997, pp. 88–111.

9. Meyer, Marc H., and DeTore, Arthur. 2001. "Creating Platform-based Approaches to New Services Development," *Journal of Product Innovation Management,* 18, pp. 188–204.

10. The 2001 10k showed annual earnings of about $150 million, and the acquisition price was set at above $2 billion—a handsome reward for Lincoln National's innovations.

11. Meyer, Marc H., and Mugge, Paul. 2001. "Make Platform Innovation Drive Business Growth," *Research Technology Management,* January-February, pp. 25–39.

56

The Elements of Platform Leadership

MICHAEL A. CUSUMANO AND ANNABELLE GAWER

With vision that extends beyond their current business operations or the technical specifications of one product, platform leaders can create an industry ecosystem greater than the sum of its parts.

You would think that a company like Intel, which in 2001 provided nearly 85% of the microprocessors for personal computers, would feel relatively secure. But companies holding the keys to popular technology don't live in a vacuum. In many cases, they are dependent not only on economic forces in the wider world but also on the research-and-development activities of partners. David Johnson, one of the directors of the Intel Architecture Labs (IAL) in Hillsboro, Oregon, goes so far as to call that reality desperate. "We are tied to innovations by others to make our innovation valuable. If we do innovation in the processor, and Microsoft or independent software parties don't do a corresponding innovation, our innovation will be worthless. So it really is a desperate situation for us."[1]

THE DESPERATION OF BEING ON TOP

Although leading companies from all industries know that business in our interconnected world has become too complex for complacency, the issues are particularly clear in the information-technology industry. There, *platform leaders* (companies that drive industrywide innovation for an evolving system of separately developed pieces of technol-ogy) are navigating more frequent challenges from *wannabes* (companies that want to be platform leaders) and *complementors* (companies that make ancillary products that expand the platform's market). To put their organizations in the best competitive position, managers need to master two tricks: coordinating internal units that play one or more of those roles and interacting effectively with outsiders playing those roles.

Intel and its computer on a chip, the microprocessor, illustrate the issues. In 2000, Intel had revenues of nearly $34 billion and net profits of more than $10.5 billion. Even after the economic downturn, with microprocessors still the core hardware component of the personal computer and increasingly in demand for new programmable devices, Intel should feel on top of the world.

But a microprocessor can do little or nothing useful by itself. It's a component in a broader platform or system. Even the PC has no value without other companies' products: operating systems, software applications, software-development tools and hardware (monitors, keyboards, storage devices, memory chips and the like). Those complementary products fueled the growth of the PC market. But Intel considers its situation desperate because it cannot be certain that its own key complementors will continue to produce market-expanding innovations as fast as Intel does. Nor can it be sure that its target platform, the personal computer, will evolve in compatible ways.

That dependency has been seen in other industries as well—for example, with the compact-disc player and the video recorder, also platform

Reprinted from *MIT Sloan Management Review*, vol. 43, no. 3, Spring 2002, pp. 51–58. Reprinted with permission.
Michael A. Cusumano is the MIT Sloan Management Review Distinguished Professor of Management at the MIT Sloan School of Management. **Annabelle Gawer** is Professor of strategy and management at INSEAD in Fontainebleu, France.

products tied to complementary products. Customers would never buy a CD player or video recorder if they couldn't get prerecorded CDs and videotapes. Platform dependency is an ongoing concern. That's why in November 2001, the mobile-phone maker Nokia decided to partner with competitor NTT's DoCoMo on software standards—and with AT&T Wireless, Motorola, Fujitsu and others to create middleware between service and mobile networks.[2] Nokia wanted more players to adopt compatible standards and potentially follow its lead.

Another Intel manager, Bala Cadambi, uses an automobile analogy. "Intel is in the business of providing the engine for the PC, just like Honda would be in the business of providing the engine for the automobile," he says. "That engine is doubling in capacity every 18 to 24 months. . . . What we really want is to ensure that the rest of the platform goes with it. This means that, if the engine gets better, the tires get better, the chassis gets better, the roads get better, you get better gas mileage. You can have navigation systems that are scalable. Everything that goes with having a better experience. *The platform around the engine limits the engine. So we want the platform—which is everything that's around the microprocessor—to be keeping pace and improving and scaling, such that the microprocessor can deliver its potential.*"[3]

Intel and Microsoft are simultaneously complementors and platform leaders for the personal computer. They are complementors in the sense that PC makers need them to make critical components. But in the main, they are platform leaders because of their influence over PC system architecture and over other companies that produce complementary products. For the same reasons, NTT DoCoMo and Palm are platform leaders.[4]

Platform leaders face three problems. First is how to maintain the integrity of the platform (the compatibility with complementary products) in the face of *future* technological innovation and the independent product strategies of other companies. A related problem is how to let platforms evolve technologically (as they must or become obsolete) while maintaining compatibility with *past* complements. A third problem is how to maintain platform

leadership. Insights from Intel as well as Microsoft, Cisco Systems, Palm and NTT DoCoMo can help companies manage innovation in industries characterized by platform technologies and complementary products that make the platform valuable to consumers.[5]

LEADERS NEED FOLLOWERS

Most platform leaders do not have the capabilities or resources to create complete systems by making all the complements themselves. They need to collaborate. The combined efforts of platform leaders and complementary innovators increase the potential size of the pie for everyone. That was true for the parties that delivered the first PCs to users. And it's true for those trying to improve the PC platform today.

Dave Ryan, director of technology marketing in the Intel Architecture Labs, is typical of managers we interviewed who focus on streamlining Intel's interactions with complementors. He says that the layers of industry groups with which Intel interacts to deliver a complete PC capability are like a stack. (See Figure 56.1.) At the base is the hardware platform. At the top is the goal (the user's awareness of the innovation). The layers appear as orderly and sequential as a stack of encyclopedias. However, interconnectedness makes for complexity, with changes occurring in one layer at the same time that changes are occurring in another.

As Ryan says, "For the user actually to see that new capability, some companies need to make new hardware. Some companies need to make changes in the operating system. There need to be new types of networking or new products developed. . . . So to get a new use or new capability to occur, we have to work simultaneously across all these layers."[6]

Platform leadership is the ability of a company to drive innovation around a particular platform technology at the broad industry level. Whether the dynamic is called "network externalities," "bandwagon effects" or "positive-feedback effects," the more people who use platform products, the more incentives there are for complement producers to

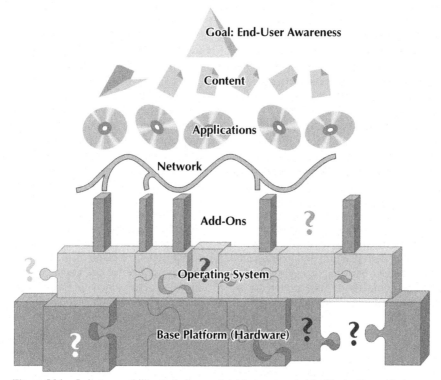

Figure 56.1. Industry capability stack. Source: Intel in-house materials. Used with permission.

introduce more complementary products, causing a virtuous cycle.[7]

Platform leaders actively solicit innovation on complementary products. But the game is complex and sometimes features fierce standards wars. One company that failed to make its platform the standard was Sony, whose Betamax lost the VCR wars. Apple's Macintosh computers, despite some resurgence, have yet to unseat Windows machines as the mass-market standard. Another platform-type product, Netscape Navigator, declined sharply from its early 90% market share after Microsoft unleashed Internet Explorer.

THE LEVERS OF PLATFORM LEADERSHIP

After analyzing Intel, Microsoft, Cisco and NTT DoCoMo, we developed some practical guidelines for managing innovation whether the innovator is a platform leader, a wannabe or a complementor.

(See "The Four Levers of Platform Leadership.") Four distinct but closely related levers of platform leadership can assist managers in both strategy formulation and implementation.

Lever One: Scope

Determining the scope of the company—that is, which complements to make in-house and which to leave to external companies—is probably the most important decision. Companies that want to become platform leaders first need to assess how dependent they are on complements. Then they need to determine how to increase demand for their platform. Palm has stimulated the external development of more than 8,500 applications for its operating system. In Japan, NTT DoCoMo has encouraged the creation of 40,000 Web sites to provide content to DoCoMo customers. Since 1993 Cisco has made more than 70 acquisitions of both complementary companies and substitute technologies that it could tie into its Internet router (a

The Four Levers of Platform Leadership

1. *Scope.* Scope comprises the amount of innovation the company does internally and how much it encourages outsiders to do. Managers of companies that are platform leaders—or that want to be (wannabes)—must weigh whether it is better to develop an extensive in-house capability to create their own complements, let the market produce complements or follow a middle road.

2. *Product Technology.* Platform leaders and wannabes must make decisions about the architecture of a product and the broader platform, if the two are not the same. In particular, they need to decide how much modularity they want, how open their interfaces should be, and how much information about both platform and interfaces to disclose to outsiders who might become complementors—or competitors. (We prefer "complementors" to the longer "developers of complementary products." See A. Brandenburger and B. Nalebuff, "Co-opetition: A Revolutionary Mindset That Redefines Competition and Cooperation" (New York: Currency Doubleday, 1997); and C. Baldwin and K. Clark, "Design Rules: The Power of Modularity" (Cambridge, Massachu-

setts: MIT Press, 2000), which is about the problem of modularity.)

3. *Relationships with External Complementors.* Managers must determine how collaborative or competitive they want relationships to be between platform producers and complementors. Platform producers also need to work on creating consensus and handling potential conflicts of interest (for example, how to behave when the move to a complementary market turns former collaborators into competitors).

4. *Internal Organization.* The right internal structure can help platform producers manage external and internal conflicts of interest. Organizational options include, first, keeping groups with similar goals under one executive or putting them in distinct departments if they have outside constituencies or potentially conflicting goals; second, addressing organizational culture and processes; third, improving internal communication of corporate strategy. Because of the ambiguity of innovative, modular industries, a company culture that encourages debate can accelerate strategy reformulation when it's needed.

---

specialized computer for sending packets of information across the Internet).

Platform producers should not develop their own complements if they lack the technical, organizational or financial capabilities to compete in the relevant markets. Microsoft did have such capabilities, and—even though its decision to proceed damaged former complementors, including Netscape, Novell, WordPerfect and Lotus—the company developed the most popular applications for Windows by itself. Intel, however, would have trouble competing with Microsoft in making commercial operating systems or software applications. Nor is Intel likely to succeed in consumer mass markets—say, with MPEG music players or digital cameras. Its strengths lie elsewhere.

Microsoft had the technical skills to move from PC-programming languages and operating systems to software applications. Still, Microsoft

generally stays with applications that mesh with platform technology such as MS Office or common database-management programs. It remains to be seen if the company can compete long term with a consumer hardware product such as the new video-game console, xBox.

Perhaps the most complex example of external scope is Cisco, which builds few end-user applications itself. Instead, it makes acquisitions when it wants to expand its product offerings' capabilities into different areas—or when it wants to pick up technologies that, when linked in hybrid networks, might substitute for its routers.

A platform leader or wannabe deciding to work with outside developers should scrutinize others' incentives and capabilities in order to exert influence over the design and production of complements. For example, platform producers can share technical information about their own products and

Advice for Complementors

Companies making products that complement platform leaders' products have placed bets on which platform producers to follow and which technical standards to support. Should a company making ancillary software for personal digital assistants use the Palm operating system, Microsoft PocketPC's system or both? Complementors try to assess who will win the war for platform leadership. They look at how actively a platform producer is lining up outsiders to support its platform with complementary applications. They investigate how openly the platform producer is providing technical information to complementary producers.

Is it possible to dance with the elephant—that is, to avoid getting crushed when a powerful platform leader decides to compete? If complementors commit resources to innovations, they should focus on products that the platform producer is unlikely to offer. They need to work at

continuous communication because changes occur rapidly. Complementors need to keep alert to a platform leader's product plans and try to get early information on a move onto their turf. They need to react quickly to demands; slow response may give a platform leader an excuse to compete with the complementor later.

Although platform leaders need complementors as a group, usually the balance of power between a platform leader and one lone complementor is tilted toward the platform leader. The trick to being a successful complementor is always to have peanuts to offer the elephant—to create products that continuously enhance the value of the core product even as the core changes.

Complementors also should identify which groups inside the platform company are likely to take the most neutral stance to promote the platform and its innovation ecosystem—and then work with them.

---

platform interfaces or send engineers to help complementors build compatible products. Intel uses what it calls a "rabbit" strategy—targeting a promising complementor and assisting it in such a visible way that other companies follow. The approach draws the attention of investors and complementors to a potentially lucrative new market and signals that the platform leader aims to stay out of the complementary market.

As Intel has done through its labs, platform producers can develop enabling technologies, such as programming interfaces and software-development kits, and share them for little or no charge to stimulate the development of complements. And sometimes platform leaders or wannabes share market information or offer complementors marketing support.

Venture investments and mergers and acquisitions also help a company influence the production of complements. Platform leaders such as Intel, Microsoft, Cisco and Palm have taken equity positions in some complementors. Intel, Microsoft

and Cisco also have acquired complementors. However, when an acquisition makes the platform leader a competitor of former partners, it can discourage other companies from becoming complementors and can mean less competition—and possibly less innovation.

There is no simple answer on whether to make complements in-house; however, platform products do need complements. Platform producers probably should have some in-house capability, not only for producing complements but also to provide constructive direction and competition for third parties.

Lever Two: Product Technology

Product architecture—both the high-level platform design and the interface designs that determine how subsystems work together—can have a profound impact on the structure of an industry and on the nature of follow-on innovation. Product architecture can determine who does what type of innovation as well as how much investment in complementary products occurs outside the platform-leader company. Modular architecture (with easily separable components) can reduce innovation costs and encourage the emergence of specialized companies. Specialists often invest heavily and creatively in complementary products and services.

Intel, Palm and NTT DoCoMo use modular architecture. Even the Microsoft and Cisco operating systems, despite their somewhat haphazard evolution, have modular characteristics that facilitate external creation of complementary products.

Modular architectures are particularly useful when the interfaces are open—that is, when the platform leader specifies publicly how to connect components to its platform. However, open disclosure aids competitors spying on the product's inner workings. That's why Intel, for example, jealously guards its microprocessor architecture, even though it is open about interfaces such as the peripheral-component-interconnect (PCI) bus and the universal-serial bus (USB), which link computers to peripherals. Similarly, Microsoft reveals detailed specifications on the Windows programming interfaces but is careful not to give away the source code (internal structure) of the Windows software platform.

Successful companies evolve the core architecture. Intel microprocessors once encountered threats from ultrafast designs that workstation producers such as Sun Microsystems, Apollo, IBM and Silicon Graphics used in their high-powered workstations—and from the superior graphical capabilities of Motorola chips in Apple Macintosh computers. But Intel evolved its microprocessor architecture to compete more effectively on speed, processing power and graphics. Microsoft evolved its architecture too, building Windows NT/2000, a high-end operating system that enabled the company to compete more effectively with Unix and Linux in the corporate server market. Palm and NTT DoCoMo keep evolving their platforms too.

Cisco's platform is essentially the internet-working operating system (IOS), which is based on open Internet communications and networking standards that Cisco did not define alone. The company has had to make its software and hardware products compatible with any new communications technology that emerges and is vulnerable to substitutes competition and specialized players. In 2001, it was not clear that Cisco would keep up with external innovations and maintain its high growth rates.

Keeping control of the architecture is a powerful barrier against companies that might offer a competing architecture with different interfaces. A competitor to Intel, for example, not only would have to invent a microprocessor with a better price-performance ratio, it also would have to rally complementors and original-equipment manufacturers to change their designs and accept the switching costs.

Microsoft is another platform leader that uses interfaces to align the interests of a coalition of companies. Although it was taken to court for antitrust violations, it retained the right to continue encouraging the platform-specific complements that create an applications barrier to entry for companies with alternative platforms.

Thus if platform producers want to stimulate the development of complementary products, they give the technical specifications of interfaces to third parties. If they want to hinder an outsider's ability to make complements (for example, if a potential complementor is competing with a preferred partner), they keep their intellectual property from that company.

Cisco relies primarily on open standards. DoCoMo is pushing for adoption of a standard for open data transmission. Intel has an open intellectual-property policy on its PCI, USB and advanced-graphics-port (AGP) interfaces. Palm licenses its Palm OS to complementors and even to competitors such as Handspring. Nokia more recently concluded that information about interfaces encourages external innovation. Nevertheless, it's a delicate balance, and disclosing too much information can be dangerous.

Finding ways to stimulate innovation involves a trade-off between secrecy and disclosure. Like a patent, secrecy is good for blocking substitute innovation. It encourages profit-seeking entrepreneurs to innovate on a stand-alone product. But disclosure is best for supporting complementary innovation.

Decisions abut product technology—architecture, interfaces and intellectual property—are critical to platform leadership. Successful companies protect their core technology but use modular architectures and disclosure of interfaces to get complementary products and services. Spending resources on design issues such as platform architecture and interfaces—or on promoting industry consensus about interface standards—can help platform producers shape their environment.

Lever Three: External Relationships
To be effective over the long term, platform leaders need to pursue two objectives simultaneously. First, they must seek consensus among key complementors about what technical specifications and standards will make platforms work with other products. Second, they must influence partners' decisions affecting how well everything works together through new product generations. Pursuing consensus and control at the same time, though essential, can be difficult, as other companies naturally fear being dictated to.

Consensus among industry players depends on one company driving the process. It must have some degree of control over interfaces between components and between the hardware platform and the software operating system. The company that leads exerts control not over others' specific choices but over the premises of choice. We call that *ecological control*. Control presupposes some degree of consensus, because leadership is possible only when others agree to follow.

Intel designed interface standards defining how the microprocessor would communicate with other components, and it developed the organizational capabilities to encourage other companies to design products. But it was a challenge. Some interfaces, though part of the PC system, were not part of the microprocessor. Thus a critical mass of key players had to agree on interface specifications for the whole product. Without agreement, an industry will not develop enough complementary and compatible products or will innovate too slowly.

Specific management processes can help a platform leader achieve consensus and maintain control at the same time. Intel's experience demonstrates the importance of a carefully thought-out balancing act of collaboration and competition that recognizes mutual dependency. Such a balancing act requires companies to trust the platform producer. But maintaining trust is difficult. It is not always clear if another company is supplier, competitor or complementor, or if today's supplier or complementor will become tomorrow's competitor. Some companies play multiple roles. For example, IBM bought Intel microprocessors but also made microprocessors that competed with Intel products.

There is a real threat to complement producers that dance with the elephant. (See "Advice for Complementors.") Although platform leaders usually avoid partners' markets, they invade often enough to make complementors wary. A platform leader is less likely to intrude into a complementor's turf if the latter can innovate in ways that the platform leader cannot.

Platform leaders should be industry enablers—helping others innovate in ever better ways around the platform. Leaders need to sacrifice short-term interests in favor of the common good. That's why Intel invested in interface standards and relinquished royalty rights for technologies that facilitated evolution of the PC as a system. (See "Ideas from Intel on Managing Platform Leadership.") Intel coordinated the efforts of hundreds of engineers in developers' forums and compliance workshops for ensuring that peripherals and other complementary products worked properly with Intel microprocessors and with one another.

Platform producers should build reputations for not impulsively stepping out of their product boundaries into complementors' territory. Intel is generally careful not to destroy partners' business models. The same cannot be said of other companies. Microsoft often prefers to crush complementors that start looking like competitors. Cisco tries to acquire them. Palm and its software licensee, Handspring, have a more complex relationship, with Handspring both a complementor to Palm

(Handspring's Visor works on the Palm operating system) and a competitor in the PDA market. Palm needs to keep developing its operating system or Handspring could switch to another one.

How do platform leaders manage external tensions? Intel perfected a gradual, low-key approach when pushing a particular agenda. It allows input from collaborating companies—and permits both sides to test the waters. Intel learned from past mistakes. With its first foray into videoconferencing, it incautiously tried to impose a new standard while there was a strong incumbent using a different standard. By giving away its cheaper alternative technology, it nearly destroyed the ability of complementors to make a profit. Intel learned to push an agenda more subtly, with managers assuring complementors that critical technical information would remain open and that there would be adequate protection of intellectual property.

By demonstrating to potential complementors that it is acting on behalf of the whole industry, a platform leader can establish credibility in those technical areas where it wants to influence future designs or standards. Thus in drafting interface specifications, Intel did not insist on complete ownership of all related intellectual property. It also protected others' intellectual property by working with only a few companies at first. Later, when the specifications were almost stable, it involved more companies in setting standards.

The balancing act is tricky: collaborating with external complementors and championing the public interest while competing with those complementors when necessary to stimulate a new complements market. The ambiguity of relationships sometimes generates tensions and conflicts of interest both sides must address. One way is through internal organization.

Lever Four: Internal Organization

A platform producer must create an internal organization that allows it to manage relationships with complementors effectively. What happens if some groups within a platform-leader company compete with complementors, while other groups need those same complementors to cooperate and adopt the

platform's technical standards? Therein lies the challenge.

Intel has some groups focus on competition with other companies, while other groups focus on consensus building with partners. Intel executives acknowledge the necessity to pursue conflicting goals—which they call "job 1" versus "job 2" or "job 3." Job 1 is selling more microprocessors, which includes encouraging external, demand-enhancing innovation on complementary products. Job 2 is to compete directly in complementary markets. Job 3 involves building new businesses that are potentially unrelated to the core microprocessor business.

Intel's top management acknowledges the conflicts among goals: Entering complementary markets means direct competition with partners but occasionally is necessary; investing heavily in new business development can be a distraction from core businesses but helps Intel diversify.

It is vital to communicate the multiple goals to the whole company and create a process for re-

---

Ideas from Intel on Managing Platform Leadership

- Protect the core technology but share interface technology.
- Sacrifice short-term interests in favor of the industry's common good.
- Do not step carelessly onto partners' turf.
- When pushing an agenda, test the waters in a low-key way.
- Help complementors protect their intellectual property.
- Separate internal groups that produce complements from those that assist complementors.
- Leverage internal processes, such as senior-management arbitration of conflicting goals.
- Communicate diligently with partners.
- Communicate diligently with internal constituencies.

solving conflicts. Intel puts up "Objectives" posters everywhere. It keeps groups with different goals separate so that outside companies can more easily entrust Intel people with confidential information.

Microsoft created separate divisions for its applications and operating-systems groups, allowing it to deal with competitors who were also complementors, such as IBM/Lotus, Netscape, Intuit and Oracle. Cisco keeps its product units relatively independent, enabling those units to work with outside companies that compete with other Cisco groups.

In general, platform leaders can appear more neutral if they establish an internal Chinese wall, with different groups playing different roles vis-à-vis third parties. But organizational design is usually not enough. Intel people rely heavily on internal processes, too, such as formal planning and off-site meetings. They also count on senior executives to arbitrate when conflicts arise among company units—and to foster an organizational culture that encourages debate and tolerates ambiguity.

Intel management understands that a platform is a complex system calling for a neutral industry broker to oversee development of the system through external collaboration. NTT devotes extensive R&D resources to studying technologies of general utility to the wireless industry as a whole. Microsoft, Cisco and Palm also recognize that their platforms contribute to larger systems. But so far Intel may be the best example of how to lead an industry by representing the interests of other companies besides itself.

MANAGERS WITH VISION

It is possible to be too platform-centric. There are other ways to compete: say, as a niche player with superior quality or service. Not every company can be the platform leader.

Sometimes platform leaders become so tied to certain technologies that they find it difficult to evolve their platforms. Intel, for example, is closely tied to the $\times 86$ microprocessor family and is unlikely to move to radically new types of computer architectures. Microsoft continues to have a Windows-centered view of the Internet and might never take full advantage of the open-standards movement. In fact, its leaders have publicly opposed the open-source concept, though Microsoft will now let some complementors view the Windows source code. Cisco depends heavily on its ability to weave multiple technologies together through its IOS software, a patchwork of code and standards that will someday outlive its usefulness. Palm is becoming a hostage to the internal architecture and external interfaces that define the Palm OS. No wonder NTT DoCoMo is partnering with Nokia. It's either collaborate or live with standards for wireless data transmission and content that others don't share.

Thus platform leaders eventually struggle with platform evolution. For some Intel groups, the platform is becoming the Internet—and new devices that run Internet software rather than use Windows and $\times 86$ chips. Microsoft is trying to reconcile traditional applications with use of the Internet as a computing platform for Web-based services. Cisco finds itself moving beyond the Internet router as a platform to software linking various types of networking equipment that communicate through Internet protocols.

Platform leaders need to have a vision that extends beyond their current business operations and the technical specifications of one product or one component. The ecosystem can be greater than the sum of its parts if companies follow a leader and create new futures together. Complementors need to understand the vision of the platform leader in their industry and make some bets on what that vision means for their own future. But it is the platform leaders, with the decisions they make, that have the most influence over the degree and kind of innovations that complementary producers create. Platform leadership and complementary innovation by outside companies are not things that happen spontaneously in an industry. Managers with vision *make* them happen.

NOTES

1. D.B. Johnson, interview with authors, Nov. 11, 1997.

2. E. Harris, "Nokia Will Open Up Parts of Its Software," *Wall Street Journal*, Tuesday, Nov. 13, 2001, sec. B, p. 11; and "DoCoMo, Nokia Form Link on Standards," *Wall Street Journal*, Thursday, Nov. 15, 2001, sec. A, p. 21.

3. B.S. Cadambi, interview with authors, Aug. 4, 1998. Emphasis added.

4. Platform leader Intel counts as complementors every company that has adopted the Windows-Intel platform for the PC and makes or sells PC hardware and peripherals (for example, Dell and Lexmark) as well as 5 million software developers who make applications (for example, Microsoft and Adobe) and networking or systems software (Microsoft, Red Hat and Netscape). NTT DoCoMo's complementors include every company that provides content to one of the 40,000 Web sites compatible with DoCoMo technology in Japan (for example, Nikkei News and Tokyo Wine News). Palm's complementors are the 145,000 external software-applications developers and 500 hardware developers (as of March 2001) that create add-on accessories: keyboards, voice recorders, digital cameras and GPS wireless-connection systems.

5. A. Gawer and M.A. Cusumano, "Platform Leadership: How Intel, Microsoft and Cisco Drive Industry Innovation" (Boston: Harvard Business School Press, 2002). Our research consisted of approximately 80 interviews with Intel managers and engineers between 1997 and 2000, as well as interviews and analyses of publicly available materials on Intel and other companies.

6. D. Ryan, interview with the authors, Aug. 4, 1998.

7. Some references we found useful: M. Katz and C. Shapiro, "Product Introduction with Network Externalities," *Journal of Industrial Economics* 40 (March 1992): 55–83; M. Cusumano, Y. Mylonadis and R. Rosenbloom, "Strategic Maneuvering and Mass-Market Dynamics: The Triumph of VHS Over Beta," *Business History Review* 66 (spring 1992): 51–94; R. Langlois, "External Economies and Economic Progress: The Case of the Microcomputer Industry," *Business History Review* 66 (spring 1992): 1–50; R. Langlois and P.L. Robertson, "Networks and Innovation in a Modular System: Lessons from the Microcomputer and Stereo Component Industries," *Research Policy* 21 (August 1992): 297–313; C. Shapiro and H. Varian, "Information Rules: A Strategic Guide to the Network Economy" (Boston: Harvard Business School Press, 1998); J. Farrell and G. Saloner, "Installed Base and Compatibility: Innovation, Product Preannouncements and Predation," *American Economic Review* 76 (December 1986): 940–955; J. Farrell and C. Shapiro, "Dynamic Competition with Switching Costs," *RAND Journal of Economics* 29 (spring 1988): 123–137; J. Farrell, H.K. Monroe and G. Saloner, "The Vertical Organization of Industry: Systems Competition Versus Component Competition," *Journal of Economics & Management Strategy* 7 (summer 1998): 143–182; and T. Bresnahan and S. Greenstein, "Technological Competition and the Structure of the Computer Industry," *Journal of Industrial Economics* 47 (March 1999): 1–40.

57

Managing Technological Innovation in Organizations

RALPH KATZ

Over the past decades a plethora of studies has convincingly demonstrated a very consistent, albeit disturbing, pattern of results with respect to the management of innovation. In almost every industry studied, a set of leading firms, when faced with a period of discontinuous change, fails to maintain its industry's market leadership in the new technological era. Tushman and O'Reilly (1997) nicely summarize this point in their research when they describe how Deming, probably the individual most responsible for jump-starting the quality revolution in today's products, highlighted this recurring theme in his lectures by showing a very long list of diverse industries in which the most admired firms rapidly lost their coveted market positions.[1] It is indeed ironic that so many of the most dramatically successful organizations become so prone to failure.

This pathological trend, described by many as the *tyranny of success*, in which winners often become losers—in which firms lose their innovative edge—is a worldwide dilemma, exemplified by the recent struggles of firms such as Xerox in the United States, Michelin in France, Philips in Holland, Siemens in Germany, EMI in England, and Nissan in Japan. The Xerox brand, for example, has penetrated the American vernacular so strongly that it has the distinction of being used both as a noun and as a verb, although the company is probably now hoping that to be "Xeroxed" won't eventually take on a different meaning. The histories of these and so many other outstanding companies demonstrate this time and time again. It seems that the very factors that lead to a firm's success can also play a significant role in its demise. The leadership,

vision, strategic focus, valued competencies, structures, policies, rewards, and corporate culture that were all so critical in building the company's growth and competitive advantage during one period can become its Achilles heel as technological and market conditions change over time. In the mid-1990s, an outside team of experts worked for five or so years with senior management on a worldwide cycle time reduction initiative at Motorola Corporation, arguably one of the world's most admired organizations at that time. In just about every workshop program they conducted across the different businesses, the consultants constantly reminded all of those managing and developing new products and services of this alarming demise of previously successful companies by referring to a notable 1963 public presentation made by Thomas J. Watson, Jr., IBM's Chairman and CEO. According to Watson:

> Successful organizations face considerable difficulty in maintaining their strength and might. Of the 25 largest companies in 1900, only two have remained in that select company. The rest have failed, been merged out of existence, or simply fallen in size. Figures like these help to remind us that corporations are expendable and that success—at best—is an impermanent achievement which can always slip out of hand.

DUALISM AND CONFLICTING ORGANIZATIONAL PRESSURES

It is important to recognize, however, that this pattern of success followed by failure—of innovation

Reprinted by permission of the author.
Ralph Katz is Professor of Management at Northeastern University's College of Business and Research Associate at MIT's Sloan School of Management.

followed by inertia and complacency—is not deterministic. It does not have to happen! Success need not be paralyzing. To overcome this tendency, especially in today's rapidly changing world, organizations more than ever before are faced with the apparent conflicting challenges of *dualism*, that is, functioning efficiently today while innovating effectively for the future. Not only must business organizations be concerned with the financial success and market penetration of their current mix of products and services, but they must also focus on their long-term capabilities to develop and incorporate what will emerge as the most customer-valued technical advancements into future offerings in a very quick, timely, and responsive manner. Corporations today, no matter how they are structured and organized, must find ways to internalize and manage both sets of concerns simultaneously. In essence, they must build internally those contradictory and inconsistent structures, competencies, and cultures that not only foster more efficient and reliable processes but that will also encourage the kinds of experiments and explorations needed to re-create the future even though such innovative activities are all too often seen by those running the organization as a threat to its current priorities, practices, and basis of success. Fortunately, there is more recent evidence, as reported in the research studies conducted by Tushman and O'Reilly (1997), of a few successful firms that have managed to do this very thing—to continue to capture the benefits of their existing business advantages even as they build their organizational capabilities for long-term strategic renewal. They are somehow able to transform themselves through proactive innovation and strategic change, moving from today's strength to tomorrow's strength by setting the pace of innovation in their businesses.

While it is easy to say that organizations should internalize both sets of concerns in order to transform themselves into the future, it is a very difficult thing to do. The reality is that there is usually much disagreement within a company operating in a very pressured and competitive marketplace as to how to carry out this dualism. Amidst

the demands of everyday requirements, decision makers representing different parts of the organization rarely agree on the relative merits of allocating resources and management attention among the range of competing projects and technical activities; that is, those that directly benefit the organization's more salient and immediate needs versus those that might possibly be of import sometime in the future.

Witness for example the experiences of Procter and Gamble (P&G) over the past five or more years. In the beginning, the analysts claimed that P&G was doing a very good job at managing its existing businesses but unfortunately was not growing the company fast enough through the commercialization of new product categories. Over the last couple of years, P&G impressively introduced a number of very successful new products (Swiffer, Whitestrips, Thermacare, and Febreze—just to name a few) that are collectively bringing in considerably more than a billion dollars in added revenue per year. The analysts, however, now claim that while P&G has managed to introduce some very exciting new products, in doing so, it took its eye off the existing brands and lost important market share to very aggressive competitors. It is not particularly surprising that these same analysts now want P&G to de-emphasize its new venture strategies and investments in order to concentrate on protecting and strengthening its bedrock major brands. The pendulum just seems to keep on swinging.

While much has been written about how important it is to invest in future innovative activities—"to innovate or die"—as management guru Tom Peters so picturesquely portrays it (Peters, 1997), there is no single coherent set of well-defined management principles on how to structure, staff, and lead organizations to accommodate effectively these two sets of conflicting challenges. Classical management theories are primarily concerned with the efficient utilization, production, and distribution of today's goods and services. Principles such as high task specialization and division of labor, the equality of authority and responsibility, and the unities of command and direction (implying that employees should have one and *only* one boss and that information should *only* flow through the formal chain of command) are all concentrating on problems of structuring work and information flows in routine, efficient ways to reduce uncertainty and facilitate control and predictability through formal lines of authority and job standardization. What is needed, therefore, are some comparable models or theories to help explain how to organize both incremental and disruptive innovative activities[2] within a functioning organization such that creative, developmental efforts not only take place but are also used to keep the organization competitive and up-to-date within its industry over generations of technological change.

While empirically tested frameworks and theories for achieving dualism may be sparse, there is no shortage of advice. In one of the more recent prominent books on the subject, for example, Hamel (2000) proposes the following ten rules for reinventing one's corporation:

1. *Set Unreasonable Expectations:* Organizations must establish bold, nonconformist strategies to avoid the typical bland, unexciting goals that most organizations use, especially if they hope to avoid behaving and thinking like managers of mature businesses or like mature managers trying to lead mature businesses. Only nonlinear innovation, according to Hamel, will drive long-term success, or, as Collins and Porras (1997) describe it, organizations that are "built to last" have *BHAGs* (i.e., Big Hairy Audacious Goals).

2. *Stretch Your Business Definition:* Companies need to define themselves by what they know (their core competencies) and what they own (their strategic assets) rather than by what they do. Senior managers must look continuously for opportunities outside of the business they manage, redefining their markets and challenging their assumptions in ways that allow them to challenge conventional wisdom and outgrow their competitors. One organization with which I worked, for example, required all of its business units to periodically redefine their markets in larger contexts. As a consequence, the associated market shares would be smaller but the areas and opportunities for both

extension and growth would now be more salient and enticing. General Electric's CEO, Jack Welch, adopted this approach whenever his division managers "gamed" his demands that their businesses be "number 1 or 2" by narrowing the definition of their markets. Welch made them redefine their markets globally so that they had no more than 10 percent and had to grow (Welch and Byrne, 2001). Medical systems, for example, went from measuring its share of the diagnostic imaging market to measuring all of medical diagnostics, including all equipment services, radiological technologies, and hospital information systems. Power systems went from having 63 percent share of a service business that focused only on supplying spares and doing repairs on GE technology to having only 10 percent share of a business that focused on total power plant maintenance.

3. *Create a Cause, Not a Business:* Revolutionary activity must have a transcendent purpose or there will be a lack of courage to commit and persist. Employees within the organization have to feel that they are contributing to something that will make a genuine difference in the lives of their customers or in society.

4. *Listen to New Voices:* If a company truly wants to remain at its business forefront, it must refrain from listening only to the old guard that is more likely to preserve its old routines and comfort levels. Instead, management must give disproportionate attention to three often underrepresented constituencies, namely, those with a youthful perspective, those working at the organization's periphery, and newcomers who bring with them fresh ideas less contaminated by historical investments and preconceived notions.

5. *Design an Open Market for Ideas:* What is required to keep an organization innovative is not the vision per se but the way the vision is implemented through at least three interwoven markets: a market for ideas, a market for capital, and a market for talent. Employees within the organization must believe that their pursuit of new framebreaking ideas are the welcomed means by which the organization hopes to sustain its success, and they must be able to give their new possibilities "air time" without rigid bureaucratic constraints and interference.

6. Offer an Open Market for Capital: Rather than designing control and budgeting processes that weed out all but the most comfortable and risk averse ideas, the organization needs to think more like a venture capitalist and permit investments in experiments and unproven markets. Individuals or teams experimenting with small investments and unconventional ideas should not have to pass the same screens and hurdles that exist within the large established business. The goal is to make sure there are enough discretionary resources for winners to emerge—not to make sure there are no losers.

7. Open Up the Market for Talent: Organizations should create an internal auction for talent across the different businesses and opportunities. The organization's professional and leadership talent cannot feel that they are locked inside moribund, mature businesses. Instead, they must have the chance to try out something new, to experiment with an idea, or to proceed with their imaginations. With more fluid boundaries inside the organization, people can search out and follow through on the most promising new ventures and ideas by voting with their feet.

8. Lower the Risks of Experimentation: Neither caution nor brash risk taking is likely to help an organization maintain its innovative vitality. Successful revolutionaries, according to Hamel, are both prudent and bold, careful and quick. They prefer fast, low-cost experimentation and learning from customers to gambling with vast sums of money in uncertain environments. And just like experienced venture capitalists, the organization needs to invest in a portfolio rather than in any single project.

9. Make Like a Cell—Divide and Divide: Using cell division as a metaphor, Hamel claims that innovation dies and growth slows when companies stop dividing and differentiating. Division and differentiation free the leadership, the professional talent, and the capital from the constraints and myopic evaluations of any single large business model. And just as important, it keeps the business units small, focused, and more responsive to their customers.

10. Pay Your Innovators Well—Really Well: To maintain entrepreneurial enthusiasm, companies need to reward those individuals taking the risks in ways that really demonstrate that they can have a piece of the action. The upside has to look sufficiently attractive for them to stick around—to create something out of nothing. Innovators should have more than a stake in the company, they should have a stake in their ideas. If you treat people like owners, they will behave like owners.

Implicit in all of this well-intentioned exhortation, as well as in so many other similar examples of advice, is the need to learn how to build parallel structures and activities that would not only permit the two opposing forces of today and tomorrow to coexist but would also balance them in some integrative and meaningful way. Within a technological environment, the operating organization can best be described as an "output-oriented" or "downstream" set of forces directed toward the technical support of problems within the business's current products and services in addition to getting new products out of development and into the marketplace through manufacturing and/or distribution. For the most part, these kinds of pressures are controlled through formal job assignments to business and project managers who are then held accountable for the successful completion of product outputs within established schedules and budget constraints, that is, for making their quarterly forecasts.

At the same time, there must be an "upstream" set of forces that are less concerned with the specific architectures, functionalities, and characteristics of today's products and services but are more concerned with all of the possible core technologies that could underlie the industry or business environment sometime in the future. They are essentially responsible for the technical health and vitality of the corporation, keeping the company up to speed in what could become the dominant and most valued technical solutions within the industry. And as previously discussed, these two sets of forces, downstream and upstream, are constantly competing with one another for recognition and resources which can either be harmful or beneficial depending on how the organization's leadership resolves the conflicts and mediates the priority differences.

If the product-output or downstream set of forces become dominant, then there is the likelihood that sacrifices in using the latest technical advancements may be made in order to meet more immediate schedules, market demands, and financial projections. Strong arguments are successfully put forth that strip the organization of its exploratory and learning research activities. Longer-term, forward-looking technological investigations and projects are de-emphasized in order to meet shorter-term goals, thereby mortgaging future technical expertise. Under these conditions, important technological changes in the marketplace are either dismissed as faddish or niche applications or go undetected for too long a period. The dilemma in these instances of course is that the next generation of new product developments either disappears or begins to exceed the organization's in-house technological capability.

At the other extreme, if the research or upstream technology component of the business is allowed to dominate development work within R&D, then the danger is that products may include not only more sophisticated and more costly technologies but also perhaps less proven, more risky, and less marketable technologies. This desire to be technologically aggressive—to develop and use the most attractive, most advanced, most clever, state-of-the-art technology—must be countered by forces that are more sensitive to the operational environment and the infrastructure and patterns of use. Customers hate to be forced to adapt to new technologies unless they feel some pressing need or see some real added value. Technology is not an autonomous system that determines its own priorities and sets its own standards of performance and benchmarks. To the contrary, market, social, and economic considerations, as so poignantly pointed out by Steele (1989), eventually determine priorities as well as the dimensions and levels of per-

formance and price necessary for successful applications and customer purchases. Such effects are vividly portrayed by the painful experiences of all three satellite consortia—Motorola's, Qualcom's, and Microsoft's—as each consortia group still tries to penetrate the cellular and wireless tele-markets with their respective Iridium, Globalstar, and Telstar systems. Interestingly enough, the $5 billion Motorola-backed bankrupt Iridium company was recently snapped up by a group of private investors for just $25 million—commercial services have been re-launched with the new investors claiming that they will be profitable by mid-2003.

A MODEL OF INNOVATION DYNAMICS IN INDUSTRY

In a previous but still seminal piece of research, Abernathy and Utterback (1978) put forth a model to capture some interesting dynamics of innovation within an industry (see Utterback, 1994, for the most thorough discussion of these issues and their implications). The model, as shown in the Figure 57.1 suggests that the rates of major innovations for both products and processes within a given industry often follow fairly predictable patterns over time and, perhaps more importantly, that product and process innovation interact with each other. In essence, they share an important tradeoff relationship.

The rate of product innovation across competing organizations in an industry or product class, as shown in Figure 57.1, is very high during the industry's early formative years of growth. It is during this period of time, labeled the *fluid* phase by Abernathy and Utterback in their model, that a great deal of experimentation with product design and technological functionality takes place among competitors entering the industry. During this embryonic stage, no single firm has a "lock" on the market. It seems that once a pioneering product has demonstrated the feasibility of an innovative concept, rival products gradually appear. As long as barriers to entry are not too high, these new competitors are inspired to enter this new emerging market with their own product variations of design

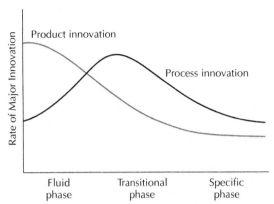

Figure 57.1. The dynamics of innovation (adapted from Utterback, 1994).

choices and features. No one's product is completely perfected and no organization has as yet mastered the manufacturing process or monopolized all of the means of distribution and sales. Furthermore, customers do not as yet have a firm conception of what an ideal product design would look like nor enough experience from which to indicate what they would want or even what they would be willing to pay for in terms of features, functions, and attributes in this new product class. In a sense, the market and the industry are in a *fluid* stage of development with all of the involved participants learning as they move along together.

One of the more illustrative examples offered by Abernathy and Utterback of the *fluid* phase lies in the early years of the automobile industry in which a bewildering variety of *horseless carriages* were designed and sold to the public. Not only were internal combustion machines being commercialized as the power source at this time, but a whole host of other electric and steam-driven cars were emerging from the workshops of dozens of other producers as viable competitive alternatives. Each manufacturer hoped to capture the allegiance of the public with its own novel new design and driver amenities. Electric or battery powered cars, in fact, were produced and sold to customers from the earliest years of the industry and were discontinued primarily because Henry Ford eschewed the battery option (influenced in part on advice he re-

ceived from Thomas Edison) and selected the internal combustion engine as the basis of his Model T design. In addition to relatively poor acceleration and the inconvenience of finding outlets for recharging, batteries became an unpopular choice simply because they were much too heavy for the mostly unpaved muddy roads that existed at that time.

According to the model, in many product class situations, this period of fluidity usually gives way to a *transitional* phase in which the rate of major innovations in the product's design decreases substantially but the rate of major innovations within the process begins to increase. It is within this period that the wide variety of different designs within the product class begins to disappear mainly because some aspects of a design standard have emerged. Most commonly, certain elements within the overall design have proven themselves in the marketplace as most desirable for satisfying the customers' needs or because certain design standards have been established either by formal agreements among major producers or by legal and regulatory constraints. As the basic form of the product becomes more predictable and consistent, the industry begins to concentrate more on improvements in the way the product is produced and how costs in the overall production and operation can be reduced.

In discussing their automobile example, Abernathy and Utterback describe how early imaginative designs of the auto age (three-, four-, and five-wheel auto designs, for example) gave way to a set of fairly standardized designs among the many competitors. As a result of this convergence, the features and basic form of the automobile achieved a reasonable degree of uniformity. In a sense, customers had developed a pretty clear understanding about what cars should look like and how they should be driven. The automobile companies developed a set of technologies and the driving public developed a set of expectations that mutually adjusted to one another over time to essentially define the basic format of the automobile. At the same time, however, substantial progress was being made in the overall ability of the firms to manufacture large quantities of cars at lower costs. By

1909, the rate of product design innovation had diminished so much that an article in the prestigious journal of *Scientific American* proudly proclaimed: "That the automobile has practically reached the limits of its development is suggested by the fact that during the past year no improvements of a radical nature have been introduced" (Scientific American, 1909). Undoubtedly, it was this kind of standardization and general stability in the overall design that probably allowed Henry Ford to devote the numerous years it took to establish successfully his assembly line manufacturing process for the Model T.

If the market for the product class continues to grow, the industry gradually passes into what Abernathy and Utterback call the *specific* phase, as shown in Figure 57.1. The researchers deliberately chose the term "specific" to emphasize the notion that during this period, the business unit strives to manufacture very specific assembled products at the highest possible rates of efficiency. During this progression, the products themselves become increasingly well-defined with product similarities greatly outpacing product differences. Customers, moreover, have grown considerably more comfortable and experienced with the product mix; consequently, they have become more demanding and price sensitive. With this increase in familiarity, customers are now more capable of assessing what they need and value and what they are willing to pay. What typically results from all this movement is the gradual commoditization of the marketplace in which the value ratio of perceived functional quality to price (i.e., the business unit's cost) essentially becomes the basis of competition. Furthermore, the interrelationships between product and process are so intertwined that changes in either one of them require corresponding changes in the other. Even small changes can therefore be difficult and expensive.

DOMINANT DESIGN

The critical supposition underlying this gradual shift from a *fluid* phase to the *transitional* and *spe-*

cific phases is the emergence of a dominant design within the industry. Retrospective studies of innovation within a product class typically reveal that at some point in time there emerges a general acceptance of how the principal components comprising the product's overall architecture interface with one another (e.g., Utterback, 1994; Anderson and Tushman, 1990). A dominant design does not necessarily incorporate the best technologies nor is it necessarily the most optimal. It is established and defined experientially by the marketplace when a particular design architecture becomes the one most preferred by the overwhelming majority of the product's purchasers. Ford's Model T; the Douglas DC-3 aircraft; Boeing's 700-series; the QWERTY keyboard; IBM's 360 mainframes and PCs; JVC's VHS format; the Sony Walkman; WINTEL-based PC's; Powerpoint, Word, and Excel software programs; and CDMA or GSM-type cell phones, for example, all accounted for upward of at least 70 to 80 percent of their respective markets at the height of their popularity.

What the emergence of a dominant design does seem to lead to, however, is a shift in concentration of an industry's innovative resources and energies away from coming up with significant design alternatives to focusing on process improvements in what has become the ensconced product architecture. To put it bluntly, it would be a very risky bet for any organization to make a significant and irretrievable capital investment in a manufacturing process dedicated to a specific architectural configuration while there is still major flux among competing designs within the industry. In creating a convergent set of engineering conventions and design choices, a dominant design also provides equipment and component suppliers within the industry a clearer and more well-defined product and technological context within which their engineers can work to improve their pieces of the overall manufacturing system.

Once a dominant design emerges, the basis of competition within the marketplace changes substantially. The industry changes from one characterized by many firms with many unique designs to one that will eventually contain many fewer firms competing with very similar product designs. A dominant design, therefore, is not a particular product or machine but a set of engineering standards and options representing an accepted pattern of design choices within the product's scope. Customers today, for example, rarely have to ask how to start, drive, or steer their cars; use their lights, windshield wipers and brakes; check their tires; or fill their gas tanks. When children's battery-operated toys are purchased, customers expect to use either the AA or AAA-standard size batteries that can be purchased in just about every retail outlet—the initial A- and B-size batteries having fallen by the wayside in most of today's product designs.[3]

More recently, scholars have begun studying the survival rate of organizations in industries as a function of their entry with respect to the dominant design. Research by Suarez and Utterback (1995) and by Christensen et al. (1998), for example, revealed that firms attempting to enter and compete in industries after dominant designs had become established faced much lower chances of success and survival. According to these scholars, the existence of a dominant design significantly restricts the engineers' freedom to differentiate their products through innovative design. As a result, there are fewer opportunities for small or entering firms to find beachheads in markets operating in a post–dominant design era. Their data on company survival further suggest that there might be a bounded "window of opportunity" for entry in fast-moving industries. In their industry samples, not only did the firms that entered after the dominant design had emerged have much lower probabilities of survival but the firms that entered too early also had reduced chances of survival. Most likely, many of these early firms had exerted efforts that built proprietary and/or specialized technological and design capabilities—which in the high turbulence of the fluid phase may have been appropriate and therefore successful—but which are now no longer sought by markets enveloped by a dominant architecture in which other factors and capabilities are more relevant and highly valued.

The emergence of a dominant design is a watershed event in that it delimits the *fluid* phase[4] in which a rich mixture of design experimentation and competition had been taking place within the product class. It is important to understand that a dominant design is not predetermined. Nor does it come about through some rational, optimal, or natural selection process. Instead, it emerges over time through the interplay of technological possibilities and market choices that are influenced and pushed by the many individual and allied competitors, regulators, suppliers, and sales channels, all of whom have their own technological, political, social, and economic agendas.

EXOGENOUS FACTORS INFLUENCING THE DOMINANT DESIGN

Complementary Assets

By understanding the many factors that can influence the establishment and persistence of a dominant design within a given industry, managers can try to enhance the long-term success of their products by making sure they are actively involved in this process rather than dismissing, resisting, or functioning apart from it. In one of the more notable articles on how companies capture value from technological innovation, Teece (1987) describes just how powerful complementary assets can be in shaping the dominant design that eventually wins in the marketplace. Complementary assets are those assets necessary to translate an innovation into commercial success. Innovative products usually embody the technical knowledge about how to do something better than the existing state of the art. However, to generate attractive financial returns and profits, this technical know-how must be linked with other capabilities or assets in order for them to be sold, used, and serviced in the market. These assets span a wide variety of nontechnical properties, including brand name and company image; supply chain logistics; distribution and sales channels; customer service and support; specialized manufacturing capabilities; deep financial pockets; peripheral products; switching costs; political, regulatory, and customer knowledge; critical real estate or institutional associations; and control over raw materials or key components.

The more a firm possesses control over these kinds of complementary assets, the more advantage it has over its competitors in establishing its product as the dominant design. While numerous PCs were available in the market long before IBM introduced its first personal computer in 1981, the IBM PC quickly became the dominant design primarily because of its exceptional brand name, service reputation, and the emergence of so much other peripheral and applications software. In controlling access to homes through their established infrastructures, the "baby Bell" companies had an enormous advantage over new DSL and other telecom start-up competitors most of whom, to their dismay, realized much too late just how precarious a situation they were in even though they may have had attractive cost-performance capabilities and governmental support as a result of industry deregulation. Not surprisingly, established telephone and cable companies, accustomed to operating as near monopolies, were not particularly eager to open their systems and infrastructure to rivals bent on offering similar services. The Verizons and Time Warners of the world threw up numerous legal and technological roadblocks to the upstart service providers.

In sharp contrast, Christensen (2001) recently concluded from his historical investigations of disruptive technologies that one of the reasons the new entrepreneurial transistor-type companies were able to successfully displace the leading vacuum tube producers was that they were in fact able to bypass the critical distribution channels that had been controlled, or at least heavily influenced, by the premier vacuum tube companies. The new transistor firms did not have to rely solely on the legion of existing appliance store outlets that had grown all too comfortable with the opportunistic profits they were deriving from the servicing and replacement of vacuum tube components and products. They were, instead, able to sell their "revolutionary" transistorized products through the many large department store chains that were also rapidly growing at that time throughout the country.

External Regulations and Standards

Government requirements and regulations can also play a significant role in defining a dominant design especially when they impose a particular standard within an industry. The FCC's (Federal Communications Commission) approval of RCA's television broadcast standard, for example, provided RCA a tremendous advantage not only in establishing its receivers as the dominant design within the industry but also in favoring its black-and-white TV strategy over the color TV strategies that were being developed and lobbied by very worthy competitors at that time.[5] Many governments around the world, either individually or collectively, try to determine standards for emerging technologies within their industries to facilitate easier and quicker product developments and to encourage increased compatibility among system components within the infrastructure of use.

The standards that a government establishes for package labeling, high definition television, telecommunications, automobile safety, or even for the content definitions of certain foods, such as ice cream, orange juice, or peanut butter, can either favor or undermine the interests and strategies of particular competitors within an industry. Governmental requirements and regulations can also be used to enhance the attractiveness of domestic producers over foreign competitors. By knowing that the U.S. government was going to require synthetic detergent producers to eliminate phosphates in order to reduce environmental pollution, Whirlpool Corporation realized that suppliers would have to augment their synthetic detergents in the future with more potent chemical additives to compensate for the banned phosphates. The company also understood that over time these more forceful chemical compounds would probably be substantially more corrosive to many of the washing machine's internal parts, including the inside pumps and drums. As a result, Whirlpool gained a distinct advantage in the industry when it quickly developed new washing machine appliance models that could be marketed to withstand the reformulated, albeit more corrosive, detergents. Whirlpool's products were offered with a complete 3-year warranty on its newly designed machines. Most of its major competitors, in comparison, found themselves backpedaling when they had to lower their equivalent warranties to only a couple of months. Obviously, such companies subsequently rushed to recover with stronger model designs that could resist the caustic characteristics of the new detergents' chemical compositions. An alternative scenario could easily unfold in 2007 when the federal government, in order to promote greater water conservation, has purportedly mandated that all washing machines sold in the U.S. will be "front" rather than "top loaders." Front load machines are exactly the kind of design that has been made and used extensively for many years throughout all of Europe while American manufacturers have concentrated predominantly on top load designs.

Organizational Strategies

The business strategy pursued by a firm relative to its competitors can also significantly influence the design that becomes dominant. The extent to which a company enters into alliances, agreements, and partnerships with other companies within its industry can substantially affect its ability to impose a dominant design. One very notable example of this in recent times is the success of JVC's VHS system over Sony's Betamax system in the VCR (video cassette recorder) industry. By establishing formidable alliances first in Japan and then in Europe and the United States, JVC was able to overcome Sony's initial market success even though it is generally acknowledged that Sony's Betamax was better technologically. To its regret, Sony deliberately chose to go it alone, relying for the most part on its own strong brand name, reputation, and movie recordings. It avoided the technical alliances and market partnerships that would ultimately make JVC's VHS product so much more attractive. Microsoft and Intel pursued similar relationships when they embarked on a strategy to make their "WINTEL" machine and its associated application programs more dominant than many alternative products. At the time, many of these competitive products were seen as technically superior and most had already been successfully commercialized, in-

cluding Lotus's 1-2-3 spreadsheet, Wordperfect's word processor, Harvard's graphics package, the Unix and OS/2 operating systems, or the Macintosh PC.

It is also quite common for companies planning to develop products in an emerging industry to send qualified participants to industrial meetings and/or professional conferences that have been convened explicitly to reach agreements on specific technical and interface standards for the common interest of all, including product developers, suppliers, and users. A great many of the technical protocols and specifications underpinning today's Internet architecture, for example, come from such organized meetings and resultant agreements. The World Wide Web, for example, is currently coordinated through a global consortia of hundreds of companies organized and led by MIT's Laboratory of Computer Science, even though Tim Berners-Lee developed the World Wide Web while he was working in Europe at CERN, the world's largest particle physics center.[6]

On the other hand, companies often disagree about the comparative strengths and weaknesses of particular technical approaches or solutions. As a result, those with similar views often band together to form coalitions that will both use and promote their technical preferences within their product class. Alternative technical camps can subsequently evolve into rival company consortia. Direct competition within the industry becomes centered around these different but entrenched innovative design choices. In replacing cables with wireless technology, for example, many companies have decided to design their new innovative products around the bluetooth wireless set of specifications originally developed by Ericsson. Many other companies, contrastingly, have chosen to base their wireless products on the IEEE 802.11 set of professional standards. The marketplace has yet to decide which if any of these will emerge as the dominant design.[7]

For a business unit to be successful over time with its innovative products, it needs to know much more than whether the innovation creates value for the targeted customers. The firm also needs to consider at least two other important factors. First, it needs to know whether it can "appropriate" the technology, that is, to what extent can it control the know-how within the innovation and prevent others from copying, using, or developing their own versions of it. And second, the firm needs to know whether it has or can secure the necessary complementary assets to commercialize the innovation in a timely and effective manner within the marketplace.

It is critical for senior management teams to realize that their businesses should be building their strategies based on the combined answers to these two important questions. Economists discovered long ago that profitability within an industry can be significantly affected by the interplay of answers to these two questions (see Figure 57.2 for a summary). In general, the more a company can protect its intellectual knowledge and capability from competitors, whether it be through patents, license agreements, technical secrecy, or some other means, the more likely the company can derive profits from its product innovations as long as the complementary assets are freely available. However, if the complementary assets (e.g., distribution channels, raw materials, specialized machinery or key components) are tightly controlled, then there are likely to be "tugs of war" or alliances between the "keepers" of the technology and the "keepers" of the complementary assets. If the intellectual property, on the other hand, cannot be appropriated, that is, the technical knowledge and capabil-

ity are widely available to all competitors, then profits are likely to be made by those firms that control the complementary assets. If both complementary assets and the intellectual property are freely available, then it will be hard for any of the major businesses operating in such an industry to secure and maintain high profitability.

CORRESPONDING CHANGES ACROSS THE PRODUCT/PROCESS MODEL

Organizational Changes
As a business unit forms around an initial new product category and becomes highly successful over time, it goes through the same transformation that any entrepreneurial enterprise experiences as it grows in size and scope. During the pioneering period of the new product industry, i.e., the *fluid* stage, the processes used to produce the new products are relatively crude and inefficient. As the rates of product and process innovation inversely shift, however, the organization has worked to take more and more costs and inefficiencies out of its processes, making production increasingly specialized, rigid, and capital intensive. Through the early 1920s, major automakers could assemble a car in four to five hours but it still required three to eight weeks to paint.[8] With ever increasing capital improvements and innovations in the painting process, especially Dupont's development of Duco lacquer, the painting time had been cut by 1930

	Complementary Assets	
	Freely Available	**Tightly Controlled**
Intellectual Property — Freely Available	Profitability Is Difficult in a Commodity Business	Holders of Complementary Assets Are Profitable
Intellectual Property — Tightly Controlled	Owners of Intellectual Property Are Profitable	Profitability Based on Power Alliances

Figure 57.2.   Profitability winners.

from more than 25 separate operations requiring many weeks to a more continuous spraying process applied over a few days. Painting had become another unskilled task and the strong, independent painters' union collapsed.

In recounting the rich example of incandescent lighting, Utterback (1994) points out that Edison's initial lighting products were made by a laborious process involving no specialized tools, machines, or craft laborers. At the time, getting the product to work was far more exciting and important to the innovators than creating an efficient volume production process. However, as companies built capacity to meet the growing market demand for incandescent lighting more cost effectively than their competitors, vast improvements in the process were gradually developed and introduced by both manufacturers and suppliers, including the use of specialized glass-blowing equipment and molds in addition to high-capacity vacuum pumps. As a result of all these process improvements, especially the development of successive generations of glass-blowing machines, the number of manufacturing steps, according to Utterback, fell dramatically from 200 steps in 1880 to 30 steps in 1920. The manufacturing process had evolved over this period into an almost fully automated continuous process. Amazingly enough, the glass-blowing Ribbon Machine which was introduced in 1926 is essentially the same device that is used today, and it still remains the most cost-effective way to produce light bulb blanks.

As organizations make these kinds of shifts, that is, from focusing on exploratory product designs to concentrating on larger-scale operational efficiencies and the production of more standardized offerings, many other important parallel changes also take place within the industry (Utterback,1994; Tushman and O'Reilly, 1997). Not only do changes in products and processes occur in the previously described inverse systematic pattern, but organizational controls, structures, and requirements also change to adjust to this pattern. In the *fluid* period of high market and technical uncertainties, for example, business leaders are more willing to take the risks required for the commer-

cialization of radical new ideas and innovations. Individuals and cross-functional areas are able to function interdependently, almost seamlessly, in order to enhance their chances of success. Formal structures and task assignments are flatter, more permeable, and more flexible—emphasizing rapid response, development, and adjustment to new information and unexpected events. The company is *organic* in that it is more concerned with the processing of information and the effective utilization of knowledge than the more rigid *mechanistic* following of formal rules, bureaucratic procedures, and hierarchical positions (Mintzberg, 1992).

As the industry and its product class evolve, the informal networks, information exchanges, and fast-paced entrepreneurial spirit that at one time had seemed so natural and valued within the organic-type firms slowly give way to those leaders increasingly skilled at and experienced in the coordination and control of large, established businesses. As operations expand, the focus of problem-solving discussions, goals, and rewards shifts away from those concerned with the introduction of new, radical innovations to those who are capable of meeting the more pronounced and complex market, production, and financial pressures facing the organization. Such demands, moreover, become progressively more immediate and interconnected as the products become more standardized, the business more successful, and the environment more predictable—all of which combine to make it even more difficult and costly to incorporate disruptive kinds of innovations. Rather than embracing potentially disruptive innovations and changes—encouraging and sponsoring explorations and new ways of doing new things—the organization seeks to reduce costs and maximize the efficiencies in its ongoing tasks and routines through more elaborate rules, procedures, and formal structures.

The real dilemma in all of this is that major changes in the environment get responded to in old ways. The organization myopically assumes that the basis by which it has been successful in the past will be the same basis by which it will be successful in the future. As the technical and market environments become increasingly stable, the growth of the enterprise relies to a greater extent on stretching its existing products and processes. The organization encourages and praises its managers and leaders for achieving consistent, steady results that predictably build on past investments and sustainable improvements. More often than not, ideas that threaten to disturb the comfortable stability of existing behavioral patterns and competencies will be seriously discouraged, both consciously and unconsciously. Contrastingly, ideas that build on the historical nature of the business's success, including its products, markets, and technical know-how, are more likely to be positively received and encouraged. A large body of research shows that when comfortable, well-run organizations are threatened, they tend to increase their commitments to their status quo—to the practices and problem-solving methodologies that made them successful in the past, not necessarily to what's needed or possible for the future (e.g., Tushman and O'Reilly, 1997; Katz, 1982). They tend to cultivate the networks and information sources that affirm their thinking and commitments rather than diligently search for information and/or alternatives that might disagree with their inveterate patterns. Without outside intervention, they become increasingly homogeneous and inward-looking, demanding increased loyalty and conformity. They hire and attract individuals who "fit in" rather than those who might think and behave in significantly new and different ways.

Industry Changes

During an industry's *fluid* stage, there is usually sufficient flux in both the technology and the marketplace that large numbers of competitors are able to enter the industry with their own product variations. In their detailed studies of the cement and minicomputer industries, for example, Tushman and Anderson (1986) show just how fragmented industries can be in this stage as each industry had more than a hundred separate firms introducing new products during the early formative years of the product's initial *fluid* cycle. Utterback (1994) draws the same conclusion when he quotes Klein's

assessment of the auto industry's competitive landscape. According to Klein (1977), there were so many initial competitors jockeying for market share leadership positions in the early 1900s that it would have been impossible to predict the top ten "winners" over the next fifteen or so years. Even today, we tend to forget that many of the initially successful PC competitors included firms that were at one time very well-known but which have now disappeared, firms such as Sinclair, Osborne, and Commodore Computer.

After the emergence of a dominant design, the industry moves toward a more commodity-type product space with many fewer surviving competitors. One would expect rapid market feedback and consumer value to be based more on features and functionality than on cost per se during the industry's *fluid* period. As markets become more stable with fewer dominant players, however, the basis of competition also shifts. Market feedback tends to be slower and direct contact with customers tends to decrease, although there are increases in the availability of industry information and statistical analyses. While incremental changes in products may stimulate market share gains during the *transitional* and *specific* stages of the product-process life cycle, they can usually be copied quickly and introduced by competitors, thereby resulting over time in much greater parity among the price, performance, features, and service characteristics of competitors' comparable products. At the same time, however, there is much greater emphasis within the industry on process innovation. Each organization tries to make the kinds of changes and improvements that benefit its particular manufacturing process, changes that are not easily copied or transferable between rival production lines that have been uniquely modified and fine-tuned. In short, price and quality become relatively stronger elements within the overall competitive equation.

As the longevity of a particular dominant design persists within the industry, the basis of competition resides more in making refinements in product features, reliabilities, packaging, and cost. In this kind of environment, the number of competing firms declines and a more stable set of efficient producers emerges. Those companies with greater engineering and technical skills in process innovation and process integration have the advantage during this period. Those that do not will be unable to compete, and ultimately, will either fail, ally together, or be absorbed by a stronger firm. It is perhaps for these reasons that Mueller and Tilton (1969), Utterback (1994), and many others contend that large corporations seldom provide their people with real opportunities or incentives to introduce developments of radical impact during this stage. As a result, changes tend to be introduced either by veteran firms or new entrants that do not have established stakes in a particular product market segment. Radical innovations and changes are usually introduced by disruptive players in small niche market segments (see Christensen, 1997), for as technological progress slows down and process innovation increases, barriers to entry in the large established markets become more formidable. As process integration proceeds, firms with large market shares, strong distribution networks, dedicated suppliers, and/or protective patent positions are also the organizations that benefit the most from product extensions and incremental improvements. It should not be particularly surprising, therefore, that in their study of photolithography, Henderson and Clark (1990) discovered that every new emerging dominant design in process innovation in their study's sample was introduced by a different firm than the one whose design it displaced. Radical, disruptive innovations that start in small market segments are too easily viewed as distractions by the more dominant firms trying to satisfy the cost and quality demands of their large customer base. These same innovations, however, are more easily seen as opportunities by both smaller firms and new entrants.

INNOVATION STREAMS AND AMBIDEXTROUS ORGANIZATIONS

Because industry forces and corresponding organizational priorities operate so differently between product and process innovation phases, Tushman

and O'Reilly (1997) argue that a business unit must redirect its leadership attention away from a particular stage of innovation and move toward managing a series of contrasting innovations if it is to survive through recurring cycles of product-process innovations. An organization has to function in all three phases simultaneously, producing streams of innovation over time that allow it to succeed over multiple cycles of technological products. Innovation streams emphasize the importance of maintaining control over core product subsystems and proactively shaping dominant designs while also generating incremental innovations, profiting from architectural innovation introductions that reconfigure existing technologies, and, most importantly, initiating its own radical product substitutes.

To demonstrate this stream of innovations capability, Tushman and O'Reilly refer to the enviable success of the Sony Walkman. Having selected the WM-20 platform for its Walkman, Sony proceeded to generate more than thirty incremental versions within the WM-20 family. More importantly, it commercialized four successive product families over a ten-year period that encompassed more than 160 incremental versions across the four families. This continuous stream of both generating incremental innovation while also introducing technological discontinuities (e.g., flat motor and miniature battery at the subsystem level) enabled Sony to control industry standards and outperform all competitors within this product class.

The obvious lesson, according to the two authors, is that organizations can sustain their competitive advantage by operating in multiple modes of innovation simultaneously. In organizing for streams of innovation, managers build on maturing technologies that also provide the base from which new technologies can emerge. These managers emphasize discipline and control for achieving short-term efficiencies while also taking the risks of experimenting with and learning from the "practice products" of the future. Organizations that operate in these multiple modes are called "ambidextrous organizations" by Tushman and O'Reilly. Such firms host multiple, internally inconsistent patterns of structures, architectures, competencies, and cultures—one pattern for managing the business with efficiency, consistency, and reliability and a distinctly different pattern for challenging the business with new experimentation and thinking.

Each of these innovation patterns requires distinctly different kinds of organizational configurations.[9] For incremental, sustained, and competence-enhancing types of changes and innovations, the business can be managed with more centralized and formalized roles, responsibilities, procedures, structures, and efficient-minded culture. Strong financial, supply chain, sales, and marketing capabilities coupled with more experienced senior leadership teams are also beneficial. In sharp contrast, the part of the organization focusing on more novel, discontinuous innovations that have to be introduced in a more fluid-type stage requires a wholly different kind of configuration—a kind of entrepreneurial, skunkworks, or start-up spirit and mentality. Such organizational entities are relatively small with more informal, decentralized, and fluid sets of roles, responsibilities, networks, and work processes. The employees themselves are usually less seasoned and disciplined—they are, however, eager to "push the state of the art" and to test "conventional wisdoms."

What all too often becomes very problematic is that the more mature, efficient, and profitable business sees its entrepreneurial unit counterpart as inefficient and out of control—a "renegade, maverick" group that violates established norms and traditions. The contradictions inherent in these two contrasting organizational configurations can easily lead to powerful clashes between the traditional, more mature organizational unit that's trying to run the established business versus the part that's trying to re-create the future. If ambidextrous organizations are to be given a real chance to succeed, the management teams need to keep the larger, more powerful business unit from trampling and grabbing the resources of the entrepreneurial entity. While the company can protect the entrepreneurial unit by keeping the two configurations physically, structurally, and culturally apart, the company also has to decipher how it wants to integrate the strategies, accomplishments, and markets of the two differentiated units. Once the entrepreneurial unit has been separated, it becomes

all too convenient for the firm to either sell or kill the smaller activity. One extremely well-known European organization sadly discovered from its own retrospective investigation that the vast majority of its numerous internal skunkworks for new products in the 1990s were eventually commercialized by competitors. Even more worrisome was the finding that none of the skunkworks products commercialized by the firm *itself* became successful businesses. Clearly the need to differentiate is critically important to foster both the incremental and radical types of innovation; but at the same time, the organization has to figure out how it is going to eventually integrate the so-called upstart unit into the larger organization so that the potential of the ambidextrous organization is not lost. The ability of senior executives and their teams to integrate effectively across highly differentiated organizational configurations and innovative activities remains the true challenge of today's successful enterprise.

ENDNOTES

1. Industries such as watches, automobiles, cameras, stereo equipment, radial tires, hand tools, machine tools, optical equipment, airlines, color televisions, etc.

2. Many other terms have been used to denote this continuum of activities, including continuous versus discontinuous; pacing versus radical; competence-enhancing versus competence-destroying; and sustaining versus disruptive innovation (See Foster and Kaplan, 2001, for the most recent publication in this area).

3. According to Tony Mazzola, a technical marketing manager at Eveready, the A battery is still used in some power tools and camcorders and the B is used for bicycles in Europe. Also, there are still AAAA, F, J, and N-size batteries but the E- and G-size have become obsolete.

4. In their descriptive model of these innovation patterns, Tushman and O'Reilly (1997) call this fluid period an "era of ferment" to denote the intense *agitation* that transpires among alternative designs and features within the product class.

5. RCA's political and VHF technical strategies, masterminded by the renowned David Sarnoff for the direct benefit of its NBC network, were pitted directly against the UHF color strategy of CBS, led by the relatively young unknown Peter Goldmark, who was hired for the expressed purpose of getting CBS into television and "the urge to beat RCA and its ruler, David Sarnoff." When the FCC unexpectedly affirmed the monochrome standards of RCA on March 18, 1947, CBS scrambled to buy, at what became a quickly inflated price, the VHF licenses it had previously abandoned. Interestingly enough, the U.S. government is now seriously reconsidering UHF as the possible standard for its broadband.

6. Neil Calder, who was the Director of Public Relations at CERN, sadly tells of the day that Tim Berners-Lee walked into his office wearing jeans and a T-shirt to tell him of his progress and latest developments. After listening in complete bafflement, Calder thanked Tim for coming and to "keep in touch." It was only much later that Calder realized that Berners-Lee had just described to him the creation of the World Wide Web. "It was the biggest opportunity I let slip by," says Calder (2002), who is now Director of Communications of the Stanford Linear Accelerator Center.

7. It is by no means clear which wireless standard will emerge as the winner or even if a single winner will emerge. The professional IEEE 802.11specifications have split into four separate 802.15 standards project groups with differing views on low-complexity and low-power consumption solutions. The older 802.11a standard, for example, has a throughput of 54Mbits/s while 802.11b has a throughput of only 11 Mbits/s. Additional company coalitions have formed to develop products based on other possible wireless standards, including the standards, known as Wi-Fi and IPv6. More recently, a new standard dubbed WiMedia has emerged claiming to be cheaper than Wi-Fi while also offering much greater range and higher speeds than Bluetooth. Clearly, wireless products are firmly in the *fluid* stage of Abernathy and Utterback's model.

8. Part of Ford's rationale for offering the Model T in "any color as long as it was black" was that black absorbed more heat than lighter colors and therefore dried significantly faster. It also lasted considerably longer, as black varnish was most resistant to ultraviolet sunlight. Interestingly enough, the Model T was originally offered in six different colors before Ford resorted to the black-only option.

9. See Tushman and O'Reilly (1997) for their original and more complete discussion of organizational congruence, culture, innovation streams, and ambidextrous organizations.

REFERENCES

Abernathy, W., and J. Utterback, 1978, "Patterns of Industrial Innovation," *Technology Review*, 80:40–47.

Anderson, P., and M. Tushman, 1990, "Technological Discontinuities and Dominant Designs: A Cyclical Model of Technological Change," *Administrative Science Quarterly*, 35:604–633.

Calder, N., 2002, Presentation at School of Engineering, M.I.T.

Christensen, C., 2001, Presentation at Sloan School of Management, M.I.T.

Christensen, C., F. Suarez, and J. Utterback, 1998, "Strategies for Survival in Fast-Changing Industries," *Management Science*, 44:1620–1628.

Collins, J., and J. Porras, 1994, *Built to Last*, New York: Harper Business.

Foster, R., and S. Kaplan, 2001, *Creative Destruction: Why Companies That Are Built to Last Underperform the Market—And How to Successfully Transform Them*, New York: Doubleday Press.

Hamel, G., 2000, *Leading the Revolution*, Cambridge, MA: Harvard Business School Press.

Henderson, R., and K. Clark, 1990, "Architectural Innovation: The Reconfiguration of Existing Product Technologies and the Failure of Existing Firms," *Administrative Science Quarterly*, 35:9–30.

Katz, R., 1982, "The Effects of Group Longevity on Project Communication and Performance," *Administrative Science Quarterly*, 267:81–104.

Klein, B., 1977, *Dynamic Economics*, Cambridge, MA: Harvard University Press,

Mintzberg, H., 1992, *Structuring in Fives: Designing Effective Organizations*, Englewood Cliffs, NJ: Prentice-Hall.

Mueller, D., and J. Tilton, 1969, "R&D Costs as a Barrier to Entry," *Canadian Journal of Economics*, 2:570–579.

Peters, T., 1997, *The Circle of Innovation*, New York: Vintage Books.

Steele, L., 1989, *Managing Technology: The Strategic View,* New York: McGraw-Hill.

Suarez, F., and J. Utterback, 1995, "Dominant Designs and the Survival of Firms," *Strategic Management Journal*, 16:415–430.

Teece, D., 1987, "Capturing Value from Technological Innovation: Integration, Strategic Partnering, and Licensing Decisions," In B. Guile and H. Brooks (eds.), *Technology and Global Industry*, Washington DC: National Academy Press.

Tushman, M., and P. Anderson, 1986, "Technological Discontinuities and Organizational Environments," *Administrative Science Quarterly*, 31:439–465.

Tushman, M., and C. O'Reilly, 1997, *Winning Through Innovation*, Cambridge, MA: Harvard Business School Press.

Utterback, J., 1994, *Mastering the Dynamics of Innovation*, Cambridge, MA: Harvard Business School Press.

Welch, J., and J. Byrne, 2001, *Straight from the Gut*, New York: Warner Books.

58

Ally or Acquire?
How Technology Leaders Decide

EDWARD B. ROBERTS AND WENYUN KATHY LIU

In their quest to develop profitable products, technology companies are constantly faced with the need to choose between alliances and acquisitions. Executives who understand where their products fit within the technology life cycle are more likely to make the right call.

There are four phases in the life cycle of a technology, and for each there are appropriate ways of partnering with outsiders. Increasingly, the challenge for managers is to recognize which phase each of their products is in and decide what kinds of external partnerships are most likely to facilitate speedy development. Each product a company is juggling may be in a different phase, and because the partnerships developed for one phase of a given technology could serve a different purpose in another phase of another technology, partnerships must be handled with care. Despite the complexity that comes with the need to manage a variety of alliances, Microsoft Corp. and others are demonstrating that it can be done successfully.

The most dramatic change in global technological innovation—the movement toward externally oriented collaborative strategies that complement internal research-and-development investments—began more than a decade ago.[1] Today companies use alliances, joint ventures, licensing, equity investments, mergers and acquisitions to accomplish their technological and market goals over a technology's life cycle. How can companies decide when

to use which form of partnership? In part, by understanding the externally focused technology-life-cycle model.[2]

THE TECHNOLOGY-LIFE-CYCLE MODEL OF ALLIANCES AND ACQUISITIONS

Understanding the role of alliances and acquisitions in the technology life cycle starts with understanding the cycle's four stages: the fluid phase, the transitional phase, the mature phase and the discontinuities phase.[3] The first three were identified in the 1970s by James M. Utterback. (See Figure 58.1.) He later added a fourth, discontinuities, stage. Each

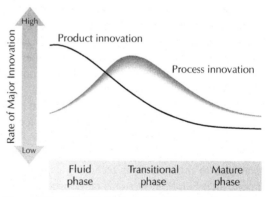

Figure 58.1. The Utterback model of the technology life cycle. (Reprinted by permission of Harvard Business School Press. From "Mastering the Dynamics of Innovation" by J.M. Utterback, Boston, 1994, p. xvii. Copyright © 1994 by the Harvard Business School Publishing Corporation, all rights reserved.)

Reprinted from *MIT Sloan Management Review*, vol. 41, no. 4, Fall 2001, pp. 26–34. Reprinted with permission.

Edward B. Roberts is Sarnoff Professor of Management at MIT's Sloan School of Management. *Wenyun Kathy Liu* is an Associate at Salomon Smith Barney in New York.

stage is shaped by changes in the character and frequency of innovations in technology-based products and processes and by market dynamics. (See Table 58.1.)

The Fluid Phase

In the fluid phase, the earliest pioneering products enter the market for that technology amid a high level of product and market uncertainty.[4] For example, CDs that use fluorescent technology are now entering the market for digital data storage, but it is too soon to tell if that technology will win out over DVDs—or other concepts. With the technology in flux, organizations seeking to increase data-storage capacity (say, the military or the movie industry) hesitate to place R&D bets on a single technology. (In an earlier time period, Exxon Enterprises found eight alternative computer-storage technologies equally attractive for investments during a fluid phase of emerging technology.)

The fluid stage also is characterized by a high rate of growth in market demand. Barriers to entry are low; companies with proprietary technologies can enter with ease. There is little brand loyalty; customers seek functionality and quality instead. Direct competition among existing companies is relatively low, so profit margins are high. The bargaining power of suppliers is low because the materials and equipment used to make the products are general in nature.

Today, as product life cycles in high-tech markets shrink, new technology needs quick acceptance. Hence managers of companies in the fluid stage should pursue aggressive outward-licensing strategies to promote their technologies. For example, after Sun Microsystems introduced SPARC (scalable processor architecture) reduced-instruction-set computing (RISC) in 1989, the company licensed it to 21 hardware manufacturers and software developers, including IBM, Novell and Toshiba. And after introducing its Java technology in 1995, Sun made 32 Java licensing agreements in two years.

Startup companies often adopt variations of traditional licensing. For instance, open-source software creators make their source code available to independent programmers who then make compatible changes and share the results—an effective launch strategy for Linux and Apache, among other software products.

To enable a new product to reach customers quickly, high-tech companies in the fluid stage also form marketing alliances with key players in their supply chain. A difference from traditional marketing alliances is the current focus on spreading products to market. Business-to-business Internet companies such as Allaire, Ariba, BroadVision, ChannelWave Software, Veritas Software, Vignette and Webridge allied with key channel players: solutions providers, application-service providers (ASPs), systems integrators, Internet-service providers (ISPs) and consulting companies.

Among the new kinds of alliances, those organized to establish standards are increasingly important. Indeed, in 1999 and 2000, most prominent computer companies participated in one standards alliance or more. (See Table 58.2.) Such alliances involve not only the promotion of the technology but also its further development—often among competitors. For example, in the high-profile Trusted Computing Platform Alliance (TCPA) of 1999, five competing computer giants joined forces to create standards for better security solutions.

During the fluid stage, well-established technology companies often acquire startups. The acquired companies get access to a wider range of resources, and the acquirer gains critical competitive technologies that would have been costly to develop in-house. Another attractive alternative is to form an R&D alliance with a startup while also making minority equity investments in it. Such strategic alliances allow established companies to keep pace with change while building high-level managerial connections and operational links that can lead to later acquisitions.

The Transitional Phase

The transitional phase of a technology life cycle starts with the emergence of a dominant design. As product and market uncertainty lessens and R&D efforts become focused on improving the dominant technology, design cycles shrink.

TABLE 58.1
Characteristics of the Four Technology Phases

	Fluid phase	Transitional phase	Mature phase	Discontinuities phase
Dynamics of the phase	■ Uncertainty in products and markets ■ High rate of product innovation and high degree of process flexibility ■ Fast-growing demand; low total volume ■ Greater importance of product functionality than brand names ■ Little direct competition	■ Appearance of dominant design ■ Increased clarity about customer needs ■ Increased process innovation ■ Importance of complementary assets ■ Competition based on quality and availability	■ Strong pressure on profit margin ■ More similarities than differences in final products ■ Convergence of product and process innovations	■ Invasion of new technologies ■ Increasing obsolescence of incumbents' assets ■ Lowered barriers to entry; new competition ■ Convergence of some marekts as new technologies emerge
Priorities	■ Development and preservation of technology (with a focus on product development and aggressive patenting) ■ Promotion of proprietary technology as industry standard	■ Realignment of technological capabilities with the dominant design ■ Continued exploration of technological opportunities ■ Pursuit of a growth strategy (through aggressive capacity building or by establishing a close relationship with suppliers and customers)	■ Cost control throughout the value chain ■ Strong customer focus ■ Lean and efficient organization	■ A need for incumbents to identify new technologies and realign core competencies ■ An option for incumbents to exit the market ■ Attackers' need to gain market recognition ■ Attackers' need to focus on product development
Strategic alliances	■ Formation of alliances to promote technology as the industry standard ■ Adoption of licensing strategies (say, open-source licensing or aggressive licensing to users) ■ Formation of marketing alliances (with key players of the supply chain or with one industry leader) ■ Formation of technology alliances with established companies, often coupled with equity investments	■ Winners' aggressive licensing to customers and to companies that lost the dominant-design battle ■ Formation of joint R&D ventures with companies in the market ■ Formation of marketing alliances; signing of supply agreements to guarantee consistent quality, price and availability	■ Formation of joint R&D ventures to share risks and costs of technology development ■ Formation of marketing alliances to attack latent markets or lure customers away from competitors ■ Manufacturing alliances to ensure availability of essential products ■ Open alliances with suppliers and customers	■ Attackers' formation of marketing alliances to gain market recognition ■ Attacker agreements to supply technology leaders ■ Incumbents' acquisition of the disruptive technology through license agreements
Mergers and acquisitions	■ Acquisitions of startups by well-established technology companies from a more mature high-tech industry ■ Corporate equity investment by well-established high-tech companies	■ Acquisitions of competitors by the winners of the dominant-technology battle ■ Acquisitions by established technology companies entering the market	■ Horizontal mergers between companies with complementary products and services ■ Divestiture of manufacturing capabilities that are not essential ■ Acquisition of technology startups making products that would be difficult to develop in-house	■ Possible equity financing for attacker from established technology companies ■ Established companies' move into new markets through acquisition of niche technology companies ■ Established companies' acquisition of enterprises that have related product capabilities ■ Divestiture of companies as priorities shift with market convergence

TABLE 58.2
Recent Standards Alliances in the Computer Industry

Date	Participants	Objective
Jan. 13, 1999	Adaptec, Compaq, Hewlett-Packard and IBM	To create a new input/ output standard
April 26, 1999	Dictaphone, eDigital, IBM, Intel, Norcom Electronics, Olympus America and Philips Electronics	To develop a standard for the way voice commands and information are transmitted and received by mobile devices
Oct. 18, 1999	Compaq, Hewlett-Packard, IBM, Intel and Microsoft	To develop security standards for hardware and software used in e-commerce
Dec. 14, 1999	Akamai Technologies, Allaire, BroadVision, Exodus Communications, Finlan Software, Network Appliance, Network Associates Novell, Open Market and Oracle	To create a standard that connects multiple Web functions on any Internet-access device
Feb. 8, 2000	3Com, Cisco Systems, Extreme Networks, Intel, Nortel Networks, Sun Microsystems and World Wide Packets	To develop standards and technology for 10-gigabit Ethernet networks

Source: Techweb, http://www.techweb.com.

During the transitional phase, industry demand grows rapidly, customers require quality products and timely delivery, and barriers to entry become even lower if the dominant design is easily accessible. Companies must realign themselves with the new standards and pursue an aggressive growth strategy.[5] To signal their commitment, they also should consider capital investment in production capacity. And to ensure product availability, they need supply and marketing agreements with customers. In the transitional phase, companies often collaborate to improve the dominant design and develop new technological extensions, features and applications through joint R&D. Typically, once they possess sufficient technological capabilities, companies of similar size join forces.

For increased market share and revenue growth, companies need to move quickly to develop or adopt the dominant design. Those with the dominant design pursue their advantage and collect royalties (as Texas Instruments has done most effectively) by aggressively licensing the product to other enterprises. Organizations that have lost the standards battle should adopt the standards quickly by getting a license to use someone else's discovery or through internal R&D. The growth potential in the transitional phase makes entry particularly attractive for companies in mature technology markets. When the PC industry was in transition, Japanese electronics giants—Hitachi, NEC and Toshiba—invested heavily to enter the business. Mature companies seek to acquire businesses that either possess the dominant design or have the capabilities needed for quick adoption of the new standards.

Growing companies that possess the dominant technology may be able to make acquisitions of their own, thanks to the financial clout of their surging stock prices. Ideally, those acquisitions would have a strong strategic objective, and target companies would possess complementary technologies or an attractive customer base.

The Mature Phase

In the mature phase, products built around the dominant design proliferate. R&D emphasis shifts from product innovation toward process innovation.

Because process innovations are inherently time-consuming and expensive, many companies form R&D alliances to share the cost and risk. Last year's alliance between Fujitsu and Toshiba to codevelop one-gigabit DRAM (dynamic random-access memory) computer chips is a good example.

The high cost and risk of internal R&D make technology acquisitions attractive, too. In some respects, an acquisition is better than an alliance, in which partners are also competitors and have equal access to the new technology. Acquisitions give the acquirer exclusive rights to the proprietary technology. Between 1993 and 2000, Cisco Systems spent roughly $9 billion buying more than 50 companies.[6] Cisco's technology acquisitions freed up important resources for an internal focus on core competencies.

During the mature stage of a technology's life cycle, the growth rate of market demand slows, but the total volume of demand expands. The once highly profitable market becomes commoditized, a direct result of cost reduction and excess capacity. There is fierce price competition and pressure on profit margins. The need to bring down cost and grow volume increases. Given the technological and capital requirements, entry is harder for outsiders. The key to surviving the mature stage is strong commitment to organizationwide improvement in efficiency. One way to reduce development cost is through cooperative alliances with suppliers—or even with competitors.

High-tech industries are notoriously cyclical, and companies that pursue manufacturing joint ventures in the mature stage have a better shot at controlling cost and guaranteeing availability and quality despite marketplace fluctuations. Marketing alliances are important, too, as competition intensifies and focusing on customers becomes critical. Marketing alliances help companies target the latent market, pursue competitors' customers and expand into new geographic markets.

In high-tech industries, horizontal mergers with complementary product lines are another popular method to reduce costs and obtain a stronger market position by offering more products and services. That was the apparent rationale when Compaq bought Digital Equipment Corp. in 1998. Compaq, facing a saturated PC market, saw DEC as its key to expanding into the then more lucrative high-end server and service markets.

In the mature stage, some companies divest noncore properties to alleviate pressures on earnings. In 1998, Texas Instruments, wishing to concentrate on its core digital-signal processing (DSP) business, sold its facilities for manufacturing DRAMs to Micron Technologies. This also is the phase during which technology companies have the highest propensity to make equity investments and acquisitions and to form alliances for R&D, marketing or manufacturing.

The Discontinuities Phase

The existing technology can be rendered obsolete by the introduction of next-generation technology (NGT), a more advanced technology or converging markets. During the discontinuities stage, the marketplace is volatile. A new market develops, taking demand away from the old market. As many previous barriers to entry (incumbents' specialized production facilities, their investments in R&D and their technology portfolios) lose their force, the likelihood of new entrants is high. The technology in the field gradually turns toward the fluid phase of a new technology life cycle. The process of technological evolution starts again.

Because technological discontinuities can render a company's competitive ability obsolete, companies must adjust business strategies. When next-generation technology increases system performance, it may either destroy or enhance company competencies.[7] If existing producers initiate the NGT (an uncommon scenario in high-tech industries), it is competence-enhancing to them. Even if the market conditions for their mainstream product deteriorate, the new technology may provide first-mover advantage. Companies that are first to increase capacity can capture near-monopoly rents. A one-month time-to-market advantage can dramatically increase the product's total profit margin. Marketing alliances and agreements to supply end users can accelerate the transition; they guarantee the new product's availability to high-end customers and alleviate the first mover's concern about uncertain market demand.

The definition of the technology's market becomes blurred as markets converge and horizontal mergers appear between attackers and incumbents. Companies with stronger financial bases become acquirers. For example, in an effort to stay ahead of the transition into IP (Internet-protocol) networks, telecommunications giant AT&T purchased cable, telecommunications, high-speed Internet and networking companies. At the same time, to focus on core competencies and avoid redundancy, AT&T made several divestitures. The company's joint R&D alliances and minority equity investments supported both capital and technology goals—without incurring the costs or risks of acquisitions. Such arrangements are popular between technology giants that have a need to collaborate. In May 1999, AT&T entered a critical alliance in which Microsoft put $5 billion into AT&T and supplied its Windows CE operating system to AT&T's set-top boxes. Similarly, Intel, having observed that networking and communications were displacing PCs as the opportunity for semiconductors, made 15 acquisitions in those fields during 1999 and 2000 alone.

COMPANIES' DECISIONS TO ALLY OR ACQUIRE

A decision to ally or acquire depends not only on company-specific competencies and needs but also on overall market development and the company's position relative to its competitors. (See Figure 58.2.) Industry structure and critical success factors change as the underlying technology evolves over its life cycle and as competitive pressures vary. Companies are more inclined to form alliances as the technology becomes better defined and as competitive pressure increases. Then the number of alliances declines in the discontinuities phase, when consolidation decreases the total number of companies in the industry. The number of M&As is often high during the transition stage because established companies acquire startups to enhance their technology portfolios. As the dominant design becomes clear and technology becomes more mature, companies increase acquisition efforts to stay ahead of the competition.[8]

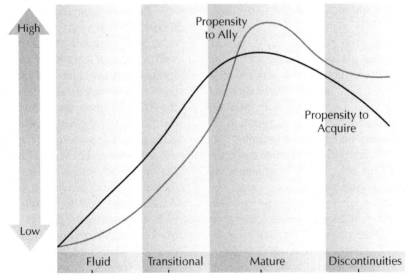

Figure 58.2. Propensity to ally or acquire.

MICROSOFT AND THE TECHNOLOGY LIFE CYCLE

Microsoft Corp. presents a good example of a company managing its external activities in sync with the underlying technological life cycle. (See Table 58.3.)

Microsoft's Fluid Phase (1975 to 1981)

When Microsoft was established in 1975 to provide software for the first personal computer, PC software was relatively new and crude; direct competition was minimal. Throughout the late 1970s, Microsoft focused on developing its PC-software products and technology portfolio. It built strategic relationships with domestic and foreign computer manufacturers, choosing licensing as its principal collaborative mechanism for attracting new customers. During its first six-year period, sales grew at an average annual rate of 165%; revenue reached $16 million by the end of 1981.[9]

Aggressive licensing of its own versions of generic software products such as Basic, Cobol, Fortran and Pascal helped Microsoft attract the attention of the most powerful mainframe-computer company, IBM. In 1980 Microsoft signed a contract to develop operating systems for IBM's first personal computer. The strategic partnership between the startup and the established manufacturer was key in making Microsoft's operating system technology (MS-DOS 1.0) part of the dominant design for PCs. IBM quickly triumphed in the personal computer market, and as other computer manufacturers started to clone the IBM PC, they adopted Microsoft's operating system, too.

Microsoft's Transitional Phase (1982 to 1987)

By 1982 Microsoft's MS-DOS was the dominant operating system for the dominant IBM PC. While Microsoft continued to improve the functionality of its operating systems, it also enhanced its efforts to develop products for specific customer needs. It broadened its technology base by adding application-software programs to its portfolio. Its main objective was to create user-friendly operating systems and software programs for PC users. As the market for PC software grew, Microsoft continued its aggressive licensing strategy. In the first 16 months that MS-DOS was on the market, it was licensed to 50 hardware manufacturers.[10] By retaining distribution rights to DOS, Microsoft benefited from the PC boom. And the continuing IBM alliance helped establish Windows as the standard operating system after MS-DOS.

Microsoft's strong stock performance after its 1986 initial public offering enabled the company to make its first acquisition in 1987—of Forethought, the developer of PowerPoint.

Microsoft's Mature Phase (1988 to 1994)

By 1988 Microsoft had developed a complete technology portfolio. It surpassed Lotus Development Corp. as the world's top software vendor. In the mature stage of its technology life cycle, Microsoft continued to develop and improve its operating systems and application software and to address PC users' needs for word processing, spreadsheets and multimedia software. Sales grew an average of 45% annually. The growth rate was significantly lower than during the fluid and transitional stages, but overall sales volume was larger.

To strengthen its dominance in the high-growth software market, Microsoft participated in strategic partnerships and made key acquisitions. From 1988 to 1994 Microsoft entered 36 joint ventures and alliances, 61% of which involved joint R&D agreements.[11] More important, because it had partnerships with all seven of the leading computer-hardware companies (and because 21 out of 36 joint ventures were exclusive partnerships), Microsoft was able to develop more-functional, user-friendly software.

Microsoft's Discontinuities Phase (1995 to 1999)

The Internet changed the industry's competitive landscape. Although it fueled the growth of the PC market, it also lowered the barriers to entry. Alternative devices appeared. From 1995 to 1999 Microsoft continued to improve the product lines that embodied its base technology. Meanwhile it began

TABLE 58.3
Microsoft and the Four Technology Phases

	Fluid phases	Transitional phases	Mature phase	Discontinuities phase
Dynamics of the phase	■ 165% growth rate ■ Poorly defined technology ■ Proliferation of products ■ Small niche market—the PC market ■ Limited competition	■ 69% growth rate ■ Establishment of Microsoft's operating system as the industry standard ■ Competition from other application-software companies ■ Market growth as technology is better defined	■ 45% growth rate ■ Enormous software sales ■ More competition in the application-software industry ■ Product upgrades and product development to meet customers' needs	■ 32% growth rate ■ Slower growth of the current technology ■ Invasion of the Internet; converging markets ■ Market invasion by new companies with new Internet technologies ■ Need for incumbent to play technological catch-up
Priorities	■ Rapid market recognition ■ Promotion of the technology as part of the dominant design for PCs	■ Pursuit of growth strategy ■ Development of application software to run on Microsoft's operating systems	■ Easy-to-use application software programs ■ Keeping pace with technology developments	■ Development of new technologies ■ Attempt to get established in the new technology area
Strategic alliances	■ Important strategic alliance with IBM ■ Aggressive licensing to commercial users and PC manufacturers	■ Continued aggressive licensing to commercial users and PC manufacturers ■ Ongoing strategic alliance with IBM	■ 36 joint ventures ■ Joint R&D ventures with PC-hardware companies ■ Joint marketing agreements ■ Continued focus on licensing	■ High level of alliance activity ■ 35 alliances, including several that had more than one objective (46% for joint R&D; 52% for joint marketing; 17% for licensing)
Mergers and acquisitions	■ None	■ Acquisition of Forethought, the development of PowerPoint ■ No other M&A activity	■ Two equity minority investments (one in competing OS, one in applications) ■ Two technology acquisitions ■ Slightly more activity than during the transitional stage	■ Much activity ■ 26 minority equity investments ■ Fifteen equity investments since 1999, all internet-related ■ Fifteen companies bought, half Internet-related, half application-software techno-

to develop Internet technologies to establish itself in the new growth market, introducing Internet Explorer 2.0 in 1995 in competition with leading Web browser Netscape Navigator. As it did with application software, Microsoft leveraged its effective operating-system monopoly, bundling Internet Explorer with Windows in the hope that people would use something preinstalled in their PCs.

Microsoft provides a textbook case of a company facing technological discontinuity. As the incumbent, the company possessed deep financial resources and a strong customer base. However, it neither pioneered the Internet nor reacted immediately to it—and consequently fell behind its attackers in the technology and product arenas. Once aware of the seriousness of the threat, Microsoft increased its alliance and acquisition efforts. From 1995 to 1999 alone, it participated in 35 joint ventures. Joint marketing agreements mounted as the market became more volatile and Microsoft strove to maintain a strong relationship with customers.

With the dawn of the Internet and mobile technologies, Microsoft bought 15 companies and, in four years, made 26 minority equity investments focusing on Internet-related technologies and application software.

THE RIGHT PARTNERSHIP ARRANGEMENT FOR THE RIGHT SHAPE

Our life-cycle model indicates that during the fluid stage companies focus on improving product functionality and gaining quick market recognition. Because Microsoft established an important strategic relationship with IBM in that stage, its operating system became the industry standard and Microsoft could proceed to an aggressive licensing strategy.

In the transitional stage, high-tech companies generally form joint R&D ventures, pursue aggressive licensing strategies to realign their technology portfolio, and sign marketing and supply agreements to guarantee consistent quality, price and availability for their customers. Microsoft continued its licensing strategy, and its strategic alliance with IBM remained instrumental to growth.

In the mature stage of the technology life cycle, companies use numerous strategic alliances and acquisitions to share the risks and costs of technology development, ensure availability of essential products and expand into latent markets. Collaborating with leading hardware companies in 36 joint ventures and alliances during its mature phase, Microsoft obtained access to the advanced technologies it needed.

The model anticipates that companies in the discontinuities stage will establish marketing and licensing agreements as well as joint R&D ventures. The phase also features a high level of product and market uncertainty, with technologies invading and markets merging. Thus, when the Internet changed the computer industry's competitive landscape, Microsoft turned increasingly to alliance efforts. Its 35 joint ventures helped maintain

customer relationships and provide comprehensive solutions to clients. As of this writing, Microsoft also has made 26 minority investments and has acquired 15 companies to address the challenge.

Clearly, Microsoft exemplifies our model of the externally focused technology life cycle. Unpublished life-cycle case studies on Cisco and Compaq—plus statistical analyses of industry data—also lend support to our framework.[12] Nevertheless, additional empirical validation from many companies and industries is needed.

Ultimately, the model raises concerns for managers. It shows that a company should use, in a timely and appropriate way, every form of business development—alliances, joint ventures, licensing, equity investments, mergers and acquisitions—in order to perform optimally over its underlying technology life cycle. But doing so requires integrated technology and market and financial planning that may be beyond most companies.

Furthermore, few companies seem to excel, with either organizational or managerial processes, in implementing even a portion of what is needed in business development. A subjective search for benchmarks finds Texas Instruments remarkably capable of carrying out profitable outbound licensing. Cisco excels in acquisitions. Intel and 3M do different but comparably effective jobs of corporate venture-capital investing. Millennium Pharmaceuticals in Cambridge, Massachusetts, has rapidly built a multibillion-dollar enterprise from alliances and joint ventures. But no one company seems to be outstanding at more than one mode of business development. The challenge is not beyond companies' reach, but in order to rise to it, managers must understand the externally focused technology-life-cycle model, think about how it applies to their own situation—and learn to use partnerships that are targeted to a particular technology-life-cycle stage.

NOTES

1. E.B. Roberts, "Benchmarking Global Strategic Management of Technology," *Research-Technology Management* 44 (March-April 2001): 25–36.

2. The authors appreciate the financial support of the global industrial sponsors of the MIT International Center for Research on the Management of Technology, as well as funding from the National Science Foundation to the MIT Center for Innovation in Product Development.

3. For a comprehensive literature review on different models of the technology life cycle, see P. Anderson and M.L. Tushman, "Technological Discontinuities and Dominant Designs: A Cyclical Model of Technological Change," *Administrative Science Quarterly* 35 (1990): 604–633. Further discussion of the model's evolution is provided by J.M. Utterback, "Mastering the Dynamics of Innovation" (Boston: Harvard Business School Press, 1994). Utterback's pioneering life-cycle work, begun in the 1970s, is best summarized by his book. Following Utterback, Tushman and Rosenkopf propose a similar technology-life-cycle model with four stages: eras of ferment, dominant designs, eras of incremental change, and technological discontinuities. See M.L. Tushman and L. Rosenkopf, "Organizational Determinants of Technological Change: Towards a Sociology of Technological Evolution," *Research in Organizational Behavior* 14 (1992): 311–347. See also R.R. Nelson and S.G. Winter, "Simulation of Schumpeterian Competition," *American Economic Review* 67 (1977): 271–276; R.R. Nelson and S.G. Winter, "The Schumpeterian Tradeoff Revisited." *American Economic Review* 72 (1982): 114–132; G. Dosi, "Technological Paradigms and Technological Trajectories: A Suggested Interpretation of the Determinants and Directions of Technical Change," *Research Policy* 11 (1982): 147–162; N. Rosenberg, "Inside the Black Box: Technology and Economics" (Cambridge: Cambridge University Press, 1982); D.J. Teece, "Profiting from Technological Innovation: Implications for Integration, Collaboration, Licensing and Public Policy," *Research Policy* 15 (1986): 285–305; D.J. Teece, "Capturing Value from Technological Innovation: Integration, Strategic Partnering and Licensing Decisions," *Interfaces* 18 (1988): 46–61; and R.R. Nelson, "Recent Evolutionary Theorizing About Economic Change, *Journal of Economic Literature* 33 (1995): 48–90.

4. W.J. Abernathy and J.M. Utterback, "Patterns of Industrial Innovation," *Technology Review* 80 (1978): 40–47.

5. R.M. Henderson, "Underinvestment and Incompetence as Responses to Radical Innovation: Evidence From the Photolithographic Alignment Equipment," *Rand Journal of Economics* 24 (1993): 248–269.

6. http://www.cisco.com/warp/public/750/acquisition.

7. M.L. Tushman and P. Anderson, "Technological Discontinuities and Organizational Environments," *Administrative Science Quarterly* 31 (1986): 439–465.

8. We define "propensity to ally" as the likelihood that a company will participate in joint ventures and alliances. Several ways for post hoc measurement of a company's propensity to ally seem plausible—for example, by examining its total number of alliances normalized by sales. We define "propensity to acquire" as the likelihood that a company will make an acquisition, perhaps measured similarly by the total number of acquisitions normalized by sales.

9. http://www.microsoft.com.

10. Ibid.

11. Data from the Securities Data Company's (SDC) joint-venture database. Many joint ventures involve multiple agreements.

12. W. Liu, "Essays in Management of Technology: Collaborative Strategies for American Technology Industries" (Ph.D. diss., MIT Department of Political Science, 2000).

INDEX

CPSIA information can be obtained at www.ICGtesting.com
Printed in the USA
BVOW08s1212240914

368100BV00005B/15/P